ACRIDINES

Edited by

R. M. Acheson

*The Department of Biochemistry
and
The Queen's College
University of Oxford*

SECOND EDITION

INTERSCIENCE PUBLISHERS

a division of

JOHN WILEY & SONS

NEW YORK • LONDON • SYDNEY • TORONTO

Copyright © 1973, by John Wiley & Sons, Inc.

All rights reserved. Published simultaneously in Canada.

No part of this book may be reproduced by any means, nor transmitted, nor translated into a machine language without the written permission of the publisher.

Library of Congress Cataloging in Publication Data:

Acheson, Richard Morrin.
Acridines.

(The Chemistry of heterocyclic compounds, 9)
Includes bibliographies.
1. Acridine. [DNLM: 1. Acridines. W1 CH364H v. 9
1973. XNLM: [QD 401 A187 1972]]

QD401.A23 1973   547'.593   72–5847
ISBN 0–471–37753–8

Printed in the United States of America

10 9 8 7 6 5 4 3 2 1

ACRIDINES

This is the second edition of the ninth volume published in the series
THE CHEMISTRY OF HETEROCYCLIC COMPOUNDS

THE CHEMISTRY OF HETEROCYCLIC COMPOUNDS
A SERIES OF MONOGRAPHS
ARNOLD WEISSBERGER and EDWARD C. TAYLOR
Editors

Contributors

R. M. Acheson, *The Department of Biochemistry and The Queen's College, University of Oxford, England*

B. A. Adcock, *Flintshire College of Technology, Flintshire, England*

N. R. Ayyangar, *National Chemical Laboratory, Poona, India*

Margaret L. Bailey, *Chemistry Department, Victoria University of Wellington, Wellington, New Zealand*

R. G. Bolton, *I.C.I. Pharmaceuticals Division Research Laboratories, Cheshire, England*

David B. Clayson, *Department of Experimental Pathology and Cancer Research, School of Medicine, Leeds, England*

A. C. R. Dean, *Physical Chemistry Laboratory, University of Oxford, England*

J. M. F. Gagan, *Department of Chemistry and Chemical Technology, Bradford University, Bradford, England*

David W. Henry, *Department of Bio-Organic Chemistry, Stanford Research Institute, Menlo Park, California*

Frank McCapra, *The Chemical Laboratory, University of Sussex, Brighton, England*

B. H. Nicholson, *Department of Physiology and Biochemistry, The University of Reading, Whiteknights Reading, England*

A. R. Peacocke, *St. Peter's College, Oxford, England*

N. R. Raulins, *Department of Chemistry, University of Wyoming, Laramie, Wyoming*

D. A. Robinson, *Molecular Pharmacology Research Unit, Medical Research Council, Cambridge, England*

J. E. Saxton, *Department of Organic Chemistry, The University of Leeds, Leeds, England*

I. A. Selby, *Pharmaceutical Division, Reckitt and Colman, Hull, England*

B. D. Tilak, *Director, National Chemical Laboratory, Poona, India*

The Chemistry of Heterocyclic Compounds

The chemistry of heterocyclic compounds is one of the most complex branches of organic chemistry. It is equally interesting for its theoretical implications, for the diversity of its synthetic procedures, and for the physiological and industrial significance of heterocyclic compounds.

A field of such importance and intrinsic difficulty should be made as readily accessible as possible, and the lack of a modern detailed and comprehensive presentation of heterocyclic chemistry is therefore keenly felt. It is the intention of the present series to fill this gap by expert presentations of the various branches of heterocyclic chemistry. The subdivisions have been designed to cover the field in its entirety by monographs which reflect the importance and the interrelations of the various compounds, and accommodate the specific interests of the authors.

In order to continue to make heterocyclic chemistry as readily accessible as possible, new editions are planned for those areas where the respective volumes in the first edition have become obsolete by overwhelming progress. If, however, the changes are not too great so that the first editions can be brought up-to-date by supplementary volumes, supplements to the respective volumes will be published in the first edition.

ARNOLD WEISSBERGER

Research Laboratories
Eastman Kodak Company
Rochester, New York

EDWARD C. TAYLOR

Princeton University
Princeton, New Jersey

Preface

In the 15 years since the publication of the first edition, there have been many developments in acridine chemistry and biochemistry. These have taken place over a broad front, too broad in fact for one person to review the field both adequately and rapidly enough to ensure up-to-date publication. I have been very fortunate in having a number of colleagues who were willing and able to find the time necessary to revise the chapters of the first edition, or to write entirely new chapters where necessary, for this new edition. Building a book of contributed chapters always presents difficulties of possible duplication and accidental omission of material that is on the borderlines of two or more chapters. A small amount of overlap has resulted, but this has been left in order to maintain continuity in the individual chapters.

All those working with acridines should be grateful that agreement has now been achieved concerning the numbering of the acridine ring. The agreed system is that employed in both editions of this book, and by *Chemical Abstracts*, and the *Ring Index*, and recommended by the International Union of Pure and Applied Chemistry.

I should like to thank most sincerely all my co-authors, all of whom have given up much leisure time in order to help. I also wish to thank The Queen's College and The University of Oxford for leave during which the manuscript was edited, and Professor C. A. Grob for the hospitality of the Institute for Organic Chemistry, The University of Basel, which was greatly appreciated.

I also thank Miss M. B. Acheson, Mrs. R. F. Flowerday, Messers P. J. Abbott, M. P. Acheson, and N. D. Wright for their help in preparing the index, and Miss P. Lloyd for typing the copy. Last, but not least, I must express my gratitude to all those working in Wiley-Interscience and associated with this volume for their advice and continuous assistance, which greatly facilitated my task.

R. M. ACHESON

March 1972
The Department of Biochemistry
and
The Queen's College
University of Oxford
Oxford, England

Contents

ACRIDINES

Second Edition

Nomenclature
and Numbering System

R. M. ACHESON

Department of Biochemistry, and The Queen's College,
University of Oxford, England

The discovery of a new basic material in the anthracene fraction of coal tar was announced by Graebe and Caro[1] in 1870. On account of its acrid smell and irritating action on the skin and mucous membrane, this new substance was called "acridin" (acris = sharp, or pungent). Apart from the addition of a terminal "e," and the replacement of the "c" by "k" in a few older papers, this name has not subsequently been changed. The general nomenclature and numbering system employed for acridine and its derivatives in this monograph is the same as that used in *Chemical Abstracts* since 1937.

At least seven different systems of numbering have been used for the simple acridine ring system, and many more methods have been proposed for fused ring systems containing the acridine nucleus. Much difficulty, therefore, arises when a literature search for acridine derivatives becomes necessary, especially in regard to early publications.

The first numbering system (1) was suggested by Hess and Bernthsen[2] but found no support. Schöpff's system (2), suggested in 1892,[3] was almost as unpopular, although a slight modification (3) was used in 1922 and occasionally later.[4]

In 1893 Graebe,[5] the discoverer of acridine, suggested a numbering system (4) based on the then accepted numbering used for anthracene, xanthene, etc. This system was generally approved at the time. In 1900, however, method

(5), which was largely ignored, was propounded by von Richter in his textbook,[6] while in the same year M. M. Richter[7] used another system (6) in his *Lexikon der Kohlenstoff-Verbindungen.*

This method of numbering gained some popularity. Borsche[8] suggested another system (7), which did not find acceptance. In 1921 Stelzner[9] extended Graebe's system to 8.

Another variation (9) suggested by Patterson,[10] which is in conformity with the *Ring Index* rules,[11] was used in a standard textbook[12] in 1926, but the numbering was changed to that of Graebe in the 1936 edition of the book. Yet another system[13] (10) appeared in 1947, adding to the confusion.

At the present time only two systems of numbering, 4 and 6, are significantly used for acridine derivatives. A record was kept during the preparation of the first edition (1956) of this monograph of the number and year of publication of all original papers available and referred to (whether mentioned subsequently in the monograph or not) using these systems. Figure 1 shows the results. From the graph it is clear that (1) the great majority of publications use Graebe's system; and (2) although an increase in the popularity of Richter's system in recent years is evident, Graebe's system is still used in the majority of publications. The *Ring Index* has adopted the latter system, which is the same as that generally used for anthracene, although it is not consistent with Patterson's rules for the numbering of cyclic compounds.

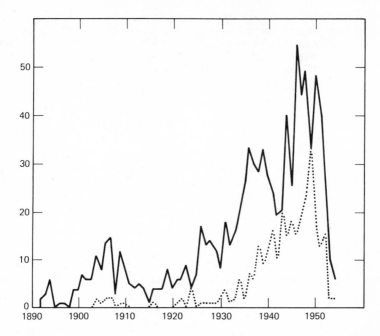

Fig. 1. The number of papers published yearly using Graebe's (———) and Richter's (......) systems.

Graebe's system of numbering is used in *Beilstein's Handbuch* and in *Chemisches Zentralblatt*. Until 1937 *Chemical Abstracts* officially used Richter's system of numbering; at that time a change was made to Graebe's system. *British Chemical Society Abstracts* uses Richter's system, which was also employed by Albert.[14] Although the British and American abstracts officially use particular methods of numbering, papers in which the numbering of acridine derivatives differs from that officially used in the abstracts may be abstracted in the original numberings. This calls for much care when using abstracts. The stage has even been reached at which both systems are used in the *same* chapter of a standard textbook.[15]

The continued use of two numbering systems for the acridine ring system added to the nomenclature difficulties, so that a decision as to which system should be used in the future became essential. This decision has recently been made (see below). The arguments put forward by Albert[14] in support of Richter's system (**6**) are:

(a) It has been used in a significant portion of the literature including almost the whole of the literature in the English language; (b) it makes no suggestion of there being more than five possible mono-derivatives whereas **4** gives rise to such names as 9-chloroacridine which suggests that there are nine; and (c) a number of medical products have been introduced in Britain and Australia in recent years not under trade names but described systematically according to **6**—e.g. 5-aminoacridine; it would be confusing to the physician if they should simultaneously be numbered in two ways, and he would be apt to think that two isomeric substances are involved.

The first statement is incorrect; the increase in popularity of Graebe's system is largely a result of its use in American publications. With regard to (b), it is necessary to point out that many accepted numbering systems involve this disadvantage, which is now unlikely to be a source of confusion. As to (c), the medical profession can be little interested in acridine nomenclature, since the systematic name "5-aminoacridine hydrochloride" (system **6**) has been officially replaced by "Aminacrine Hydrochloride B.P.," and the trade names for this substance, e.g., "Acramine," "Dermacrine," and "Monacrin," are widely used. Incidentally, in spite of his support for Richter's system, Albert has, in fact, found it convenient to use Graebe's system in one section (on carbazimes) in the first edition of his own monograph.[14]

From the evidence discussed above, it is clear that Graebe's system (**4**), which is employed throughout this monograph, is used in the large majority (ca. 75.3%) of the papers published before 1955 in which a numbering system is required, and in the more important abstract journals.

Richter's system (**6**) offers no particular advantage to compensate for its minority position and should therefore be completely dispensed with in the future. This view has also been taken by the Union of Pure and Applied Chemistry[16] and now appears to have been generally accepted.[17]

The carbon atoms common to both rings can bear substituents only when the acridine ring is suitably reduced, and in this situation it becomes necessary to number these positions. Albert[17] employs Stelzner's extension (**7**) of Graebe's system for these carbon atoms. This use is unfortunate and should be discontinued. It is inconsistent with the generally agreed rules for the numbering of such atoms, and it is inconsistent with current *Chemical Abstracts* practice. The atoms common to both rings should be numbered as shown in structure **11**.

11

Many arbitrary numberings, as well as generic numberings, have been used for fused ring systems containing the acridine nucleus. Generic numbering is built up from the numberings of the constituent ring systems regarded as being fused together, subscripts being used to differentiate between the figures. In this system **12** is 1′,2′-1,2-benzacridine, or 1,2-benzacridine, since this simplification leads to no ambiguity. This type of numbering is much used in the German literature. A similar system, based on Richter's method of numbering acridine, is also largely used in *British Chemical Abstracts*. It was often used in *Chemical Abstracts* until 1937, when the *Ring Index* numbering was adopted for condensed acridines. In order to minimize the use of numbers and subscripts, the sides of the parent ring system have been lettered $a, b, c, \ldots$ (13) starting from position 1. Thus 1,2-benzacridine (12) becomes benz[a]-acridine, and 14 is 3'-methylbenz[a]acridine.

12   13   14

Although generic numbering for comparatively simple, fused ring systems has the advantage of being easily worked out from the numbering of the constituent ring systems, it has the disadvantage of requiring the use of many figures and brackets. For this reason, and in order to conform to current usage in *Chemical Abstracts*, the arbitrary *Ring Index*[10,18] numbering of condensed acridine derivatives has been adopted here. 14 is then known as 4-methyl-benz[a]acridine. The main disadvantage in using an arbitrary numbering system for polycyclic compounds is that similar compounds may have substantially different numberings. This, however, is a small price to pay for the convenience offered by a generally accepted, *and used*, numbering system. In view of the variety of current methods of numbering employed for condensed acridines, as much care is necessary when making a literature search for particular compounds of this type as in the case of simple acridine derivatives.

A variety of unsystematic and misleading names have been used in the past for most condensed acridines, an example being benz[a]acridine, which has also been referred to as β-chrysidine, β-phenonaphthacridine and 1,2-naphthacridine. Misleading nomenclature of the latter type is hardly ever used in current literature and should not be revived. The only remaining example of unsystematic nomenclature in the acridine series officially used by *Chemical Abstracts* until about 1957 is that of "carbazime" for 2,9-dihydro-2-iminoacridine. Obsolete names, as well as the *Chemical Abstracts* names and numberings used in this monograph for condensed acridines, are given in many cases when the individual compounds are discussed.

In the naming of a substituted acridine derivative, there are still two points to be considered. First, the lowest possible numbers should be chosen for the substituents; a full discussion of this is available in *Chemical Abstracts*[19]

TABLE I. Numbering Systems used for Acridine and the Benzacridines

Compound and *Ring Index* number[18]	Numbering used here and in *Chem. Abstr.* from *1937*	Alternative (minority) used in *Chem. Abstr.* before *1937*
Acridine R.R.I. 3523 (R.I. 1973)[10]		
Benz[a]acridine R.R.I. 5144 (R.I. 2735)[10]		
Benz[b]acridine R.R.I. 5140 (R.I. 2731)[10]		
Benz[c]acridine R.R.I. 5148 (R.I. 2737)[10]		
7(H)-Benz[kl]acridine R.R.I. 5564 (R.I. not listed)[10]		
1, 7(H)-Pyrid[3,2,1-de]acridine R.R.I. 5104 (R.I. 2712)[10]		

Note: The *Chemical Abstracts* numberings for more complicated condensed compounds are given in the sections in which the compounds are discussed.

6

and in International Union of Pure and Applied Chemistry publications.[20] Since the acridine ring can be numbered in either direction, **15** may be called either 9-chloro-6-dimethylamino-7-fluoro-4,5-dimethylacridine or 9-chloro-3-dimethylamino-2-fluoro-4,5-dimethylacridine, the latter being "correct." Such a compound, however, may be found "incorrectly" numbered in *Chemical Abstracts*. Second, substituents should be placed in alphabetical order, regardless of their number or position in the molecule, and compound radical names should be treated as a unit according to their first letter. Agreement on this issue has been reached between American and British workers.[21] For instance **15** should be called 9-*c*hloro-3-*d*imethylamino-2-*f*luoro-4,5-dimethylacridine, and not 9-chloro-4,5-dimethyl-3-dimethylamino-2-fluoro-acridine. An order of preference for the last radical in the naming of complex organic compounds is also used by *Chemical Abstracts*: onium compound, acid, acid halide, amide, imide, amidine, aldehyde, nitrile, isocyanide, ketone, alcohol, thiol, amine, imine, ether, sulfide, sulfoxide, sulfone, etc. On this basis **16** should be known as 2-(6-chloro-2-methoxy-9-acridylamino)-ethanol.

15

16

However, the heterocyclic part of this molecule is generally considered more important than the aliphatic part, so that the compound is called 6-chloro-9-(2-hydroxyethylamino)-2-methoxyacridine in this monograph and usually in *Chemical Abstracts*, where it may also be indexed under its alternative name.

The system of numbering approved by the *Ring Index* and by the International Union of Pure and Applied Chemistry, used in this monograph and officially used in *Chemical Abstracts* for acridine and the benzacridines, and alternative numberings used in *Chemical Abstracts* before 1937, are given in Table I.

References

1. C. Graebe and H. Caro, *Chem. Ber.*, **3**, 746 (1870).
2. W. Hess and A. Bernthsen, *Chem. Ber.*, **18**, 689 (1885).
3. M. Schöpff, *Chem. Ber.*, **25**, 1980 (1892). .
4. C. H. Browning, J. B. Cohen, R. Gaunt, and R. Gulbransen, *Proc. Roy. Soc.*, *Ser. B*, **93**, 329 (1922).

5. C. Graebe and K. Lagodzinski, *Justus Liebigs Ann. Chem.*, **276,** 35 (1893).
6. V. von Richter, *Organic Chemistry*, 3rd American ed., from the 8th German ed., 1900, p. 559.
7. M. M. Richter, *Lexikon der Kohlenstoff-Verbindungen*, Voss, Hamburg, and Leipzig, 1900, p. 24.
8. W. Borsche, *Justus Liebigs Ann. Chem.*, **377,** 70 (1910).
9. R. Stelzner, *Literatur-Register der organischen Chemie*, Vieweg and Sohn, Braun-schwieg, **3,** 59 (1921).
10. A. M. Patterson, *J. Amer. Chem. Soc.*, **47,** 556 (1925).
11. A. M. Patterson and L. T. Capell, *The Ring Index*, 1st ed. Reinhold, New York, 1940.
12. J. Schmidt, *Organic Chemistry*, 1st English ed., transl. by H. G. Rule, Gurney and Jackson, London, 1926, p. 619.
13. B. Pullman, *Bull. Assoc. Franc.*, *Etude Cancer*, **34,** 245 (1947).
14. A. Albert, *The Acridines*, 1st ed., Arnold, London, 1951, p. xi; A. Albert, "The Acridines," in *Heterocyclic* Compounds, Vol. 4, Elderfield, Ed., Wiley, New York, 1952, p. 491.
15. H. Ing, in *Organic Chemistry*, Vol. 3, Chap. V., H. Gilman, Ed., Wiley, New York, 1953.
16. *International Union of Pure and Applied Chemistry*, *Nomenclature of Organic Chemistry*, Sections A and B, 2nd ed., Butterworths, London, 1966, p. 58.
17. A. Albert, *The Acridines*, 2nd ed., Arnold, London, 1966, p. xi.
18. A. M. Paterson, L. T. Capell, and D. F. Walker, *The Ring Index*, 2nd ed., American Chemical Society, Washington, D.C., 1960.
19. *Chem. Abstr.*, **39,** 5867 (1945).
20. *International Union of Pure and Applied Chemistry*, *Nomenclature of Organic Chemistry*, Section C, Butterworths, London, 1965.
21. "Editorial Report on Nomenclature," *J. Chem. Soc.*, 3699 (1950).

CHAPTER I

Acridines

N. R. RAULINS

Department of Chemistry,
University of Wyoming,
Laramie, Wyoming

1. Historical Introduction and the Formulation of Acridine

It has been one hundred years since Graebe and Caro[1] announced the isolation of acridine in their report in the *Berichte*, "Wir geben derselben den Namen Acridin wegen der scharfen und beissenden Wirkung, die sie auf die Haut ausübt." The new basic material was isolated from the anthracene fraction of coal tar by extraction with dilute sulfuric acid, followed by precipitation as its dichromate. This compound, assigned the empirical formula, $C_{12}H_9N$, could be only incompletely characterized because of the small amount of material available. However, its appearance, melting point, steam volatility, stability, and ability to form a variety of well-crystallized salts were duly noted. From this stimulus have come a vast body of research and the varied useful applications of acridine and its derivatives known today.

Early experiments suggested the molecular formula $(C_{12}H_9N)_2$ for acridine.[2] Later its structure was considered[3] to be **1**, partly because alkaline permanganate oxidation gave acridinic acid,[4] a quinoline dicarboxylic acid, which on successive decarboxylation gave a quinoline monocarboxylic acid and quinoline. However, Riedel showed that this structure was not possible,[5] as the quinoline monocarboxylic acid obtained on degradation was identical with quinoline-3-carboxylic acid. He suggested that the earlier analyses were in error, that acridine had the molecular formula $C_{13}H_9N$ and was better represented by **2**. This was, in fact, proved correct by the

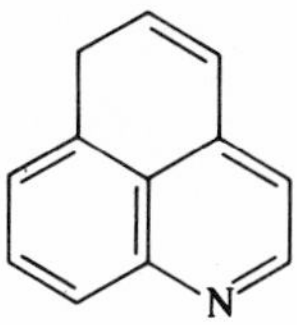

1 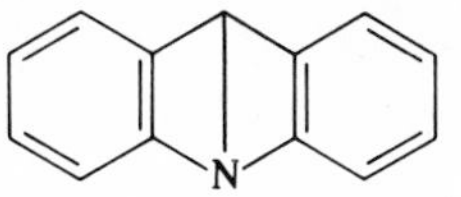   2 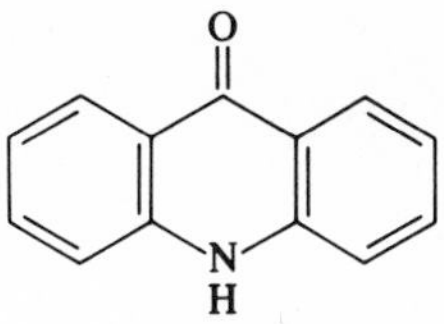   3

synthesis of acridine, although in poor yield, from N-formyldiphenylamine and zinc chloride;[6] 9-phenylacridine had been synthesized a few years before,[7] but its structure had not been recognized. The formation of 9-acridanone (3), both by the oxidation of acridine[3] and by the sulfuric acid cyclization of diphenylamine-2-carboxylic acid,[8] supported Riedel's view.

Following the early observation[9] that acridine reacted with sulfurous acid, methods for estimation[10] and extraction[11, 12] of acridine using aqueous sodium bisulfite have been devised. Acridine has been purified by its phosphate,[13] and can be estimated by titration in aqueous ethanol with sulfuric acid using phenolphthalein as the indicator,[14] or gravimetrically as the picrate[15] or perchlorate.[16] Spectrophotofluorimetric methods have recently been used for the detection and estimation of acridine in the airborne particulates of urban atmospheres.[17]

The methods used for the synthesis of acridines will be discussed in the next section.

The numbering system used for acridine in *Chemical Abstracts*, as shown in **4**, will be used throughout this discussion.

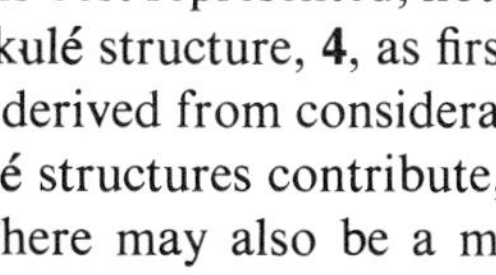

4

The structure of acridine is best represented, not with a centric bond, **2**, but in terms of an ordinary Kekulé structure, **4**, as first suggested by Hinsberg.[18] A more complete picture is derived from consideration of a resonance hybrid to which all possible Kekulé structures contribute, as well as the centric and the others shown here.[19] There may also be a minor contribution from **8**, where the electrons are unpaired and ready to participate in homolytic reactions. Such a representation is in accord with the optical exaltation,[20, 21] the highly conjugated ultraviolet (uv) absorption spectrum (which is considered in Chapter X), the diamagnetic susceptibility,[22] and the resonance energy. The resonance energy of 106 kcal m^{-1} determined from combustion data[23] has been replaced by a value of 84 ± 3.0 kcal m^{-1} from the work of

5   6

7   8

Jackman and Packham.[24] This lower value is preferred because it is derived from bond energy values obtained from lithium aluminum hydride heats of hydrogenation.

A dipole moment of 2.09 D has been reported for acridine.[25] This suggests a shift in emphasis on the contributing structures from that implied in the earlier value[26] of 1.95 D. Leroy and his colleagues recorded an experimentally determined ionization potential of 7.78 eV for acridine.[27a] The calculated value is 7.59 eV. The n ionization potential[27b] has been found to be 2.2 eV larger than the π ionization potential, even though the $n \rightarrow \pi^*$ and $\pi \rightarrow \pi^*$ transitions have the same energy (3.3 eV). This implies charge redistribution, associated with $n \rightarrow \pi^*$ transitions resulting from the coulombic attraction between the promoted electron and the hole it leaves behind. The molar refraction[28] was found to be 64.3.

The early calculations of π electron densities made by Longuet-Higgins and Coulson[29] by the molecular orbital method and by Pullmann,[30] using the valence bond method, have been greatly expanded by more sophisticated applications of the principles of quantum chemistry to heterocyclic molecules. To the atomic spectroscopic data[31] and proton chemical shifts[32] have been added many semiempirical parameters that have been successfully used to gain a picture of electron distribution, electron densities,[33] bond orders,[34] and reactivity indices[35] of acridines.

The X-ray diffraction studies of Phillips and his colleagues[36a,b] have provided detailed information about the crystal structure of two of the crystalline forms of acridine known as acridine III and II (see Section 3). The weighted, mean bond lengths of the two crystallographically distinct molecules in acridine II agree with those in acridine III. These two molecules exhibit significant departures from planarity, the central ring of one being the "chair" and that of the other, the "boat" form. Acridine III, monoclinic, has polar molecules arranged in antiparallel pairs, distorted slightly from planar in ways suggestive of molecular interactions. Bond lengths and angles have been calculated. The unit cell of acridine III, $Z = 4$, has the dimensions, $a = 11.375$, $b = 5.988$, $c = 13.647$ Å, $\beta = 98°58'$. It is a modification of the anthracene structure.

2. Methods of Preparation of Acridines

There is no general method of synthesis that can be used for most acridines. The frequently used syntheses appear to be those proceeding via the 9-acridanone or 9-chloroacridine. The 9-acridanones are readily reduced to acridans, which can be oxidized to acridines. The 9-chloro compounds can be reduced to acridines. Both 9-acridanones and 9-chloroacridines are readily available from the cyclization of diphenylamine-2-carboxylic acids (Chapter III). The most direct potentially general method is perhaps the cyclization of diphenylamine-2-aldehydes and ketones, but these substances are difficult to prepare. Bernthsen's synthesis, a variation of which was used in the original preparation of acridine itself, involves the combination of diphenylamines and carboxylic acids under vigorous conditions and is useful for the synthesis of 9-substituted acridines. A similar reaction with formic acid gives 3-amino-acridines or 3,6-diaminoacridines according to the conditions employed. A related synthesis, in which the carboxylic acids are replaced by aldehydes, initially gives acridans, easily oxidized to the acridines. A small number of acridines have also been obtained from dehydrogenation, cyclodehydrogena-tion, and other reactions that will be considered here.

A. Preparation of Acridines from 9-Acridanones or 9-Chloroacridines

Excluding the zinc dust distillation of 9-acridanones and their reactions with Grignard and similar reagents, there are no reports of useful one-stage reductions of a 9-chloroacridine, prepared from the 9-acridanone with phosphorus oxychloride (Chapter III), and only two of 9-acridanones (p. 20 and Ref. 58b) to the corresponding acridine.

The reason for this failure is that it is much more difficult to reduce a 9-acridanone or a 9-chloroacridine to the acridine than it is to reduce the acridine to the acridan. Consequently, the reduction of a 9-acridanone (or 9-chlororacridine) gives largely the acridan in excellent yield under the proper conditions. As the quantitative oxidation of an acridan to the acridine is easily carried out, this two-stage conversion is a very valuable, frequently used procedure. Table I lists the monosubstituted acridines that have been prepared in these two-stage processes; they are representative of a much

TABLE I. Conversion of 9-Acridanones and 9-Chloroacridines to the Corresponding Acridines

Compound reduced	Procedure	Yield of acridine (%)	Ref.
9-Acridanone	Na and AmOH; CrO_3	80	53,65
	Zn dust distillation	100	8
9-Acridanone-2-carboxylic acid	Al/Hg and EtOH; $FeCl_3$	75	108
9-Acridanone-4-carboxylic acid	Al/Hg and NaOH (aq); $FeCl_3$	80	108
9-Acridanone-2-sulfonic acid	Na/Hg and water; $FeCl_3$	75	109
1-Amino-9-acridanone	Na/Hg and NaOH(aq); $FeCl_3$	70	84
	Al/Hg and EtOH	0	67
2-Amino-9-acridanone	Na/Hg, EtOH and CO_2; $FeCl_3$	85	67
	Al/Hg and EtOH; $FeCl_3$	75	67
3-Amino-9-acridanone	Na/Hg, EtOH and CO_2; $FeCl_3$	70	67
	Na/Hg and NaOH (aq); hot air	54	110
4-Amino-9-acridanone	Na/Hg, EtOH, $NaHCO_3$	46 (impure)	64
2-Aminomethyl-9-acridanone	Na/Hg, EtOH and CO_2; $FeCl_3$	50	108
2-Bromo-9-acridanone	Toluenesulfonhydrazide method	49	48
2-Bromo-9-chloroacridine	Toluenesulfonhydrazide method	52	49
4-Bromo-9-chloroacridine	Toluenesulfonhydrazide method	88	49
4-Bromo-9-chloro-1-ethylacridine	H_2 and Raney nickel, then H_2 and $Pd/SrCO_3$; CrO_3	39[a]	111
9-Chloroacridine	H_2 and Raney nickel; CrO_3	70	53, 55
	Toluenesulfonhydrazide method	73	44
	Hydrazine hydrate; O_2, Pt	55	50

(Table Continued)

TABLE I. (Continued)

Compound reduced	Procedure	Yield of acridine (%)	Ref.
4-Chloro-9-acridanone	Na/Hg, EtOH and CO_2; product isolated as the acridan	?	19
2-Chloro-9-acridanone 10-oxide	Na/Hg and NaOH (aq)	?	112
9-Chloro-2-cyano-acridine	Toluenesulfonhydrazide method	65	44
	Hydrazine hydrate; O_2, Pt	35	50
9-Chloro-2-ethylacridine	H_2 and Raney nickel; CrO_3	58	111, 78
9-Chloro-3-ethylacridine	H_2 and Raney nickel; CrO_3	?	111
9-Chloro-4-ethylacridine	H_2 and Raney nickel; CrO_3	53, 24	111, 78
9-Chloro-1-methoxyacridine	Toluenesulfonhydrazide method	80	47
9-Chloro-2-methoxyacridine	H_2 and Raney nickel; CrO_3	79	113
9-Chloro-4-methoxyacridine	Toluenesulfonhydrazide method	70	47
	Hydrazine hydrate; O_2, Pt	40	50
9-Chloro-1-methylacridine	H_2 and Raney nickel; CrO_3	14	78
9-Chloro-2-methylacridine	H_2 and Raney nickel; CrO_3	70	78
9-Chloro-3-methylacridine	H_2 and Raney nickel; CrO_3	37	78
9-Chloro-4-methylacridine	H_2 and Raney nickel; CrO_3	80	78
9-Chloro-2-nitroacridine	Toluenesulfonhydrazide method	52	44
9-Chloro-3-nitroacridine	Toluenesulfonhydrazide method	53	44
9-Chloro-4-*n*-Propylacridine	H_2 and Raney nickel; CrO_3	16	78

(Table Continued)

15

TABLE I. (Continued)

Compound reduced	Procedure	Yield of acridine (%)	Ref.
3, 9-Dichloro-acridine	Toluenesulfonhydrazide method	71	44
	KOH, H_2 and Raney nickel; CrO_3	47	79
4-Ethoxy-9-acridanone	Na and AmOH; air?	50	114
2-Ethoxycarbonyl-amino-9-chloroacridine	Toluenesulfonhydrazide method	40^b	44
2-Hydroxy-9-acridanone	Na and EtOH; $FeCl_3$ (trace), air	60	89 .
2-Methoxy-9-acridanone	Na/Hg, EtOH and CO_2; HNO_2	?	92
	Na and EtOH; CrO_3	?	115
3-Methoxy-9-acridanone	Na/Hg, EtOH and CO_2; HNO_2	?	92
4-Methoxy-9-acridanone	Na/Hg, EtOH and CO_2; CrO_3	?	92
1-Methyl-9-acridanone	Na and AmOH; $FeCl_3$	?	88
2-Methyl-9-acridanone	Na and AmOH; $FeCl_3$	?	88
	Zn distillation	?	116
3-Methyl-9-acridanone	Na and AmOH; $FeCl_3$	?	88
4-Methyl-9-acridanone	Na and AmOH; $FeCl_3$	?	88
	Zn distillation	80	40
	Na/Hg and EtOH; boil with $C_6H_5NO_2$	?	103
10-Methyl-9-acridanone	Na and AmOH	?	88
	Zn and AcOH	?	102
	Na/Hg and water	?	90
2-Nitro-9-acridanone	Al/Hg and EtOH; $FeCl_3$	75	67
3-Nitro-9-acridanone	Al/Hg and EtOH; $FeCl_3$	70	67
	Na/Hg, EtOH and CO_2	70	67

(Table Continued)

TABLE I. (Continued)

Compound reduced	Procedure	Yield of acridine (%)	Ref.
4-Nitro-9-acridanone	Al/Hg and EtOH; $FeCl_3$	65	67
	Na/Hg, EtOH and CO_2; $FeCl_3$	70	59, 67
10-Phenyl-9-acridanone	Na and AmOH; I_2	?	65

[a] of 1-ethylacridine. [b] of 2-aminoacridine.

larger group of derivatives. In the discussion that follows, the exceptions to this general scheme, (1) and (2), are considered first; then the details of the oxidation-reduction methods are reviewed.

(1) *9-Substituted Acridines*

The many reactions of 9-chloroacridines that depend upon the ease of replacement of the halogen atom in nucleophilic substitution are dealt with in connection with the halo compounds (Section 5.B) and the amino derivatives prepared in these reactions (Chapter II). 9-Acridanone, on treatment with methyl magnesium iodide, gave a mixture of 9-methylacridine (44%) and 9,9-dimethylacridan (17%),[37] the latter compound probably being produced from the Grignard reagent and the first formed 9-methylacridine. Much 9-acridanone was recovered from the reaction mixture. In a similar reaction, only 10–15% of 2-methoxy-9-methylacridine was obtained from 2-methoxy-9-acridanone.[38] Phenyl magnesium bromide and 9-acridanone gave some 9-phenylacridine (19%),[39] but very much better yields (92%) of this substance, **9**, and its 3-methyl derivative were obtained from the 9-acridanones and phenyl lithium.[39] Two equivalents of the latter reagent were required, unless the potassium salt of the acridanone was used in the reaction. No 2-methoxy-9-methylacridine could be obtained from 2-methoxy-9-acridanone and methyl lithium under a variety of conditions.[38]

9-Substituted acridines have also been prepared from acridine by reaction with Grignard reagents or aryl metals and subsequent oxidation of the resulting acridans; this is discussed in Chapter V.

(2) *Zinc Dust Reduction of 9-Acridanones*

The reduction of 9-acridanone to acridine by distillation with zinc dust was first carried out by Graebe and Caro in 1880.[3] Other examples of the reaction, which gives variable yields and is only suitable for the synthesis of acridines that can be distilled from the reaction mixture, have been reported.[8, 40a—c, 41] The reaction tends to become uncontrollable on a large scale and is of little use for preparative purposes.

(3) *Acridines from 9-Chloroacridines with Hydrazides and Hydrazine*

In 1885 Escales showed that benzenesulfonhydrazide was decomposed by warm, aqueous sodium hydroxide into benzene, nitrogen, and benzenesulfinic acid.[42] Much later, McFadyens and Stevens used a variant of this reaction for the conversion of acyl chlorides to the corresponding aldehydes by reaction with p-toluenesulfonhydrazide, followed by decomposition of the complex hydrazide with aqueous sodium carbonate.[43]

$$RCOCl \rightarrow RCONHNHSO_2C_6H_4Me \xrightarrow{Na_2CO_3 \text{ (aq), HCl (aq)}}$$
$$RCHO + N_2 \uparrow + MeC_6H_4SO_2H$$

Since 9-acridanone is amidic in structure (Chapter III), it is not unexpected that the corresponding chloro compound, 9-chloroacridine, reacts with p-toluenesulfonhydrazide to give the corresponding acridyl derivative, **10**. Treatment of the latter with sodium hydroxide gave acridine (**11**) and sodium p-toluenesulfinate in a similar decomposition to that of the Escales benzene derivative.[44]

$$+ \text{MeC}_6\text{H}_4\text{SO}_2\text{H}$$

If sodium carbonate replaces sodium hydroxide in the sulfonhydrazide decomposition, 2- and 3-nitro- and 2-cyano-acridine can be prepared from the corresponding 9-chloroacridines.[44] This is of great interest because no other procedure is available for preserving the nitro and cyano groups during the complete synthesis sequence. It probably proceeds through the formation of the 9-diazoacridan, followed by the loss of nitrogen and migration of the hydrogen atom, in an analogous way to that in the decomposition of the acid p-toluenesulfonhydrazides.[43, 45] The reaction has been applied successfully to the preparation of four of the methoxynitroacridines[46] in yields of about 44%. The monosubstituted acridines[44–49] that have been prepared in this fashion are listed in Table I.

In 1965 Albert reported the preparation of acridines from 9-chloroacridines through their conversion to 9-hydrazinoacridines.[50] These can be oxidized with oxygen in the presence of platinum and a trace of sodium hydroxide at room temperature. Table I lists the monosubstituted acridines prepared by this method, which is considered especially valuable for hot alkali-sensitive compounds.

(4) *Reduction of 9-Acridanones or 9-Chloroacridines to Acridans, Followed by Oxidation to Acridines*

This is undoubtedly the most frequently used procedure for the conversion of 9-acridanones or 9-chloroacridines to acridines; the reduction and oxidation stages are considered separately. The reduction of 9-acridanone probably involves an initial dihydro product, 12, which is dehydrated and immediately reduced to acridan.

Little is known of the mechanism of the oxidation of acridan to acridine, which may vary with the reagent employed. During the oxidation, highly colored acridan-acridine complexes are sometimes formed; these as well as more details on acridan preparation are to be found in Chapter V.

12

(a) REDUCTION PROCEDURES. Although the catalytic reduction of 9-acridanone in tetralin over a copper chromite catalyst at 190° under 120 atm hydrogen to acridan has been reported,[51] the catalytic reduction of acridones cannot be effected under mild conditions (cf. Ref. 52). 9-Chloroacridines which do not contain nitro groups can, however, be reduced by hydrogen over Raney nickel at room temperature and pressure to the corresponding acridans in excellent yield,[53, 54] provided that pyridine or potassium hydroxide is present to remove the hydrogen chloride formed in the reaction. A small quantity of 9,9'-biacridyl is also formed. Ethanol and ethanol-benzene are satisfactory as solvents, but methanol causes some hydrolysis of 9-chloroacridine to 9-acridanone.[55] A careful investigation of the reaction has shown that reduction does not take place in the absence of alkali,[55] and that an external supply of hydrogen is essential; the hydrogen adsorbed on the catalyst alone will not effect reduction to acridan. This reduction method, as well as lithium aluminum hydride reduction, has been used successfully for reactions leading to 4,5-dimethyl- and 1,4,5,8-tetramethylacridines.[56]

The reduction of 9-acridanone to a mixture of acridan, acridine, and biacridyl with sodium amalgam was first mentioned without practical details in 1893.[57] A generally useful reagent for the reduction of acridanones, as well as of amino-9-acridanones, employs the sodium amalgam[58a, b] along with hot 90% alcohol and carbon dioxide,[59, 60] aqueous sodium hydroxide,[61, 62, 63] or aqueous ethanol and sodium bicarbonate[64] at 60–70°. (In the reduction of 1-methoxy-8-amino-9-acridanone, the acridine was formed directly.) An early reduction that yielded principally acridan used a large excess of sodium metal in amyl alcohol.[65] This combination was too drastic when amino or nitro groups were present in the molecule[59, 61, 65] and was thought responsible for partial hydrolysis of alkoxyl groups in other experiments.[66] Amalgamated aluminum foil and boiling water or 90% ethanol has given satisfactory results with a number of amino- and nitro-9-acridanones.[67] The same amalgamated aluminum foil with 4% methanol-KOH was used by Brockmann and Mux-feldt in the preparation of 2,6-dimethylacridine.[41] The reaction is accelerated by alkali and has the advantage of not requiring large quantities of mercury. The aluminum amalgam cannot be used in the presence of substrates such as 1-aminoacridine, which form nonreducible chelated complexes with the aluminum. Aluminum amalgam in aqueous ethanol reduced 1,2,3,4-tetra-fluoro-9-acridanone directly to the fluorinated acridine.[68]

Although 9-acridanone is not affected by aluminum powder and concentrated sulfuric acid,[69] some 10-methyl derivatives give the corresponding acridans and other products (Chapter VI) with zinc and acetic acid[70] or hydrochloric acid.[71] Stannous chloride and hydrochloric acid have been used for the reduction of nitro-9-acridanones to the corresponding aminoacridanones,[72, 58b] but did not effect further reduction. Wechter's attempts to reduce 9-acridanone with diborane met with little success.[73]

(b) Oxidation of Acridans to Acridines. Many acridans are so easily oxidized that the acridine, and not the acridan, is obtained on working up the 9-acridanone reduction product. The results of this sort of accidental aerial oxidation tend to be variable, and it is much better to ensure proper oxidation with a suitable reagent. A large number of oxidizing agents has been used for this purpose.

Graebe and Caro,[2] the discoverers of acridine, were the first to oxidize acridan to acridine by passing its vapor through a red hot tube (dehydrogenation?), by heating it to 100° in concentrated sulfuric acid (cf. Ref. 74) and by treatment with chromic acid. Of these procedures only the latter is satisfactory,[53] and also gives good results with many derivatives of acridan.[41, 56, 59, 75, 76] Chromic acid also oxidizes 10-methylacridans to the corresponding acridinium salts[77] and is an excellent reagent for the oxidation of acridans not possessing sensitive substituents (e.g., NH_2 groups).[78, 79]

Aeration of a hot, finely divided suspension of an acridan in aqueous alkali, such as that obtained after a sodium amalgam reduction, will cause oxidation to the acridine, provided that the original acridan does not contain electron-attracting groups. Substituents of this sort prevent oxidation by air, which is facilitated by amino and hydroxy groups. The reaction is best carried out with hot reagents; 4-amino-1-methylacridan, which is stable indefinitely at room temperature,[59] gave an excellent yield of the acridine on aeration in hot $2N$ sodium hydroxide.[80]

Ferric chloride was first used in 1896 for the oxidation of 3,6-bisdimethyl-aminoacridan to the corresponding acridine[81]; it has since been employed successfully for the oxidation of many amino[67, 82-84] and other acridans.[85-87] A common procedure is to add an excess of ferric chloride to an acidified suspension or solution of the acridan from an amalgam reduction. The reaction is rapid with aminoacridans, but in the presence of electron-attracting groups oxidation only takes place on boiling. Cold ferric chloride also oxidizes cold alcoholic solutions of 1-, 2-, 3-, and 4-methylacridans to the acridines.[88] Boiling an acid suspension of hydroxyacridans with a trace of ferric chloride, presumably to catalyze atmospheric oxidation, gave the acridines.[89] Ferric chloride also oxidized 2,7-diamino-10-methylacridan to the corresponding acridinium chloride.[90] A number of aminoacridans have been acetylated, oxidized with chromic[91] or nitrous acid,[92] and hydrolyzed to the corresponding aminoacridines; but this protection procedure is unnecessary and, in fact, a disadvantage, since the yield of acridine is invariably much less than that obtained by the direct oxidation of the acridan with ferric chloride. Ferric chloride is undoubtedly the best reagent for the oxidation of aminoacridans.

Silver nitrate in aqueous alcohol has been employed for the oxidation of acridan to acridine.[93, 94] The only examples of its subsequent use are the almost quantitative oxidation of 4-methoxyacridan[95] and of 9-phenylacridan.[96] Chlorine and bromine[97] are also useful in the oxidation of acridans. For instance, 10-phenylacridan gives the 10-phenylacridinium halides on treatment in alcoholic solution with iodine or in benzene solution with chlorine.[65] The yields are so good that quantitative determinations of the amount of acridan in acridine-acridan mixtures have been made by oxidation with excess iodine in a sodium acetate buffer, followed by titration of the remaining iodine.[54] A mixture of 10-methyl-9-acridanone and 10-methylacridinium chloride is obtained from 10-methylacridan and $N,2',4',6'$-tetrachlorobenzanilide in benzene.[98] Stoichiometric amounts of triphenylmethyl perchlorate in acetic acid dehydrogenated both acridan and 9,10-dimethylacridan to the corresponding acridinium perchlorates, each in 97% yield.[99]

Nitric acid has been used for the oxidation of 2,7-dimethylacridan to the acridine,[100, 101] and of 10-methyl-[102] and 2,10- and 3,10-dimethylacridan to the corresponding acridinium salts.[77] This reagent does not attack 1,10- and 4,10-dimethylacridan, which are, however, easily oxidized by chromic acid.[77]

Nitrous acid has been used as an oxidizing agent in a few instances,[91, 92, 97, 101] and some acridans have given the acridines on boiling with nitrobenzene.[103, 104]

Jackson and Waters examined a radical reaction using the dimethylacetonitrile radical to abstract hydrogen from acridan, which could then transfer a hydrogen atom to a suitable acceptor (polynitroaromatic).[105] This does not, however, constitute a good preparative method. Pratt and McGovern found manganese dioxide oxidation a very useful method for converting acridan to acridine under reflux conditions in apparatus equipped with a Bidwell-Sterling water trap.[106] The role of pyridoxal hydrochloride in certain biosynthetic schemes has been studied.[107] Pyridoxal was found to be a useful catalyst, although possibly not a required one, in the oxidation of 9-aminomethylacridan to acridine and formaldehyde.

B. The Bernthsen Reaction and Its Modifications

This reaction was one of the earliest used for the synthesis of acridines and consists in heating a mixture of an aromatic or aliphatic carboxylic acid with a diphenylamine and zinc chloride (1.5–3.0 moles) in the absence of a solvent to 200–270°. The yields are variable; formic acid gives particularly poor results.

The reaction was a development of the earliest synthesis of acridine (13), designed to provide a proof of structure, from N-formyldiphenylamine and zinc chloride.[6]

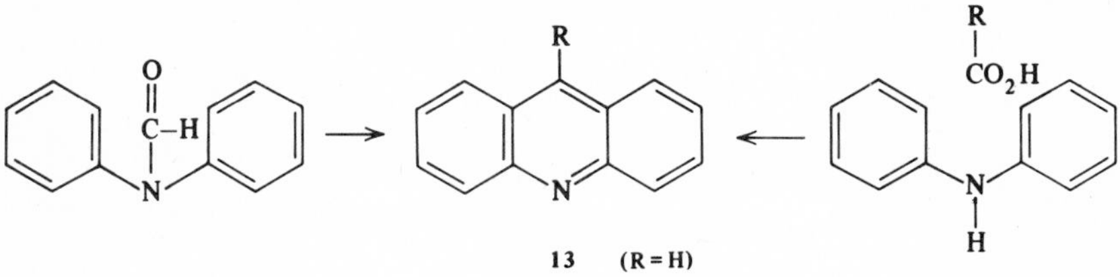

13   (R = H)

The very poor yield obtained was not improved when the formyl derivative was replaced by diphenylamine and formic, or oxalic, acid,[117] although acetic and benzoic acids gave reasonable yields of the 9-substituted acridines.[117] Subsequently, a large number of acridines (Table II) and benzacridines have been made by this method from a variety of carboxylic acids and diphenylamines. The temperature, time of reaction, and quantity of zinc chloride present are of importance in obtaining the optimum yields and vary with each product. Many hours are often allowed for the reaction.

Mixtures of acridines might be formed from 3-substituted diphenylamines, but there is little evidence on this point. Few reactions of this type appear to have been carried out (Table II). From 3-hydroxydiphenylamine and benzoic[118] and 4-hydroxybenzoic[119] acids, the only products reported were the 3-hydroxyacridines (14), cyclization taking place para to the hydroxyl.

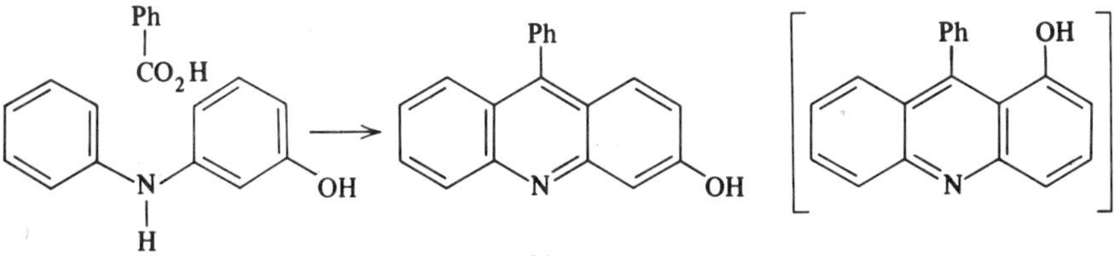

14

TABLE II. Mono- and disubstituted Acridines Prepared by Bernthsen's Reaction and Popp's Modification[a]

Acridine	mp(°C)	Yield (%)	Substituents in diphenylamine used	Acid or acid component	Ref.
Unsubstituted	111	Very low	—	Formic acid	117
		Very low	—	Oxalic acid	117
		7.5	—	Chloroform	139
2-Amino-9-phenyl-	204	8-10	4-Amino-	Benzoic acid	129
9-(4-Aminophenyl)-	270-272	24	—	4-Aminobenzoic acid[a]	121
2-Benzamido-9-phenyl-	246	?	4-Benzamido-	Benzoic acid	132
9-Benzyl-	173	50	—	Phenylacetic acid	140-142
	170-173	11	—	Phenylacetic acid[a]	121
9-(4-Bromophenyl)-	234	?	—	4-Bromobenzoic acid	143
	239-240	6	—	4-Bromobenzoic acid[a]	121
9-Butyl-	?	?	—	Valeric acid	144
9-isoButyl-	38-39	15	—	Isovaleric acid	140
9-tertButyl-	62	20	—	Trimethylacetic acid	140
9-(2-Carboxyphenyl)-	>315 darkens 293	40-50	—	Phthalic anhydride	61, 145
9-(2-Carboxyphenyl)-3-phenylamino-[b]	>300	?	3-Phenylamino-	Phthalic anhydride	146
2-Chloro-9-methyl-	124-125	71.5	2-Chloro-	Acetic acid	156
9-(4-Chlorophenyl)-	>270	29	—	4-Chlorobenzoic acid	147
2,9-Dimethyl-	122-123	?	4-Methyl-	Acetic acid	148
3,9-Dimethyl-	89.5-90	45.5	3-Methyl-	Acetic acid	156

Compound	M.P.	Substituent	Yield	Acid/Solvent	Ref.
4, 9-Dimethyl-	53.5-54	2-Methyl-	51.2	Acetic acid	156
9-(2,4-Dimethylphenyl)-	159	—	40	2,4-Dimethylbenzoic acid	149
9-(2,5-Dimethylphenyl)-	176	—	46	2,5-Dimethylbenzoic acid	149
9-Ethyl-	116	—	?	Propionic acid	150, 157
9-(1-Ethylpropyl)-	81	—	30	2-Ethylbutyric acid	140
9-Heptyl-	59	—	20	Octanoic acid	140
9-Heptadecyl-	69-70	—	?	Stearic acid	151
	66-68	—	trace	Stearic acid[a]	121
3-Hydroxy-9-(4-hydroxy-phenyl)-	350	3-Hydroxy-	13	4-Hydroxybenzoic acid	119
9-(2-Hydroxyphenyl)-	289-290	—	1.5	2-Hydroxybenzoic acid	152
9-(3-Hydroxyphenyl)-	366-367	—	82	3-Hydroxybenzoic acid	152
9-(4-Hydroxyphenyl)-	355-356 (dec)	—	27	4-Hydroxybenzoic acid	152
2-Hydroxy-9-phenyl-	275 (Sinters)	4-Hydroxy-	Low	Benzoic acid	129
3-Hydroxy-9-phenyl-[b]	135 (Labile red form)	3-Hydroxy-	27-34	Benzoic acid	118
	264 (Stable yellow form)				
2-Methyl-	110-112	4-Methyl-	Low	Formic acid	148
		4-Methyl-	Low	Chloroform	148
9-Methyl-	114		55	Acetic acid	117
	117-118		?	Acetic acid	153
			70	Acetic anyhdride	133

(Table Continued)

TABLE II. (Continued)

Acridine	mp(°C)	Yield (%)	Substituents in diphenylamine used	Acid or acid component	Ref.
	117-118	40	—	Acetic anhydride	38
		?	—	Acetic anhydride	135
		?	—	Acetic acid[c]	123
	118-118.5	?	—	Acetic acid	156
9-(2-Methylphenyl)-	212	50	—	2-Methylbenzoic acid	149
9-(3-Methylphenyl)-	165	63	—	3-Methylbenzoic acid	149
9-(4-Methylphenyl)-	189-190	40	—	4-Methylbenzoic acid	149
	188-189	18	—	4-Methylbenzoic acid[a]	121
9-[3-(p-Methylphenyl)-propyl]-	98	30	—	4-(p-Methylphenyl)-butyric	140
2-Methyl-9-phenyl-	135-136	37	4-Methyl-	Benzoic acid	148
9-Methyl-3-phenyl-amino-[b]	215-216	76	3-Phenylamino-	Acetic acid	146
-9-Pelargonic acid	207-208	31		Ethyl sebacyl chloride	154
9-Pentadecyl-	65	?	—	Palmitic acid	150
9-Phenyl-	184	48	—	Benzoic acid	117
		10	—	Benzonitrile	6
	184-185	48	—	Benzoic acid[a]	121
		20	—	Benzoic acid[a]	126
9-Phenyl-3-phenyl-amino-[b]	196-197	76	3-Phenylamino-	Benzoic acid	146
3-Phenylamino-[b]	175-176	"Good"	3-Phenylamino-	Formic acid	146

26

9-(1-Phenylethyl)-	138	30	—	2-Phenylpropionic acid	140
9-(2-Phenylethyl)-	102	30	—	3-Phenylpropionic acid	140
9-(3-Phenylpropyl)-	104	25	—	4-Phenylbutyric acid	140
	101-103	8	—	4-Phenylbutyric acid[a]	121
-9-Propionic acid	300	10	—	Succinic acid	151, 155
9-Propyl-	72-75	?	—	Butyric acid	150
9-isoPropyl-	oil	20	—	isoButyric acid	140
9-(3-Pyridyl)-	118	10	—	Nicotinic acid	127
9-Undecyl-	45	20	—	Lauric acid	140
-9-Valeric acid	265-269 (dec)	12	—	Ethyl adipyl chloride	154

[a] Polyphosphoric acid as catalyst and solvent.

[b] Alternative formulation is not rigorously excluded.

[c] 730°F, SiO_2-Al_2O_3 catalyst.

The constitution of the product in the latter case was proved by an independent synthesis.[119] Only one product was isolated from each of a number of similar condensations leading to benzacridines,[120] but the unlikely possibility that the products might have alternative structures was not considered in these cases.

Popp has suggested the use of polyphosphoric acid in place of zinc chloride in these cyclization reactions.[121] With a considerable reduction in reaction time and a lower reaction temperature, a number of the 9-substituted acridines have been prepared from both aliphatic and aromatic acids and diphenylamine. The reactants, in the proportion of 1 mole of acid to 2 moles of amine, were heated at 200° for 15 min in a large volume of polyphosphoric acid, which served as both catalyst and solvent medium. Although little effort was made to alter conditions to maximize the yield, the products formed include 9-(p-aminophenyl)acridine, which had not been prepared successfully by the zinc chloride method. These results are included for comparison in Table II. Birchall and Thorpe prepared 2,7-dibenzoyl-9-phenylacridine in 78% yield from benzoic acid, 4,4'-dibenzoyldiphenylamine, and polyphosphoric acid.[122] The patent literature details an additional example of a change in the acidic cyclization catalyst. When a vaporized mixture of diphenylamine (171 parts) in acetic acid (240 parts) was passed over the cracking-type SiO_2-Al_2O_3 catalyst at 730°F, 9-methylacridine was obtained.[123] Since no yields are given, the efficiency of this procedure for 9-alkylacridines cannot be compared with the other Bernthsen-like experiments.

The mechanism of the reaction has not been completely investigated. There is no evidence for the supposition of Hollins that it proceeds through N-acylation.[124] The replacement of diphenylamine by its N-acetyl derivative in the reaction did give 9-methylacridine, but the yield was not increased.[125] A small amount of 9-phenylacridine was among the products obtained when N-benzoyldiphenylamine was heated for 1 hr at 130° in polyphosphoric acid.[122] It appears more likely that acylation takes place first at position 2 of the diphenylamine. The resulting ketone, 15, can then cyclize under the influence of the acidic zinc chloride.

This same reaction sequence has been suggested in the 9-phenylacridine preparation in which polyphosphoric acid is the catalyst. This catalyst serves to produce the benzoyl carbonium ion that aroylates the diphenylamine at position 2, prior to cyclization. Additional benzoylation at the 4 and 4′ positions of diphenylamine has been observed in this reaction.[122, 126] In any event, a common, acid-catalyzed, ketone cyclization-dehydration scheme for all Bernthsen-like acridine preparations appears plausible.

When volatile acids are employed, sealed reaction vessels are essential. Reactants containing relatively sensitive groups cannot be used because of the high temperatures necessary; no acridines have been obtained from attempted condensations involving picolinic acid,[127] 3-nitrobenzoic acid,[128] 4-nitrobenzoic acid,[128, 129, 121] and 2,2′-dimethoxy-4,4′-dinitrodiphenylamine.[130] As previously mentioned, the polyphosphoric acid cyclization scheme, involving a somewhat lower temperature and shorter reaction time, has yielded the acridine from 4-aminobenzoic acid,[121] which was not obtainable with zinc chloride and the usual conditions.[129]

The product of the acetic acid-diphenylamine reaction is occasionally an addition compound of 9-methylacridine and diphenylamine, mp 92–94°, which has been mistaken for the pure acridine,[117, 131] mp 118°. 9-Methylacridine, on crystallization from ethanol, is also stated to form an alcoholate, mp 98°, which loses ethanol on standing in air.[131] When benzoic acid and 4-benzamidodiphenylamine undergo the reaction, 2-amino-, 2-hydroxy-, as well as the expected 2-benzamido-9-phenylacridine and a high molecular weight product, possibly a biacridine, are formed.[132] In the Popp modification of the synthesis for 9-phenylacridines,[121] the excess benzoic acid present is considered responsible for small amounts of uncyclized diaroyldiphenylamine, as well as some 2,7-dibenzoyl-9-phenylacridine.[122, 126]

An improvement in the synthesis is reported if the acids are replaced by their anhydrides[120, 133–136]; this modification is much used in the preparation of benzacridines,[137] which are dealt with in Chapter VII. It appears, however, that in at least one instance[133] the 9-methylacridine reported in 70% yield and having a melting point of 94–96° is, in reality, the previously mentioned yellow complex of diphenylamine and 9-methylacridine.[138] In experimental repetition of this work, a 40% yield of the 9-methylacridine was obtained.[38]

The acid has been replaced by the corresponding trichloro compound in a few cases; the best preparation of acridine (7.5%) from diphenylamine is that achieved by heating it with chloroform and zinc chloride or aluminum chloride.[117, 139] 9-Phenylacridine has been prepared in a similar way from benzotrichloride,[117] and by heating benzonitrile with diphenylamine hydrochloride[7] at 230–250°; lower temperatures in the last case gave *N,N*-diphenylbenzamidine. In Table II are recorded data for Bernthsen and modified Bernthsen preparations of mono- and disubstituted acridines.

C. Preparation of Acridines from *m*-Phenylenediamines and Formic (or Oxalic) Acid

This method of synthesis is undoubtedly the best available for the preparation of symmetrically substituted 3,6-diaminoacridines. These acridines can be obtained, free from isomers, in excellent yield in one reaction from accessible starting materials. The synthesis is not in general suitable for the preparation of asymmetrically substituted diaminoacridines, but 3-aminodiphenylamine (*N*-phenyl-*m*-phenylenediamine) will give 3,6-bisphenylaminoacridine or 3-aminoacridine, according to the conditions. These two types of reactions are now treated separately.

(1) *3,6-Diaminoacridines*

The preparation of the dyestuff, acridine yellow (**17**) by heating 2,4-diaminotoluene (**16**) with oxalic acid, glycerol and zinc chloride was first reported[158, 159] in 1890.

16   17

The constitution of this dyestuff was shown to be **17** by an unambiguous synthesis from 2,2',4,4'-tetraamino-5,5'-dimethyldiphenylmethane (**18**) by cyclization to the corresponding acridan, followed by oxidation with ferric chloride.[82]

18

Acridine orange, 3,6-bisdimethylaminoacridine (19), has also been prepared by these two routes.[81, 160]

The valuable antiseptic proflavine, 3,6-diaminoacridine (20), has been made by the oxalic or formic acid method for a considerable time,[161] and the reaction has been examined in detail.[162, 163]

The synthesis is best carried out, in the case of oxalic acid, by heating with m-phenylenediamine (2 moles) and zinc chloride (1.3 moles) to 155° in sufficient glycerol to enable efficient stirring. Higher temperatures reduce the yield, which is about 60%. The effect of variations in the procedure will now be considered.[162]

The glycerol can be replaced by ethylene glycol, 1,2-propylene glycol, or sorbitol, but not by 1,3-propylene glycol or phenylethyl alcohol. Therefore formic acid is an intermediate in the synthesis, as the decarboxylation of oxalic acid is facilitated only by 1,2-diols at the reaction temperature.

In the absence of zinc chloride, no acridine was formed. The N-(3-amino-phenyl)oxamic acid obtained is not an intermediary in the formation of 3,6-diaminoacridine. Increasing the amount of zinc chloride from zero increased the yield of 3,6-diaminoacridine to a maximum of 60%. No appreciable reaction took place when the zinc salt was replaced by aluminum chloride, hydrogen chloride, or stannous chloride, while the best yield of acridine in the presence of calcium chloride was 30%. 3-Aminoformanilide (21) has been isolated from a zinc chloride condensation not allowed to proceed to completion. One function of the zinc chloride is therefore to catalyze the decarboxylation of the oxalic acid; the anilide is later shown to be an intermediate in the synthesis. When formic acid replaces the oxalic acid in a zinc chloride-catalyzed reaction, acridine formation is negligible. This results from a lack of protons, which are essential for the reaction; oxalic acid is a very much stronger acid than formic acid. The addition of hydrogen chloride, in slight excess of the quantity required to convert the diamine into its monohydrochloride, restores the yield of acridine to 60%. Now the omission of the zinc chloride only reduces the yield to ca. 50%. The use of a 1,2-diol as a solvent is also now unnecessary.

In summary, the reaction is acid-catalyzed and proceeds via the formation of formic acid if this substance is not used as a starting material. If oxalic acid is used as a reactant, the main function of the glycerol and the zinc chloride is to effect decarboxylation to formic acid. Zinc chloride often has a slight, but beneficial, effect on the condensation.

Formic acid is now almost always used in preference to oxalic acid for several reasons. Zinc chloride can usually be omitted from the reaction without a great lowering of the yield, and this greatly simplifies the isolation of the acridine. Frothing, due largely to the evolution of carbon dioxide, is eliminated, and the formation of oxanilides is precluded.

The reaction is best carried out by heating the diamine (1 mole), hydrochloride acid (1–1.3 moles) and formic acid (1–2 moles), in sufficient glycerol to ensure homogeneity, to 155–175° for about an hour. In only one case, that of 2,6-diaminotoluene, is zinc chloride reported essential for acridine formation.[164] A list of acridines prepared by this method is given in Table III.

The mechanism of the reaction has been investigated in some detail.[162, 163] The first stage is undoubtedly the production of 3-aminoformanilide (21). This compound was isolated when the reaction was not allowed to go to completion; when treated with m-phenylenediamine under the usual reaction conditions, it gave a larger yield of 3,6-diaminoacridine than could be obtained from formic acid and the diamine. The other products obtained from incomplete reactions were bis-N-(3-aminophenyl)formamide, 3-aminophenyl-amino-2′,4′-diaminophenylmethanol (22), and 2,2′,4,4′-tetraaminobenz-hydryl ether (25), in 1, 14, and 6% yield, respectively.[163]

Bis-N-(3-aminophenyl)formamide is not a reaction intermediate, since with hydrochloric acid and glycerol at 155° it gave only 30% of 3,6-diamino-acridine, along with much polymer which is not formed in the normal cyclization.

The second substance, 22, is an intermediate, as it gave a 75% yield of 3,6-diaminoacridine with hot hydrochloric acid in glycerol. The "aldehyde ammonia" structure assigned to the compound is consistent both with its easy hydrolysis to m-phenylenediamine by moist air or $1N$ hydrochloric acid, and with its synthesis from m-phenylenediamine and formic acid in slowly distilling toluene.

The third substance, 25, is not a reaction intermediate and gave bis-3,6-diamino-9-acridanylether (27) with only 40% of 3,6-diaminoacridine (26) on treatment with glycerol and hydrochloric acid. Its formation can be explained if 2,2′,4,4′-tetraaminobenzhydrol (23) is an intermediate in the synthesis, although none could be isolated. Compound 23 has been made by reduction of the corresponding benzophenone; on treatment with one equivalent of hydrochloric acid, it formed a crimson anhydrosalt (cf. 24) that gave the benzhydrol ether, 25, on standing. However, on heating with two equivalents of hydrochloric acid in glycerol, it gave a quantitative yield of 3,6-diaminoacridine (26).

TABLE III. Acridines from Formic Acid and *m*-Phenylenediamines

Product	mp(°C)	Yield (%)	*m*-Diamine	Ref.
3-Amino-6-diethyl-amino-2-methyl-acridine (and acridine Yellow, 25%)	216-217	14	2,4-Diaminotoluene and 3-hydroxydi-ethylaniline	160
3-Amino-2,7-di-methyl-6-methyl-aminoacridine (and acridine yellow, 35%)	264	20	3-Dimethylamino-4-methylaniline and 5-formamido–2-methylaniline	160
3-Amino-6-hydroxy-acridine (with 3,6-dihydroxyacridine)	—	67 (Total)	3-Aminophenol	160
3,6-Diaminoacridine (proflavine)	288 (corr)	63	*m*-Phenylenediamine	163
3,6-Diaminoacridine-1,8-dicarboxylic acid	>365	65	3,5-Diaminobenzoic acid	160
3,6-Diamino-2,7-dichloroacridine	>300 (dec)	35	2,4-Diaminochloro-benzene, with zinc chloride	160
3,6-Diamino-2,7-diethoxyacridine	238	20	2,4-Diaminophene-tole	160
3,6-Diamino-2,7-dimethoxyacridine	244 (Sealed)	20	2,4-Diaminoanisole	160
3,6-Diamino-1,8-dimethylacridine	294-295 (Sealed)	20	3,5-Diaminotoluene	160
3,6-Diamino-4,5-dimethylacridine	170	57	2,6-Diaminotoluene	164
3,6-Diamino-2,7-dimethylacridine (acridine yellow)	325	75	2,4-Diaminotoluene, oxalic acid and zinc chloride	160
2,7-Dimethyl-3,6-bismethylamino-acridine	308-309 (Sealed)	20	4-Amino-2-dimethyl-aminotoluene	160
3,6-Bisdimethyl-aminoacridine (acridine orange)	181-182	60	3-Aminodimethyl-aniline	160
3,6-Bisphenylamino-acridine	>365	40	3-Aminodiphenyl-amine	160

The mechanism of the reaction can therefore be written:

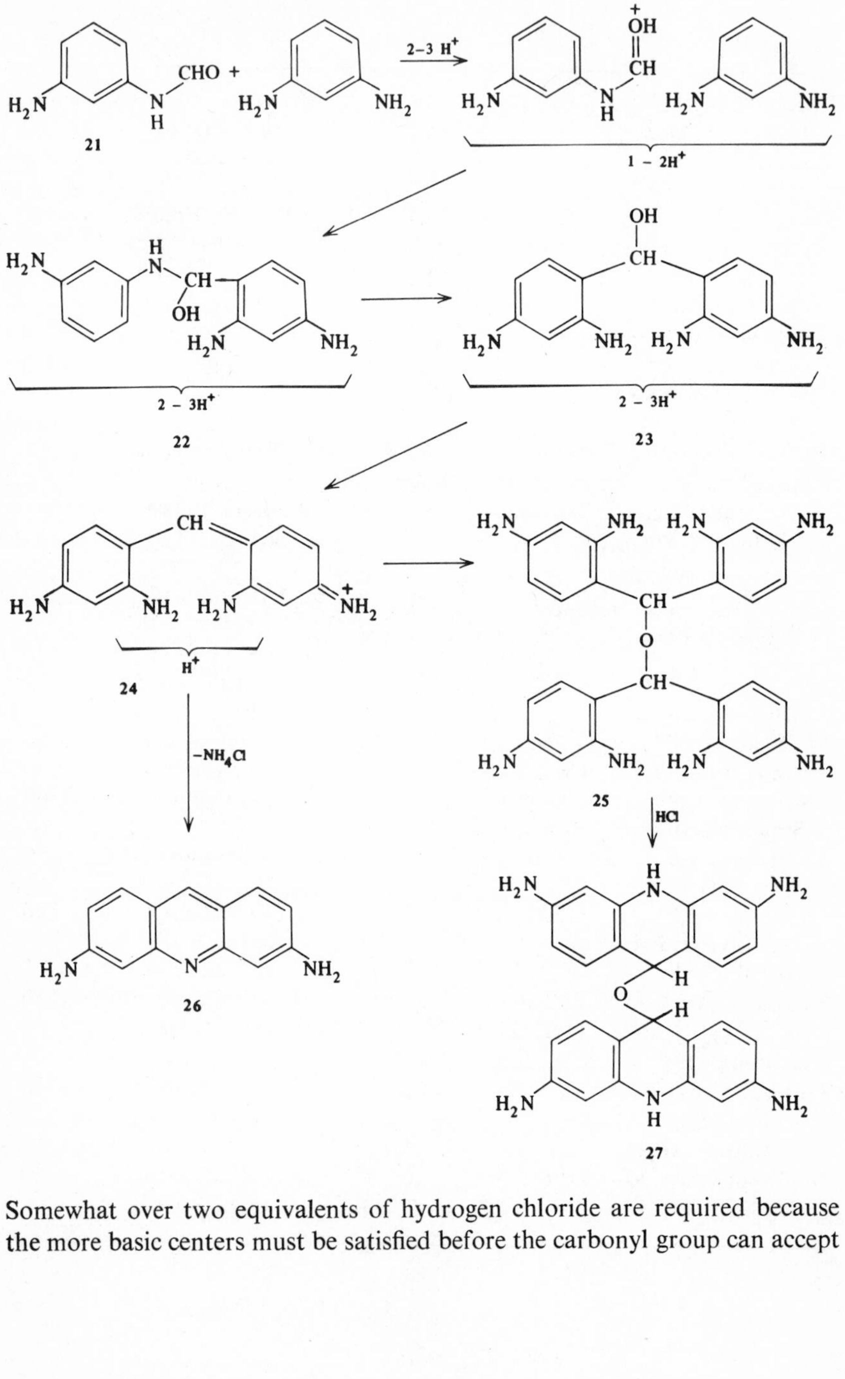

Somewhat over two equivalents of hydrogen chloride are required because the more basic centers must be satisfied before the carbonyl group can accept the proton needed to start the reaction. Increasing the amount of acid much above that required to bring about reaction should clearly hinder the condensation by deactivating the nucleus being attacked by the carbonium ion; this is in agreement with the experimental results. The postulate that the reacting species are **28** and **29** does not account for the reaction being acid-catalyzed at all,[163] as both reactants are merely deactivated by their respective charges.

28

29

If the suggestion were correct, appreciable reaction should (but apparently does not) take place after a little more than 1 mole of hydrochloric acid has been added to 2 moles of the diamine in the reaction mixture. Nothing is known about the details of the third stage in the scheme, which probably proceeds via hydrolysis to the aldehyde and *m*-diamine, followed by recombination, or possibly by dehydration and rearrangement. Two equivalents of acid are required for the cyclization of tetraaminobenzhydrol, since the first will form the mesomeric mono-ion with no appreciable charge on either of the amino groups at position 2 required to induce cyclization. It is of interest that each stage of the reaction, with the possible exception of the final cyclization, produces an intermediate more basic than its predecessors, as in the benzidine rearrangement.

A number of acridines (Table III) have been made by this reaction, which gives only one acridine with a single *m*-diamine. The reaction always takes place at the most activated positions (4 or 6) of the diamine. If such a position is not available, as in the case of 1,3-diamino-4,6-dimethylbenzene, no acridine is formed.[160] Other amines, apart from *m*-phenylenediamines and certain aminophenols, do not undergo the reaction. Mixtures are always formed with the aminophenols. The formic acid used in the condensation cannot be replaced by acetic or benzoic acids, but phthalic acid reacts with *m*-phenylene-diamine at 220° to give 3,6-diamino-9-(2-carboxyphenyl)acridine[165]; the synthesis of flaveosin (Section 2.F) is also of interest in this connection. When mixtures of *m*-diamines are used, the product is almost invariably a mixture of the two symmetrical cyclization products, the asymmetric product being formed, if at all, in small yield. An examination of a number of reactions of the latter type has shown that many patents erroneously claim the formation of such asymmetric acridines.[160] 3-Aminodimethylaniline undergoes the formic acid reaction normally to give 3,6-bisdimethylaminoacridine (acridine orange), but 5-amino-2-methyldimethylaniline is reported to give 2,7-dimethyl-3,6-bismethylaminoacridine, with the loss of two *N*-methyl groups.[160] Carbon monoxide reacts with *m*-phenylenediamine at about 3000 atm pressure to give a brown powder that may be a polymeric 3,6-diaminoacridine derivative.[166]

(2) *3-Aminoacridines*

When 3-aminodiphenylamine (*N*-phenyl-*m*-phenylenediamine) (**30**) was used in the formic acid synthesis,[160] some 3-aminoacridine (**33**) was formed, as well as the normal product, 3,6-bisphenylaminoacridine (**31**). Further investigation showed that reducing the amount of hydrogen chloride present from the usual 2.3 mole to 0.75 mole mole^{-1} of amine gave a maximum yield (60%) of the 3-aminoacridine, as well as 10% of the normal product.[80] Further reduction in the quantity of hydrochloric acid progressively reduced the yield of acridines, only traces of which were formed in its absence. When a large amount of hydrochloric acid is present, the primary amino group will be in the form of a cation and thus will not activate the nucleus. Under these conditions the attack of the formic acid, or formylated amine on the 4-position of the diphenylamine leading to **31**, is in agreement with the normal substitution reactions of diphenylamines. Reducing the amount of hydrochloric acid below one equivalent allows the ordinary activating effect of the primary amino group to take precedence. The intermediate in the production of 3-aminoacridine is almost certainly **32**. It has never been isolated from the reaction, but diphenylamine-2-aldehydes cyclize quantitatively to the corresponding acridines under milder acidic conditions than required by the present reaction. The reaction is usually carried out by heating the reactants in glycerol to 155–175° for 1 hr.

The formic acid cannot be replaced by acetic or benzoic acid. N-Phenyl-N-3-aminophenylformamide (**34**) is not an intermediate, as it gives lower yields of acridines than 3-aminodiphenylamine.

3-Nitrodiphenylamine and 4-aminodiphenylamine did not undergo the reaction, but 3-hydroxyacridine was formed from 3-hydroxydiphenylamine. 2,3′-Diaminodiphenylamine is reported to cyclize to a benziminazole.[80] 3-Amino-6-methyldiphenylamine, in which the methyl group blocks the reactive position, does not give an acridine.[167] Comparatively few acridines (Table IV) have been prepared by this method, which is very suitable for the synthesis of 3-aminoacridines.

TABLE IV.   Acridines from 3-Substituted Diphenylamines and Formic Acid [a]

Acridine	mp(°C)	Yield (%)	Amine
3-Aminoacridine	216	60	3-Aminodiphenylamine
9-Aminobenz[a]acridine	264-265 (Sealed)	45	N-(3-Aminophenyl-2-naphthylamine
10-Aminobenz[c]acridine	200 (Sealed)	40	N-(3-Aminophenyl)-1-naphthylamine
7-Amino-3-dimethylamino-acridine	242 (Sealed)	70	4-Amino-3′-dimethyl-aminodiphenylamine
3,6-Diaminoacridine	276 (Sealed)	55	3,3′-Diaminodiphenyl-amine
3,7-Diaminoacridine	350 (Sealed)	60	3,4′-Diaminodiphenyl-amine
3-Dimethylamino-acridine	183	60	3-Dimethylaminodi-phenylamine
3-Hydroxyacridine	285	35	3-Hydroxydiphenyl-amine

[a] From Ref. 80.

D. Acridines and Acridans from Aromatic Amines and Aldehydes and Similar Syntheses

Acridans, often oxidized deliberately or unintentionally to the acridines, have been obtained by the cyclization of 2,2'-diaminodiphenylmethanes. The required 2,2'-diaminodiphenylmethanes can be obtained by nitration and reduction of appropriate 4-substituted compounds. More frequently, the syntheses are begun with the aromatic amines and aldehydes and carried through the cyclization and oxidation procedures without the isolation and purification of the intermediate diphenylmethanes and acridan.

Typical of the relatively small number of acridines obtained by the first method is the one whose preparation was used to provide a proof of structure for acridine orange (**37**). [81, 168] The cyclization of 2,2'-diamino-4,4'-dimethyl-aminodiphenylmethane (**35**) was carried out by heating with hydrochloric acid to 140°, and the resulting acridan, **36**, oxidized with ferric chloride. Proflavine, 3,6-diaminoacridine, has been obtained from diphenylmethane and its 4,4'-diamino derivative by similar methods in satisfactory yield;[83, 169, 170] if the cyclization is carried out at a higher temperature, 170°, the mineral acid catalysis assists in the replacement of the amino by hydroxyl groups and 3,6-dihydroxyacridine is obtained (See Refs. 83 and 171; also earlier papers.)

Since a number of the compounds prepared by this method are of interest for their biological importance (Chapters XIV–XVIII) and as acridine dyes (Chapter VIII), only representative examples are include in Table V.

Aromatic amines can react with formaldehyde under neutral or alkaline conditions to form a variety of products (cf. Ref. 172) but if the amine is substituted in the *para* position and heated with formaldehyde and hydrochloric acid, the initial reaction gives a diphenylmethane such as **35**, which cyclizes at about 140° in the acid solution, to the acridan. Oxidation to the acridine is best effected with ferric chloride, although air oxidation during the workup has been noted. These condensations, cyclizations, and oxidations are the fundamental reactions for the other preparations included in Table V. The reaction worked well with p-toluidine,[100, 101] 2,4-diaminotoluene,[82,173] and 3-amino-N,N-dimethylaniline.[81] For the preparation of 9-methylacridines, the formaldehyde has been replaced by acetaldehyde.[174] Better yields have been obtained with benzaldehyde[100] and other aromatic amines, 2,4-diaminotoluene,[175] and p-anisidine.[87] In addition to the aldehydes tabulated, m- and p-nitrobenzaldehyde[100] and furfural[176] also undergo the reactions. Although the time required may be a matter of several hours, further improvements in the method have resulted from proper combinations of amine

TABLE V. Acridines from Diphenylmethanes or Aromatic Amines and Aldehydes

Acridan or (by oxidation) acridine	Diphenylmethane or aldehyde + amine	Cyclization acid and oxidation method	Ref.
3,6-Diamino-	2,2',4,4'-tetraamino-diphenylmethane	HCl; FeCl$_3$	83
3,6-Diamino-2,7-dimethyl-	Formaldehyde + 2,4-diaminotoluene	HCl; −	173
	2,2',4,4'-tetraamino-5,5'-dimethyldiphenylmethane	HCl; FeCl$_3$	82
3,6-Diamino-2,7-dimethyl-9-phenyl-	Benzaldehyde + 2,4-diaminotoluene · HCl	PPA (+NaOH)	175
3,6-Diamino-2,7,9-trimethyl-	Acetaldehyde + 2,4-diaminotoluene	PPA; dil H$_2$SO$_4$	175
3,6-bisDimethylamino-	2,2'-Diamino-4,4'-dimethylamino-diphenylmethane	HCl; FeCl$_3$	169
3,6-Dihydroxy-	2,2',4,4'-tetraamino-diphenylmethane	HCl at 170°	171
2,7-Dimethyl-	Formaldehyde + p-toluidine · HCl	FeCl$_3$ HNO$_3$; NaNO$_2$	100 101
2,7-Dimethyl-9-phenyl-	Benzaldehyde + p-toluidine	HCl; FeCl$_3$	100
2,4,5,7-Tetramethyl-9-phenyl-	Benzaldehyde + m-xylidine	HCl; FeCl$_3$	100

and amine hydrochloride,[87, 175] cyclizations with polyphosphoric acid,[175] and a change in oxidizing agent.[101] 2-Naphthylamine reacted very easily with formaldehyde in boiling ethanolic hydrochloric acid[177–179] to give good yields of the dibenzacridine (38), while 1-naphthylamine gave no acridine.

Half of the 2-naphthylamine in this condensation has been replaced by 2-naphthol,[97] and many similar reactions have been carried out. When the formaldehyde was replaced by benzaldehyde and more vigorous conditions employed, both 1- and 2-naphthylamines gave the appropriate benzacridines.[180, 181]

The syntheses just described led only to symmetrically substituted acridines, but it was found that when 2,4-diaminotoluene and methylenebis-2-naphthol and hydrochloric acid, or 2-naphthol and 2,2′, 4,4′-tetramino-5,5′-dimethyldiphenylmethane were heated,[182] the product was 9-amino-10-methylbenz[a]acridine (39). A similar synthesis of 3-amino-2,7-dimethylacridine from 2,4-diaminotoluene and *p*-toluidine, as well as other examples of this reaction are known.[171]

A large number of unsymmetrically substituted polyaminodiphenylmethanes have been prepared.[183] Recent extensions of the Ullmann-Fetvadjian acridine cyclizations[136, 137, 184, 185] with paraformaldehyde, 1- and 2-naphthol, and selected primary aromatic amines provide more examples of unsymmetrical substitution in the complex benzacridine products.

The aldehydes in the symmetrical condensations have been replaced, apparently without advantage, by the corresponding dihalogen compounds. For example, 3,4-dimethylaniline, on heating with methylene chloride or iodide, gave 2,3,6,7-tetramethylacridine (see Ref. 186, and earlier papers); ethylidene dichloride and benzal chloride have also been used.[180]

Among other modifications of this synthesis method is the preparation of 6-hydroxy-2-methyl-9-phenylacridine (**40**) from 2,4-dihydroxybenzhydrol and *p*-toluidine by the elimination of water.[187]

No xanthene was isolated from the reaction.[188]

2-Aminobenzyl alcohol reacted with 3-hydroxydiethylaniline and 2-naphthol at about 200° and gave the corresponding acridines,[189] doubtless through the formation of the acridans. A related reaction from the preparative work of Baezner[190a-c] and his colleagues is

E. Cyclization of Diphenylamine-2-aldehydes and Ketones

A small number of acridines has been synthesized by the cyclization of the corresponding diphenylamine-2-aldehydes and -ketones (Table VI). The preparation of the aldehydes has not been thoroughly investigated, presumably because of some serious difficulties in handling them and because there are other excellent synthetic pathways to most of these acridines via the corresponding acridanones (Section 2. A and Chapter III).

TABLE VI. Acridines from Diphenylamine Aldehydes and Ketones. (The intermediate aldehyde or ketone was not isolated in all cases.)

Acridine	Yield (%)	Prepared from cyclization of the product from	Ref.
Unsubstituted	?	2-Aminobenzaldehyde and iodobenzene	192
	80	Diphenylamine-2-carboxy-(ω-p-toluenesulfonyl) hydrazide	80
2-Amino-5,7-dinitro-9-phenyl-	85	2-Chloro-3,5-dinitrobenzo-phenone, 1,4-phenylene-diamine	209
2-Amino-7-nitro-9-phenyl-	90	2-Chloro-5-nitrobenzo-phenone and 1,4-phenyl-enediamine	197
6-Chloro-2-methoxy-9-methyl-	37	2-Amino-4-chloroaceto-phenone and 4-bromo-anisole	38
3-Chloro-9-methyl-	3	2,4-Dichloroaceto-phenone and aniline	200
2-Chloro-4-nitro-	?	2-Chlorobenzaldehyde and 4-chloro-2-nitroaniline	193
2-Chloro-4-nitro-9-phenyl-	?	2-Chlorobenzophenone and 4-chloro-2-nitroaniline	199
2,7-Dibenzoyl-9-phenyl-	17	N-Benzoyldiphenylamine	122
2,9-Dimethyl-	?	4-Bromotoluene and 2-aminoacetophenone	195
2,9-Dimethyl-7-nitro-	77	2-Chloro-5-nitroaceto-phenone and p-toluidine	198
2,6-Dimethyl-4-nitro-9-phenyl	?	2-Bromo-4-methylbenzo-phenone and 4-methyl-2-nitroaniline	199
2,4-Dinitro-	?	2-Chlorobenzaldehyde and 2,4-dinitroaniline	192
	30	2,4-Dinitrobromobenzene and 2-aminobenzalde-hyde	84
2,4-Dinitro-5-hydroxy-9-phenyl-	85	2-Chloro-3,5-dinitrobenzo-phenone and 2-amino-phenol	209
4,5-Dinitro-2-methoxy-9-phenyl-	?	2-Bromo-3-nitrobenzo-phenone and 4-methoxy-2-nitroaniline	130

(Table Continued)

TABLE VI. *(Continued)*

Acridine	Yield (%)	Prepared from cyclization of the product from	Ref.
2,4-Dinitro-6-methyl-9-phenyl-	?	2-Bromo-4-methylbenzophenone and 2,4-dinitroaniline	199
2,4-Dinitro-7-methyl-9-phenyl-	?	2-Chloro-5-methylbenzophenone and 2,4-dinitroaniline	199
2,4-Dinitro-9-phenyl-	?	2-Chlorobenzophenone and 2,4-dinitroaniline	199
	90	2-Chloro-3,5-dinitrobenzophenone and aniline	209
2-Ethoxy-9-phenyl-	?	4-Bromoethoxybenzene and 2-aminobenzophenone	195
2-Methoxy-	?	4-Bromoanisole and 2-aminobenzaldehyde	195
4-Methoxy-	?	2-Bromoanisole and 2-aminobenzaldehyde	195
5-Methoxy-2-nitro-9-phenyl-	60	2-Chloro-5-nitrobenzophenone and 2-aminoanisole	197
9-*p*-Methoxyphenyl-2-nitro-	70	2-Chloro-4'-methoxy-5-nitrobenzophenone and aniline	197
2-Methyl-	?	4-Bromotoluene and 2-aminobenzaldehyde	210
4-Methyl-	?	2-Bromotoluene and 2-aminobenzaldehyde	210
9-Methyl-	?	Bromobenzene and 2-aminoacetophenone	195
9-Methly-1,2-benz[*a*]-	80	1-Acetyl-2-hydroxynaphthalene and aniline	184
2-Methyl-4-nitro-	?	2-Chlorobenzaldehyde 4-methyl-2-nitroaniline	192
	?	4-Chloro-3-nitrotoluene and 2-aminobenzaldehyde	192
9-Methyl-2-nitro-	?	4-Bromonitrobenzene and 2-aminoacetophenone	195

(Table Continued)

TABLE VI. *(Continued)*

Acridine	Yield (%)	Prepared from cyclization of the product from	Ref.
	80	2-Chloro-5-nitroaceto-phenone and aniline	195, 198
2-Methyl-4-nitro-9-phenyl-	?	2-Chlorobenzophenone and 4-methyl-2-nitroaniline	199
2-Methyl-9-phenyl-	?	4-Bromotoluene and 2-aminobenzophenone	195
2-Nitro-	?	4-Nitrobromobenzene and 2-aminobenzal-dehyde	210
3-Nitro-	50?	Bromobenzene and 2-amino-4-nitrobenz-aldehyde	211
4-Nitro-	?	2-Chlorobenzaldehyde and 2-nitroaniline	192
	60	2-Nitrobromobenzene and 2-aminobenzal-dehyde	196, 210
2-Nitro-3-chloro-7-methoxy-9-methyl-	?	3-Nitro-4-chloroaceto-phenone and 4-methoxyaniline	200
2-Nitro-3-chloro-9-methyl-	?	3-Nitro-4-chloroaceto-phenone and aniline	200
2-Nitro-3,7-dichloro-9-methyl-	?	3-Nitro-4-chloroaceto-phenone and 4-chloroaniline	200
2-Nitro-9-phenyl-	80	2-Chloro-5-nitrobenzo-phenone and aniline	197
4-Nitro-9-phenyl-	?	2-Chlorobenzophenone and 2-nitroaniline	199
	?	2-Bromonitrobenzene and 2-aminobenzo-phenone	195
9-Phenyl-	?	Iodobenzene and 2-aminobenzophenone	199
	9	*N*-Benzoyldiphenylamine	122

(1) *Aldehyde Cyclizations*

Diphenylamine-2-aldehydes can be made in two ways by an extension of the Ullmann reaction:

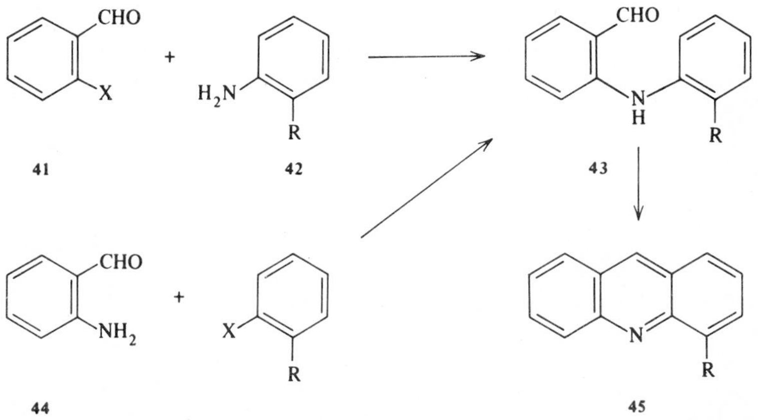

When preliminary experiments showed that 2-chlorobenzaldehyde (type **41** structure) reacted with 1-aminoanthraquinone to give a secondary amine (subsequently cyclized to a complex acridine),[191] its reaction with 2-nitroaniline (**42**, R = NO$_2$) in the presence of copper and sodium carbonate was examined.[192] The reaction product was not isolated, but, after treatment with hot concentrated sulfuric acid, it gave 4-nitroacridine (**45**, R = NO$_2$). In a similar way, acridines were prepared from 4-chloro-2-nitroaniline, 4-methyl-2-nitroaniline, and 2,4-dinitroaniline, the aldehyde being isolated in the last instance. The reaction only took place with amines that did not form anils. The following amines could not be converted to acridines by this method: aniline, and its 3- and 4-nitro; 2-, 3-, and 4-chloro; 2,4-dichloro; 2,4,6-trichloro; 2-chloro-4-nitro; 2-chloro-4,6-dinitro; and 2,6-dichloro-4-nitro derivatives; and 1-nitro-2-naphthylamine. 2-Chloro-5-nitrobenzaldehyde is also stated not to react with 2,4-dinitroaniline.[192, 193]

The alternative synthesis starting with 2-aminobenzaldehyde (**44**) is much more widely applicable, since the reaction can be carried out with a variety of aromatic halogen compounds. However, the 2-aminobenzaldehyde is not very stable, although it is easily made.[194]

The first example of this reaction was the synthesis of acridine.[192] Iodobenzene and 2-aminobenzaldehyde were refluxed in nitrobenzene or naphthalene with sodium carbonate and copper, the solvent removed by steam, and the ether soluble portion of the product cyclized without further purification. Sulfuric acid in acetic acid (15% by volume) has been used,[195] but better results appear to be obtained with sulfuric acid alone[80] at 100°. Other similar reactions are recorded in Table VI. It is not necessary to isolate the intermediate aldehyde in pure form before cyclization, although this has been done in the cases of 2'-nitrodiphenylamine-2-aldehyde (**43**, R = NO$_2$)[196] and 2'-nitro-4'-chlorodiphenylamine-2-aldehyde.[192]

The mechanism of the formation of the diphenylamine aldehyde is probably similar to that of the formation of diphenylamine-2-carboxylic acids by the Ullmann reaction (Chapter III). The data for the preparation of 2,4-dinitroacridine (Table VI) emphasize the importance of the influence of the nature and position of substituent groups on the ease of the diphenylamine formation.

Diphenylamine-2-aldehydes have also been obtained from the 2-carboxylic acids by the McFadyen-Stevens decomposition of their *p*-toluenesulfonhydrazides by dilute alkali. One acridine preparation in Table VI is of this type. However, the adverse effect of electron-attracting groups *ortho* and *para* to the carboxyl group very much limits this acid reduction to aldehyde followed by cyclization scheme. The cyclization of the diphenylamine-2-carboxylic acid to the chloroacridine, followed by treatment with *p*-toluenesulfonhydrazide and decomposition with alkali (as discussed in Section 2.A) seems a workable preparation method.

(2) *Ketone Cyclizations*

Diphenylamine-2-alkyl or aryl ketones have been prepared by methods similar to those for the aldehydes; on treatment with acids, they cyclize to the corresponding 9-substituted acridines.

One of the first examples of the reaction was the formation of 2-benzoyl-4-nitrodiphenylamine from 2-chloro-5-nitrobenzophenone and aniline by heating with potassium carbonate. The ketone was cyclized to 2-nitro-9-

phenylacridine by sulfuric acid in acetic acid.[197] This acridine has also been prepared directly,[197] in excellent yield, from the benzophenone and aniline by heating them with sodium acetate to 180°. However, better results are claimed in the two-stage preparation for certain acridines.[198]

If the formation of the diphenylamine involves a comparatively unreactive halogen atom, the use of a copper catalyst, as in the Ullmann reaction, becomes essential.[199] Both copper powder and copper (I) iodide have been used in some diphenylamine preparations.[200] Best results are obtained, of course, when the halogen is activated. In a somewhat related benzacridine synthesis, Buu-Hoï[184] and his collaborators used a trace of iodine in the reaction between aniline and 1-acetyl-2-hydroxynaphthalene. The wide applicability of the synthesis when ketones are the starting materials (Table VI), in contrast to the use of aldehydes, is doubtless because of the difficulty with which the ketones form anils.

The alternative synthesis starting from an o-aminoketone has only been carried out with 2-aminobenzophenone, and with 2-aminoacetophenone[201] and its 4-chloroderivative.[38] These amines have been condensed with a small number of aromatic halogen compounds in the presence of copper and a base, and the resulting ketone cyclized by acid in the usual way. 6-Chloro-2-methoxy-9-methylacridine is best prepared from the corresponding acetophenone according to Table VI but has also been obtained in poor yield from 5-chloro-4'-methoxydiphenylamine-2-carboxyl chloride by successive treatment with diazomethane, hydriodic acid, and acetic-sulfuric acid.[38]

The cyclizations of these aldehydes and ketones have been brought about most frequently by heating them in acetic acid-sulfuric acid mixtures. The reactions almost certainly proceed through the addition of a proton to the carbonyl group, followed by cyclization and dehydration, as in the case of the anthracenes.[202]

This mechanism is supported by the fact that the uv absorption maximum of diphenylamine-2-aldehyde moves toward the visible when acid is added to its solution. If the proton reacted with the nitrogen, the shift should be toward the ultraviolet.

The Fries rearrangement of N-benzoyldiphenylamine, catalyzed by polyphosphoric acid, yielded a mixture of diphenylamine, 9-phenylacridine and benzoyl derivatives of these compounds.[122] Although it appears not to be an especially good synthetic method for 9-phenylacridine or 2,7-dibenzoyl-9-phenylacridine, it is of interest here, since it is postulated to proceed via the formation of 2-benzoyldiphenylamine and seems to involve carbonium ion intermediates similar to those suggested for the other acridines considered above.

A similar cyclization,[203, 204] which proceeds in the presence of aqueous alkali, is that of the N-phenylisatins (47) to the corresponding acridine-9-carboxylic acids (48), presumably via:

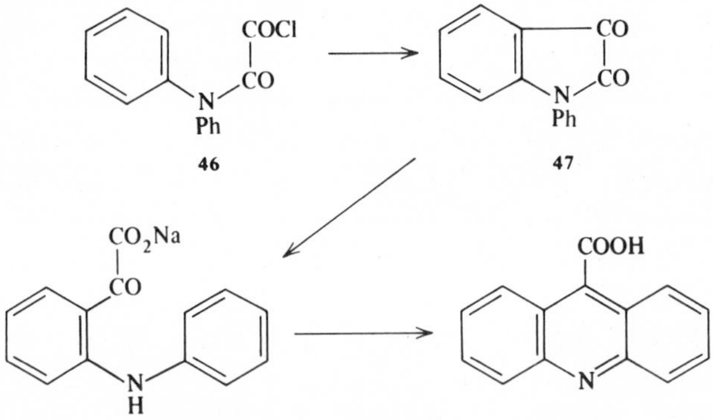

The reaction has, in recent years, been studied further, since the acridine-9-carboxylic acids are easily decarboxylated to the acridines. The early work of Stollé mentions the preparation of acridine-9-carboxylic acid.[203] This preparation and that of the 2-chloro derivative were documented by the work of Friedländer and Kunz.[204] One preparation of acridine-9-carboxylic acid[205] was carried out by heating sodium N-phenylisatinate to 250°. The N-phenylisatins have been prepared by the oxidation of the indoxyls with ferric chloride[204] and from the diphenylamines with oxalyl chloride.[56, 203] The syntheses of 4,5-dimethylacridine and 1,4,5,8-tetramethylacridine were begun with the refluxing of the appropriate ditolyl- and dixylyl-amines with oxalyl chloride and aluminium chloride in carbon disulfide solution.[56] Compounds of general structure 46 were formed first. Then the purified isatins were converted, in good yield, into the acridine-9-carboxylic acids (structure type 48) after 12 hr of reflux with 10% potassium hydroxide solution. Fifteen minutes of heating of the dry acids sufficed to effect decarboxylation.

The isatin ring has also been built onto 2-phenylaminonaphthalene[206] by reaction with ethyl mesoxalate at 180°. The product, ethyl 2,3-dihydro-3-hydroxy-2-keto-1-phenylbenz[e]indole-3-carboxylate, on hydrolysis in the absence of air, followed by alkaline air oxidation, gave **49**. The isatin, **49**, was converted with aqueous alkali[206] into benz[a]acridine-12-carboxylic acid (**50**).

This acid, **50**, has also been prepared from benzocoumarindione (**51**) by its reaction with aniline to form **52** which yielded the acridine upon treatment with acetic acid.[207] The direct formation of acid **50** when the coumarindione, **51**, is boiled with aniline in acetic acid has also been carried out. These results do not invalidate the possibility of **52** or even **49** as intermediates in the reaction.

An interesting synthesis of 1,3-dinitroacridine has been carried out from the reaction of 2,4,6-trinitrotoluene with 4-nitrosodimethylaniline.[208] The first-formed anil, **53** (R = $-C_6H_4N(Me)_2$), when boiled with aniline in ethanol, loses nitrous acid to yield **54**. This last compound on heating with acetic acid cyclizes to the acridine **55**, with the loss of p-aminodimethylaniline.

F. Other Methods of Preparing Acridines

It is difficult to categorize completely all acridine syntheses in terms of starting materials, reaction conditions, or products formed. All of those that can be described as related to phosphorus oxychloride cyclizations of diphenylamine-2-carboxylic acids, which were prepared in the Ullmann reaction, are considered in Chapter III. A very large number of 9-acridanones and chloroacridines have been prepared by this method. Other reactions that have been used successfully to form products at the acridine oxidation level will be dealt with here. Many of these are not recommended as synthetic methods: they simply yielded some acridine as a product. The final section comprises a miscellaneous group.

(1) *Dehydrogenations Giving Acridines*

In comparison with the oxidation of 9,10-dihydroacridines or acridans (Section 2.A), only a small number of more highly reduced acridines have been dehydrogenated successfully, often in rather poor yields. Palladium on charcoal in boiling diphenyl ether dehydrogenated the 5-, 6-, and 7-methoxy-1,2,3,4-tetrahydroacridines to the corresponding acridines in 15, 65, and 80% yields, respectively.[47] The dehydrogenation of 5-phenyl-1,2,3,4-tetrahydroacridine has been carried out, using palladium on charcoal.[212] Masamune and Homma obtained acridine as well as partially dehydrogenated products in a series of studies in which dihydro-, tetrahydro- and two octahydro-acridines were placed in a hydrogen stream in the presence of 30% palladium on charcoal.[213, 214] In the presence of 10% palladium on charcoal at 210–220° 1,4-dimethyl-5,6-dihydrobenz[c]acridine was dehydrogenated in 25% yield.[215]

Dehydrogenations using lead oxide, zinc, selenium, and chloranil, although stated to give bad results, have been used in a number of cases.[47] Acridine has been obtained from 1,2,3,4-tetrahydroacridine and its 9-carboxylic acid by distillation over hot litharge in a carbon dioxide stream,[216] and from 1,2,3,4-tetrahydro-9-acridanone[217] and some other reduced acridines[218, 219] on distillation over zinc. Selenium dioxide in ethyl acetate gives mainly the acridines from 1,2,3,4-tetrahydroacridine and its 2-methyl derivative.[220] Selenium itself catalyzes the formation of 3-methylacridine (57) from both 6-methyl-1,2,3,4-tetrahydroacridine (56) and 3-methyl-1,2,3,4-tetrahydroacridine (58).[221]

56   57   58

More recent applications of selenium dehydrogenation of perhydroacri- dines[222] and tetrahydroacridine[223] have been reported, but there is no indica- tion that yields of acridines are improved. Some reduced benzacridines have also been dehydrogenated by selenium.[221] Yields of 5 to 17% of acridine have been obtained when tetra- and octa-hydroacridines were heated in boiling xylene with 5–10% excess of chloranil.[224] Redmore[225] used chloranil in the dehydrogenation of diethyl acridan-9-phosphonate.

Decarboxylation and distillation with lead dioxide[226] has transformed appropriately substituted 1,2,3,4-tetrahydroacridine-9-carboxylic acids into 2- and 3-methylacridine; this reagent also aromatized some reduced benz- acridines.[227] Chloranil in boiling xylene or cumene, and even bromine water, dehydrogenated certain benzacridines when hot lead oxide caused decomposi- tion.[120] The distillation of 3,9-dimethyl-1,2,3,4-tetrahydroacridine with zinc under hydrogen effected dehydrogenation to 3,9-dimethylacridine[228]; 4- methylene-1,2,3,4-tetrahydroacridine similarly became 4-methylacridine.[229] Other examples of dehydrogenations giving acridines may be found in Chapter V.

(2) *Cyclodehydrogenations Giving Acridines*

A series of reactions of this type has been carried out under vigorous con- ditions. On the whole, the results are poor. A little acridine was obtained by distillation of 2-aminodiphenylamine (**59**) through a red hot tube.[230] 9- Phenylacridine (**61**) has been prepared in poor yield by heating 2-amino- triphenylcarbinol (**60**) (from methyl anthranilate and phenyl magnesium bromide), alone or with picric acid.[231]

59   60

61, R = H
62, R = NH$_2$

In 1957, Petyunin, Panferova, and Konshin[232a-e]; made extensive studies of this method and applied it to the preparation of a large number of halo- phenylacridines. The best yields were obtained when the halogens were present in the methyl anthranilate ring, rather than in the Grignard reagent. The tri-

TABLE VII.   Preparation of 9-Phenylacridines by the
              Cyclodehydrogenation of Triphenylcarbinols

9-Phenylacridine	mp(° C)	Yield (%)	Triphenylcarbinol	Ref.
Unsubstituted	182-185.5	77	2-Acetamido-	232b
		87-97	2-Amino-	232b
2-Bromo-	145-146	53	2-Amino-5-bromo-	232c
2-Chloro-	150-151.5	72	2-Amino-5-chloro-	232c
3-Chloro-	151.5-152	53.4	2-Amino-4-chloro-	232c
2-Chloro-4-bromo-	228-229	79	2-Amino-3-bromo-5-chloro-	232d
2-Bromo-4-chloro-	214-215	55-66	2-Amino-3-chloro-5-bromo-	232d
2,4-Dibromo-	232-233	63-66	2-Amino-3,5-di-bromo-	232d
2,4-Dichloro-	213-213.5	53-61	2-Amino-3,5-dichloro-	232d

phenylcarbinol cyclizations were somewhat more successful with the 2-amino compounds than with their benzoyl or acetyl derivatives. Reasonable yields of products were obtained after refluxing the triphenylcarbinol in nitrobenzene for 15 to 30 min. Representative structures are summarized in Table VII. A number of 9-*p*-chlorophenylacridines have been prepared by the same method in yields of 31–63%.[232e]

An early synthesis of chrysaniline (**62**) was obtained from 2,4,4'-triamino-triphenylmethane by oxidation with arsenic pentoxide.[233] In the course of establishing the structures of acridine derivatives, recent workers obtained 3-amino-9-phenylacridine upon heating (160°) 2,4'-diaminotriphenylmethane with the same syrupy arsenic acid.[234] 2-Methyldiphenylamine, passed through a hot tube, was partially converted into acridine,[235] but oxidation with litharge with or without cupric oxide was found to be a much more satisfactory procedure with up to 25% yield.[236] 2-*o*-Tolyl-1-naphthylamine on heating with sulfur at 220° gave the corresponding benzacridine (25%), but better results (40–45%) were obtained with litharge.[237] Rieche and Moeller[238] have prepared methylacridines by passing the vapors of aromatic amines, diluted with nitrogen, over aluminum oxide at 440°: *o*-toluidine → 10% yield of 4,5-dimethylacridine; *p*-toluidine → 10% of 2,7-dimethylacridine; *p*-xylidine → 1,4,5,8-tetramethylacridine (**63**).

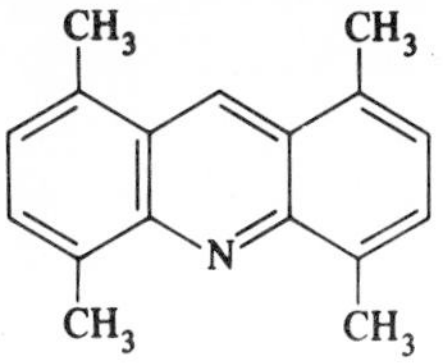

63 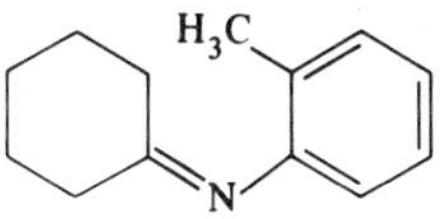　　　　　　　　64 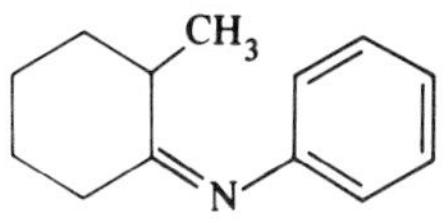　　　　　　　　65

Dehydrogenation of **64** and **65** at 550° over a copper-chromium catalyst gave 15% and 29% of acridine, and 78% and 92% of hydrogen in the evolved gas, respectively.[239]

(3) Acridines from 2-Aminobenzaldehyde and Phenols

Salicylaldehyde on heating with aniline to 260° for some time gave only 0.5% of acridine,[240] while slightly better results were claimed when the heating process was carried out in the presence of phosphorus pentoxide.[241] Phloroglucinol and 2-aminobenzaldehyde, in an almost complementary reaction, were found to give 1,3-dihydroxyacridine in the presence of sodium hydroxide.[242] This reaction was fully investigated later,[243] the reactants being allowed to stand in aqueous solution at room temperature for 5 days at definite pH values, 4-13. A maximum yield of 90% was observed at pH 8. With excess 2-aminobenzaldehyde, phloroglucinol gives condensed acridines (Chapter VII). The reaction between 1,2,3,5-tetrahydroxybenzene (**66**) and 2-aminobenzaldehyde gave a minute yield of the acridine at pH 8, but good yields were obtained when the reactants were heated in the presence of one equivalent of alkali. The product was 1,2,3-trihydroxyacridine (**67**). None

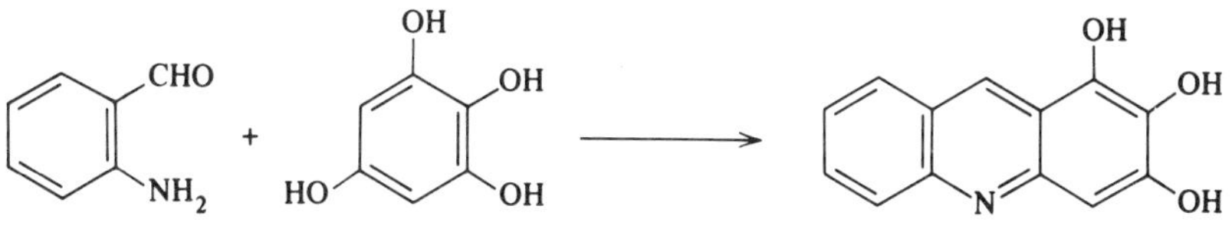

66　　　　　　　　67

of the isomeric 1,2,4-trihydroxyacridine appeared to be formed.[243] 2-Aminobenzaldehyde gave no acridines with phenol, salicyclic acid, resorcinol, catechol, β-resorcyclic acid, quinol, pyrogallol, or 1- or 2-naphthol. It is clear that a considerable proportion of the phenolic reactant must be in the tautomeric keto form, as in the similar reaction with isatin, if cyclization is to take place.

Refluxing isatin in alkaline ethanolic solution with phloroglucinol gave 1,3-dihydroxyacridine-9-carboxylic acid (**68**) in excellent yield.[244]

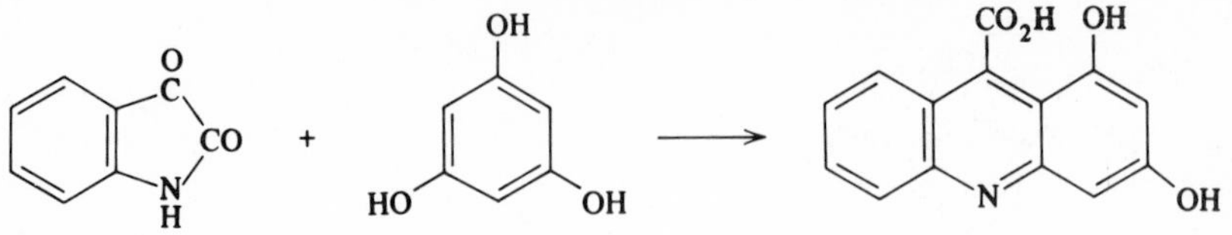

68

Resorcinol gave very little acridine under these conditions, but in aqueous alkali at 120° a 76% yield of 3-hydroxyacridine-9-carboxylic acid was obtained. Phenol, catechol, and hydroquinone did not appear to undergo the reaction.

When the terpene ketone, oxocineole, was condensed with 2-aminobenzaldehyde in the presence of NaOH, and the product dissolved in concentrated H_2SO_4 and aged a day before it was submerged in ice water, 1-isopropyl-4-methylacridine (**69**) was isolated as its bisulfate salt.[245]

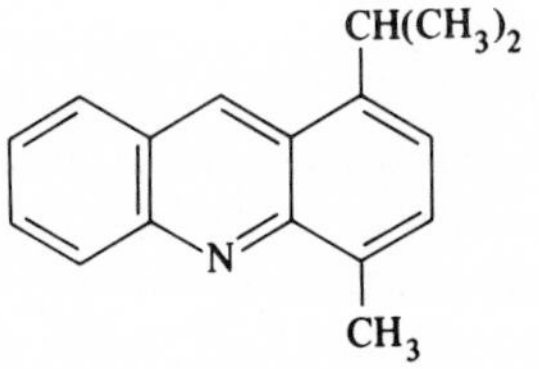

69

(4) Other Acridine Syntheses

The following are reactions that yield acridines but, as stated earlier, are not necessarily useful synthetic methods. There is only one account of the conversion of a 2,2'-azodiphenylmethane to the corresponding acridine from its treatment with boiling stannous chloride and hydrochloric acid.[169] The reaction may proceed by reduction to the hydrazo compound, **70**, and cyclization in a manner analogous to that of the Fisher Indole synthesis.

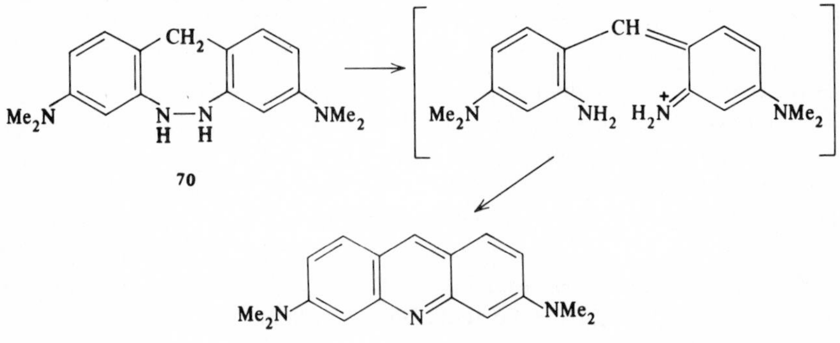

70

Poor yields of 9-methylacridine have been obtained when 2,3-bisphenyl-aminobutane sulfate was heated with aniline[246]; 2,7,9-trimethylacridine was obtained as a by-product from the reaction of 2,3-dichlorobutane with *p*-toluidine.[174] 2,4,5,7-Tetramethylacridine was formed[247] in poor yield from *o*- and *p*-toluidine hydrochlorides in methanol at 270°. Hot concentrated hydrochloric acid converted the iminostilbene, **71**, into 9-methylacridine by an appropriate sequence of ionic, rearrangement steps.[248]

71

The distillation of benzylaniline (**72**) through a red hot tube (700°) is stated to give acridine[249]; no yields are given in this and related syntheses. Pyridine and aluminum oxide at about 400° are also alleged to give some acridine.[250] The distillation of benzal-2-aminophenol (**73**) over zinc is also said to give a little acridine.[251] Benzal-1- and -2-naphthylamines gave benz[*c*]acridine and benz[*a*]acridine, respectively, on distillation through a red hot iron tube filled with pumice.[252, 253]

72   73

The obsolete dyestuff, flaveosin (**75**) was made in the reaction of phthalic anhydride with 3-diethylaminoacetanilide in the presence of acetic anhydride. The first product, **74**, on boiling with 20% HCl, was converted into the dyestuff.[254]

74   75

Heating fluorescein (76) with aqueous ammonia to 180–200° gave 3,6-diamino-9-(2-carboxyphenyl)acridine (77) in the same way that

76   77

pyrones give pyridines.[255] Acridine was obtained as a by-product in the preparation of 1-methylcarbazole from 78 with hot calcium oxide.[256] The zinc dust distillation of 2,2′diaminobenzophenone (79) gave some acridine, doubtless through the formation of 9-acridanone as the latter compound is formed from the diamine and zinc chloride.[257]

78   79

2-Nitrobenzophenone, on heating with ferrous oxalate, appears to give some acridine,[258] but attempts to cyclize 2-aminobenzophenone have not been very successful.[259]

Another obsolete dyestuff, rhenonine (81), can be prepared in very poor yield from 4,4′-bisdimethylaminobenzophenone, m-phenylenediamine, and zinc chloride [254] at 200°, and the fully methylated dyestuff in a similar way way from 3-aminodimethylaniline.[260] The anil 80 is considered to be an intermediate in the reaction.[254]

80   81

As aniline and *p*-phenylenediamine give anils that do not cyclize, it appears that activation of the *ortho* position to the anil nitrogen is necessary for acridine formation. Bisacetoxyphenylisatin (**82**) on oxidation with chromic acid gave **83**, which on treatment with concentrated sulfuric acid lost carbon dioxide with the formation[119] of a mixture of 3-hydroxy-9-(4-hydroxyphenyl)-acridine (**84**) and 2-aminophenylbis(4-hydroxyphenyl)methane (**85**).

Two anthranils have been pyrolyzed to 9-acridanones which, with POCl$_3$, yield haloacridines. Compound **86**[261] yielded 2,6,9-trichloroacridine (**87**),

and Kwok and Pranc[262] obtained a mixture of **89** and **90** from 3-(*p*-methoxyphenyl)-5-chloro-anthranil (**88**).

Acridine was among the products obtained when 9-acridanthione was desulfurized during a 13-hr reflux period in dimethylformamide and ethanol in the presence of activated Raney cobalt catalyst.[263]

Evidence for the formation of a dehydroquinoline intermediate was furnished by the formation of a small amount of acridine when a furan solution of 2-chloro-3-bromoquinoline was shaken with lithium amalgam for 5 days at room temperature.[264]

3. General Properties of Acridine

Acridine itself, although the parent compound of many useful and widely differing substances, is of comparatively little importance. It may be of interest that the irritating physiological effects that inspired its name, a lachrymatory effect and skin irritation, seem to be minimized when methyl substituents are in the ring positions 4 and 5, adjacent to nitrogen.[56]

Acridine is a very pale yellow solid that sublimes at ca. 100°, boils[2] at 345–346°/760 mm, and crystallizes well from ethanol or benzene. It dissolves in approximately 6 parts by weight of ethanol, 5 parts of benzene, and 16 parts of ether. It is sparingly soluble in water and is volatile in steam.[2] It is slightly soluble in both liquid ammonia and liquid sulfur dioxide.[265]

Five polymorphic forms of acridine are known. They differ considerably in stability and crystal habit.[266] Some of these have been examined in detail:[36,267] I is a hydrated form,[268] which loses water upon heating, yielding II. Among the anhydrous forms, the one designated as II has a density (determined by flotation) of 1.28 and is formed as prisms or elongated needles, stable in the range 20–105°, mp 109°. Form III (density, 1.27), recrystallized from aqueous alcohol at 20° as hexagonal flat plates, is converted into II at about 45°, while IV, density 1.20, sublimes slowly even at 20° and is transformed entirely into II at 70°. The fifth polymorph, the β-acridine of the notation of Herbstein and Schmidt,[266] accounts for the 110–111° melting acridine obtained from cooling benzene or alcohol solutions of analytically pure material.[268] Acridine is triboluminescent.[269]

Acridine is classed as a "π-electron-deficient" heterocycle,[270] and, as such, would be expected to show poor fluorescence properties. It is nonfluorescent in the crystalline state or when dissolved in benzene or ethyl acetate.[270] However, its blue fluorescence in dilute aqueous or aqueous alcohol solution is one of its most characteristic properties. The irradiation[271] of a 10^{-3} M solution of acridine in benzene with nitrogen laser pulses ($\lambda = 337.1$ nm) gave only one excited species with an absorption peak at 440 nm and a lifetime of approximately 100 μsec. The fluorescence was of extremely short duration (0.5 μsec), so that detection of the short-lived species was not possible. This 440-nm absorption, probably due to the π-π^* triplet, may however suggest a relationship to the fluorescence spectrum having a main peak 400–480 nm with principal summit at 440 nm,[272] even though the exact relationship and significance has not been established.[273] The violet-blue fluorescence of the weak base acridine, as well as the range of fluorescence colors (blue → orange) observed for aminoacridines and other derivatives is distinct from a green fluorescence color of the acridinium ion. Further, Förster and Weller have said that the electronic excitation of acridine makes it a much stronger base (Refs. 274a; b, as reported in Ref. 270), so that its enhanced hydrogen-bonding capacity in the excited state is responsible for still another green fluorescence at approximately pH 10.35. Despite the difficulties that surround the interpretation of various fluorescences, it has been possible to use a number of acridines as fluorescent indicators. (See Refs. 275a-c, and earlier papers.) Certain acridine derivatives have been identified by means of their characteristic fluorescences.[276] Quantitative determinations of the fluorescence of specific acridine derivatives have been made the basis of methods used for estimating streptomycin[277] and penicillin.[278]

Variation in the electrical conductivity of acridine has been studied in the 25–80° temperature range.[279] From the conductivity plot, an activation energy value of 0.337 eV was obtained. The resistivity at 25° is about 2.5 × 10^{13} ohm-cm. Studies on the electrical conductivity, σ, determined in the temperature range, 50–250°, gave a σ(liquid)/σ(solid) of 83 for acridine at its melting point.[280]

Potentiometric titration (see also Chapter II, p. 110) indicates that acridine, pK_a 5.60[281, 108] is a stronger base than quinoline, pK_a 4.94, or pyridine, 5.23[282, 283] in water; this has been discussed from a theoretical standpoint[284,285] but an inaccurate value (4.3)[281] was taken for the pK_a of acridine, in the latter instance. More recent measurements, potentiometric and photometric (in water-ethanol), are in agreement with the thermodynamic constant, pK_a = 5.60 in water at 20° for the acridinium cation.[286] The studies of Reynaud have shown that the pK value for acridine is more sensitive to solvent effects than that of quinoline or pyridine.[287]

Another approach to "basicity" evaluation for π-deficient N-heteroaromatics, such as acridine, has been found in the ir (meaning infrared) study of intermolecular hydrogen-bonded complexes with a variety of donor molecules.[288, 289] A $\Delta\nu_{OH}$ of 520 cm^{-1} is recorded for the phenol-acridine complex in carbon tetrachloride solution.

Acidity constants for excited states of acridine have been considered, both in theoretical studies[290, 291] and in experimental investigation.[292] When the triplet state was populated in a flash photolysis experiment, pK_T = 5.6, very close to pK_G (ground state) was obtained. The pK_S (first excited singlet) value is 10.6, in good agreement with the quantum chemical calculations.[290] This greatly enhanced basicity in the singlet state[292] is attributed to a much increased importance of structure **91**. Structure **92** is considered the major contributor to the triplet state.

91   **92**

Many of the spectra of acridines are to be considered in later chapters, but it may be noted here that the best electron spin resonance spectrum[293] for the mononegative ion of acridine was obtained by controlled electrolysis of a $10^{-3}M$ acridine solution in pyridine at 10°. The investigators of the magnetic

TABLE VIII. Molecular Compounds and Complexes of Acridine

Ratio of Acridine to 2nd component	Second component	C. T. Band (λ_{max} in nm)	mp(°C)	Ref.
1:1 ?	Diphenylamine		84-86	153
1:1	2,4,7-Trinitrofluorenone	441 (Shoulder)	163-165	303, 305
1:1	Chloranil	497		305
1:1	2,4-Dinitrophenol		162-163	306
1:1	4,4'-Dichlorodibenzene-sulfonamide		195-196	307
1:1	Tetrachlorophthalic anhydride	404		308
1:1	2-Nitroindan-1,3-dione		183	309
?	4-Nitrosodimethylaniline		52.1	310
1:2	Phenol		87	311
3:2	Phenol		101	311
1:1	Catechol		144.5	311
2:1	Resorcinol		179.5	311
2:1	Hydroquinone		209.5	311
2:3	2-Naphthol		135	311
1:1	Phenanthrene		66.2-66.5	312
1:1	9,9-Dimethylacridan		135-136	37
1:1	Quinone		212-214 (dec)	313
2:1	N_2O_4		25° (dec)	297
1:1	Tetracyanoethylene	460		295
1:1	Tetrakis(dimethylamino)-ethylene	462		296
1:1	IBr		168-171	299
1:1	ICl_3		212-213	299
1:1	CuI		335 (dec)	298
2:1	2,5-Dihydroperoxy-2,5-dimethylhexane		104-105	300
2:1	$CoCl_2$		264	314
2:1	$CoBr_2$		272	314
2:1	CoI_2		280	314
1:1	CrO_3		194-195 (dec)	315
1:1	Diisobutyl-9,10-dihydro-acridyl aluminum		192 (dec)	316
3:1	$H_3[Cr(NCS)_6]$		280 (dec)	317
2:1	$Pd(SCN)_2$		218-220	318

(Table Continued)

TABLE VIII. (Continued)

Ratio of Acridine to 2nd component	Second component	C. T. Band (λ_{max} in nm)	mp(°C)	Ref.
4:1	$H_2V_{10}O_{28}$		180-260 dec range	319
1:1	AlI_3		?	320
1:1	$VOCl_3$		?	321
2:1	H_2OsBr_6		?	322

rotatory dispersion of acridine found a Cotton effect of opposite sign to that of aliphatic ketones and approximately 400 times as intense.[294]

A large number of combinations of acridine with both inorganic and organic substances have been described as molecular compounds or complexes. Most of those that are properly identified as acridinium salts will be considered in Chapter V. Among the charge transfer complexes are many for which only spectral evidence has been recorded and a 1:1 molar relationship is assumed.[295] Although acridine is usually the donor molecule, its acceptor properties have been noted in its interaction with tetrakis(dimethylamino)-ethylene.[296] When acridine is added to an ethereal solution of dinitrogen tetroxide at −75°, a yellow solid complex is formed which decomposes fairly rapidly at room temperature.[297] In contrast with this unstable, but isolable, complex are the brick-red cuprous iodide complex,[298] the interhalogen-amine compounds,[299] and the 2:1 complex of acridine and 2,5-dihydroperoxy-2,5-dimethylhexane.[300] In some cases a single definite composition for the molecular compound has not readily been established.[153] The rhenium complex[301] and the calcium ammonia complex[302] fall into this category. For many others the molar ratio has been clearly determined.[303] X-ray crystal analysis has been done on the 1:1:1 acridine-cytosine-water complex.[304] Table VIII lists representative examples of this very heterogeneous group, characterized as molecular compounds and complexes. In some cases, the charge transfer bands are listed as evidence for complex formation.

In the presence of certain metallic ions and ammonium thiocyanate, acridine gives highly crystalline complexes, some of which can be utilized in microanalysis.[323] In tests on the feasibility of the use of anthracene oils as nuclear reactor coolants, liquid acridine has been subjected to radiolytic decomposition.[324] Acridine and acridan are among the substances identified as causing damage to the leaves of the plants near tar works.[325] Dogs fed on acridine excrete 2-hydroxy-9-acridanone.[326] Acridine is metabolized by rabbits to 9-acridanone and 2-hydroxy-9-acridanone.[327]

4. Chemical Properties of Acridine

Acridine is a base. It effectively dehydrochlorinates 3-chloropropanonitrile to acrylonitrile in refluxing ethanol;[328] its reactions with maleic and fumaric acids, which give acridinium hydrogen maleate and diacridinium fumarate, respectively, emphasize the importance of acid pK_a to product formation.[329] Hydrogen-bonding studies involving carboxylic acids have confirmed the existence of a tautomeric equilibrium involving protonated base and associated base.[330] A variety of spectrophotometric studies have yielded data on acid-base equilibria and acridine conjugate acid acidity.[331a, b] The very large number of amine salts and quaternary salts of acridine are, along with reduced acridines, the subject matter of Chapter V.

Perbenzoic acid oxidation of acridine occurs more readily than that of pyridine or quinoline. The high reactivity suggests a relationship between the nucleophilicity of nitrogen and its π electron density[332] as calculated by LCAO MO. The products of the oxidation, carried out in chloroform solution, are acridine 10-oxide, a small amount of 2-(2-hydroxyanilino)benzoic acid and a major amount of 2-(2-hydroxyanilino)benzaldehyde.[333] When 3-chloroperbenzoic acid is the oxidizing agent, the additional product, **93**, is observed.

93

10-Methyl-9-acridanone is the product when p-benzoquinone oxidizes 10-methylacridinium hydroxide.[334] Ozonation of acridine, followed by alkaline hydrogen peroxide oxidation, yields largely quinoline-2,3-dicarboxylic acid with small amounts of 9-acridanone as well.[335]

The nature and behavior of the radical anion produced by the addition of alkali metals to tetrahydrofuran solutions of acridine has been studied by means of esr and uv spectroscopy.[336] When acridine is refluxed with toluene and t-butyl peroxide for 9 days under nitrogen, 9,10-dibenzylacridan and 9-benzylacridine are formed.[337] Another reaction of acridine with radicals is that reported by Norman and Radda.[338]

Acridine participates in many addition chemical reactions. The tendency toward an ionic stepwise process may be related to the relatively low electron density[29] at C-9, calculated, 0.695. The nucleophilic reaction that leads to an

$$CH_2-CH_2-CO_2Me$$

$$[CO_2Me-CH_2-CH_2-CO_2]_2 \quad + \quad Acridine \xrightarrow{\Delta}$$

acridan product is followed, in many instances, by an oxidation process. By such a sequence the addition of triphenylsilyl lithium leads,[339] after oxidation, to 9-triphenylsilylacridine. Acridine reacts with sodamide, giving 9-amino-acridine (Chapter II), with sodium diethyl phosphonate[340] to form an acridan oxidized by chloranil to diethyl 9-acridinephosphonate. A number of 9-substituted acridines[341] have been prepared by the room temperature addition of active methylene compounds to acridine, followed by lead tetraacetate or manganese dioxide oxidation of the acridan addition compound. For example,

$$Acridine + CH_2(CN)_2 \xrightarrow[\text{room temperature}]{EtOH} \xrightarrow[\text{Pyridine}]{MnO_2}$$

However, it has not been possible to prepare[342] Reissert compounds (see also Section 5.H). Benzyne adds to acridine at positions 4 and 10 to give mixtures.[343, 344] One experiment of Wittig and Niethammer [345] gave 4-phenyl-acridine (2%) rather than the anticipated addition across positions 9 and 10. The reactions of acridines with dimethylketene are also addition reactions.[346a, b] Here it is thought that the first formed 10-acylacridinium ion is attacked by a second molecule of ketene or some solvated species acting as the nucleophile. Certainly, methanol participation is indicated.

$$Acridine + Me_2C=C=O \xrightarrow{MeOH} \xrightarrow{Me_2C=C=O}$$

When acridine in carbon tetrachloride is saturated with dry chlorine gas and exposed to sunlight, it adds the maximum number of chlorines.[347] The other reactions of acridine—with o-xylylene,[348] alkyl and aryl Grignard reagents,[349] acetone and NH_4SCN,[350] and dimethylsulfoxide with sodium hydride[351]—may also proceed through acridan or addition intermediates although they result, ultimately, in the products of nucleophilic attack at C-9.

Calculated π electron densities, from molecular orbital theory, are used to predict the position of attack by electrophilic, nucleophilic or radical reagents.[352] For acridine the frontier electron density for electrophilic substitution is as shown in **94**.

94

The predicted position of electrophilic attack is then at position 4. The observed nitration products are the 4-nitro, 2-nitro, and 2,4-dinitro derivatives. The agreement between calculated and experimental, according to Fukui, is "almost satisfactory."[352] Electrophilic bromination[48] occurs most readily at position 2.

The reduction of acridine has been accomplished by heating it under reflux with sodium carbonate in ethylene glycol,[353] by treatment with hydrogen with a special Raney nickel[354] or ruthenium catalyst,[355] arylthiols,[356] lithium aluminum hydride,[357] polarographic methods,[358a, b] triethylammonium formate and Raney nickel catalyst,[359] and lithium-ammonia solutions.[360] The reactions and the biacridan, di-, tetra-, and octa-hydro products formed are discussed in Chapter V.

In the presence of sunlight, acridine, but not 9-phenylacridine, gave a yellow photodimer, which decomposed on melting (276°) to acridine.[361] This same yellow dimer was observed by Kellman in 1957 when acridine was irradiated with γ rays from a Co^{60} source or uv radiation supplied by a mercury vapor lamp.[362] Since that time a vast body of literature[363, 364] has accumulated as a result of efforts to determine the products formed from photolysis,[365-367] the role of the solvent in the reaction,[366,368] the importance of oxygen,[369-371] temperature,[372] radiation,[373-375] concentration, [365] and the nature of the reactive species[376-383] participating in the photochemical (radical) reduction of acridine. Steady illumination,[371] as well as flash techniques,[384] have been used. Although the evidence indicates incomplete agreement at some points, [385] others have been clearly established. In various types of solvents, triplet states of acridine have the ability to abstract hydrogen from the solvents,[386] with the result that three radical processes yield acridan, solvent-coupled product, and 9,9'-diacridan.

Flash photolysis experiments[387] on 5×10^{-5} M solutions of acridine in the 90–300 K temperature range have shown that the primary photoreaction starts in the $^1(n - \pi^*)$ state, which is filled with an energy of activation of 0.035 eV from the excited $^1(\pi-\pi^*)$ singlet state. The primary reaction is the addition of a hydrogen atom from the solvent. Acridan is transferred into a highly excited triplet state; in different primary reactions, different radicals are formed.[387, 388] Evidence for transient isomeric acridine semiquinones, **95** and **96**, has been presented.[389]

Either ionic or radical pathways may explain the formation of a number of 9-alkylacridans (convertible to 9-alkylacridines) when acridine is irradiated in the presence of an equimolar quantity of an aliphatic carboxylic acid.[386] A suggested sequence is as follows:

When acridine is irradiated with 600-kV electrons at 0.5 mA, it polymerizes.[390] Its radiolytic stability appears to be greater than that of smaller condensed ring systems studied. The irradiation of acridine in a nuclear reactor[391] with a neutron flux of approximately 5×10^{12} neutrons per cm^2 yielded acridine containing C^{14} as well as some anthracene-9-C^{14}. Studies on C^{14} distribution in neutron-irradiated acridine showed concentrations at position 9, implying that pi electron distribution plays an important role in this type of reaction.[392] Wilzbach tritiation of crystalline acridine has been carried out.[393] Data presented suggest that radical processes are important in the Wilzbach labeling.

5. Substituted Acridines

Few direct substitution reactions have been extensively investigated. In most cases, substituents have been introduced into the benzenoid rings prior to completion of the acridine synthesis. The exceptions are the large number of reactions that occur at the 9-position but, even here, many proceed via the formation of the appropriate acridan, to be followed by an oxidation step. In Table IX are recorded a representative, but not exhaustive, list of monosubstituted acridines from widely differing sources. Nitration gave a mixture of nitroacridines. Both addition and substitution products are formed in halogenation reactions (see also Chapter V, p. 445). In conjunction with the discussion of 9-acridanones in Chapter III, some halogenated products are considered. Since aminoacridines, 9-acridanones, acridinium salts and biacridines all are dealt with in later chapters, attention is focused here on others of the principal types of substituted acridines. Considerations of space prevent any but brief summaries concerning these compounds.

The sulfonation of acridine has been little investigated,[2] but at 180° fuming sulfuric acid appears to give 9,9'-biacridine-tri- and tetra-sulfonic acids.[415] Acridine-2-sulfonic acid (Table IX) has been obtained by the reduction of 9-acridanone-2-sulfonic acid. Other sulfur compounds, mercaptans,[109] thioethers,[412] and thioesters,[412] have been prepared from 9-acridanthiones (Chapter III, pp. 239–240). The 9-triphenylsilyl compound has been investigated for possible use as a high-temperature antioxidant containing silicon.[339] Among the phosphorus compounds that have been studied is the 9-acridinyldiphenylphosphine (Table IX).[403] Redmore[225] and Scheinkman and his colleagues[416] have provided two approaches to the preparation and study of 9-acridinephosphonic esters and acid. A numer of acridinearsonic[92,417] and acridinestibonic acids,[418] as well as isothiocyanatoacridines,[409] have been prepared from the corresponding aminoacridines. Some arsenic (see ref. 419, and earlier papers) and antimony[420] analogues of acridine are known. The more important substituted acridines are now considered individually.

TABLE IX.   Acridine and Its Monosubstituted Derivatives

Acridine	mp($^\circ$C)	Appearance, solvent, and other properties	Ref.
Unsubstituted	110	Yellow needles from EtOH or benzene; picrate, mp 261-262° (dec)	2, 394
3-Acetyl-	135-136.5	Lemon yellow prisms after vacuum distillation (125-130°/0.1 mm)	76
9-Acetyl-	109	Brown crystals from CCl$_4$	395
9-Acetylmethyl-	144-146	Recrystallized from benzene; phenylhydrazone, mp 235-237°	396
9-Acridinecarbox-aldehyde	150 (146-7)	Golden yellow needles, EtOH, MeOH; oxime, mp 247° (dec)	395 397-399
2-Acridinecarboxylic acid	Dec at 270	Buff crystals, aq cellosolve; amide, mp 243°	108
4-Acridinecarboxylic acid	189-190	Yellow crystals, aq EtOH	108
9-Acridinecarboxylic acid	289-290	Yellow prismatic needles, much EtOH; amide, mp 263-264°	12
Acridine 10-oxide	169	Fine yellow needles, petroleum ether	400-402
2-Acridinesulfonic acid	>360	Golden yellow flat needles, dilute HCl; amide, mp 258°	109
9-Acridinyldiphenyl-phosphine	192	From toluene; 9-acridinyl-diphenylphosphine sulfide, mp 212°	403
2-(9-Acridinyl)-ethanol	155-155.5	Pale yellow, 75% EtOH, then benzene	135
9-Anilino-	224	Rhombic yellow leaflets, alcohol; hydrochloride, mp 205-207°	404
9-Benzoyl-	217.5	Yellow needles from EtOH; deep violet fluorescence, EtOH	39, 140
1-Bromo-	110-111	Yellow needles from petroleum ether	49
2-Bromo-	175-175.5	Yellow prisms from aq EtOH	48, 49

(Table Continued)

TABLE IX. *(Continued)*

Acridine	mp(°C)	Appearance, solvent, and other properties	Ref.
3-Bromo-	137-138	Pale yellow plates, light petroleum	49
4-Bromo-	107-108	Crystals from light petroleum	49
9-Bromo-	116	Yellow-brown needles, amm. EtOH; picrate, mp 212-213°	405
2-Chloro-	170	Pale rose crystals from benzene; mercurichloride, mp 245-247°	112
3-Chloro-	129 or 134	Pale yellow crystals, two forms, benzene	44, 79
4-Chloro-	79 or 90-91	Needles or prisms, petroleum ether	59
9-Chloro-	119-120	Almost colorless crystals, EtOH or benzene	406-408
2-Cyano-	209	Pale yellow crystals, chlorobenzene	44, 50
9-Cyano-	186	Yellow needles, EtOH	12
Diethyl 9-acridine-phosphonate	95-96	Yellow from benzene-hexane; 9-phosphonic acid, mp >300°, soluble in base, orange-yellow	225
1-Ethyl-	89-90	Pale yellow prisms, petroleum ether	111
2-Ethyl-	77-78.5	Colorless prisms, vacuum sublimation	78, 111
3-Ethyl-	90-91.5	Colorless prisms, vacuum sublimation at $110°\ 0.4\ mm^{-1}$	111
4-Ethyl-	37-38.5	Colorless prisms, vacuum sublimation at $100°\ 0.4\ mm^{-1}$	111
9-Ethyl-	110-111	Yellow needles from light petroleum; picrate, mp 203°	516
1-Hydroxy-	ca. 250 (dec)	Yellow needles, aq EtOH; cation orange, anion red	196
2-Hydroxy-	ca. 279 (dec)	Yellow needles, aq EtOH	89, 195
3-Hydroxy-	285 (Sealed)	Microscopic yellow needles	146, 196

(Table Continued)

TABLE IX. *(Continued)*

Acridine	mp(°C)	Appearance, solvent and other properties	Ref.
4-Hydroxy-	117	Pale yellow needles, aq EtOH; picrate, mp 216°	95
9-Hydroxy-	360	See acridanone (Chapter III)	
9-Iodo-	169	Yellow-brown needles, EtOH; picrate, mp 204°	405
1-Isothiocyanato-	122-123	–	409
2-Isothiocyanato-	167	–	409
3-Isothiocyanato-	174	–	409
1-Methoxy-	122	Small yellow plates, light petroleum; methosulfate, mp 216-217° (dec)	47
2-Methoxy-	105	Orange-yellow needles, light petroleum; methosulfate, mp 227° (dec)	47, 79
3-Methoxy-	91	Pale yellow needles, light petroleum; methosulfate, mp 260° (dec)	47
4-Methoxy-	134	Pale yellow opaque needles, cyclohexane; methosulfate, mp 170°	47, 50
9-Methoxy-	65	Yellow needles from light petroleum; monohydrate, aq MeOH, mp 103°	410, 411
1-Methyl-	98-99	Yellow needles, aqueous EtOH	78, 88
2-Methyl-	134	Yellow needles from EtOH	78, 88
3-Methyl-	122-123	Pale yellow plates, aq EtOH	78, 88
4-Methyl-	89	Pale yellow needles, aq MeOH	78, 88, 40a
9-Methyl-	117-119 (118-118.5)	Long needles from 75% EtOH, benzene	156, 133, 135
9-Methylthio-	113	From aqueous acetone	412
1-Nitro-	154	Flat yellow needles, PhCl; no fluorescence in EtOH	211
2-Nitro-	215.5	Yellow plates from PhCl	211
3-Nitro	183	Yellow needles from EtOH; no fluorescence in EtOH	211

(Table Continued)

TABLE IX.   *(Continued)*

Acridine	mp(°C)	Appearance, solvent and other properties	Ref.
4-Nitro-	167	Silvery leaflets from benzene	211
9-Nitromethyl-	156-157	Pale yellow, from benzene	341
9-Phenoxy-	127-128	Crystallized from benzene	413
9-Phenyl-	184, 185	Yellow prisms from ethanol	117
	bp 403-404°		121
			6,126
9-Triphenylsilyl-	287-289	Bright yellow crystals, benzene-petroleum ether	339
9-Vinyl-	85-87	Pale yellow crystals, petroleum ether	414

A. Nitroacridines

The reaction of acridine with nitric acid (d 1.45) alone was carried out at a very early stage,[2] and the two products obtained were later shown to be the 2- and 4-nitroacridines.[210] A careful reinvestigation of the nitration showed that much disubstitution took place unless the reaction was carried out in sulfuric acid solution, when, with one equivalent of nitric acid, the ratio of 1-, 2-, 3-, and 4-nitroacridines formed was 5:130:1:25, respectively.[211] (More recent simple MO calculations of pi energies by Dewar and Maitlis[421] gave a predicted order of reactivity that is in qualitative agreement with Lehmstedt's findings reported here.) Separation of the mono-nitro isomers proved difficult, especially as both 1- and 2-nitroacridine formed mixed crystals with 4-nitro-acridine. Controlled oxidation, however, converted the 2- and 4-nitroacridines to the corresponding 9-acridanones without affecting their isomers, thus achieving separation.[211, 422] The 2-, 3-, and 4-nitroacridines, as well as 2,4-dinitroacridine, have been synthesized from diphenylamine-2-aldehydes (Table VI).

The largest group of known nitroacridines contains a chloride atom at position 9; these compounds have been synthesized from the appropriate diphenylamine-2-carboxylic acids by the action of phosphorus halides. Among those recently prepared are the 1-nitro and 3-nitro-9-chloroacridines[423] and the chloromethoxynitroacridines prepared by Steck and his co-workers.[424] Maksimets and Sukhomlinov[46] used the *p*-toluenesulfonhydrazide reaction (Section 2.A) as a method of removing the 9-chloro substituents from their chloromethoxynitroacridines, leaving the nitro group intact. The major transformation of these compounds has been their conversion to various amino derivatives for bactericidal therapeutic testing. This matter is considered in a later chapter.

Most of the simple nitroacridines are yellow crystalline compounds (Table IX). All of them have been reduced to the corresponding amino compounds,[425] which are discussed in Chapter II.

B. Halogen-substituted Acridines

Early experiments in the halogenation of acridine in chloroform solution gave inconclusive results.[426] Later work[48] on the bromination showed that in carbon tetrachloride an addition compound, 10-bromoacridinium bromide (Chapter V, p. 445), was formed, while, in acetic acid, substitution took place giving a mixture of 2-bromoacridine and 2,7-dibromacridine. Heating acridine with sulfur chloride gave a mixture containing some 9-chloroacridine.[427] Acridine did not react with N-bromosuccinimide under ordinary conditions,[428] but in boiling carbon tetrachloride, in the presence of benzoyl peroxide, a number of products were formed.[429] These included two monobromo and three dibromoacridines, 9-succiniminoacridine (Chapter II, p. 120) and several of its bromo derivatives, and 9-acridanone. The positions of the bromine atoms in most of these compounds are unknown.

The 9-chloroacridines, by virtue of the reactivity of the chlorine atom, are by far the most important chloroacridines. They are usually yellow solids that tend to irritate the skin and mucous membrane. Special care must be taken, on an industrial scale, with 6,9-dichloro-2-methoxyacridine, an intermediate in the synthesis of atebrin, since it sensitizes the skin of certain workers and the dust is liable to cause extensive dermatitis.[430] 9-Chloroacridine has been used recently in a new spectrophotometric determination of small quantities of primary aromatic amines.[431]

9-Chloroacridines are best made from the corresponding 9-acridanones[432] or diphenylamine-2-carboxylic acids[433] by treatment with phosphorus oxychloride (Chapter III, p. 144). Some of the 9-acridanones have been obtained by the pyrolysis of suitably substituted anthranils[261, 262]; others have been treated with chlorine gas in glacial acetic acid or carbon tetrachloride solution so that, following the phosphorus oxychloride reaction, trichloro-[434] and pentachloro-acridines[435] have been obtained. These compounds are yellow crystalline solids that are very soluble in chloroform but can often be recrystallized from benzene and petroleum ether; very pure specimens may be obtained from the latter solvent after chromatography on alumina. Their solutions are fluorescent. Partial hydrolysis with acids may give red compounds, consisting of mixed crystals of the 9-chloroacridine hydrochloride and the corresponding 9-acridanone.[436a, b] Different 9-chloroacridines sometimes form mixed crystals.[110] 9-Bromoacridine, prepared from 9-acridanthione, red phosphorus, and bromine, gave 9-iodoacridine with sodium iodide in ethanol.[405]

The hydrolysis of 9-chloroacridines to 9-acridanones is usually effected by boiling with dilute acid, but the ease of hydrolysis depends on the substituents present (of. Chapter III, p. 146). In one series studied,[437] the reactivities of 2- and 3-substituted derivatives were found to be in the order $CH_3O <$ $CH_3 < H < NO_2$. It is not possible to make a complete quantitative comparison of hydrolysis rate data from earlier studies[408] with the more recent work.[437–442] Different temperatures, solvent concentrations, and reaction conditions have been used and their effects vary. However, all are in general agreement that electron-donating substituents seem to retard the rate slightly, while electron-withdrawing substituents enhance the rate. The hydrolysis of 9-chloroacridine in 80% acetic acid is, according to Ledochowski, a pseudo-first-order, two-step consecutive reaction.[440]

The intermediate, **97**, as the dihydrate, has been isolated. Table X summarizes the effect of the position of a methyl substituent on the rate constants that have been determined by the extended time ratio method. The effects are relatively small and are, possibly, attributable to small changes in electron density at carbon 9. Ethanolic solutions of 9-chloro-, 9-bromo- and 9-iodo-acridine have been irradiated with 365-nm light.[443] Charge transfer complexes between the acridans and unreacted acridines were observed. The hydrolysis rate of these acridinium ions to the 9-acridanone in the dark was 100 times that of the acridines themselves.[444] Hydrolysis in light in the presence of oxygen was also 100 times as fast as that in the dark. In the light, in the absence of oxygen, some as yet unidentified products were formed.

TABLE X.   Rates of Hydrolysis of 9-Chloroacridine Derivatives[442]

	Acridine	$k_1 \times 10^3$ sec^{-1}	$k_2 \times 10^3$ sec^{-1}
Temperature, 70°C	9-Chloro-	6.3	0.28
Solvent, 80% aqueous acetic acid	9-Chloro-2-methyl-	2.3	0.135
k_2/k_1 ratio = 0.047 ± 0.006	9-Chloro-3-methyl-	3.3	0.15
	9-Chloro-4-methyl-	7.5	0.3

9-Chloroacridines can react with a wide variety of compounds to give 9-substituted acridines. The most examined reaction is with amines and is discussed in Chapter II (p. 116). The conversion of 9-chloroacridines to acridines has been treated in Section 2.A of this chapter. For the reactions of 9-chloroacridines leading to biacridines, consult Chapter VI (p. 519), to 9-acridanones, 9-acridanthiones[432] and a variety of phenylamino-, as well as dialkylaminophenyl-acridines, Chapter III. 9-Chloroacridine was found to react with the sodium derivatives of phenylacetonitrile,[445] malonitrile,[445] ethyl cyanoacetate,[445] diethyl malonate,[446] phenol (see Ref. 447, where the starting compound was 9-chloro-2-methoxy-8-nitroacridine) and ethyl acetoacetate[395] to give the corresponding 9-acridyl derivatives, and with alkyl cyanides and alkoxides. With silver or potassium isothiocyanate and a dihaloacridine, the isothiocyanate group[448] appears only at position 9. In addition to the phosphines[403] and phosphonic acid derivatives,[225, 449] phosphorus and sulfur compounds, such as O,O-diphenyl-S-(2-ethoxy-6-nitro-9-acridyl)phosphorodithioate (98), have been prepared.[450]

98

Most of the known bromo- and chloro-acridines not containing a 9-chlorine atom (Table IX) have been obtained from the corresponding 9-acridanones or 9-chloroacridines. Certain di- and poly-halogen derivatives of 9-phenylacridine have been obtained[232d, e] when the appropriate triphenylcarbinols were heated in nitrobenzene (Table VII). The 1-bromo and 3-bromo compounds were prepared, however, from the corresponding nitro-acridines by the reaction sequence: reduction, diazotization, thermal decomposition of the mercury (II) bromide complexes of the diazonium salts.[49] All of these compounds are yellow solids that are weaker bases than acridine but usually form salts with acids. Those of the 9-phenylacridine group that have three or more halogens do not form picrates.[232e] Little is known of the relative reactivity of the chlorine atoms in the known 2-, 3-, and 4-chloroacridines. 4-Chloroacridine does not react[59] with ethylamine at 160° or with diethylaminoethylamine[451] and was prepared by the sodium amalgam reduction of 4-chloro-9-acridanone. 2-Chloroacridine was similarly obtained from 2-chloroacridanone 10-oxide.[112] A modification of conditions has made possible the previously unsuccessful[44] selective dehalogenation of 3,9-dichloro-acridine.[79] When this compound was hydrogenated using Raney nickel paste and an equivalent of potassium hydroxide in methanol solution, the major product was 3-chloroacridan. This was readily oxidized with chromic acid to the 3-chloroacridine. Apparently, the presence of the methoxy function as a deactivating influence has made such a preparation of 2-methoxy-6-chloro-acridine from 2-methoxy-6,9-dichloroacridine easier.[54, 79] However, a trend toward less stability and greater reactivity for chlorine bonded at 3- than for that bonded at 4-, 1-, or 2- is shown in other reactions. The 3-chloro compound is largely dehalogenated in the zinc and acetic acid reduction of all the monochloro-9,10-diphenylacridinium salts.[452] The 3-anilino derivative is obtained when 3-chloro-9,10-diphenylacridinium chloride reacts with boiling aniline. This behavior, even though a variety of reaction types is considered, is generally consistent with Fukui's frontier electron density predictions relative to nucleophilic reagent attack.[352]

Mention should be made of some fluoro derivatives of acridine that have now been prepared. Joshi and Bakel have used the diphenylamine 2-carboxylic acid with phosphorus oxychloride approach to obtain 2-fluoro and 2,7-difluoro compounds for testing for fungicidal properties.[453] Among their studies concerned with carcinogenic properties (Chapter XVII), Thu-Cuc, Buu-Hoï, and Xuong have reported the synthesis of alkylfluoroacridines by a Bernthsen-type method.[136] 1,2,3,4-Tetrafluoroacridine has been prepared via an anthranil intermediate formed in the pyrolysis of 2,3,4,5,6-pentafluoro-2'-nitrodiphenylmethanol[68] [Cf. Section 2.F(4)]. The possibility of the effect of fluorine on pharmacological properties stimulated this study.[68]

C. Acridine N-Oxides

Acridine 10-oxide (or N-oxide), erroneously called "acridol," was originally obtained from the sodium amalgam reduction of acridanone 10-oxide[400] and as a by-product of the reaction of 2-nitrobenzyl chloride, benzene, and aluminum chloride[454]; its correct structure was recognized somewhat later.[112, 401, 402] Much of the information on acridine 10-oxides has now been organized and systematized.[455] The oxide is best prepared from perbenzoic acid[401, 456] or *m*-chloroperbenzoic acid[333, 457] and acridine. It is a yellow solid, mp 169°, dissolving in ethanol to give a solution with green fluorescence; it forms a hydrochloride. Like acridine, this 10-oxide forms a green complex with acridan and hydrochloric acid.[402] The uv absorption spectrum[458] (Chapter X) and high resolution nuclear magnetic resonance (nmr) spectrum[459] (Chapter XII) have been examined and its k-electronic structure has been calculated.[460] The esr of the anion radical of acridine 10-oxide has been studied and the spin density distribution determined.[461] The dipole moments, determined for a number of acridine 10-oxides, are all higher than those of the free bases.[462]

Acridine 10-oxide undergoes a number of typical reactions: (1) acridan is formed with sodium amalgam,[400] (2) 9-chloroacridine with phosphorus pentachloride,[400] (3) 9-cyanoacridine with potassium cyanide,[402] (4) acridine with sodium bisulfite,[402] (5) 9-acridanone with boiling acetic anhydride,[400] (6) 10-hydroxy-9-acridanone with potassium permanganate,[410] and (7) 10-hydroxy-9-acridanone and 9-acridanone with ozone in methanol.[335] It has been proposed that the polarographic reduction of acridine 10-oxide involves an intermediate with the structure of an unsaturated hydroxylamine that partially dehydrates to the deoxygenated base.[463] A variety of products has been obtained in photolysis experiments on acridine 10-oxides. These appear to change with the solvent, concentration, temperature, presence of oxygen, and 9-substitution of the acridine 10-oxide.[464] Among the products that have been reported are **99**,[465] **100**,[466] and **101**.[467, 468]

99   100   101

Three recent investigations[457, 469, 470] have been concerned with the mechanism of the acridine 10-oxide reaction with acetic anhydride. The process exhibits pseudo first order kinetics and, at least in the presence of excess acetic anhydride, seems to involve the rate determining formation of a cyclic species such as **102**.

The intramolecular nature of the rearrangement has been established. Some

102

details on solvent involvement have been considered.[469, 470] Acridine 10-oxide has been found to undergo electrophilic bromination and nitration[456] at position 9. Dipole moment measurements have been used in explaining the electron distribution in the 10-oxide system.

A number of acridine 10-oxides have been obtained as by-products in the synthesis of acridanones from phenylanthranils (Chapter III). 6-Chloro-9-hydroxy-3-nitro-acridine 10-oxide[471] (tautomeric with the corresponding 10-hydroxy-9-acridanone) was formed in the condensation of chlorobenzene with 2,4-dinitrobenzaldehyde. Some eight different 9-substituted acridine 10-oxides have been prepared by the perbenzoic acid oxidation of the corresponding acridines.[401, 472, 473] Similar oxidation of 6,9-dichloro-2-methoxyacridine was carried out,[474, 475] where the ultimate goals were aminoacridine 10-oxides. Atebrin formed a di-N-oxide with 3% hydrogen peroxide.[476]

The oxidation of acridine in benzene with perbenzoic acid[477] gave a 5–7% yield of a substance isomeric with 10-hydroxy-9-acridanone, and into which it was converted by warm alcoholic alkali. The substance was insoluble in alkali, and on melting (178°) or on crystallization from pyridine gave largely 9-acridanone. There were no spectral data available with which to confirm the 9,10(9)-epidioxyacridan structure assigned. Oxidation of acridine in ether with 3-chloroperbenzoic acid has resulted in the isolation of an insoluble, compound[333] of mp 175° which may be related to the previously mentioned substance. As indicated in Section 4, it has been assigned the N,N'-oxydi-9-acridanone structure, **93**, and it is converted to 10-hydroxy-9-acridanone upon recrystallization from acetic acid.

It is perhaps appropriate to note that there is spectral evidence for the existence of both 10-oxide and 10-hydroxy compounds under certain conditions.[478a,b]

D. Hydroxyacridines

All the monohydroxyacridines (Table IX) have been obtained by the hydrochloric acid hydrolysis of the corresponding aminoacridines (Chapter II), by the dealkylation of the alkoxyacridines, and by other methods discussed earlier in this chapter. They are all very sparingly soluble, yellow crystalline substances; with the exception of 9-hydroxyacridine, which exists largely if not entirely in the tautomeric form of 9-acridanone (Chapter III), they are basic and form normal phenolic derivatives. Tautomerism is not possible with 2- and 4-hydroxyacridine, but there is considerable spectral evidence for its existence in 1- and 3-hydroxyacridines (Chapter X). Additional evidence for the relative importance of O-H vs. N-H bonding in these compounds has been accumulated in further studies of pH dependence in uv and ir absorption spectra,[479, 480] comparison with the spectral properties of methyl derivatives,[481] and ionization constant measurements.[482] Solutions of 1-hydroxyacridine in benzene and in 20% aqueous ethanol or aqueous pyridine are yellow and blue, respectively,[196, 244] the latter color suggesting the presence of **103**, since 10-methyl-1-keto-1,10-dihydroacridine[483] is blue.

103   104

4-Hydroxyacridine (**104**) has a much lower melting point and greater solubility than its isomers; this is a common property of 4-substituted acridines and is presumably because hydrogen bonding considerably reduces intermolecular association. This isomer forms chelate compounds with a larger number of metals,[95, 484, 485] than does 5-hydroxy-1,2,3,4-tetrahydroacridine. Recent consideration of substituted acridines as potential terdentate ligands has added a knowledge of protonation order [58b] and comparative chelating ability of 4,5-dihydroxy-, 4,5-diamino-, and 4-amino-5-hydroxyacridines.[486] An interesting property of 3-hydroxyacridine and 3-hydroxy-9-phenylacridine is their thermochromism.[487a, b] The color change, yellow to red, observed when the crystals are pressed or rubbed, is attributed to tautomerism in the solid state, where proton transfer occurs via "defect sites." This lactim ⇆ lactam equilibrium is also found in the following specifically oriented hydroxyphenylacridine condensation products:

Lactim   Lactam

yellow   red

4-Hydroxyacridine is best prepared by the demethylation of 4-methoxy-acridine,[95] but has been obtained from the reduction of 4-hydroxy-9-acridanone with sodium in ethanol,[114] a method used for a number of dihydroxy-acridines.[89] 3,6-Dihydroxyacridine has been obtained by heating 3,6-diamino-acridine with 45% sulfuric acid to 195°. The nitric acid oxidation of 2,4-dihydroxyacridine gave acridinic acid.[488]

E. Alkoxy- and Aryloxy-acridines

The 9-alkylthioacridines are discussed in Chapter III (pp. 238–240). The 9-alkoxyacridines have all been prepared from the 9-chloroacridines and solutions of sodium in the appropriate boiling alcohol.[410, 411, 489, 490] 9-Methoxyacridine crystallized from aqueous methanol as the monohydrate;[411] evaporation of its benzene solution gave the anhydrous substance. 9,9-Di-phenoxyacridan hydrochloride[491] was obtained from 9-chloroacridine and phenol at 100°, and on treatment with water or dilute alkali, gave 9-phenoxy-acridine. Quite an extensive series of 9-phenoxyacridines, substituted both in the acridine and phenoxy portions of the molecules, have now been prepared.[447, 492, 493] In these cases, with the substituted 9-chloroacridine and phenol as the starting materials, the yields of ethers are very good. Polaro-graphic reductions of 9-phenoxyacridine and its 10-oxide have been carried out [358a]; the half-wave potentials are quite similar. The dipole moment measurements show a markedly higher value for the 10-oxides than for the 9-phenoxyacridines and 2-methoxy-9-phenoxyacridines.[494] Ledochowski and Wojenski[495] have found the hydrolysis of 9-phenoxyacridine (in 80% acetic acid) to 9-acridanone to be a pseudo-first-order reaction. 9-Methoxyacridine, as well as some substituted 2-methoxyacridines, are converted into their

10-oxides by perbenzoic acid (Section 5.C). 9-Methoxyacridine and 9-phenoxyacridine did not react with hydrogen cyanide, but both 9,9-diphenoxyacridan and 9-methoxyacridine hydrate gave 9-cyanoacridine[411]; 9-methoxyacridine hydrate also formed an addition compound with sodium bisulfite.[410] This has been considered to suggest that 9-methoxyacridine hydrate is, in fact, the acridan, **105**. However, the uv absorption spectra of the hydrate in aqueous methanol, and of the anhydrous compound in cyclohexane, are almost identical.[496] They show typical acridine, and not acridan, absorption. The hydrate is therefore not present to any great extent as the acridan in solution. Displacement of the methoxyl group by anionoid reagents probably proceeds via the acridinium salt, **106**; this reaction is discussed with special reference to 9-aminoacridines in Chapter II, Section 3.A.

No 9,9'-biacridyl ethers appear to have been obtained pure, although 9-(9-acridylthio)acridine is known (Chapter III, p. 240). 3,3',6,6'-Tetraamino-9,9'-biacridanyl ether was obtained as a by-product in the synthesis of 3,6-diaminoacridine (Section 2.C), and larger quantities were obtained when hydrochloric acid was not added to the reaction mixture. Air oxidation gave 3,6-diamino-9-acridanyl 3,6-diamino-9-acridyl ether; some further oxidation appeared to take place. Some compounds analogous to the last named ether are known.[160]

1-, 2-, 3-, and 4-Alkoxyacridines have been prepared by many of the general methods available for the synthesis of acridines (Table IX). For its use in the preparation of a fluorescent indicator, 9-chloro-4-ethoxyacridine was obtained by the diphenylamine-2-carboxylic acid cyclization method.[497] The alkoxyacridines can be hydrolyzed to the corresponding hydroxyacridines in good yield by boiling hydrobromic[95] or hydriodic acid.[484] The use of this method, however, tends to result in the formation of 9-acridanones if it is used with alkoxy-9-chloroacridines. Aluminum chloride, which has been used for the demethylation of methoxy-10-methylacridinium salts[77] and 4-methoxy-9-acridanone (Chapter III, p. 218), gave excellent results in cases in which it was desired to retain the 9-chlorine atom.[498] 3-Methoxy-, but not 1-, 2- or 4-methoxy-bisacridinium salts, are demethylated on treatment with alkali (Chapter VI, p. 522).

F. Alkyl- and Aryl-acridines

Most of the 9-substituted compounds in this class have been prepared by Bernthsen's reaction (Section 2.B; cf. Section 2.A). The direct reduction of acridine-9-aldehyde to 9-methylacridine by zinc and formic acid is worthy of special mention.[395] The acridinium salts are discussed in Chapter V, Section 1. The quaternary salts have been the products in some recently studied 9-aminophenylacridine preparations.[499a,b] The substituted 9-phenylacridines prepared by cyclodehydrogenation procedures are discussed in Section 2.F(2). The susceptibility of C-9 to nucleophilic attack as discussed in Section 4, has been used to great advantage in the preparation of a variety of 9-alkyl-[349, 500, 501] and also 9-aryl-acridines.[349] To the rather specialized 9-methylacridine preparations referred to in Section 2.F(4) may be added its formation from 9-diazoacetylacridine[502] and the nucleophilic alkyl substitution with methylphenylsulfone[503] thought to proceed as follows:

Acridine + CH$_3$SO$_2$Ph $\xrightarrow[\text{HMPT}]{\text{NaH}}$ $\longrightarrow$ 9-methylacridine (60%)

HMPT = Hexamethylphosphoric triamide

Phenylcarbene may be responsible for the formation of a small amount of 9-benzylacridine when benzaldehyde tosylhydrazone is added to acridine and sodium methoxide in dry decalin.[504]

With the exception of the 9-substituted derivatives, the alkyl- and aryl-acridines have properties rather similar to those of acridine. The introduction of alkyl substituents causes a small change in the light absorption toward the visible and a reduction in solubility. The basic strength is increased slightly, except when the alkyl groups are in position 4 and/or 5, when the addition of a proton is sterically difficult. This is illustrated by a comparison of the

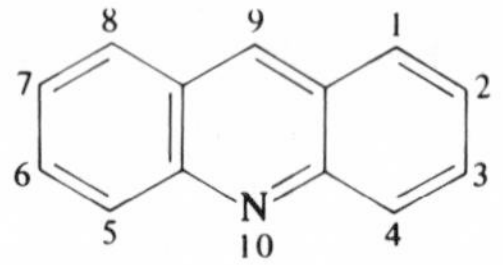

pK_a's of acridine and its 4-methyl and 4,5-dimethyl derivatives in 50%
aqueous ethanol. They are 4.11, 3.95, and 2.88, respectively.[108] 9-Phenylacri-
dine is less basic and less soluble than acridine. Its dipole moment, 2.49 D,
is slightly larger than that of acridine.[456]

(1) *Alkylacridines*

9-Alkylacridines are of particular interest because of the great reactivity
of the α-hydrogen atoms of the alkyl group, due to the electron-attracting
properties of the heterocyclic nitrogen atom. They may be compared with
the 2- and 4-methylpyridines and methylquinolines, where the methyl group
is similarly activated. The methiodides of these alkyl pyridines and quinolines
are more reactive,[398] as would be expected on electronic grounds, than the
parent tertiary amines, toward electron-deficient centers; it is, therefore,
remarkable that 9,10-dimethylacridinium iodide (in contrast to the tertiary
base) is reported not to react with 4-dimethylaminobenzaldehyde, 4-nitroso-
dimethylaniline, and other compounds.[125] It failed, also, to react with 2-
iodoquinoline methiodide and potassium hydroxide in absolute ethanol, but
after the addition of water the expected dyestuff, **107**, was formed. These
unexpected failures may occur because of solubility problems or the conversion of the methiodide into the unreactive 9-ethoxy-9,10-dimethylacridan;
the presence of water would facilitate the ionization of the latter compound
and thus cause reactivation. 9,10-Dimethylacridinium methosulfate reacted
normally in acetic acid (no acridan present here) with aldehydes to form the
9-styryl-10-methylacridinium salts,[505] and with diazonium salts to form the
hydrazones[506] such as **108**.

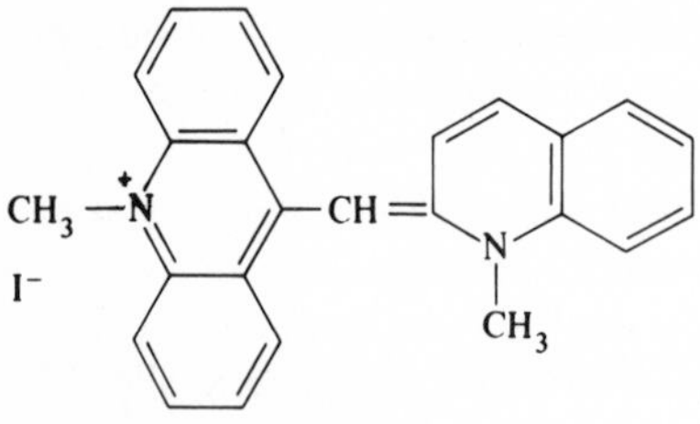

107

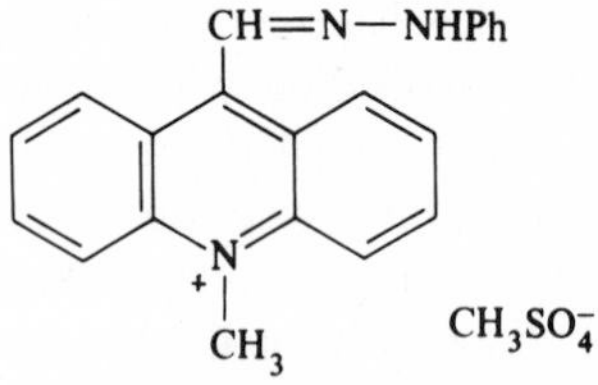

108

The alkylacridines, like acridine itself, form molecular compounds with ease. 9-Methylacridine is reported to form complexes with diphenylamine,[153] 1,3,5-trinitrobenzene,[507] and 9,9-dimethylacridan.[37] 9-Phenylacridine complexes with 1 or 2 moles of picric acid,[508] 1,3,5-trinitrobenzene,[507] trinitroresorcinol,[509] and trinitrophloroglucinol.[509] Both 4,5-dimethyl- and 1,4,5,8-tetramethyl-acridine form 1:1 complexes with 1,3,5-trinitrobenzene, although they do not complex with boron trifluoride, possibly because of methyl interference.[56] 9-Heptadecylacridine stabilizes mineral oils and increases their lubricating powers.[510]

In the presence of an ethyl dichloroaluminum catalyst, in benzene, 9-vinylacridine polymerizes to benzene-soluble and benzene-insoluble polymers, both of which no longer show vinyl absorption in the ir region.[511]

The reactions of 9-methylacridine, the only well-known acridine of this class, will now be considered in detail. It may be prepared by Bernthsen's reaction (Section 2.B), from 2-aminoacetophenone and bromobenzene (Section 2.D), or by a procedure summarized at the beginning of this section.

(a) WITH OXIDIZING AGENTS. Vigorous oxidation with alkaline potassium permanganate gave quinoline-2,3,4-tricarboxylic acid.[117] Boiling dilute nitric acid (d 1.2) had no effect, while more concentrated nitric acid (d 1.33) gave a trinitroacridine-9-carboxylic acid.[117] The effect of chromic and acetic acids does not appear to have been examined, but the oxidations of 9-benzylacridine to 9-acridyl phenyl ketone (59%), benzaldehyde (38%), and 9-acridanone[39]; and of 9-(4-nitrobenzyl)acridine to 9-acridyl 4-nitrophenyl ketone, and 4-nitrobenzaldehyde by this reagent are of interest.[142] 9-Methylacridine was oxidized by selenium dioxide to acridine-9-aldehyde,[512] while 9-benzylacridine and the 12-benzylbenz[a]acridine in xylene similarly gave 9-acridyl phenyl ketone,[140] and 12-benzoylbenz[a]acridine, respectively.[513] Chromyl chloride did not oxidize 9-methylacridine to acridine-9-aldehyde.[397]

(b) WITH ALDEHYDES. 9-Methylacridine[395, 514] reacted rapidly with formaldehyde in boiling ethanol to give 2-(9-acridyl)ethanol (109), easily oxidized to acridine-9-carboxylic acid by chromic acid; with excess formaldehyde, 110 was formed. If dimethylamine and hydrochloric acid were also present, a Mannich reaction, giving 9-(2-dimethylaminoethyl)acridine, took place.[515, 516] When 9-methylacridine was refluxed in methanol with piperidinomethanol[414] (from formaldehyde and piperidine) pale yellow crystals of 9-(2-N-piperidinoethyl)acridine were obtained. 9-Methyl- and 9-ethyl-acridine react with formaldehyde and 4-aminobenzenesulfonamide to give 111 (R = H and CH$_3$).[517]

The only other aliphatic aldehyde that appears to have been condensed with 9-methylacridine is chloral,[397] which reacted in refluxing benzene to give 1-(9-acridyl)-3,3,3-trichloropropane-2-ol. This, on boiling with alkali, was

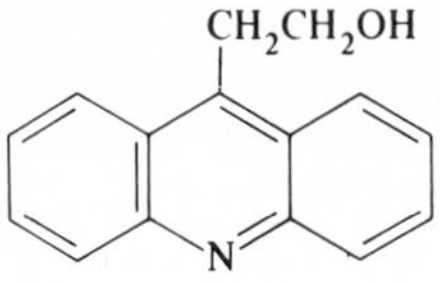

CH₂CH₂OH

(structure)

109

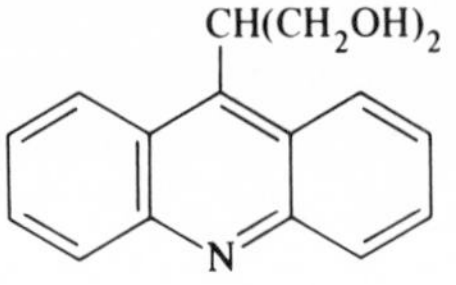

CH(CH₂OH)₂

(structure)

110

RCH—CH₂NHSO₂—⟨phenyl⟩—NH₂

(structure)

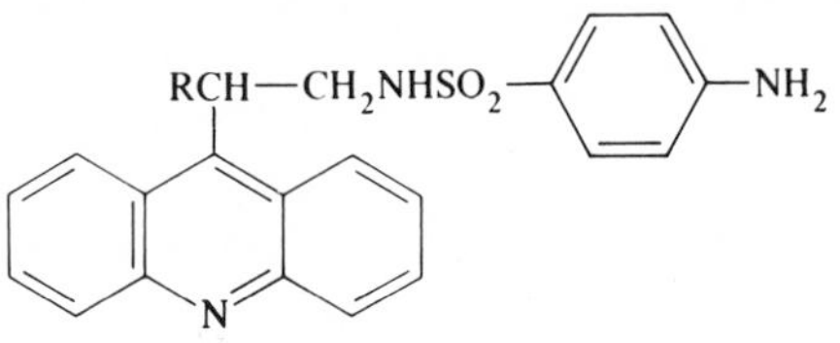

111

both hydrolyzed and dehydrated to 3-(9-acridyl)acrylic acid, which was also oxidized by permanganate to acridine-9-aldehyde and reduced to 3-(9-acridyl)propionic acid.

Benzaldehyde[518] and its 4-nitro derivative [519] with 9-methylacridine at 100° gave the corresponding ethanols, **112**, while 3-nitrobenzaldehyde gave the styrene, **113**, (R = 3-nitrophenyl).[520] Heating with zinc chloride caused dehydration of **112** (R = 4-nitrophenyl) to the corresponding **113**.

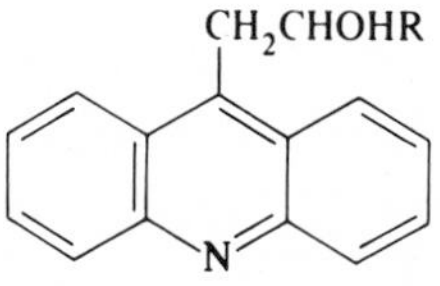

CH₂CHOHR

(structure)

112

CH=CHR

(structure)

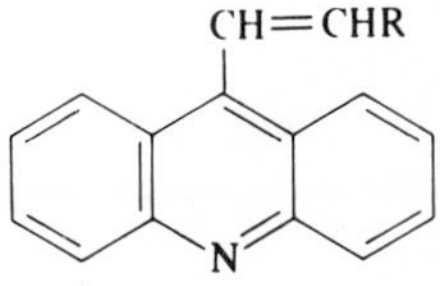

113

This was made,[519] as were the corresponding derivatives from 3-[519, 520] and 4-nitrobenzaldehyde[520] and 4-dimethylaminobenzaldehyde,[520] directly from the aldehyde, the acridine, and zinc chloride. It is thought that some interaction between hydroxyl and nitro groups may have been responsible for the fact that the 2-nitrobenzaldehyde condensation yielded the alcohol, **112** (R = 2-nitrophenyl), which could not be dehydrated to the styrene. The nitrostyrenes have been reduced to amines, which are discussed in Chapter II, Section 5. Others of the styrene products were obtained in the zinc chloride-catalyzed condensation of 9-methylacridine with p-hydroxybenzaldehyde[521] and with p-methoxybenzaldehyde.[521] Similarly, 4 hr of reflux with acetic anhydride brought about the condensation with p-bis(β-chloroethyl)amino-benzaldehyde to give the styrene expected.[522, 523]

(c) WITH AROMATIC NITROSO COMPOUNDS. A number of workers[395, 398,524, 525] investigated the reaction of 9-methylacridine with a series of 4-nitroso-(mono- and di-)alkylanilines and concluded that the products were anils because on hydrolysis all gave acridine-9-aldehyde in good yield. This was supported by the observation that the product of the reaction of acridine-9-aldehyde and 4-aminodimethylaniline had, after much purification, the same melting point as the anil. However, the analytical data and general properties of the alleged anils indicated that all were mixtures. A number of subsequent investigations have attempted to clarify this matter.

The product from 9-methylacridine and 4-nitrosodimethylaniline in the presence of piperidine at 115°, or boiling ethanol, proved to be a molecular complex of the nitrone, **114**, and the anil, **115**, in 3:1 ratio.[38]

114   **115**

2-Methoxy-9-methylacridine with 2 moles of the nitroso compound gave 82% of the corresponding nitrone.[38] This is in agreement with the earlier observation[524] that different products were obtained if one, or two, equivalents of *p*-nitroso-*N,N*-diethylaniline were condensed with the acridine. The constitution of the nitrone was confirmed by its reduction with sulfur dioxide[38] to the anil. (This is similar to amine oxide reductions.)

At the same time, other investigators condensed nitrosobenzene and its 4-dimethylamino, 4-diethylamino, and 4-ethylamino derivatives with 9-methylacridine in ethanol in the presence of sodium carbonate.[526] The benzene soluble fraction of the 4-nitrosodimethylaniline condensate was identical with the violet anil, **115**, mp 248°, synthesized from acridine-9-aldehyde and 4-aminodimethylaniline, and gave a quantitative yield of the aldehyde with cold 10% aqueous hydrochloric acid. A benzene insoluble fraction crystallized from chlorobenzene to give the carmine nitrone, **114**, mp 243°, which was hydrolyzed quantitatively to acridine-9-aldehyde with boiling dilute hydrochloric acid. With molar equivalent quantities of reactants, the proportions of the anil to the nitrone product varied with the individual nitroso compounds. With nitrosobenzene,[526] as with 4-methylnitrosobenzene,[527] the nitrone was the principal product. In each case it could also be synthesized from acridine-9-aldehyde and the appropriate *N*-phenylhydroxylamine for comparison purposes. The nitrone, **116**, was isomerized by sunlight to the amide, **117**, obtainable by an independent synthesis from acridine 9-carboxylic acid treated with thionyl chloride, and then with aniline.

                                    116   **117**

More recently, Tsuge and his colleagues[156, 157, 528a,b] have reinvestigated a number of the condensations and added some new ones. The results from the use of different amount of mineral acid have been compared with those in which a carbonate was the catalyst. A variety of 4-substituted nitrosobenzenes has been condensed with methylacridines and halo-9-methylacridines, as well as 9-ethylacridine. Their results show that the ratio of anil to nitrone product formed appears to be related not only to the structures and proportions of reactants, but also to the condensation catalyst and the oxidizing ability of *p*-nitrosodimethylaniline (formed in some of the reactions). With 9-ethyl-acridine, there is evidence for products of structures **118**, **119**, and **120**, separable with difficulty.[528b]

      118   **119**

                  120

The Russian workers reported the formation of anil **121** from 9-methyl-acridine as shown here.[522, 523] Both groups describe hydrolysis to 9-acridine-carboxaldehyde as part of the evidence for the anil structure.

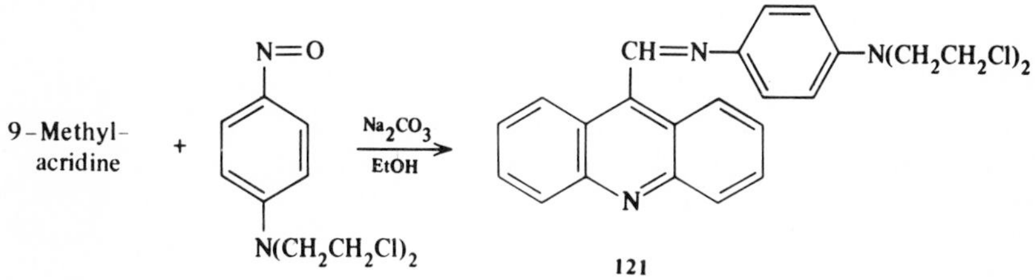

9-Methyl-
acridine   +

$$N=O$$

$$N(CH_2CH_2Cl)_2$$

$$\xrightarrow[\text{EtOH}]{Na_2CO_3}$$

$$CH=N--N(CH_2CH_2Cl)_2$$

121

However, the discrepancy between the melting points for the supposedly identical anils in the two reports suggests a continuing difficulty, associated with the isolation and identification of products from such condensations.

(d) WITH DIAZO COMPOUNDS. 9-Methylacridine reacted with many techni-cally used diazo compounds to give the corresponding azo compounds, **122**, which were identical with the appropriate hydrazones prepared from acridine-9-aldehyde.[529]

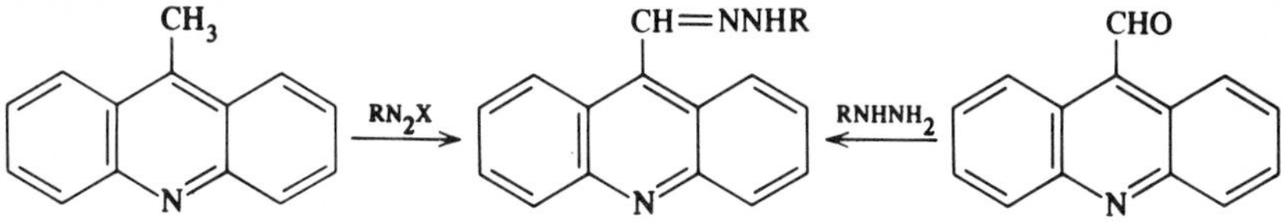

$$CH_3 \qquad\qquad CH=NNHR \qquad\qquad CHO$$

$$\xrightarrow{RN_2X} \qquad\qquad \xleftarrow{RNHNH_2}$$

122

9,10-Dimethylacridinium methosulfate behaved similarly.[506] Sodium hydro-sulfite converted **122** into 9-aminomethylacridine, **126**; (see also p. 128). Comparative studies have shown 9-methylacridine to be more readily reactive in these diazo coupling processes than its 10-oxide.[530] Kinetic measurements over a range of pH values, for the coupling of 9-methylacridine with p-nitrobenzenediazonium chloride have shown that the base enters the reaction as a salt, ionizing to form the zwitterion, **123**, which reacts with the diazo cation.[531]

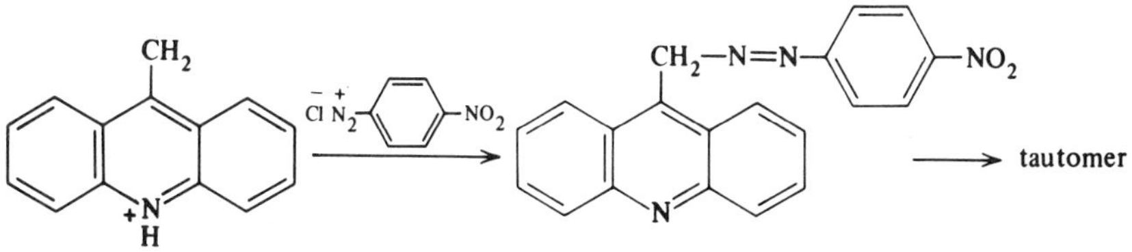

$$CH_2 \qquad\qquad\qquad CH_2-N=N--NO_2$$

$$\xrightarrow{\overset{-}{Cl}\,\overset{+}{N_2}--NO_2}$$

$$\xrightarrow{} \text{tautomer}$$

123

(e) WITH ORGANOMETALLIC AND OTHER REAGENTS. 9-Methylacridine is reported, by successive reaction with phenyl lithium and phenylmethylcyanamide to give 9-acridylacetonitrile in 70% yield.[532] Isomeric adducts, **124** and **125**, are formed when 9-methylacridine reacts with dimethylketene.[533]

124   **125**

9-Methylacridine undergoes the King reaction[534] with pyridine and iodine, forming 1-(9-acridylmethyl)pyridinium iodide. Radical monobromination of 9-methylacridine with N-bromosuccinimide converts it into 9-bromomethylacridine,[501] much used in the synthesis of compounds required in studies of tumor inhibition.[535] The base-catalyzed reaction of 9-methylacridine with dicarboxylic acid anhydrides[536] is

(2) *Arylacridines*

Comparatively little is known about the chemical properties of 9-phenyl-acridine, which is most conveniently prepared by Bernthsen's reaction (Section 2.B). It is a crystalline solid, mp 184°. It is stable to distillation, but, if this is carried out in the presence of zinc, decomposition to acridine, benzene, and tar is reported to take place.[219] 9-Phenylacridine was unaffected by alkaline permanganate, but with sulfuric acid oxidation gave 4-phenylquinoline-3-carboxylic acid and 4-phenylquinoline-2,3-dicarboxylic acid.[117] The bromination of 9-phenylacridine was examined; it appeared that both substitution and addition of bromine may have taken place.[147, 415] These results were inconclusive and provided no evidence regarding the position of the entering bromine atoms. Further experiments, however, have shown that bromine reacted with 9-phenylacridine in carbon tetrachloride to give 10-bromo-9-phenylacridinium bromide.[48] Di- and tri-nitro derivatives of 9-phenylacridine and a disulfonic acid have been prepared by direct substitution,[117] and the nitration of the 10-methylacridinium perchlorate gave a mononitro derivative.[537] There appears to be no direct evidence in support of the notions of 3,6-disubstitution in chlorination and nitration, and of sulfonation in the benzene ring. In this connection, it is of interest to be reminded that nitration of 9-benzylacridine gave 9-(4-nitrobenzyl)acridine, proved by its oxidation to 4-nitrobenzaldehyde by chromic acid.[142]

The phenyl group in 9-phenylacridine can only be weakly conjugated with the acridine system because of the proximity of the neighboring hydrogen atoms in the two-ring systems. It is possible that rotation about the single bond joining the rings is restricted, and that the rings are most of the time at about 90° to each other. There is, however, little evidence on this point.

G. Acridine Aldehydes and Ketones

The comparatively few known compounds of this type all have the aldehyde or ketone group attached to position 9 of the acridine ring. They are yellow compounds which give blue-violet fluorescent solutions, form salts with acids and have typical aldehyde and ketone reactions.

(1) *Aldehydes*

9-Acridinecarboxaldehyde may be prepared in a number of ways, but the most convenient is from 9-methylacridine and 4-nitrosodimethylaniline, followed by hydrolysis of the resulting nitrone (Section 5.F). It has been prepared in the lithium aluminum hydride reduction of ethyl 9-acridine-carboxylate.[399] When 9-hydroxymethylacridan is shaken in benzene over-night with manganese dioxide, golden yellow needles of the aldehyde may be recrystallized from aqueous methanol. Although it is only sparingly soluble in water, the yellow aqueous solution exhibits a blue fluorescence. It forms an oxime,[398] a series of substituted phenylhydrazones,[397, 529] a number of anils,[398, 538] (see also Section 5.F), and Schiff bases with aliphatic amines.[395] Several of the phenylhydrazones have been reduced to 9-aminomethylacridine (126), characterized as its benzoyl derivative.[529] The aldehyde also reacted with nitromethane in the presence of sodium methoxide to give 127, which like the Schiff bases could not be reduced to the corresponding amine.[395] The con-densation product of hippuric acid[539] and 9-acridinecarboxaldehyde was by a sequence of steps converted into β-(9-acridinyl)alanine. A normal base-catalyzed reaction took place between 9-acridinecarboxaldehyde and aceto-phenone,[538] as well as with some heterocyclic compounds containing activated methyl groups,[540] giving the corresponding styrenes, 128 (R = COPh, etc.).

126   127   128

9-Vinylacridine[541] and a number of 1-(9-acridinyl)-2-alkylethylenes (cis and trans) and trans 9-styrylacridine[542] have been prepared under modified Wittig conditions.

9-Acridinecarboxaldehyde reacted with methyl, ethyl, and n-propyl magnesium halides to give the corresponding carbinols, which by successive treatment with hydrobromic acid (60%) at 150–170° and secondary amines were converted into the corresponding aminoacridines.[395] 2-Methoxy-9-acridinecarboxaldehyde also reacted normally with Grignard reagents.[38] 9-Acridinecarboxaldehyde did not give a positive reaction with Schiff's reagent but was oxidized by hot ammoniacal silver nitrate to the correspond-ing acid.[397] Oxidation also took place slowly in air and more rapidly in aqueous acid. The aldehyde was reduced to 9-methylacridine by zinc and formic acid,[395] as previously mentioned.

(2) Ketones

9-Acridinyl methyl ketone (129) has been prepared by the oxidation of 1-(9-acridinyl)ethanol with chromic acid.[395] The ketone behaves normally with amyl nitrite and sodium ethoxide, giving 130.

129   130

The bromoethyl and chloromethyl 9-acridinyl ketones have been prepared from acridine-9-carboxyl chloride by a reaction with diazomethane, followed by treatment of the 9-acridinyl diazomethyl ketone with hydrobromic and hydrochloric acids.[407] This appears to be the only 9-acridinyl alkyl ketone that has been studied.

Large numbers of ketones in which the carbonyl function is one carbon removed from the acridine ring are known. 9-Acridinylacetone has been made by the alkylation of sodium ethyl acetoacetate with 9-chloroacridine with subsequent hydrolysis and decarboxylation.[395] Sheppard and Levine[543] have prepared ketones in this series by acylation of the lithium salt of 9-methylacridine with esters. In a more generally useful process, sodium salts of active methylene compounds with acridine form acridans that may be readily oxidized to the desired acridinylmethyl ketones as follows:

Acridine + (RCOCH$_2$)$^-$Na$^+$ ⟶ [Pb(OAc)$_4$] ⟶

This same synthetic approach has been used by Hayashi and Nakura.[500] The reaction products included compounds such as **131**, mp 186°.

131

The methods just mentioned have also been applied to the synthesis of aryl acridinylmethyl ketones.[500, 543] Information on 9-acridinyl phenyl ketones is limited. The preceding section included the statement that 9-benzyl-acridine and 9-(4-nitrobenzyl)acridine had been oxidized to the corresponding acridinyl phenyl ketones. 9-Acridinyl phenyl ketone has also been prepared from 9-cyanoacridine and phenyl lithium.[39] Benzaldehyde and its 2-chloro and 4-nitro derivatives have been heated (160°) under nitrogen for 5 hr with 9-acridinecarboxylic acid to bring about its decarboxylation with the formation, in fair yield, of the corresponding phenyl ketones.[544]

H. Cyanoacridines

9-Cyanoacridine, a bright yellow solid, mp 181°, is most conveniently obtained from 9-chloroacridine with potassium or sodium cyanide in methanol[407, 545] at 120°, or in the presence of cuprous cyanide at atmospheric pressure.[415] Some 9-methoxyacridine may be formed at the same time.[545] 9-Methoxyacridine hydrate, but not 9-methoxyacridine, is reported to give 9-cyanoacridine with hydrogen cyanide.[411] When 3,9-dichloroacridine in anhydrous methanol under nitrogen atmosphere was treated with sodium cyanide and heated for 4.5 hr in a steel reactor, 3-chloro-9-cyanoacridine was obtained.[546]

Acridine reacted with potassium cyanide in ethanol to form 9-cyanoacridan, which was oxidized by air to 9-cyanoacridine.[39] A variation is to treat acridine with anhydrous hydrogen cyanide and benzoyl chloride, or with aqueous potassium cyanide, followed by benzoyl chloride.[547] There is some evidence for the formation of benzaldehyde in the latter reaction, which suggests that 10-benzoyl-9-cyanoacridan may be an intermediate compound.

2-Cyanoacridine,[44] obtained from 9-chloro-2-cyanoacridine by the p-toluenesulfonhydrazide method (Section 2.A) is the only known cyanoacridine without substituents at position 9; this method of synthesis should be applicable to its isomers. The new dehalogenation method of Albert for 9-chloro-2-cyanoacridine has been applied,[50] as noted earlier. Here the treatment of the 2-cyano-9-hydrazinoacridine with oxygen in the presence of a platinum catalyst and sodium hydroxide furnished a 35% yield of 2-cyano-acridine.

9-Cyanoacridine has been hydrolyzed to the acid, and hydrogenated to 9-cyanoacridan.[395] It reacts with aryl metals; with ammonium hydrogen sulfide it gives the corresponding thioamide[12]; with N,N-di-n-butylamino-magnesium bromide, the N,N-dibutylamidine is formed in excellent yield.[548] 9-Amino-6-cyanoacridine and its 2-methoxy derivative with hydroxylamine in boiling aqueous alcoholic pyridine gave the 6-aminoisonitrosomethylacrid-ines.[549, 550] 9-Cyanoacridine with nitric and sulfuric acids was converted to 9-cyano-4-nitroacridine, the constitution of which was proved by hydrolysis and decarboxylation to the known 4-nitroacridine.[415]

I. Acridine Carboxylic Acids

The acridinecarboxylic acids, few of which are known, are all pale yellow, high melting solids that form salts with acids and bases. The free acids exist partly as zwitterions and are very sparingly soluble in most solvents. The solutions are fluorescent. Most of the acids are decarboxylated to the corresponding acridines on heating above their melting points[86, 12, 397] or in nitrobenzene.[415] Attempts to decarboxylate 3,6-diamino-1,8-acridinedi-carboxylic acid failed.[160]

4-Acridinecarboxylic acid (132), prepared by reduction of the 9-acridanone (Table I), is much lower melting (mp 189°) and generally more soluble than 2-acridinecarboxylic acid, mp 270° (dec) (Tables I and IX), or 9-acridine-carboxylic acid, mp 290° (dec). These differences in properties are undoubtedly a result of intramolecular interaction, as shown in 132, which precludes the intermolecular association which takes place with the isomers.

132

9-Phenyl-2-acridinecarboxylic acid has been obtained[148] by the oxidation of 2-methyl-9-phenylacridine with chromic and sulfuric acids. The direct oxidation of other alkylacridines has not yet been carried out, but the chromic sulfuric acid oxidation of the nitrone, 114, obtained from 9-methylacridine and 4-nitrosodimethylaniline, or of 2-(9-acridinyl)ethanol (109) gave excellent yields of 9-acridinecarboxylic acid. However, this acid is more conveniently obtained by the sulfuric acid (100°) hydrolysis of 9-cyanoacridine to the amide, followed by treatment with nitrous acid.[12] Saggiomo and Weinstock used

90% sulfuric acid for the hydrolysis of 3-chloro-9-cyanoacridine to 3-chloro-9-acridinecarboxylic acid.[546] The 9-carboxylic acid has also been made from N-phenylisatin (Section 2.E). Very recently, there has appeared a patented preparation for some methyl and chlorine substituted 4,9-acridinedicarboxylic acids, which begins with the appropriately substituted isatin salt and potassium o-chlorobenzoate.[551] Methyl substituted 2-, and 4-acridinecarboxylic acids have been prepared by the usual cyclization and acridanone reduction methods.[86] One new synthesis of 9-phenyl-2-acridinecarboxylic acid began with benzilic acid p-carboxanilide ethyl ester.[552]

9-Acridinecarboxylic acid was not esterified by ethanol in the presence of sulfuric or hydrochloric acids,[195] as might have been expected from its structural resemblance to 2,6-dimethylbenzoic acid. However, esters were easily made by converting the acid to the acid chloride hydrochloride with thionyl chloride, followed by treatment with the appropriate alcohol.[195, 553] The amide is readily obtained in the reaction of the same acid chloride hydrochloride with ammonia.[502] The acid chloride also reacted with diazomethane to give the diazoketone. Certain 9-carboxylic acids [Section 5.G(2)] have been decarboxylated in the presence of benzaldehydes with the production of 9-benzoylacridines.[544]

References

1. C. Graebe and H. Caro, *Chem. Ber.*, **3**, 746 (1870).
2. C. Graebe and H. Caro, *Justus Liebigs Ann. Chem.*, **158**, 265 (1871).
3. C. Graebe and H. Caro, *Chem. Ber.*, **13**, 99 (1880).
4. G. Koller and E. Strang, *Monatsh. Chem.*, **50**, 48 (1928).
5. C. Riedel, *Chem.Ber.*, **16**, 1609 (1883).
6. A. Bernthsen and L. Bender, *Chem. Ber.*, **16**, 1802 (1883).
7. A. Bernthsen, *Justus Liebigs Ann. Chem.*, **192**, 1 (1878).
8. C. Graebe and K. Lagodzinski, *Chem. Ber.*, **25**, 1733 (1892).
9. C. Graebe, *Chem. Ber.*, **16**, 2828 (1883).
10. V. I. Khmelevskii and I. Ya. Postovskii, *J. Appl. Chem. (USSR)*, **17**, 463 (1944).
11. E. Wirth, German Patent 440,772; *Friedl.*, **15**, 342 (1925).
12. K. Lehmstedt and E. Wirth, *Chem. Ber.*, **61**, 2044 (1928).
13. K. H. Engel, U.S. Patent 2,408,975; *Chem. Abstr.*, **41**, 998 (1947).
14. V. I. Khmelevskii and I. I. Ovchinnikova, *Org. Chem. Ind. (USSR)*, **7**, 626 (1940).
15. A. Bolliger, *Analyst* (London), **64**, 416 (1939).
16. V. Cordier, *Monatsh. Chem.*, **43**, 525 (1923); *Sitzungsber.*, **81**, 477 (1922).
17. E. Sawicki and C. R. Engel, *Mikrochim. Acta*, 91 (1969).
18. O. Hinsberg, *Chem. Ber.*, **38**, 2800 (1905).
19. A. Albert, *The Acridines*, 2nd ed. Arnold, London, 1966, p. 236.
20. K. von Auwers and R. Kraul, *Chem. Ber.*, **58**, 543 (1925).
21. D. V. van Krevelen, L. Blom, and H. A. G. Chermin, *Nature*, **171**, 1076 (1953).
22. N. P. Buu-Hoï and A. Pacault, *J. Phys. Radium*, [8], **6**, 33 (1945).
23. J. B. Willis, *Trans. Faraday Soc.*, **43**, 97 (1947).

24. L. M. Jackman and D. I. Packham, *Proc. Chem. Soc.*, 349 (1957).
25. C. W. N. Cumper, R. F. A. Ginman, and A. I. Vogel, *J. Chem. Soc.*, 4518 (1962).
26. E. Bergmann, L. Engel, and H. Meyer, *Chem. Ber.*, **65**, 446 (1932).
27. (a) G. Leroy, C. Aussems, and F. van Remoortere, *Bull. Soc. Chim. Belges*, **77**, 181 (1968); (b) M. A. El-Bayoumi and O. S. Khalil, *J. Chem. Phys.*, **47**, 4863 (1967).
28. J. Eisch and H. Gilman, *Chem. Rev.*, **57**, 525 (1957).
29. H. C. Longuet-Higgins and C. A. Coulson, *Trans. Faraday Soc.*, **43**, 87 (1949).
30. B. Pullmann, *Bull. Soc. Chim. Fr.*, 533 (1948).
31. A. Veillard and G. Berthier, *Theor. Chim. Acta*, **4**, 347 (1966).
32. P. J. Black, R. D. Brown, and M. L. Hefferman, *Aust. J. Chem.*, **20**, 1305 (1967).
33. A. F. Terpugova, *Spectroscopiya, Metody Primenenie, Akad. Nauk SSR, Sibirsk. Otd.*, 105 (1964); *Chem. Abstr.*, **62**, 121d (1965).
34. A. T. Amos and G. G. Hall, *Mol. Phys.*, **4**, 25 (1961).
35. R. Zahradnik and J. Koutecky in *Advances in Heterocyclic Chemistry*, Vol. 5, A. R. Katritsky, ed., Academic Press, New York, 1965, pp. 69 ff.
36. (a) D. C. Phillips, *Acta Crystallogr.*,9, 237 (1956). (b) D. C. Phillips, F. R. Ahmed, and W. H. Barnes, *Acta Crystallogr.*, **13**, 365 (1960), discussed by D. J. Sutor in *Ann. Rep. Chem. Soc.*, **57**, 491 (1961).
37. W. L. Semon and D. Craig, *J. Amer. Chem. Soc.*, **58**, 1278 (1936).
38. T. D. Perrine and L. J. Sargent, *J. Org. Chem.*, **14**, 583 (1949).
39. K. Lehmstedt and F. Dostal, *Chem. Ber.*, **72**, 804 (1939).
40. (a) C. Graebe and J. Locher, *Justus Liebigs Ann. Chem.*, **279**, 275 (1894); (b) C. Graebe and S. Kahn, *Justus Liebigs Ann. Chem.*, **279**, 268 (1894); (c) R. A. Reed, *J. Chem. Soc.*, 679 (1944).
41. H. Brockmann and H. Muxfeldt, *Chem. Ber.*, **89**, 1379 (1956).
42. R. Escales, *Chem. Ber.*, **18**, 893 (1885).
43. J. S. McFadyen and T. S. Stevens, *J. Chem. Soc.*, 584 (1936); W. R. Bamford and T. S. Stevens, *J. Chem. Soc.*, 4735 (1952).
44. A. Albert and R. Royer, *J. Chem. Soc.*, 1148 (1949).
45. C. Niemann and J. T. Hayes, *J. Amer. Chem. Soc.*, **65**, 482 (1943).
46. V. P. Maksimets and A. K. Sukhomlinov, *Zh. Org. Khim.*, **1**, 1137 (1965); *Chem. Abstr.*, **63**, 11495h (1965).
47. G. K. Hughes, N. K. Matheson, A. T. Norman, and E. Ritchie, *Aust. J. Sci. Res.*, **A5**, 206 (1952).
48. R. M. Acheson, T. G. Hoult, and K. A. Barnard, *J. Chem. Soc.*, 4142 (1954).
49. G. S. Chandler, R. A. Jones, and W. H. F. Sasse, *Aust. J. Chem.*, **18**, 108 (1965).
50. A. Albert, *J. Chem. Soc.*, 4653 (1965).
51. von Schuckmann, Frames 3306-8, Publishing Boards Microfilm PB 82023 on I. G. Farbenindustrie, U.S. Dept. of Commerce (1942).
52. A. Albert and E. Ritchie, *J. Proc. Roy. New South Wales*, **74**, 77, 373 (1940).
53. A. Albert and J. B. Willis, *J. Soc. Chem. Ind.*, **65**, 26 (1946).
54. A. L. Tarnoky, *Biochem. J.*, **46**, 297 (1950).
55. A. M. Grigorovski and V. S. Fedorov, *J. Appl. Chem.* (*USSR*), **21**, 259 (1948).
56. M. S. Newman and W. H. Powell, *J. Org. Chem.*, **26**, 812 (1961).
57. C. Graebe and K. Lagodzinski, *Justus Liebigs Ann. Chem.*, **276**, 35 (1893).
58. (a) M. Schöpff, *Chem. Ber.*, **27**, 2316 (1894); (b) A. Corsini and E. J. Billo, *J. Inorg. Nucl. Chem.*, **32**, 1241 (1970).
59. G. R. Clemo, W. H. Perkin, and R. Robinson, *J. Chem. Soc.*, **124**, 1751 (1924).
60. A. A. Goldberg and W. Kelly, *J. Chem. Soc.*, 102 (1946).
61. M. T. Bogert, A. D. Hirschfelder, and P. G. T. Lauffer, *Collect. Czech. Chem. Commun.*, **2**, 383 (1930).

62. F. R. Bradbury and W. H. Linnell, *Quart. J. Pharm. Pharmacol.*, **15**, 31 (1942).
63. F. R. Bradbury and W. H. Linnell, *J. Chem. Soc.*, 377 (1942).
64. E. Koft and F. H. Case, *J. Org. Chem.*, **27**, 865 (1962).
65. F. Ullmann and R. Maag, *Chem. Ber.*, **40**, 2515 (1907).
66. S. M. Sherlin, G. I. Braz, A. Ya. Yakubovich, E. I. Vorobeva, and F. E. Rabinovich, *Zh. Obshch. Khim.*, **8**, 884 (1938); *Chem. Abstr.*, **33**, 1330 (1939).
67. A. Albert and E. Ritchie, *J. Soc. Chem. Ind.*, **60**, 120 (1941).
68. P. L. Coe, A. E. Jukes, and J. C. Tatlow, *J. Chem. Soc.*, C, 2020 (1966).
69. A. Eckert and R. Pollak, *Monatsh., Chem.*, **38**, 11 (1917).
70. K. Lehmstedt and H. Hundertmark, *Chem. Ber.*, **62**, 1065 (1929).
71. P. Ehrlich and L. Benda, *Chem. Ber.*, **46**, 1931 (1913).
72. A. Albert and W. H. Linnell, *J. Chem. Soc.*, 88 (1936).
73. W. J. Wechter, *J. Org. Chem.*, **28**, 2935 (1963).
74. J. Boes, *Apoth. Zt.*, **30**, 406 (1915).
75. J. N. Graves, G. K. Hughes, and F. Lions, *J. Proc. Roy. Soc. New South Wales*, **71**, 255 (1938).
76. L. J. Sargent, *J. Org. Chem.*, **22**, 1494 (1957).
77. S. Nitzsche, *Chem. Ber.*, **76**, 1187 (1943).
78. R. G. Bolton, Ph.D. Thesis, University of Oxford, England, 1970.
79. V. S. Federov, *Zh. Obshch. Khim.*, **26**, 591 (1956); *Chem. Abstr.*, **50**, 13929h (1956).
80. A. Albert, *J. Chem. Soc.*, 1225 (1948).
81. J. Biehringer, *J. Prakt. Chem.*, **54**, 217 (1896).
82. F. Ullmann and A. Marič, *Chem. Ber.*, **34**, 4307 (1901).
83. L. Benda, *Chem. Ber.*, **45**, 1787 (1912).
84. A. Albert and W. H. Linnell, *J. Chem. Soc.*, 22 (1938).
85. P. Kränzlein, *Chem. Ber.*, **70**, 1785 (1937).
86. G. Stapleton and A. I. White, *J. Amer. Pharm. Assn. (Sci. Ed.)*, **43**, 993 (1954).
87. H. Lund, P. Lunde, and F. Kaufmann, *Acta Chem. Scand.*, **20**, 1631 (1966).
88. R. A. Reed, *J. Chem. Soc.*, 679 (1944).
89. W. H. Linnell and R. E. Stuckey, *Quart. J. Pharm. Pharmacol.*, **13**, 162 (1940).
90. K. Lehmstedt and H. Hundertmark, *Chem. Ber.*, **64**, 2386 (1931).
91. W. H. Perkin and G. R. Clemo, British Patent 214,756; *Chem. Abstr.*, **18**, 2715 (1924).
92. S. M. Sherlin, G. J. Braz, A. J. Jakubowitsch, E. L. Worobjowa, and A. P. Ssergejef, *Justus Liebigs Ann. Chem.*, **516**, 218 (1935).
93. A. Bernthsen and L. Bender, *Chem. Ber.*, **16**, 1971 (1883).
94, C. Graebe and V. Kaufmann, *Justus Liebigs Ann. Chem.*, **279**, 281 (1894).
95. H. Irving, E. J. Butler, and M. F. Ring, *J. Chem. Soc.*, 1489 (1949).
96. J. J. Lingane, C. G. Swain, and M. F. Fields, *J. Amer. Chem. Soc.*, **65**, 1348 (1943).
97. E. R. Blout and R. S. Corley, *J. Amer. Chem. Soc.*, **69**, 763 (1947).
98. S. L. Cosgrove and W. A. Waters, *J. Chem. Soc.*, 907 (1949).
99. W. Bonthone and D. H. Reid, *Chem. Ind.* (London), 1192 (1960).
100. F. Ullmann, *Chem. Ber.*, **36**, 1017 (1903).
101. O. Dimroth and R. Criegee, *Chem. Ber.*, **90**, 2207 (1957).
102. H. Decker and D. Dunant, *Chem. Ber.*, **39**, 2720 (1906).
103. K. Lehmstedt, W. Bruns, and H. Klee, *Chem. Ber.*, **69**, 2399 (1936).
104. H. Gilman and D. A. Shirley, *J. Amer. Chem. Soc.*, **72**, 2181 (1950).
105. R. A. Jackson and W. A. Waters, *J. Chem. Soc.*, 4632 (1958).
106. E. F. Pratt and T. P. McGovern, *J. Org. Chem.*, **29**, 1540 (1964).
107. H. M. Whitlock and G. A. Digenis, *Tetrahedron Lett.*, 1521 (1964).
108. A. Albert and R. Goldacre, *J. Chem. Soc.*, 706 (1946).

109. K. Matsumura, *J. Amer. Chem. Soc.*, **57**, 1533 (1935).
110. A. Albert and W. H. Linnell, *J. Chem. Soc.*, 1614 (1936).
111. L. J. Sargent, *J. Org. Chem.*, **19**, 599 (1954).
112. I. Tanasescu and E. Ramontianu, *Bull. Soc. Chim. Fr.*, [5], **1**, 547 (1934).
113. A. Albert, *The Acridines*, 2nd ed., Arnold, London, 1966, p. 13.
114. K. Matsumura, *J. Amer. Chem. Soc.*, **49**, 810 (1927).
115. W. Borsche, F. Runge and W. Trautner, *Chem. Ber.*, **66**, 1315 (1933).
116. C. Graebe and S. Kahn, *Justus Liebigs Ann. Chem.*, **279**, 270 (1894).
117. A. Bernthsen, *Justus Liebigs Ann. Chem.*, **224**, 1 (1884).
118. F. Kehrmann and Zd. Matusinsky, *Chzm. Ber.*, **45**, 3498 (1912).
119. S. Inagaki, *J. Pharm. Soc. Jap.*, **58**, 961 (1938).
120. N. P. Buu-Hoï, *J. Chem.Soc.*, 792 (1946).
121. F. D. Popp, *J. Org. Chem.*, **27**, 2658 (1962).
122. J. M. Birchall and D. H. Thorpe, *J. Chem. Soc.*, *C*, 2071 (1967).
123. W. E. Erner, U.S. Patent 3,019,227; *Chem. Abstr.*, **56**, 12864h (1962).
124. C. Hollins, *Synthesis of Nitrogen Ring Compounds*, Benn, London, 1924, pp. 67, 81, 92.
125. F. M. Hamer, *J. Chem. Soc.*, 995 (1930).
126. B. Staskun, *J. Org. Chem.*, **29**, 2856 (1964).
127. A. H. Cook, I. M. Heilbron, and A. Spinks, *J. Chem. Soc.*, 417 (1943).
128. S. Meeker, *Textile Colourist*, **49**, 447 (1927).
129. A. Bernthsen and W. Hess, *Chem. Ber.*, **18**, 689 (1885).
130. R. Robinson and M. L. Tomlinson, *J. Chem. Soc.*, 1524 (1934).
131. H. Decker, *Chem. Ber.*, **38**, 2493 (1905).
132. F. Kehrmann and A. Stepanoff, *Chem. Ber.*, **41**, 4133 (1908).
133. A. E. Porai-Koshits, and A. A. Kharkharov, *Bull. Acad. Sci. URSS Classe Sci. Chim.*, 243 (1944); cf. *Chem. Abstr.*, **39**, 1631 (1945).
134. N. P. Buu-Hoï and J. Lecocq, *C. R. Acad. Sci.*, Paris, **218**, 792 (1944).
135. S. M. H. van der Krogt and B. M. Wepster, *Rec. Trav. Chim. Pays-Bas*, **74**, 161 (1955).
136. T. Thu-Cuc, N. P. Buu-Hoï, and N. D. Xuong, *J. Chem. Soc.*, *C*, 87 (1966).
137. D. C. Thang, E. K. Weisburger, Ph. Mabille, and N. P. Buu-Hoï, *J. Chem. Soc.*, *C*, 665 (1967).
138. A. Albert, *The Acridines*, 2nd ed., Arnold, London, 1966, p. 93.
139. O. Fischer and G. Körner, *Chem. Ber.*, **17**, 101 (1884).
140. N. P. Buu-Hoï and J. Lecocq, *Rec. Trav. Chim. Pas-Bas*, **64**, 250 (1945).
141. H. Decker and T. Hock, *Chem. Ber.*, **37**, 1564 (1904).
142. E. H. Huntress and E. N. Shaw, *J. Org. Chem.*, **13**, 674 (1948).
143. A. E. Dunstan and J. A. Stubbs, *Chem. Ber.*, **39**, 2402 (1906).
144. A. Bernthsen and J. Traube, *Chem. Ber.*, **17**, 1510 (1884).
145. H. Decker and T. Hock, *Chem. Ber.*, **37**, 1002 (1904).
146. E. Besthorn and W. Curtman, *Chem. Ber.*, **24**, 2039 (1891).
147. A. E. Dunstan and T. P. Hilditch, *J. Chem. Soc.*, **91**, 1659 (1907).
148. A. Bonna, *Justus Liebigs Ann. Chem.*, **239**, 55 (1887).
149. A. Schmid and H. Decker, *Chem. Ber.*, **39**, 933 (1906).
150. A. Volpi, *Chem. Ber.*, **24R**, 912 (1891).
151. A. Volpi, *Chem. Ber.*, **25R**, 940 (1892)
152. E. Landauer, *Bull. Soc. Chim. Fr.*, **31**, 1083 (1904).
153. O. Blum, *Chem. Ber.*, **62**, 881 (1929).
154. E. Graef, J. M. Fredericksen, and A. Burger, *J. Org. Chem.*, **11**, 257 (1946).
155. H. Jensen and L. Howald, *J. Amer. Chem. Soc.*, **48**, 1988 (1926).

156. O. Tsuge, M. Nishinohara, and M. Tashiro, *Bull. Chem. Soc. Jap.*, **36**, 1477 (1963).
157. O. Tsuge, M. Nishinohara, and K. Sadano, *Bull. Chem. Soc. Jap.*, **38**, 2037 (1965).
158. Ciba, French Patent 203,467 (1890).
159. Leonhardt and Co., German Patent 67,126; *Friedl.*, **3**, 290, (1890–1894).
160. A. Albert, *J. Chem. Soc.*, 244 (1947).
161. Poulenc, Frères, and Mayer, British Patent 137,214; French Patent 509,610 (1919); U.S. Patent 1,419,474; German Patent 347,819; *Friedl.*, **14**, 799 (1921–1925).
162. A. Albert, *J. Chem. Soc.*, 121 (1941).
163. A. Albert, D. K. Large, and W. Kennard, *J. Chem. Soc.*, 484 (1941).
164. A. Albert and D. Magrath, *J. Soc. Chem. Ind.* (London), **64**, 30 (1945).
165. German Patent 141,356; *Friedl.*, **7**, 316 (1902–1904).
166. G. D. Buckley and N. H. Ray, *J. Chem. Soc.*, 1152 (1949).
167. A. Albert, *The Acridines*, 2nd ed., Arnold, London, 1966, p. 99.
168. K. L. Moudgill, *J. Chem. Soc.*, 1506 (1922).
169. J. Duval, *Bull. Soc. Chim. Fr.*, [4], **7**, 527 (1910).
170. M. Schöpff, *Chem. Ber.*, **27**, 2316 (1894).
171. F. Ullmann and R. Fitzenkam, *Chem. Ber.*, **38**, 3787 (1905).
172. J. G. Miller and E. C. Wagner, *J. Amer. Chem. Soc.*, **54**, 3698 (1932).
173. R. Meyer and R. Gross, *Chem. Ber.*, **32**, 2352 (1899).
174. G. T. Morgan and W. J. Hickinbottom, *J. Chem. Soc.*, **123**, 97 (1923).
175. H. R. Snyder and M. S. Konecky, *J. Amer. Chem. Soc.*, **80**, 4388 (1958).
176. S. A. Mahood and C. R. Harris, *J. Amer. Chem. Soc.*, **46**, 2810 (1924).
177. R. Möhlau and O. Haase, *Chem. Ber.*, **35**, 4164 (1902).
178. G. T. Morgan, *J. Chem. Soc.*, **73**, 536 (1898).
179. J. H. Reed, *J. Prakt. Chem.*, **35**, 298 (1887).
180. A. Senier and P. C. Austin, *J. Chem. Soc.*, **89**, 1388 (1906).
181. F. Ullmann and A. Fetvadjian, *Chem. Ber.*, **36**, 1027 (1903).
182. F. Ullmann and E. Naef, *Chem. Ber.*, **33**, 913 (1900).
183. J. H. Billman, D. G. Thomas, M. Hedrick, G. Schrotenboer, D. K. Barnes, J. Nemec, P. Trix, and E. Cleland, *J. Org. Chem.*, **11**, 773 (1946).
184. N. P. Buu-Hoï, R. Royer, and M. Hubert-Habert, *J. Chem. Soc.*, 1082 (1955).
185. N. P. Buu-Hoï, M. Mangane, and P. Jacquignon, *J. Heterocycl. Chem.*, **7**, 155 (1970).
186. A. Senier and A. Compton, *J. Chem. Soc.*, 1623 (1909).
187. O. Dischendorfer, *Monatsh. Chem.*, **48**, 543 (1927).
188. F. G. Pope and H. Howard, *J. Chem. Soc.*, **97**, 78 (1910).
189. F. Ullmann and C. Baezner, *Chem. Ber.*, **35**, 2670 (1902).
190. (a) C. Baezner, *Chem. Ber.*, **37**, 3077 (1904); **39**, 2650 (1906); (b) C. Baezner and A. Gardiol, *Chem. Ber.*, **39**, 2623 (1906); (c) C. Baezner and J. Gueorguieff, *Chem. Ber.*, **39**, 2438 (1906).
191. G. Kalischer and F. Mayer, *Chem. Ber.*, **49**, 1994 (1916).
192. F. Mayer and B. Stein, *Chem. Ber.*, **50**, 1306 (1917).
193. F. Mayer and I. Levis, *Chem. Ber.*, **52**, 1641 (1919).
194. L. I. Smith and J. W. Opie, *Org. Syntheses*, **28**, 11 (1948).
195. H. Jensen and F. Rethwisch, *J. Amer. Chem. Soc.*, **50**, 1144 (1928).
196. A. Albert and B. Ritchie, *J. Chem. Soc.*, 458 (1943).
197. F. Ullmann and H. W. Ernst, *Chem. Ber.*, **39**, 298 (1906).
198. W. Sharp, M. M. J. Sutherland, and F. J. Wilson, *J. Chem. Soc.*, 344 (1943).
199. F. Mayer and W. Freund, *Chem. Ber.*, **55**, 2049 (1922).
200. M. J. Sacha and S. R. Patel, *J. Indian Chem. Soc.*, **34**, 821 (1957); *Chem. Abstr.*, **52**, 9127g (1958).

201. J. C. E. Simpson, *J. Chem. Soc.*, 646 (1945).
202. C. K. Bradsher, *Chem. Rev.*, **38**, 494 (1946).
203. R. Stollé, *J. Prakt. Chem.*, **105**, 137 (1922).
204. P. Friedländer and K. Kunz, *Chem. Ber.*, **55**, 1597 (1922).
205. R. Stollé, R. Bergdoll, M. Luther, A. Auerhahn, and W. Wacker, *J. Prakt. Chem.*, **128**, 1 (1930).
206. J. Martinet and A. Dansette, *Bull. Soc. Chim. Fr.*, **45**, 101 (1929).
207. K. Saftien, *Chem. Ber.*, **58**, 1958 (1925).
208. S. Secareanu, *Chem. Ber.*, **64**, 837 (1931).
209. F. Ullmann and J. Broido, *Chem. Ber.*, **39**, 356 (1906).
210. H. Jensen and M. Friedrich, *J. Amer. Chem. Soc.*, **49**, 1049 (1927).
211. K. Lehmstedt, *Chem. Ber.*, **71**, 808 (1938).
212. G. Wittig and K. Niethammer, *Chem. Ber.*, **93**, 944 (1960).
213. T. Masamune and G. Homma, *Nippon Kagaku Zasshi*, **77**, 1766 (1956); *Chem. Abstr.*, **53**, 5234d (1959).
214. T. Masamune and G. Homma, *J. Fac. Sci., Hokkaido Univ.*, Ser. III, **5**, 64 (1957); *Chem. Abstr.*, **52**, 14581h (1958).
215. J. L. Adelfang and N. H. Cromwell, *J. Org. Chem.*, **26**, 2368 (1961).
216. W. Borsche, *Chem. Ber.*, **41**, 2203 (1908).
217. H. Tiedkte, *Chem. Ber.*, **42**, 621 (1909).
218. D. Vorländer and F. Kalkow, *Justus Liebigs Ann. Chem.*, **309**, 356 (1899).
219. D. Vorländer and O. Strauss, *Justus Liebigs Ann. Chem.*, **309**, 375 (1899).
220. W. Borsche and H. Hartmann, *Cher. Ber.*, **73**, 839 (1940).
221. V. A. Petrow, *J. Chem. Soc.*, 693 (1942.)
222. T. Masamune, G. Homma, and M. Ohno, *J. Fac. Sci., Hokkaido Univ.*, Ser. III, **5**, 59 (1957); *Chem. Abstr.*, **52**, 10998e (1958).
223. G. E. Hall and J. Walker, *J. Chem. Soc.*, C, 2237 (1968).
224. T. Masamune, T. Saito, and G. Homma, *J. Fac. Sci., Hokkaido Univ.*, Ser. III, **5**, 55 (1957); *Chem. Abstr.* **52**, 11062h (1958).
225. D. Redmore, *J. Org. Chem.*, **34**, 1420 (1969).
226. W. Borsche and W. Rottsieper, *Justus Liebigs Ann. Chem.*, **377**, 101 (1910).
227. W. Braun and H. Wolff, *Chem. Ber.*, **55**, 3675 (1922).
228. W. Borsche, H. Schmidt, and R. Tiedkte, *Justus Liebigs Ann. Chem.*, **377**, 79 (1910).
229. C. Mannich and B. Reichert, *Arch. Pharm.* **271**, 116 (1938).
230. O. Fischer and H. Schütte, *Chem. Ber.*, **26**, 3085 (1893).
231. A. Baeyer and V. Villiger, *Chem. Ber.*, **37**, 3191 (1904).
232. (a) P. A. Petyunin, N. G. Panferova, and M. E. Konshin, *Zh. Obshch. Khim.*, **26**, 2050 (1956); *Chem. Abstr.*, **51**, 5078e (1957). (b) P. A. Petyunin and N. G. Panferova, *Zh. Obshch. Khim.*, **26**, 3191 (1956); *Chem Abstr.*, **51**, 8745e (1957). (c) P. A. Petyunin and M. E. Konshin, *Zh. Obshch. Khim.*, **27**, 475 (1957); *Chem. Abstr.*, **51**, 15522h (1957). (d) *Zh. Obshch. Khim.* **27**, 1558 (1957); *Chem. Abstr.*, **52**, 3764b, (1958). (e) *Zh. Obshch. Khim.*, **28**, 974 (1958); *Chem. Abstr.*, **52**, 17267e (1958).
233. O. Fischer and G. Körner, *Chem. Ber.*, **17**, 203 (1884).
234. P. J. Grisdale, J. C. Doty, T. H. Regan, and J. L. R. Williams, *J. Org. Chem.*, **32**, 2401 (1967).
235. C. Graebe, *Chem. Ber.*, **17**, 1370 (1884).
236. A. Y. Yakubovich and M. Nevyadomski, *J. Gen. Chem. (USSR)*, **18**, 887 (1948).
237. F. Ullmann and A. La Torre, *Chem. Ber.*, **37**, 2922 (1904).
238. A. Rieche and R. Moellér, *J. Prakt. Chem.*, **15**, 44 (1961); *Chem. Abstr.*, **57**, 9702b (1962).

239. C. Hansch, F. Gschwend, and J. Bamesberger, *J. Amer. Chem. Soc.*, **74**, 4554 (1952).

240. R. Möhlau, *Chem. Ber.*, **19**, 2451 (1886).

241. H. Blau, *Monatsh. Chem.*, **18**, 123 (1897).

242. J. Eliasberg and P. Friedländer, *Chem. Ber.*, **25**, 1752 (1892).

243. G. K. Hughes and E. Ritchie, *Aust. J. Sci. Res.*, **A4**, 423 (1951).

244. W. H. Linnell and K. L. Sharp, *Quart. J. Pharm. Pharmacol.*, **21**, 58 (1948).

245. G. Minardi and P. Schenone, *Ann. Chim.* (Rome), **49**, 702 (1959); *Chem. Abstr.*, **53**, 22055g (1959).

246. G. T. Morgan, W. J. Hickinbottom, and T. V. Barker, *Proc. Roy. Soc.* (*Ser. A*), **110**, 502 (1926).

247. D. H. Hey, *J. Chem. Soc.*, 1581 (1931).

248. P. Rumpf and R. Reynaud, *Bull. Soc. Chim. Fr.*, 2241 (1961).

249. H. Meyer and A. Hofmann, *Monatsh. Chem.*, **37**, 681 (1916).

250. J. Ploquin, *C. R., Acad. Sci., Paris*, **224**, 481 (1947).

251. A. Pictet and H. J. Ankersmit, *Justus Liebigs Ann. Chem.*, **266**, 138 (1891).

252. C. Graebe, *Justus Liebigs Ann. Chem.*, **335**, 122 (1904).

253. A. Pictet and S. Ehrlich, *Justus Liebigs Ann. Chem.*, **266**, 153 (1891).

254. E. Grandmougin and A. Lang, *Chem. Ber.*, **42**, 4014 (1909).

255. R. Meyer and O. Oppelt, *Chem. Ber.*, **21**, 3376 (1888).

256. F. Ullmann, *Justus Liebigs Ann. Chem.*, **332**, 82 (1903).

257. W. Staedel, *Chem. Ber.*, **27**, 3362 (1894).

258. H. C. Waterman and D. L. Vivian, *J. Org. Chem.*, **14**, 289 (1949).

259. C. Graebe and F. Ullmann, *Chem. Ber.*, **27**, 3483 (1894).

260. E. Grandmougin and A. Lang, *Chem. Ber.*, **42**, 3631 (1909).

261. I. Tanasescu, C. Anghel, and A. Popescu, *Studia Univ. Babes-Bolyai Ser. Chemia*, **8**, 141 (1963); *Chem. Abstr.*, **61**, 13279e (1964).

262. R. Kwok and P. Pranc, *J. Org. Chem.*, **33**, 2880 (1968).

263. G. M. Badger, N. Kowanko, and W. H. F. Sasse, *J. Chem. Soc.*, 440 (1959).

264. T. Kauffmann and K. Udluft, *Angew. Chem., Int. Ed. Engl*, **2**, 45 (1963).

265. F. de Carli, *Gazz. Chim. Ital.*, **57**, 347 (1927).

266. F. H. Herbstein and G. M. J. Schmidt, *Acta Crystallogr.*, **8**, 399 (1955).

267. D. C. Phillips, *Acta Crystallogr.*, **7**, 649 (1954).

268. A. Albert, *The Acridines*, 2nd ed., Arnold, London, 1966, p. 235.

269. M. Trautz, *Z. Phys. Chem.* (Leipzig), **53**, 59 (1905).

270. E. J. Bowen, N. J. Holder, and G. B. Woodger, *J. Phys. Chem.*, **66**, 2491 (1962).

271. L. Lindquist, *C. R. Acad. Sci., Paris, Ser. C*, **263**, 852 (1966); *Chem. Abstr.*, **66**, 70835n (1967).

272. A. Albert, *The Acridines*, 2nd ed., Arnold, London, 1966, p. 206.

273. D. Bertrand, *Bull. Soc. Chim. Fr.*, **12**, 1019 (1945).

274. (a) Th. Förster, *Z. Electrochem.*, **54**, 42 (1950); (b) A. Weller, *Z. Electrochem.*, **61**, 956 (1957).

275. (a) J. A. Radley and J. Grant, *Fluorescence Analysis in Ultraviolet Light*, 3rd ed., Chapman and Hall, London, 1939; (b) Débériré, *Ann. Chim. Anal.*, **19**, 290 (1937); (c) H. Goto and Y. Kakita, *Nippon Kagaku Zasshi*, **64**, 575 (1943).

276. K. Nikolic and L. Sablic, *Arh. Farm.*, **17**, 121 (1967); *Chem. Abstr.*, **69**, 30159a (1968).

277. V. C. Jelineck and G. E. Boxer, *J. Biol. Chem.*, **175**, 367 (1948).

278. J. V. Scudi and V. C. Jelineck, *J. Biol. Chem.*, **164**, 195 (1946).

279. J. K. D. Verma and M. D. Aggarwal, *Indian J. Pure Appl. Phys.*, **7**, 771 (1969); *Chem. Abstr.* **72**, 16488n (1970).

280. P. K. Mitskevich and M. I. Bashmakova, *Zh. Fiz. Khim.*, **38**, 1606 (1964); *Chem. Abstr.*, **61**, 7817a (1964).

281. A. Albert, S. D. Rubbo, and R. Goldacre, *Nature*, **147**, 332, (1941).
282. A. Albert, R. Goldacre, and J. Phillips, *J. Chem. Soc.*, 2240 (1948).
283. H. T. S. Britton and W. G. Williams, *J. Chem. Soc.*, 796 (1935).
284. D. A. Brown and M. J. S. Dewar, *J. Chem. Soc.*, 2410 (1953).
285. H. C. Longuet-Higgins and C. A. Coulson, *J. Chem. Soc.*, 971 (1949).
286. R. Reynaud and P. Rumpf, *Bull. Soc. Chim. Fr.*, 1805 (1963).
287. R. Reynaud, *Bull. Soc. Chim. Fr.*, 4597 (1967).
288. G. Sellier and B. Wojtkowiak, *J. Chim. Phys.*, **65**, 936 (1968).
289. L. Joris and P. von R. Schleyer, *Tetrahedron*, **24**, 5991 (1968).
290. R. Cetina, D. V. S. Jan, F. Peradejordi, O. Chalvet, and R. Daudel, *C. R. Acad. Sci., Paris, Ser. C.*, **264**, 874 (1967).
291. N. Tyutyulkov, F. Fratev, and D. Petkov, *Theor. Chim. Acta*, **8**, 236 (1967); *Chem Abstr.*, **67**, 108075a (1967).
292. G. Jackson and G. Porter, *Proc. Roy. Soc., Ser. A*, **260**, 13 (1961).
293. H. G. Hoeve and W. A. Yeranos, *Mol. Phys.*, **12**, 597 (1967).
294. B. Briat and M. Le Liboux, *C. R. Acad. Aci., Paris, Ser. C*, **262**, 803 (1966).
295. P. Machmer and J. Duchesne, *C. R. Acad. Sci., Paris, Ser. C*, **260**, 3775 (1965).
296. P. R. Hammond and R. H. Knipe, *J. Amer. Chem. Soc.*, **89**, 6063 (1967).
297. D. A. Davenport, H. J. Burkhardt, and H. H. Sisler, *J. Amer. Chem. Soc.*, **75**, 4175 (1953).
298. A. U. Malik, *J. Inorg. Nucl. Chem.*, **29**, 2106 (1967).
299. R. D. Whitaker, *J. Inorg. Nucl. Chem.*, **26**, 1405 (1964).
300. A. A. Oswald and F. Noel, U.S. Patent 3,236,850; *Chem. Abstr.*, **64** 14045d (1966).
301. B. J. Brisdon and D. A. Edwards, *J. Chem. Soc., D*, **278** (1966).
302. A. R. Utke and R. T. Sanderson, *J. Org. Chem.*, **29**, 1261 (1964).
303. M. Orchin, L. Reggel, and E. O. Woolfolk, *J. Amer. Chem. Soc.*, **69**, 1225 (1947).
304. E. Shefter, *Science*, **160**, 1351 (1968).
305. P. R. Hammond, *Nature*, **201**, 922 (1964).
306. P. I. Petrovich. *Zh. Prikl. Khim.*, **32**, 353 (1959); *Chem. Abstr.*, **53**, 17031h (1959).
307. F. Runge, H. J. Engelbrecht, and H. Frank, *Chem. Ber.*, **88**, 533 (1955).
308. S. K. Chakrabarti, *Spectrochim. Acta, A*, **24**, 790 (1968).
309. G. Wanag and A. Lode, *Chem. Ber.*, **70**, 547 (1937).
301. R. Kremann and O. Wlk, *Monatsh. Chem.*, **40**, 57 (1919).
311. R. Kremann and F. Solvak, *Monatsh, Chem.*, **41**, 5 (1920).
312. P. Pascal, *Bull. Soc. Chim. Fr.*, **29**, 644 (1921).
313. G. Castelfranchi, *Chem. Abstr.*, **45**, 5694 (1951).
314. J. R. Allan, D. H. Brown, R. H. Nuttall, and D. W. A. Sharp, *J. Inorg. Nucl. Chem.*, **26**, 1895 (1964).
315. H. H. Sisler, N. El-Jadir, and D. H. Busch, *J. Inorg. Nucl. Chem.*, **16**, 257 (1961).
316. W. P. Neumann, *Justus Liebigs Ann. Chem.*, **667**, 1 (1963).
317. R. Ripan, I. Gainescu, and C. Varheliji, *Rev. Roum. Chim.*, **11**, 1051 (1966); *Chem. Abstr.*, **66**, 91240e (1967).
318. P. Spacu and D. Camboli, *Analele Univ. Bucuresti, Ser. Stünt. Nat.*, **14**, 101 (1965); *Chem. Abstr.*, **66**, 7694w (1967).
319. I. Lukacs, C. Strusievici, and C. Liteanu, *J. Prakt. Chem.*, **28**, 209 (1965); *Chem. Abstr.*, **63**, 4256d (1965).
320. E. Ya. Gorenbein, G. G. Rusin, and E. P. Skorobogat'ko, *Zn. Neorg. Khim.*, **14**, 516 (1969); *Chem. Abstr.*, **70**, 92738t (1969).
321. H. L. Krauss and G. Gnatz, *Chem. Ber.*, **95**, 1023 (1962).
322. P. Spacu and C. Gheorghiu, *Z. Anal. Chem.*, **174**, 340 (1960); *Chem. Abstr.*, **55**, 1289i (1961).

323. A. Langer, *Mikrochemie*, **25**, 71 (1938).
324. E. P. Wien, *Atomkernenergie*, **13**, 294 (1968); *Chem. Abstr.*, **69**, 72775d (1968).
325. R. Ewart, *Chem. Abstr.*, **13**, 1332 (1919).
326. H. Fühner, *Arch. Exptl. Path. Pharmakol.*, **51**, 391 (1904).
327. M. Kumasaka, *Nichidai Igaku Zasshi*, **19**, 3726 (1960); *Chem. Abstr.*, **61**, 7467h (1964).
328. S. A. Heininger, *J. Org. Chem.*, **22**, 704 (1957).
329. C. D. Hurd and R. Kotani, *J. Chem. Soc.*, *C*, 2655 (1967).
330. J. Nasielski and E. Vander Donckt, *Spectrochim. Acta*, **19**, 1989 (1963).
331. (a) T. Jasinski and H. Smagowski, *Rocz. Chem.*, **41**, 363, 791 (1967); *Chem. Abstr.*, **67**, 57693r, 90325v (1967). (b) J. D'Souza and A. Bruylants, *Bull. Soc. Chim. Belges*, **74**, 591 (1965).
332. J. Foucart, J. Nasielski, and E. Vander Donckt, *Bull. Soc. Chim. Belges*, **75**, 17 (1966).
333. R. M. Acheson and B. Adcock, *J. Chem. Soc.*, *C*, 1045 (1968).
334. T. Hase, *Acta Chem. Scand.*, **18**, 1806 (1964).
335. E. J. Moriconi and F. A. Spano, *J. Amer. Chem. Soc.*, **86**, 38 (1964).
336. S. Niizuma, M. Okuda, and M. Koizumi, *Bull. Chem. Soc. Jap.*, **41**, 795 (1968).
337. W. A. Waters and D. H. Watson, *J. Chem. Soc.*, 253 (1957).
338. R. O. C. Norman and G. K. Radda, in *Advances in Heterocyclic Chemistry*, Vol. 2, Academic Press, New York, 1963, p. 155.
339. H. Gilman and G. D. Lichtenwalter, *J. Org. Chem.*, **23**, 1586 (1958).
340. D. Redmore, *J. Org. Chem.*, **34**, 1420 (1969).
341. F. Kröhnke and H. L. Honig, *Justus Liebigs Ann. Chem.*, **624**, 97 (1959).
342. W. E. McEwen and R. L. Cobb, *Chem. Rev.*, **55**, 516 (1955).
343. G. A. Reynolds, *J. Org. Chem.*, **29**, 3733 (1964).
344. B. H. Klanderman, *Tetrahedron Lett.*, 6141 (1966).
345. G. Wittig and K. Niethammer, *Chem. Ber.*, **93**, 944 (1960).
346. (a) S. A. Procter and G. A. Taylor, *J. Chem. Soc.*, 5877 (1965); (b) S. A. Procter, G. A. Taylor, and T. Wood, *Chem. Ind.* (London), 1019 (1969).
347. A. Cattaneo, *Farmaco, Ed. Sci.*, **12**, 930 (1957); *Chem. Abstr.*, **52**, 11850g (1958).
348. K. Sisido, K. Tani, and H. Nozaki, *Tetrahedron*, **19**, 1323 (1963).
349. E. Hayashi, S. Ohsumi, and T. Maeda, *Yakugaku Zasshi*, **79**, 967 (1959); *Chem. Abstr.*, **53**, 21947b (1959).
350. E. Hayashi, *Yakugaku Zasshi*, **79**, 969 (1959); *Chem. Abstr.*, **53**, 21934d (1959).
351. H. Nozaki, Y. Yamamoto, and R. Noyori, *Tetrahedron Lett.*, 1123 (1966).
352. K. Fukui, T. Yonezawa, C. Nagata, and H. Shingu, *J. Chem. Phys.*, **22**, 1433 (1954).
353. A. Albert and G. Catterall, *J. Chem. Soc.*, 4657 (1965).
354. G. M. Badger and W. H. F. Sasse, *J. Chem. Soc.*, 616 (1956).
355. A. A. Ponomarev, A. S. Chegolya and V. N. Dyukareva, *Khim. Geterotsikl. Soedin.*, 239 (1966); *Chem. Abstr.*, **65**, 2219c (1966).
356. H. Gilman, J. L. Towle, and R. K. Ingham, *J. Amer. Chem. Soc.*, **76**, 2920 (1954).
357. R. E. Lyle and P. S. Anderson, in *Advances in Heterocyclic Chemistry*, Vol. 6, Academic Press, New York, 1966, p. 78.
358. (a) O. N. Nechaeva and Z. V. Pushkareva, *Zh. Obshch. Khim.*, **28**, 2693 (1958); *Chem. Abstr.*, **53**, 9229c (1959). (b) L. B. Radina, Z. V. Pushkareva, N. M. Voronina, and N. M. Khvorova, *Zh. Obshch. Khim.*, **30**, 3480 (1960); *Chem. Abstr.*, **55**, 23528d (1961).
359. K. Ito, *Yakugaku Zasshi*, **86**, 1166 (1966); *Chem. Abstr.*, **66**, 75899w (1967).
360. A. J. Birch and H. H. Mantsch, *Aust. J. Chem.*, **22**, 1003 (1969).
361. W. R. Orndorff and F. K. Cameron, *Amer. Chem. J.*, **17**, 658 (1895).
362. A. Kellmann, *J. Chim. Phys.*, **54**, 468 (1957); *Chem. Abstr.*, **51**, 16112f (1957).

363. H. Wolter, *Meded. Vlaam. Chem. Ver.*, **31**, 172 (1969); *Chem. Abstr.*, **72**, 89416x (1970).
364. V. Zanker, F. Mader, and W. Körber, *Angew. Chem. Int. Ed. Eng.*, **3**, 388 (1964).
365. V. Zanker and P. Schmid, *Z. Phys. Chem.* (Frankfurt am Main), **17**, 11 (1958).
366. F. Mader and V. Zanker, *Chem. Ber.*, **97**, 2814 (1964).
367. H. Goth, P. Cerutti, and H. Schmid, *Helv. Chim. Acta*, **48**, 1395 (1965).
368. V. Zanker, E. Erhardt, and J. Thies, *Ind. Chim. Belge*, **32**, 24 (1967); *Chem. Abstr.*, **70**, 87538y (1969).
369. S. Kato, S. Minagawa, and M. Koizumi, *Bull. Chem. Soc. Jap.*, **34**, 1026 (1961).
370. V. Zanker and E. Erhardt, *Ber. Bunsenges. Phys. Chem.*, **72**, 267 (1968); *Chem. Abstr.*, **69**, 26527b (1968).
371. S. Niizuma, Y. Ikeda, and M. Koizumi, *Bull. Chem. Soc. Jap.*, **40**, 2249 (1967).
372. A. Kellmann, *J. Chim. Phys.*, **56**, 574 (1959); *Chem. Abstr.*, **53**, 21088c (1959).
373. A. Weller, *Z. Elektrochem.*, **61**, 956 (1957); *Chem. Abstr.*, **52**, 4297i (1958).
374. M. Giurgea, V. Topa, and S. Haragea, *J. Chim. Phys.*, **58**, 705 (1961); *Chem. Abstr.*, 56. 13698e (1962).
375. T. Shida and A. Kira, *Bull. Chem. Soc. Jap.*, **42**, 1197 (1969).
376. A. Kira, Y. Ikeda, and M. Koizumi, *Bull, Chem. Soc. Jap.*, **39**, 1673 (1966).
377. A. Kellmann and J. T. Dubois, *J. Chem. Phys.*, **42**, 2518 (1965).
378. A. Kira, S. Kato, and M. Koizumi, *Bull. Chem. Soc., Jap.* **39**, 1221 (1966).
379. A. Niizuma and M. Koizumi, *Bull. Chem. Soc., Jap.*, **41**, 1090 (1968).
380. E. Vander Donckt and G. Porter, *J. Chem. Phys.*, **46**, 1173 (1967).
381. A. Kellmann, *J. Chim. Phys.*, **63**, 936 (1966); *Chem. Abstr.*, **66**, 45971w (1967).
382. F. Wilkinson and J. T. Dubois, *J. Chem. Phys.*, **48**, 265 (1968).
383. K. Nakamaru, S. Niizuma, and M. Koizumi, *Bull. Chem. Soc. Jap.*, **42**, 255 (1969).
384. A. Kira and M. Koizumi, *Bull. Chem. Soc. Jap.*, **40**, 2486 (1967).
385. M. Koizumi, Y. Ikeda, and T. Iwaoka, *J. Chem. Phys.*, **48**, 1869 (1968).
386. R. Noyori, M. Kato, M. Kawanisi, and H. Nozaki, *Tetrahedrom*, **25**, 1125 (1969).
387. V. Zanker and G. Prell, *Ber. Bunsenges. Phys. Chem.*, **73**, 791 (1969); *Chem. Abstr.*, **70**, 130670v (1969).
388. M. Koizumi, Y. Ikeda, and H. Yamashita, *Bull. Chem. Soc. Jap.*, **41**, 1056 (1968).
389. A. Kira and M. Koizumi, *Bull. Chem. Soc. Jap.*, **42**, 625 (1969).
390. E. de Gorski and G. de Gaudemaris, *C. R. Acad. Sci., Paris*, **253**, 2965 (1961).
391. A. P. Wolf and R. C. Anderson, *J. Amer. Chem. Soc.*, **77**, 1608 (1955).
392. M. Forys, *Radiochim. Acta*, **7**, 13 (1967).
393. J. L. Garnett and S. W. Law, *Aust. J. Chem.*, **20**, 1875 (1967).
394. R. M. Acheson and M. L. Burstall, *J. Chem. Soc.*, 3240 (1954).
395. O. Eisleb, *Med. Chem. Abhandl. Med.-Chem. Forschungstatten I. G. Farbenindustrie*, **3**, 41 (1936); *Chem. Abstr.*, **31**, 5802 (1937).
396. E. Hayashi, *Yakugaku Zasshi*, **79**, 969 (1959); *Chem. Abstr.*, **53**, 21934d (1959).
397. A. Bernthsen and F. Muhlert, *Chem. Ber.*, **20**, 1541 (1887).
398. A. Kaufmann and L. G. Vallette, *Chem. Ber.*, **45**, 1736 (1912).
399. A. Campbell and E. N. Morgan, *J. Chem. Soc.*, 1711 (1958).
400. A. Kliegl and A. Fehrle, *Chem. Ber.*, **47**, 1629 (1914).
401. A. Kliegl and A. Brösamle, *Chem. Ber.*, **69**, 197 (1936).
402. K. Lehmstedt and H. Klee, *Chem. Ber.*, **69**, 1155 (1936).
403. K. Issleib and L. Bruesehaber, *Z. Naturforsch., B*, **20**, 181 (1965); *Chem. Abstr.*, **63**, 628b (1965).
404. P. A. Petyunin, M. E. Konshin, and N. G. Panferova, *Khim. Geterotsikl Soedin., Akad. Nauk Latv. SSR*, 257 (1965); *Chem. Abstr.*, **63**, 11496d (1965).

405. A. Edinger and W. Arnold, *J. Prakt. Chem.*, **64**, 471 (1901).

406. A. Albert and B. Ritchie, *Org. Syntheses*, **22**, 5 (1942).

407. G. I. Braz and T. V. Gortinskaya, *J. Gen. Chem.* (*USSR*), **10**, 1751 (1940).

408. O. J. Magidson and A. M. Grigorovski, *Chem. Ber.*, **66**, 866 (1933).

409. P. Kristian, K. Antos, A. Hulka, P. Nemec, and L. Drobnica, *Chem. Zvesti*, **15**, 333 (1961); *Chem. Abstr.*, **55**, 27322a (1961).

410. K. Lehmstedt, *Chem. Ber.*, **68**, 1455 (1935).

411. N. S. Drozdov and O. M. Cherntzov, *J. Gen, Chem.* (*USSR*), **14**, 181 (1944).

412. M. Ionescu, I. Goia, and H. Mantsch, *Acad. Rep. Populare Romine, Filiala Cluj, Stud. Cercet. Chim.*, **13**, 95 (1962); *Chem. Abstr.*, **60**, 5456h (1964).

413. D. J. Dupré and F. M. Robinson, *J. Chem. Soc.*, 549 (1945).

414. T. D. Perrine, *J. Org. Chem.*, **25**, 1516 (1960).

415. K. Lehmstedt, *Chem. Ber.*, **64**, 1232 (1931).

416. A. K. Scheinkman, G. V. Saniodenko, and S. N. Baranov, *Zh. Obshch. Khim.*, **40**, 700 (1970); *Chem. Abstr.*, **73**, 14931y (1970).

417. N. S. Drozdov, *J. Gen. Chem.* (*USSR*), **6**, 1641 (1936).

418. G. J. O'Donnell, *Iowa State College J. Sci.*, **20**, 34 (1945).

419. R. E. Davies, H. T. Openshaw, F. S. Spring, R. H. Stanley, and A. R. Todd, *J. Chem. Soc.*, 295 (1948).

420. G. T. Morgan and G. R. Davies, *Proc. Roy. Soc., Ser. A*, **143**, 38 (1933).

421. M. J. S. Dewar and P. M. Maitlis, *J. Chem. Soc.*, 2521 (1957).

422. K. Lehmstedt, *Chem. Ber.*, **60**, 1370 (1927).

423. French Patent, 1,458,183; *Chem. Abstr.*, **68**, 39493s (1968).

424. E. A. Steck, J. S. Buck, and L. T. Fletcher, *J. Amer. Chem. Soc.*, **79**, 4414 (1957).

425. T. Sasa, *J. Soc. Org. Synthet. Chem. Jap.*, **12**, 183 (1954); *Chem. Abstr.*, **51**, 2780fg (1957).

426. A. Senier and P. C. Austin, *J. Chem. Soc.*, **85**, 1196 (1904).

427. A. Edinger and W. Arnold, *J. Prakt. Chem.*, **64**, 182 (1901).

428. N. P. Buu-Hoï, *Justus Liebigs Ann. Chem.*, **556**, 1 (1944).

429. H. Schmid and W. E. Leutenegger, *Helv. Chim. Acta*, **30**, 1965 (1947).

430. R. W. Watrous, *Brit. J. Ind. Med.*, **4**, 111 (1947).

431. J. T. Stewart, T. D. Shaw, and A. B. Ray, *Anal. Chem.*, **41**, 360 (1969).

432. M. Ionescu, H. Mantsch, and I. Goia, *Acad. Rep. Populare Romine, Filiala Cluj, Stud. Cercet. Chim.*, **12**, 135 (1961); *Chem. Abstr.*, **57**, 12432i (1962).

433. A. S. Samarin and A. G. Lebekov, *Tr. Tomskogo Gos. Univ., Ser. Khim.*, **154**, 253 (1962); *Chem. Abstr.*, **60**, 4110f (1964).

434. M. Ionescu and I. Goia, *Acad. Rep. Populare Romine, Filiala Cluj, Stud. Cercet. Chim.*, **10**, 335 (1959); *Chem. Abstr.*, **55**, 533g (1961).

435. M. Ionescu, I. Goia and I. Felmeri, *Acad. Rep. Populare Romine, Filiala Cluj, Stud. Cercet. Chim.*, **8**, 351 (1957); *Chem. Abstr.* **54**, 4587h (1960).

436. (a) O. M. Cherntzov and N. S. Drozdov, *J. Gen. Chem.* (*USSR*), **9**, 1373 (1939); (b) N. S. Drozdov and N. S. Leznova, *J. Gen. Chem.* (*USSR*), **5**, 690 (1935).

437. A. Ledochowski, E. Zylkiewicz, and F. Muzalewski, *Rocz. Chem.*, **42**, 445 (1968); *Chem. Abstr.*, 69, 85854d (1968).

438. G. Illuminati, G. Marino, and O. Piovesana, *Ric. Sci. Rend., Sez. A*, **4**, 437 (1964); *Chem. Abstr.*, **61**, 14504a (1964).

439. A. Ledochowski, *Rocz. Chem.*, **40**, 2015 (1966).

440. A. Ledochowski, *Rocz. Chem.*, **41**, 717 (1967).

441. A. Ledochowski, *Rocz. Chem.*, **41**, 1255 (1967).

442. A. Ledochowski, *Rocz. Chem.*, **41**, 1561 (1967).

443. V. Zanker and W. Fluegel, *Acta. Chim. Acad. Sci. Hung.*, **40,** 45 (1964); *Chem. Abstr.*, **61,** 9056f (1964).

444. V. Zanker and W. Fluegel, *Z. Naturforsch., B.*, **19,** 376 (1964); *Chem. Abstr.*, **61,** 10559h (1964).

445. A. A. Goldberg and W. Kelly, British Patent 600,354; *Chem. Abstr.*, **42,** 7797 (1948).

446. E. N. Morgan and D. J. Tivey, British Patent 789,696; *Chem. Abstr.*, **52,** 13806f (1958).

447. K. G. Yekundi and S. R. Patel, *Chem. Ber.*, **90,** 2448 (1957).

448. P. Kristian, *Chem. Zvesti*, **23,** 371 (1969); *Chem. Abstr.*, **72,** 21584v (1970).

449. G. M. Kosolapoff, *J. Amer. Chem. Soc.*, **69,** 1002 (1947).

450. V. G. Pesin, I. G. Vitenberg, and A. M. Khaletskii, *Zh. Obshch. Khim.*, **34,** 2769 (1964); *Chem. Abstr.*, **61,** 14663e (1964).

451. G. R. Clemo and W. Hook, *J. Chem. Soc.*, 608 (1936).

452. M. Gomberg and D. L. Tabern, *J. Amer. Chem. Soc.*, **48,** 1345 (1926).

453. K. C. Joshi and S. C. Bakel, *J. Indian Chem. Soc.*, **38,** 877 (1961); *Chem. Abstr.*, **56,** 15483e (1962).

454. K. Drechsler, *Monatsh. Chem.*, **35,** 533 (1914); *Sitzungsber.*, **73,** 51 (1914).

455. I. Goia, *Stud. Cercet. Chim.*, **14,** 155 (1966); *Chem. Abstr.*, **65,** 2219d (1966).

456. R. M. Acheson, B. Adcock, G. M. Glover, and L. E. Sutton, *J. Chem. Soc.*, 3367 (1960).

457. J. H. Markgraf and C. G. Carson, *J. Org. Chem.*, **29,** 2806 (1964).

458. G. A. Dima and P. Pogangeanu, *Bull. Sect. Sci. Acad. Roumanie*, **22,** 19 (1939).

459. H. Mantsch, W. Seifert, and V. Zanker, *Rev. Roum. Chim.*, **12,** 1137 (1967); *Chem. Abstr.*, **69,** 48105q (1968).

460. H. Mensch, W. Seifert, and V. Zanker, *Rev. Roum. Chim.*, **14,** 125 (1969); *Chem. Abstr.* **71,** 21606h (1969).

461. T. Kubota, K. Nishikida, H. Miyozaki, K. Iwatani, and Y. Oishi, *J. Amer. Chem. Soc.*, **90,** 5080 (1968).

462. Z. V. Pushkareva, L. V. Varyukshina, and Z. Yu. Kokoshko, *Dokl. Akad. Nauk SSSR*, **108,** 1098 (1956); *Chem. Abstr.*, **51,** 21i (1957).

463. G. Anthoine, J. Nasielski, E. Vander Donckt, and N. Vanlautern, *Bull. Soc. Chim. Belges*, **76,** 230 (1967).

464. C. Kaneko, S. Yameda and M. Ishikawa, *Chem. Pharm. Bull.*, (*Tokyo*), **17,** 1294 (1969); *Chem. Abstr.*, **71,** 70517p (1969).

465. H. Mantsch and V. Zanker, *Tetrahedron Lett.*, 4211 (1966).

466. M. Ishikawa, C. Kaneko, and S. Yamada, *Tetrahedron Lett.*, 4519 (1968).

467. V. Zanker, W. Seiffert, and G. Prell, *Justus Liebigs Ann. Chem.*, **723,** 95 (1969); *Chem Abstr.*, **71,** 38089q (1969).

468. W. Seiffert, H. H. Mantsch, V. Zanker, and H. H. Limpach, *J. Mol. Struct.*, **5,** 227 (1970).

469. J. H. Markgraf and M-K. Ahn, *J. Amer. Chem. Soc.*, **86,** 2699 (1964).

470. S. Oae, S. Kozuka, Y. Sakaguchi, and K. Hiramatsu, *Tetrahedron*, **22,** 3143 (1966).

471. I. Tanasescu and Z. Frenkel, *Stud. Cercet. Chim.*, **4,** 227 (1956); *Chem. Abstr.*, **51,** 10527d (1957).

472. I. Tanasescu and E. Ramontianu, *Bull. Soc. Chim. Fr.*, **3,** 2009 (1936).

473. M. Ionescu, I. Goia, and H. Mantsch, *Rev. Roum. Chim.*, **11,** 243 (1966); *Chem. Abstr.*, **65,** 3713e (1966).

474. L. V. Varyukhin, O. N. Nechaeva, and Z. V. Pushkareva, *Metody Polucheniya Khim. Reaktivov Preparatov, Gos. Kom. Sov. Min. SSSR Khim.*, 84 (1964); *Chem. Abstr.*, **64,** 15837e (1966).

475. L. B. Radina, K. V. Aglitskaya, A. I. Cherkasova, and Z. V. Pushkareva, *Zh. Obshch. Khim.*, **34**, 1543 (1964); *Chem. Abstr.*, **61**, 5753g (1964).
476. F. Linsker and M. T. Bogert, *J. Amer. Chem. Soc.*, **68**, 192 (1946).
477. K. Lehmstedt and H. Klee, *Chem. Ber.*, **69**, 1514 (1936).
478. (a) M. Ionescu, H. Mantsch, and I. Goia, *Chem. Ber.*, **96**, 1726 (1963); (b) M. Ionescu, A. R. Katritsky, and B. Ternai, *Tetrahedron*, **22**, 3227 (1966).
479. H. H. Perkampus and Th. Rössel, *Z. Electrochem.*, **62**, 94 (1958); *Chem. Abstr.*, **52**, 9760h (1958).
480. S. F. Mason, *J. Chem. Soc.*, 4874 (1957).
481. S. F. Mason, *J. Chem. Soc.*, 5010 (1957).
482. S. F. Mason, *J. Chem. Soc.*, 674 (1958).
483. S. Nitsche, *Angew. Chem.*, **52**, 517 (1939).
484. D. H. Freeman and F. Lions, *J. Proc. Roy. Soc. New South Wales*, **74**, 520 (1940).
485. A. Albert and D. Magrath, *Biochem. J.*, **41**, 534 (1947).
486. A. Corsini and E. J. Billo, *J. Inorg. Nucl. Chem.*, **32**, 1249 (1970).
487. (a) A. G. Cairns-Smith, *J. Chem. Soc.*, 182 (1961); (b) N. Campbell and A. G. Cairns-Smith, *J. Chem. Soc.*, 1191 (1961).
488. A. Konopnicki and E. Sucharda, *Rocz. Chem.*, **7**, 183 (1927).
489. H. J. Barker, J. H. Wilkinson, and W. G. H. Edwards, *J. Soc. Chem. Ind.* (London) **66**, 411 (1947).
490. R. O. Clinton and C. M. Suter, *J. Amer. Chem. Soc.*, **70**, 491 (1948).
491. N. S. Drozdov and O. M. Cherntzov, *J. Gen. Chem.* (*USSR*), **5**, 1576 (1935).
492. S. P. Acharya and K. S. Nargund, *J. Karnatak University*, **6**, 33 (1961); *Chem. Abstr.* **59**, 7489h (1963).
493. A. G. Lebekov and A. S. Samarin, *Khim. Geterotsikl. Soedin.*, 838 (1969); *Chem. Abstr.*, **72**, 111271v (1969).
494. Z. V. Pushkareva and O. N. Nechaeva, *Zh. Obshch. Khim.*, **28**, 2702 (1958); *Chem. Abstr.*, **53**, 9229g (1959).
495. A. Ledochowski and S. Wojenski, *Rocz. Chem.*, **44**, 43 (1970).
496. R. M. Acheson, M. L. Burstall, C. W. Jefford, and B. F. Sansom, *J. Chem. Soc.*, 3742 (1954).
497. G. N. Kosheleva, *Zh. Obshch. Khim.*, **26**, 2567 (1956); *Chem. Abstr.*, **51**, 5081a (1957).
498. W. W. Carlson and L. H. Cretcher, *J. Amer. Chem. Soc.*, **70**, 597 (1948).
499. (a) O. N. Chupakhin, V. A. Trofimov, and Z. V. Pushkareva, *Dokl. Akad. Nauk SSSR*, **188**, 376 (1969); *Chem. Abstr.*, **72**, 3341j (1970). (b) O. N. Chupakhin, V. A. Trofimov, and Z. V. Pushkareva, *Khim. Geterotsikl. Soedin.*, 954 (1969); *Chem. Abstr.*, **72**, 111270u (1970).
500. E. Hayashi and T. Nakura, *Yakugaku Zasshi*, **87**, 570 (1967); *Chem. Abstr.*, **67**, 64228w (1967).
501. A. Campbell, C. S. Franklin, E. N. Morgan, and D. J. Tivey, *J. Chem. Soc.*, 1145 (1958).
502. G. I. Braz and S. A. Kore, *Zh. Obshch. Khim.*, **23**, 868 (1953); *Chem. Abstr.*, **48**, 3979d (1954).
503. H. Nozaki, Y. Yamamoto, and T. Nisimura, *Tetrahedron Lett.*, 4625 (1968).
504. H. Nozaki, M. Yamabe, and R. Noyori, *Tetrahedron*, **21**, 1657 (1965).
505. R. Wizinger and H. Wenning, *Helv. Chem. Acta*, **23**, 247 (1940).
506. R. Wizinger and B. Cyriax, *Helv. Chim. Acta*, **28**, 1018 (1945).
507. B. J. Sudborough, *J. Chem. Soc.*, **109**, 1339 (1916).
508   H. Bassett and T. A. Simmonds, *J. Chem Soc.*, 416 (1921).
509. R. Schmidt, *Chem. Abstr.*, **38**, 5753 (1944).

510. N. V. de Bataafsche Petroleum, *Chem. Abstr.*, **35**, 1626 (1941).
511. Y. Seoka, K. Tanikawa, H. Hirata, S. Kusabayashi, and H. Mikawa, *Chem Commun.*, 652a (1969).
512. L. Monti, *Atti. Accad. Naz. Lincei Rend., Cl. Sci. Fis. Mat. Nat.*, **24**, 145 (1936).
513. N. P. Buu-Hoï, *J. Chem.Soc.*, 670 (1949).
514. A. W. Homberger and H. Jensen, *J. Amer. Chem. Soc.*, **48**, 800 (1926).
515. L. Monti and M. Procopio, *Gazz. Chim. Ital.*, **63**, 724 (1933).
516. N. Fisher, C. S. Franklin, E. N. Morgan, and D. J. Tivey, *J. Chem. Soc.*, 1411 (1958).
517. L. Monti and V. Felici, *Gazz. Chim. Ital.*, **70**, 375 (1940).
518. K. Friedländer, *Chem. Ber.*, **38**, 2840 (1905).
519. W. Sharp, M. M. J. Sutherland, and F. J. Wilson, *J. Chem. Soc.*, 5 (1943).
520. A. E. Porai-Koshits, P. A. Solodovnikoff, and M. V. Troitzki, *Z. Farben Ind.*, **6**, 291, (1907).
521. N. F. Kazarinova and I. Ya. Postovskii, *Zh. Obshch. Khim.*, **27**, 3325 (1957); *Chem. Abstr.*, **52**, 9126i (1958).
522. Z. V. Pushkareva, Z. P. Penyugalova, and R. Kh. Batulina, *Zh. Obshch. Khim.*, **34**, 2475 (1964); *Chem. Abstr.*, **61**, 9463e (1964); *J. Gen. Chem.*, *USSR*, **34**, 2492 (1964).
523. I. I. Chizhevskaya, L. I. Gapanovich, and N. N. Khovratovich, *J. Gen. Chem. USSR*, **34**, 4118 (1964).
524. A. E. Porai-Koshits, V. Auschkap, and N. K. Amsler, *J. Russ. Phys. Chem. Soc.*, **43**, 518 (1910).
525. E. Bergmann and L. Haskelberg, *J. Chem. Soc.*, 1 (1939).
526. L. Chardonnens and P. Heinrich, *Helv. Chim. Acta*, **32**, 656 (1949).
527. N. S. Drozdov and E. V. Yavorskaya, *Zh. Obshch. Khim.*, **30**, 3421 (1960); *Chem. Abstr.*, **55**, 18728h (1961).
528. (a) O. Tsuge and M. Nishinohara, *Bull. Chem. Soc. Jap.*, **38**, 2034 (1965); (b) O. Tsuge, M. Nishinohara, and K. Sadano, *Bull. Chem. Soc. Jap.*, **38**, 2037 (1965).
529. A. E. Porai-Koshits and A. A. Kharkov, *Bull. Acad. Sci. URSS, Classe Sci. Chim.*, 143 (1944).
530. I. M. Mishina and L. S. Efros, *J. Gen. Chem. USSR*, **32**, 3185 (1962).
531. I. M. Mishina and L. S. Efros, *Zh. Obshch. Khim.*, **34**, 2358 (1964); *Chem. Abstr.*, **61**, 14813f (1964).
532. H. Lettré, P. Jungmann, and J. C. Salfield, *Chem. Ber.*, **85**, 397 (1952).
533. (a) S. A. Procter and G. A. Taylor, *Chem. Commun.*, 569 (1965); (b) S. A. Procter and G. A. Taylor, *J. Chem. Soc., C*, 1937 (1967).
534. F. Kröhnke and K. F. Gross, *Chem. Ber.*, **92**, 22 (1959).
535. (a) W. Rzeszotarski and Z. Ledóchowski, *Rocz. Chem.*, **38**, 1631 (1964); *Chem. Abstr.*, **60**, 14472c (1964). (b) *Rocz. Chem.*, **39**, 93 (1965); *Chem. Abstr.*, **63**, 4256a (1965).
536. M. Milosev and B. Aleksiev, *Chem. Ber.*, **102**, 2869 (1969).
537. R. J. W. Le Fevre and J. Pearson, *J. Chem. Soc.*, 482 (1933).
538. L. Monti, *Gazz. Chim. Ital.*, **69**, 749 (1939).
539. L. B. Radina and Z. V. Pushkareva, *Zh. Obshch. Khim.*, **31**, 2362 (1961); *Chem. Abstr.*, **56**, 3447f (1962).
540. W. L. Glen, M. M. J. Sutherland, and F. J. Wilson, *J. Chem. Soc.*, 1484 (1936); 654 (1938).
541. O. Tsuge, A. Torii, and T. Tomita, *Nippon Kagaku Zasshi*, **90**, 1263 (1969); *Chem. Abstr.*, **72**, 78837j (1970).
542. O. Tsuge, T. Tomita, and A. Torii, *Nippon Kagaku Zasshi*, **89**, 1104 (1968); *Chem. Abstr.*, **70**, 96595s (1969).

543. C. S. Sheppard and R. Levine, *J. Heterocycl. Chem.*, **1,** 67 (1964).
544. A. I. Gurevich, *Zh. Obshch. Khim.*, **28,** 322 (1958); *Chem. Abstr.*, **52,** 13725d (1958).
545. G. I. Braz, *J. Gen. Chem.* (*USSR*), **11,** 851 (1941).
546. A. J. Saggiomo and J. Weinstock, U.S. Patent 3,016,373; *Chem. Abstr.*, **57,** 2201f (1962).
547. K. Bauer, *Chem. Ber.*, **83,** 10 (1950).
548. E. Lorz and R. Baltzly, *J. Amer. Chem. Soc.*, **70,** 1904 (1948).
549. A. A. Goldberg and W. Kelly, *J. Chem. Soc.*, 637 (1947).
550. A. A. Goldberg and W. Kelly, British Patent 595,401; *Chem. Abstr.*, **42,** 4204 (1948).
551. K. Sada and K. Nishide, Japanese Patent 7,006,816; *Chem. Abstr.*, **72,** 132565k (1970).
552. P. A. Petyunin, A. F. Soldatova, and A. K. Sukhomlinov, *Khim. Geterotsikl. Soedin.*, 702 (1969); *Chem,. Abstr.* **72,** 31578k (1970).
553. B. Samdahl and C. F. Weider, *Bull. Soc. Chim. Fr.*, [5], **2,** 2008 (1935).

CHAPTER II

Aminoacridines

B. ADCOCK

Flintshire College of Technology,
Flintshire, Wales

The stimulus for many investigations of the aminoacridines has come primarily from a recognition of their important biological activity. The valuable antibacterial properties of 3,6-diaminoacridine (proflavine) were discovered by Browning in 1913, and the antimalarial properties of some substituted 9-aminoacridines (including atebrin) were described by Mauss and Mietzsch in 1930. More recently, the carcinostatic activity of several 9-aminoacridines has attracted a good deal of attention. The aminoacridines also display a great variety in their physical and chemical behavior. The properties of the 9-aminoacridines, in particular, differ considerably from those of the isomers; for this reason, the two groups are considered separately.

The synthesis of an aminoacridine most often involves the cyclization of an appropriately substituted diphenylamine-2-carboxylic acid, prepared by the Ullmann reaction, to the corresponding 9-acridanone or 9-chloroacridine. These may be reduced to the acridine in a variety of ways, or the 9-chloroacridine may be converted to a 9-aminoacridine.

1. Basic Strengths and Solution Properties

The base strengths of large numbers of aminoacridines have been determined,[1-6] usually in connection with studies of biological activity. Most determinations have been by potentiometric titration, but other techniques used include absorptiometry, solubility, and partitions (with ether) on buffer solutions containing the acridine. These results have been very successfully correlated with antibacterial activity.

The first pK_a values in water at 20° of the five monoaminoacridines are shown in Table I, that of acridine under the same conditions[2] is 5.6. Second pK_a values for 1-, 2-, 3-, 4-, and 9-aminoacridine are 0.6, 1.1, approximately −1.4, 0.5, and less than −2, respectively.[2, 7] The base strengths of many aminoacridines with low solubility in water have been measured in 50% ethanol. Often, the pK_a values in this solvent are of the order of 0.5 units lower than in water; however, in some cases, unusually large differences can arise, so that caution must be exercised in comparing basicities measured in the two solvents. For example, the pK_a of 4,5-dimethylacridine is 1.68 units greater in water than in 50% ethanol, a difference attributed to differential solvation of the neutral species and the cation.[2] The effect of temperature on pK_a values has also been studied.[8]

9-Aminoacridine is by far the most basic monoaminoacridine, pK_a 9.99. This is because it can form a very stable resonant ion (1), such as formed from guanidines and amidines.[9] Spectroscopic evidence shows that in most cases the first proton does, indeed, attack the ring nitrogen atom. The addition of a second proton to the 9-aminoacridinium ion takes place only in strong acids, giving the nonresonant dication (2).

1　　　　　　　　　　　　　　　　　　　2

Further evidence for the existence of (1) comes from a study of the first and second dissociation constants of a series of 9-diethylaminoalkylaminoacridines (3, where n is 2–8).

3

For aliphatic diamines a plot of ΔpK against the distance between the nitrogen atoms agrees closely with the curve calculated from the mutual interaction of the two ionic centers. In the case of the acridines (3), the calculated distance between the ionic centers was found to be intermediate between the distance from the diethylamino group to the 9-amino nitrogen and that from the diethylamino group to the ring nitrogen atom.

A similar type of resonance is possible for both 1-aminoacridine and 3-aminoacridine, pK_a's 6.04 and 3.04, respectively, but only the latter is appreciably more basic than acridine. This is because of a contribution to the stability of the monocation from a *para*-quinonoid structure (4) in the case of 3-aminoacridine but little contribution from the less stable *ortho*-quinonoid structure (5) in the case of the 1-isomer.

2-Aminoacridine and 4-aminoacridine cannot exhibit such stabilization of the ion and in consequence are weaker bases, pK_a's 5.88 and 4.40, respectively. The fact that the latter is even less basic than acridine is attributed[1] to hydrogen bonding (6), a hypothesis supported by the ir spectrum.[10]

The introduction of substituent groups into an aminoacridine usually affects the basicity in a predictable fashion, electron-withdrawing groups tending to diminish, and electron donors increasing, the basic strength. However, Albert has drawn attention to the fact that a methoxyl group at position 2 does not exhibit the base-strengthening mesomeric effect expected, but rather the base-weakening inductive effect predominates.[7] Steric hindrance to the approach of a proton toward the heterocyclic nitrogen atom by groups at positions 4 and 5 reduces the basic strength. For example, 4,5-dimethylacridine has a pK_a of 4.56 (in water) compared with acridine, pK_a 5.6. However, the introduction of a single methyl group at position 4 slightly increases the base strength to pK_a 5.65. Steric hindrance to coplanarity of the amino group at position 9 also reduces basicity, and 9-dimethylaminoacridine is a weaker base than the parent amine and its monomethyl derivative,[2,7] resonance such as in 1 being possible only to a much more limited extent. An additional amino group at positions 1 or 2 causes a small increase in basicity, but at positions 3 or 9 it increases the base strength considerably because of the additional resonant structures possible. The introduction of a second amino group at position 4 results in a small diminution in basicity, and 4,5-diaminoacridine has a pK_a of only 4.12. Spectral evidence indicates that in this case the proton has attacked an exocyclic nitrogen atom.[7]

A number of workers have investigated the correlation of the basicity of the aminoacridines with calculated parameters, such as the charge on the nuclear nitrogen atom, the charge on the carbon adjacent to the nuclear nitrogen in the corresponding carbanion, and the first ionization potential.[11,12,13] Major contributions to the energy of protonation (ΔE) from the π electron energy change accompanying protonation (ΔE_π) and from solvation energy changes (ΔE_{solv}) have also been computed, and used with some success to predict base strengths of the aminoacridines.[14,15] For details of the calculations, the reader is referred to the original literature.

The R_F values for the monoaminoacridines and for some other amino derivatives on paper using ammonium chloride solution or butanol/acetic acid mixtures as the solvent have been tabulated.[16] Other workers have also studied the chromatographic properties of a variety of aminoacridines on paper[17−20] or using thin-layer techniques.[21−23] The differences between observed R_F values on wet, buffered paper and those calculated from partition data have been used to calculate adsorption potentials for a number of aminoacridines.[24] The partition of aminoacridines between olive oil[25] and water at pH 7, between castor oil[26a,b] and water at pH 7.3, and between benzene[24] and various buffer solutions has been examined. On the whole, the aminoacridines have little effect on the surface tension of water, compared with aliphatic amines of similar molecular weight. Of the simple monoaminoacridines, 4-aminoacridine has the largest effect, being the only one to depress the surface tension more efficiently than acridine itself. The effect decreases as the pH diminishes and the acridines are converted to their salts.[25] Both the aminoacridines and their salts associate in solution; there is considerable evidence, from the conductivity of aqueous 9-aminoacridine hydrochloride and also from deviations from Beer's law for solutions of monoaminoacridines, that micelles are formed.[27] The association phenomena have been examined spectroscopically in various solvents by Levshin[28, 29] and others.[30a,b] The adsorption of 9-aminoacridine on charcoal[31] and on albumin[32] has been investigated.

2. 1-, 2-, 3-, and 4-Aminoacridines

The general properties and best methods of preparation are listed in Table I. In most cases, the preparation involves the reduction of an amino-9-acridanone or a nitro-9-acridanone, followed by reoxidation of the intermediate aminoacridan. The amines are all very soluble in ethanol and pyridine, but only 4-aminoacridine dissolves in light petroleum, probably because intramolecular hydrogen bonding is important in this case.

In general, the amino groups in these four aminoacridines behave in the normal aromatic fashion. Spectral studies confirm the presence of the amino structure,[43a,b,c] but reveal that where a group with sufficiently strong electron-attracting properties is attached to the 3-amino group (as in 3-benzene-sulfonamidoacridine), the tautomeric imino form (**7**) becomes important.[44] Methyl iodide attacks the primary amino groups in preference to the ring-nitrogen giving mixtures that are difficult to separate,[45] but 4-aminoacridine was converted to 4-(2-diethylaminoethylamino)acridine by 2-diethylamino-ethyl chloride.[46] All give monoacetyl derivatives with acetic anhydride at 100°; these derivatives, with the exception of 4-acetamidoacridine, reacted with methyl p-toluenesulfonate at 180° to give the 10-methylacridinium salts.[45] Treatment of 4-p-toluenesulfonamidoacridan with methyl sulfate and alkali, followed by hydrolysis, gave 4-methylaminoacridine, oxidation having taken place.[46]

All the aminoacridines with aromatic aldehydes gave anils, the 1- and 2-isomers quantitatively but the 3- and 4-isomers in poorer yield.[45] Hydrolysis with concentrated hydrochloric acid at 180° gave the corresponding hydroxy-acridines,[45] but alkali had little effect. Catalytic reduction over Raney nickel resulted in the formation of aminoacridans,[45] and in each case the first reduction stage in controlled potential coulometric reduction was shown to involve one electron only.[47] On boiling the aminoacridines with aniline in the presence of one equivalent of hydrochloric acid, the corresponding anilino-acridines were obtained.[45] Little work has been done on the nitration of aminoacridines, but nitration of 2-acetamidoacridine gave a red inhomo-geneous material, and nitration of 1-acetamidoacridine gave an orange crystalline solid that was not investigated further.[48] The 1-, 2-, and 3-amino-acridines diazotized normally and coupled with phenols. The 4-isomer,

    7   **8**

TABLE I. Monoaminoacridines

Positions of amino group	Appearance, solvent and mp(°C)	pK_a^2 in H_2O at 20°	Fluorescence color of base in EtOH; salts in H_2O	mp(°C) of acetyl derivatives	Color of diazo solution	Best Method of Synthesis	Ref.
1-	Small yellow needles ex. aq EtOH, mp 180°	6.04	Orange	None	230	Yellow	Reduction of 1-amino-9-acridanone	33-37
2-	Fine yellow needles ex. EtOH, mp 213·4°	5.88	Intense yellow-green	None	230	Orange	Reduction of 2-amino-9-acridanone or 2-nitro-9-acridanone	34, 35, 38
3-	Large yellow crystals, mp 218°, ex. EtOH	8.04	Intense yellow green	Intense yellow	229	Yellow	3-Aminodiphenylamine and formic acid, or reduction of 3-amino-9-acridanone	34-38
4-	Red-brown needles ex. petroleum ether (bp 60/80°), mp 105-106°	4.40	None	None	117	Buff ppt.	Reduction of 4-nitro-9-acridanone	35, 39, 40

114

| 9- | Yellow cryst. ex. acetone mp 234° | 9.99 | Green-blue | Green or blue-violet at great dilution | 266(mono), 164 (di) | — | 9-Chloroacridine and ammonium carbonate in phenol | 35, 41, 42 |

however, gave a buff precipitate, a reaction reminiscent of that of 8-amino-quinoline[49] with nitrous acid and suggested[45] to be caused by the formation of **8**. Successful replacement of amino groups by way of the diazonium salts is rather rare, and attempts to produce 3-hydroxy compounds from derivatives of 3-aminoacridine failed.[50] However, replacement by the arsonic acid group,[38] bromine[36] and hydrogen[50] have been reported, although in low yield. The repetition[38] of early work, which claimed the formation of 3-aminoacridine by partial diazotization of 3,6-diaminoacridine followed by reduction with ethanol,[51] was not successful; but 3-aminoacridine was formed from 3,6-diaminoacridine by monoacetylation, followed by diazotization, reduction with hypophosphorous acid, and hydrolysis.[52] 3-Aminoacridine did not react with carbon disulfide alone; in the presence of a large excess of triethylamine, the triethylammonium salt of 3-acridylthiocarbamate was produced.[53] Isothiocyanatoacridines were obtained from 1-, 2-, and 3-amino-acridines and thiophosgene, but the 4-isomer failed to react.[37] 2-Amino-acridine reacted normally with cyanamide, forming 2-guanidinoacridine.[54]

The resonance energies for the aminoacridines have been calculated, to-gether with the energies of the highest occupied and lowest free molecular orbitals and the electronic charge distribution.[55]

3. 9-Aminoacridines

A. Preparation

The first preparations of 9-aminoacridines were from 9-chloro-, 9-ethoxy-or 9-phenoxy-acridines by heating with amines, or suitable derivatives, at about 140° in the presence of copper salts.[56] Yields were usually quite good and replacement of the ammonia with urea or ammonium carbonate avoided the use of pressure vessels.

Although phenol had been used as a solvent for the reaction of 9-chloro-acridine with a number of amines,[57] its importance in facilitating the reaction was not generally recognized until the work of Magidson and Grigorovskii.[58] Reaction in phenol can usually be carried out at 100–120°, the phenoxy-acridinium ion (**11**; X = OPh) almost certainly being an intermediate. Sub-sequently, other workers[59–62] used the method; Drozdov and Cherntzov re-ported a 94% yield of 9-aminoacridine from 9-chloroacridine and ammonium chloride in phenol,[63] although it has been noted that yields are variable because of the poor solubility of ammonium chloride. 9,9-Diphenoxyacridan was isolated from one such reaction.[61]

Best results are usually obtained when the 9-chloroacridine is condensed with a free amine[64–67] (see also earlier papers) or a 9-phenoxyacridine with the amine hydrochloride[67] in phenol at 100–120°. The product may be isolated as the hydrochloride on diluting the reaction mixture with ether or acetone, or as the free base by the action of aqueous alkali. 9-Aminoacridine itself is most conveniently prepared from 9-chloroacridine and ammonium carbonate in phenol and may be isolated as the free base[68] or the hydrochloride[42] with equal ease. Although the reaction is acid-catalyzed, the presence of an excess is harmful. The more reactive 9-chloroacridines[69] give better yields at temperatures in the region of 60°. A very large number of substituted 9-aminoacridines were prepared in this way for testing as antimalarials during World War II[70]; more recently, many more have been synthesized as potential carcinostatic drugs (see Ref. 71 and earlier papers) as well as for testing for other useful biological properties,[72–76] by the same route.

Following the recognition of the acid catalysis of the above condensations, it was found that 6,9-dichloro-2-methoxyacridine reacted with aromatic amines or morpholine,[77] but not with ethylamine, in aqueous acid to yield the 9-aminoacridines. The rate of formation of the aminoacridine was increased by increasing the acid concentration, but so also was the rate of hydrolysis of the chloroacridine to the 9-acridanone. The reaction was suggested to proceed by initial protonation of the ring nitrogen, facilitating nucleophilic attack[77]; this can be a useful method when it is necessary to avoid the use of phenol.[78] It has also been used successfully in the preparation of 9-phenoxyaminoacridines from aminophenols, the aminophenol often being liberated in situ by hydrolysis of an N-acetylaminophenol.[79a,b]

Other solvents have also been used for the reaction of 9-chloroacridines with amines. 6,9-Dichloro-2-methoxyacridine does not appear to react with 2-aminobenzyldiethylamine in phenol, but does so in boiling toluene.[80] The reverse is usually found to be more true for primary amines.[81] Atebrin, for example, is formed from this chloroacridine and the appropriate diamine in 92% yield in phenol, but in less than 25% yield in toluene.[80] Methylcellosolve at reflux temperature has also been used as a solvent with some success.[82] An interesting case of a change of solvent bringing about the formation of a different product occurs in the reaction of 9-chloroacridine with aminophenols. In aqueous acids[79] or in phenol,[83] the products are phenoxyaminoacridines (9), whereas in refluxing methanol the aminophenoxyacridine (10) is formed.[84]

9   10

A kinetic study has been made of the reaction between 9-chloroacridine and piperidine in toluene, and the energy and entropy of activation calculated from the second-order rate constants.[85]

9-Acridanthione and its S-esters or ethers react with primary or secondary amines in hot phenol to give the corresponding 9-aminoacridines.[86,87] The addition of inorganic compounds, such as lead acetate, copper acetate, and silver nitrate, to the reaction mixture has been found to be beneficial by some workers.[88] The same reactions carried out in butanol gave inferior yields, but the addition of acetic acid improved the yield by a factor of 15.[89] These methods have been little used, since the thiones are usually obtained from 9-chloroacridines. However, since acridines fused with sulfur can give good yields of the 9-thiones,[90] the method does afford a means of introducing an amino group into a previously unoccupied 9-position.[91]

9-Alkoxyacridines (prepared from 9-chloroacridine and sodium alkoxides) also react readily with primary amines in alcohol under acid catalysis,[92] 9-aminoacridine being formed in 87% yield from 9-methoxyacridine and ammonium chloride.[93] The latter reaction in aqueous ethanol at 50° is probably second order (K = ca. 0.185), the rate of reaction with methylamine hydrochloride under the same conditions (K = ca. 1.6) was too fast to be measured.[93] The same authors showed that when 6-chloro-2-methoxy-9-(−)-(1-methylheptyloxy)acridine reacted with benzylamine, the eliminated alcohol retained its optical activity.[93]

The exchange of amino groups can occur in a reversible reaction. 6-Chloro-9-(4-diethylamino-1-methylbutylamino)-2-methoxyacridine, for example, may be prepared from the 9-aminoacridine with 4-amino-1-diethylaminopentane at 170°. The addition of acetic acid shortens the reaction time and increases the yield nearly fivefold.[94] Similarly, 9-aminoacridine with boiling aniline in the presence of one equivalent of hydrochloric acid gives 9-anilinoacridine[45]; primary aliphatic amines are reported to condense with 9-acetamido- and 9-benzamido-acridine with a loss of acetamide or benzamide.[95]

All the above acid-catalyzed reactions probably occur by the following route:

A compound, rather similar in structure to **12**, 9,9-diphenoxyacridan, has been isolated as an intermediate in one such reaction[61]; also, acridans, such as 9-amino-10-methyl-9-phenylacridan, are well known. The use of pyridine as a leaving group in a similar reaction, starting from 1-(9-acridyl)pyridinium chloride, has also been described.[96]

Amides derived from diphenylamine-2-carboxylic acid chlorides and amines cyclize under the influence of phosphoryl chloride to give very good yields of 9-aminoacridines[62, 97−99]. This method has been particularly favored by Indian workers for the preparation of dihalogenated 9-aminoacridines when other methods have sometimes given poor yields. (See Ref. 100 and earlier

X = Cl, OPH, OR, SR, etc.   11

12

papers.) However, attempts to prepare 9-*tert*-butylaminoacridine by the cyclization of *N*-(diphenylamine-2-carbonyl)*tert*-butylamide failed because the amide was stable toward all the cyclization reagents tried.[101] 9-Phenoxyacridine with *tert*-butylammonium chloride in phenol, and also 9-chloroacridine, *tert*-butylamine, and phenol in a sealed tube, merely gave 9-aminoacridine, the tertiary butyl group having been lost. 9-*sec*-Butylaminoacridines are readily obtained under similar conditions, so that the difference in behavior must be ascribed to the extra steric strain and the extra stability associated with the leaving *tert*-butyl cation.[101] A similar difficulty in accommodating a *tert*-butyl group at position 9 is shown by the dealkylation of 9-*tert*-butyl-3,6-bis-dimethylaminoacridan on attempted oxidation to the acridine.[102]

Methods of less preparative value include the acid reduction of 1-(9-acridyl)-2-phenylhydrazines, 9-acridylhydrazines, or 1,2-di-(9-acridyl)hydrazines,[103–105] the decomposition of 9-azidoacridines[106] and the Hofmann degradation of the amide.[107]

9-Phenylaminoacridine may be prepared by the action of isocyanatobenzene on acridine 10-oxide in dimethylformamide, carbon dioxide being evolved at room temperature. The reaction presumably involves a 1,5-dipolar cycloaddition, with the possible intermediacy of **13**.[108]

Sodamide attacks acridine in dimethylaniline at 150° with the evolution of hydrogen and the formation of the sodium salt of 9-aminoacridine, which gives 9-aminoacridine (72%) with ice water.[109] This sodium salt[110] is also formed in good yield when the adduct from acridine and sodium hydrazide is allowed to lose ammonia at 60°. Low yields of 9-aminoacridine were obtained from acridine and potassium or barium amides in liquid ammonia,[111]

13

while the action of N-bromosuccinimide gave mixtures containing 9-succini-minoacridine.[112] A few 9-aminoacridines have been made from 9-chloro-acridines and the silver salt of imidazole.[113]

Steric hindrance prevents the 9-amino group from being coplanar with the acridine ring in 9-dialkylaminoacridines and, in consequence, they are less stable (see below). At the same time, reactions giving the 9-dialkylamino-acridines tend to be slower than the corresponding reactions leading to primary and secondary 9-acridylamines, again as a result of steric factors. Thus attempts to prepare 9-dialkylaminoacridines from 9-alkoxyacridines and salts of secondary amines resulted in much hydrolysis to the 9-acri-danones, a difficulty encountered with reactions in phenol solution.[93] How-ever, good yields can be obtained if the reactants are thoroughly dry and long reaction times are used.[67] 9-Dimethylaminoacridine was prepared in 46% yield from 9-phenoxyacridine and dimethylamine hydrochloride in phenol at 120° for 9 hr, while the 9-diethylamino analogue was better obtained by reacting 9-chloroacridine with diethylamine and phenol in a sealed tube at 120° for 15 hr.[67] 9-Dialkylaminoacridines have also been obtained, for ex-ample, from 6,9-dichloro-2-methoxyacridine and methylaminoethanol with-out a solvent[114] at 110°, from 3-methoxy-9-phenoxyacridine and 2-diethyl-aminoethylmethylamine[62] at 150–180° for 20 hr, and from 6,9-dichloro-2-methoxyacridine and 1,2,3,4-tetrahydroquinoline in phenol in the presence of potassium carbonate[115] at 100°.

Polish workers have prepared 9-(3-dimethylaminopropylamino)-1-nitro-acridines labeled with ^{14}C, either on C-9 of the ring[116] or on C-1 of the side chain,[117] in connection with their studies of the carcinostatic properties of the compound. In each case, the preparation was by way of the 9-chloroacridine.

B. Properties

The general properties of 9-aminoacridine are given in Table I. This is by far the most basic of the monoaminoacridines and its chemical behavior is very different from that of its isomers. For the latter reason, and because of difficulties that arose in interpreting some spectroscopic data, the question of whether 9-aminoacridine exists as the amino form (14) or as the imino tautomer (15) was a controversial one.

14   15

However, it is now accepted that 9-aminoacridine is an amino compound (14) and this conclusion is supported by the ir spectra of the compound and its deuterated derivatives[118–120a,b] (Chapter XI, p. 666) by uv spectra[118, 121] (Chapter X, p. 639) by low-temperature absorption spectra[122] and by nmr spectra[123] (Chapter XII, p. 696).

The resonance energy of 9-aminoacridine, together with the energies of the highest occupied and lowest free molecular orbitals and the electronic charge distribution, have been calculated by Pullman.[55] Heats of combustion for 9-aminoacridine and its nitrate have been determined,[124] as have the solubilities of a number of aminoacridines,[125] and X-ray diffraction measurements have shown 9-aminoacridine hemihydrate to be orthorhombic with the rare space group Fddd.[126] The dipole moments of many 9-aminoacridines have been measured in benzene and found to be unusually large.[127a,b] Half-wave potentials have also been recorded for the polarographic reduction of some aminoacridines,[128] and the first reduction stage for the parent compound has been shown to involve two electrons.[47]

9-Aminoacridine with acetic anhydride at 105°[129] or with the calculated quantity of acetic anhydride in pyridine[130] gave a monoacetyl derivative (16), while with boiling acetic anhydride a diacetyl compound (17) was formed.[129]

16   17

One acetyl group may be removed from the diacetyl compound by boiling alcoholic 1.5% hydrochloric acid,[130] or 9-acridanone may be obtained with boiling 0.5 M sulfuric acid. Under the latter conditions the monoacetyl derivative is merely hydrolyzed to 9-aminoacridine, implying the direct elimination of the diacetamido group from **17**. That acetylation does indeed occur on the exocyclic nitrogen atom was shown by quaternization of the ring nitrogen with methyl sulfate followed by hydrolysis,[129] when both **16** and **17** gave the known 9-amino-10-methylacridinium salt. The structure of 9-diacetylaminoacridine has also been confirmed by uv[131] and ir[44] absorption measurements. When a strongly electron-withdrawing group is attached to the amino-nitrogen, as in 9-trichloroacetamidoacridine or 9-benzenesulfonamidoacridine, the compound exists as the imino tautomer (**18**), but 9-acetamidoacridine prefers the amino form.[44]

$X = COCCl_3$ or SO_2Ph

18

1-Trifluoromethyl-9-aminoacridine did not form a diacetyl derivative[125] and 9-(4-acetamidobenzenesulfonamido)-6-chloro-2-methoxyacridine on crystallization from ethanol[132] hydrolyzed to 6-chloro-2-methoxy-9-acridanone. 9-Aminoacridine reacted with chloroacetyl chloride[133] to yield 9-chloroacetamidoacridine, which with diethylamine in boiling toluene gave 9-diethylaminoacetamidoacridine. In contrast, 6-chloro-2-methoxy-9-chloroacetamidoacridine was reported to be inert toward boiling diethylamine.[134] Attempted acylations of 9-aminoacridine with acid chlorides of complex boron-containing acids or of 1-substituted acetylene-2-carboxylic acids failed.[135]

9-Aminoacridine reacts readily with methyl[45] or ethyl[136] iodide, but unlike its isomers it is attacked at the ring nitrogen, giving 9-amino-10-alkylacridinium iodides. However, propyl iodide, methyl 11-iodoundecanoate or methyl 6-iodohexanoate with the aminoacridine in boiling toluene gave 9-aminoacridinium iodide as the only isolable material.[101] The same product was formed from 9-aminoacridine and 1,10-diiododecane, together with a small yield of a deep purple-red material that was not investigated further.[137] Ethyl chloroformate in nitrobenzene gave 6-chloro-9-ethoxycarbonylamino-2-methoxyacridine frcm the 9-aminoacridine by attack on the amino group.[138]

9-Aminoacridines[139] do not diazotize normally, but 9-aminoacridine with nitrosyl chloride in concentrated sulfuric acid yields a solution that gives an intense red color with alkaline 2-naphthol.[45]

9-Aminoacridine reacted normally with boiling aniline[45] in the presence of one equivalent of hydrochloric acid to give 9-anilinoacridine, and with carbon disulfide in ethanol to give the dithiocarbamate ester,[53] but it did not yield anils with aromatic aldehydes.[45]

Oxidation with lead dioxide, silver oxide, potassium ferricyanide, or potassium permanganate, converts 9-aminoacridine to 9,9′-azoacridine (19), catalytic hydrogenation of which gives 1,2-di-(9-acridyl)hydrazine (20).[140]

19   **20**

This compound is said to be identical with the higher melting-point product from hydrazine and 9-chloroacridine in boiling ethanol[141]; although there is a considerable difference in the melting points recorded for the two products, the uv spectra are identical. The mechanism of the oxidation of **19** to **20** in nonaqueous media has been studied electrochemically.[140] The mono-(9-acridyl)hydrazines are useful in that they are oxidized to acridines in moderate yields by the action of oxygen at 25° in the presence of platinum and a trace of alkali. Thus they afford a route from 9-chloroacridines to acridines when alkali-sensitive groups are present[141] and the more usual method of heating the 9-(β-4-toluenesulfonhydrazino)acridine in molar sodium hydroxide solution at 100° is inapplicable.[142]

The nitration of 9-methoxycarbonylaminoacridine in sulfuric acid at 0–5° gave a good yield of 9-amino-2,7-dinitroacridine with a small amount of 9-amino-2,4,7-trinitroacridine[69]; sulfonation gave a 9-aminoacridinedisulfonic acid, which was not identified, although it is presumably the 2,7-derivative.[41] Yellow alcoholic solutions of 9-amino-2-nitroacridine[45] or its 7-ethoxy derivative[58] turn violet on the addition of alkali. The color is thought to be a result of the resonance stabilized ion **21**, since the 3-nitro derivative, where such an ion cannot be formed, does not exhibit the color change.

21

9-Aminoacridines do not appear to give 9-aminoacridine 10-oxides with peroxyacids directly,[143, 144] and these compounds are obtained either by way of the 9-chloroacridine 10-oxide[72, 144, 145] or from 9-phenoxyacridine 10-oxides.[146, 147]

Sufficiently bulky substituents on the 9-amino group can prevent co-planarity with the acridine nucleus and hence cause a diminution of stability. Thus, in the series 9-aminoacridine, 9-methylaminoacridine, and 9-dimethyl-aminoacridine, there is a general decline in stability toward hydrolysis. The alkaline hydrolysis to 9-acridanone has been studied kinetically by Kalatzis,[148] and the energies and entropies of activation calculated. The deamination of all three is first order with respect to the amine but zero order with respect to the hydroxide ion concentration. The enhanced rate of hydrolysis of 9-methylaminoacridine relative to that of 9-aminoacridine is the result of a drop in the energy of activation (72.4 kJ mol^{-1} compared with 78.2 kJ mol^{-1}), the entropy of activation in fact falling slightly. 9-Dimethylaminoacridine however not only has a reduced energy of activation for the hydrolysis (69 kJ mol^{-1}) but also a much increased entropy of activation ($+21.8$ JK^{-1} mol^{-1} compared with 13.8 JK^{-1} mol^{-1} for 9-aminoacridine). The same increased ease of hydrolysis on passing along the series was observed by earlier workers who obtained yields of 11, 24, and 68% of 9-acridanone after boiling the three amines with $5M$ potassium hydroxide for 2 hr.[45] Although 9-amino-acridine is more stable toward acid hydrolysis than its isomers (10% yield of 9-acridanone after treatment with concentrated hydrochloric acid at 180° for 2 hr[45]), the acid hydrolysis of 9-dialkylaminoacridines is particularly easy, aqueous solutions of their hydrochlorides rapidly depositing the appropriate 9-acridanone on standing at room temperature.[59, 62] The energy of activation for the hydrolysis of 9-(3-dimethylaminopropylamino)-1-nitroacridine dihydrochloride in dilute aqueous solution is 88 ± 17 kJ mol^{-1}; in general, the longer the side chain at position 9, the greater the rate of hydrolysis.[149] The loss of stability associated with increased steric hindrance is also reflected in the results of Peck,[150] who found that 6-chloro-9-(2-hydroxyethyl)-methylamino-2-methoxyacridine (22) reacted with glycols at 115° to yield ethers (23).

MeNCH$_2$CH$_2$OH   OCH$_2$CH$_2$OH

OMe   OMe

Cl   Cl

22   23

A less bulky substituent at position 9 (e.g., morpholino) reacted with ethylene glycol in a similar way but only at higher temperatures.

9-Methylaminoacridine with nitrous acid forms 9-nitrosomethylamino-acridine, which, unlike secondary aromatic nitrosamines, is completely hydrolyzed in very dilute acidic solutions to the original acridine.[151] Hydrogen sulfide in alcoholic ammonia attacks 9-methylaminoacridine, but not 9-aminoacridine, to give 9-acridanthione in good yield.[152]

4. Synthesis and Properties of Atebrin

Although atebrin[153] was commercially available as an antimalarial in 1932, the original patents did not disclose its structure, which was not revealed until late in 1933 as 6-chloro-9-(4-diethylamino-1-methylbutyl)amino-2-methoxyacridine (26) dihydrochloride dihydrate.[154] Shortly afterward, the results of degradative work on the commercial material were published.[155, 156] The original synthesis[57] was subsequently somewhat improved [59, 66]; although the importance of atebrin, particularly during the World War II, when the major sources of quinine were lost, stimulated a great deal of work on its production, the best method was essentially the original, described below. An American synthesis[157] of atebrin by this method has been broadly described, as has the detailed German synthesis used at Eberfield during the war.[158] The overall yield of atebrin by the latter route from the potassium salt of the diphenylamine-2-carboxylic acid is 75%.

2,4-Dichlorobenzoic acid was the key material for the American synthesis and was obtained from 2,4-dichlorotoluene by chlorination to 2,4-dichloro-benzotrichloride, followed by hydrolysis. The dichlorotoluene was made by chlorination of 4-chlorotoluene, or by Sandmeyer reactions on the appropriate aminochlorotoluenes, or diaminotoluenes. 2,4-Dichlorobenzotri-chloride was also obtained from toluene-4-sulfonyl chloride,[159] by chlorination in the presence of a catalyst such as tellurium chloride to 3-chloro-4-methyl-benzenesulfonyl chloride, followed by further chlorination at 200°. The Ullmann reaction between 2,4-dichlorobenzoic acid and p-anisidine was performed in water, the potassium salt of 24 being then directly converted by phosphoryl chloride in chlorobenzene to the 9-chloroacridine (25), which can cause severe dermatitis.

24

Condensation[57, 160] with 4-amino-1-diethylaminopentane (29) is best performed in phenol at 100–115°, cooling being necessary to control the temperature in large-scale work.[161] The use of toluene as a solvent gives poor
results.[80] Isolation of atebrin is accomplished by making the reaction mixture
basic under benzene,[160] followed by extraction from the benzene into aqueous
acetic acid and precipitation of the dihydrochloride by the addition of concentrated hydrochloric acid. The benzene extraction can be avoided with a
slight loss of product by precipitating atebrin free-base under carefully controlled conditions.[161]

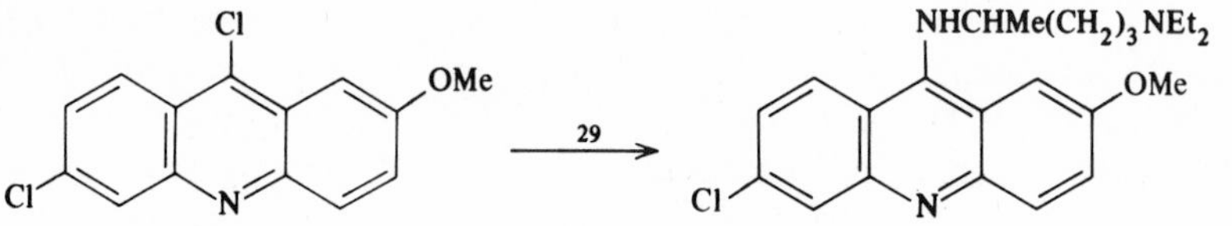

A number of methods are available[156, 162, 163] for the synthesis of 4-amino-
1-diethylaminopentane (29), that from 2-chlorotriethylamine hydrochloride
being quite satisfactory.[158, 164, 165] Reaction with ethyl acetoacetate in the
presence of two equivalents of sodium gives the ester (27), which on vigorous
hydrolysis with hydrochloric acid gives 1-diethylaminopentane-4-one (28).
This on catalytic reduction in methanolic ammonia is converted to the
diamine (29) in excellent yield:

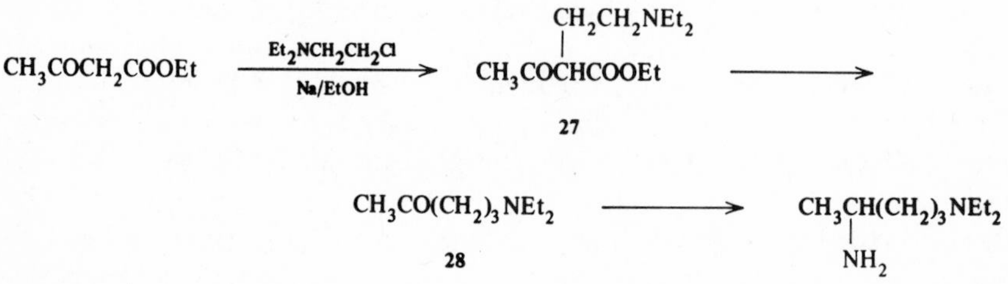

Atebrin labeled in various ways has been prepared by modifications of
standard routes. For example, atebrin labeled with tritium in the side chain[166]
has been obtained from the chloroacridine (25) and the appropriately labeled
diamine (29). Atebrin with the 6-chloro group replaced by ^{131}I has also been
described.[167] The synthesis starts from 6-amino-2-methoxy-9-acridanone and
continues by way of the diazonium salt to 6-iodo(^{131}I)-2-methoxy-9-acridanone,
which is converted to the 9-chloroacridine by phosphoryl chloride. Atebrin

associated with ^{131}I in a more ambiguous way has been obtained by treating atebrin free-base with iodine monochloride,[168] followed by sodium iodide (^{131}I). Chromatography of the product showed that it consisted, in fact, of two components.

Atebrin exists in optically active forms, since there is an asymmetrically substituted carbon atom in the side chain. The first resolution was via the α,α'-bromocamphorsulfonic acid salts[169]; this has been disputed[170, 171] but also confirmed.[172] The salts of 3-bromo-(+)-camphor-7-sulfonic acid[170] or, better, (+)-4,4′,6,6′-tetranitrodiphenic acid can also be separated.[171] The oily free bases[171] had $[\alpha]_D^{16} + 205° \pm 5°$ and $[\alpha]_D^{17} - 197° \pm 5°$ in ethanol, and the hydrochlorides had mp 244–245° (dec), $[\alpha]_D^{10} +388° \pm 7°$ and $[\alpha]_D^{14} - 379° \pm 6°$ in water, respectively.

The hydrolysis of atebrin (26) dihydrochloride by boiling with water for 60 hr gave 6-chloro-2-methoxy-9-acridanone and 4-amino-1-diethylamino-pentane,[173] but decomposition with concentrated hydrochloric acid[59] at 120–125° cleaved the side chain to give a mixture of 9-amino-6-chloro-2-methoxyacridine and 9-amino-6-chloro-2-hydroxyacridine. The rates of hydrolysis of atebrin in aqueous buffer solutions under physiological conditions have been investigated polarographically,[174] and it was concluded that of the several mechanisms consistent with the kinetic evidence, the reaction most probably proceeded by the attack of a hydroxyl ion on the acridine dication. This is a similar mechanism to that put forward for the analogous reaction of atebrin with hydrogen sulfide in alkali at room temperature, when the thione (30) is obtained.[152] A similar reaction occurs with both ethanethiol and thiophenol in pyridine or with thioglycollic acid in acetic acid.[175] The side chain in atebrin is also replaced in low yield by the action of aminoacids

30

31

in buffer solutions of pH 6.9 at 37° over a period of 10 days.[175] *N*-(9-Acridinyl)amino acids are formed, which are more readily obtained from the corresponding 9-phenoxyacridine and amino acids in hot phenol. Cysteine reacts with 2 moles of atebrin at pH 7.1 and 37° to give *N,S*-bis(6-chloro-2-methoxy-9-acridinyl)cysteine (**31**).[175]

Calculations of the electronic charge distribution and the energies of the highest occupied and lowest empty molecular orbitals have been made for atebrin and some related molecules.[176]

Some compounds closely related to atebrin have important physiological properties. Among these are 2-amino-3-chloro-6-methoxy-9-(4-diethylamino-1-methylbutyl)aminoacridine (**32**) and 3-chloro-9-(4-diethylamino-1-methyl-butyl)aminoacridine 10-oxide (**33**), which are both powerful antimalarials.

These are synthesized by routes closely related to that described above. The former compound is prepared from 2,4-dichloro-5-nitrobenzoic acid by an Ullmann condensation with *p*-anisidine and conversion to the 9-chloroacridine with phosphoryl chloride.[177] After condensation of the 9-chloroacridine with 4-amino-1-diethylaminopentane in phenol, the synthesis is concluded by stannous chloride reduction of the nitro group. The acridine **33** is readily obtainable from 3,9-dichloroacridine, which is oxidized by peroxybenzoic acid to the 10-oxide before condensation with 4-amino-1-diethylamino-pentane.[72]

5. Aminoalkyl- and Aminoaryl-Acridines

A number of monoaminoacridines in which the amino group is not directly attached to the nucleus are known. 9-Aminomethylacridine has been prepared by sodium hydrosulfite reduction of acridine-9-aldehyde phenylhydrazone[178] and from 9-bromomethylacridine[179] by way of the hexamethylenetetrammonium salt. A more unusual method reported is from acridine and "methylolbenzamide" or "methylolchloroacetamide," followed by hydrolysis.[180] 9-Bromomethylacridine also reacts readily with other amines under mild conditions[181, 182] and with diethyl acetamidomalonate in ethanolic sodium ethoxide to give, after hydrolysis, 9-(2-amino-2-carboxyethyl)-acridine.[183] This latter compound may also be prepared starting from acridine-9-aldehyde, using the azlactone synthesis.[183] 9-Methylacridine serves as a convenient starting point for the preparation of other 9-aminoalkylacridines, in that it reacts with formaldehyde in boiling ethanol, giving 9-(2-hydroxy-ethyl)acridine, [184, 185] readily converted to 9-(2-halogenoethyl)acridine and thence to amino compounds.[181, 186] 9-Methylacridine also undergoes Mannich reactions with, e.g., dimethylamine or 2-amino-1-phenyl-propane and formaldehyde in the presence of hydrochloric acid, giving 9-(2-dimethylaminoethyl)acridine[180] and 9-(1-methyl-2-phenylethylamino-ethyl)acridine, respectively.[187] The red enamine, 9-(2-dimethylaminovinyl)-acridine, is also obtained from 9-methylacridine[188] in 67% yield by the action of bis-dimethylamino-*tert*-butyloxymethane at 160°. 2-Chloro-9-dimethyl-aminopropylacridine has been produced,[189] together with 2-chloro-9,9-bis-(dimethylaminopropyl)acridan, by the action of 3-dimethylaminopropyl-magnesium bromide on 2,9-dichloroacridine. 2-Aminomethylacridine[2], 4-aminomethylacridine[190] and substituted 4-aminomethylacridines[191] have been synthesized by standard methods. The amino groups in these compounds, where studied, have shown normal chemical properties.

9-(4-Dialkylaminophenyl)acridines have long been used as derivatives for the characterization of the high-melting 9-acridanones, from which they are obtained by boiling with dialkylanilines and phosphoryl chloride.[192, 193, 194] They may also be obtained from 9-chloroacridines and dialkylanilines in the presence of aluminium chloride.[194, 195] More recently, similar compounds have been reported to be obtained by the action of aniline or dimethylaniline on acridine in the presence of sulfur.[196, 197] The behavior of these compounds and also that of 9-aminostyrylacridines, obtained from 9-methylacridine and an appropriate aldehyde,[198, 199] toward acids is of interest. For example, 9-(4-dimethylaminophenyl)acridine base (**34**) is yellow; it gives a violet solution of the monocation (**35**; R = H) in acetic acid or in the correct amount of dilute aqueous-alcoholic hydrochloric acid, and the color reverts to yellow when additional hydrochloric acid is added. Similar colors are given by, e.g., 9-(4-dimethylaminophenyl)-10-methylacridinium chloride (**35**; R = Me), which is deep blue[200]; by 9-(4-aminophenyl)-10-methylacridinium iodide, which is violet[197]; and by compounds in which the 4-amino groups are replaced by other donor groups in the 2- or 4- but not 3-positions of the 9-phenyl ring.[201, 202] The deepening of color has been attributed to resonance of the type **35** ↔ **36**.[200]

When initial attack on the heterocyclic nitrogen does not occur, as in the weakly basic 2,4,5,7-tetrabromo-9-(4-dimethylaminophenyl)acridine, deeply colored cations are not produced.[194] The pK_a value of 9-(4-dimethylamino-phenyl)acridine[2] in 50% ethanol at 20° is 4.4, compared with 4.11 for acridine[2] under the same conditions and about 4.4 for dimethylaniline,[203] indicating that in the ground state the monocation is not stabilized to any great extent by resonance of the type **35** ↔ **36**. The two ring systems cannot, in fact, lie in the same plane as required by structure **36** because of overlapping hydrogen atoms. Some authors[194, 201, 202, 204] have attributed the spectral changes to an intramolecular charge transfer effect; an electron, initially localized on the donor group, passing to the positively charged acridine ring system during light absorption. The dependence of the spectrum upon the type of donor group present, the solvent, and the temperature have all been suggested to be consistent with this theory. However, Albert has preferred the explanation that the color of the monocation is caused by a resonance effect with a sig-nificant contribution from the quinonoid system in an excited state.[205] The reversion to yellow on the addition of more acid to a solution of the mono-cation indicates the formation of dications (**37**) with spectra almost identical with that of the cation of acridine.

37

9-(4-Aminostyryl)acridine and 9-(4-dimethylaminostyryl)acridine are yellow and, similarly, give purple and blue monocations; the 3-aminostyryl analogue gives only a red hydrochloride forming yellow solutions.[198] The pK_a values for these compounds show that the monocations have little or no additional resonance above that of the free bases.[2]

6. Diamino- and Triamino-Acridines

Fourteen of the 20 theoretically possible diaminoacridines are known, as well as a large number of their derivatives. They have all been synthesized by standard methods; some of their properties, together with key references to their preparation, are given in Table II. The chemical properties of the diaminoacridines can be approximately predicted from the known properties of the two monoaminoacridines with the amino groups in analogous positions and therefore will not be considered in detail.

Diaminoacridines on diazotization often give bright colors, varying from violet to orange, of unknown significance. The dark red solution formed by diazotizing 6,9-diamino-2-ethoxyacridine has been studied spectroscopically[213]; it was concluded that only the 6-amino group participates in the diazotization. 3,6-Diaminoacridine appears to form 3-aminoacridine-6-diazonium chloride with 1 mole of nitrous acid, and 3-amino-6-iodoacridine[51] and 3-amino-acridine-6-arsonic acid[38] have been obtained by this route, although in small yield. However, the reported conversion[51] of 3,6-diaminoacridine to 3-amino-acridine by partial diazotization and treatment with alcohol has been disputed.[38] A better method involves monoacetylation, followed by diazotization of the unprotected 6-amino group and reduction.[52] A number of diamino-acridines have been deaminated by treating their bisdiazonium salts with alcohol[50, 214, 215] and 3,6-di-iodoacridine has been obtained from the bis-diazonium salt and potassium iodide.[51] Attempts to convert[50] acridine-3,6- and -2,7-bis-diazonium salts[209] into the hydroxyacridines were not successful. 3,6-Diaminoacridine, however, is converted to the dihydroxy compound by hydrolysis[50] with 45% sulfuric acid at 195°; 2,7-diaminoacridine is best hydrolyzed with hydrochloric acid.[209] A partial hydrolysis of 2,7-diamino-3,6-dimethylacridine to 2-amino-3,6-dimethyl-7-hydroxyacridine by 10% sulfuric acid at 200° for 10 hr has been claimed.[216]

3,6-Diaminoacridine (proflavine) has important antiseptic properties and is best prepared by heating m-phenylenediamine with formic acid and zinc chloride in glycerol.[210] It will undergo the Skraup reaction[217] and forms mono-anils with benzaldehyde[217, 218] and some other aldehydes.[218, 219] 3,6-Diamino-10-methylacridinium salts also form anils with aromatic aldehydes[220] and aliphatic and aromatic ketones.[221] The formation of 3,6-diisothiocyanato-

TABLE II. Diaminoacridines

Positions of amino groups	Solvent, appearance and mp($^\circ$C)	Fluorescence colors in EtOH	aq HCl	Ref.
1,5-	Orange-brown needles ex. aq. EtOH, mp 194-196	?	?	206
1,6-	Brownish-orange needles ex. H_2O, mp 200-201	None	None	207
1,7-	Chocolate prisms ex. aq. pyridine, mp 254-256	?	?	208
1,9-	Orange needles ex. benzene/light petroleum, mp 147-148	Weak green	Weak green	69
2,4-	Orange feathery cryst. ex. EtOH, mp 225 (dec., cor)	None	None	33
2,5-	Orange needles ex. H_2O, mp 126 (Sealed,cor)	None	None	33
2,6-	Glittering golden leaves ex. aq. alc. pyridine, mp 345-346 (Sealed)	Yellow-green	None	208
2,7-	Pale yellow needles ex. H_2O, mp 318 (cor)	Yellow-green	None	209
2,9-	Orange crystals ex. PhCl mp 245 (Sealed)	?	?	142
3,5-	Felted orange needles ex. EtOH, mp 248-250	None	None	206
3,6- (proflavine)	Long yellow-brown needles ex. H_2O mp 277 (Sealed)	Yellow-green	Yellow-green	50, 210
3,9-	Yellow needles ex. H_2O, mp 141 (cor)	Intense green	Yellow-green	103, 211

(Table Continued)

TABLE II.   *(Continued)*

Positions of amino groups	Solvent, appearance and mp(°C)	Fluorescence colors in		Ref.
		EtOH	aq HCl	
4,5-	Orange-brown needles ex. aq. EtOH, mp 182	?	?	206, 212
4,9-	Yellow-brown crystals ex. aq. Me₂CO, mp 178-179	Green	Green	69

acridine from the diamine by the thiophosgene route has been recorded,[222] but 2,9-, 3,9-, and 4,9-diaminoacridines failed to give even the monoiso-thiocyanate. The treatment of 3,6-diaminoacridine with potassium thio-cyanate in acetic acid, followed by bromine, results in thiocyanation in positions 4 and 5. The product has been further converted to pentacyclic systems.[223] Neither 3,6-diaminoacridine nor 6,9-diamino-2-ethoxyacridine gave 10-oxides with peroxybenzoic acid, and both gave unidentifiable products

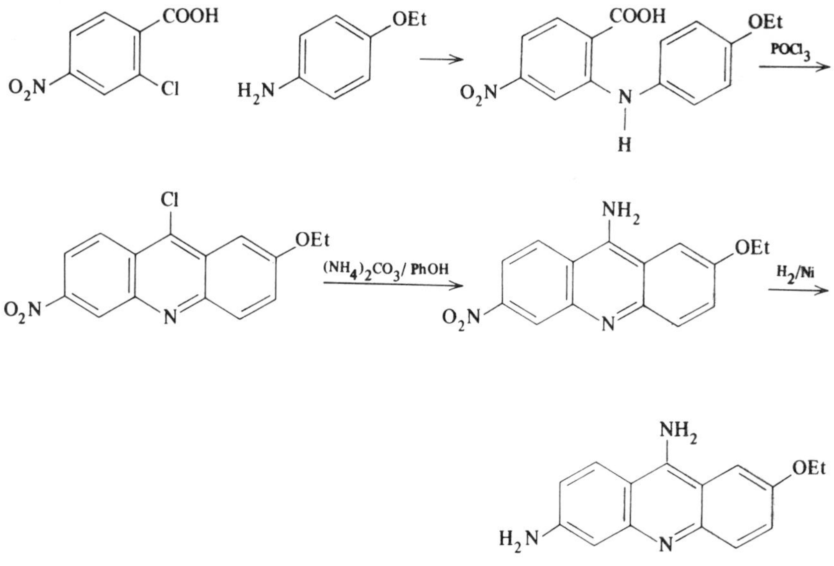

38

with peroxyacetic acid. The 10-oxides were obtained, however, after acetylation of the amino groups.[144]

6,9-Diamino-2-ethoxyacridine (**38**), first prepared[224] in 1920, is a potent antibacterial and is supplied as the dihydrochloride or as the much more soluble lactate under the name rivanol.

A synthesis following the general scheme of the original patents has been published and is given in outline above.[225] The overall yield is approximately 40%. Rivanol is said to give the 6,9-diphthalimido derivative with phthalic anhydride in acetic acid[226]; like 9-aminoacridine, it gave the 10-methylacridinium iodide with methyl iodide.[227] The analytical uses of the halides and pseudo halides,[228] the complex metallocyanides,[229] and the nitroprusside[230] of 6,9-diamino-2-ethoxyacridine have been studied as a means of determining both the acridine and also the ions. Rivanol has also been employed as a useful precipitating agent in the fractionation of proteins.[231]

The electron density at the ring nitrogen, and the energy of the highest occupied molecular orbital and that of the lowest empty molecular orbital, have been calculated for a series of diaminoacridines in an effort to correlate these parameters with pK_a values and the therapeutic activity.[232]

2,7,9-Triaminoacridine has been prepared by reduction of 9-amino-2,7-dinitroacridine but is of little practical importance.[69]

References

1. A. Albert and R. J. Goldacre, *J. Chem. Soc.*, 454 (1943).
2. A. Albert and R. J. Goldacre, *J. Chem. Soc.*, 706 (1946).
3. J. H. Wilkinson and I. L. Finar, *J. Chem. Soc.*, 759 (1947).
4. J. L. Irvin, R. W. McQuaid, and E. M. Irvin, *J. Amer. Chem. Soc.*, **72,** 2750 (1950).
5. J. L. Irvin and E. M. Irvin, *J. Amer. Chem. Soc.*, **72,** 2743 (1950).
6. J. D'Souza and A. Bruylants, *Bull. Soc. Chim. Belges*, **74,** 591 (1965).
7. A. Albert, *J. Chem. Soc.*, 4653 (1965).
8. A. Albert, R. J. Goldacre, and J. Phillips, *J. Chem. Soc.*, 2240 (1948).
9. N. V. Sidgwick, *The Organic Chemistry of Nitrogen*, 3rd ed., I. T. Millar and H. D. Springall, eds. Oxford University Press, Oxford, 1966, p. 438.
10. L. N. Short, *J. Chem. Soc.*, 4584 (1952).
11. T. Nakajima and A. Pullman, *J .Chim. Phys.*, **55,** 793 (1958).
12. J. J. Elliott and S. F. Mason, *J. Chem. Soc.*, 2352 (1959).
13. B. Pullman, *C. R. Acad. Sci., Paris*, **255,** 3255 (1962).
14. F. Peradejordi, *C. R. Acad. Sci., Paris*, **258,** 1241 (1964).
15. F. Peradejordi, *Cahiers Phys.*, **17,** 158, 393 (1963).
16. A. Albert, *The Acridines*, 2nd ed., Arnold, London, 1966, p. 150.
17. D. Ratajewicz and A. Waksmundzki, *Chem. Anal.* (Warsaw), **10,** 361 (1965).
18. A. M. Luly and K. Sakodynsky, *J. Chromatogr.*, **19,** 624 (1965).
19. S. M. Deshpaude and R. R. Upadhyay, *Curr. Sci.*, **36,** 374 (1967).

20. S. M. Deshpaude and R. R. Upadhyay, *Curr. Sci.*, **37**, 73 (1968).
21. W. Kamp, W. J. M. Onderberg, and W. A. van Seters, *Pharm. Weekblad*, **98**, 993 (1963).
22. J. E. Gill, *J. Chromatogr.*, **26**, 315 (1967).
23. L. A. Roberts, *J. Assn. Offic. Anal. Chemists*, **49**, 837 (1966).
24. D. Ratajewicz, *Chem. Anal.* (Warsaw), **11**, 929 (1966); **12**, 517 (1967).
25. A. Albert, R. J. Goldacre, and E. Heymann, *J. Chem. Soc.*, 651 (1943).
26. (a) D. L. Hammick and S. F. Mason, *J. Chem. Soc.*, 348 (1950); (b) J. Cymerman and A. A. Diamantis, *J. Chem. Soc.*, 1619 (1953).
27. R. J. Goldacre, in A. Albert, *The Acridines*, 2nd ed., Arnold, London, 1966, p. 156.
28. L. V. Levshin, *Sov. Phys. Doklady*, **1**, 296 (1956).
29. L. V. Levshin, *Termodinam. Stroenie Rastvorov, Akad. Nauk SSSR, Otdel. Khim. Nauk Khim. Fak., Moskov. Gosudarst. Univ., Trudy Soveshchaniya, Moscow*, **1958**, 285 (1959); *Chem. Abstr.*, **55**, 7036 (1961).
30. (a) G. R. Haugen and W. H. Melhuish, *Trans. Faraday Soc.*, **60**, 386 (1964); (b) A. Olszowski and Z. Ruziewicz, *Bull. Chem. Soc. Jap.*, **37**, 1883 (1964).
31. D. E. Weiss, *Nature*, **162**, 372 (1948).
32. D. J. R. Laurence, *Biochem. J.*, **51**, 168 (1952).
33. A. Albert and W. H. Linnell, *J. Chem. Soc.*, 22 (1938).
34. K. Matsumura, *J. Amer. Chem. Soc.*, **61**, 2247 (1939).
35. A. Albert and B. Ritchie, *J. Soc. Chem. Ind.*, **60**, 120 (1941).
36. G. S. Chandler, R. A. Jones, and W. H. F. Sasse, *Aust. J. Chem.*, **18**, 108 (1965).
37. P. Kristian, *Chem. Zvesti*, **15**, 333 (1961).
38. S. M. Scherlin, G. I. Bras, A. I. Jakubovich, E. I. Vorob'ova, and A. P. Sergeev, *Justus Liebigs Ann. Chem.*, **516**, 218 (1935).
39. G. R. Clemo, W. H. Perkin, and R. Robinson, *J. Chem. Soc.*, **124**, 1751 (1924).
40. K. Lehmstedt, *Chem. Ber.*, **60**, 2413 (1927).
41. K. Lehmstedt, *Chem. Ber.*, **64**, 1232 (1931).
42. A. Albert and B. Ritchie, *Org. Syntheses*, **22**, 5 (1942).
43. (a) D. P. Craig and L. N. Short, *J. Chem. Soc.*, 419 (1945). (b) N. H. Turnbull, *J. Chem. Soc.*, 441 (1945). (c) P. Kristian, *Chem. Zvesti*, **15**, 815 (1961).
44. A. I. Gurevich and Yu. N. Sheinker, *Z. Fiz. Khim.*, **36**, 734 (1962).
45. A. Albert and B. Ritchie, *J. Chem. Soc.*, 458 (1943).
46. G. R. Clemo and W. Hook, *J. Chem. Soc.*, 608 (1936).
47. F. P. Wilson, C. G. Butler, P. H. B. Ingle, and H. Taylor, *J. Pharm. Pharmacol.* **12**, *Suppl.*, 220T (1960).
48. K. Lehmstedt, *Chem. Ber.*, **71**, 808 (1938).
49. R. P. Dikshoorn, *Rec. Trav. Chim. Pays-Bas*, **48**, 517 (1929).
50. L. Benda, *Chem. Ber.*, **45**, 1787 (1912).
51. E. Grandmougin and K. Smirous, *Chem. Ber.*, **46**, 3425 (1913).
52. R. F. Martin and J. H. Y. Tong, *Aust. J. Chem.*, **22**, 487 (1969).
53. W. O. Foye, D. H. Kay, and P. R. Amin, *J. Pharm. Sci.*, **57**, 1793 (1968).
54. R. Royer, *J. Chem. Soc.*, 1665 (1949).
55. B. Pullman, *C. R. Acad. Sci.*, **255**, 3255 (1962).
56. Meister, Lucius and Bruning, German Patent 360,421; 364,034; 364,032; 367,084; Swiss Patent 94,950; *Chem. Abstr.*, **18**, 1130 (1924).
57. F. Mietzsch and H. Mauss, German Patent 553,072; *Chem. Abstr.*, **26**, 4683 (1932).
58. O. Yu. Magidson and A. M. Grigorovskii, *Chem. Ber.*, **66**, 866 (1933).
59. O. Yu. Magidson and A. M. Grigorovskii, *Chem. Ber.*, **69**, 396 (1936).

60. N. S. Drozdov and O. M. Cherntzov, *J. Gen. Chem.* (*USSR*), **14**, 181 (1944).

61. N. S. Drozdov and O. M. Cherntzov, *J. Gen. Chem.* (*USSR*), **5**, 1576 (1935).

62. R. R. Goodall and W. O. Kermack, *J. Chem. Soc.*, 1546 (1936).

63. N. S. Drozdov and O. M. Cherntzov, *J. Gen. Chem.* (*USSR*), **5**, 1736 (1935).

64. A. Albert, *The Acridines*, 2nd ed., Arnold, London, 1966, p. 306.

65. W. Huber, R. K. Bair, W. Boehme, S. C. Laskowski, M. Jackman, and R. O. Clinton, J. Amer. Chem. Soc., **67**, 1849 (1945).

66. D. S. Breslow, H. G. Walker, R. S. Yost, J. C. Shivers, and C. R. Hauser, *J. Amer. Chem. Soc.*, **68**, 100 (1946).

67. D. J. Dupré and F. A. Robinson, *J. Chem. Soc.*, 549 (1945).

68. J. E. R. Falk, U S. Patent 2,488, 904; *Chem. Abstr.*, **44**, 2042 (1950).

69. A. Hampton and D. Magrath, *J. Chem. Soc.*, 1008 (1949).

70. F. Y. Wiselogle, *A Survey of Antimalarial Drugs*, 1941–1945, University of Michigan Press, Ann Arbor, Mich., 1946.

71. A. Ledochowski, C. Wasielewski, G. Blotny, J. F. Biernat, J. Peryt, B. Horowska, and B. Stefanska, *Rocz. Chem.*, **43**, 1083 (1969).

72. E. F. Elslager, R. E. Bowman, F. H. Tendick, D. J. Tivey, and D. F. Worth, *J. Med. Pharm. Chem.*, **5**, 1159 (1962).

73. K. C. Joshi and S. C. Bakel, *J. Indian Chem. Soc.*, **38**, 877 (1961).

74. O. N. Nechaeva and Z. V. Pushkareva, *Trudy Ural. Politekh. Inst. S. M. Kirova*, No. 81, 36 (1959); *Chem. Abstr.*, **55**, 9399 (1961).

75. E. A. Steck, U.S. Patent 2,762,806; *Chem. Abstr.*, **51**, 4444 (1957).

76. E. F. Elslager, F. W. Short, and F. H. Tendick, *J. Heterocycl. Chem.*, **5**, 599 (1968).

77. C. K. Banks, *J. Amer. Chem. Soc.*, **66**, 1127 (1944).

78. J. H. Burckhalter, F. H. Tendick, E. M. Jones, P. A. Jones, W. F. Holcomb, and A. L. Rawlins, *J. Amer. Chem. Soc.*, **70**, 1363 (1948).

79. (a) V. I. Stavrovskaya, *J. Gen. Chem.* (*USSR*), **25**, 177 (1955); (b) V. K. Mehta and S. R. Patel, *J. Indian Chem. Soc.*, **43**, 235 (1966).

80. D. M. Hall and E. E. Turner, *J. Chem. Soc.*, 694 (1945).

81. K. Kitani, *J. Chem. Soc. Jap.*, **75**, 396 (1954).

82. R. M. Peck, R. K. Preston, and H. J. Creech, *J. Org. Chem.*, **26**, 3409 (1961).

83. E. F. Elslager and F. H. Tendick, *J. Med. Pharm. Chem.*, **5**, 1153 (1962).

84. A. G. Lebekhov and A. S. Samarin, *Khim. Geterotsikl. Soedin.*, 838 (1969).

85. K. R. Brower, J. W. Way, W. P. Samuels, and E. D. Amstutz, *J. Org. Chem.*, **19**, 1830 (1954).

86. Mitsui Chemical Industry Co., Japanese Patent, 163,130; *Chem. Abstracts*, **42**, 6382 (1948).

87. S. J. Das Gupta, *J. Indian Chem. Soc.* **20**, 137 (1943).

88. K. Kitani, *J. Chem. Soc. Jap.*, **75**, 480 (1954).

89. K. Kitani, *J. Chem. Soc. Jap.*, **75**, 482 (1954).

90. A. Edinger and W. Arnold, *J. Prakt. Chem.*, **64**, 182 (1901).

91. E. F. Elslager, *J. Org. Chem.*, **27**, 4346 (1962).

92. K. Kitani, *J. Chem. Soc. Jap.*, **75**, 475 (1954).

93. H. J. Barber, J. H. Wilkinson, and W. G. H. Edwards, *J. Soc. Chem. Ind.*, **66**, 411 (1947).

94. K. Kitani, *J. Chem. Soc. Jap.*, **75**, 477 (1954).

95. S. Kuroda, *J. Pharm. Soc. Jap.*, **64**, 59, 69 (1944); *Chem. Abstr.*, **45**, 2949 (1951).

96. J. Peryt, E. Zylkiewicz, and A. Ledochowski, *Rocz. Chem.*, **43**, 623 (1969).

97. N. S. Drozdov, *J. Gen. Chem.* (*USSR*), **8**, 1192 (1938); *Chem. Abstr.*, **33**, 4251 (1939).

98. E. A. H. Friedheim and E. Bergmann, British Patent 587,050; *Chem. Abstr.*, **42,** 621 (1948).

99. N. B. Ackerman, D. K. Haldorsen, F. H. Tendick, and E. F. Elslager, *J. Med. Chem.*, **11**, 315 (1968).

100. A. G. Munshi, P. M. Dholkia, and K. S. Nargund, *J. Indian Chem. Soc.*, **35,** 130 (1958).

101. R. M. Acheson and C. W. C. Harvey, in press.

102. B. L. van Duuren, B. M. Goldschmidt, and H. H. Seltzman, *J. Chem. Soc.*, *B*, 814 (1967).

103. Meister, Lucius, and Brüning, German Patents 364,031; 364,033; Swiss Patents 93,439; 93,752; 94,363; 94,625; 94,982, 96,608-9; *Chem. Abstracts*, **18**, 1131 (1924).

104. Lucius and Brüning, *Friedl.*, **14**, 807 (1924).

105. G. Cauquis and G. Fauvelot, *Bull. Soc. Chim. Fr.*, **8**, 2014 (1964).

106. Meister, Lucius, and Brüning, German Patent 364,035-6; *Chem. Abstr.*, **18**, 1131 (1924).

107. K. Lehmstedt and F. Dostal, *Chem. Ber.*, **72**, 804 (1939).

108. B. Adcock, in press.

109. K. Bauer, *Chem. Ber.*, **83**, 10 (1950).

110. T. Kauffmann, H. Hacker, and H. Mueller, *Chem. Ber.*, **95**, 2485 (1962).

111. F. W. Bergstrom and W. C. Fernelius, *Chem. Rev.*, **12**, 163 (1933).

112. H. Schmid and W. E. Leutenegger, *Helv. Chim. Acta*, **30**, 1965 (1947).

113. S. I. Lur'e, M. G. Kuleshova, and N. K. Kochetkov, *J. Gen. Chem. (USSR)*, **9**, 1933 (1939).

114. R. M. Peck, *J. Org. Chem.*, **28**, 1998 (1963).

115. F. G. Holliman and F. G. Mann, *J. Chem. Soc.*, 34 (1945).

116. B. Stefanska, J. F. Biernat, J. Zielinski, T. Uminski, and A. Ledochowski, *Rocz. Chem.*, **42**, 2187 (1968).

117. J. F. Biernat, B. Stefanska, E. Jereczek-Morawska, T. Uminski, and A. Ledochowski, *Rocz. Chem.*, **9**, 1749 (1969).

118. S. F. Mason, *J. Chem. Soc.*, 1281 (1959).

119. N. Bacon, A. J. Boulton, R. T. C. Brownlee, A. R. Katritzky, and R. D. Topsom, *J. Chem. Soc.*, 5230 (1965).

120. (a) Yu. N. Sheinker and E. M. Peresleni, *Doklady Akad. Nauk SSSR*, **131**, 1366 (1960); (b) *Fiz. Prob. Spektroskopii Akad. Nauk SSSR*; *Materialy 13-go Trinadtsatogo Soveshch.*, *Leningrad*, **1**, 437 (1960); *Chem. Abstr.*, **59**, 12303 (1963).

121. A. K. Sukhomlinov, *Zh. Obshch. Khim.*, **28**, 1038 (1958).

122. V. Zanker and A. Wittiwer, *Z. Phys. Chem. (Frankfurt)*, **24**, 183 (1960).

123. J. P. Kokko and J. H. Goldstein, *Spectrochim. Acta*, **19**, 1119 (1963).

124. J. B. Willis, *Trans. Faraday Soc.*, **43**, 97 (1947).

125. J. H. Wilkinson and I. L. Finar, *J. Chem. Soc.*, 32 (1948).

126. H. A. Rose and A. J. van Camp, *Acta Cryst.*, **9**, 824 (1956).

127. (a) Z. Yu. Kokoshko and Z. V. Pushkareva, *Zh. Obshch. Khim.*, **24**, 877 (1954), and earlier papers. (b) Z. V. Pushkareva and O. N. Nechaeva, *Zh. Obshch. Khim.*, **28**, 2702 (1958).

128. O. N. Nechaeva and Z. V. Pushkareva, *Zh. Obshch. Khim.*, **28**, 2693 (1958).

129. J. H. Wilkinson and I. L. Finar, *J. Chem. Soc.*, 115 (1946).

130. A. M. Grigorovski, *C. R. Acad. Sci. URSS*, **53**, 229 (1946).

131. E. J. King, M. Gilchrist, and A. L. Tarnoky, *Biochem. J.*, **40**, 706 (1946).

132. A. K. Choudhury, P. Das-Gupta, and U. Basu, *J. Indian Chem. Soc.*, **14**, 733 (1937).

133. V. Ettel and J. Neumann, *Chem. Listy*, **51**, 1906 (1957).

134. M. Neeman, *J. Chem. Soc.*, 1079 (1956).
135. M. A. Davis and A. H. Soloway, *J. Med. Chem.*, **10**, 730 (1967).
136. I. S. Ioffe and N. A. Selezneva, *Zh. Obshch. Khim.*, **31**, 50 (1961).
137. E. P. Taylor, *J. Chem. Soc.*, 5048 (1952).
138. M. P. Gerchuk, P. G. Arbuzova, and I. A. Kel'manskaya, *J. Gen. Chem. (USSR)*, **11**, 948 (1941).
139. U. P. Basu and S. J. Das-Gupta, *J. Indian Chem. Soc.*, **16**, 100 (1939).
140. G. Cauquis and G. Fauvelot, *Bull. Soc. Chim. Fr.*, **8**, 2014 (1964).
141. A. Albert, *J. Chem Soc.*, 4653 (1965).
142. A. Albert and R. Royer, *J. Chem. Soc.*, 1148 (1949).
143. R. M. Acheson, B. Adcock, G. M. Glover, and L. E. Sutton, *J. Chem. Soc.*, 3367 (1960).
144. A. M. Khaletskii, V. G. Pesin, and Chou Tsin, *Zh. Obshch. Khim.*, **28**, 2821 (1958).
145. E. F. Elslager and F. H. Tendick, *J. Med. Pharm. Chem.*, **5**, 1149 (1962).
146. B. Stefanska and A. Ledochowski, *Rocz. Chem.*, **42**, 1973 (1968).
147. N. M. Voronina, Z. V. Pushkareva, L. B. Radina, and N. V. Babikova, *Zh. Obshch. Khim.*, **30**, 3476 (1960).
148. E. Kalatzis, *J. Chem. Soc.*, *B*, 96 (1969).
149. A. Ledochowski and E. Zylkiewicz, *Rocz. Chem.*, **43**, 291 (1969).
150. R. M. Peck, *J. Org. Chem.*, **28**, 1998 (1963).
151. E. Kalatzis, *J. Chem. Soc. B*, 273 (1967).
152. R. S. Asquith, D. L. Hammick, and P. L. Williams, *J. Chem. Soc.*, 1181 (1948).
153. H. Mauss and F. Mietzsch, *Chem. Ber.*, **69**, 641 (1936).
154. H. Mauss and F. Mietzsch, *Klin. Wochenschr.*, **12**, 1276 (1933).
155. G. V. Chelintzev, I. L. Knunyantz, and Z. V. Benevolenskaya, *C. R. Acad. Sci. URSS*, (N.S.), **1**, 63 (1934).
156. I. L. Knunyantz, G. V. Chelintzev, Z. V. Benevolenskaya, E. D. Osetrova, and A. I. Kursanova, *Bull. Acad. Sci. URSS, Classe Sci. Math. Nat.*, 165 (1934).
157. A. E. Sherndahl, *Chem. Eng. News*, **21**, 1154 (1943).
158. L. W. Greene, *Amer. J. Pharm.*, **120**, 39 (1948).
159. W. Davies, N. H. Oliver, and B. W. Wilson, *Aust. J. Sci. Res.*, **A5**, 198 (1952).
160. W. A. Cowdrey and A. G. Murray, U.S. Patent 2,410,406; *Chem. Abstr.*, **41**, 2088 (1947).
161. R. G. Jones, G. L. Shaw, and J. H. Waldo, *Ind. Eng. Chem.*, **37**, 1044 (1945).
162. P. C. Guha, P. L. N. Rao, and T. G. Verghese, *Curr. Sci.*, **12**, 82 (1943).
163. W. Reppe, BIOS Final Rep. No. 353.
164. (a) O. Eisleb and G. Ehrhart, German Patent 551,436; *Friedl.*, **19**, 1473 (1932); (b) W. Schulemann, F. Schonhofer, and A. Wingler, U.S. Patent 1,747,531; German Patent 486,079; 486,771; *Friedl.*, **16**, 2684, 2694 (1927-9).
165. D. S. Breslow, R. S. Yost, H. G. Walker, and C. R. Hauser, *J. Amer. Chem. Soc.*, **66**, 1921 (1944).
166. Y. M. Young, F. Wild, and I. Simon-Reuss, *Brit. J. Cancer*, **19**, 370 (1965).
167. H. Medenwald, *Med. Chem., Abhandl. Med.- Chem. Forschungsstatten Farbenfabriken Bayer*, **5**, 206 (1956); *Chem. Abstr.*, **55**, 8403 (1961).
168. L. J. Anghileri, *Arg., Rep., Com. Nacl. Engergia At., Inform.*, No. 124 (1964); *Chem. Abstr.*, **62**, 5254 (1965).
169. G. V. Chelintzev and E. D. Osetrova, *J. Gen. Chem. (USSR)*, **10**, 1978 (1940).
170. F. A. Bacher, R. P. Buhs, J. C. Hetrick, W. Reiss, and N. R. Trenner, *J. Amer. Chem. Soc.*, **69**, 1534 (1947).
171. B. R. Brown and D. L. Hammick, *J. Chem. Soc.*, 99, (1948).

172. N. S. Drozdov, *Trudy Kafedry Biokhim. Moskov. Zootekh. Inst. Konevodstra*, 28 1944; *Chem. Abstr.*, **41**, 763 (1947).
173. F. Mietzsch, H. Mauss, and G. Hecht, *Indian Med. Gas.*, **71**, 521 (1936).
174. D. L. Hammick, S. F. Mason, and G. W. Meacock, *J. Chem. Soc.*, 4745 (1952).
175. F. Wild and J. M. Young, *J. Chem. Soc.*, 7261 (1965).
176. J. A. Singer and W. P. Purcell, *J. Med. Chem.*, **10**, 754 (1967).
177. N. N. Dykhanov, G. A. Gorlach, and V. P. Sergovskaya, *Med. Prom. SSSR*, **14**, No. 6, 22 (1960).
178. A. E. Porai-Koshits and A. A. Kharkharov, *Bull. Acad. Sci. URSS*, 243 (1944); *Chem. Abstr.*, **39**, 1631 (1945).
179. W. Rzeszotarski and Z. Ledochowski, *Rocz. Chem.*, **37**, 1631 (1963).
180. L. Monti and M. Procopio, *Gazz. Chim. Ital.*, **63**, 724 (1933).
181. N. Fisher, C. S. Franklin, E. N. Morgan, and D. J. Tivey, *J. Chem. Soc.*, 1411 (1958).
182. W. Rzeszotarski and Z. Ledochowski, *Rocz. Chem.*, **39**, 93 (1965).
183. L. B. Radina and Z. V. Pushkareva, *Zh. Obshch Khim.*, **31**, 2362 (1961).
184. O. Eisleb, *Med. Chem., Abhandl. Med.-Chem. Forschungsstatten, I. G. Farbenindustrie*, **3**, 41 (1936); *Chem. Abstr.*, **31**, 5802 (1937).
185. A. W. Homberger and H. Jensen, *J. Amer. Chem. Soc.*, **48**, 800 (1926).
186. H. Jensen and L. Howland, *J. Amer. Chem. Soc.*, **48**, 1988 (1926).
187. N. V. Koninklijke, Pharmaceutische Fabrieken Voorheen Brocades-Stheeman and Pharmacia, Dutch Patent 6,608,741; *Chem. Abstr.*, **67**, 11438 (1967).
188. H. Bredereck, G. Simchen, and R. Wahl, *Chem. Ber.*, **101**, 4048 (1968).
189. A. Marxer, *Helv. Chim. Acta*, **49**, 572 (1966).
190. W. Gruszecki and E. Borowski, *Rocz. Chem.*, **42**, 733 (1968).
191. W. Gruszecki and Z. Ledochowski, *Rocz. Chem.*, **41**, 393 (1967).
192. F. Ullmann, *Chem. Ber.*, **40**, 4796 (1907).
193. H. Gilman and D. A. Shirley, *J. Amer. Chem. Soc.*, **72**, 2181 (1950).
194. R. M. Acheson and M. J. T. Robinson, *J. Chem. Soc.*, 484 (1956).
195. N. S. Drozdov, *J. Gen. Chem. (USSR)*, **6**, 219 (1936).
196. O. N. Chupakhin, V. A. Trifimov, and Z. V. Pushkareva, *Khim. Geterotsikl. Soedin.*, 954 (1969).
197. O. N. Chupakhin, V. A. Trifimov, and Z. V. Pushkareva, *Doklady Akad. Nauk SSSR*, **188**, 376 (1969).
198. A. Porai-Koschitz, P. Ssolodownikow, and M. Troitzki, *Z. Farben Ind.*, **6**, 291 (1907); *Chem. Zentr.*, XI, ii, 1527, (1907).
199. I. I. Chizhevskaya, L. A. Gapanovich, and N. N. Khovratovich, *Zh. Obshch. Khim.*, **34**, 4059 (1964).
200. K. Gleu and A. Schubert, *Chem. Ber.*, **73**, 757 (1940).
201. V. Zanker and A. Reichel, *Z. Elektrochem.*, **63**, 1133 (1959).
202. V. Zanker and A. Reichel, *Z. Elektrochem.*, **64**, 431 (1960).
203. M. Mizutani, *Z. Phys. Chem.* (Leipzig), **116**, 350 (1925).
204. V. Zanker and G. Schiefele, *Z. Elekrochem.*, **62**, 86 (1958).
205. A. Albert, *The Acridines*, 2nd ed., Arnold, London, 1966, p. 185.
206. A. A. Goldberg and W. Kelly, *J. Chem. Soc.*, 595 (1947).
207. A. Albert and W. H. Linnell, *J. Chem. Soc.*, 88 (1936).
208. A. A. Goldberg and W. Kelly, *J. Chem. Soc.*, 102 (1946).
209. M. T. Bogert, A. D. Hirschfelder, and P. G. Lauffer, *Collect. Czech. Chem. Commun.* **2**, 383 (1930).
210. A. Albert, D. K. Large, and W. Kennard, *J. Chem. Soc.*, 484 (1941).

211. A. Albert and W. H. Linnell, *J. Chem. Soc.*, 1614 (1936).

212. E. R. Klein and F. N. Lahey, *J. Chem. Soc.*, 1418 (1947).

213. P. B. Ivakhnenko and K. N. Bagdasarov, *Aptech. Delo*, **14**, 38 (1965); *Chem. Abstr.*, **62**, 12427 (1965).

214. C. Haase *Chem. Ber.* **36**, 588 (1903).

215. R. Meyer and R. Gross *Chem. Ber.* **32**, 2352 (1899).

216. L. Cassella, German Patent 121,686; *Friedl.*, **6**, 487 (1901).

217. L. Schmid and O. Friesinger, *Sitzber. Akad. Wiss. Wien.*, *IIb*, **156**, 76 (1947).

218. L. Schmid and O. Friesinger, *Monatsh.*, **77**, 76 (1947).

219. W. L. Glen, M. M. J. Sutherland and F. J. Wilson, *J. Chem. Soc.*, 1484 (1936).

220. (a) M. Brockmühl and O. Sievers, U.S. Patent 2,145,070; *Chem. Abstracts*, **33**, 3536 (1939); (b) O. Sievers, *Chem. Zentr.*, 752 (1935).

221. G. W. Raiziss and B. C. Fisher, U.S. Patent 1,670,740; *Chem. Abstr.*, **22**, 2439 (1928).

222. P. Kristian, K. Antos, A. Hulka, P. Nemek, and L. Drobnica, *Chem. Zvesti*, **15**, 730 (1961).

223. A. Fravolini, G. Grandolini, and A. Martani, *Ann. Chim.* (Rome), **58**, 533 (1968).

224. Lucius and Brünning, German Patent 360,421; 364,033; *Friedl.*, **14**, 804 (1925).

225. A. Albert and W. Gledhill, *J. Soc. Chem. Ind.*, **61**, 159 (1942); Brit. I O S Final Rep. No. 766.

226. G. Vanags and A. Veinbergs, *Chem. Ber.*, **75** 1558 (1942).

227. A. Weizmann, *J. Amer. Chem. Soc.*, **69**, 1224 (1947).

228. I. C. Dragulescu and I. Florea, *Acad. Rep. Populare Romine, Baza Cercet. Stiint. Timisoara, Stud. Cercet. Stiinte Chim.*, **9**, 205, 219 (1962); *Chem. Abstr.*, **59**, 9330 (1963).

229. I. C. Dragulescu and I. Florea, *Acad. Rep. Populare Romine, Baza Cercet. Stiint. Timisoara, Stud. Cercet. Stiinte Chim.*, **9**, 227 (1962), *Chem. Abstr.*, **59**, 9331 (1963); **10**, 31 (1963), *Chem. Abstr.*, **61**, 3898 (1964).

230. I. C. Dragulescu and I. Florea, *Acad. Rep. Populare Romine, Baza Cercetari Stiint. Timisoara, Stud. Cercet. Stiinte Chim.*, **9**, 247 (1963); *Chem. Abstr.*, **59**, 9331 (1963).

231. E. D. West and M. J. West, U.S. Patent 3,382,227; *Chem. Abstr.*, **69**, 30105 (1968).

232. N. E. Sharpless and C. L. Greenblatt, *Exp. Parasitol.*, **24**, 216 (1969).

CHAPTER III

9-Acridanones

J. M. F. GAGAN

Department of Chemistry and Chemical Technology,
University of Bradford, England

141

9-Acridanone (2), commonly known as "acridone," occupies a place between the fully aromatic acridine system (1), and the bridged diphenylamine system of acridan (3). By invoking the principle of vinylogy, the relationship between the three molecules 1, 2, and 3 is seen to parallel that of imine to amide to amine in the aliphatic series.

This midway position and the ease of synthesis of an extensive range of substituted 9-acridanones accounts for the importance of 9-acridanone in acridine chemistry, for it may readily be converted by a number of reliable methods to either of the other systems.

That 9-acridanone is a stable system is indicated by the high melting points (often without decomposition) of many of its derivatives, the powerful reagents used in synthetic procedures, and its frequent isolation after the vigorous oxidation of complex acridine-based molecules.[1] Certain of its N-substituted derivatives form a group of alkaloids (Chapter IV, p. 381), well distributed in the *Rutaceae* of Australasia. 9-Acridanone itself is also present to a small extent in the polluted urban atmosphere.[2]

Aspects of 9-acridanone chemistry elaborated elsewhere in this volume include dyestuffs (Chapter VIII), pharmaceuticals (Chapters II, XVI, XVII, and XVIII), the spectroscopic properties of 9-acridanone (Chapters X and XIII) and the reduced 9-acridanones (Chapter V).

In this chapter, discussion centers on the synthetic routes to 9-acridanones, including the details of several important intermediates, the structure of the 9-acridanone molecule, the chemical behavior of 9-acridanones in general, and the idiosyncracies of a number of particular 9-acridanones. Many of the known 9-acridanones, and the 9-chloroacridines, which are so often their precursors, are listed in the tables at the end of the chapter. 9-Acridanthiones and 9-acridanselenones are also described.

1. Preparation of 9-Acridanones

All other methods of 9-acridanone synthesis are subordinate to the cyclization of appropriate diphenylamine-2-carboxylic acids (described as N-phenyl-anthranilic acids by *Chemical Abstracts*), which is effected by a number of acidic reagents (Tables II, III, and VII). The acids (4) are easily prepared in great variety (Table IV), and the yields obtained in the cyclization step are frequently excellent. A number of other general methods are available, but they have the serious disadvantage of a multiplicity of steps. Of these, the Lehmstedt-Tanasescu route via phenylanthranils (Table V), the Sternbach group cyclization of 2-amino-2'-fluorobenzophenones (Table VI) and the oxidation routes from acridines, seem to be the most useful.

A. The Cyclization of Diphenylamine-2-carboxylic Acids

(1) Cyclization with Phosphorus Oxychloride

This most effective reagent was discovered independently by Runge (Dissertation, Gottingen, Germany, 1922 but not reported until 1933 in Borsche, Runge, and Trautner[3]) and Lesnianski and Dziewonski,[4] and has been used extensively by Russian chemists and many others. A mixture of phosphorus oxychloride and pentachloride had, however, been used to convert 9-acridanone to 9-chloroacridine as early as 1893.[5]

In a typical cyclization procedure,[6] the diphenylamine-2-carboxylic acid is refluxed with 5 times its weight, or more, of phosphorus oxychloride until solution is complete and evolution of hydrogen chloride has ceased. Although chlorobenzene is commonly used in industry as a solvent, excess oxychloride not only performs this function, but also facilitates the reaction.[7] The danger of a sudden explosive start is avoided by adding a few drops of sulfuric acid to the reaction mixture. The product under these conditions is a 9-chloro-acridine, which may be quantitatively hydrolyzed to the corresponding 9-acridanone, in dilute aqueous hydrochloric acid at 100°. Phosphorus oxychloride has also been used directly on crude Ullmann reaction products, from which diphenylamine-2-carboxylic acids were not easily isolated, and gave low yields of 9-chloroacridines.[8]

9-Acridanones may be produced directly by using a limited quantity of the oxychloride (63% yield for the parent compound),[9] and although they may be separated from 9-chloroacridines by chromatography on alumina,[10] it is more satisfactory, in practice, to use an excess of the reagent and hydrolyze the 9-chloroacridines (87% yield for parent).[9] Cyclization of ethyl diphenylamine-2-carboxylate also seems to give 9-acridanone directly.[11]

9-Acridanone thus appears to be the first formed product, but its subsequent conversion to 9-chloroacridine seems possible without the intervention of the enol (5), as 10-methyl-9-acridanone gives 9-chloro-10-methylacridinium dichlorophosphate with phosphorus oxychloride under the same conditions.[12]

It is not necessary to postulate the formation of an acid chloride, as phosphorus oxychloride does not usually convert acids to their chlorides, and cyclizes diphenylamine-2,4'-dicarboxylic acid to 9-acridanone-2-carboxylic acid without doing so. A more probable intermediate is a mixed anhydride-like complex (6), formed by the elimination of hydrogen chloride between the reagents, which on protonation will break down to give a carbonium ion. A similar intermediate complex (7) from the 9-acridanone and phosphoryl chloride would allow the ready substitution of chlorine at position 9.

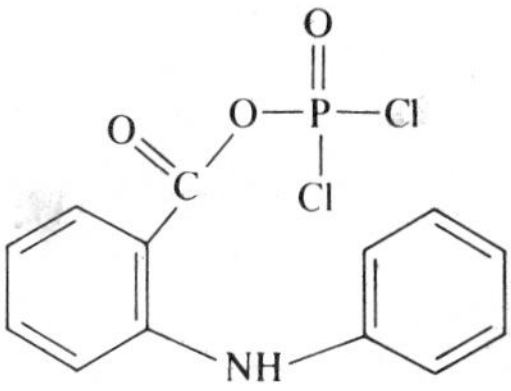

6

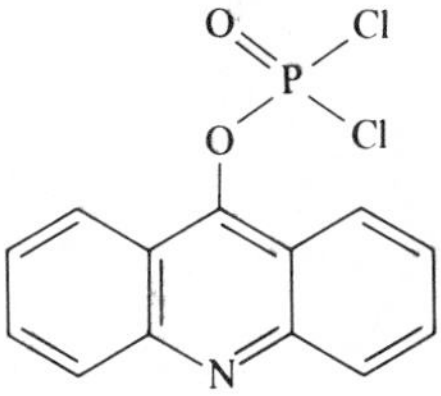

7

Further evidence that the first step is an intramolecular acylation, comes from the observations that diphenylamine-2-carboxyamides give 9-amino-, rather than 9-chloro-acridines, on cyclization[13]; and the ratio of 9-acridanone to 9-chloroacridine is higher (26:56) when the products obtained in the cyclization have methyl groups in both positions *peri* to the carbonyl group than (6:89) when hydrogen atoms occupy these positions.[10] Also, if nitro substituents are present, deactivating the carbonyl group toward complex formation, the cyclization reaction stops at the 9-acridanone stage.

The separation of crystalline 9-chloroacridinium dichlorophosphates (8) when toluene is used as the solvent,[14] supports the following equation for this cyclization.[12, 15]

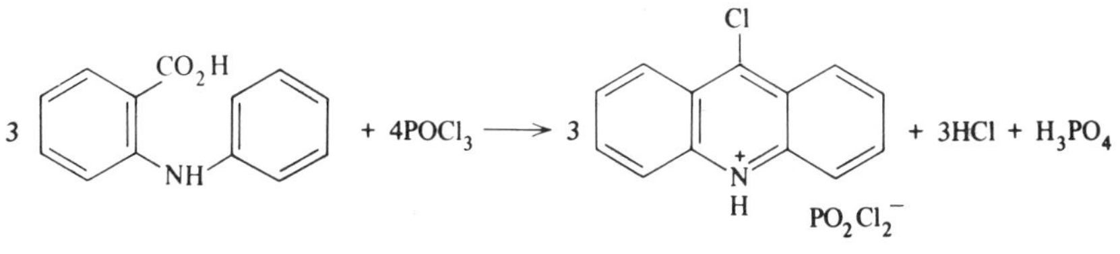

8

However, an alternative scheme has been suggested,[16] in which the product is not a dichlorophosphate, although the author acknowledges that the salt (8) is a primary product of this reaction.

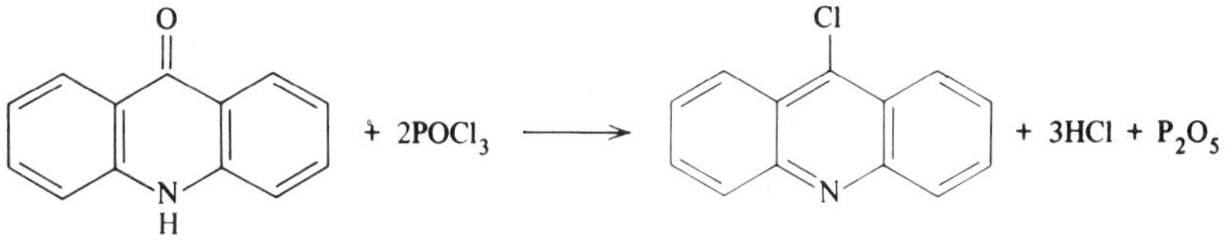

Other interpretations of these reactions are not excluded by the experimental results.

Phosphorus oxychloride, in contrast to sulfuric acid, does not hydrolyze alkoxy or cyano groups,[17] and carboxylic acid groups are unaffected. Hydroxyl and amino groups in the diphenylamine-2-carboxylic acids should be protected before cyclization is attempted. The former group may be protected as its methyl ether, although if more than one alkoxy group is present, polyphosphoric acid is a more satisfactory cyclization reagent.[18] The latter group may be protected by acetylation, but the use of the urethane, prepared from the amine with ethyl chloroformate, is claimed to give better results.[19] These protecting groups may be removed by refluxing the 9-acridanone with constant boiling point hydrobromic acid. An alternative route to amino-substituted 9-acridanones is sometimes available via the phosphorus oxychloride cyclization and subsequent stannous chloride reduction of the corresponding nitro compounds.

Diphenylamine-2-carboxylic acids reluctant to cyclize and 9-acridanones not converted to 9-chloroacridines in refluxing phosphorus oxychloride, may often be persuaded to react by employing a high-boiling inert solvent, e.g., chlorobenzene,[20] xylene,[21, 22, 23] or nitrobenzene,[24] to raise the reaction temperature. However, the use of too high a temperature for the evaporation of the phosphorus oxychloride, after the cyclization of 5-nitrodiphenylamine-2-carboxylic acid, resulted in the formation of 3,9-dichloroacridine, in which the nitro group had been replaced by a chlorine atom. When petroleum ether was used to wash away the excess oxychloride, 9-chloro-3-nitroacridine was obtained.[25]

With one nitro group present, cyclization is usually straightforward, but conversion to the 9-chloroacridine may require vigorous conditions; e.g., 4-nitro-9-acridanone needs a 10-hr reflux with phosphoryl chloride in nitrobenzene.[24] A second nitro group may prevent reaction completely.[21, 26] Hydrogen bonding between the 10-hydrogen atom and the *peri* nitro group may partly account for the difficulty,[23] but electronic factors appear to be paramount. Sulfonic acid[27] and chloro[20] substituents may also be responsible for failure under normal conditions.

A cyclization reagent of phosphorus oxychloride and pentachloride together in formic acid is reported to give a 99% yield of 9-chloroacridine.[28]

9-Chloroacridines are often extremely sensitive to acid hydrolysis, and even a pure sample will suffer autocatalyzed decomposition to the 9-acridanone and hydrochloric acid if open to the air. In isolating 9-chloroacridines, therefore, the minumum contact with aqueous solutions is desirable, and these must always be kept alkaline to phenolphthalein if 9-acridanone formation is to be avoided. Unless electron-donating groups are present, decomposition of the glassy solid left after evaporating excess phosphorus oxychloride, or its solution in chloroform,[29, 30] must also be carried out in the presence of ice, although a more rapid separation of the 9-chloroacridine takes place if temperatures up to 50° may be used. If 9-acridanones are formed, they co-precipitate with the 9-chloroacridines or collect at the interface of the organic and aqueous layers, making separation and purification difficult.

The increased ease of hydrolysis, or other nucleophilic attack,[31, 32] at the 9-position of 9-chloroacridine in comparison with the 4 position of 4-chloro-pyridine and quinoline has been ascribed to annellation, which leads to a decrease in the charge density at the site of attack[33] and an increased possi-bility of delocalization of charge in the transition state.[32] The much faster rate of hydrolysis of 9-chloroacridine when compared with 9-chloro-10-nitro-anthracene has been attributed to steric inhibition of resonance by the *peri* hydrogens in the latter compound.[34] As expected, electron-withdrawing sub-stituents promote nucleophilic attack, while electron-donating groups have the reverse effect.   Surprisingly, *N*-oxide formation appears to reduce the reactivity of the chlorine atom at 9,[35] and only poor yields of 9-alkylamino-acridine 10-oxides are obtained in substitution reactions.[36]

The considerable catalytic effect of acid is shown in the reaction between 9-chloroacridine and piperidine, where in ethanolic solution, in which piperi-dine hydrochloride is soluble, a much enhanced reaction rate is observed in comparison with that found for reaction in toluene.[32] Better yields are also found for a similar reaction in phenol compared with xylene, or a no-solvent reaction.[37] The hydrolysis of the 9-chloroacridinium cation takes place 100 times faster than that of the base, but curiously, in the presence of light and oxygen the hydrolysis rate of the 9-chloroacridine is much increased over the dark reaction rate.[38]

However, 6,9-dichloro-2-hydroxyacridine is not hydrolyzed on pouring into ice-cold concentrated hydrochloric acid; instead the acridine hydrochloride precipitates in 95% yield,[39] and 9-chloro-2-ethoxy- and -7-methoxy-acridines are only converted to 9-acridanones under conditions that cause hydrolysis of the alkoxy groups.[40] 6,9-Dichloro-2-methoxyacridine may be nitrated with mixed acid to the 4- and 7-nitro derivatives without hydrolysis[41]; alcoholic sodium hydroxide is the hydrolyzing solvent used with 6,9-dichloro-2-(2-hydroxyethoxy)acridine to avoid the cleavage of the alkoxy group that occurs under acid conditions.[39]

Nevertheless, the facile replacement of a 9-halogen, or alkoxy group, from a protonated acridine, is exploited in the synthesis of 9-alkylaminoacridines valuable as antiseptics[42]; 9-arylaminoacridines with local anesthetic proper-ties[872]; and 9-hydrazinoacridines, which are useful pesticides.[873] Phenol is frequently used as a solvent[43, 874] and catalyst, as the intermediate 9-phenoxy compounds may be isolated. Indeed, aminophenols react with 9-chloro-acridine to give 9-aminophenoxyacridine hydrochlorides, rather than the alternative 9-hydroxyanilinoacridine hydrochlorides.[875] The addition of acetic acid also improves both the rate and the yield in these substitutions,[44] in which primary amines are more effective than secondary amines.[43] The reaction between 9-chloroacridine and primary amines has been suggested as suitable for colorimetric analysis.[45, 46]

Early studies of the kinetics of 9-chloroacridine hydrolysis[29] have been extensively continued by a Polish group. In aqueous acetic acid, hydrolysis occurs in a pseudo-first-order, two-step consecutive reaction according to the scheme below,[49] which was first shown for 9-chloroacridine.[48] The actual steps in the mechanism are said to be protonation at the nitrogen atom, followed by the addition of hydroxyl at position 9, and the subsequent loss of hydrogen chloride.[38] The intermediate addition compounds (9), some of which precipitated from concentrated solutions of the 9-chloroacridine,[49] had been discovered earlier, when 10% aqueous hydrochloric acid was used as a hydrolysis medium.[50, 51]

1-, 2-, 3-, and 4-Methoxy-, methyl-, and nitro-9-chloroacridines were later examined (See also Chapter I, Section 2.A.(3)). For the 2- and 3-substituted compounds, the reactivity increased as methoxyl was replaced successively by methyl, hydrogen, and nitro groups, as expected; the biggest increases in rate were noted for the 2- and 4-nitro compounds.[53] The 9-alkylamino series also indicated that the dimethylamino group was more deactivating than methoxyl.[54] The ratio k_1/k_2 was found to be constant through the series of substituents,[55] but it varied for different temperatures and concentrations of acetic acid.[49]

The hydrolyses of 9-phenoxyacridine[876] and 9-alkylaminoacridine hydrochlorides[54] also follow pseudo-first-order kinetics, although much slower reaction rates have been observed in the latter case.[56] The reaction of 9-

chloroacridine with ethoxide anion, however, has been shown to be of second order.[32]

The yields of 9-acridanone, or 9-chloroacridine, from the cyclization of diphenylamine-2-carboxylic acids, are with very few exceptions excellent (up to 90%) and by-products of the cyclization are practically unknown. Traces of the 10-(9-acridinyl)-9-acridanone (10) have been reported in 6,9-dichloro-2-methoxyacridine produced this way,[57] and it probably arises by a similar route to that shown in the synthetic sequence below.[58] The yield of 10 in the last step is only 5%. The anhydride appears to be the essential intermediate, as the acid chloride of the diphenylamine-2-carboxylic acid forms only the expected 9-acridanone on heating.[57]

The parent compound of 10 may also be prepared by heating 9-acridanone and 9-chloroacridine[58] at 300°.

Amides of the diphenylamine-2-carboxylic acids are cyclized by phosphorus oxychloride to 9-aminoacridines.[59, 60] Those derived from 4-[61—63] and 5-[64]amino-1-diethylaminopentane, 5-amino-1(N)-piperidinopentane, ethanolamine, urethane, piperazine, and ethylenediamine[66, 67] have all been cyclized this way, and excellent yields have been obtained. The amide of triphenyl-amine-2-carboxylic acid also undergoes this cyclization in 97% yield. This route is especially useful for the synthesis of 3-substituted-9-alkylamino-acridines, for an amide with a large *N*-substituent may give mainly this compound, when the diphenylamine-2-carboxylic acid would have given a mixture of the 1- and 3-substituted 9-acridanones.[68] Hydrolysis of the 9-amino compound could then give the 3-substituted 9-acridanone.

These results confirm the proposed mechanism for the cyclization, since the phosphoryl moiety in the intermediates (11) or (12) is expected to be a better leaving group than the amine.

11   12

(2) *Cyclizations with Sulfuric or Phosphoric Acid*

These strong mineral acids are convenient for laboratory use and have the advantage of giving 9-acridanones directly, but the products are often less pure than those obtained by the hydrolysis of 9-chloroacridines. Sulfuric acid was the first reagent used for the cyclization of diphenylamine-2-carboxylic acids to 9-acridanones[69] and still finds frequent use, including, e.g., the assay[877, 878] of the pharmaceuticals mefenamic and flufenamic acids (p. 328). Polyphosphoric acid (20–60 parts by weight) has been used less frequently,[70, 140] but it was the reagent of choice for the Brockmann group in their extensive studies on depeptidoactinomycin analogues.[71]

The cyclization temperatures with these reagents seldom need to be higher than 100°, and reaction times rarely exceed 2 hr. These conditions, however, seem to be much more critical when sulfuric acid is used, and optimum conditions vary considerably for each diphenylamine-2-carboxylic acid. After cyclization, the hot reaction mixture is poured into boiling water, and made just alkaline, to precipitate the 9-acridanone as a granular solid.

Although kinetic studies are lacking for the cyclization of diphenylamine-2-carboxylic acids, the mechanism is clearly analogous to the acid-catalyzed cyclization of 2-acyldiphenylmethanes to anthracenes, where a carbonium ion is the postulated intermediate.[72] The catalytic effect of acid is shown by the way the cyclization is retarded on dilution, but the trace of acid needed to initiate the phosphoryl chloride cyclization may only be required to catalyze the complex formation, a task subsequently assumed by the liberated hydrogen chloride.

The following mechanism accounts qualitatively for the effect of substituents on the course of the reaction:

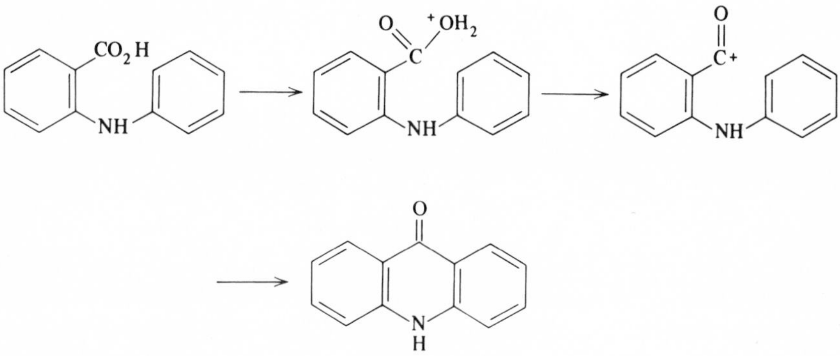

Sulfonation, but not phosphorylation, accompanies cyclization, accounting for the drop in yield if the acid is too concentrated or too vigorous conditions are employed. With diphenylamine-2-carboxylic acid, a 9-acridanone yield of 98% after 30 min at 85° drops to 35% after 90 min at 100°. A further 59% of the acid is converted to 9-acridanonemonosulfonic acids, of which a high proportion is the 2-sulfonic acid.[27] To ensure that the product is 9-acridanone-2-sulfonic acid, fuming sulfuric acid should be used.[73–75] If sulfonation is to be avoided, diluted sulfuric acid, which retards both reactions but favors cyclization more than sulfonation; or a low proportion of sulfuric acid, which becomes progressively and significantly more dilute as the cyclization proceeds, may give good results.

Electron-donating substituents in the acridanone rings facilitate sulfonation; even the yield of 4-nitro-9-acridanone falls from an initial 90% after 15 min at 100° as the reaction continues,[76] showing that a single nitro group is not sufficiently deactivating to prevent sulfonation.

Another serious limitation of sulfuric acid is the effect it has on substituents. Amines need not be protected,[77, 78] indeed, N-acetyl groups are hydrolyzed under cyclization conditions, but cyano groups are hydrated to amides and alkoxy groups are usually dealkylated. Methoxy groups are slightly more resistant than ethoxy groups,[79] but only 28% of 6-chloro-2-methoxy-9-acridanone could be recovered after cyclization of the appropriate diphenylamine-2-carboxylic acid.[80] Polyphosphoric acid does not cleave ether groups,[71, 81] nor does it hydrolyze and desulfonate the trifluoromethylsulfonyl group.[82]

An effective use of polyphosphoric acid is found in the synthesis of acronycin.[83] A three-carbon side chain is required in the 4-position of the 9-acridanone, and this is produced as the amide link is hydrolyzed prior to cyclization.

Sulfuric acid, in contrast to phosphorus oxychloride, gave 9-acridanone, rather than 9-anilinoacridine, when used to cyclize the anilide of diphenylamine-2-carboxylic acid, but the yield was low. Presumably, the amide was first hydrolyzed, as N-acyl groups are also removed under these conditions.[13]

(3) Cyclizations with Other Reagents

A mixture of phosphorus oxychloride and sulfuric acid in chlorobenzene has been used to cyclize 50-lb batches of diphenylamine-2-carboxylic acid to 9-acridanone,[84] but the precise advantages of this reagent have not been defined.

No advantage in yield is gained by using phosphorus pentachloride in a refluxing inert solvent, instead of phosphorus oxychloride, as the cyclization reagent; nuclear chlorination may occur.[85] However, 9-acridanones with 1- and 4-methoxy or ethoxy substituents and N-acetyl groups have been formed, without hydrolysis, using phosphorus pentachloride in benzene.[86, 87] The reaction temperature may influence the nature of the product, for 3',4'-dimethoxydiphenylamine-2-carboxylic acid gave the corresponding 9-acridanone in carbon disulfide but the 9-chloroacridine in benzene.[3]

Thionyl chloride, again usually in an inert solvent, has been little used as a cyclization reagent, and is said, like phosphorus trichloride, to convert diphenylamine-2-carboxylic acids only to the 9-acridanones. However, it will convert 9-acridanones to 9-chloroacridinium chlorosulfites,[88] and under pressure cyclizes 5-chloro-4'-methoxydiphenylamine-2-carboxylic acid to 6,9-dichloro-3-methoxyacridine hydrochloride,[80, 89] as does phosgene.[89] The sulphonamido group of 2'-nitro-4'-sulfonamidodiphenylamine-2-carboxylic acid survives intact when this acid is cyclized by thionyl chloride in a dichlorobenzene-dimethylformamide solvent mixture[90] at 160°.

The cyclization may be carried out in two stages if the acid chloride is first prepared, and then cyclized. Thionyl chloride (1 mole) in dichloromethane, with a trace of dimethylformamide, is the recommended reagent[91] if 9-acridanone formation is to be avoided during the first step, but excellent yields of acid chloride are given by phosphorus pentachloride.[80] The cyclization step is carried out by heating the acid chloride alone,[92, 93] in a refluxing solvent[80, 92, 93] or in a phenol melt.[80] In some cases much decomposition took place. In one instance, a nitro group blocking the carbon atom needed for cyclization was eliminated with the formation of 13.

The reaction of the acid chloride with pyridine is alleged to give N-phenyl-anthranylpyridinium (14); with concentrated sulfuric acid, this substance gave a mixture of 9-acridanone and diphenylamine-2-carboxylic acid.[58]

13   14

Aluminium chloride, used earlier and still occasionally employed,[95] is unnecessary in this cyclization[96, 97] and may cause decomposition, demethylation of methoxy groups,[98] or reaction with the solvent. The isolation of 9-phenylacridine, as well as 9-acridanone, after treating diphenylamine-2-carboxylic acid chloride with aluminium chloride in benzene,[99] led to the development of a useful synthesis of 9-substituted acridines, from 2-acyl- and 2-aroyl-diphenylamines. Such an intermediate was presumably formed by the reaction of the acid chloride with the benzene.

Phosphorus pentoxide in toluene gives much poorer yields in the cyclization of diphenylamine-2-carboxylic acid than reagents discussed above,[21] and has had little use.[100] In xylene, however, 100% cyclization of 5-phenyltriphenylamine-2-carboxylic acid is claimed,[101] although sulfuric acid gave poor results, and phosphorus pentachloride and aluminium chloride caused chlorination.

Hydrochloric acid [102] and boric acid[103] at high temperatures have been used to prepare polynuclear 9-acridanones, thereby avoiding sulfonation. Acetic acid alone did not bring about the cyclization of octachlorodiphenylamine-2-carboxylic acid, but with zinc chloride present 5'-anilino-2'-methyldiphenyl-amine-2- carboxylic acid does cyclize.[105] Chlorosulfonic acid, used earlier for cyclization in dyestuff syntheses,[106, 107] has been shown to give a 90% yield of 6-chloro-2-methoxy-9-acridanone, where sulfuric acid gave 72% of the demethylated product.[80] Attempted chlorosulfonation of 2'-nitro-[108] or 2',4'-dinitro-diphenylamine-2-carboxylic acid[109] gave good yields of 9-acridanones with a chlorosulfonyl group probably in the 7 position, which could be charac-terized as sulfonamides[109] or phenylsulfonic esters.[108]

The action of acetyl chloride or acetic anhydride and perchloric acid on 5-chloro-4'-methoxydiphenylamine-2-carboxylic acid, did not give the ex-pected N-acetyl acid but 6-chloro-2-methoxy-9-acridanone and its N-acetyl derivative, respectively.[110] Benzoyl chloride,[111] or better p-nitrobenzoyl chloride[112] in a high-boiling solvent has also been used to effect cyclization to some complex 9-acridanones in good yield.[113] The 9-acridanone may be formed from the N-acyldiphenylamine-2-carboxylic acid (Section 1.A(5)) or from a mixed anhydride.

Some 3,5-diamino-9-acridanone was isolated as a by-product of the oxida-tive cyclization of 3',6-diaminodiphenylamine-2-carboxylic acid to 6-amino-phenazine-1-carboxylic acid, in refluxing nitrobenzene.[114] A carbonium ion mechanism would not appear to be operative in this case.

(4) *Alternative Cyclizations*

A given 9-acridanone may always, in theory, be obtained by the cyclization of two isomeric diphenylamine-2-carboxylic acids, as in the example shown. 2-Iodo-[115] and 2-iodo-7-methoxy-9-acridanones[116] have also been prepared from both isomers. In practice one of the routes always gives better results, and in some cases only one isomer undergoes cyclization. For example, 3-nitrodiphenylamine-2-carboxylic acid fails to cyclize to 1-nitro-9-acridanone in contrast to the behavior of the 3′-nitro isomer.

Both 3′- and 5-dimethylaminodiphenylamine-2-carboxylic acids were cyclized with phosphorus oxychloride to 9-chloro-3-dimethylaminoacridine, but the former gave a yield of only 30% after 3 hr, compared with the 60% yield obtained after half an hour from the latter.[117] Some of the 3′-dimethyl-amino acid may have been lost in the concurrent formation of 9-chloro-1-dimethylaminoacridine, emphasizing the point that if 1- or 3-substituents are required in the 9-acridanone nucleus, they must be present as 3- or 5-substituents in that ring of the starting material bearing the carboxylic acid group to ensure a specific synthesis.

The cyclization of 3′-substituted diphenylamine-2-carboxylic acids with no substituents in position 2′ or 6′ usually gives rise to a mixture of 1- and 3-substituted 9-acridanones. The results of a number of investigations of this type of cyclization are collected in Table VII.

This possibility was first recognized in the example given.[118] Although only 1-nitro-9-acridanone was isolated[119] and shown to differ from the nitro-acridanone prepared from 5-nitrodiphenylamine-2-carboxylic acid, later investigators obtained a small amount of 3-nitro-9-acridanone from the same reaction.[30]

The primary determining factor of the direction of cyclization appears to be the nature of the 3′-substituent. Electron-withdrawing groups (chloro, nitro, and trifluoromethyl) favor the formation of 1-substituted 9-acridanones, possibly because their increased negative charge attracts the carbonium ion to the vicinity of the ortho-carbon atom. Electron-donating groups with lone pairs of electrons (amino, dimethylamino, and alkoxy) favor the formation of the 3-substituted isomers, since the activated position para to the substituent is not sterically hindered. A methylenedioxy group seems to activate its para position more effectively than a methoxy group.[120] The 3′-methyl group, whose inductive electron donation is more effective at the ortho position, finishes mainly in the 1-position of 9-acridanone,[97, 121] but a series of 3′,4′-dialkyl and cycloalkyldiphenylamine-2-carboxylic acids appear to give 80% yields of 2,3-disubstituted 9-acridanones on cyclization in sulfuric acid, in contradiction to the former findings.[122] The 3′-fluoro and 3′-bromo substituents contrast with 3′-chloro group in favoring the 3-substituted 9-acridanone. Mesomeric electron donation by the former, and steric hindrance by the latter, could account for this preference.

An interesting consequence of the directing powers of 3′-substituents is that contrary to earlier claims,[118] either 1- or 3-amino-9-acridanones can be made in good yield from 3′-nitrodiphenylamine-2-carboxylic acid by cyclization followed by reduction, or reduction followed by cyclization, respectively.[124] Exclusive 1-substitution has also been achieved by cyclizing the 6′-bromo-3′-substituted acid and subsequently removing the bromine by catalytic hydrogenation.[125]

The formation of a mixture of 9-acridanones when flufenamic acid (3′-trifluoromethyldiphenylamine-2-carboxylic acid) is cyclized in sulfuric acid fortunately does not diminish the value of this assay technique because differences in uv and fluorescence spectra of the isomers are negligible.[877]

The influence of the cyclization reagent on the proportions of the isomers has not been fully investigated, but replacing sulfuric acid by phosphorus oxychloride is said not to alter the ratio.[30]

The choice of reagent has a significant effect on the mode of cyclization of 5-nitrodiphenylamine-2,2′-dicarboxylic acid (15); a molecule offering another type of alternative cyclization. After cyclization with phosphorus oxychloride, the product isolated was 5-carboxy-3-nitro-9-acridanone (16),[126] identified after decarboxylation by comparison with authentic 3-nitro-9-acridanone, whereas phosphorus pentachloride in nitrobenzene gave the alternative acridanone (17).[127]

Reduction and decarboxylation of the latter gave a compound identified as 1-amino-9-acridanone. The possibility remains, however, that both reagents gave isomer mixtures, and the two groups both lost each others product during their identification procedures.

Fractional crystallization and differential hydrolysis of the 9-chloroacridines have been used most frequently for the separation of isomers produced in these ambiguous cyclizations. Chromatographic methods appear not to have been used, and only one case of countercurrent separation is recorded.[128] The 1-substituted 9-chloroacridines are generally more soluble and more readily hydrolyzed than their 3-substituted isomers, but there are exceptions.[129, 130] Short hydrolysis times give the best separation, with half an hour being sufficient for the 1- and 3-methyl-, and only 5 min for the 1- and 3-methoxy-9-chloroacridines.[23] This technique did not separate the 1- and 3-ethyl analogues, however.[125]

In a number of cases,[879] heating a mixture of 1- and 3-substituted 9-chloroacridines in pyridine allows separation. The 3-isomer remains undissolved, and the 1-isomer may be precipitated from the filtrate, as a pyridine complex, by the addition of ether. Treatment of this complex with phosphoryl chloride regenerates the 9-chloroacridine.

Differences in solubility between the pairs of 9-chloroacridines or the 9-aminoacridine hydrochlorides (in which the 1-isomer again shows enhanced solubility[121]) have even been used to allocate structures to the isomers isolated after cyclization. The general principle, that 1-substituted 9-chloro-[30, 130] and -9-amino-acridines[121] have lower melting points than the corresponding 3-substituted compounds has been similarly employed. These criteria are unreliable if used alone,[20, 131, 132] and there seem to be possibilities for nmr spectroscopy in this area (see Chapter XII, p. 687).

(5) Cyclization of N-Substituted Diphenylamine-2-carboxylic Acids

There are few examples of this type of cyclization, but they indicate that this route to 10-substituted 9-acridanones is of general application. N-Methyl-diphenylamine-2-carboxylic acid cyclizes with either phosphoryl chloride[12] or sulfuric acid,[133] and triphenylamine-2-carboxylic acid gives a good yield of 10-phenyl-9-acridanone with sulfuric acid.[134] However, N-methyl-2'-nitro-diphenylamine-2-carboxylic acid was reluctant to cyclize in sulfuric acid,[135] in contrast to the unmethylated acid.[136] N-Phenyldiphenylamine-2-carboxylic acids, with substituents in the benzene rings, offer alternative modes of cyclization, and a mixture of the isomeric products was found in the following example:[134]

Phosphorus pentachloride with aluminium chloride[137] and phosphorus pentoxide[101] have also been used to cyclize triphenylamine-2-carboxylic acids.

Several N-benzoyldiphenylamine-2-carboxylic acids lose benzoic acid on heating to 270° or less,[111] to give the corresponding 9-acridanones in good yield. The claim for base catalysis in this reaction has not been substantiated. Esters of these acids, which are the usual products of the thermal rearrangement of imino ethers,[138] do not react readily, but they may be converted into the corresponding 9-acridanones by combined hydrolysis and cyclization in hot sulfuric[139] or phosphoric acid.[140] Mechanistic studies have shown that on fusing the esters (18) without solvent, 9-acridanones are produced by an intermolecular reaction. As the solvent is introduced, a second intramolecular mechanism appears that eventually predominates in dilute solution.[112]

It is interesting to note that if N-benzoyldiphenylamine-2-carboxylic acids are treated with phosphorus oxychloride, followed by ammonia, the product is not a 9-chloroacridine but a 2-phenyl-4(3H)-quinazolinone (19).[141, 142]

Esters of 2-carboxydiphenylamine-N-acetic acid (20) have been cyclized both to 9-acridanones[143] and to 10-methyl-9-acridanones.[144]

R = NO$_2$, NH$_2$

20

R = H

The determining factor seems to be the substituent at position 4, as treatment of the ester (R = nitro) with 1% aqueous potassium hydroxide is sufficient to eliminate the acetic acid residue,[143] whereas 9-acridanone-10-acetic acid can be isolated after cyclization of the ester **20** (R = H) with sulfuric acid.

(6) Cyclization Failures and Difficult Cyclizations

Diphenylamine-2-carboxylic acids may fail to cyclize to 9-acridanones because they are too deactivated or too hindered for the intramolecular condensation to take place, or because a more favorable reaction path is open to them.

This alternative is frequently decarboxylation, which 5-aminodiphenylamine-2-carboxylic acids undergo readily,[25, 145] although their esters are stable.[146] The 3-acetamido acid also gave only 5% of 1-aminoacridine with sulfuric acid,[147] while its 3'-nitro analogue, in many attempted cyclizations, gave no 9-acridanone; only starting material and 3-amino-3'-nitrodiphenylamine were recovered.[148] For this reason, it is usually preferable to prepare amino-9-acridanones by cyclization and reduction of the appropriate nitrodiphenylamine-2-carboxylic acids, most of which are stable under these conditions.

3-Nitrodiphenylamine-2-carboxylic acid also decarboxylates easily,[147, 149,150] but if the nitro groups are in the other phenyl ring, as with the 2'-methyl-4'-nitro acid,[105] decarboxylation probably takes place because the cyclization is inhibited. The 2'-nitro acid gave the acid chloride with phosphoryl chloride,[23] suggesting that a phenylamino substituent *ortho* to the carboxyl group is not an effective as a *para* amino group in promoting decarboxylation.

Steric hindrance probably accounts for the resistance of N-benzoyl-2-*tert*-butyldiphenylamine-2-carboxylic acid toward hydrolysis and cyclization.[151]

An alternative product in the attempted cyclization of 2'-aminodiphenylamine-2-carboxylic acids[26, 143] is the 7-membered ring lactam **21** (R = H), which may also be prepared by heating **20** (R = NH$_2$). This product forms an N-nitroso derivative but not a diazonium compound with nitrous acid.[136] A similar compound, **21** (R = NH$_2$) almost certainly is the product of the vigorous reduction of 2′,4′-dinitrodiphenylamine-2-carboxylic acid,[26, 69] as it is unlike authentic 2,4-diamino-9-acridanone, and the proposed structure, **22** (R = NH$_2$),[26] based on the unsubstantiated **22** (R = NO$_2$),[7] is doubtful both on steric and on basicity considerations. Acids having 2′-secondary amino groups do not form lactams. The analogous lactone, depsazidone (**23**), may also be obtained, by treating 2′-hydroxydiphenylamine-2-carboxylic acid with thionyl chloride and pyridine.[152]

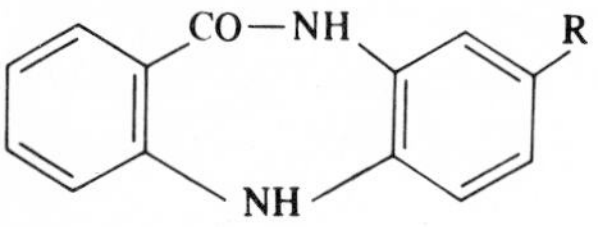

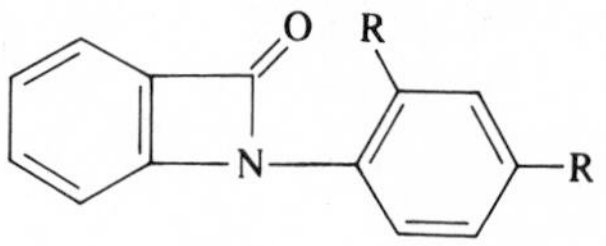

In some cases the incorrect choice of reagent may lead to difficulties. Sulfuric acid cyclizes 2′,4′- and 2′,5′-dichlorodiphenylamine-2-carboxylic acids to 9-acridanones,[118, 153] but with phosphorus pentachloride in chloro-benzene the acid chlorides are isolated. In the latter case, nuclear chlorination also occurs.[85] The cyclization shown below is effected by sulfuric acid,[154] but not by phosphorus oxychloride, since prior hydrolysis of the imide seems to be necessary.

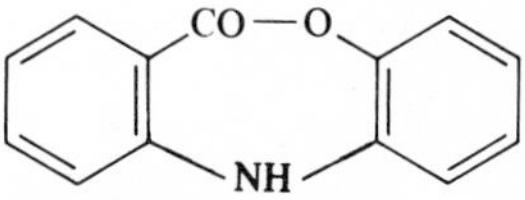

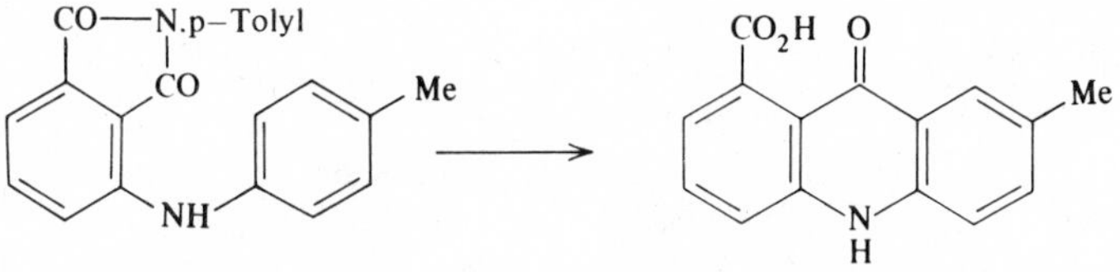

B. Preparation of Diphenylamine-2-carboxylic Acids

A review of the preparation of these compounds is included because they are essential intermediates in the synthesis of many 9-acridanones and acridines. In earlier literature, these acids were made as intermediates, but in recent years they have acquired importance under the title "phenamic (or fenamic) acids," as antipyretic and anti-inflammatory drugs in the treatment of rheumatic complaints. A number of fenamic esters[880, 881] and the cyclic analogue, acridine-4-carboxylic acid,[882] show similar activity. Although this has increased research on alternative synthetic routes, the majority of the diphenylamine-2-carboxylic acids described in the literature, and included in Table IV, have been made by the Ullmann reaction. *Chemical Abstracts* now uses the name *N*-phenylanthranilic acids for these compounds, which alters the numbering system used throughout this chapter, and shown on diagram **24**, to that shown on **25**.

24   **25**

(1) *The Ullmann Reaction*

The basis of this reaction is the substitution of the halogen atom of a halobenzene by a primary or secondary Section 1.B(1)(d)(iii) phenylamine to give a diphenylamine. Clearly, two routes are possible to the diphenylamine-2-carboxylic acids, according to whether the carboxyl group occupies a position *ortho* to the halogen, or the amino group in the starting materials. Amino- and halo-benzenes with *ortho* amido or ethoxycarbonyl groups may also be employed.[155, 156]

Early examples of both types of reaction depended for success on the activation of the halogen atom by electron-withdrawing groups in *ortho* and *para* positions,[69, 157] and a little sodium acetate could often be added with advantage.[109, 158] Ullmann's discovery that the addition of a small proportion of powdered copper to the reaction mixture made possible the facile substitution of nonactivated halogen atoms,[159] removed this serious limitation to the scope of the reaction and allowed the preparation of the parent acid from type 1[159] and type 2[160, 161] reactions in 85 and 95% yields, respectively. Diphenylamine-2-arsonic acids may be made in a type 2 reaction under identical conditions.[162]

The usual procedure is to reflux a mixture of the amine and the halide for a few hours in a solvent, which may be removed by steam distillation, in the presence of a copper-containing catalyst; and a base, which absorbs the liberated hydrogen halide. An excess of the amine component is preferred, and an alkali salt of the acid is frequently used. The product may precipitate as the alkali salt on cooling, or as the acid on acidification, or may be leached

Type 1

Type 2

from the residues with aqueous alkali. Residual halobenzoic or anthranilic acid may be removed by washing the product with boiling water, in which diphenylamine-2-carboxylic acids are almost insoluble, or they remain in solution when the required acid is precipitated from acetone by the addition of water.

(a) THE CATALYST. Since Ullmann's use of mechanically divided copper, the catalytic effect of many other preparations of copper, cuprous chloride, and bromide; and cupric oxide, carbonate, acetate, and sulfate, alone and in admixture, has been demonstrated. The spongy copper, precipitated by zinc powder from aqueous copper sulfate,[163, 164] and that left after treating Devarda's alloy with 30% aqueous sodium hydroxide are highly recommended.[165–167] They should be freshly prepared or kept under light petroleum and they may be activated before use by treatment with iodine.[86, 164] The addition of a little potassium iodide may,[131] or may not,[168] increase the yields. Salts of iron, nickel, platinum, and zinc had a weaker catalytic effect; those of manganese and tin were ineffective.

More catalyst is required if nonactivated amines and halides are used, but the quantity needed is still small. Weights in excess of 3% of the halogen compound may even reduce yields because of tar formation. There appears to be a lower limit, too, as a 10-fold increase of catalyst from $1 \times 10^{-4}\%$ by weight brought about a 36-fold increase in the yield.[118] This contrasts with the similar Gomberg reaction, between bromobenzene and 4-acetamidotoluene, where, without the activation of the *ortho* carboxyl group, at least 2 g of copper is needed per mole of reactant.[169]

The true catalyst appears to be copper in the cuprous form[170] because the activity of metallic copper is much reduced in the absence of air,[169, 171] and cuprous iodide is effective in the presence of potassium iodide;[161] these are conditions under which cupric ions cannot exist. In the Gomberg reaction above, a red powder, formed from copper and the amide only when air is present, appears to be the catalyst.[169] It may be a substituted cuprammonium salt.

The suggestion of a free-radical course for this reaction[172] has little support,[173] and it is more probably ionic in character. A satisfactory mechanism must explain not only why cuprous ions catalyze both types of reaction, but also why only *ortho* halogen groups are replaced, even when 2,4-dichlorobenzoic acid reacts with the strongly nucleophilic anisole;[174] and why *meta* and *para* aminobenzoic acids are so reluctant to react with bromobenzene, when the *ortho* isomer reacts readily.[175] The conclusion that the cuprous salt of the acid component plays a decisive part seems inescapable.[176] In the case of the *ortho* halobenzoic acid, a copper chelate (26) has been suggested as the reactive species.[177] The coordinate bond between the halogen and the copper atom will enhance the polarization of the carbon-halogen bond and facilitate nucleophilic attack by the amine at the ring carbon atom. For reactions of type 2, an intermolecular coordinate bond between copper atom and halogen (27) will bring the electron-depleted ring carbon atom into proximity with the amino group of the anthranilic acid.

26

27

Surprisingly, methyl salicylate and salicylamide have been shown to catalyze Gomberg and Ullmann reactions in the absence of copper or its salts.[178]

(b) THE BASE. One equivalent of base is essential to remove the hydrogen halide liberated as the Ullmann condensation proceeds, or reaction slows and decarboxylation tends to occur.[118] Improved yields are often found if the acid component is introduced as the dry, powdered potassium or sodium salt,[147] or a further equivalent of base is added as an acid-binding agent, even though it is unlikely that the anion of the acid plays an important part in the substitution. Up to a third equivalent also appears to have a beneficial effect, possibly by neutralizing local acid concentrations or by-product acid, but yields fall off rapidly when further base is added.[177]

Ammonia has been superseded as the base,[69] first by excess amine,[159] and then by potassium carbonate, which remains the reagent of choice in most cases.[147, 179] Sodium carbonate is not always so satisfactory, possibly because the sodium salt of the product acid is more soluble, but it is cheaper. Cupric carbonate has been used as the combined catalyst and base,[180] and calcium carbonate has found a special use in the preparation of diphenylamine-sulfonic acids.[181, 182] Calcium hydride has recently come into use,[91, 180] and sodium acetate is useful if alkali-sensitive substituents are present.[183] N-Ethylmorpholine[180] has proved a suitable base to give a homogeneous reaction mixture when nonhydroxylic solvents are employed, but a general suggestion that a tertiary amine might be chosen to act as the base, solvent, and temperature control in this reaction has not been implemented.[184]

Potassium hydroxide, which dissolves in alcoholic solvents, is less effective than potassium carbonate, and the efficiency of the latter base falls in aqueous alcoholic solutions. This is suggested as evidence for the participation of halogen to copper coordinate bonds that would be disrupted by the alkali.[177]

(c) THE SOLVENT. If possible the reaction mixture should remain fluid and homogeneous throughout, as immiscibility of the components and precipitation of product on the surface of undissolved reagents can be causes of low yields or failure in the Ullmann reaction. To achieve this end, a refluxing alcoholic solvent is generally employed, of which amyl alcohol (bp 130°) is the usual choice, unless its boiling point is unsuitable. Water formed when the acid is neutralized may slow the reaction[177] or poison the catalyst,[169] so that it is frequently azeotropically distilled before[147] or during[185] the reaction. However, a trace of water is said to be necessary for the substitution to proceed,[147] and water has even been used as solvent under both atmospheric[186] and higher pressures.[187, 188] If the amine is sufficiently soluble, yields may be high,[168, 189] particularly if the alkali salt of the diphenylamine-2-carboxylic acid is less soluble than that of the substituted benzoic acid.

A wide variety of other solvents has also been used. Nitrobenzene,[190] with a little water added,[191] is better than glycerol[4] if a high reaction temperature is required, but it may cause oxidation. Bromo- or chloro-benzene may be used as alternatives.[192] The use of excess amine as solvent in type 1 condensations leads to lower yields than when amyl alcohol is employed[193]; hydrocarbon solvents, in which salts are insoluble, give poor yields, unless the components are activated by their substituents.[3] Dimethyl sulphoxide,[180], dimethylformamide,[182] dimethylacetamide,[194, 195] hexamethylphosphoramide[8] and diglyme[196, 197] have also been used; but direct comparisons with established solvents, and in many cases even yields, are lacking.

The use of esters of the acid components, as an aid to forming homogeneous reaction mixtures, remains largely unexplored.[114, 198, 199]

(d) EFFECT OF SUBSTITUENTS ON THE CONDENSATION

(i) *Type 1 Reaction; 2-Halobenzoic Acids and Anilines.* All four 2-halobenzoic acids have been used successfully in this reaction. This is expected from the mechanism proposed, as those halogens less willing to undergo nucleophilic displacement are the ones whose electrons are more readily polarized for the formation of a coordinate bond to the catalyst. Few direct comparisons of their respective reactivities have been made, but 2-chlorobenzoic acid has been used most frequently, possibly because of its lower cost. Bromine seems to be easier to replace than chlorine,[200] and triphenylamine derivatives are sometimes formed.[201] 2-Fluorobenzoic acid, now available from the oxidation of 2-fluorotoluene, might be tried in difficult cases,[170] since the substituted fluoro acids that have been used condensed without a catalyst[202, 203]; but success is not guaranteed.[114] The zwitterionic compound from 2-iodobenzoic acid and benzene, (2-phenyliodonio)benzoate (28), has been used,[204] as well as the acid itself.[187]

28   29

The ease of substitution and the yield of diphenylamine-2-carboxylic acid are greatly increased when electron-attracting groups are in positions *ortho* and *para*, but not *meta*, to the halogen atom; conversely, electron-donating groups in these positions retard the reaction. 2-Chloro-3,5-dinitrobenzoic acid (29, X = Cl, R = R' = NO$_2$)[171] or its methyl ester[205] gave high yields of diphenylamine-2-carboxylic acids or esters in no-catalyst reactions, and the introduction of a 5-nitro group into 2-chlorobenzoic acid increased the yield of product acid from 85 to 95% in the copper-catalyzed condensation with aniline. Even 2-bromoisophthalic acid (29; X = Br, R = CO$_2$H, R' = H), despite steric hindrance, reacted readily with 2,3-dimethylaniline.[195] However, the yield of substituted diphenylamine-2-carboxylic acid dropped from 50% to 20% when 2-chloro-5-methoxy (29; X = Cl, R = H, R' = MeO), rather than 2-chlorobenzoic acid, was used in reaction with 3-chloroaniline.[176]

Although both 2- and 6-halogen atoms in a substituted benzoic acid may be replaced by anilino groups under Ullmann conditions,[150, 206] there has never been an indication that substitution of a halogen more remote than *ortho* occurs, whether fluorine,[207] chlorine,[150] bromine,[208] or iodine.[209]

Low yields in the Ullmann reaction are also related to the low nucleophilic activity, or basic strength of the component amine, due to electron-withdrawing groups in the ring. *Ortho* substituents are particularly deactivating. Even 2-chloro-5-nitrobenzoic acid gave yields of only 13, 9, and 17%, respectively, in reaction with o-,[247] m-,[147] and p-nitroanilines,[210] although the introduction of a further nitro group at position 3 of the benzoic acid increased the yields into the 60–70% range.[205]

The corresponding introduction of electron-donating substituents into the aniline ring leads to the increased facility of reaction, as shown by the drop in optimum reaction temperature from 180° with *p*-nitroaniline,[25] through 100° for *p*-toluidine to only 60° for *p*-aminophenol.[147] Steric hindrance possibly accounts for the enhanced reactivity of anilines with *para*, rather than *ortho*, substituents.[211] Unfortunately, activating substituents also encourage oxidation of the anilines and dehalogenation of the acid components. *Meta*-[212] and *para*-[213, 214] phenylenediamine, or better their monoacetyl derivatives,[215] are sufficiently activated to react with excess 2-halobenzoic acid to give the bis-compounds **30** and **31**.

30   31

When chloroanilines are used in Ullmann condensations, the addition of "dinitrobenzene," with which they form a complex, is said to improve the smoothness of reaction.[216]

(*ii*) *Type 2 Reaction: Anthranilic Acids and Halobenzenes.* This is a less popular route to diphenylamine-2-carboxylic acids than the type 1 reaction, but a sufficient number of examples is available in the literature to show that the same electronic considerations apply in both cases. The amino group of anthranilic acid is even more sensitive to deactivation by electron-withdrawing substituents than the amino group of a substituted aniline in the type 1 reaction; therefore, such substituents are preferably introduced in the halobenzene ring, where their effect in *ortho* and *para* positions is activating. Bromobenzenes are generally preferred to chlorobenzenes, unless activating substituents are present,[217] and iodobenzenes have also found use, especially in alkaloid syntheses.[218–221] *o*-Chloroiodobenzene with anthranilic acid gave a 73% yield of 2'-chlorodiphenylamine-2-carboxylic acid, but none of the 2'-iodo compound,[222] thereby demonstrating the greater reactivity of the heavier halogen. 2,4-Dinitrofluorobenzene is the only fluorobenzene tried in reaction with anthranilic acid, and the already high yields obtained[223] without the use of a catalyst[224] are shown to be improved if potassium fluoride is added to the reaction mixture.[225] Bis-compounds may also be made by this route, using *p*-dibromobenzene.[213, 226]

(*iii*) *N-Substituted Amino Components*. The expectation, on electronic
grounds, that *N*-alkylamines would be more reactive, and *N*-acylamines less
reactive in the Ullmann reaction, is not realized in practice. Steric crowding
at the amino nitrogen atom may be the decisive factor.

In type 1 reactions, *N*-methylaniline reacted with neither 2-chloro-[97] nor
2-chloro-5-nitro-benzoic acids,[227] under conditions where aniline would have
given almost quantitative yields of the diphenylamine-2-carboxylic acids.
Yet 2-chlorobenzoic acid reacted with *N*-phenylglycine to give the tertiary
amine **32**,[144] and *N*-(*o*-chlorobenzoyl)aniline reacted with the anilide of 2-
chlorobenzoic acid to give compound **33** in 90% yield.[13]

32   **33**

More examples of type 2 reactions of *N*-substituted anthranilic acids are
recorded. *N*-Methylanthranilic acid fails to react with iodobenzene[97] but
condenses with 2,4-dinitrochlorobenzene without the need for a catalyst.[228]
Both *N*-methyl- and *N*-ethyl-anthranilic acids react with a number of *o*-
bromonitrobenzenes,[135, 229] but the yields are somewhat lower than in equivalent reactions with anthranilic acid itself. These 2′-nitrodiphenylamine-2-
carboxylic acids are reduced by sodium dithionite[229] or ammoniacal ferrous
sulfate,[135] to the corresponding amines in excellent yield. *N*-(2-Carboxyphen-
yl)glycine,[143] and *N*-acetylanthranilic acid[82] both lost their side chains, while
undergoing the Ullmann reaction; diphenylamine-2-carboxylic acids without
N-substituents could be isolated from the reaction mixture.

Many examples of the reaction of diphenylamine-2-carboxylic acid (*N*-
phenylanthranilic acid) with halobenzenes in the presence of copper show
good yields,[134, 137, 230] but iodobenzenes[137, 231] seem to be better phenylating
agents than bromobenzenes.[232] Traditional *N*-alkylation methods, such as the
treatment of diphenylamine-2-carboxylic acids with methyl sulfate[233] or alkyl
halides in basic solution, [234–237] appear to be more satisfactory than condensing *N*-alkylamines under Ullmann conditions. The reductive condensation of
aldehydes onto the secondary amine has also proved successful.[238]

(*iv*) *Comparison of Type 1 and Type 2 Reactions*. As similar experimental
conditions are employed for both reactions, the choice of a route to a diphenyl-
amine-2-carboxylic acid will depend on the availability of starting materials,
and the electronic factors discussed above. Type 1 reactions are almost
always preferred.

Comparative studies shown in Example 1[239] and Example 2[240] illustrate the general principle that unless every electronic factor is favorable, a type 2 reaction will furnish less product and more difficulty than a type 1 reaction.

Example 1

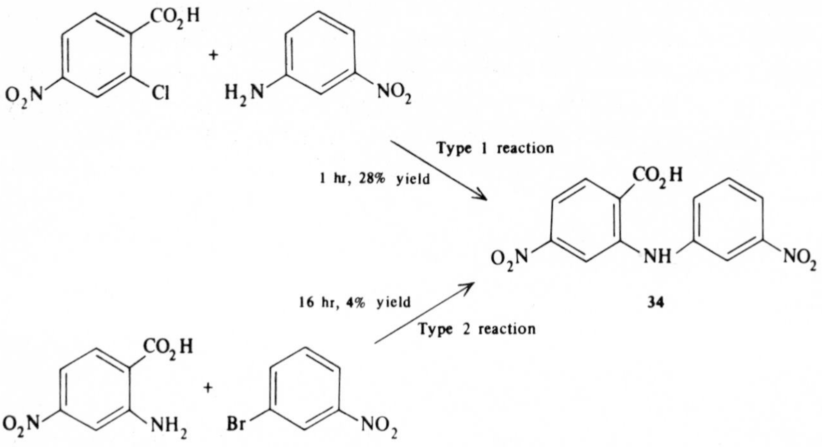

Example 2

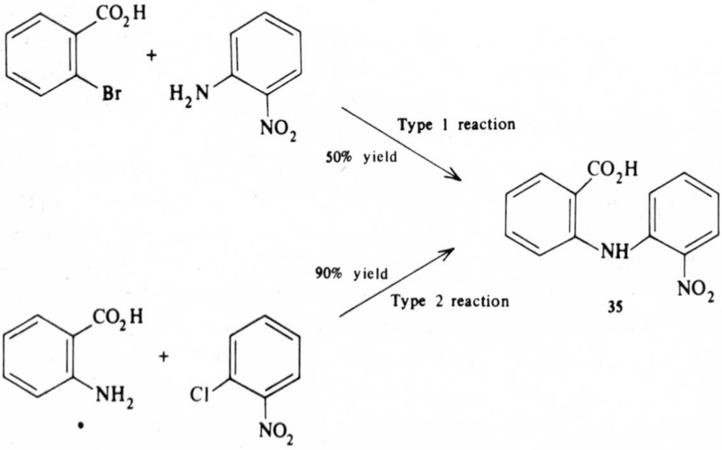

A single, mild deactivating influence can destroy the viability of a type 2 reaction, and even the apparently favored reaction between methyl 2,6-diaminobenzoate and *o*-iodonitrobenzene failed.[199] In Example 1, leading to

34, the *meta*-nitro groups have complementary activating and deactivating powers, but the deactivation of the anthranilic acid amino group leads to almost complete failure in the type 2 reaction. 3-Chloronitrobenzene does not react at all. In Example 2, even the powerful *ortho* nitro-group deactivation of the aniline does not have so devastating an effect on the result of the type 1 reaction. Other direct comparisons confirm this general rule.[132, 241]

The preparation of 2',6'-dichlorodiphenylamine-2-carboxylic acid by either route is equally satisfactory (type 1, 74%; type 2, 73%)[242] but the advantage of the one-step synthesis over the 5-step Chapman rearrangement Section 1.B(2)(a) is clearly demonstrated in this paper. Although the average yield in each step was well over 80%, an overall yield of only 39% was obtained.[242]

(*v*) *Limitations of the Ullmann Reaction.* Other examples of Ullmann reactions giving a zero yield of diphenylamine-2-carboxylic acid, due to deactivation of the type considered in the last section, have been recorded. Failure in type 2 reactions is common,[139, 199, 210, 243] and cyano[17] and sulfonic acid[244] or amide[27] groups have a strongly deactivating influence on anilines in type 1 reactions. 2-Chloro-6-nitrobenzoic acid is also unwilling to condense with anilines.[148, 171, 245] Difficulties of isolation and purification of the product may also be the cause of low yields.[114, 158, 171]

The part played by steric hindrance is sometimes difficult to disentangle from deactivation effects, as in the unsuccessful reactions of 2-chlorobenzoic acid with 2-amino-4,4'-dinitrodiphenyl,[22] and 2,4,6-tribromoaniline.[246] However, it must be responsible for the decreasing yields as the alkyl group in 2-alkylanilines increases in size from methyl (35%) to ethyl (15%)[8] and for the failure of the reaction with 2-propyl- and 2-butyl-anilines.[151] A by-product found in these reactions was 2-cyclohexyloxybenzoic acid, formed by the copper-catalyzed nucleophilic attack of the solvent, cyclohexanol, on the 2-chlorobenzoic acid.[8]

The dark colors of Ullmann reaction mixtures, which have been ascribed to the formation of chelate complexes,[177] are also partly caused by the oxidation of the amines, often at the expense of the halogen compound. 2,6-Dichlorobenzoic acid has been completely dehalogenated, giving only benzoic and diphenylamine-2-carboxylic acids in reaction with aniline or dimethylaniline,[150] and 4-nitrobenzoic acid is a regular by-product, when 2-chloro-4-nitrobenzoic acid is a reaction component.[25, 239, 243, 247, 248] 3-Acetamido-2-chloro-,[147] 2-chloro-3,5-dibromo-,[139] and 2,4-dichloro-benzoic acids also suffer dehalogenation.[131, 249, 250] High temperatures favor this side reaction, for when the solvent was changed from amyl alcohol (bp 130°) to isopropanol (bp 80°) in the reaction between *p*-phenylenediamine and 2-chloro-3-nitrobenzoic acid, the almost quantitative yield of 4-nitrobenzoic acid dropped to 30%, and a 59% yield of the expected diphenylamine-2-carboxylic acid was obtained.[147] The use of monoacetyl p-phenylenediamine prevented dehalogenation,[158] even at high temperatures.[147]

Aminodiphenylamine-2-carboxylic acids have a tendency to decarboxylate during preparation,[25, 145, 239, 883] as well as during attempted cyclization (Section 1.A(6)), but both their esters and their hydrochlorides are stable. Even if the corresponding nitro acid, from the Ullmann reaction, is reduced at room temperature,[240] decarboxylation may occur.[105]

Chloronitrobenzoic acids sometimes undergo self-condensation to biphenyl-dicarboxylic acids,[239] such as 36, isolated from a reaction mixture of 2-chloro-5-nitrobenzoic acid and 3-aminodiethylaniline.[191] Compound 37 was another unusual by-product, in the reaction between 2,6-dichlorobenzoic acid and 4-anisidine.[251] A minor product of the reaction between 2-chloro-5-nitrobenzoic acid and p-phenylenediamine, from the analysis, could be 38.[247]

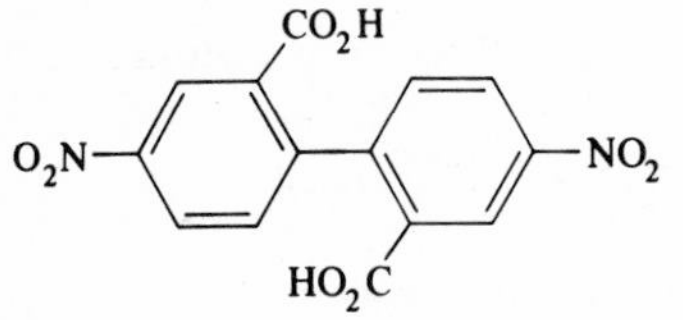

36

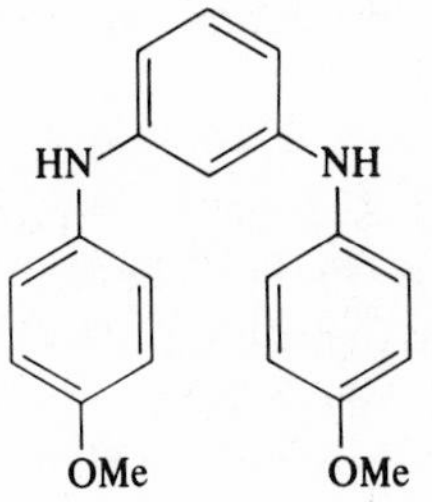

37

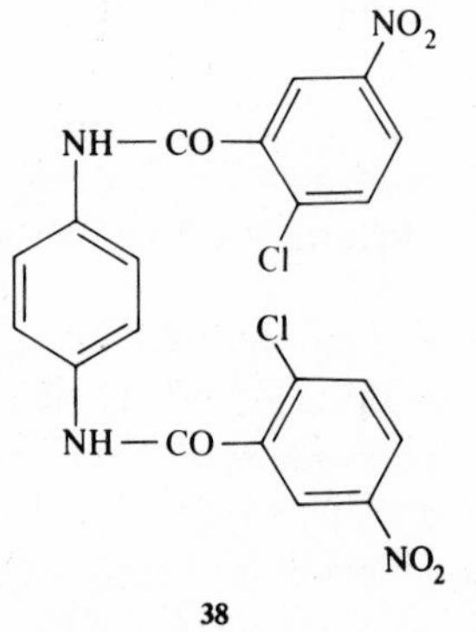

38

This same acid with 2-nitroaniline gave some 5-nitrosalicylic acid by hydrolysis of its reactive chloro group.[247]

If an o-dinitrohalobenzene is used in Ullmann reactions, one of the nitro groups may be substituted instead of the halogen. In a type 1 reaction, aniline condensed with 2-chloro-4,5-dinitrobenzoic acid to give 5-chloro-2-nitro-diphenylamine-4-carboxylic acid, and 5'-chloro-4-methyl-2'-nitrodiphenyl-amine-2-carboxylic acid was obtained in the type 2 reaction between 5-methylanthranilic acid and 3,4-dinitrochlorobenzene.[217] Both 4-nitro and 2-fluoro groups are replaced if 4,5-dinitro-2-fluorobenzoic acid is heated to 190° with an excess of aniline.[203] 2,4,5-Trinitrophenol also condensed with anthranilic acid to give 5'-hydroxy-2',4'-dinitrodiphenylamine-2-carboxylic acid in a similar reaction.[252]

The yield in the preparation of 2'-hydroxydiphenylamine-2-carboxylic acid is only 38%, but the 2'-methoxy analogue, available in much higher yield, could not be demethylated.[152] 5-Chloro-4'-methoxy-2'-nitrodiphenylamine-2-carboxylic acid gave poor yields in a type 1 reaction,[253] (despite claims to the contrary[254]), and starting materials for the type 2 reaction are not easily accessible.[255] However, it may be obtained by nitrating the 5-chloro-4-methoxy acid, which is readily available from the Ullmann condensation.[253] Bromination of the methyl ester of diphenylamine-2-carboxylic acid in acetic acid gave the 2',4,4'-tribromoderivative.[140]

(2) *Other Methods of Preparing Diphenylamine-2-carboxylic Acids*

(a) CHAPMAN REARRANGEMENT OF IMINO ETHERS. Jamison and Turner's application[138] of the Chapman diphenylamine synthesis for the preparation of diphenylamine-2-carboxylic acids is shown below.

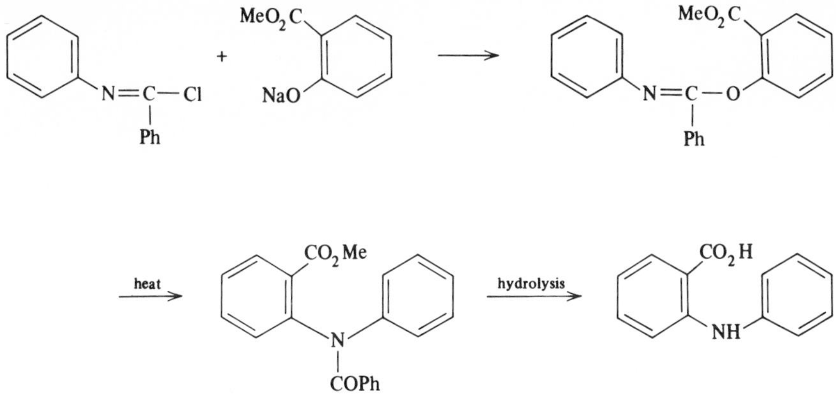

Iminochlorides are readily obtained by treating benzanilides with phosphorus pentachloride or thionyl chloride, and certain salicylic acids; for example, the 5-chloro and 3,5-dibromo derivatives, are much more accessible than the corresponding 2-chlorobenzoic acids. The rearrangement is exothermic at about 270°, and the N-benzoyl ester is hydrolyzed to the required acid in almost quantitative yield, and free from the tarry contaminants often associated with Ullmann products.

The electronic requirements of the rearranging molecule parallel those of the Ullmann reaction, in that higher yields and a smoother substitution are obtained if electron-withdrawing substituents are present in the migrating nucleus and absent from the aniline ring.[256] Compounds 39[257] and 40[151] do not undergo rearrangement, showing that steric hindrance about the migration terminus can cause the reaction to fail.

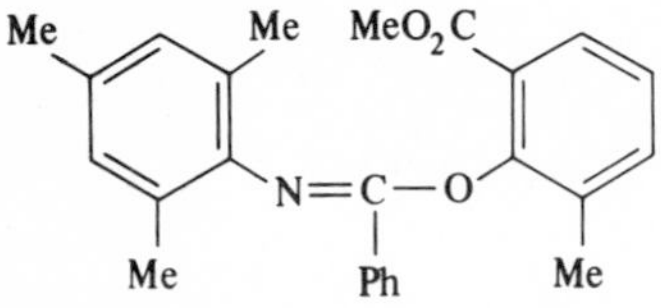

39

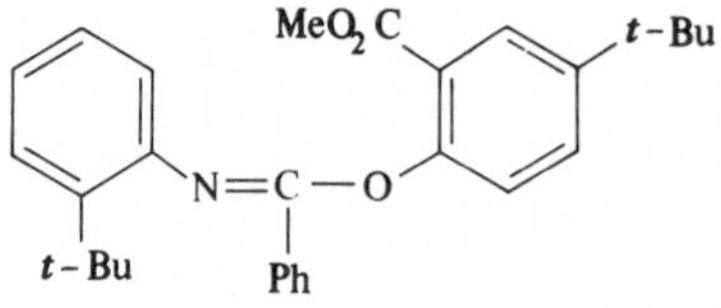

40

The possibilities of the Chapman rearrangement have been widely explored by the Parke, Davis group in their search for an economic synthesis of mefenamic acid (2′,3′-dimethyldiphenylamine-2-carboxylic acid). Some of their interesting variations are shown on the next page.[258–262]

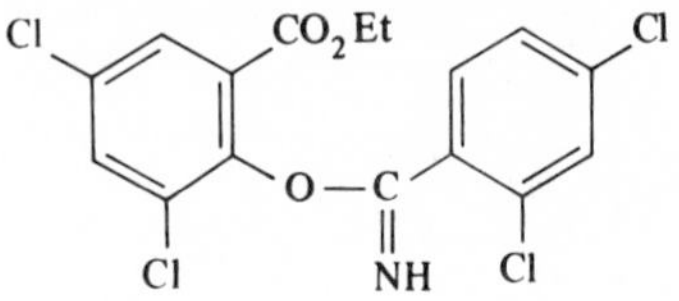

41

The diphenylamine-2-carboxylic acid is obtained by hydrolysis, but unfortunately no yields are given for these ingenious routes. Compound 41 is also said to give 2,4,5,7-tetrachloro-9-acridanone in 9% yield after heating to 260°, and then treating with sulfuric acid. The intermediate was not isolated.[263]

(b) SMILES REARRANGEMENT OF AMIDES. This rearrangement, which is really an internal nucleophilic substitution reaction, has been used in a few cases to prepare diphenylamine-2-carboxylic acids. A typical example with the proposed mechanism is shown on page 174.[264]

A strongly electron-withdrawing group must be present in an *ortho* or *para* position to the displaced group, as the phenyl and *m*-nitrophenyl ethers do not undergo this reaction; neither do disubstituted amides. Instead, hydrolysis of the amide linkage takes place.

When 2-bromo-2′,3′-dimethylbenzanilide reacts in a high-boiling solvent, in the presence of a base and a trace of water, to give mefenamic acid, a similar reaction must occur, in which the bromine atom, activated by the *ortho*

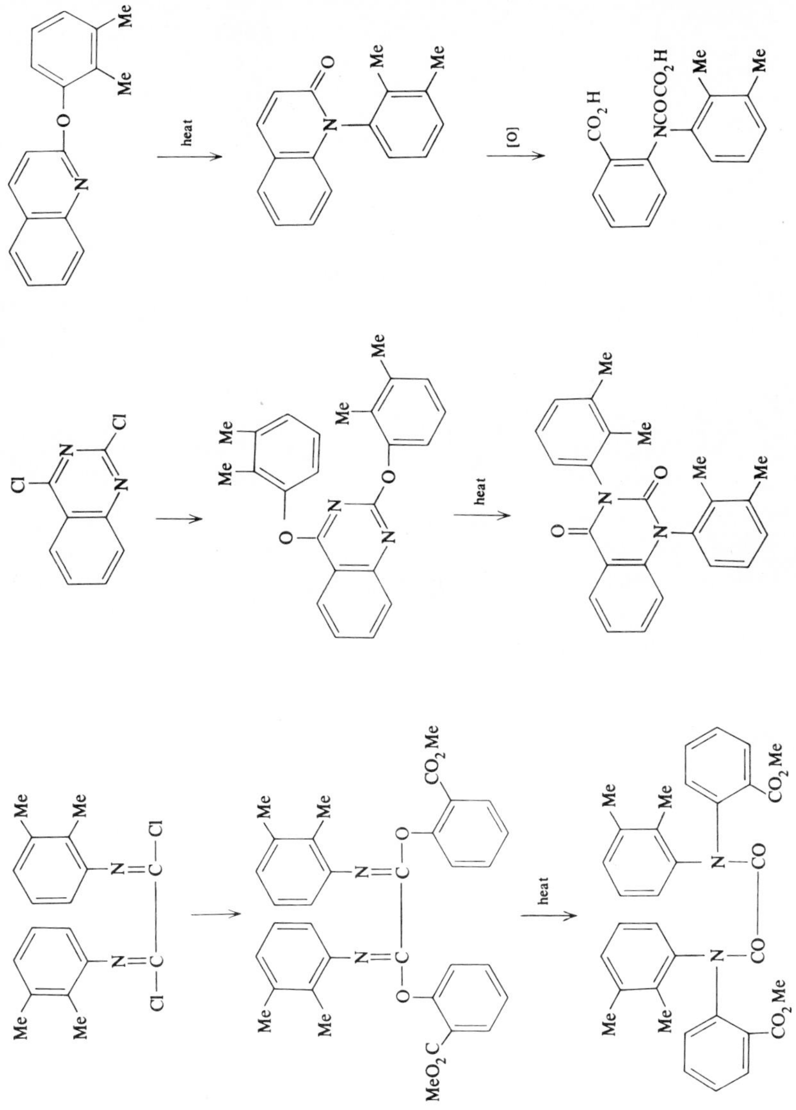

heat
[O]
heat
heat

carbonyl group is displaced by the anilino group of the amide.[265] Such an explanation also accounts for the by-products **42** and **43** in the reaction shown, which takes place in N-methylpyrrolidone with sodium hydride present.[266]

X = Halogen; R = SO$_2$NMe$_2$

(c) CONDENSATIONS OF ANTHRANILIC ACIDS AND QUINONES. Much of the early work on this reaction is of poor quality and it has aroused little interest recently.[267-270] Yields are generally low and the products badly characterized.

N-Methyl-, and N-ethyl-anthranilic acid react similarly,[884] and methyl and 2,3-dimethoxyquinones have also been used. In a refluxing solvent, the 2,3-positions of the quinone must be blocked to stop the reaction at stage **44**; even with excess quinone, the only pure product obtained was **45**.[271] However, **44** was the product after reaction in aqueous acetic acid at room temperature.[272, 884] Reductive methylation of **44** gave methyl 2′, 5′-dimethoxydiphen-

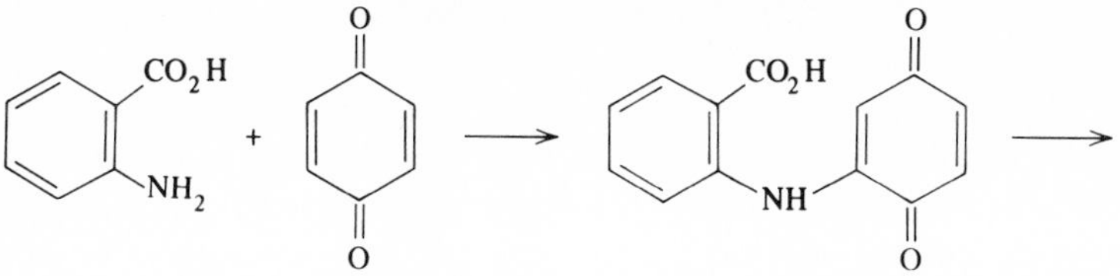

44

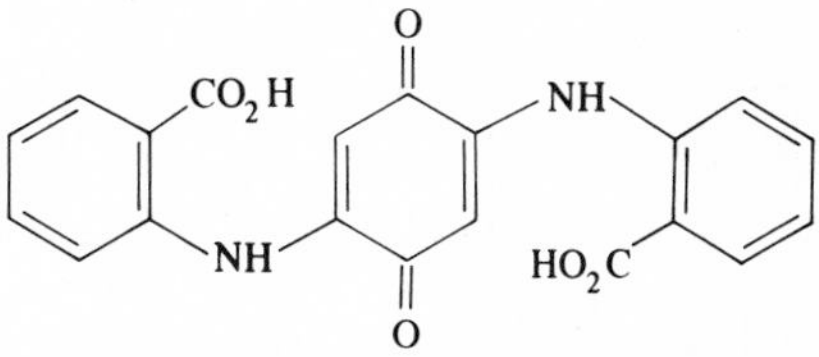

45

ylamine-2-carboxylate,[272] and treatment of **45** with diazomethane gave a product identical to the ester prepared from methyl anthranilate and quinone.[271]

Cyclization of **45** in hot concentrated sulphuric acid presumably gave **46**,[271] rather than **47**,[268] since the product is insoluble in aqueous alkali. The N-methyl analogue is better characterized.[271]

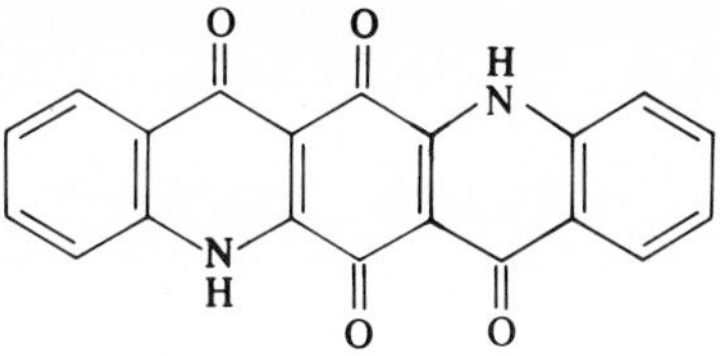

46 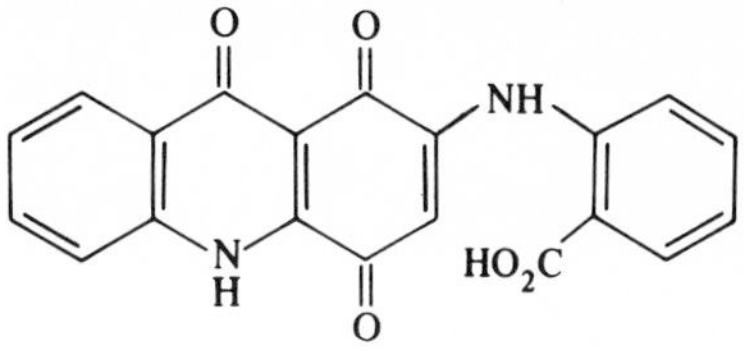   **47**

(d) DIRECT CARBOXYLATION. Only a 14% yield of N-methyldiphenylamine-2-carboxylic acid is obtained if the tertiary amine is treated successively with n-butyl lithium, carbon dioxide, and dilute acid.[273] However, 2-bromo-diphenylamine gives better yields with this treatment, or with an alkyl magnesium bromide and extra magnesium,[274] and 2,2'-dibromodiphenylamine gives 84% of the dicarboxylic acid.[275] 2'-Mercaptodiphenylamine-2-carboxylic acids may also be obtained from N-ethyl- or N-phenyl- (but not N-benzyl-) phenothiazines on treatment with lithium followed by solid carbon dioxide.[276]

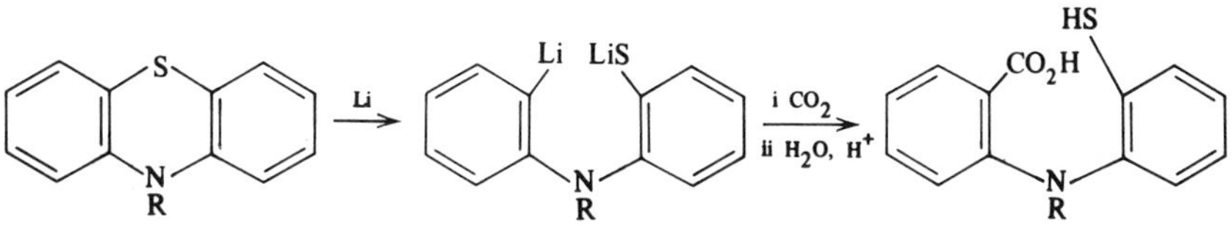

Attempts to carboxylate diphenylamines with oxalyl chloride or carbonyl chloride in the presence of aluminium chloride all failed.[272] The product of the former reaction is N-phenylisatin (48).[297]

An unusual carboxylation reaction takes place when the anilide of diphenylamine N-carbamic acid is heated with potassium carbonate in a stream of carbon dioxide. The major product is diphenylamine-4-carboxylic acid, but a little of the 2-isomer is also formed.[278]

(e) OXIDATION METHODS. The oxidation of N-phenylisatin, 48, may be carried out with peracetic acid[262] or alkaline hydrogen peroxide,[277] and the diphenylamine-2-carboxylic acid obtained in 80% yield. The hydrolysis and oxidation of compound 49 also gives a high yield of the same acid,[277] as does the permanganate oxidation of N-phenylquinaldinium chloride.[279]

48   49

(f) MISCELLANEOUS METHODS. An interesting ring opening and recyclization took place when 2-(2-bromo-4-methylanilino)-5,7-dibromotropone (50) was treated with sodium methoxide.

50

Diazotization of anthranilic acid by amyl nitrite in dry tetrahydrofuran gives the phenoxy ester (50A) and tetramethyleneglycol diphenyl ether. Benzyne, formed by the loss of carbon dioxide and nitrogen from the internal diazonium salt, cleaves the solvent molecule to form a zwitterion, which reacts with undiazotized anthranilic acid and another benzyne molecule to give 50.[885]

3,4,5,6-Tetrachloroanthranilic acid gives both octachlorodiphenylamine-2-carboxylic acid (20%)[236] and octachloro-9-acridanone (6–10%)[104, 236] on diazotization. These two products of the reaction of benzyne with anthranilic acid must be formed independently, as the diphenylamine-2-carboxylic acid is not cyclized under these conditions.[104]

PhO(CH$_2$)$_4$OCO   PhO(CH$_2$)$_4$OCO

50A

Anthranilic acid with 2-nitroanisidine in alkaline solution gave some 2'-methoxydiphenylamine-2-carboxylic acid[280]; esters of diphenylamine-2-carboxylic acid may be converted to the equivalent thionic acids with phosphorus pentasulphide.[886] Direct mono- and di-halogenation of diphenylamine-2-carboxylic acids is possible by using N,N-dihalobenzenesulfonamides.[887] Although the positions of substitution are not specified, 2'-, 4'-, and 6'-positions should be the most reactive sites for electrophilic attack.

C. Other Methods of 9-Acridanone Synthesis

The hydrolysis of 9-amino-, 9-alkoxy- and 9-alkylthio-acridines, although a general route to 9-acridanones, has little preparative importance, since most 9-substituted acridines are made from 9-chloroacridines, which are themselves readily hydrolyzed to 9-acridanones. The rapid and quantitative acid hydrolysis of 9-alkoxyacridines, in contrast to the slower hydrolysis of the 9-amino analogues,[281] has been used in kinetic studies of the conversion of 9-alkoxy- to 9-amino-acridine. The 9-acridanone, precipitated on acidification, was determined gravimetrically.[282] 9-Methoxyacridine may also be demethylated by thiophenol,[283] or dimethyl azodicarboxylate.[284] Aqueous alkali,[281, 285] or even aqueous alcohol,[23] is a better reagent for hydrolyzing 9-aminoacridines; also, secondary,[281] or especially tertiary,[62, 174, 282] amino groups hydrolyze more readily than the primary amine function, possibly because of the steric inhibition of resonance between the ring and the 9-amino group. The presence of electron-withdrawing substituents in the ring also facilitates these hydrolyses.[24]

9-Acetylaminoacridines, but not the unacetylated amino compounds, are converted to 9-acridanones in high yield after a long period of reflux with decamethylene diiodide in alcohol,[286] and 9-salicylaminoacridine, on boiling for some hours in alcoholic alkali, gave not only 9-acridanone but bis(9-acridinyl)amine.[287]

(1) *Rearrangement of Phenylanthranils to 9-Acridanones* *(the Lehmstedt-Tanasescu Reaction)*

A series of related reactions in which 9-acridanones are obtained from 2-nitrodiphenylmethane, 2-nitrobenzophenone, phenylanthranil, or 2-nitrobenzaldehyde and a substituted benzene has been much studied, probably more for its intrinsic interest than as a potential synthetic route.[288, 289] Two distinct types of reaction appear to be involved.

A scheme for the first type, in which acetic anhydride,[290] or concentrated sulfuric or phosphoric acid is the reaction medium,[291, 292] and nitrous acid has a catalytic role,[293, 294] is shown below:

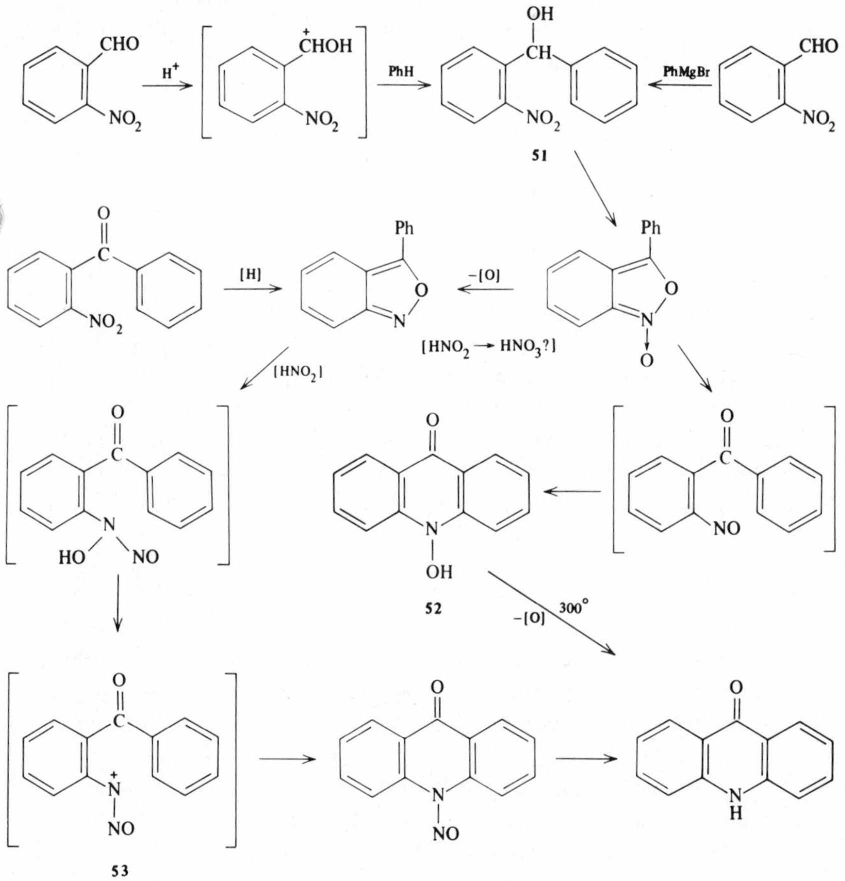

The postulation of a benzhydrol (**51**) intermediate from the attack of a carbonium ion on benzene explains the orientation of substituents[130] and makes unnecessary the controversial supposition that a tautomer of *o*-nitrobenzaldehyde takes part. (See Ref. 295 and references quoted therein.) The participation of benzhydrols is demonstrated by the fact that an independently synthesized benzhydrol gives 70% of 9-acridanone on treatment with sulfuric and nitrous acids.[296] The benzhydrol does not appear to be oxidized to benzo-phenone[297] because benzophenones are rarely found as by-products.[298, 299]

10-Hydroxy-9-acridanone (52) is always a product of this reaction, and the yields obtained (10–65%, according to the starting materials) are not affected if nitrous acid is absent. This indicates that the 10-hydroxy compound is pro-duced by a separate route from the 9-acridanone; the need for much higher temperatures than those achieved during the reaction,[297, 300] to bring about deoxygenation of the 10-hydroxy compound, confirms this conclusion. The status of the 2-nitrosobenzene as an intermediate on this route is in doubt, since a prepared sample of this compound did not give 9-acridanones in sulfuric acid.[301]

3-Phenylanthranils are the precursors of 9-acridanones in all these reac-tions, and other syntheses of them are described later. Nitrous acid is essential only for the rearrangement step; without it, phenylanthranils and products of the N-oxide route are obtained. A trace is sufficient, as enough is formed from the interaction of sulfuric acid and 2,4-dinitrobenzaldehyde[288, 302] to take the reaction to the 9-acridanone stage. The rearrangement takes place readily, even at $-10°$, but higher temperatures (ca. 40°) are recommended for com-plete reaction.[130] The sole evidence for the intermediate 53 is the isolation of a corresponding diazo compound, expected as a product of its reduction, but N-nitroso groups are known to be removed by strong acids. The nature of the reducing agent remains unresolved, but it seems unlikely to be nitrous acid, in view of the tiny amounts present in the reaction mixture. Both hydrogen bromide and quinol have been shown to act as reducing agents when they take part in this reaction.[303]

2-Nitro-[304] and 2,4-dinitro-benzaldehydes[305] condensed satisfactorily with benzyl alcohol,[298] benzene, toluene, and halobenzenes but failed to react in sulfuric acid with benzoic acid, and cyano-[305] dimethylamino-, methoxy-,[305] or nitro-benzene.[306] The failure, in some cases, is caused by the deactivation of the benzene nucleus by an electron-withdrawing substituent, in others by the susceptibility of the substituent to attack by sulfuric acid. In contrast, the use of polyphosphoric acid allowed a clean reaction of o-nitrobenzaldehydes with anisole. The facts that nitrous acid is not necessary in this modification, and that 3-phenylanthranil is also produced,[292] suggest that the N-oxide route is followed, and the 10-hydroxy-9-acridanone formed is subsequently de-oxygenated.[291] If fuming nitric acid is used to rearrange 3-phenylanthranil, nitro-9-acridanones are formed.[307]

Anthranil with p-amino-[304] and p-methoxy-phenyl[308] substituents at position 3 could not be rearranged to 9-acridanones, suggesting that electron-donating substituents inhibit the reaction. N-Benzoylphenylanthranils also failed to react.[304]

The rearrangement of 3-(2,4-dimethoxyphenyl)anthranil on heating did not give the expected 1,3-dimethoxy-9-acridanone but its 2,4-isomer.[309] This reaction, together with the pyrolyses of 3-phenylanthranil,[310] 2,3,4,5,6-pentafluoro-2′-nitrobenzhydrol,[311] and 2-nitrodiphenylmethane,[302] comprise a group of reactions of the second type, for which a nitrene intermediate, 54, is suggested. Again, the 3-phenylanthranil is the vital intermediate.

The higher electron density at position 1′ when the 4′ substituent is methoxyl encourages attack to take place there, rather than at position 2′, which is the site of attack when alkyl or halogen substituents are present. An ionic mechanism may be written for these reactions, but as the fluorinated derivative reacts in the same way as the unsubstituted anthranil,[311] the free radical explanation is preferred.

A spiro-intermediate is also proposed to account for the azoxybenzene 56, formed when 3-(2,4-dimethoxyphenyl)anthranil is treated with nitrous acid.[309] Des-methoxy 55 is a resonance form of 53.

The irradiation of 3-phenylanthranils with uv light, which normally leads to azepines (57), gives 9-acridanones if position 7 is substituted. A nitrene intermediate is also proposed for this low-yield reaction.[312]

(2) *Preparation of 3-Phenylanthranils*

The chemistry of the anthranils (now described as 2,1-benzisoxazoles by *Chemical Abstracts*) has been reviewed up to 1965.[295] Many of the general methods of preparation mentioned are suitable for the preparation of 3-phenylanthranils (58). These include the cyclization of benzophenones with *ortho* nitrogen functions, or their reduced derivatives; the addition of the 3-phenyl group to a benzaldehyde with an *ortho* nitrogen group; and the insertion of the 3-carbon atom, as well as the phenyl group, into an isoxazole ring built onto a nitrobenzene.

Two complementary syntheses of the first type are the reduction of 2-nitro-benzophenones and the oxidation of 2-aminobenzophenones.

Excellent yields are found on the reduction route if sodium sulfide is used,[313] but tin in hydrochloric[314] or acetic[310] acids, and zinc in methanolic ammonium chloride, [315] have also been employed. A nitrene intermediate has been proposed[888] to account for the formation of 3-phenylanthranil and the 2-aminobenzophenone isolated as a by-product, when triethyl phosphite reacts with 2-nitrobenzophenone.[889,890]

Persulfuric acid is an effective oxidizing agent,[310, 316] in the alternative synthesis, but pertrifluoracetic acid yields the further oxidized 2-nitrobenzophenone.[316] Sodium hypobromite gives a 58% yield of anthranil from 2-aminobenzophenone in this reaction.[317]

The Hofmann degradation of 4'-chlorobenzophenone-2-carboxyamide (59), or its cyclic tautomer (60), and the cyclization of o-nitrobenzhydrol (61) with thionyl chloride in chloroform, which gives 5-chloro-3-phenylanthranil (62),[318] are reactions of a similar type.

The introduction of chlorine at position 5 during the latter synthesis is probably the chlorination of a benzene ring under the mildest conditions known.

Treatment of 2-aminobenzophenone with nitrous acid gave only 15% of 3-phenylanthranil; the major product was fluorenone.[317] However, treatment of the diazonium compound with sodium azide gave o-azidobenzophenone (64), which cyclized to 3-phenylanthranil with a loss of nitrogen on heating in decalin.[309, 319] This was predicted, as the azido compound had been postulated as an intermediate in the formation of 3-phenylanthranil from the oxime[320] of o-aminobenzophenone (63).

2-Nitrodiphenylmethanes form 3-phenylanthranils on heating,[302] by an internal oxidation-reduction process initiated by nucleophilic attack at methylene by the adjacent nitro group.[321] The yields are poor because isomerization to 9-acridanones takes place at the high tempreatures required.

63   **64**

The second type of anthranil synthesis, in which 2-nitro- or 2,4-dinitro-benzaldehyde condenses with aromatic compounds under acid conditions, has been described earlier as part of the route to 9-acridanones. 3-Substituted anthranils have been made this way using benzene,[289, 297, 298, 322] toluene,[305, 322] fluoro-,[305] chloro-,[130, 292, 300] bromo-,[292, 300] or iodo-benzenes[88] in the presence of concentrated sulfuric acid. Aniline, or its sulfate,[323] reacts with 2-nitro-benzaldehyde or 5-chloro-2-nitrobenzaldehyde under the influence of zinc chloride,[324, 325] but if phosphorus oxychloride is used, the reaction takes the course shown.[326] Some triphenylmethane derivatives, as well as the 2:1 condensation product (**65**), are also found in the former reactions.

65

Phenols and dimethylaniline give 3-arylanthranils with nitrobenzaldehyde if hydrogen chloride[327] in acetic acid[328–330] or ether,[303] or hydrogen bromide[303] is used as the condensing acid. A chlorine atom is usually introduced at position 5 during this reaction,[323, 325, 326] unless a substituent is already present there, when substitution at position 7 may take place.[303] The mixture of chlorinated and nonchlorinated anthranils found when aniline reacts with 2-nitro-benzaldehyde in these conditions may be separated because the 5-chloro compound is less soluble in hydrochloric acid.[331] Bromination is less frequent and only occurs in the 5 position. The participation of the hydrogen halide, presumably by a mechanism similar to the equation **61 → 62**, as a benzhydrol may be isolated from this reaction,[332] avoids the need for the unexplained reduction step in the **51 → 53** sequence (p. 178).

At the next higher level of oxidation, 2-nitrobenzoic acid reacts with trifluoroacetic anhydride, boron trifluoride, and benzene to give 3-phenyl-anthranil 1-oxide.[333]

The third type of synthesis is exemplified by the reaction of benzyl cyanide with 4-chloronitrobenzene in methanolic sodium hydroxide[334] to form **67** (X = Cl) in 97% yield. The use of potassium hydroxide as the base,[308, 335, 336] is said to give poorer yields,[334] and no improvement was found if sodium methoxide was employed.[337] The isolation of the intermediate potassium salt (**66**), which is then treated with dilute acid, is considered to be a superior technique[336]; substituted benzyl cyanides also take part in this reaction.[338]

A *para*-substituted nitrobenzene must be used, or 4α-cyanobenzylidene-2,4-cyclohexadien-1-one oximes (**68**) are isolated,[335, 338, 339] and if pyridine is used as the solvent, nucleophilic substitution and reduction takes place to give a mixture[308, 340] of **69** and **70**. The reaction also fails if the *para* substituent[308] is electron-donating (X = Me or OMe).

(3) *Oxidation of Acridines*

No general, high-yield method is yet available for the direct oxidation of acridines to 9-acridanones, but even so, this reaction was valuable in establishing the structure of acridine. The principle of vinylogy indicates that this is a process comparable to the oxidation of a Schiff's base to an amide, for which ozone is the sole, and inefficient, reagent.[341]

Poor yields of 9-acridanone have been obtained in the oxidation of acridine by calcium hypochlorite in the presence of cobalt,[342] by bromine and sodium methoxide,[343] by potassium permanganate,[344] by N-bromosuccinimide in the presence of perbenzoic acid,[345] and by a mixture of chromic and acetic acids.[5] 10,10'-Biacridan-9-onyl (71) was also formed in the last of these reactions, unless a low proportion of oxidizing agent was used.[5]

A route via the oxidative hydrolysis of 9-acridanthione, which is produced in excellent yield from acridine and sulfur at 190°, is, however, highly successful with acid ferricyanide, or better, sodium hypochlorite. This method was used in the purification of carbon-14 labeled acridine to constant radioactivity.[346]

The nucleophilic attack of ozone on acridine is solvent-dependent, giving 2.5% of 9-acridanone in methylene chloride but less than 0.1% in methanol. The major product is quinoline-2,3-dicarboxylic acid. When acridine 10-oxide was used, a maximum yield of 40% was obtained, provided the ozone was in excess. The suggested mechanism also shows why 10-hydroxy-9-acridanone, tautomeric with 9-hydroxyacridine 10-oxide, predominates if less ozone is available.[347]

Acridine and dimethyl acetylenedicarboxylate in ether gave the 9-acridanones **73** with *cis* and *trans* 10-substituents. In alcoholic solution the 9-alkoxyacridan **74** is formed, suggesting that aerial oxidation of the first-formed ylid, **72**, leads to the 9-acridanone.[348]

Alkaline peroxide or catalytic hydrogenation cleaves the 10-side chain.

Substituted acridines are more readily oxidized. 2- and 4-Nitroacridines are more rapidly oxidized than the 1- and 3-isomers, but all give good yields of the corresponding nitro-9-acridanones, with chromic acid.[349, 350] 3-Chloroacridine is similarly oxidized.[351]

9-Substituents are particularly susceptible to oxidation. 9-Allylacridine gives 9-acridanone on treatment with dichromate in acetic acid,[352] and 9-diethylaminoacridine gives a mixture of 9-acridanone, nitro-9-acridanone, and 9-ethylaminoacridine with nitric acid, showing that dealkylation precedes, and oxidation accompanies, nitration.[353]

9-Methyl-,[354] but not 9-ethyl-acridine[355] is oxidized to 9-acridanone by *p*-nitrosodimethylaniline in hydrochloric acid. An intermediate 9-carboxaldehyde group seems to result from the hydrolysis of the first-formed anil (**75**) or anil-oxide (**76**).

Peroxide oxidation of 9-carboxaldehyde[354] or 9-carboxylic acid[356] groups also gives 9-acridanones, suggesting that these reactions are analogues of the Dakin hydroxylation reaction.

The mechanism of the rearrangement of acridine 10-oxide to 9-acridanone in acetic anhydride,[307, 357] has recently been investigated.[358-360] The pseudo-unimolecular kinetics favor a solvent-separated, ion-pair process, within which a concerted intramolecular mechanism fits the thermodynamic parameters.[360] This contrasts with the intermolecular ionic route found for the rearrangement of pyridine 1-oxide to 2-pyridone under these conditions.[361] Two possible reaction paths, involving **77** and **78**, are consistent with the results of some oxygen-18 labeling experiments, outlined below.

The path via **78** is favored when a large amount of solvent is used, suggesting that the mode of solvation affects the conformation of the rearranging ion. As the reaction rate is not altered when 9-deuteroacridine 10-oxide is used, the rate-determining step must be the formation of the bridged intermediate.[359]

The metabolic oxidation of acridines has been studied frequently by Japanese workers. A common product is 2-hydroxy-9-acridanone,[362] excreted as glucuronate[363] or sulphate,[891] and formed in the kidney by the action of 9-acridanone dehydrogenase[362–364a,b] on 9-acridanone.[365, 366] Both liver and kidney contain acridine dehydrogenase,[367] which acts to give 9-acridanone[363] from acridine, its 10-oxide and quaternary salts, and also 9-substituted acridines, provided that the 9-substituent is easily removed.[367–369] The latter enzyme is present in man, and rabbit,[363] cat, and rat but not in the dog[193] or mouse.[367]

Reduced 9-acridanones, especially the readily available 1,2,3,4-tetra-hydro-9-acridanone (Chapter V, p. 478) may be dehydrogenated to 9-acridanones[370] by treatment with hot concentrated sulfuric acid,[892] sulfur in quinoline,[371, 372] or on heating in dry air at 280°;[373] but better yields are obtained if copper powder is added and the temperature is raised to 360°.[372, 374] A quantitative dehydrogenation is said to occur when 9-acridanone is sublimed from a mixture of litharge, pumice, and the tetrahydro compound, in an atmosphere of carbon dioxide.[375]

(4) *Cyclization of 2,2'-Disubstituted Benzophenones*

A 2-fluoro group in benzophenone is readily displaced by nucleophilic reagents.[376] This observation has been exploited in an intramolecular condensation reaction to give both 9-acridanones and their 10-substituted derivatives. Dimethylformamide is the recommended solvent, but 2-ethoxyethanol was needed with *N*-*p*-tosyl-2-amino-2'-fluorobenzophenone to avoid the loss of the tosyl group. It is also essential to add slightly more than one equivalent of potassium carbonate.

2-Amino-2'-fluorobenzophenones were originally prepared as intermediates in the synthesis[377] of benzo-1,4-diazepin-2(1*H*)-ones (**79**), and may be regenerated by the acid hydrolysis of these diazapinones or the alkaline hydolysis of the related diazepinium salts.[378]

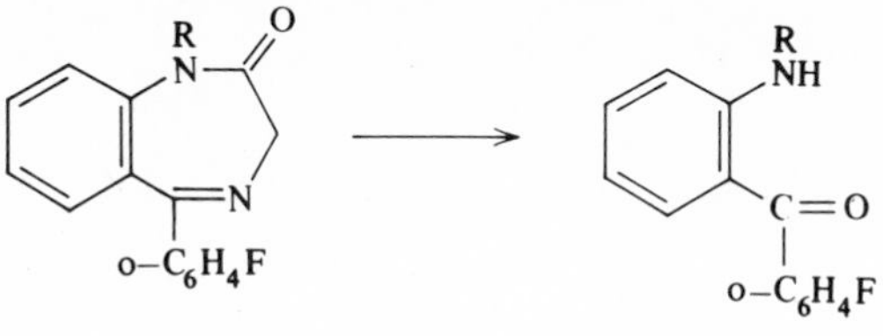

79

These diazepinones may be nitrated before hydrolysis,[379] and compounds **79** (R = H) are readily N-alkylated,[380, 381] giving a range of aminofluorobenzophenones on hydrolysis.

The more direct synthetic route[377, 382, 383] is the reaction of o-fluorobenzoyl chloride with an arylamine, preferably having its *para* position blocked, at about 200°, in the presence of zinc or aluminium chloride. The o-fluorobenzoic acid, obtained together with the required ketone, should be recycled for optimum yield.[384] A chlorine atom *para* to the amino group may be catalytically reduced out of the benzophenone system, leaving the 2'-fluoro group intact[385]; a methyl group in this position may successively be converted to carboxy, amido, and cyano groups without destroying the rest of the molecule.[386]

An alternative synthesis of highly fluorinated aminobenzophenones is possible from the heating of polyfluorotriphenylcarbinols or benzophenones with aqueous ammonia in an autoclave.[387]

Although yields lie in the range 80–100% when 2'-fluoro compounds are used, the corresponding 2'-chloro- and 2'-hydroxy-2-aminobenzophenones do not undergo this cyclization,[122] except as shown on this page and p. 190.

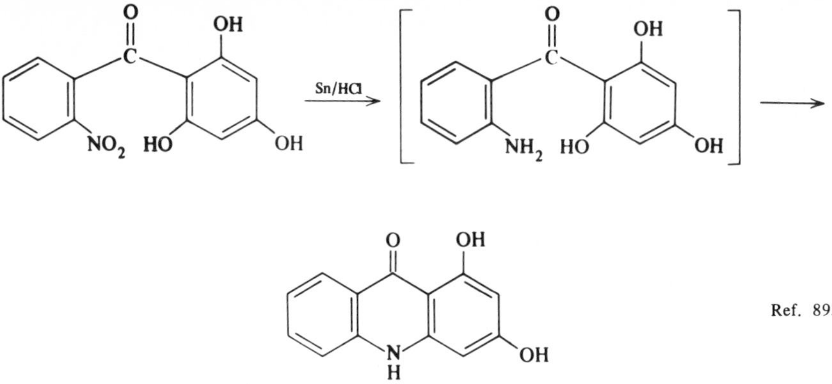

Ref. 893

Hot concentrated hydrochloric acid or zinc chloride is said to cyclize 2,2'-diaminobenzophenone to 9-acridanone,[390] but a repeat of these experiments showed yields of 0 and 7%, respectively.[391] However, high-temperature cyclization in 100% phosphoric acid gave an almost quantitative yield of 9-acridanone, as well as a 76% yield of 3,7-dichloro-9-acridanone from 2,2′-diamino-4,4′-dichlorobenzophenone,[391] a reaction previously reported to fail.[392] 2,2′,4,4′-Tetraaminobenzophenone failed to cyclize under these conditions, even though the reductive cyclization of the corresponding tetranitro compound with stannous chloride in hydrochloric acid [393, 394] is said to be the best route to 3,7-diamino-9-acridanone. These are examples in a series of reactions in which the acridine system may be obtained at the acridan, acridine, or 9-acridanone level of oxidation, by the elimination of ammonia from 2,2′-diamino-diphenylmethanes, -benzhydrols or -benzophenones, respectively.

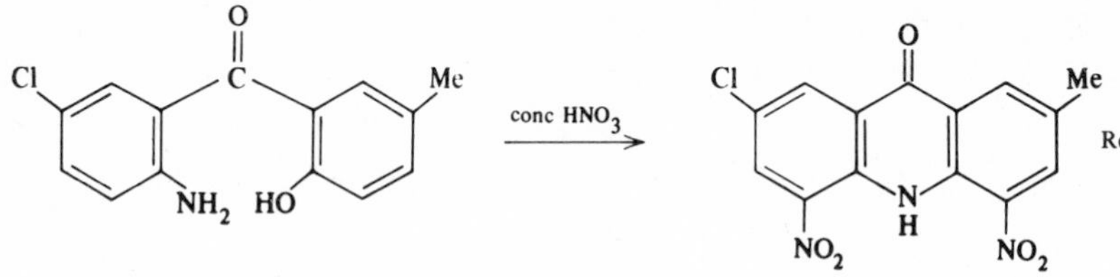

Cyclization of 2-amino-2′-methoxybenzophenones is known only in the benzacridanone series.[388] The reaction shown below, however, took place on heating the benzophenone with ammonia or a primary amine.

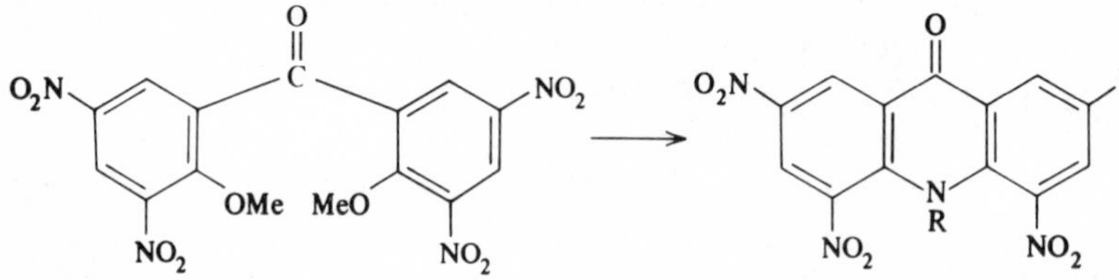

Cyclodehydrogenation to 9-acridanones is said to occur on heating 2-aminobenzophenone with litharge[395] at 350–360° (50%) and on treating 2-amino-3′-hydroxybenzophenone with manganese triacetate or potassium persulphate.[893] With zinc chlorides as cyclizing reagent, 2-aminobenzophenones gave only traces of 9-acridanones. The major products were diphenyldibenzodiazocines, and benzanilides were also formed by rearrangement.[894] Similarly, only enough 9-acridanone for chromatographic identification was obtained when diphenylamine-2-carboxaldehyde was distilled from zinc dust.[396]

(5) Miscellaneous Laboratory Routes

Both thermolysis[397] and uv irradiation of 3-phenylbenzotriazin-4-one (80) in aprotic solvents[398] give 9-acridanone, although some phenanthridone was also obtained in the former reaction. Ionic[398] and free radical[397] mechanisms have been suggested for these reactions

There is some evidence for both pathways because, during irradiation, β-naphthol will trap the diazonium salt **81** and water or ethanol will add to the ketene **82** to give diphenylamine-2-carboxylic acid or its ester[398]; while thermolysis in paraffin gives a benzanilide as a major product, due to free radical hydrogen abstraction from the solvent.[397]

N-p-Chloro-[399] and p-iodo-[400] phenylbenzimino-5-methoxy-2-methoxycar-bonylphenyl ethers (**83**), typical starting materials for the Chapman rearrange-ment (Section 1.B(2)(a)) are reported to give acridanones directly on heating to 320°. Concentrated sulfuric acid at 200° converts similar imino ethers to 9-acridanones,[401] but poor yields are found in both reactions.[263] Salicylic acid, on heating with phosphorus trichloride and m-toluidine, gave a low yield of a methyl-9-acridanone,[402] later identified as the 1-substituted compound.[372] The anilide was possibly formed first, as salicylanilides (**84**) are known to undergo rearrangement and cyclization to 9-acridanones,[403] together with some decarboxylation to diphenylamines.

9-Acridanone is found as a by-product in the hydrolysis of 9-acridinyl-acetamide with methanolic potash, and in the reaction of 9-bromoacetyl-acridine with secondary amines.[404] Heating calcium anthranilate to 340° and treating the product with hydrochloric acid at 170° also gave a little 9-acridanone.[405]

(6) Biosynthesis

Acridine alkaloids (see also Chapter IV, p. 379), more properly named 9-acridanone alkaloids,[893] as the 9-keto group is always present, also display with few exceptions[900] the characteristics of 1,3-oxygen substitution and the absence of substituents from the left-hand ring. 10-Methylation is another common but not essential feature.

The agreed source of the left-hand ring is a metabolic equivalent of anthranilic acid, which may arise from shikimic acid[893] or tryptophan[895, 897] as tritiated material, is incorporated solely into this part of the molecule.[406] Carbon-14 studies[896] suggest that the right-hand ring is assembled from three acetate units,[409, 895] rather than from preformed phloroglucinol[408] or a simple monosaccharide[901]; but whether the route lies via N-arylation[407] or N-acylation,[897] the intermediate formation of a quinolone[897] or an o-amino-benzophenone,[893] remains to be decided.

Whether the initial cyclization product is an acridine,[407, 896] subsequently oxidized to a 9-acridanone, is not known, but it is generally considered that both O- and N-methylation, and the introduction of further hydroxyl groups, takes place after cyclization. Nevertheless, N-methylanthranilic acid has been shown to be both an alkaloid precursor and a plant metabolite.[897] Other metabolic by-products of these routes are rare,[407] but labeled 1,3-dihydroxy-acridine, and 1,3-dimethoxy-10-methyl-acridinium sulfate and -9-acridanone have been converted to arborinine (2,3-dimethoxy-1-hydroxy-10-methyl-9-acridanone) in the plant.[896] This result further suggests that hydroxylation at C_2 occurs after N-methylation and the oxidation of the acridine to the 9-acridanone, even though the activity of labeled 1,2,3-trihydroxy-10-methyl-9-acridanone is not found in the isolated alkaloids.[897]

In vitro analogues are available for some of these reactions. The condensation of o-aminobenzaldehyde with phloroglucinol gives excellent yields of 1,3-dihydroxyacridine under physiological conditions,[408, 898] but the reaction of anthranilic acid with phloroglucinol in the presence of zinc chloride is less satisfactory.[410-412] The 9-acridanone that results may be purified by conversion into the diacetyl compound, however, and recovered by a quantitative alkaline hydrolysis.[899] o-Aminobenzophenones may be cyclized to 9-acridanones by oxidative coupling or intramolecular dehydration.[893]

D. Synthesis of 10-Substituted 9-Acridanones

The principal routes to 10-substituted acridanones, other than the cyclization of *N*-substituted diphenylamine-2-carboxylic acids (Section 1.A(5)) are the oxidation of 10-alkylacridinium salts or acridans, and the *N*-alkylation of acridanones or 9-substituted acridines.

(1) *Oxidation*

The oxidation of 10-alkylacridinium salts and acridans can usually be carried out easily and with good yield, in contrast to the difficult oxidation of acridines. Chromic acid oxidizes 10-methylacridinium acetate in acetic acid to 10-methyl-9-acridanone in almost quantitative yield,[227, 413] but the more common oxidizing agent is alkaline ferricyanide.[408, 413–415] As 10-methyl-9-hydroxyacridan is very rapidly oxidized in air,[416] the true reaction with the latter reagent may be the oxidation of the 10-substituted acridinium hydroxide in the form of a pseudobase.[417, 418]

Only in the absence of air, or other oxidizing agents,[227] do acridinium hydroxides disproportionate on heating to a mixture of 9-acridanone and acridan.[419–421] However, 10-methylacridinium hydroxide with a dilute solution of alkali in aqueous acetone gives the diacridinium ether **84a**, which decomposes to 10-methylacridan and 10-methyl-9-acridanone on warming with acetic anhydride and a trace of hydrochloric acid.[902]

Photochemical stability and stability to oxidizing agents run parallel, as acridinium hydrochloride is not oxidized on irradiation with uv light, in an alcoholic solution saturated with oxygen, whereas 10-methylacridinium chloride gives 10-methyl-9-acridanone. In chloroform solution, disproportionation occurs, as both 10-methyl-9-acridanone and 10-methylacridan are obtained.[422]

10-Methyl-9-acridanone has also been prepared by the aerial oxidation of 9-benzylidene-10-methylacridan (85), formed by the treatment of 9-benzyl-10-methylacridinium chloride with a base,[420] and by the hydrogen peroxide oxidation[423] of 9-cyano-10-methylacridan (86), formed from potassium cyanide and 10-methylacridinium chloride.[424] These reactions further illustrate the easy nucleophilic attack possible at position 9, when the acridine nitrogen bears a positive charge. Acridan 86, in the presence of gaseous oxygen and cyanide ions, gives an 80–90% yield of 10-methyl-9-acridanone and cyanate ions in a chemiluminescent reaction.[903] (See Chapter IX, p. 622.) A 9-cyano-9-peroxy anion, and the four-membered ring anion 86a may be intermediates. Alternatively, the prior formation of 9-cyanoacridanyl dimer or bis-(9-cyanoacridanyl)-9-peroxide could account for the results.[903]

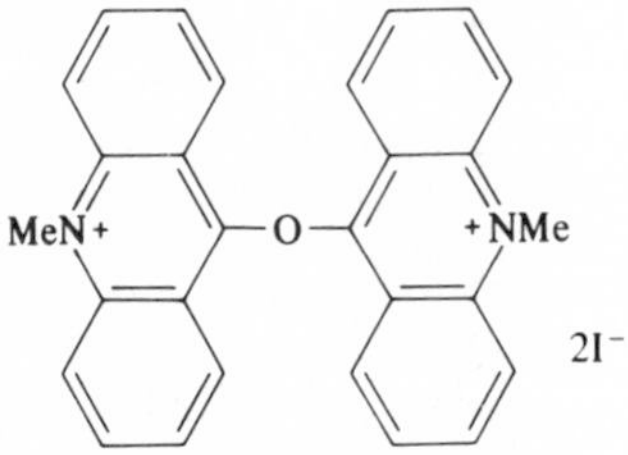

84a

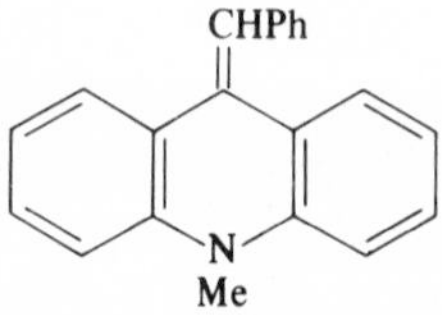

85

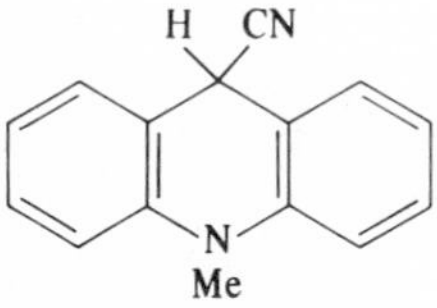

86

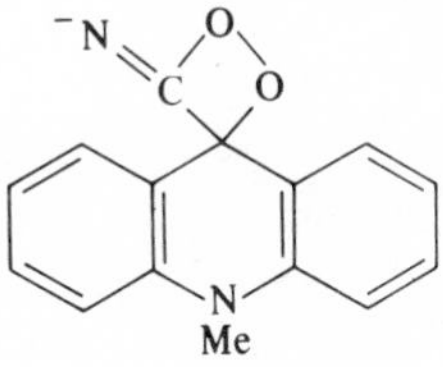

86a

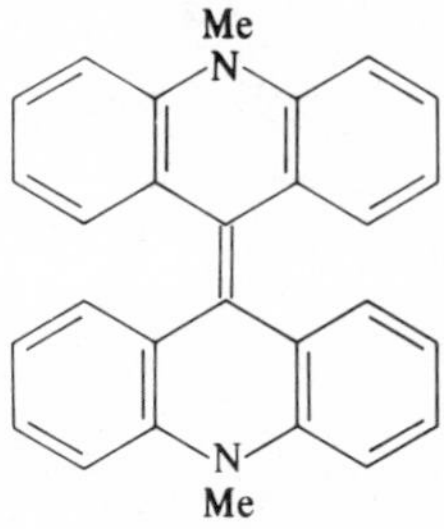

87

The oxidation of 10,10'-dimethyl-9,9'-$\Delta^{9,9'}$-biacridan to 10-methyl-9-acridanone with singlet oxygen, ozone, or bromine in alkaline ethanolic hydrogen peroxide is a chemiluminescent reaction.[425] Some of the intermediate radicals, in the oxidation of lucigenin (the dinitrate of the 9,9'-biacridinium salt corresponding to Structure 87) have been detected by esr spectroscopy.[426]

Potassium permanganate,[416] nitrous acid, and nitrosyl chloride,[427] have been used as oxidizing agents to give 10-methyl-9-acridanones from 10-methylacridinium salts with 9-alkyl substituents. Potassium ferricyanide, but not hydrogen peroxide, readily oxidizes the methyl quaternary salts of 9-methoxycarbonylacridine to 10-methyl-9-acridanone.[899] As acridine-9-

carboxylic acid is the product of a Pfitzinger reaction between the cheap reagents isatin and phloroglucinol,[899] this may prove to be a useful general route to 10-alkyl-9-acridanones.

Although acridine methiodide is said to give a poor yield of 10-methyl-9-acridanone on oxidation,[428] 10-phenylacridinium iodide may readily be oxidized in alkaline solution to 10-phenyl-9-acridanone.[429]

A satisfactory route to 10-aryl-9-acridanones is the dehydrogenation of the corresponding, readily available tetrahydro-9-acridanones. (Chapter V, Section 6). Copper powder at high temperatures dehydrogenates 10-p-tolyl-[430] and 10-p-ethylphenyl-1,2,3,4-tetrahydro-9-acridanones,[374] and sulfur in quinoline is an effective reagent for their o- and p-chlorophenyl[371] and p-bromophenyl[431] analogues.

(2) 10-Alkylation

The sodium or potassium salts of 9-acridanones may be prepared by evaporating a solution of the 9-acridanone in excess alcoholic alkali,[5] or by the azeotropic separation of water or alcohol from a suspension of the 9-acridanone and a base in boiling decalin[432] or xylene.[433] These salts react with many alkyl halides,[434] including benzyl chloride[435] and allyl iodide,[436] and the alkylating agents methyl[437, 438] or ethyl[434, 438] sulfate, and methylbenzenesulfonate[439] to give 10-alkyl-9-acridanones.

Yields are usually high, but isopropyl bromide gave only 4% of the 10-isopropyl-9-acridanone,[434] possibly because of steric interference with the course of the reaction. Failures experienced with 4-methoxy-9-acridanones[218] have been ascribed to hydrogen bonding, but 2- and 3-nitro-5-methoxy-9-acridanones have been methylated via their potassium salts.[439]

10-Methylation may be carried out in a single step by using methyl iodide with sodium methoxide in refluxing methanol,[440] or with potassium carbonate in boiling acetone,[414, 441] and cyclopropylmethyl bromide also alkylates 9-acridanone under the latter conditions.[82] 10-(2-Diethylaminoethyl)-9-acridanone[442] and many similar 9-acridanones[443–445] have been made from 9-acridanone and a suitable alkyl chloride in the presence of sodamide. With ethanolic potash as the base, however, much of the starting 9-acridanone is recovered,[446] but with sodium hydroxide in dimethyl sulfoxide, an alkyl halide needs less than 5 min to alkylate a 9-acridanone.[447]

Two more unusual routes to 10-substituted 9-acridanones are shown. The formation of 10-vinyl-9-acridanone (88) needs the high temperatures and pressures of an autoclave, but 92–95% yields are achieved.[448, 449] The benzyne reaction gives 10-phenyl-9-acridanone only as a by-product, but the binuclear product 89 is suggested as the precursor since it is readily oxidized to the 9-acridanone.[450]

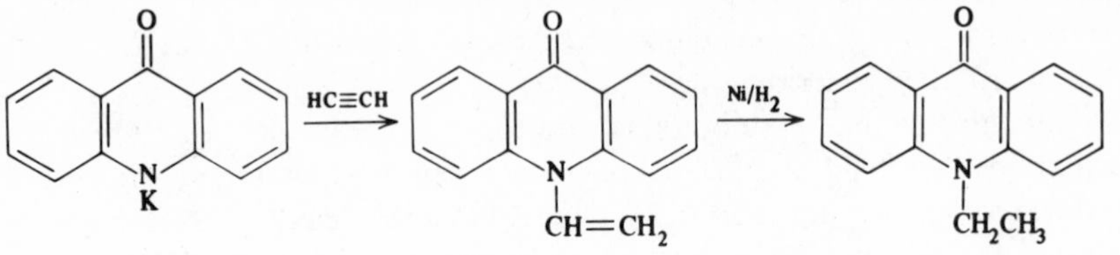

88

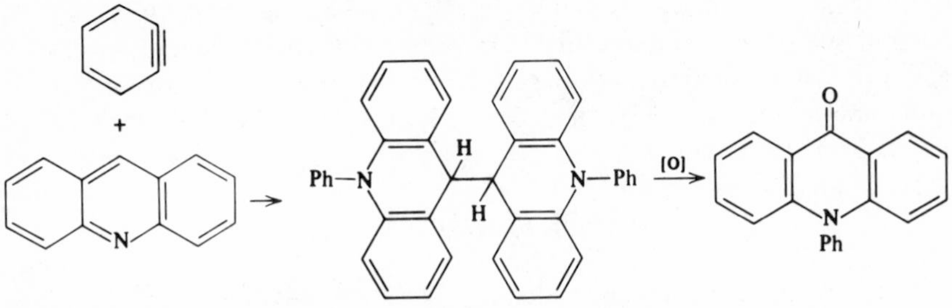

89

A further satisfactory synthesis of 10-methyl-9-acridanones takes advantage of the easy hydrolysis of 9-chloro, alkoxy, or amino groups from an acridine with a positive charge on the nitrogen atom. The 9-substituted acridine is first 10-methylated with methyl sulfate,[97] iodide,[281] or p-toluenesulfonate[451, 904]; the 10-methylacridinium salt is then refluxed in strong alkali to give the 9-acridanone.

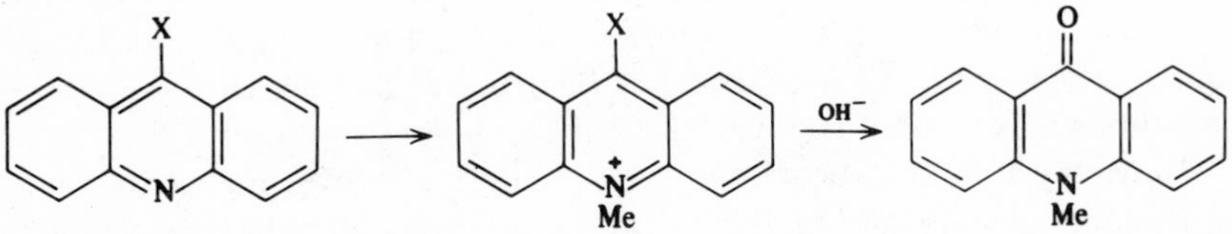

X = Cl, PhO, NH$_2$

Certain methoxylated 9-aminoacridines are converted to the corresponding 9-acridanones by this procedure,[218] rather than the 10-methylated compounds, possibly due to preferential methylation of the exocyclic nitrogen atom.

9-Amino-10-ethylacridinium iodide gives a pseudobase (**90**), when treated with 10% aqueous potash. In concentrated sulfuric, or ethanolic hydrochloric acid, or on heating to 150°, the base eliminates ammonia to give 10-ethyl-9-acridanone. However, with more dilute aqueous acid, 9-amino-10-ethyl-acridinium chloride is the product.[452]

9-Chloroacridine with trimethyloxonium borofluoride gives the 10-methyl-acridinium borofluoride. The hydrolysis of this compound is not recorded, but with 9-acridanimine it forms a 9,9′-nitrogen-bridged binuclear compound.[453]

90

9-Methoxyacridine rearranges to 10-methyl-9-acridanone[454] on heating to 200°, presumably via an intermolecular methylation-demethylation pathway. In the presence of methyl iodide[97] the reaction proceeds more readily, and its scope has been much increased by this simple expedient.[218, 414, 455] 4-Methoxy-quinoline undergoes a similar rearrangement.[456]

2. Structure, Properties, and Reactions of 9-Acridanone and Its Derivatives

A. General Properties and Structure

9-Acridanone was first obtained at 1880[344] and named acridone in 1892.[157] A number of syntheses of the parent compound are included in Table I, of which the cyclization of diphenylamine-2-carboxylic acid is the most satisfactory. 9-Acridanone has a high melting point and shows unusual stability, in that it distills unchanged at atmospheric pressure. It is almost insoluble in common solvents but it may be recrystallized from a large volume of acetic acid or amyl alcohol, or, better, on a large scale, from m-cresol or a mixture (5:12.5) of aniline and acetic acid. Purification may also be achieved by sublimation. 9-Acridanone is one of the most fluorescent substances known, showing a blue-violet fluorescence in water or alcohol that changes to green on the addition of alkali. This property has proved valuable in the estimation of trace quantities of 9-acridanones.[877]

9-Acridanone is a feeble base (pK_a − 0.32) and a very weak acid, but the excited acridanone molecule is shown by fluorescence spectroscopy to have much stronger basic properties.[457−459]

The hydrochloride, which precipitates on cooling from a solution of 9-acridanone in hot concentrated hydrochloric acid, and the potassium salt present in a solution of 9-acridanone in ethanolic potassium hydroxide, are both immediately hydrolyzed by water. Unlike the other hydroxyacridines, which dissolve in 0.1N sodium hydroxide solution, 9-acridanone is undissolved, even by strong aqueous alkali.[460]

When subjected to paper, or polyamide thin layer, chromatography, 9-acridanone requires strongly acid eluting solvents, or concentrated ammonia solution to show appreciable R_F values.[461, 462] Acridine spots run much more freely.[463]

As its potentiometric titration curve shows[464] no inflection between pH 2 and 12, 9-acridanone cannot be represented as 9-hydroxyacridine (91). The tautomeric keto formulation (92) is also unsatisfactory, since generally neither 9-acridanone[5] nor 10-substituted 9-acridanones[442] react with hydroxylamine or phenylhydrazine. These characteristic carbonyl group derivatives must be made from 9-chloroacridines, except for the single example (93), which appears to form an oxime directly.[465]

91   **92**   **93**

A calculation of the keto:enol ratio from a comparison of the basic strengths of 9-acridanone and 9-methoxyacridine[466] shows that the proportion of **91** is only 1 in 10^8. Despite the small proportion of the hydroxytautomer, 9-acridanone may be converted into 9-chloroacridine as readily as 4-pyridone, where the enol:keto ratio is 1:2200, is converted to 4-chloropyridine. The importance of the form **91** in the reactions of acridanone is clear, in that both cation **94** and anion **95** are derived from it.[440, 467]

94   **95**

Protonation is expected to occur on the oxygen rather than nitrogen, by analogy with 4-pyridone,[468] and because oxygen is known to tolerate a negative charge better than nitrogen. Evidence from uv spectroscopy confirms the predominance of the keto form **92**, as the spectrum of 9-acridanone is very similar to that of 10-methyl-9-acridanone, but differs from that of 9-methoxy-acridine.[469] The nmr spectrum also shows N—H to ring proton coupling, not possible in 9-hydroxyacridine.[470]

9-Hydroxyacridine has never been isolated, as the only compound claimed to have this form[471] was subsequently shown to be the tautomeric 10-hydroxy-9-acridanone.[50, 454, 472] Nor has it been detected in nonpolar solvents,[473] which are known to favor hydroxy, rather than keto, tautomers.[460] Even in such compounds as 4-hydroxyacridine (**96**), where the enol form might have been favored by hydrogen bonding,[474] and the complex **97**, where a proto-tropic shift would have allowed further coordination through the nitrogen lone pair,[232] spectroscopic evidence indicates that the keto form of the 9-acri-danone is maintained.

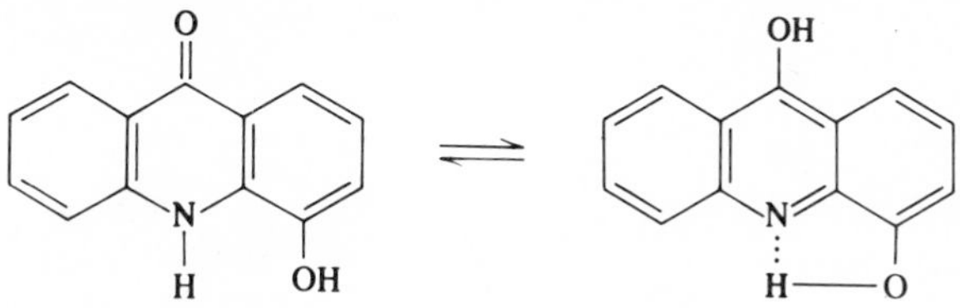

96

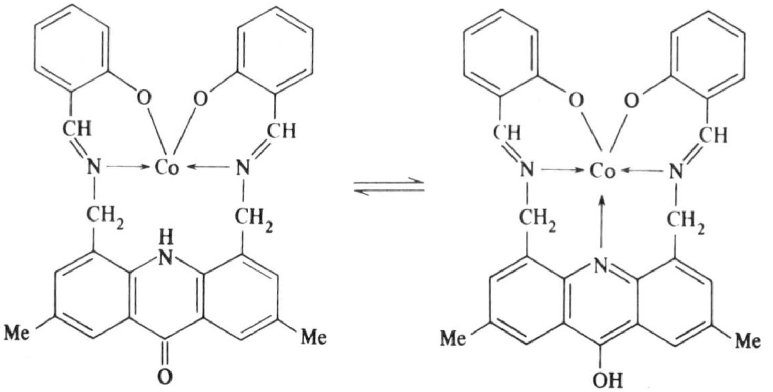

97

In contrast, 9-aminoacridine has been shown to exist predominantly in the amino form rather than as the imino tautomer ($10^3:1$), by ir,[475, 476] uv,[1, 47'] and nmr[470] spectroscopic data, ionization constant calculations[475, 478] and dipole moment measurements.[479] (See Chapter II, p. 121.) Earlier erroneous conclusions favoring the imino form are possibly the result of neglecting steric interaction,[469, 480–482] between substituents on the exocyclic nitrogen and the *peri* hydrogen atoms. However, stabilization of the imino form is achieved by replacing the imine hydrogen atom with a benzenesulfonyl-,[483] or trifluoro- or trichloro-acetyl group.[484] Acetylation does not bring about this stabilization.[485]

The high melting point and the solubility characteristics of 9-acridanone indicate that it is a highly associated substance. 9-Acridanones with 10-substituents, or groups in positions 4 or 5, melt at lower temperatures than the parent compound, suggesting that intermolecular hydrogen bonding (**98**) is normally present, but the fixed position of the hydrogen atoms required by "mesohydric tautomerism"[486] is now unacceptable.[487] 9-Acridanones with proton-donating substituents in the 1-position, e.g., the *nor*-9-acridanone alkaloids (Chapter IV, Section 2.D), and 9-acridanones with proton-accepting substituents in the 4-position, may show altered physical properties (cf. Chapter XI, p. 670) and chemical reactivity because of intramolecular hydrogen bonding.

98   **99**

9-Acridanone certainly has a strong tendency to form hydrogen bonds, as its changing uv spectrum in mixtures of acetic acid and ethanol shows that acridanone both associates with, and breaks up, acetic acid dimers.[488] The more effective hydrogen bonds, however, seem to be made through the oxygen rather than the nitrogen atom, as the appearance of free radical complexes in the 10-ethyl-9-acridanone:isobutyric acid system has been ascribed to the lowering of the activation energy of hydrogen atom transfer along a strong carbonyl-hydroxyl hydrogen bond.[489]

This highly polarized nature of the acridanone molecule has been expressed in the formulation **99**[454]; this canonical form is considered an important contributor to the resonance hybrid.[464] Absorptions in the ir spectrum show that the carbonyl-group stretching frequency of 9-acridanone (see Chapter XI, p. 668) is lower than that expected even for an associated amide,[490] in common with other 2- and 4-pyridones,[491] and the characteristic frequency of the tertiary *N*-methyl group, which vanishes on quaternization, is very weak in the spectrum of 10-methyl-9-acridanone.[492]

The similarity of the uv spectrum[493] of 9-acridanone to those of the 9-

acridanone and 9-aminoacridine cations,[1] and the dipole moment of 10-methyl-9-acridanone (3.5 D),[494] which is appreciably higher than that of benzophenone (2.95 D) and contrasts with the zero dipole moment of 9-chloroacridine,[479] also lend support to the significance of structure **99**.

B. Chemical Properties of 9-Acridanone

(1) *Reduction*

Distillation with zinc dust is the only general one-step process known [see Chapter I (p. 13) for the two known exceptions] for the reduction of 9-acridanones to acridines.[81, 157, 344, 495–497] The reaction may be violent, and the product, if allowed to distill from the reaction mixture, is contaminated with acridan and unreduced 9-acridanone.[122] If left in the vessel, acridan is not found, as temperatures above its dehydrogenation temperature are reached. Only alkyl groups are certain to survive; hydroxy, alkoxy, and halogen substituents are always lost.[498]

Fortunately, a number of satisfactory two-stage processes are available, most of which involve the reduction of the 9-acridanone to the acridan and its reoxidation to the acridine. A large excess of sodium in amyl alcohol,[429] which is better than ethanol,[499] gives an 85% yield of acridan from 9-acridanone and is also effective with methyl-9-acridanones.[497] Sodium in [O-²H]butan-1-ol has been used to reduce 9-acridanone to 9,9-dideuteroacridan, which had an isotopic purity of over 95%.[905] If sodium amalgam in alkali,[5, 146, 210, 499, 500] or 90% ethanol, [18, 136, 147, 394] is used, a smaller proportion of sodium is needed, and it is the preferred reducing agent for nitro- and amino-9-acridanones.[130, 500] Halogen substituents survive this reduction,[18] but bimolecular products, such as **100** are sometimes obtained.[5, 18, 499]

100   101

Aluminium amalgam[81, 124] is said to be even more effective than sodium amalgam, but it fails with 1-amino-9-acridanone, because an aluminium chelate compound is formed.[124]

9-Acridanone, in contrast to anthraquinone, is not reduced by aluminium in concentrated sulfuric acid,[501] and tin and hydrochloric acid only reduces nitro-9-acridanones and their 10-methyl derivatives to amino-9-acridanones. [145, 239] Neither is catalyic reduction over Raney nickel any more successful with nitro-9-acridanones, [145] but copper chromite under forcing conditions effectively catalyzes the hydrogenation of 9-acridanone to acridan.[502] Diborane in tetrahydrofuran converts 9-acridanone to a mixture of acridine and acridan,[503] but the usefulness of this method is severely limited by the insolubility of 9-acridanone.

Acridan may be oxidized to acridine by air,[429] bromine,[504] nitrous[504, 505] or nitric acids,[506] silver nitrate,[507] and dichromate in acetic acid,[81, 136, 483, 508] but ferric chloride is the recommended reagent.[40, 122, 497]

Two, very good, three-stage reduction processes are also known. 9-Acridanone is readily converted to 9-chloroacridine, which is reduced to acridan, by hydrogen over Raney nickel,[351, 509, 510] in an alkaline medium. Subsequent oxidation gives acridine.[483] However, reduction with lithium aluminium hydride is said to give acridine directly in 75% yield.[10] Alternatively the 9-chloroacridine may be treated with p-toluensulfonylhydrazine, and alkaline hydrolysis of the 9-acridinyl p-tosyl hydrazine (Chapter I (100a)), gives acridine, nitrogen, and p-toluenesulfinic acid.[19, 414, 511, 906] Although this sequence constitutes a reduction, reducing conditions are not employed; this allows reducible cyano and nitro groups to survive into the product.

(2) Oxidation

The oxidation of 9-acridanone or its 10-methyl derivative with dichromate in hot acetic acid gave 10,10'-di-9-acridanone, together with carbon dioxide, in the latter case.[5] Under the same conditions, 4-methylacridine also gives compound 101, mixed with 9-acridanone. The 4-methyl group had been eliminated from both products, probably via the decarboxylation of the corresponding carboxylic acids.[495] Compound 101 was split into 2 moles of acridan on reduction with sodium amalgam in ethanol and gave acridine on distillation from zinc dust. It was soluble in organic solvents, but much less so in alcohols, acids, or alkalis. No substituted derivatives are known.

9-Acridanone is oxidized by ozone in a chemiluminescent reaction, which has been suggested as a very sensitive method of detection and estimation of amounts[512] down to 2 ng. Acridine alkaloids (Chapter IV, p. 384) are oxidized to 4-quinolone-3-carboxylic acid by nitric acid. The susceptible hydroxylated ring is broken down, and the 2-carboxyl group of the first-formed quinolone-dicarboxylic acid decarboxylates under the conditions of the reaction.[406, 441]

(3) Electrophilic Substitution—Halogenation

Chlorination of 9-acridanone in acetic acid gave an inseparable mixture of chlorinated products,[513] but with carbon tetrachloride as solvent a 9% yield of 2,7-dichloro-9-acridanone was precipitated, leaving the mixture of mono-chloro-9-acridanones in solution.[514] The 2,7-dichloro-9-acridanone was further chlorinated to 2,4,5,7-tetrachloro-9-acridanone in 98% yield[401, 514] and gave 4,5-dibromo-2,7-dichloro-9-acridanone when treated with bromine.[206] When antimony pentachloride with a trace of iodine was used as the chlorinating agent, both 9-acridanone and its 10-methyl derivative gave octachloro products.[437]

Bromination gives excellent yields of either 2,7-dibromo- or 2,4,5,7-tetra-bromo-9-acridanone (102), according to the conditions,[18, 139, 206] and not 2,3-dibromo-9-acridanone, as previously suggested.[513] This reaction is not easily stopped at the mono- or tri-bromo stages, but if the red compound (presumably 103), formed from 10-methyl-9-acridanone and bromine in chloroform at room temperature, is refluxed in the same solvent, 2-bromo-10-methyl-9-acridanone is formed, uncontaminated with 2,7-dibromo-9-acridanone.[515]

102

103

The more nucleophilic solvent, methanol, breaks down the complex to the free base, which is much more susceptible to electrophilic attack and rapidly brominates at positions 2 and 7.[516] 2,7-Dibromo-10-methyl-9-acridanone is the normal product of mild bromination, but more vigorous reaction conditions led to the formation[139] of the demethylated product (102). 2-Bromo-9-acridanone is readily converted to the 2,7-dibromo compound,[517] and a hexabromo-9-acridanone results from the treatment of 102 with boiling bromine in the presence of aluminium bromide.[139] If chlorine was passed through a solution of 2,7-dibromoacridine at 100°, 2,7-dibromo-4,5-dichloro-9-acridanone was obtained,[206] but the iodine of 2-iodo- or 2,7-diiodo-9-

acridanone was sometimes replaced by chlorine under similar conditions,[907] giving eventually the 2,4,5,7-tetrachloro compound.

9-Acridanone with iodine monochloride gave mixtures of 2-iodo- and 2,7-diiodo-9-acridanone, but extending the reaction time led to 4-chloro- and 4,5-dichloro-2,7-diiodo-9-acridanone.[908] 2,7-Dichloro-9-acridanone gave only the 2,4,7-trichloro compound in this reaction,[908] suggesting that with a deactivated 9-acridanone nucleus only the more reactive chlorine molecule, formed by disproportionation, can bring about electrophilic substitution.

Many of these structures have been checked by independent synthesis of the halogenated 9-acridanones from the corresponding diphenylamine-2-carboxylic acids.

(4) *Electrophilic Substitution—Nitration*

The nitration of 9-acridanone at 30° is reported to give a mixture of 2-nitro- (85%) and 4-nitro- (15%) 9-acridanones,[505] but other workers have been unable to repeat these experiments.[518] Great care is needed to obtain mono nitro products, even if just 1 mole of nitric acid is used. Further nitration of 2-nitro-9-acridanone gave mainly 2,7-dinitro-9-acridanone with a small proportion of 2,5-dinitro- and 2,4,7-trinitro-9-acridanone,[505] the last of these structures being confirmed by synthesis.[24] 4-Nitro-9-acridanone, on the other hand, is said to give only 2,5-dinitro-9-acridanone,[343, 505] but 2,4,5,7-tetranitro-9-acridanone is also claimed as a product of this reaction.[519] The 1-nitro- and 3-nitro-9-acridanones are also further nitrated in the 7-position.[147]

Just as in the halogenation series, the introduction of a nitro group into one of the rings of 9-acridanone sufficiently deactivates the ring to ensure that a second electrophilic substituent enters the other ring.

(5) *Electrophilic Substitution—Sulfonation*

This reaction corresponds to the nitration of 9-acridanone, in that 9-acridanone-2-sulfonic acid is the major product and a little of the 4-sulfonic acid is also obtained.[73] The allocated structures were confirmed by converting the sulfonic acids into hydroxy-9-acridanones, which were compared with authentic specimens. Hot oleum (80%) was required to introduce sulfonic acid groups into both rings; they entered at positions 2 and 7, as expected.[73]

(6) *Reactions at Position 9*

9-Bromoacridine is formed when bromine is added to a suspension of 9-acridanone in refluxing phosphorus tribromide.[520] 9-Acridanone gave almost a quantitative yield of 9-phenylacridine with two equivalents of phenyl lithium,[149] but only a 20% yield with phenyl magnesium bromide. With methyl magnesium iodide, the major product was 9-methylacridine,[521] but with excess Grignard reagent a little 9,9-dimethylacridan was obtained.[522, 523] A better route to this compound is the acid cyclization of the tertiary alcohol obtained from diphenylamine-2-carboxylic esters with methyl magnesium iodide.[247] Methyl lithium did not attack position 9[521] but did react with an ester group in a side chain, during a stage in the synthesis of acronycin.[412] (See Chapter IV, p. 419.)

9-Acridanone may not be 0-alkylated, and 9-alkoxyacridines must be prepared from 9-chloroacridines and alkali alkoxides. Neither is it 0-acylated with acetic anhydride and sodium acetate,[5, 157] in contrast to 4-pyridone. The hemiacetal **104** (R = H, R' = Me) is said to be formed when sodium methoxide reacts with 9-chloroacridine. It has a uv spectrum very similar to that of 9-methoxyacridine,[469] and so is probably a nonbonded hydrate, despite its reaction with hydrogen cyanide to give 9-cyanoacridine, which is not shown by the 9-methoxy compound.[524]

The acetal **104** (R,R' = Ph) is formed when 9-chloroacridine is heated to 100° with 10 parts of phenol. It instantly eliminates phenol if treated with dilute alkali,[524, 525] and reflux with dilute hydrochloric acid regenerates 9-acridanone.[525]

The typical carbonyl reagents, hydrazine,[526] thiosemicarbazide,[527] and aniline[281] do not react with 9-acridanone, but the tautomeric N-(9-acridinyl) derivatives may be made from 9-chloroacridine, or 9-chloroacridinium chlorophosphite[431] and the appropriate reagent. A low yield of 1,2-di-(9-acridinyl)hydrazine is also obtained in the first of these reactions. Acid hydrolysis of 9-anilinoacridine (**105**), also made from 9-aminoacridine and aniline hydrochloride,[281] gave 9-acridanone and aniline, the result expected from the hydrolysis of anil **106**.

| 104 | 105 | 106 |

A useful range of derivatives with lower melting points than the high, and correspondingly unreliable, values given by many 9-acridanones, is available from the reaction of a 9-acridanone with phosphorus oxychloride and a dialkylaniline. Under a more vigorous application of these conditions, acridines give the same derivatives,[528] and in the presence of sulfur, acridine, and aniline give the compound **107** (R = H).[529]

Even 4-aminodiphenyl undergoes the latter reaction,[909] in which a fast complex formation step is followed by a slow dehydrogenation.[910] As the reaction fails if the *para* position of the aniline is blocked (as in *N,N*-dimethylamino-*p*-toluidine) or strongly deactivated (as in *N,N*-dimethyl-*m*-nitraniline),[140] and as identical compounds may be made from acridine and 4-(*N,N*-dialkylamino)-phenyl lithium, followed by oxidation,[530] they are clearly 9-(4-dialkylamino-phenyl)acridines (**107**).[95]

A mixture of 9-acridanones is better separated at the 9-chloroacridine stage,[139] as a mixture of these dialkylamino derivatives may form mixed crystals, which are resistant even to chromatographic separation.

C. Substituted 9-Acridanones

N-Substituted 9-acridanones are prepared by direct substitution and oxidation or hydrolysis of 10-alkylacridinium salts or acridans. *C*-Substituents are usually introduced before cyclization or by further electrophilic or nucleophilic substitution. Groups may also be removed from the 9-acridanone rings by decarboxylation, desulfonation, and dehalogenation, or modified by additional chemical processes, such as oxidation, hydrolysis, and diazotization. The photoionization potential of 9-acridanone is lowered by *N*-alkylation, indicating that electrons are supplied to the ring by the alkyl group.[531] The mass spectra of many 9-acridanones have been measured, (see Chapter XIII, p. 709); they can provide valuable information regarding the substituents at positions 1, 2, and 4, and occasionally 10, when used as an aid to structure determination.[532, 533] Nuclear magnetic resonance spectroscopy has not proved so useful, but features such as unsubstituted benzene rings and hydroxyl groups *peri* to the 9-carbonyl group are readily detected[534] (see Chapter XII, p. 698). The fluorescence spectra of several substituted 9-acridanones have also been examined and discussed.[370, 535, 536]

Aspects of the chemistry of the more important groups of substituted 9-acridanones are considered below.

(1) Alkyl- and Aryl-acridanones

10-Methyl-9-acridanone was until recently the only *N*-alkyl compound available by direct alkylation,[537] possibly due to steric hindrance between the entering alkyl group and the hydrogen atoms *peri* to the nitrogen. It is more soluble in ethanol than 9-acridanone, but by contrast its solubility is not increased by the addition of potassium hydroxide. The dilution of a solution in 10*N* hydrochloric acid does not precipitate the hydrochloride but 10-methyl-9-acridanone itself.

On reduction with sodium amalgam[227] or zinc in aqueous alkali,[432] 10-methyl- and several other 10-alkyl-9-acridanones give 10-alkylacridans. However, if zinc in acetic acid,[538, 539] or preferably in ethanolic hydrogen chloride,[424, 434, 540] is used as a reducing agent for 10-methylacridanone, 10,10'-dimethyl-$\Delta^{9,9'}$-biacridan (**108**; R = Me) is formed, together with an acridinium salt and 10-methylacridan. If the reaction is prolonged, the yield of acridan increases. Bimolecular compounds are typical of acid reductions of 10-substituted 9-acridanones, as similar compounds (**108**; R = Et,[540] or Bu[434, 436]) have also been obtained. As a pinacol-type reduction with magnesium and magnesium iodide also gives these bimolecular products,[540, 541] the reaction scheme on the next page is suggested.

The intermediate pinacols (**109**) have not been isolated, but either lucigenin analogues (**110**)[538] (see Chapter IX, p. 616) or $\Delta^{9,9'}$-biacridans (**108**) may be obtained.[540] 10-Methyl-9-acridanones with methoxy,[541, 542] methyl,[542] and amino[424] substituents, but only melicopine among the alkaloids,[543] behave the same way (Chapter IV, p. 391). 10-Methyl-9-chloroacridinium dichlorophosphate, made from the 9-acridanone and phosphorus oxychloride,[12] or a mixture of oxy- and penta-chlorides,[544] also gives bimolecular reduction products.[493, 542]

A chlorosulfite analogue [545] of the dichlorophosphate salt may be intermediate in the preparation of 10-methyl-9-acridanimine from thionyl chloride and aqueous[546] or liquid ammonia.[547] Alternatively, a reaction path via a geminal dichloroacridan may be followed,[12] as many ketones form dichloro derivatives with this reagent; this route is also proposed when oxalyl chloride[548] is used in place of thionyl chloride. 10-Methyl-9-chloroacridinium dichlorophosphate also reacts with ammonia,[549] or aniline[541, 550] to give imines, and with formyl hydrazine[33] to give the hydrazone 111, all of which revert to 10-methyl-9-acridanone on treatment with dilute acid.[541, 546] However, perchloric acid hydrolyzes only the formyl group of 111 to precipitate the 10-methyl-9-acridanone hydrazone perchlorate.[33] With hydrazine hydrate,[550] these 9-chloroacridinium salts give the azine (112; R = Me, or Et); the dialkyl derivative of the tautomeric form of the diacridinylhydrazine obtained from the 9-chloroacridine salt (Section 2.B(6)).

10-Methyl-9-chloroacridinium salts yield 9-(4-dialkylaminophenyl)-10-methylacridinium chlorides with N,N-disubstituted anilines, which form the corresponding, unstable 9-hydroxyacridans (113) on basification[15, 551]; the ethoxy analogues or pyridine complexes of these pseudobases are more easily isolated.[551a,b]

Similar products to 113 are also made by the action of aryl[552] or alkyl[553] Grignard reagents on 10-methyl-9-acridanone. Acridols from the former reaction are converted by acids into 9-aryl-10-methylacridinium salts,[137, 435, 552] and those from the latter reaction may readily be dehydrated and hydrogenated to 9,10-dialkylacridans.[82, 554, 555] Phenyl lithium also yields a 9-phenyl-acridol.[435]

Lithium metal gave the organometallic product 114 with 10-methyl-9-

111   **112**   **113**

acridanone, which condensed with dichlorodiphenylmethane to form the oxide **115**.

114   **115**

The nitration of 10-methyl-9-acridanone gives 2-nitro- and 2,7-dinitro-10-methyl-9-acridanones just like the parent 9-acridanone,[227] and for the bromination of 10-methyl-9-acridanone see Section 2.B(3).

Preparations of 10-ethyl-[5, 157, 434] and 10-benzyl-9-acridanones[60] are similar to those used for the 10-methyl compound. 10-Vinyl-9-acridanone (Section 1.D(2)) forms a 9-chloro dichlorophosphate salt, and in somewhat poorer yield, a chlorosulfite salt, both of which react readily with aqueous alcoholic primary amine solutions[88] to give the imines (**116**). These imines, however, hydrolyze easily in water, without the need for acid catalysts. Reduction of the substituted 10-vinyl-9-acridanone (**72**) with zinc and hydrochloric acid, gives $\Delta^{9,9'}$-biacridan **117**, in which the side chain is also hydrogenated.[348]

The expected reduction products, 10-phenylacridan[429] and 10,10'-diphenyl-Δ-9,9'-biacridan[371, 540] respectively, are found when 10-phenyl-9-acridanone is reduced in alkaline or acid media, but yields with substituted 10-phenyl derivatives under the latter conditions may be low.[374, 430] Phenyl magnesium bromide with 10-phenyl-9-acridanone gives a 9-phenylcarbinol,[429, 556] but when 10-(p-bromophenyl)-9-chloroacridinium dichlorophosphate is treated with this reagent, and then nitric acid, a bimolecular lucigenin (cf. **110**) analogue results.[431] The treatment of 9,10-diphenylacridol with sodium formate and formic acid reduces it to the acridan[101]; during treatment of this compound with zinc, the acridyl radical **118** may be extracted into an organic layer.[137]

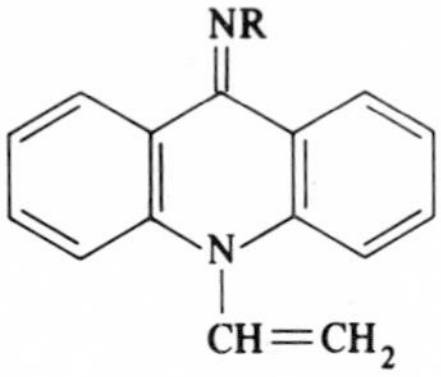

116

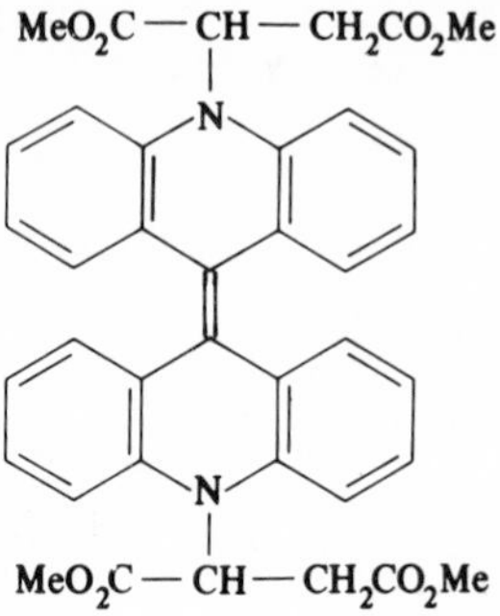

117

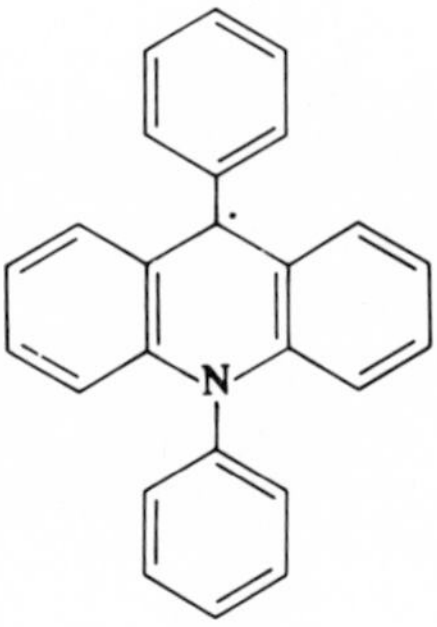

118

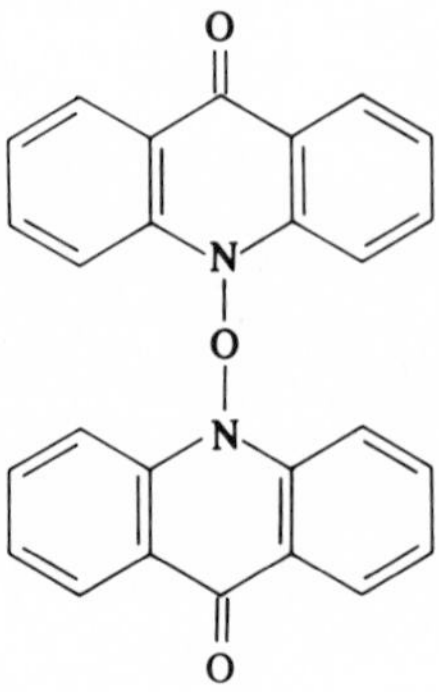

119

(2) *10-Hydroxy-9-acridanone*

These compounds, also known as acridone *N*-oxides, or 9-hydroxyacridine *N*-oxides, have been prepared by the oxidation of acridine oxides with permanganate,[454] alkaline hydrogen peroxide,[347] or ozone (Section 1.C(3)).[347] The alkaline hydrolysis of 9-chloro-[557] and 9-methoxy-[520, 558] 9-acridanone 10-oxides, made by the action of perbenzoic acid in chloroform on the corresponding acridines,[558–562] the acid hydrolysis of 9-bromo- and 9-nitro-acridine 10-oxides,[400] and the condensation of *o*-nitrobenzaldehyde and benzene in sulfuric acid, have also been employed.[307]

Alternatively, the peroxide oxidation product of acridine, originally thought to be 9,10(9*H*)-epidioxyacridan,[563] but now believed to be 10,10′-bisacridanone ether (**119**),[396] may be hydrolyzed to 10-hydroxy-9-acridanone with acids or bases.[563] Only a 9% yield of **119** is obtained when acridine is treated with 3-chloroperbenzoic acid in ether,[396] as acridine 10-oxide and 2-(2-hydroxyanilino)benzaldehyde are also produced. The uv spectrum of **119** is very similar to that of 10-methyl-9-acridanone,[396] and its ir spectrum shows absorption previously assigned to the N—O stretching frequency of *N*-oxides.[564]

Like 9-acridanone, 10-hydroxy-9-acridanone shows low solubility in acids, water, and organic solvents. However, it dissolves in aqueous sodium hydroxide to give a sodium salt, which crystallizes as the solution is concentrated. Reduction with sodium amalgam gave a mixture of acridine 10-oxide, acridine, and acridan,[307] while heating the 10-hydroxy compound alone,[558] in nitrobenzene[305, 565] or in tetralin[291, 305] caused decomposition to 9-acridanone and oxygen. The methyl derivative formed when 10-hydroxy-9-acridanone was treated with dimethyl sulfate and alkali,[307] or diazomethane,[558] differed from 9-methoxyacridine 10-oxide (**120**), and is most probably 10-methoxy-9-acridanone (**121**). No compounds of structure **122**, the remaining possibility, have been made, but it seems less likely to undergo hydrolysis to methanol and 10-hydroxy-9-acridanone than **121** and is not an expected product of the diazomethane reaction.

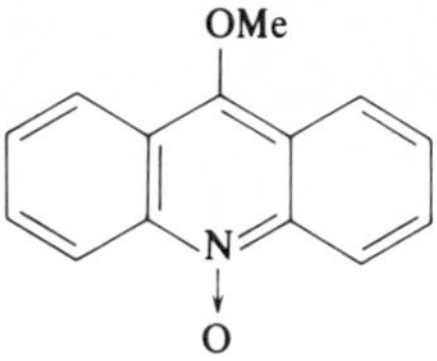

120 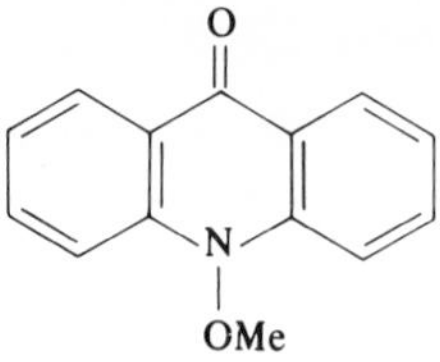   121 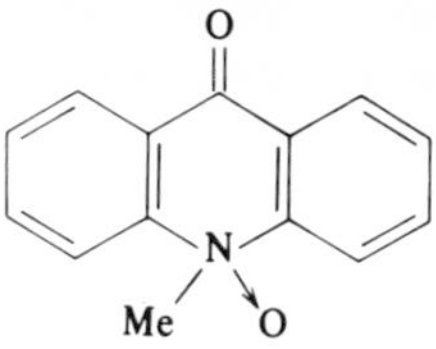   122

The product formed when this methyl derivative reacts with phenyl lithium[566] shows behavior comparable to that of 9-hydroxy-10-methyl-9-phenylacridan (**123**), for in both compounds the hydroxy group is replaced by alkoxy in refluxing alcohol[418, 567] and both give acridinium salts with acids. This suggests that the product is **124**, and establishes structure **121** for the methylated compound. The action of benzoyl chloride on 9-acridanone 10-oxide, similarly, gives the 10-benzoylated derivative **125**.

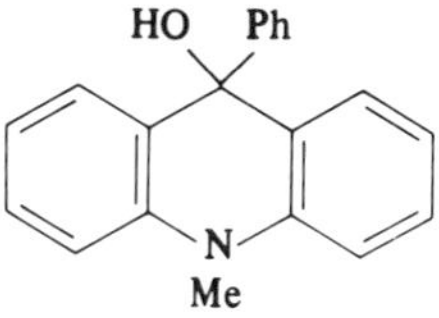

123 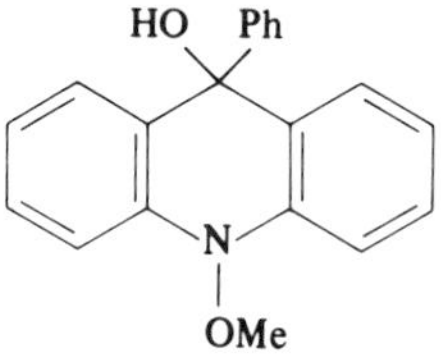   124 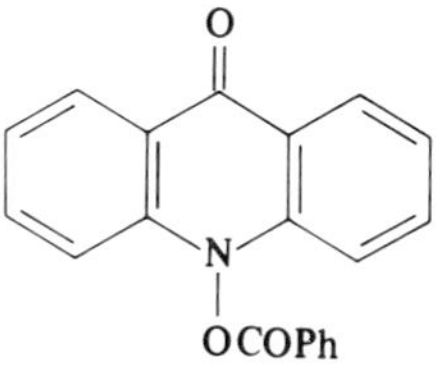   125

Both **120** and **121** give 9-acridanone and formaldehyde when heated above their melting points. In refluxing ethanolic hydrogen chloride, 10-hydroxy-,[454] 10-methoxy-[454] and 10-benzoyloxy-9-acridanones[559] all give 2-chloro-9-acridanone; if a 2-chloro substituent is already present, a dichloro derivative, probably 2,7-dichloro-9-acridanone, results. The intermediate formation and rearrangement of a 10-chloro-9-acridanone could be the pathway for these reactions.[566] Direct halogenation gives 2,7-dichloro- and 2,7-dibromo-10-hydroxy-9-acridanones,[18, 139] not the 2,3-dihalogen compounds previously reported,[513] and 10-hydroxy-3-nitro-9-acridanone[18] has been shown to brominate at position 7. Attempts to prepare 9-(p-dimethylaminophenyl) derivatives of 10-hydroxy-9 acridanones, with phosphoryl chloride and dimethylaniline, always resulted in the loss of oxygen from the ring nitrogen atom, and the isolation of the corresponding 9-acridanone derivatives.[18, 568] This deoxygenation reaction was used to establish the positions of electrophilic attack on the 10-hydroxy-9-acridanone.

The possibility of different structures for the methyl derivatives is one aspect of the controversy over the structure of 10-hydroxy-9-acridanone, which has only recently been resolved by applications of spectroscopy. Structures **126**[357, 569] and **127**[558, 570, 911] both had their champions, but both sides made the same mistake in assuming that the structure of the *N*-oxide and its methyl derivative would be the same. In the crystalline state, to judge from the ir spectrum of a nujol paste, and in carbon tetrachloride solution, the molecules exist in a strongly hydrogen-bonded 9-hydroxyacridine 10-oxide structure (**126**),[564] as *N*-oxide bands, shown by the sodium salt **128** and both 9-chloro- and 9-methoxy-acridine 10-oxide, are present.[564]

These bands are absent from the spectra of 10-methoxy- and 10-benzoyloxy-9-acridanone. In the slightly more polar solvent chloroform, the ir spectrum indicates that the 10-hydroxy-9-acridanone tautomer (**127**) is present to a substantial extent[396]; in alcholic solution, the uv spectra of the 10-hydroxy- and 10-methoxy-9-acridanones are very similar, but quite different from that of the sodium salt of the former.[571, 572] This establishes the structure of the sodium salt to be **128**. However, the spectra of the neutral species of the unmethylated compound and the two methyl derivatives **120** and **121** at pH 4.8, and those of their conjugate acids in 15N sulfuric acid, are all so similar that no conclusion can be made as to which structure predominates.[573] Estimates of the ionization constants of **126** and **127** show them to be almost the same,[573] suggesting that there will be approximately equal quantities of the two species present under these conditions. The high proportion of N-oxide form (**126**) is said to be a result of the stabilization of canonical forms **129** and **130** by annellation.

(3) Halo-9-acridanones

Although cyclization of halogen-substituted diphenylamine-2-carboxylic acids, and the electrophilic substitution of 9-acridanones (Section 2.B(3)) are the usual routes to halo-9-acridanones, 2,7-dibromo-9-acridanone has also been made by a Sandmeyer reaction on diazotized 2,7-diamino-9-acridanone.

The usual syntheses of 9-acridanones involve electrophilic ring closure, but when 1,2,3,4-tetrafluoro-9-acridanone was required, the high deactivation of the fluorinated ring was expected to preclude cyclization. The pyrolysis of 2′-nitro-2,3,4,5,6-pentafluorobenzhydrol (**131**) was found to be an effective route, however, giving a mixture of 3-pentafluorophenylanthranil and 1,2,3,4-tetrafluoro-9-acridanone, in which the proportion of acridanone increased as the temperature was raised.[311] A nitrene intermediate is suggested in this reaction (p. 180).

2,4,5,7-Tetrahalo-9-acridanones have lower melting points than the 2,7-dihalogen compounds; this is interpreted as a screening effect of the halogen atoms on the hydrogen at position 10, which diminishes the strength of the hydrogen bonding in the crystal (cf. Chapter XI, p. 670).[206] Neither do they exhibit the strong molecular association in solution shown by di- and tri-halo-9-acridanones, even though infrared spectra indicate that the polarity of the C—O bond increases with increasing halogenation.[907]

The reduction of 2-bromo-9-acridanone with sodium in ethanol led to the removal of the bromine and the formation of acridan.[226] However, sodium amalgam in water reduced 9-acridanone-2- or -3-halo-7-sulfonic acids to the corresponding acridines,[574] and in ethanol reduced 2,7-dibromo-9-acridanone to a mixture of 2,7-dibromo-acridine and -acridan without affecting the halogen atoms.[18] 2,4,7-Tribromo- and 2,4,5,7-tetrabromo-9-acridanones gave only 9-chloroacridines when treated with phosphorus oxychloride and dimethylaniline, but in the presence of aluminium chloride 9-(p-dialkylaminophenyl) derivatives could be prepared.[139, 575]

Halo-9-acridanones are nitrated[150] and sulfonated[574] in the ring that does not contain the halogen substituent. The usual site for electrophilic attack is position 7.

A characteristic reaction of 9-acridanones with 1-halogen substituents, is nucleophilic substitution, which is more effective if electron-withdrawing substituents are present in the same ring. A number of 9-acridanones with chloro or bromo groups at position 1, give 1-amino-9-acridanones in yields of about 50% when refluxed with an aliphatic amine,[576, 577] piperidine,[153, 578] or aniline.[150] Other halogen atoms at positions 3, 4, or 6 are not replaced. 1-Chloro-4-methyl-9-acridanone was unchanged after refluxing in piperidine,[153] but the chlorine atom in 1-chloro-4-nitro-9-acridanone could readily be displaced by aryl and alkyl thiols[217]; that in 1-chloro-2,4-dinitro-9-acridanone was so labile that it was replaced by a hydroxy group when the compound was boiled with 7% aqueous potash.[252]

The 3-chloro group may also be replaced, but the much more vigorous conditions of an autoclave are required.[300, 579] The substitution of 3-bromo groups is slightly easier, but poor yields were obtained when a 3-iodo group was replaced.[580] After 5 hr at about 200° in the presence of strong aqueous ammonia and a little copper sulfate, even 2,7-dibromo-9-acridanone underwent nucleophilic substitution to 2,7-diamino-9-acridanone.[18]

A consideration of the canonical forms 132 and 133 suggests a theoretical explanation for these observations, but 10-hydroxy-9-acridanone for which similar forms (134) may be drawn, is found to give low yields in nucleophilic substitution reactions.[579, 580]

132   133   134

The bromine atom in the side chain of 4-bromomethyl-9-chloroacridine is also readily replaced by amines, to give 4-aminomethyl-9-chloroacridines, 4-aminomethyl-9-acridanones or their salts.[581]

(4) Nitro-9-acridanones

In addition to the cyclization of nitrodiphenylamine-2-carboxylic acids and the nitration of 9-acridanones (Section 2.B(4)), nitro-9-acridanones can be made by the chromic acid oxidation[350] of the corresponding acridines (Section 1.C(3)).

The mononitro-9-acridanones are more acidic than 9-acridanone, dissolving appreciably in aqueous, and very easily in alcoholic, alkali.[350] As the number of nitro groups increases, so does the acidity. 2,4-Dinitro-9-acridanone is soluble in hot aqueous sodium hydroxide, but not carbonate, while 2,4,7-trinitro- and 2,4,5,7-tetranitro-9-acridanone dissolve in the latter reagent with the evolution of carbon dioxide.[227] Alkali salts, which are precipitated from strong solutions in aqueous alkali, may be sparingly soluble in water; a 0.5% solution of 2,4,5,7-tetranitro-9-acridanone in 1% aqueous lithium carbonate solution has been suggested as a useful precipitating agent for potassium ions.[519]

Nitro-9-acridanones are readily N-methylated by methyl iodide in alcoholic alkali,[227] but 4-nitro-9-acridanone does not give the corresponding 9-chloro-acridine with phosphorus oxychloride. Even after 8 hr in refluxing xylene with phosphorous pentachloride, only a 20% yield was obtained.[23] As other nitro-9-acridanones are also reluctant to undergo this reaction, association between the 4-nitro group and the N—H group in the ring cannot be held entirely responsible for this failure.

A nitro group at position 1 will undergo nucleophilic substitution when refluxed with piperidine,[153, 578] and 2- and 3-nitro-9-acridanones are both brominated in position 7,[18] and not in the 3- and 2-positions, respectively, as formerly reported.[513] Cold oleum converts 3-nitro-9-acridanone to 6-nitro-9-acridanone-2-sulfonic acid in 97% yield,[582] and 4-nitro-9-acridanone is chlorosulfonated at an unspecified position in the unsubstituted carbocyclic ring.[912]

(5) Amino-9-acridanones

A valuable route to these materials is through the reduction of the corresponding nitro compounds, which are usually more easily prepared by standard methods than the amino-9-acridanones. The most frequently used reducing agent is stannous chloride in hot concentrated hydrochloric acid,[124, 130] but sodium dithionite,[583] ammonium sulfide,[136] and sodium sulfide,[118, 119] are possible alternatives. Catalytic hydrogenation over Raney nickel has also proved effective.[145, 584] More powerful reduction gives aminoacridans[394]; indeed, this is the normal route to aminoacridines,[585] e.g., when 3-aminoacridine-7-sulfonic acid, or sulfonamide, is prepared by the sodium or aluminium amalgam reduction of the corresponding 3-nitro-9-acridanone[586] followed by ferric chloride oxidation of the acridan.

Reduction of the nitro group may be accompanied by the elimination of a halogen atom. 3-Amino-9-acridanone was the product from the reduction of both 6-bromo-[300] and 6-iodo-3-nitro-9-acridanone[580] with zinc and calcium chloride in aqueous ethanol, and only 5% of the product retained the chloro group, when 2-chloro-7-nitro-9-acridanone was treated with the same reagent.[401]

9-Acridanones with amino groups at positions 2 or 4 shown an unusual solubility in aqueous alkali[26, 148, 587] which has not been explained. 1- And 3-amino-9-acridanones do not show this property.

The diazotization and replacement of amino groups in the 9-acridanone ring proceeds satisfactorily and has been of particular use in the synthesis of 9-acridanones that are not easy to prepare directly. 3-Cyano-9-acridanone,[588] and the 9-acridanone-arsonic and -stibonic acids[589] fall into this category, but bromo[300, 517, 590] and chloro groups[249, 300, 392, 514, 591] have also been introduced in this way, using both mono- and diamino-9-acridanones. A step in the preparation of a labeled atebrin analogue was the replacement of the amino group of 6-amino-2-methoxy-9-acridanone with radioactive iodine,[584] via diazotization (Chapter II, p. 126). A diazonium group has also been replaced by hydrogen on reflux in ethanol, reduced to a hydrazino group with stannous chloride,[582] and coupled with a reactive aromatic nucleus[18] to form dyestuffs.[86, 87]

4-Amino-9-acridanone, on treatment with nitrous acid, cyclized to 6-triazolo[de]-9-acridanone (135),[150] which was reduced, with ring opening, to give 4-hydrazino-9-acridanone.[136] When boiled with acetic anhydride, 4-amino-9-acridanone gave the similar imidazolo-9-acridanone (136).

Diazotization and hydrolysis is not a satisfactory way to introduce hydroxyl groups into the 9-acridanone ring.[393] An amino group is better replaced by vigorous acid hydrolysis. Dilute sulfuric acid at 195°[297] and concentrated hydrochloric acid at 220°[227] were both effective for the substitution of amino groups at position 2 by hydroxyl.

Amino-9-acridanones, which have an amino group at position 2 or 4, oxidize easily, while their isomers do not. The product from the aerial oxidation of 2-anilino-9-acridanone was the quinonoid compound 137, which was reduced back to 2-anilino-9-acridanone by hydroquinone.[913] 1-Anilino- and 1-(p-tolylamino)-9-acridanones exhibit thermochromism, a yellow form changing to a red form, with the same melting point, on heating.[252] 4-Amino- and 4,6-diamino-9-acridanones take part in Skraup reactions, even though the corresponding aminoacridines do not. The product in the first instance is 138.

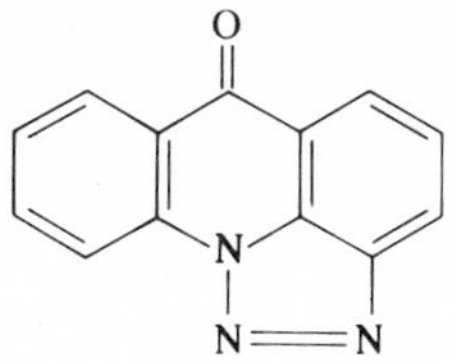

135

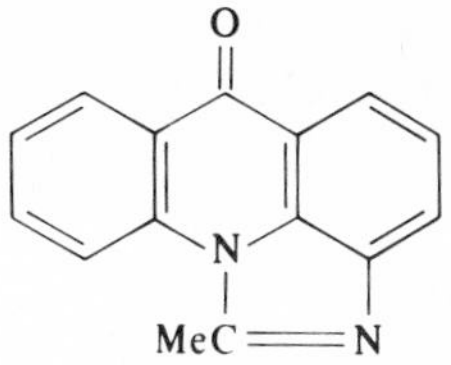

136

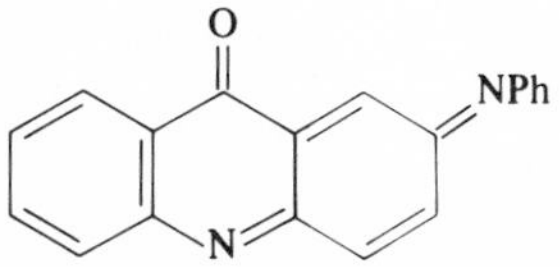

137

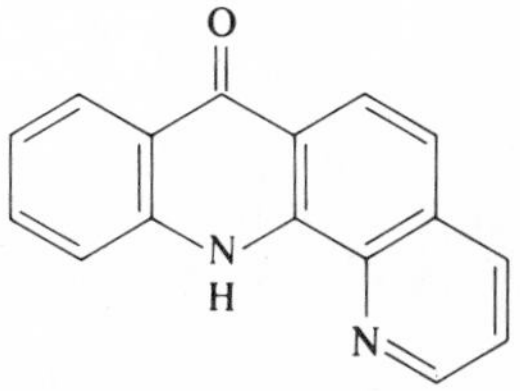

138

(6) *Hydroxy- and Alkoxy-9-acridanones*

Most naturally occurring 9-acridanones have hydroxy, methoxy, or methylenedioxy functions in the right-hand ring; much of the chemistry of hydroxy- and alkoxy-9-acridanones, is a result of synthetic or degradative studies on the acridine alkaloids (see also Chapter IV, Section 2.D). Synthetic compounds in this group have been made from diphenylamine-2-carboxylic acids, where the use of sulfuric acid usually gives simultaneous cyclization and dealkylation, from amino-9-acridanones by hydrolysis, and from sulfonic acids by fusion with potassium hydroxide.

1-Hydroxy- and 1-methoxy-9-acridanones show contrasting behavior with their isomers. 1-Hydroxy-10-methylacridanone is insoluble in refluxing aqueous alkali, has no fluorescence in solution, and gives a deep green color with ferric chloride; 1-hydroxy-9-acridanone is similar. 2-, 3-, and 4-Hydroxy-9-acridanones do not have sharp melting points; are soluble in dilute alkali; dissolve better in polar solvents but less well in nonpolar solvents than the 1-hydroxy compounds,[543] giving fluorescent solutions; and show yellow ferric chloride colors. The methylation of 1-hydroxy-9-acridanone is more difficult, but the demethylation of 1-methoxy-9-acridanone is very much easier to achieve than are the equivalent reactions with the other isomers. These differences have been attributed to intramolecular hydrogen bonding,[914] which is only possible with 1-substituents (e.g., 139). Certainly, 2-, 3-, and 4-hydroxy-

9-acridanones show an absorption in the infrared, allocated to free O—H stretching, which is not present in the spectrum of 1-hydroxy-9-acridanones[543] (Chapter XI, Section 2).

139   140

The enhanced basic properties of the 1-methoxy-9-acridanones could also be caused by the stabilization of the cation by an intramolecular hydrogen bond (140).[592, 593] Some 10-methyl-hydroxy- and -alkoxy-9-acridanones are basic enough to form stable picrates,[227] and 1- and 4-hydroxy- and -alkoxy-9-acridanones give yellow colors with pyroboric acetic anhydride (Dimroth's reagent), triphenylborate or a mixture of boric and citric acids (Wilson's reagent).[594] These colors may be caused by the presence of chelated boric esters, such as 141 and 142.

141   142

The demethylation of methoxy-9-acridanones has been the subject of careful study.[414] The nonselective reagent, constant boiling point hydrobromic acid, demethylates all methoxy groups after a few hours at reflux temperature,[414] and a trace of stannous chloride is said to aid reaction in difficult cases.[595] Anhydrous aluminium chloride, which demethylated 4-methoxy-9-acridanone,[118] and deethylated 6,9-dichloro-2-ethoxyacridine without hydrolyzing the 9-chloro group,[39] may be of more general application, and so may a mixture of hydriodic acid and acetic anhydride, which demethylated 3-methoxy-9-acridanone.[596]

The presence of a 10-methyl group, or further alkoxy or hydroxy substituents on the 9-acridanone molecule, greatly eases the hydrolysis of 1-methoxy groups. The alkaloids lose this methyl group easily in acid conditions, and

1,3-,[455] but not 1,4-dimethoxy-9-acridanone,[414] is converted to the 1-hydroxy-methoxy-9-acridanone after a 1-hr reflux in N-hydrochloric acid.[455] However, 1-methoxy-10-methyl-9-acridanone gave only a 25% yield of demethylated product after a solution in $2N$ ethanolic hydrochloric acid had been boiled for 8 hr.[414] The hydrolysis of 9-chloroacridines is noticeably faster than the demethylation of 1-methoxy-9-acridanones, as the action of refluxing 1% hydrochloric acid on 9-chloro-1-methoxyacridine for half an hour, gave the 9-acridanone without affecting the methoxy group at all.[441] The fusion of the hydrochloride salts of the 1-methoxy-9-acridanones is also an effective demethylation procedure, [414, 597] but like other methods employing hydrochloric acid, it leaves 2-, 3-, and 4-methoxy groups untouched.[441] The 2-methoxy group of arborinine (2,3-dimethoxy-1-hydroxy-10-methyl-9-acridanone) was eventually demethylated after a long period of refluxing in concentrated hydrochloric acid.[598]

Other reagents which have been used successfully in the demethylation of 1-methoxyacridanones include sodium hydride in dimethylformamide,[533] boron trichloride in methylene chloride,[83] which is known to be a selective demethylating reagent for methoxy groups *ortho* or *peri* to a carbonyl group,[599] and methyl magnesium iodide, which surprisingly, left the 9-acridanone oxo-group unchanged.[412]

Alkaline demethylating agents are less effective. Even after reflux for 21 hr in aqueous ethanolic sodium hydroxide, only a partial hydrolysis of the 3-methoxy group of 10-methyl-2,3,4-trimethoxy-9-acridanone was achieved, despite the presence of activating groups.[600] Ethoxy substitution occurred at position 3, as well as demethylation at positions 1- and 3- of 10-methyl-1,2,3,4-tetramethoxy-9-acridanone, when ethanolic potassium hydroxide was employed.

Nevertheless, alcoholic potassium hydroxide will displace a methylene-dioxy group from a 9-acridanone ring and leave adjacent hydroxy and alkoxy groups. With 2,3-[533, 601] and 3,4-[543, 602] methylenedioxy groups, the alkoxy group, which could be methoxy, ethoxy, isopropoxy, benzyloxy or 3,3-di-methylallyloxy (prenyloxy), was always found in the 3-position, with the hydroxy group in position 2 or 4, respectively. Position 3 is the expected site of nucleophilic attack (p. 392). Well-characterized products could not be obtained after the acid hydrolysis of methylenedioxy groups.[414]

A 1-hydroxyl group was removed completely from the 9-acridanone ring, by first treating the compound with *p*-toluenesulfonyl chloride and then reducing out the tosyl group with freshly prepared Raney nickel.[600]

The most useful general technique for alkylating hydroxyl groups in the 9-acridanone ring is to treat the hydroxy-9-acridanone with an alkyl halide and potassium carbonate in dry acetone. By the use of this method, methyl,[441, 455, 536] ethyl and prenyl ethers[533] have been prepared in good yields.

Methyl sulfate requires these anhydrous conditions to methylate 1-hydroxy-9-acridanone,[534] but hydroxyl groups at other positions are methylated by this reagent in aqueous alkali.[393, 536] Alternatively, sodium methoxide and an alkyl halide in chlorobenzene may be used to form 2- and 4-alkoxy groups,[603] and diazomethane will methylate a hydroxy group at position 3,[600] but not at position 1.[534] The former reagent,[603] in contrast to the anhydrous alkyl halide potassium carbonate mixture,[441] does not alkylate the ring nitrogen atom as well as the hydroxyl group oxygen atoms, while methyl sulphate in dimethylformamide requires sodium hydride to bring about both *N*- and *O*-alkylation.[899] Methyl sulphate with potassium carbonate, a reagent also suitable for the methylation of hydroxydiphenylamine-2-carboxylic acids prior to cyclization,[11] produces only *O*-methylation.[899]

If 1,3-dihydroxy-10-methyl-9-acridanone is treated with 3-chloro-3-methylbut-1-yne, potassium carbonate and sodium iodide in dimethyl-formamide, and the temperature is kept below 52°, the propargyl ether (**143A**) is obtained. Increasing the temperature or heating the product in diethylani-line gives noracronycin (**144, R = OH**).[899, 914]

The yield, which approaches 90%, and the absence of any linear isomer from the reaction mixture make this the best available synthetic route to acronycin (**144; R = OMe**). By contrast the pyridine-catalyzed condensation of 1,3-dihydroxy-9-acridanone with 3-hydroxyisovaleraldehyde dimethyl acetal gives only a 32% yield of a 3:1 mixture of bisnoracronycin and its linear isomer.[915]

If the acetylenic bond of the ether (**143A**) is partly reduced using a Lindlar catalyst, the resulting allyl ether undergoes a Claisen rearrangement to compound (**143**), which may be oxidatively cyclized with chloranil.[899, 914]

The reduction of a number of hydroxy-9-acridanones with sodium in alcohol has given the corresponding acridines.[40] 1,4-Dihydroxy-, and 2- or 4-hydroxy-9-acridanones are readily oxidized to 9-acridanone-1,2- or 1,4-quinones[595]; methoxy derivatives may undergo demethylation and oxidation to form quinones if treated with nitric or nitrous acids.[604] Two general observations may be made. First, if a 4-hydroxyl group is present, a 1,4-quinone results; while the presence of a 2-hydroxyl group ensures that a 1,2-quinone is the product. Second, a methylenedioxy group is attacked preferentially (to give adjacent keto and phenolic groups), rather than an alkoxy group suffering demethylation and oxidation. For example, the 1-hydroxy-4-methoxy-2,3-methylenedioxy system gave on oxidation the 3-hydroxy-4-methoxy-1,2-quinone, and the 1-hydroxy-2-methoxy-3,4-methylenedioxy system the 2-methoxy-3-hydroxy-1,4-quinone.[543] 3,4-Quinones have not been found.

The presence of hydroxy and methoxy groups activates the ring toward electrophilic substitution. The Friedel-Crafts reaction of 3,3-dimethylallyl chloride in refluxing trifluoracetic acid with a zinc chloride catalyst is an alternative route to compound 143.[83] A second molecule of halide will attack position 2.[412] 3-Hydroxy-9-acridanone also reacts with dimethylallyl phosphine to give compound 144 (R = H). 1,3-Dihydroxy-9-acridanone was mono-nitrated even in hot dilute nitric acid, but the product differed from the mono-nitro compound isolated from a nitration reaction using concentrated nitric acid.[916]

Bromination has been thoroughly studied in the alkaloid series (see Chapter IV, Section 4.B). Early results showed that with bromine in acetic acid, a mixture of 9-acridanones brominated at position 7 and/or demethylated at positions 1 and 3, and 9-acridanone-1,4-quinones was produced.[605] With bromine in alcoholic solvents, more complex products could be isolated.[600]

These were eventually formulated as tribromide salts,[606] in which addition rather than substitution had taken place and the alcohol had also participated. In some cases, all the bromine entering the ring had subsequently been replaced by alkoxy groups. A hypobromite group, at 1- or 9-, according to whether a hydroxyl or methoxyl group was originally present at position 1, is also a feature of these molecules. On alkaline hydrolysis, a 1,9-diketo compound is produced, and treatment with acid gives a 9-acridanone-1,4-quinone.[606]

The sequences for normelicopicine (145) and normelicopidine (146) are shown, but others are very similar.[606]

When the 4-position was not occupied, the first formed salt on treatment with alkali, and heating gave 4-bromo-9-acridanones and 1-oxo compounds corresponding to those above, but no quinones.[607] In one case, bromine entered the 4-position directly.[607]

When chloroform was used as a solvent,[515] 7-bromo-9-acridanones were the normal products; but intermediate tribromide salts with hypobromite characteristics could also be isolated. Hydrogen bromide, liberated in the ring bromination step, demethylated methoxyl groups at position 1.[515] The salt **147**, on treatment with 1% aqueous caustic soda, showed a ring contraction[515] to the cyclopentenoquinolone **148**. This compound had previously been isolated after treating the alkaloid **149** with hydrochloric acid and sodium nitrite.[543]

The site of bromination in the addition reactions depends on the attack of the nucleophile, methanol, on the intermediate bromonium ion. It is the stabilization of this ion by charge delocalization among the oxygen atoms that allows addition, rather than substitution, to take place. A further influence is the relief of strain in the methylenedioxy ring when the sp^2 hybridized carbon atoms of the aromatic ring are converted to sp^3.[516]

A sequence for the complete dealkylation of evoxanthine (150) after successive treatment with bromine in methanol, alkali, and acid, and then warming the product (151) in dilute alkali, is shown.[607]

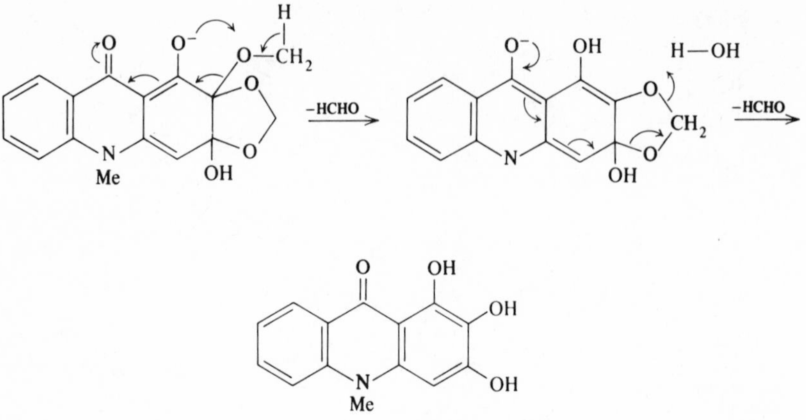

Attack of the solvent alcohol on the stable sigma-complex **152** is suggested to explain the formation of the ethers found in this reaction.[516]

(7) *9-Acridanone Carboxylic Acids*

A few 9-acridanonecarboxylic acids have been made by cyclizing diphenyl-amine-2-carboxylic acids, but a more effective general method could be the hydrolysis of cyano-9-acridanones, produced from the corresponding amines by diazotization and copper-catalyzed substitution reactions. 9-Acridanone-3-carboxylic acid has been made this way.[588] The potassium hydroxide fusion of vat dyes of the naphth[2,3-*c*]acridan-5,8,14-trione structure (**153**) may also have wider applications as a synthetic route.[588]

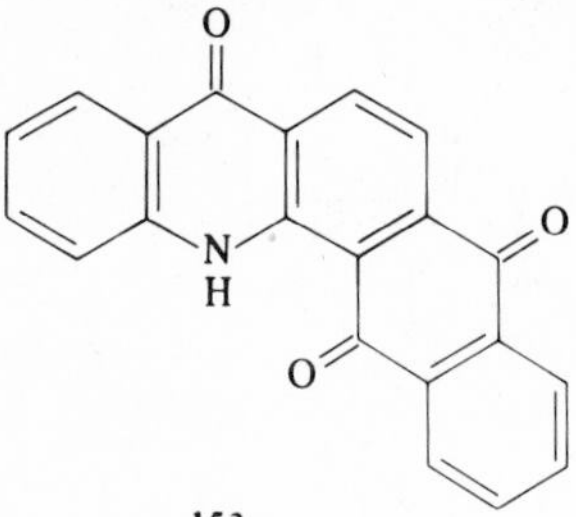

153

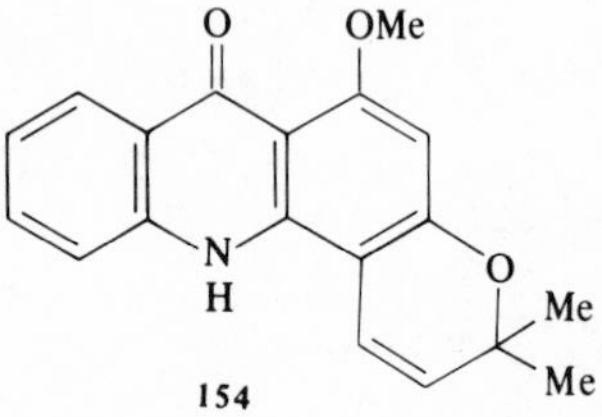

154

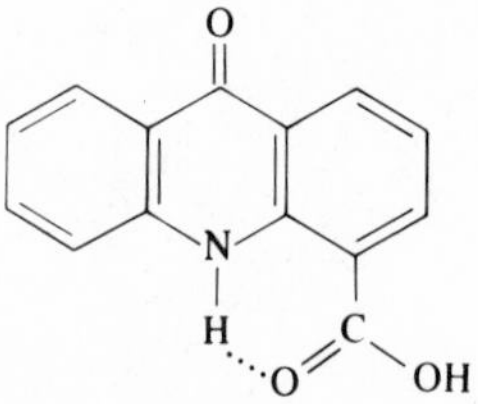

155

3-Hydroxy-1-methoxy-10-methyl-9-acridanone-4-carboxylic acid is obtained by ozonolysis and further oxidation of the alkaloid acronycin (**154**).[455, 597] 9-Chloroacridines with carboxylic acid groups are rarely obtained, as the 9-acridanone forms readily by internally catalyzed hydrolysis.

Methyl esters are best prepared by treating the acid with diazomethane[455] or its sodium salt with methyl sulfate.[188] The 4-carboxylic acid gave only 20% of the ester after reflux in methanolic sulfuric acid, possibly because of intramolecular hydrogen bonding (**155**). Extensive intermolecular hydrogen bonding[486] seems to be present in the crystals of 9-acridanone-2- and -3-carboxylic acids, as their melting points, and those of their methyl esters, lie well above the corresponding values for the 4-substituted isomers.

9-Acridanone-4-carboxylic acid[343] is nitrated at position 7, a further example of a deactivating substituent directing electrophilic reagents to the position *para* to the nitrogen in the other ring. Decarboxylation of the 4-acid occurs on heating with silver.[75, 126, 127, 150, 582]

(8) *9-Acridanone Sulfonic Acids*

A few sulfonated 9-acridanones have been prepared by cyclizing sulfonamidodiphenylamine-2-carboxylic acids, and a number from the unintentional or planned sulfonation of diphenylamine-2-carboxylic acids during cyclization with sulfuric or chlorosulfonic acids.

These sulfonic acids may be converted to acid chlorides,[574] esters, and amides[27, 574] by standard procedures. The treatment of 9-acridanone-2-sulfonic acid with phosphorus oxychloride and dimethylaniline gave the comparatively low melting 2-(9-*p*-dimethylaminophenylacridinyl) *p*-dimethylaminophenyl sulfone,[73] and these compounds may be generally useful as derivatives for characterization.

9-Acridanone-2-sulfonic acid gives fluorescent solutions,[93] and the quenching of this fluorescence has been the subject of detailed examination.[608]

The sulfonic acids can be desulfonated to the corresponding 9-acridanones by refluxing 30% sulfuric acid,[30] for about 3 hr, and can be converted to the equivalent hydroxy-9-acridanones by fusion with potassium hydroxide. The 2- and 4-sulfonic and 2,7-disulfonic acids all gave good yields in the latter reaction.[27, 252, 393] The reduction of 9-acridanone-2-sulfonyl chloride to biacridinyl-2,2'-disulfide and acridine-2-thiol has been reported.[27]

(9) *Acridanone Quinones*

If oxygen is passed through a methanolic solution of 2-hydroxyacridine in the presence of a secondary aliphatic amine and cupric acetate, a 4-dialkyl-aminoacridine-1,2-quinone is formed.[609, 610] This unlikely reaction, which has parallels in the quinoline series, gives surprisingly good yields.

Derivatives of 9-acridanone-1,2- (156) and -1,4- quinones (157) are known, but no 9-acridanone-3,4-quinones (158) have been identified.

156   157   158

The rules for the formation of quinones via the oxidation of hydroxy- or methoxy-9-acridanones have been discussed above and an example of quinone formation by bromination and hydrolysis given on p. 222. Nitric acid, ferric chloride, and chloranil have been used to oxidize 1,4-dihydroxy-9-acridanones to the 9-acridanone-1,4-quinones and their efficiency decreases in that order.[595] Nitric acid is also the most effective reagent with 1,4-dimethoxy-9-acridanones. To oxidize 4-hydroxy-9-acridanone to the 1,4-quinone, or 2-hydroxy-9-acridanone to the 1,2-quinone, potassium nitrodisulfonate is the recommended reagent.[595]

Additional methoxy groups in positions 2 or 4 of acridanonequinones are much more easily demethylated than such groups in the parent 9-acridanone. Reflux in aqueous sodium carbonate is sufficient for the hydrolysis of these methoxy groups, but the 3-methoxy compound still needs to be heated with constant boiling-point hydrobromic acid.[604]

Both types of quinone also form acetates with acetic anhydride and pyridine, or on reductive acetylation. Reduction of the 10-methyl-9-acri-

159

danone-1,4-quinone monoacetal (159) with sodium borohydride, follows the course shown.[605]

Derivatives of both 156 and 157 may be reduced to the corresponding quinols with phenylhydrazine, or further oxidized, first to 1,2,3,4-tetraoxo-9-acridanone, which may be isolated as the dihydrate,[604] and eventually to a 4-quinolone-2-carboxylic acid.[593]

Acridine quinones are listed in Table VIII.

(10) Actinomycinol, or De(s)peptidoactinomycin

During the 1950s, 2,5-dihydroxy-3,6-dimethyl-9-acridanone-1,4-quinone, the most important of the acridanonequinones, attracted the attention of groups in Nottingham and Göttingen, who gave to it the names that head this section. The bright red actinomycins are a group of antibiotics, possessing the same chromophore, but differing in the composition of the two peptide chains that each carries. Their chemistry has been reviewed.[611, 612a,b] All actinomycins are hydrolyzed in refluxing aqueous barium hydroxide to actinomycinols,[613-615] which although they have ill-defined melting points,

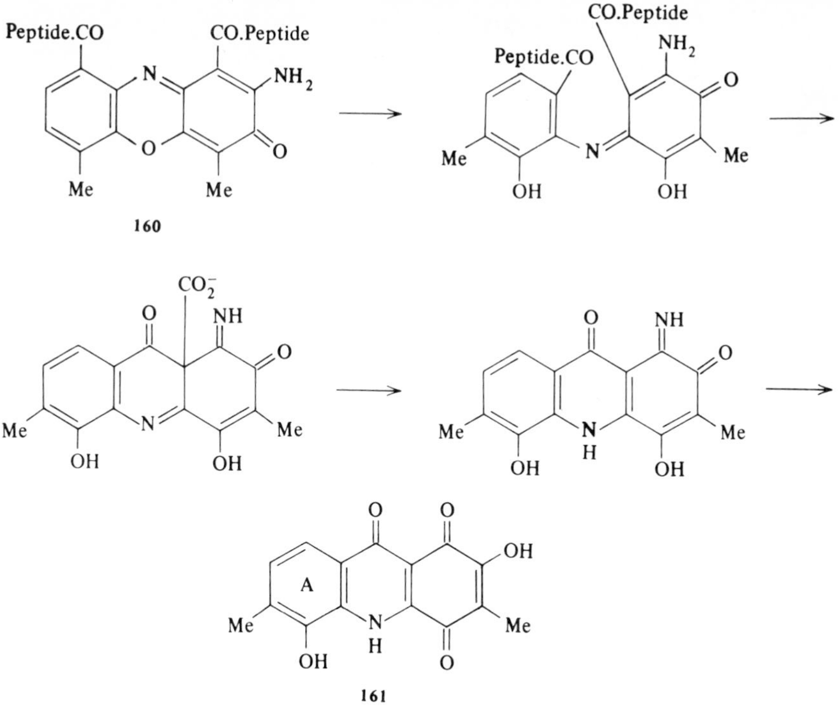

160

161

have been shown to be identical by ir spectroscopy.[616] Actinomycinol derivatives are listed in Table IX.

The natural chromophore is not an acridanonequinone, but a phenoxazinone (160), which is converted to actinomycinol (161) in the basic medium, possibly by the pathway shown on the previous page.[272, 611]

A hydrolysis of the oxygen bridge is followed by the loss of the peptide chains and a Dieckmann-like cyclization. Decarboxylation and hydrolysis of the 1,2-quinonimine then gives a tautomer of 161, which undergoes protoctropic rearrangement to actinomycinol. Ammonia was given off as the reaction proceeded,[617] but carbon dioxide was presumably absorbed by the alkaline solution.

The structure of actinomycinol was determined independently by both groups,[119, 618, 619] who announced their results on the same day.[611] A significant degradation sequence was the oxidation of actinomycin with alkaline hydrogen peroxide,[618, 620] and the hydrolysis of the peptide-containing product, to the benzoxazolone 162, in which ring A of actinomycinol remains intact. Spectral comparisons and redox potential measurements were also of value,[71, 272, 617, 619] but the final proof of structure was furnished by synthesis. The two synthetic routes to actinomycinol diacetate, which is the derivative of choice for purposes of characterization, are shown.

Actinomycinol is stable to alkali and heat, and forms an alkali metal salt.[621] Its titration curve shows three inflections corresponding to pK values of

Brockmann and Muxfeld [119, 498]– 27% overall yield.

Hanger, Howell, and Johnson [272]

4.8, 8.2, and 12.1, which were allocated,[81] respectively, to the 2-hydroxy group (its position adjacent to the quinone carbonyl accounts for its strongly acid nature), the amidelike N—H group at position 10, and the phenolic group at position 6. However, 9-acridanone is known to be an extremely weak acid, and a decreasing order of acidity from 2-hydroxy, through 6-hydroxy to 10-NH, seems preferable.

With methyl iodide and silver oxide, actinomycinol is said to form the trimethyl ether (163), which is subsequently hydrolyzed to a monomethyl ether with $2N$ sodium carbonate solution.[81, 621] Since the monomethyl ether retains the inflections in its titration curve, corresponding to the two higher acid values, it is said to be the 6-methylated 9-acridanonequinone.[81] This example of O-methylation at position 9 is unique; the possibility that 10-methylation has occurred in the first step, giving (164; R = Me), and that the 10-methyl group has survived hydrolysis, must be considered,

especially as this agrees with the revised order of acidity. Unfortunately, the
uv spectrum of the trimethyl ether, which should resolve this problem, is not
recorded. In comparable reactions,[616] acetic anhydride, or benzoyl chloride,
in pyridine converts actinomycinol to a triacetate or tribenzoate, given the
structure (164; R = COMe or COPh).[81]

Reduction with hydrogen iodide and red phosphorus,[81] sodium dithionite,
or stannous chloride gave dihydroactinomycinol,[621] in which the quinone
system must be hydrogenated, as a dihydropentaacetate (166; R = COMe)
may be obtained. 9(0)-Acetylation had not taken place, as the pentaacetate
had a uv spectrum similar to those of 9-acridanones but differing significantly
from the spectra of acridines.[81] In contrast, with acetic anhydride and zinc
in acetic acid, actinomycinol gives a dihydrodeoxytetraacetate, whose acri-
dinelike spectrum indicates that it has the structure 167.

In confirmation of structure 161, actinomycinol yields 1 mole of iodoform
when oxidized with hypoiodite solution, which must come from the methyl
group in the quinone ring, and 3,6-dimethylacridine sublimes from a heated
mixture of actinomycinol and zinc dust.[81] It is also nitrated by sodium nitrate
in hydrochloric acid,[81] presumably at position 7. The formation of the
phenazine derivative 168 shows that one hydroxyl group is adjacent to a
carbonyl group in the ring.[81, 272]

166   167

168

The intermediate 165 was expected to be identical with the dihydrotetra-
methyl ether prepared from actinomycinol, but the synthetic compound had a
melting point 15° below that of the product from the natural material.[272]

Details of actinomycinol and a number of its derivatives appear in Table IX.

(11) *Dihydroacridinones Other Than 9-Acridanone*

There are three other ways to introduce a keto group into the outer rings of the acridine system, and examples of each class are known (Table X). The term "isoacridones" has been used to describe those compounds retaining a hydrogen atom, an alkyl or an aryl group, at position 10, but with the keto group at 1 (**169**) or 3 (**170**); such compounds are tautomers of 1- and 3-hydroxyacridine.

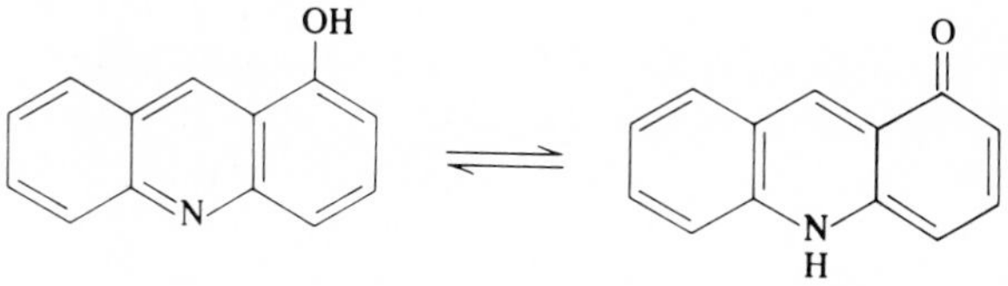

169

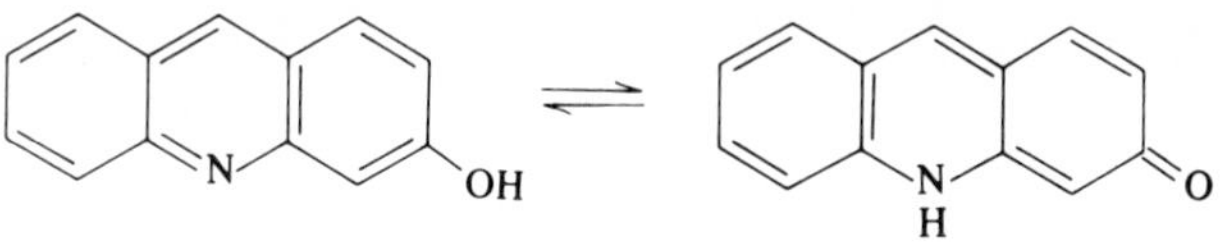

170

Alternatively, the hydrogen atom or a suitable substituent may, in theory, be at some other position in the ring, which is usually position 9, as in the "2-carbazons" (**171**). In fact, no compounds of type **171** (R = H) are known, and the analogous "4-carbazons" remain undiscovered.

The largest category, comprising compounds in which the keto group has been introduced into a reduced acridine in place of a methylene group, is discussed in Chapter V (p. 433).

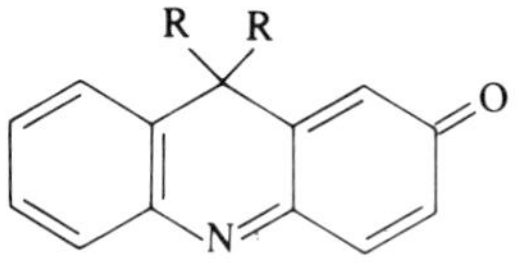

171

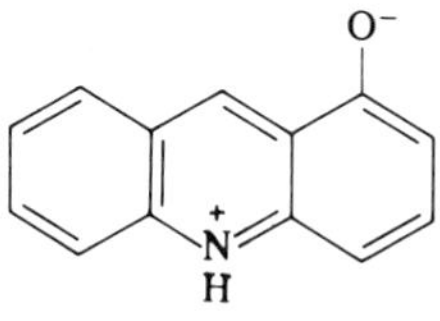

172

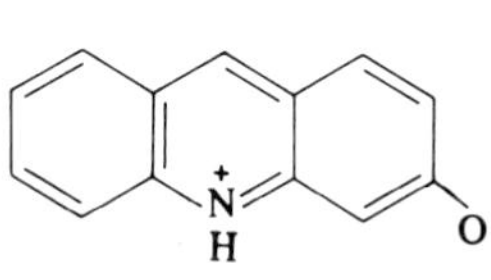

173

In dioxan, the uv spectra (see also Chapter X, p. 645) of the 1-, 2-, 3-, and 4-hydroxyacridines are similar, and like those of the corresponding methoxyacridines.[1] In aqueous alcohol, the spectra of the 2-and 4-hydroxy compounds remain the same, but those of the 1- and 3-hydroxyacridines change as the color of their solutions changes from yellow to blue.[281,622] This suggests the presence of keto tautomers, as solutions of 10-methyl-1,10-dihydroacridin-1-one are also blue.[623] As the dielectric constant increases, the tautomeric equilibrium moves to favor the keto forms,[1, 460] which are stabilized by resonance with the zwitterions 172 and 173.[457] Calculations of the tautomer ratio, by comparison of ionization constants[624] and extinction coefficients of the uv spectrum in alcohol,[625] show that in 170 the 3-keto compound predominates, but in 169 the 1-hydroxyacridine is the major component. The yellow 3-hydroxy-9-phenylacridine even becomes red on grinding,[626] due to the formation of some 3,10-dihydro-9-phenylacridin-3-one, which may be detected spectroscopically,[627] and the equilibrium 174 may be followed by observing the color change of yellow to red on heating.[487]

174

Ultraviolet spectroscopy shows that these tautomers are distinct compounds, not a single resonance hybrid,[487] thus destroying the suggestion of "mesohydric tautomerism" in 9-acridanone derivatives,[471] which allowed only one, intermediate, position for the hydrogen atom in the hydrogen-bonded system.

Amido-hydroxyimino tautomerism is not possible for the 2- and 4-hydroxy-9-acridines, but there are indications that a type of tautomerism is present,[473, 624] possibly between uncharged (175) and zwitterionic (176) forms, rather than structures with hydrogen at another position (171, R = H).

175   176

A direct route to 3,10-dihydroacridin-3-ones[628] is the condensation of *m*-dihydroxybenzenes with dicarboxylic acids and urea. Yields of 50% of **177** (*n* = 0) and **177** (*n* = 2) were obtained when oxalic and succinic acids were used, although a by-product (**178**) was also isolated in the former reaction. *o*- and *p*-Diphenols are also stated to take part in this reaction, even though "isoacridone" formation is not possible from these materials.[628]

10-Methyldihydroacridinones are made by treating 1- and 3-hydroxyl-10-methylacridinium salts with alkali. In contrast to 9-acridanone, they are strong bases with deep colors, [730] and are moderately soluble in water.[629] Under these conditions[629] or on heating in a dry atmosphere[630] 2- and 4-hydroxy-10-methylacridinium salts form betaines (**179**).

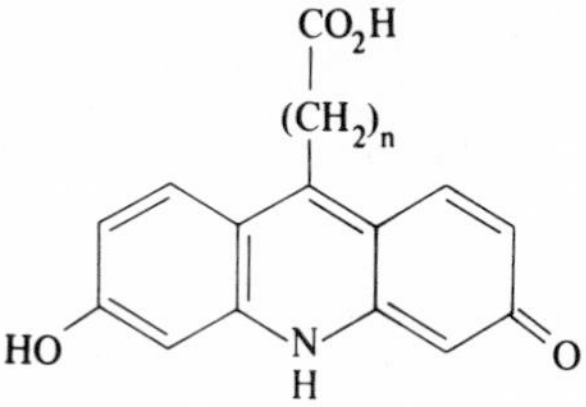

177

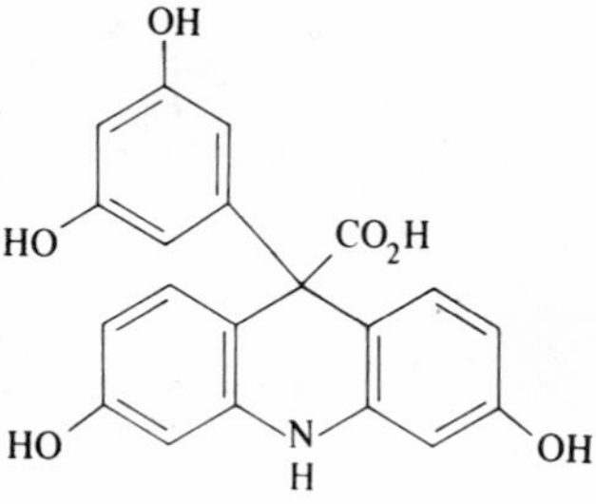

178

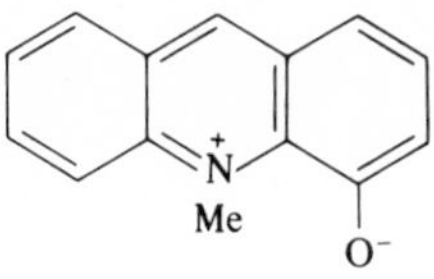

179

Nitration of 10-acetyl- or 10-methyl-9,9-disubstituted acridans[631, 632] gave a series of derivatives with nitro groups in the 2-, 4-, 5- or 7-positions, which were subsequently reduced to their amino analogues. Oxidation of 2-amino-9,9-disubstituted acridans by air or ferric chloride[633] yielded 9,9-disubstituted-2,9-dihydro-2-iminoacridines or "carbazims" (**180**), which may readily be reduced back to the leuco bases.[883] *N*-Substituents, which survive the nitration, may be removed during the oxidation step or by treatment with hydrobromic acid.[632] 7-Amino-2-carbazims may also be made by the action of a Grignard reagent, followed by dilute acid and ferric chloride, on the ester of 4,4'-diaminodiphenylamine-2-carboxylic acid.[883] Ferric chloride oxidation of the 2-dimethylaminoacridan hydrochloride gives the salt **181**.[634, 635]

Mild alkaline hydrolysis of carbazims with aqueous ethanolic sodium hydroxide[636] or carbonate,[637] or even boiling water alone, yields the corresponding "carbazons," or 2,9-dihydroacridin-2-ones (171), which may also be made by the oxidation of a 2-hydroxyacridan.[638] The 7-amino group of a carbazon is replaced by hydroxyl after more vigorous hydrolysis with aqueous sodium hydroxide.[636, 639]

Members of both classes are soluble in organic solvents, may be isolated as crystalline salts,[633] and exhibit a range of colors in solution.[636, 640, 917] All carbazons have additional hydroxy or amino substituents, and the imino-amino and keto-enol tautomeric possibilities, which may be realized in a structure such as 182 as the nature of the solvent is varied, account for this polychromatic behavior.[640]

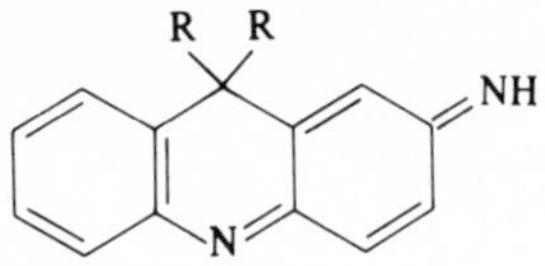

180 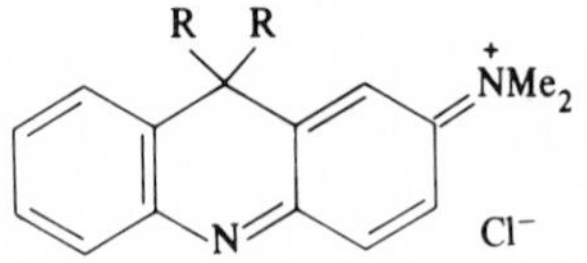   181

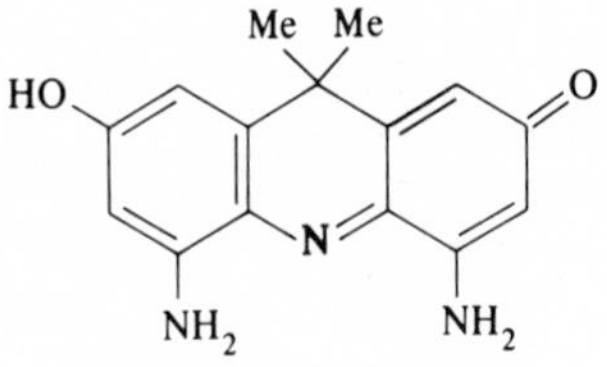

182

The resemblance of the carbazons and carbazims to the indamine and azine dyes (e.g., methylene blue) has been pointed out, but they are much too unstable for use as dyestuffs.

However, a "dyestuff," which occurs as a by-product (10%) in the Liebermann test for phenols[918, 919] has recently been shown to be the carbazon[917] 183a (R = H), and not a hydrated indophenol (183b), as previously suggested.[920] This compound (183a; R = H) and analogues such as 183a (R = Me) are made by treating a mixture of 183b and a phenol (unsubstituted in the para position) with 90% sulfuric acid, or directly and in better yield from phenol and nitrosyl sulfuric acid.[917] The uv spectra of compound 183a and the reduction product 183c are similar, and like those of other carbazons, and their nmr spectra indicate either that a rapid tautomeric equilibrium is established between keto and phenolic groups, or a median form 183d exists in solution.[917]

The lucigenin analogue 184 on alkaline hydrolysis gives the binuclear 10-methyl-3,10-dihydroacridin-3-one 185. The 3,3'-dimethoxy compound is regenerated by treatment of 185 with dimethyl sulfate.

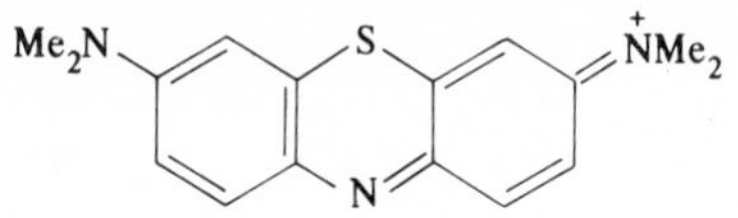

183

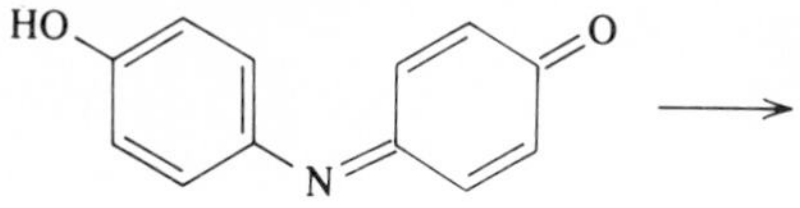

183b

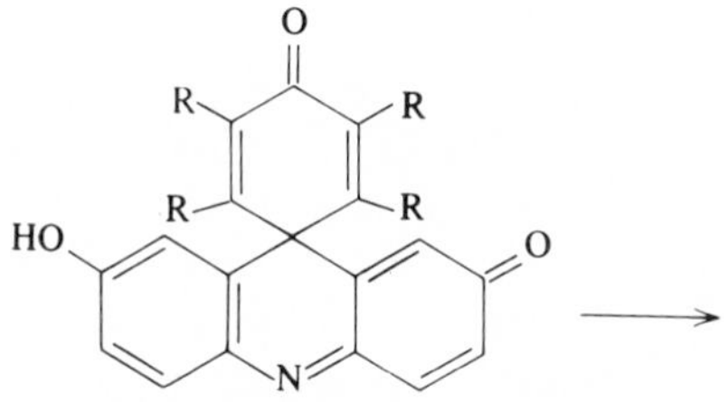

183a

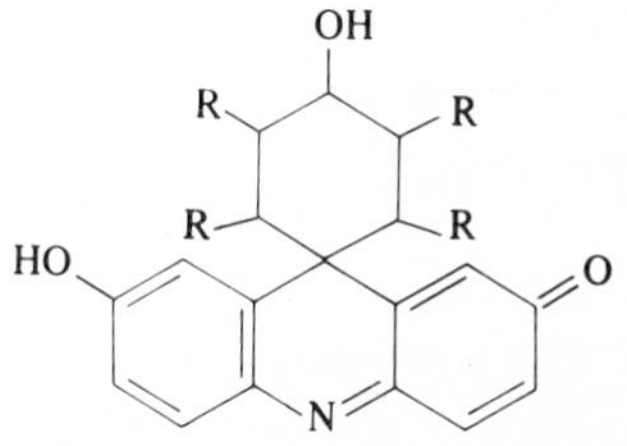

183c

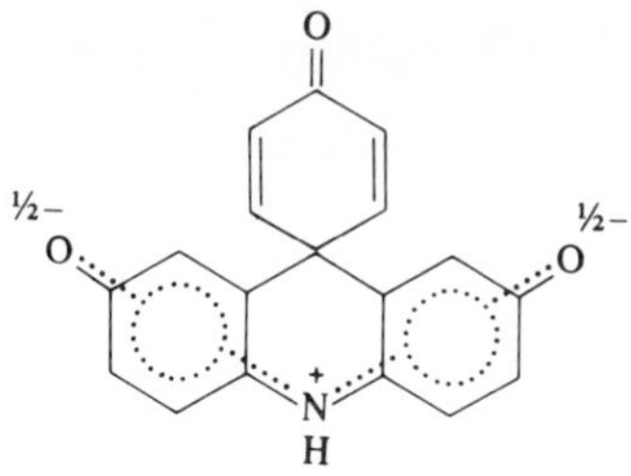

183d

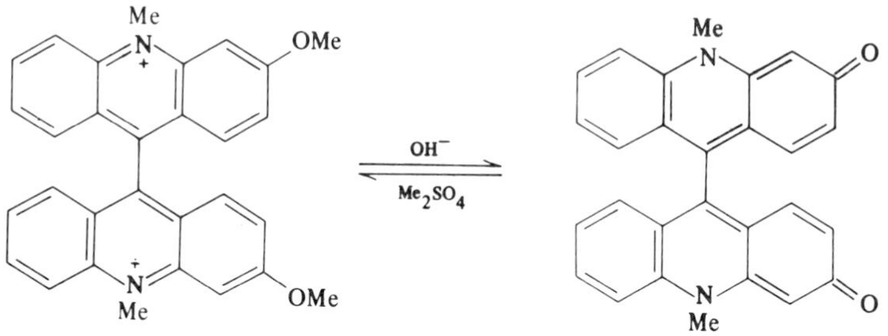

184

185

The corresponding 1-, 2-, and 4-methoxy compounds must be refluxed in hydrobromic acid for demethylation, and the conversion of the 1-methoxy compound is especially difficult, even with this reagent.[493]

D. 9-Acridanthione and 9-Acridanselenone

(1) *9-Acridanthiones*

9-Acridanthione may also be found described in the literature as thio-acridone, or acridine-9-thiol (see also Table XI). Its chemistry is similar in many ways to that of 9-acridanone, as will become apparent in the following review.

The early method of preparation of 9-acridanthione, from 9-acridanone and phosphorus pentasulfide,[136, 641] or a mixture of sulfur and phosphorus, at high temperatures is made more satisfactory by using refluxing pyridine as the solvent. However, the 93% yield after 45 min seems to drop if longer reaction times are employed.[642, 643] Heating acridine and sulfur together at about 190° is a suitable way of making small batches of 9-acridanthione,[644] as in the preparation of ^{14}C-9-acridanthione from neutron-irradiated acridine.[346] Benz[c]- and dibenz[cd]-acridines do not react with sulfur in this way, supporting the suggestion that an N-sulfide is an intermediate.[645]

The most frequently used preparative technique is the replacement of the 9-chloro group of 9-chloroacridines by the thiol group. Sodium sulfide[206, 908] and sodium hydrogen sulfide[646−649] in ethanol and calcium polysulfide in water[283] has each been used, and halo-, methyl-, and methoxy-9-acridanthiones have been prepared this way. 9-Chloroacridine[353, 650] also reacts with thiourea and potassium xanthogenate in phenol to give 9-acridanthione.

Solutions of 9-aminoacridines in ethanolic ammonia,[648] or better pyridine,[651] give up to 90% yields of 9-acridanthiones when treated with hydrogen sulfide at room temperature. Aminoacridine hydrochlorides react more readily in these mildly basic solvents, but the use of acetic acid or alcoholic sodium hydroxide as the solvent inhibits the reaction. Therefore, it seems that the presence of both H^+ ions to protonate the ring nitrogen, and SH^- ions for nucleophilic attack at position 9, are required.

Substituents in the acridine ring do not appear to influence the course of the reaction, but it fails with the amines **186** (R = H, $COCH_3$, Ph, and *m*-nitrophenyl).[648] Electron-withdrawing substituents presumably discourage protonation at position 10, but reasons for failures with unsubstituted amino and anilino groups are not clear. The acridine **186** (R = *p*-nitrophenyl) does react with hydrogen sulfide, but the yields are poor. Protonation at position 10 of the ion **187**, which has no counterpart from 9-(*m*-nitrophenylamino)-acridine, may be involved. Certain aminoacridine hydrochlorides, such as atebrin (Chapter II, p. 127) also react with thiols, under these conditions,[651] but the reaction is slower, possibly because the conformation required for proton transfer **(188)** is less easily accessible than when the sulfydryl proton is available **(189)**.

186   189

187   188

The acridyl group, introduced by treatment with a 9-chloroacridinium salt and cleaved with hydrogen sulfide, has been suggested as a suitable protecting group for the primary amino function.[651]

Acridine 10-oxide undergoes intramolecular rearrangement to 9-acridanone in acetic anhydride (see p. 76), but in acetyl sulfide, an intermolecular reaction takes place, producing only 9-acridanthione.[921] This suggests that O-acetylation is followed by an attack at C_9 by the thioacetate anion, which is a more powerful nucleophile than the acetate ion; subsequently, acetate and acetyl groups are lost.[921]

The treatment of 9-chloro-2-ethoxy-6-nitroacridine with potassium diphenylphosphorodithioate [(PhO)$_2$PSSK] gave a quantitative yield of the corresponding acridanthione, but the diisopropyl analogue gave an 82% yield of the 9-acridyl phosphorodithioate (**190**).[652] Sodium dimethyldithiocarbamate similarly gave 9-acridinyl dimethyldithiocarbamate.[647]

The use of either potassium diphenyl phosphorothiolate,(PhO)$_2$POSK,[653] or ammonium diethyl phosphorothionate [(EtO)$_2$PSONH$_4$][654] in this reaction gives the bimolecular sulfide **191** in better than 95% yield.

190

191

This compound is also formed when 2-ethoxy-6-nitro-9-thiocyanatoacridine is treated with a mixture of diethyl and triethyl phosphites.[922] A similar sulfide is formed from 9-acridanthione and 9-chloroacridine, and it decomposes to 9-acridanone and 9-acridanthione on reflux with hydrochloric acid.[646]

9-Acridanthione is more acidic than 9-acridanone, just as thioamides are more acidic than amides; nevertheless, it does not dissolve in aqueous sodium carbonate.[644] It may be recrystallized from 0.5N sodium hydroxide, and its hydrochloride is hydrolyzed by water. Unlike 9-acridanone, its solutions are not fluorescent, and unlike many thio compounds it does not have an unpleasant smell. It shares with 9-acridanone a low solubility in common solvents.

The melting points of N- and S-alkyl derivatives of 9-acridanthione are well below that of the unsubstituted compound, suggesting that the intermolecular hydrogen bonding, observed in 9-acridanone (p. 200), is also present with the thio analogue. This association has been confirmed in a nitrobenzene solution of 9-acridanthione, and shown to be absent for its S-benzoyl derivative in naphthalene.[655] The weaker nature of the hydrogen bonding, expected when sulfur replaces oxygen,[206] is indicated by the lower melting points of the 9-acridanthiones, when compared with their oxygen analogues.

Acridine-9-thiols (**192**) are not known, as the differently colored 6-chloro-2-methyl-9-acridanthiones, believed to be tautomers,[650] have been shown by ir spectroscopy, X-ray powder photographs[656] and thermogravimetric analysis[657] to be different crystal forms, one of which includes a molecule of water of crystallization. Comparisons of the uv spectrum of 9-acridanthione in a variety of solvents, with those of 10-methyl-9-acridanthione and 9-methylthioacridine, show that in neutral media, thione **193**, and thiol **192** exist together,[923] but structure **193** is the preponderant tautomer.[469] In acid solution, however, the 9-thiolacridinium ion is present.[923] The ir evidence suggests that the crystal is constructed of associated molecules in the thione form, but the large dipole moment (5.2 D),[494] indicates that the dipolar canonical form **194** makes a significant contribution to the resonance hybrid.

192   193   194

A number of 9-acridanthiones form complexes of definite melting point, with metallic ions, which being insoluble in water or alcohols are suitable for gravimetric,[658] as well as spectroscopic,[659] analysis. Such compounds are obtained from the ions copper(II), mercury(II),[647, 658, 659] palladium(II), platinum(IV),[656, 657, 659] silver(I) and gold(III),[650] and the thiol form **192**.

Two features account for most of the differences in the behavior of 9-acridanthione, when compared with 9-acridanone. The polarizable sulfur atom is more open to electrophilic attack than oxygen, and it is readily replaced by oxygen during reactions in aqueous media.

In contrast to 9-acridanones, 9-acridanthiones form S-esters with acetic anhydride,[660] or even glacial acetic acid alone,[661] and with benzoyl chloride[662, 663] and p-toluenesulfonyl chloride in the presence of alkali.[660] Alkyl halides in alcoholic sodium hydroxide[645, 662] or phenol solution,[664] or methyl sulfate in alkali, give S-ethers with 9-acridanthione, some of which have also been prepared by the complementary reaction between 9-chloroacridines and the sodium salts of alkyl thiols.[665] Proof of S-alkylation was furnished by the isolation of 9-acridanone and the respective thiol after the acid hydrolysis of 9-benzylthio- and 9-methylthio-acridines, made from 9-acridanthione.[644, 663] N-Alkylation with aminoalkyl chloride hydrochlorides and sodamide is possible for 9-acridanthiones, just as in the 9-acridanone series.[445]

9-Acridanthione is converted to 9-acridanone by sulfuric acid[663] at 200°, or by oxidation with selenium dioxide in acetic acid,[666] or with alkaline sodium hypochlorite or potassium ferricyanide,[642, 644, 667] This, together with the formation of 9-acridanthiones from acridines and sulfur, constitutes an oxidation of acridine to 9-acridanone,[346] which is not otherwise easily achieved (p. 184). It is possibly a technique of limited applicability. Attempts to oxidize 9-phenylthioacridines to sulfones also resulted in the formation of 9-acridanones,[668] but 9-methylthioacridine was oxidized not to the 10-oxide, but to 9-methylsulfinylacridine 195 with perbenzoic acid.[669] The oxidation of 9-acridanthione with iodine and potassium iodide in aqueous ethanolic sodium hydroxide gives 9,9'-biacridinyldisulfide 196.[669]

195   196

By treating 9-acridanthione with a mixture of phosphorus pentachloride and oxychloride, or bromine and red phosphorus, the expected 9-haloacridine was obtained, but a 9-iodo compound could not be made this way.[663] If the phosphorus halides are not present in excess, however, some 9,9'-biacridinyl sulfide is also produced.[646] 9-(p-Dimethylaminophenyl)- and 9-(p-diethyl-aminophenyl)-acridines, prepared as derivatives of 9-acridanones, are similarly obtained from 9-acridanthiones.

9-Acridanthione,[642] in contrast to 5,6-dihydrophenanthridine-6-thione, re-sists desulfurization; after 13 hr in refluxing ethanolic dimethylformamide with Raney cobalt, only low yields of acridine (6%) and 9,9'-biacridinyl (25%) were obtained.[670] With copper at 280° for 2 hr, the respective yields of these two materials[671] were 25 and 60%.

Unlike 9-acridanone, 9-acridanthione does react directly with primary and secondary amines in phenol to give 9-aminoacridines. A variety of inorganic reagents, of which zinc chloride was the most effective, has been added in this reaction, to improve yields, but the use of the less acidic solvent, butanol, had the opposite effect.[660] These reactions therefore seem to be acid-catalyzed substitutions, comparable to the replacement of the chlorine of 9-chloro-acridine. 9-Alkylthioacridines also undergo this reaction but not as readily as their alkoxy analogues.[283] Again, inorganic reagents facilitate the substitution, possibly through the formation of complexes, of which one (with cuprous chloride) has been isolated.[283] These reactions have little preparative value,[645] as substitution of the more readily available 9-chloroacridines is usually preferred.

The 9-thioketo group survives neither reduction nor electrophilic substitution. 4-Nitro-9-acridanthione was reduced to 4-aminoacridan by sodium amalgam in aqueous bicarbonate,[136] and nitration[663] and chlorination[649] of 9-acridanthione gave nitro- and chloro-9-acridanones, respectively.

The normal syntheses of 10-substituted 9-acridanones, via N-substitution and oxidation of 10-alkylacridinium salts, have no part in the 9-acridanthione series. However, a convenient route to 10-substituted 9-acridanthiones is available by treating 10-alkyl-9-chloroacridinium dichlorophosphate, prepared from the 10-alkyl-9-acridanone and phosphoryl chloride, with alcoholic sodium hydrogen sulfide[541] or aqueous sodium thiosulfate.[550] The probable course of the latter reaction is shown.

Neither 10-methyl- nor 10-ethyl-9-acridanone gives its thioanalogue with refluxing phosphorus pentasulfide, but a 65% yield of 10-phenyl-9-acridanthione is obtained under these conditions.[672] 9-Methylthioacridine does not rearrange to 10-methyl-9-acridanthione if heated alone at 200°,[550] but the addition of methyl iodide has not been tried in this case (cf. p. 197). The 10-methyl compound may, however, be made by heating 10,10′-dimethyl-Δ-9,9′-biacridan with sulfur at 240° in an atmosphere of carbon dioxide.[550] The proposed mechanism[673] for the reaction of 10-substituted 9,9-dichloroacridans with potassium xanthogenate is shown.[440] Thiourea reacts similarly, via a thiouronium salt, to give 10-substituted acridanthiones.

10-Methyl-9-acridanthione is easily converted to 10-methyl-9-methylthio-acridinium iodide in refluxing methyl iodide[643]; this salt gives a 10-methyl-9-sulfonylhydrazone with the corresponding sulfonylhydrazine.[643] 10-Phenyl-9-acridanthione gave neither bimolecular nor desulfurated compounds on heating with copper.[672]

Acridine-9-thiol 10-oxide, prepared from 9-chloroacridine 10-oxide and sodium sulfide, [669] is believed, from ionization data, to exist as approximately equal proportions of the tautomers **197** and **198**.[573]

197   198

It is readily *S*-methylated with diazomethane or alkaline solutions of methyl iodide or sulfate,[669] and gives the oxide of the disulfide (**196**) when oxidized with iodine in alkaline potassium iodide solutions.[669]

(2) 9-Acridanselenone

This compound has attracted little interest in the 70 years since it was first prepared (see also Table XI). A yield of 80% is obtained in the reaction between 9-chloroacridine and sodium hydrogen selenide in alcoholic solution,[646] but heating 9-acridanone with selenium fails to give 9-acridanselenone. Its structure appears to be analogous to that of 9-acridanthione. The replacement of the oxygen atom in 9-acridanone by sulfur to give 9-acridanthione, and by selenium to give 9-acridanselenone, results in a characteristic deepening in color and a lowering of the melting point.

9-Acridanselenone is not soluble in dilute acids, but hydrolysis to 9-acridanone takes place in refluxing alcoholic sodium hydroxide. Like 9-acridanthione, 9-acridanselenone is alkylated at the exocyclic atom, rather than at the ring nitrogen atom, when treated with methyl iodide or benzyl chloride in the presence of sodium ethoxide, as shown by the hydrolysis of these alkyl derivatives to 9-acridanone in refluxing alcoholic hydrochloric acid.

10-Substituted 9-acridanselenones have been made by treating 10-alkyl-9-chloroacridinium chlorophosphates with potassium selenosulfate, which gave better yields than potassium hydrogen selenide.[550] They are deeply colored compounds, like the parent 9-acridanselenone, in contrast to the pale 9-alkylselenylacridines.

TABLE I. 9-Acridanone and Derivatives

9-Acridanone, mp 356-358; 362-365 (dec) recrystallized from acetic acid (glacial or aq), amyl acetate, aniline acetate, benzyl alcohol, *m*-cresol, 95% ethanol, isoamyl alcohol, or sublimes

Preparation	Yield %	Ref.
Hypochlorite oxidation of 9-acridanthione	100	346
Cyclization of 2-amino-2'-fluorobenzophenone	99	674
Cyclization of 2,2'-diaminobenzophenone	98	391
Cyclization of diphenylamine-2-carboxylic acid; H_2SO_4;	91-96	535, 536, 675
$POCl_3$ and hydrolysis	63-83	9, 29, 469
Hydrolysis and cyclization of methyl *N*-benzoyldiphenylamine-2-carboxylate; PPA	88-93	140
Decamethylene diiodide on 9-acetamidoacridine	84	286
Rearrangement of acridine 10-oxide	82	358
Lehmstedt-Tanasescu reaction	40-90	288, 292, 294 301, 333, 357
Heat on *N*-benzoyldiphenylamine-2-carboxylic acid	70	111
p-Nitrosodimethylaniline and HCl on 9-methylacridine	64	354
Lead oxide oxidation of 2-aminobenzophenone	50	395
Ozone on 10-hydroxy-9-acridanone	40	357
Irradiation of 3-phenylbenztriazin-4-one	38	398
Dehydrogenation of 1,2,3,4-tetrahydro-9-acridanone	27	372
Dehydrogenation of 2-aminobenzophenone or diphenylamine 2-carboxaldehyde	Trace	396, 594
Hydrolysis of *N*-(9-acridinyl)acetamide	?	404
Oxidation of 9-allylacridan	?	352
Alkaline ferricyanide on 9-acridanthione	?	642

(Table Continued)

Table I   (Continued)

Derivatives of 9-Acridanone, mp (°C)	Yield %	Ref.
9-(*p*-Aminophenyl) 269;		529
9-(*p*-Dimethylaminophenyl) 290;		529, 528, 575, 642
9-(*p*-Diethylaminophenyl) 197, 239;		95,258
10-Mercuriphenyl 238-239;		924
9-Thiosemicarbazone 205;		
Tosylhydrazone hydrochloride infusible		
10-Hydroxy 256-257, ex. EtOH, HOAc,		
Lehmstedt-Tanasescu		301,520
Alkaline H_2O_2 on acridine 10-oxide		307, 347
9-Chloroacridine 119-120 $POCl_3$	65-99	6, 9, 28, 676
10-oxide 209 (dec), ex. dioxan	80-90	35, 677, 678
9-Bromoacridine 116, ex. petrol	?	128, 663
10-oxide 174	80-90	677
9-Iodacridine 10-oxide 145, ex. acetone	80-90	677

TABLE II. *C*-Substituted 9-Acridanones

Note: All 9-acridanones and 9-chloroacridines containing one or two *types* of substituent and further-substituted compounds that direct the reader to references not mentioned elsewhere in the text are included in this table. Polysubstitution with a single functional group qualifies as one *type* of substituent.

Substituents on 9-acridanone	mp (°C)	Recrystallizing solvents and derivatives, mp (°C)	Yield (%)	Cyclizing reagent or conditions	Ref.
2-Acetamido-6-chloro	(b) 242–243			POCl₃	244
2(4-Acetamidobenzene-sulfonamido)-7-methoxy	(b) 243–244			POCl₃	679*
1-Amino	(a) 290–292	MeOH	3–5	H₂SO₄	147
2-Amino	(a) 300–302	EtOH, AmOAc; acetyl > 350,[642] benzoyl > 350,[642] urethane 355 (dec)[19]	93–95	H₂SO₄	76, 535, 536
				Redn of NO₂	589
	(b)	Urethane 183[19]			
3-Amino	(a) 300–302	EtOH, EtOAc-petrol	90	H₂SO₄	25, 124, 147
		H₂O	30	L-T	300
			17	Redn of NO₂	580
4-Amino	(a) 355 (dec)	EtOH	100	Redn of NO₂	118, 585

*The cited reference contains details of further compounds of a similar nature.

(a) Indicates 9-acridanone;
(b) Indicates 9-chloroacridine;
L-T: the Lehmstedt-Tanasescu synthesis (Section 1.C (1)).

(Table Continued)

245

Table II. (Continued)

Substituents on 9-acridanone	mp (°C)	Recrystallizing solvents and derivatives, mp (°C)	Yield (%)	Cyclizing reagent or conditions	Ref.
3-Amino-6-bromo	(a) 345-350	aq EtOH	37	L-T	300
2-Amino-7-chloro	(a) 317	EtOH		Redn of NO_2	401
3-Amino-6-chloro	(a) >300	aq EtOH	70	Redn of NO_2	130
7-Amino-1-chloro-4-nitro	(a) 300	o–$C_6H_4Cl_2$		H_2SO_4	217
2-Amino-1,4-diethoxy	(a)	Acetyl, diazonium salt		PCl_5	86, 87
2-Amino-1,4-dimethoxy	(a)	Acetyl		PCl_5	86, 87
4-Amino-3, 6-dimethyl-1, 2, 5-trimethoxy	(a) 183-188 (dec)			Redn of NO_2	119*
6-Amino-2-dimethylamino	(a) 312-314		94	Redn of NO_2	147
1-Amino-2, 4-dinitro	(a) 319	$PhNO_2$		Subs of 1-Cl	252
6-Amino-2-methoxy	(a) 290 (dec)	$PhNO_2$		Redn of NO_2	584
2-Amino-6-nitro	(a) 370-375 (dec)		90	H_2SO_4	147
2-Amino-7-nitro	(a)		86	Redn of NO_2	78
3-Amino-5-nitro	(a) 314-316		87	H_2SO_4	148
3-Amino-6-nitro	(a) > 400			H_2SO_4	147
			8	NH_3 on 6-I	580
			16	NH_3 on 6-Br 10-oxide	579
			20	NH_3 on 6-Br	579
			3.5	NH_3 on 6-Cl	579
6-Amino-1-nitro	(a) 366-368		80	H_2SO_4	147

246

Table II. (Continued)

Substituents on 9-acridanone	mp (°C)	Recrystallizing solvents and derivatives, mp (°C)	Yield (%)	Cyclizing reagent or conditions	Ref.
6-Amino-2-nitro	(a)		86-93	H_2SO_4	77
7-Amino-1-nitro	(a) 340-342		94	H_2SO_4	147
2-Amino-7(?)-sulfonic acid	(a)		88	H_2SO_4	147
3-Amino-7(?)-sulfonic acid	(a) >400		31	H_2SO_4	147
4-Aminoacetic acid		Me ester 237-238, nitroso cpd 180(dec); Et ester 195, nitroso cpd 185-186		Br-acetic acid on 4-NH_2	136
2-Aminomethyl	(a) 340			H_2SO_4	680
4-Aminomethyl	(a) 212-216	aq EtOH-NH_3; HCl 295 (dec) acyl derivatives	61	NH_3 on 4-CH_2Br-9-Cl	681
4-(p-Aminophenyl)	(a) >300	EtOH; HCl salt, dehydro 280		Red[n] of NO_2	22
2-Anilino	(a) 303-305			H_2SO_4	913
1-Anilino-2,4-dinitro	(a) 290-291	Xylene-PhNO_2		Substitution of 1-Cl	252*
1-Anilino-4-methyl	(a) 140-160				105
2-Arsonic Acid	(a) >400	HOAc; 9-DMP 230-232,[682]			589

(Table Continued)

247

Table II. (Continued)

Substituents on 9-acridanone	mp (°C)	Recrystallizing solvents and derivatives, mp (°C)	Yield (%)	Cyclizing reagent or conditions	Ref.
1-Bromo	(a) >310	HOAc; 9-DMP 266-267	15	H_2SO_4	590
2-Bromo	(a) 389 (dec)	HOAc, AmOAc	64	H_2SO_4	226, 246, 536
			72	L-T	308
	(b) 136	9-DMP 243-245[139,683]	100	POCl$_3$	128, 139
3-Bromo	(a) >350	9-DEP 212[683]		L-T	288, 317, 925
		aq HOAc, PhH	40	L-T	590
		DMP 231-232		Diazonium	
	(b) 170-171			POCl$_3$	684
4-Bromo	(a)			H_2SO_4	590
	(b) 98-99	9-DMP 297	92	POCl$_3$	128, 685
2-Bromo-6-chloro	(a) 414-416	Petrol, aq EtOH-NH$_3$	87	L-T	308
6-Bromo-2-chloro	(a)	HOAc; DMP 236-237		L-T	568
2-Bromo-3,4-dimethyl	(b) 175-176	10-Oxide		POCl$_3$	686
3-Bromo-7-dimethylamino	(b) 199-202			POCl$_3$	687
4-Bromo-1-ethyl	(b) 121-122		70	POCl$_3$	125
2-Bromo-7-fluoro	(b) 135		66	POCl$_3$	208
2-Bromo-7-iodo	(a) 360				
1-Bromo-2-methoxy	(b) 216-217			POCl$_3$	688
1-Bromo-7-methoxy	(b) 142-145			POCl$_3$	123
3-Bromo-7-methoxy	(b) 184			POCl$_3$	689
3-Bromo-2-methoxy	(b) 192		25	POCl$_3$	123
6-Bromo-2-methoxy	(b) 158-160			POCl$_3$	690, 691
1-Bromo-4-nitro	(a) 252-254			H_2SO_4	217, 692*

Table II. (Continued)

Substituents on 9-acridanone	mp (°C)	Recrystallizing solvents and derivatives, mp (°C)	Yield (%)	Cyclizing reagent or conditions	Ref.
2-Bromo-6-nitro	(a) >360		56	Bromination	18, 557
	(b) 236-237	PhH	75-85	POCl$_3$	200, 693
2-Bromo-7-nitro	(a) >370	DMP 284	70-80	POCl$_3$, Bromination	683
3-Bromo-6-nitro	(b) 231-232		68-70	POCl$_3$	18, 64, 694
	(a) >300 (210-212)		43-100	L-T	297, 300, 695
4-Bromo-1-nitro	(a) 305		55	H$_2$SO$_4$	578
2-Bromo-7-sulfonamido	(a) 300			Sulfonation, PCl$_5$, NH$_3$	574
2-(n-Butoxy)-6-nitro	(b) 159-160	Petrol	69	POCl$_3$	14
2-(n-Butyl)	(a) 231-232		64	POCl$_3$	8, 164
	(b) 49-50		70		
2-sec-Butyl	(b) Liquid		58	POCl$_3$	151
2-tert-Butyl	(b) 69-70	Petrol	64	POCl$_3$	151
4-sec-Butyl	(a)		10	POCl$_3$	8, 164
	(b) 68-70				
2-sec-Butyl-6-nitro	(a) 335	Toluene, EtOH	63	POCl$_3$	151
	(b) 149-150	Petrol	55		
2-Carboxy	(a) >370	Me ester 339		H$_2$SO$_4$	118, 188
3-Carboxy	(a) >360	Me ester 319-320		POCl$_3$	121
				Hydrolysis of CN	588

(Table Continued)

249

Table II. (Continued)

Substituents on 9-acridanone	mp (°C)	Recrystallizing solvents and derivatives, mp (°C)	Yield (%)	Cyclizing reagent or conditions	Ref.
4-Carboxy	(a) 325	AmOAc, EtOH Me ester 172	90 76	H_2SO_4 $POCl_3$-xylene	118, 188, 536 21
				Alkaline H_2O_2 on 9-CO_2H	356
4-Carboxy-5-chloro	(a) 352			Alkaline H_2O_2 on 9-CO_2H	356
5-Carboxy-1-chloro	(a) 336			Alkaline H_2O_2 on 9-CO_2H	356
5-Carboxy-2-chloro	(a) 330			Alkaline H_2O_2 on 9-CO_2H	356
5-Carboxy-3-chloro	(a) 360			Alkaline H_2O_2 on 9-CO_2H	356
5-Carboxy-1,4-dichloro	(a) 334			Alkaline H_2O_2 on 9-CO_2H	356
2-Carboxy-1,3-dimethoxy	(b)	Cyclohexane; Me ester 155-156	98	$POCl_3$	455
5-Carboxy-2,4-dimethyl	(a) 354			Alkaline H_2O_2 on 9-CO_2H	356
7-Carboxy-2,3-dimethyl	(a) 500 (dec)		88	$POCl_3$	200
2-Carboxy-7-methyl	(a) 500 (slow dec)		86	$POCl_3$	200

250

Table II. (Continued)

Substituents on 9-acridanone	mp (°C)	Recrystallizing solvents and derivatives, mp (°C)	Yield (%)	Cyclizing reagent or conditions	Ref.
4-Carboxy-5-methyl	(a) 315			Alkaline H_2O_2 on 9-CO_2H	356
5-Carboxy-2-methyl	(a) 343		84	$POCl_3$ Alkaline H_2O_2 on 9-CO_2H	200 356
5-Carboxy-1-trifluoro-methyl	(a) 345			Alkaline H_2O_2 on 9-CO_2H	356
1-Chloro	(a) 360 (b) 116-117	HOAc; DMP 252-253	34	$POCl_3$	30 30, 121
2-Chloro	(a) 398	HOAc, aq DMF, $PhNO_2$, AmOAc; N-tosyl 172[674] N-oxide, [297,299] DMP 265[299] DEP 236-238[308]	90	H_2SO_4 L-T Heat	85, 246, 536 288, 308, 299 138
3-Chloro	(b) 147 (a) 390	aq, EtOH Aniline; N-oxide 167-169[351,559] [560,562,697] DMP 239[288,925]	95-100 90 45 60	Sternbach $POCl_3$ L-T Diazonium Heat N-benzoyl Hydrolysis of 9-PhO	674 649, 696, 906 331 591 111 545

251

(Table Continued)

Table II. (Continued)

Substituents on 9-acridanone	mp (°C)	Recrystallizing solvents and derivatives, mp (°C)	Yield (%)	Cyclizing reagent or conditions	Ref.
4-Chloro	(b) 169-170	aq EtOH	94	$POCl_3$	30, 121, 331, 698
	(a) 360	AmOAc, HOAc, MeOH-CH_2Cl_2; DMP 279[454]		H_2SO_4	246, 536
	(a)			Irradn of Ph-anthranil	312
1-Chloro-4-(2-diethyl-amino)ethoxy	(a) 115	EtOH		PPA	699*
4-Chloro-1-(3-diethyl-aminopropyl)amino	(a)	HCl 238-239		Substitution of 1-Cl	596
6-Chloro-2,3-dimethoxy-7-nitro	(b) 240-241			$POCl_3$	41*
6-Chloro-2-dimethyl-amino	(a) >300		96	H_2SO_4, $POCl_3$	21
	(b) 206-208		65	$POCl_3$	147
			35	$POCl_3$-xylene	21
1-Chloro-2,4-dinitro	(a) 266	$PhNO_2$		$POCl_3$, H_2SO_4	252
2-Chloro-4,5-dinitro-7-methyl	(a) 250	EtOH		HNO_3 on amino-hydroxy-benzophenone	329
2-Chloro-5,7-dinitro	(a) 350		75	H_2SO_4	205
4-Chloro-5,7-dinitro	(a) 345		75	H_2SO_4	205
6-Chloro-2-ethoxy	(b) 162-163		76	$POCl_3$	39, 174, 690
6-Chloro-2-ethyl	(b) 102			PCl_5	700

252

Table II. (Continued)

Substituents on 9-acridanone	mp (°C)	Recrystallizing solvents and derivatives, mp (°C)	Yield (%)	Cyclizing reagents or conditions	Ref.
6-Chloro-2-fluoro	Unstable		50	$POCl_3$	701
6-Chloro-2-hydroxy	(a) (b) 222	EtOH; HCl	72 95	H_2SO_4 Dealkylation	80 39, 691
6-Chloro-2-(2-hydroxy-ethoxy)	(a) >300	EtOH		Hydrolysis 9-Cl	39
2-Chloro-7-iodo	(a) 365 (b) 224-225	HOAc PhH		$POCl_3$	688
1-Chloro-6-methoxy	(a) 310 (b) 168-170	HOAc PhH		$POCl_3$	702
1-Chloro-5-methoxy	(a) (b) 183-184	HCl 294-295	75	$POCl_3$	703
2-Chloro-6-methoxy	(a) >300			$POCl_3$, heat imino-ether	399, 702
2-Chloro-7-methoxy	(b) 195-196 (a) >300 (b) 212-213	PhH, CHCl$_3$-hexane PhH, hexane	20 100	L-T Heat Hydrolysis 9-PhO $POCl_3$	309 138 525 703, 704, 705 706
3-Chloro-2-methoxy	(b) 198		7	L-T $POCl_3$	309 707

(Table Continued)

253

Table II. (Continued)

Substituents on 9-acridanone	mp (°C)	Recrystallizing solvents and derivatives, mp (°C)	Yield (%)	Cyclizing reagents or conditions	Ref.
3-Chloro-5-methoxy	(b) 189-191		70-93	POCl$_3$	66, 706
3-Chloro-6-methoxy	(b) 222-223	PhH, EtOH	35-54	POCl$_3$	702, 706, 708
5-Chloro-2-methoxy	(b) 157-158		85	POCl$_3$	251, 706
			84	POCl$_3$-PhCl	20
5-Chloro-3-methoxy	(a) 358-359	HOAc		POCl$_3$	709
6-Chloro-1-methoxy	(b) 178-180	EtOH-PhH			
	(b) 183-185				
6-Chloro-2-methoxy	149-150	EtOH	36	POCl$_3$	706, 708
	(a) 364-366	MeOH; 10-oxide 236[35], 559,562,678 697,710	61	POCl$_3$	42,50,690
		Acetyl 274-276 (dec)[110]	5-7	Cyclization of acid chloride	57
			99	SOCl$_2$-PhH	80
			62-73	Heat *N*-benzoyl	111
				Hydrolysis 9-NH$_2$	56
	(b) 164		87	POCl$_3$	20,67,174, 703,711,712
7-Chloro-1-methoxy	(b) 178-180	PhH	20	POCl$_3$	702
8-Chloro-2-methoxy	(b) 182		24-37	POCl$_3$-PhCl	20,713
1-Chloro-7-methoxy-4-methyl	(b) 153-154	CHCl$_3$	70	POCl$_3$	714

Table II. (Continued)

Substituents on 9-acridanone	mp (°C)	Recrystallizing solvents and derivatives, mp (°C)	Yield (%)	Cyclizing reagents or conditions	Ref.
2-Chloro-4-methoxy-1-methyl	(a) 193-195	PhCl		PPA	699*
6-Chloro-2-methoxy-4-nitro	(b) 272-273		80-90	$POCl_3$	254
6-Chloro-2-methoxy-7-nitro	(b) 211	Dichloroethane		$POCl_3$	715, 716
1-Chloro-4-methyl	(a) 298		75	H_2SO_4	153
2-Chloro-6-methyl	(b) 200			PCl_5	700
3-Chloro-5-methyl	(b) 146-147		81	$POCl_3$	676
3-Chloro-6-methyl	(b) 206	Acetone; 10-oxide 172-174	20	$POCl_3$	559,562,697, 717
4-Chloro-5-methyl	(a) 197-198		86	Heat	711
6-Chloro-2-methyl	(b) 148-150		40	$POCl_3$	174,700
6-Chloro-2-methylthio	(b) 182-183		73	$POCl_3$	59,718
1-Chloro-4-nitro	(a) 250-251			H_2SO_4	153, 217
	(b) 247-248		95	$POCl_3$	150
1-Chloro-6-nitro	(a)	Pyridine	95	$POCl_3$	130
1-Chloro-7-nitro	(a) >300	DMP 236	20	L-T	130
	(b) 201-202	PhH	25-78	$POCl_3$	130, 719
2-Chloro-6-nitro	(a) >380		86	PCl_5-$AlCl_3$	293,500
	(b)			$POCl_3$	926
2-Chloro-7-nitro	(a) 415		80	$POCl_3$	401,500,720

(Table Continued)

255

Table II. (Continued)

Substituents on 9-acridanone	mp (°C)	Recrystallizing solvents and derivatives, mp (°C)	Yield (%)	Cyclizing reagents or conditions	Ref.
3-Chloro-5-nitro	(b) 208-209		75		720,721,722
	(a) 312		75	H_2SO_4	723
	(b) 205		85	$POCl_3$	723, 926
3-Chloro-6-nitro	(a) 300	DMP 250,[724]	40	L-T	130,297
		DEP 236,[724]			
		10-oxide 200 [130,562]		$POCl_3$	5,697,926
4-Chloro-1-nitro	(a) 320		64	H_2SO_4	578
5-Chloro-2-nitro	(a) 320		90	Acid chloride-AlCl$_3$	500
5-Chloro-3-nitro	(a) 340		75	PCl_5-PhH	500
6-Chloro-2-nitro	(b) 227	PhH	22	$POCl_3$	130, 719
3-Chloro-2-phenyl	(b) 164		100	$POCl_3$	22
6-Chloro-2-phenyl	(b) 240-242	PhH; 10-oxide 209-211[562]	83	$POCl_3$	559, 697
2-Chloro-7-sulfonamido	(a) 300			Sulfonation PCl$_5$-NH$_3$	574
6-Chloro-2-sulfonamido	(a) 300			Sulfonation PCl$_5$-NH$_3$	574
6-Chloro-1-trifluoromethyl	(b)			$POCl_3$	68
3-Chloro-6-trifluoromethyl	(b) 159-160	PhH	70	$POCl_3$	68
1-Cyano	(b)			$POCl_3$	17
2-Cyano	(a) >360				
	(b) 186		100	$POCl_3$	17, 23
3-Cyano	(a) >360	aq pyridine	90	Diazonium	23, 588

Table II. (Continued)

Substituents on 9-acridanone	mp (°C)	Recrystallizing solvents and derivatives, mp (°C)	Yield (%)	Cyclizing reagents or conditions	Ref.
6-Cyano-2-ethoxy	(b) 192-194		100	POCl$_3$	17
6-Cyano-2-methoxy	(b) 224-226		80	POCl$_3$	725
6-Cyano-2-methoxy	(b) 228-230		77-100	POCl$_3$	17,718
7-Cyano-2-methoxy	(b) 270-272		96	POCl$_3$	17
1,6-Diamino	(a) 302-304		79	Redn	147
1,7-Diamino	(a) 314-316		74	H$_2$SO$_4$ and redn	147
2,4-Diamino	(a) 222-223, 300			Redn	147
2,6-Diamino	(a) 358-360	aq pyridine	51-84	H$_2$SO$_4$	77, 147
			33	Redn	147
2,7-Diamino	(a) 350-352		54	Redn	24, 726
	300-310 (dec)		66	NH$_3$ on dibromo	18
3,6-Diamino	(a) 368-370		41	H$_2$SO$_4$	147
			97	Redn of tetranitro-benzophenone	393,394
			54	NH$_3$ on di-Cl	579
			56	NH$_3$ on di-Br	579
4,5-Diaminomethyl-2,7-dimethyl		di-HBr		H$_2$SO$_4$	232
1,3-Dibromo	(a) >300	HOAc		POCl$_3$	63
1,4-Dibromo	(a) 232-233		55	H$_2$SO$_4$	153
2,6-Dibromo	(a) >350	DMP 231-232[590]		Diazonium, L-T	317
2,7-Dibromo	(a) 438	EtOH; DMP 273, N-oxide		Bromination	18, 206

(Table Continued)

257

Table II. (Continued)

Substituents on 9-acridanone	mp (°C)	Recrystallizing solvents and derivatives, mp (°C)	Yield (%)	Cyclizing reagents or conditions	Ref.
3,6-Dibromo	(b) 219, 249-250	PhH, PhH-EtOH	100	POCl₃	18,139,511,649
	(a)			L-T	300
2,7-Dibromo-4,5-dichloro	(a) 273	HOAc		Halogenation	206
4,5-Dibromo-2,7-dichloro	(a) 292	HOAc		Halogenation	206
1,3-Dibromo-7-methoxy	(a) >300	HOAc		POCl₃	63
2,7-Di-tert-butyl	(b) 124-125	Petrol	61	POCl₃	151
1-(2-Di-n-butylamino-ethyl)amino	(a) 159	aq EtOH; HCl.H₂O 198-200	39	Substitution of 1-Cl	576*
1,3-Dichloro	(a) 310	Pyridine	85-88	H₂SO₄	576, 577
1,4-Dichloro	(a) 268		64	H₂SO₄	153
1,6-Dichloro	(b) 144-146	Heptane	25	POCl₃	727
2,4-Dichloro	(a) >360			H₂SO₄	246
2,6-Dichloro	(a) 422-424	HOAc; DMP 245-247[568], 10-oxide[568]	96	L-T	308, 323
2,7-Dichloro	(b) 201-203			POCl₃	691, 707
	(a) 434	HOAC; DMP 263	9	POCl₃, chlorination of diazonium	206,401,514
3,5-Dichloro	(b) 207	PhH-EtOH		POCl₃	649
3,6-Dichloro	(b) 177-178	PhH		POCl₃	727
	(a) >360	aq pyridine; 10-oxide 236-238[559,562,697,717]	76	Cyclization of diaminobenzo-phenone	371

Table II. (Continued)

Substituents on 9-acridanone	mp (°C)	Recrystallizing solvents and derivatives, mp (°C)	Yield (%)	Cyclizing reagents or conditions	Ref.
	(b) 223-224	Heptane, PhCl	55	L-T	392
3,7-Dichloro	(b) 201-203	DMP 240	22-74	$POCl_3$	68, 392, 707 727
1,2-Dichloro-7-methoxy	(b) 179	Acetone		L-T	331
1,3-Dichloro-2-methoxy	(b) 159	Toluene		$POCl_3$	728
2,3-Dichloro-7-methoxy	(b) 192-193	Toluene		$POCl_3$	729
3,7-Dichloro-2-methoxy	(a)			$POCl_3$	705, 728
1,6-Dichloro-4-methyl	(a)			$POCl_3$	707
1,2-Dichloro-4-nitro	(a)		95	H_2SO_4	576
1,6-Dichloro-6-nitro	(a) 257-258			H_2SO_4	217, 692*
1,7-Dichloro-4-nitro	(a) 300-301	o-$C_6H_4Cl_2$		H_2SO_4	217, 692*
4-(2-Diethylamino)ethoxy	(a) 62	PhH-petrol; 2HCl 236-238, MeBr 229-230		H_2SO_4	217, 692* 603*
2-(2-Diethylamino)ethoxy-6-nitro	(b) 159-160			PCl_5	730
4-Diethylaminomethyl	(b) 79-80		90	NH_3 on 4-CH_2Br	581*
2,7-Difluoro	(b) 179			$POCl_3$	208
1,3-Dihydroxy	(a) 370, 345 (dec)	PhH, acetone, acetone-EtOAc; 3-acetyl 286-287, 1,3-diacetyl	20-30	Anthranilic acid and phloroglucinol	412, 604 899

(Table Continued)

Table II. (Continued)

Substituents on 9-acridanone	mp (°C)	Recrystallizing solvents and derivatives, mp (°C)	Yield (%)	Cyclizing reagents or conditions	Ref.
1,4-Dihydroxy	(a) 300 (dec), 282	205-207, benzoyl 295-297; HBr	70	Hydrolysis of OMe	595
3,6-Dihydroxy	(a) >320		85	H$_2$SO$_4$ on diamino	393
1,4-Dihydroxy-3,4-methylene-dioxy (Norxanthevodine)	(a) 272-273			Heat HCl salt of MeO	441
Di-(2-hydroxyethyl)-aminomethyl	(a) 182-183		55		681*
2,7-Diiodo	(a) 350-355			POCl$_3$	688
	(b) 224-225	PhH-EtOH			688
1,3-Dimethoxy	(a) 286-287 (dec)		90	PPA	731
				Methylation	604
	(b) 162			POCl$_3$	597
1,4-Dimethoxy	(a) 221-222		92	PPA	414,731
	(b) 148-149	With chlorination products	60-73	POCl$_3$	414,731,732
				PCl$_5$	98
2,3-Dimethoxy	(a) 298		85	PPA	731
				PCl$_5$-CS$_2$	3
	(b) 186-187		79	POCl$_3$	931
				PCl$_5$-PhH	3

260

Table II. (Continued)

Substituents on 9-acridanone	mp (°C)	Recrystallizing solvents and derivatives, mp (°C)	Yield (%)	Cyclizing reagents or conditions	Ref.
2,4-Dimethoxy	(a) 260-261		99	PPA	731
	(b) 189-190	PhH-hexane	50	Pyrolysis of Ph-anthranil	309
2,7-Dimethoxy	(a) >300		78	POCl$_3$	731
	(b) 228-229			POCl$_3$	733
3,4-Dimethoxy	(a) 225	Toluene	80	POCl$_3$	733
			100	PPA	731
3,6-Dimethoxy	(a) >320		77	PCl$_5$-CS$_2$	508
	(b) 184			Methylation	393
				POCl$_3$	393
4,5-Dimethoxy	(a)			PCl$_5$	98
2,3-Dimethoxy-l-hydroxy (xanthoxoline)	(a)	Acetyl 238-241		Hydrolysis	441
1,3-Dimethoxy-4-propionic acid	(a)	EtOH, CHCl$_3$; Me ester 244-245	83	PPA	83*
1,2-Dimethoxy-3,4-methylenedioxy	(a) 243-244	EtOH	92	POCl$_3$	220, 221
	(b) 218-219 (dec)	EtOH	88	POCl$_3$	220,221
1,4-Dimethoxy-2,3-methylenedioxy (xanthevodine)	(a) 217-218	EtOH; HCl 137 (dec) Picrate 145-146	99	POCl$_3$	441
	(b) 187-188	EtOH-DMF	88		220, 221, 734

(Table Continued)

261

Table II. (Continued)

Substituents on 9-acridanone	mp (°C)	Recrystallizing solvents and derivatives, mp (°C)	Yield (%)	Cyclizing reagents or conditions	Ref.
1,2-Dimethoxy-6-nitro	(b) 246-248		72	$POCl_3$	14
1,4-Dimethoxy-6-nitro	(a) 265-266		94-95	L-T, PPA	927
	(b) 192-193	aq dioxan	96	$POCl_3$	927
2,3-Dimethoxy-6-nitro	(b) 252-253 (dec)	10-oxide 265 (dec)[36]	70-72	$POCl_3$	14,243,735
1,2-Dimethyl	(a) 297	aq HOAc	80	H_2SO_4	122
	(b) 136-137	Cyclohexane		$POCl_3$	686
1,4-Dimethyl	(b) 112-114		80	$POCl_3$	732
2,3-Dimethyl	(a) 310	HOAc	80	H_2SO_4	122
				$POCl_3$	686
2,4-Dimethyl	(a) 307	Tosylhydrazone HC1, infusible	85	H_2SO_4	507
	(b) 114	PhH	71	Heat	138
2,6-Dimethyl	(a) 324-326			$POCl_3$	19, 507
	(b) 128		90	PPA	81
				PCl_5	700
2,7-Dimethyl	(a)>300	HOAc, dioxan	55	H_2SO_4	232,736
4,5-Dimethyl	(a) 234-235	PhH-petrol	6	$POCl_3$	10, 23
	(b) 149-151	Petrol	89		10, 23
3,6-Dimethyl-4-hydroxy	(a)		88	Hydrolysis	81
3,6-Dimethyl-4-methoxy	(a) 260-261		77	PPA	81
3,6-Dimethyl-4-nitro-1,2,5-trimethoxy	(a) 195-197		82	PPA	119*, 498*

Table II. (Continued)

Substituents on 9-acridanone	mp (°C)	Recrystallizing solvents and derivatives, mp (°C)	Yield (%)	Cyclizing reagents or conditions	Ref.
2-Dimethylamino	(a) 292	EtOH, AmOAc	95	H_2SO_4	171,536
			92	$POCl_3$	21
3-Dimethylamino	(b) 158-160		75-98	$POCl_3$	117, 147
	(b) 118-119		30	3'-Me_2N acid, $POCl_3$	117
			60	5-Me_2N acid, $POCl_3$	117
4-Dimethylamino	(b) 116-118		75	$POCl_3$	117
2-Dimethylamino-6-nitro	(b) 278-282 (impure)		97	$POCl_3$	147
1,3-Dinitro	(a)		94	$POCl_3$	26
1,5-Dinitro	(a) 323-324		98	H_2SO_4	148
1,6-Dinitro	(b) 200-203		90	$POCl_3$	147
1,7-Dinitro	(a)		73	H_2SO_4	147
2,4-Dinitro	(a) >350	Pyridine	90	$POCl_3$-xylene	737
			25	Heat acid chloride in $PhNO_2$	94
2,6-Dinitro	(a)	2-Ethoxyethanol, aq DMF	80	$POCl_3$; H_2SO_4	94,738
2,7-Dinitro	(a) >370		97-100	$POCl_3$	25
	(360)		55	Sternbach	379,674
				H_2SO_4	78
			90	Nitration	726

(Table Continued)

263

Table II (Continued)

Substituents on 9-acridanone	mp (°C)	Recrystallizing solvents and derivatives, mp (°C)	Yield (%)	Cyclizing reagents or conditions	Ref.
3,5-Dinitro	(a) 318-320		85	Hydrolysis of 9-NH_2	24
3,6-Dinitro	(b) 247-249		10	$POCl_3$	25
3,8-Dinitro	(b) 200-203		90	$POCl_3$	148, 239
4,5-Dinitro	(a) 266-268			$POCl_3$	239
			100	H_2SO_4	148
2,4-Dinitro-1-hydroxy	(a) 294	$PhNO_2$, pyridine		Hydrolysis of 1-Cl	252
2,4-Dinitro-1-methoxy	(a) 239	HOAc		Substitution of 1-Cl	252
5,7-Dinitro-2-sulfon-amido	(a) >300		95	NH_3 on acid chloride	109
5,7-Dinitro-2-sulfonyl Cl	(a) 272-276 (dec)		92	$ClSO_3H$	109
2,4-Dinitro-1-(p-toluidino)	(a) 305	$PhNO_2$-xylene		Substitution of 1-Cl	252*
2-Ethoxy	(a) 320 (dec)		92	Heat acid chloride	93
	(b) 146-147		50	H_2SO_4 with 36% 2-OH	739
4-Ethoxy	(a)	$PhNH_2$-HOAc	95	$POCl_3$	40
			79	$POCl_3$	740
			89		740
2-Ethoxy-7-iodo	(b) 124-125			$POCl_3$	741
2-Ethoxy-5-methoxy	(b) 170-172		95	$POCl_3$	40
2-Ethoxy-7-methoxy	(b) 164		95	$POCl_3$	40
2-Ethoxy-6-nitro	(b) 175	10-oxide[745] 204-6(dec)		$POCl_3$	742-744
2-Ethoxy-7-nitro	(b) 189-191, (a) 378		92	$POCl_3$	4

Table II. (Continued)

Substituents on 9-acridanone	mp (°C)	Recrystallizing solvents and derivatives, mp (°C)	Yield (%)	Cyclizing reagents or conditions	Ref.
2-Ethyl	(a) 258-261		65	POCl$_3$	164
	(b) 87-89		74-79		125
3-Ethyl	(b) 51-53		100	POCl$_3$	125
4-Ethyl	(b) 68-70		55-91	POCl$_3$	8,125,164
2-Ethyl-6-nitro	(a) >300				746
2,3-Ethylenedioxy-6-nitro	(b) 299-300	xylene, PhCl	87	POCl$_3$	14.747
1-Hydroxy	(a) 260 (dec)	EtOH		Hydrolysis of 1-OMe	414
	320 (dec)				
2-Hydroxy	(a) 345-350 (dec)	EtOH, triacetylglycerine	86	H$_2$SO$_4$	3,193,536
				Hydrolysis of 2-OMe	414
3-Hydroxy	(a) 324-325	MeOH		Hydrolysis of 3-OMe	414,596
	295 (dec)				
4-Hydroxy	(a) 300 (dec)	AmOAc, aq HOAc	50-91	H$_2$SO$_4$	193
	190		74	Hydrolysis of 4-OMe	414,536
1-Hydroxy-3-methoxy or 3-Hydroxy-1-methoxy	{(a) 203			Methylation	604
	{(a) 252			Methylation	604
1-Hydroxy-4-methoxy	(a) 220			Hydrolysis of 1-OMe	414

265

(Table Continued)

Table II. *(Continued)*

Substituents on 9-acridanone	mp (°C)	Recrystallizing solvents and derivatives, mp (°C)	Yield (%)	Cyclizing reagents or conditions	Ref.
4-Hydroxy-3-methoxy	(a)			Hydrolysis of 4-OMe	595
1-Hydroxy-2,3-methylene-dioxy (norevoxanthidine)	(a) 327 (dec) 283-284	EtOH		Hydrolysis of 1-OMe	120, 441
				Bromination and hydrolysis	607*
1-Fluoro	(b) 106		40	POCl$_3$	187
2-Fluoro	(a) 364	AmOAc, Sublimes 360		H$_2$SO$_4$	536,748
	(b) 137		100	POCl$_3$	187
3-Fluoro	(a) 372		45	L-T	153, 925
	(b) 151		60	POCl$_3$	187
4-Fluoro	(b) 142		100	POCl$_3$	187
2-Fluoro-1-nitro	(b) 155 (dec)		48	POCl$_3$	208
2-Fluoro-3-nitro	(b) 265 (dec)		24	POCl$_3$	208
3-Fluoro-6-nitro	(a)		40	L-T	288, 305
6-Fluoro-2-methoxy	(a) 345-347 (dec)			POCl$_3$	725
	(b) 166-168		87		67,725
2-Iodo	(a) >360	HOAc		POCl$_3$	115
	(b) 166-167	PhH-EtOH	75	POCl$_3$	115
4-Iodo	(b) 114-116	Cyclohexane	55	POCl$_3$	685
2-Iodo-6-methoxy	(b) 208-210	PhH-acetone	70	Heat on iminoether	400

Table II. (Continued)

Substituents on 9-acridanone	mp (°C)	Recrystallizing solvents and derivatives, mp (°C)	Yield (%)	Cyclizing reagents or conditions	Ref.
2-Iodo-7-methoxy	(a) 360	HOAc		POCl$_3$	116
	(b) 166	EtOH, PhH			116
6-Iodo-2-methoxy	(a) >270	EtOH	84	Diazonium	584,690
	(b) 164-165				584,690
2-Iodo-5-methyl	(b) 178			POCl$_3$	209
2-Iodo-7-methyl	(b) 151-152		100	POCl$_3$	690,749
3-Iodo-6-nitro	(a) 286-287		19-83	L-T	580
1-Methoxy	(a) 346-348	HOAc	83	PCl$_5$-PhH	97,307
	(b) 146-147	EtOH, cyclohexane		PCl$_5$	97
	125-127				
2-Methoxy	(a) 290-292	AmOAc, aq HOAc	24	POCl$_3$	708,750
			71	POCl$_3$	50,51
				H$_2$SO$_4$,	179,536
				Hydrolysis of	525
	(b) 154	N-oxide 200-201[35,559,678]		9-PhO	
				POCl$_3$	97,733,750
3-Methoxy	(a) 290	aq HOAc		H$_2$SO$_4$	751,
	(b) 170-171	PhH	36-78	POCl$_3$	30,97,414,
					708,750,879
4-Methoxy	(a) 295-296	AmOAc, CH$_2$Cl$_2$,	75	H$_2$SO$_4$	118,193,536

(Table Continued)

267

Table II. (Continued)

Substituents on 9-acridanone	mp (°C)	Recrystallizing solvents and derivatives, mp (°C)	Yield (%)	Cyclizing reagents or conditions	Ref.
		aq HOAc		Irradn of Ph-anthranil	312
2-Methoxy-5-chloro	(b) 128-130		60	POCl$_3$	97, 750
2-Methoxy-5,7-dinitro	(b) 157		85	POCl$_3$	752
	(a) 268		65	H$_2$SO$_4$	205
1-Methoxy-3-methyl	(a) >360			POCl$_3$	11
2-Methoxy-6-methyl	(b) 159-160			PCl$_5$	700
4-Methoxy-3-methyl	(a) 249	HBr	82	PPA	595
1-Methoxy-2,3-methylenedioxy (evoxanthidine)	(a) 319-320 (dec)	EtOH	40-76	POCl$_3$	120,219,441, 734
3-Methoxy-1,2-methylenedioxy	(b) 177-178	PhH	65-71	POCl$_3$	120, 219
	(b) 218-220	MeOH	17	POCl$_3$	120
4-Methoxy-1,2-methylenedioxy	(b) 240 (dec)			POCl$_3$	414
1-Methoxy-6-nitro	(a) > 360	aq DMF	97	POCl$_3$	753
	(b) 223 (dec)	EtOH-PhH	15		753
2-Methoxy-1-nitro	(a) 339	Dichloroethane	91	POCl$_3$	241
2-Methoxy-3-nitro	(b) 195-196				241
	(b) 205-206		73	POCl$_3$	241
2-Methoxy-4-nitro	(b) 205-206			POCl$_3$	255

Table II. *(Continued)*

Substituents on 9-acridanone	mp (°C)	Recrystallizing solvents and derivatives, mp (°C)	Yield (%)	Cyclizing reagents or conditions	Ref.
2-Methoxy-6-nitro	(a) 320, 360	Pyridine, aq DMF	96	POCl$_3$	584,744,753
	(b) 216-218	Pyridine		POCl$_3$	439,584,725, 742,744,754
2-Methoxy-7-nitro	(a) 395,350		92-95	POCl$_3$	4,50,439,755
				Cyclization of acid chloride	928
				Hydrolysis of 9-PhO	525
	(b) 226		65	POCl$_3$	744
3-Methoxy-6-nitro	(a) > 360	aq DMF	96	POCl$_3$	753
			100	L-T	305
	(b) 233	PhH	58	POCl$_3$	753
4-Methoxy-1-nitro	(a) 281		85	H$_2$SO$_4$	732
	(b) 173-174		55	POCl$_3$	50
5-Methoxy-2-nitro	(a) >300		92-95	POCl$_3$-xylene	50
	(b) 265-267			POCl$_3$	439
					755
5-Methoxy-3-nitro	(a).>360	aq DMF		POCl$_3$	439,753
	(b) 223	CHCl$_3$	82	POCl$_3$	753
7-Methoxy-1-nitro	(b) 247	Toluene	8	POCl$_3$	132
2-Methoxy-6-sulfonamido	(b) >240 (dec)		100	POCl$_3$	244
2-Methoxy-7-sulfonamido	(b) 230 (dec)		80-90	POCl$_3$	244
2-Methoxy-7-sulfon-diethylamido	(b)184-186			POCl$_3$	244

(Table Continued)

269

Table II. (Continued)

Substituents on 9-acridanone	mp (°C)	Recrystallizing solvents and derivatives, mp(°C)	Yield (%)	Cyclizing reagents or conditions	Ref.
2-Methoxy-7-sulfon-dimethylamido					756
1-Methoxy-x-sulfonicacid	(a) 240-245 (dec)			H_2SO_4	307
2-Methoxy-7-sulfon-phenylamide	(b) 207-208			$POCl_3$	244
2-Methoxy-7-sulfonyl-(p-sulfonamidophenyl)-amino	(b 212-215	aq acetone		$POCl_3$	757
4-Methoxy-1-(1,1,3,3-tetramethyl)butyl	(b) 123-125			$POCl_3$	758
1-Methyl	(a) 318	Formamide, aq EtOH	35	Dehydrogenation of tetra-H	97,372
	(b) 95-96	PhH		$POCl_3$	97,392
2-Methyl	(a) 339	PhNO$_2$, AmOAc.EtOH; DMP 231-232[82]	83-85	$POCl_3$	164,496,682
				H_2SO_4	496,536
				Dehydrogenation of tetra-H	372
				Degrad 9-$CONH_2$	286
	(b) 119-121			$POCl_3$	164
3-Methyl	(a) 344-345	Formamide, pyridine	11-22	Dehydrogenation of tetra-H	372

Table II. (Continued)

Substituents on 9-acridanone	mp (°C)	Recrystallizing solvents and derivatives, mp(°C)	Yield (%)	Cyclizing reagents or conditions	Ref.
4-Methyl	(b) 125	aq EtOH	90	L-T	294, 925
	(a) 347	AmOAc, formamide, HOAc, EtOH; sublimes		POCl₃	97,100,129
		335		H₂SO₄	495,536
			88	Heat acid chloride	93
			22	Dehydrogenation of tetra-H	372
2-Methyl-5,7-dinitro	(b) 96-97		57	POCl₃	164
4-Methyl-5,7-dinitro	(a) 330		30	POCl₃	51,97,164
6-Methyl-2-methylthio	(a) 246		70	H₂SO₄	205
2-Methyl-6-nitro			45	H₂SO₄	205
	(a) >300			POCl₃	59
	(b) 199-200	PhH	92	POCl₃	746
2-Methyl-7-nitro	(b)		63	POCl₃	693
3-Methyl-6-nitro	(a) >300	DMP ca. 200; EtOH	Low	L-T	575
4-Methyl-1-nitro	(a) >300			H₂SO₄	297, 305
	(b) 160-167		69	POCl₃	136
4-Methyl-5-nitro	(a) 226		70	H₂SO₄	732
4-Methyl-6-nitro	(b) 187-188	PhH	94	POCl₃	680
4-Methyl-1-isopropyl	(a) 208-209		59	POCl₃	693
	(b) 75		10	POCl₃	8,164
					8,164

(Table Continued)

271

Table II. (Continued)

Substituents on 9-acridanone	mp (°C)	Recrystallizing solvents and derivatives, mp (°C)	Yield (%)	Cyclizing reagents or conditions	Ref.
2-Methyl-7-sulfon-dimethylamido	(b)			$POCl_3$	756
2-Methyl-3,5,7-trinitro	(a) 253	HOAc-PhH		H_2SO_4	759
2-Methyl-4,5,7-trinitro	(a) 320	HOAc		H_2SO_4	760
2-(1-Methyl-n-butyl)	(a) 182-185		57	$POCl_3$	8,164
	(b) Liquid		78		8,164
2,3-Methylenedioxy	(a) 350 (dec)	EtOH; HCl 296-298	94	$POCl_3$	219
	(b) 180-181	aq EtOH	98		219
2,3-Methylenedioxy-6-nitro	(b) 300-301	PhCl	63	$POCl_3$	14,747
2-Methylthio	(a) 260-261	PhNO2; DMP 280-281[30]	79	Acid chloride-A1Cl3	190,696
1-Nitro	(a) 362-364	10-oxide 300 (dec)[561]	61	H_2SO_4	147
	(b) 150-151	PhH-EtOAc	25	$POCl_3$	150,761,762
2-Nitro	(a) >370	PhNO2, aq DMF	83	$POCl_3$	124,505,674
		Tosyl hydrazone, HCl infusible.[19]	87	Sternbach	737
	(b) 203-204	DMP 225[350]		$POCl_3$	744,762
3-Nitro	(a) >400	HOAc, toluene	98	$POCl_3$	147
				L-T	357
		DMP 255[306]	15	H_2SO_4	147

272

Table II. (Continued)

Substituents on 9-acridanone	mp (°C)	Recrystallizing solvents and derivatives, mp(°C)	Yield (%)	Cyclizing reagents or conditions	Ref.
	(b) 214	PhH-EtOAc; 10-oxide 233-234[35,559,678]	20-100	POCl$_3$	147,744,761 762
4-Nitro	(a) 260-262	PhNO$_2$	67-91	H$_2$SO$_4$	118,136,147
	(b) 199-200	PhH	97	POCl$_3$	74
1-Nitro-4-carboxylic acid?	(a) 333		20-78	POCl$_3$	762
6-Nitro-4-carboxylic acid?	(a) 331-333		96	PCl$_5$-PhNO$_2$	75
4-Nitro-2-sulfonamido			82	POCl$_3$	150
				SOCl$_2$	90
6-Nitro-2-sulfonic acid	(a) 325-352 (dec)		78	f. H$_2$SO$_4$, POCl$_3$	582
4-Nitro-7-sulfonyl chloride				Chlorosulfonation	108
4-(p-Nitrophenyl)	(a) 262-263		65	POCl$_3$	22
Octachloro	(a) 370	PhCl	10	Benzyne	104
				H$_2$SO$_4$	104
	(b) 259			SbCl$_5$ chlorination	437
2-Phenyl	(b)		100	POCl$_3$	23
4-Phenyl	(b)		100	POCl$_3$	23
4-N-Piperidylmethyl	(b) 117-118			Amine on 4-CH$_2$Br	581*
4-n-Propyl	(b) 60-61		9	POCl$_3$	8,164
4-isoPropyl	(a) 252-253		65	POCl$_3$	8,164
	(b) 108-109		6		8,164
2-Sulfonic acid	(a) 318	aq HCl; DMP 264-265[73]	98	H$_2$SO$_4$	27,763

(Table Continued)

Table II. (Continued)

Substituents on 9-acridanone	mp (°C)	Recrystallizing solvents and derivatives, mp (°C)	Yield (%)	Cyclizing reagents for conditions	Ref.
4-Sulfonic acid	(a) 268	TrifluoroMe ester		PPA	82,696
				$POCl_3$	73
2-Stibonic acid	(a) >400	aq HCl	2	H_2SO_4	27
				Diazonium	589
2,4,5,7-Tetrabromo	(b) 255	PhH		$POCl_3$	511,649
2,4,5,7-Tetrachloro	(a) 253	PhH, toluene	98	Chlorination	401,514
				H_2SO_4 on N-benzoyl	
1,2,3,4-Tetrafluoro	(b) 196	PhH-EtOH	3-9	$POCl_3$	263,649
	(a) 360 (dec)	Sublimes	32	Pyrolysis of anthranil	311
	(b) 182-183	PhH-petrol			311
1,2,4,8-Tetrahydroxy		HBr		Hydrolysis	595
1,3,4,5-Tetrahydroxy		HBr		Hydrolysis	595
1,2,3,4-Tetramethoxy	(a) 202-203			$POCl_3$	218
1,2,4,5-Tetramethoxy	(a) 203		87	PPA	71
1,2,4,6-Tetramethoxy	(a) 193-194		47	PPA	71
1,2,4,7-Tetramethoxy	(a) 195		94	PPA	71
1,2,4,8-Tetramethoxy	(a) 216		89	PPA	71
1,3,4,5-Tetramethoxy	(a) 188		76	PPA	71
1,3,4,6-Tetramethoxy	(a) 253-254		76	PPA	71
1,3,4,7-Tetramethoxy	(a) 252		92	PPA	71
1,3,4,8-Tetramethoxy	(a) 235-237		45	PPA	71
1,4,5,8-Tetramethyl	(a) 220-221	PhH	26	$POCl_3$	10

274

Table II. (Continued)

Substituents on 9-acridanone	mp (°C)	Recrystallizing solvents and derivatives, mp (°C)	Yield (%)	Cyclizing reagents for conditions	Ref.
2,4,7-Tribromo	(b) 172	Petrol	56		10
1,4,6-Trichloro	(a) 232-233	HOAc		POCl₃	140
2,4,7-Trichloro	(a) 242-243		72	H₂SO₄	576,
	(a) 410			Halogenation	908
1,2,6-Trichloro-4-nitro	(a) 287-288	o-C₆H₄Cl₂		H₂SO₄	217, 692*
1-Trifluoromethyl	(a) 360		92	POCl₃	121
3-Trifluoromethyl	(a) 360			POCl₃	121
	(b) 128		84		121
1,2,4-Trihydroxy		HBr		Hydrolysis	595
1,4,6-Trihydroxy	(a) 200 (dec)	HBr	58	Hydrolysis	595
1,2,3-Trimethoxy	(a) 195-197	aq EtOH		POCl₃	218
	(b) 146-147	PhH-petrol			441
1,2,4-Trimethoxy	(a) 221-223			POCl₃	218
1,3,4-Trimethoxy	(a) 162-163			POCl₃	218
1,4,5-Trimethoxy	(a) 258		70	PPA	71
1,4,6-Trimethoxy	(a) 216-218		55	PPA	71
1,4,7-Trimethoxy	(a) 227-229		83	PPA	71
1,4,8-Trimethoxy	(a) 203		75	PPA	71
2,3,4-Trimethoxy	(a) 208-209			POCl₃	218
2,4,7-Trinitro	(a) 277		26	Nitration	24

275

TABLE III. 10-Substituted 9-Acridanones

10-Substituent; other substituents	mp (°C)	Recrystallizing solvents and derivatives, mp (°C)	Yield (%)	Cyclizing reagent or conditions	Ref.
10-Acetyl; 6-chloro-2-methoxy	301-303 (dec) a	aq MeOH		Acetylation	110
10-(9-Acridyl)	383-384	Pyridine	30	H_2SO_4	58
10-Allyl	136-137	aq EtOH		Alkylation	436
10-Benzyl	180-181			Alkylation	435
10-(p-Bromophenyl)	279-280		72	Dehydrogenation	431
10-(n-Butyl)	100		57	Alkylation	434
10-iso Butyl	178-180		12	Alkylation	434
10-Cyclopropylmethyl 2-trichloromethyl				Alkylation	696
2-trifluoromethyl				Alkylation	82
10-(9-(3-Chloroacridinyl))	327-330	Pyridine	30	H_2SO_4	58
10-(o-Chlorophenyl)	214-216		32	Dehydrogenation	549*
4-chloro		Toluene	74	PCl_5-$AlCl_3$	137
10-(p-Chlorophenyl)	279-280		70	Dehydrogenation	371
2-chloro	>270	Xylene		PCl_5-$AlCl_3$	137
10-(2-Diethylamino)ethyl		HCl 234-235	58	Alkylation	443*
1-nitro		HCl 240-241	60	Alkylation	445*

*The cited reference contains details of further compounds of a similar nature.

276

Table III. (Continued)

10-Substituent; other substituents	mp (°C)	Recrystallizing solvents and derivatives, mp (°C)	Yield (%)	Cyclizing reagent or conditions	Ref.
10-2(Dimethylamino)-ethyl	112	HCl 213-214	94	Alkylation	432*
1-methoxy		HCl 253-254	63	Alkylation	764*
2-methoxy		HCl 251-252	59	Alkylation	764*
3-methoxy		HCl 264-266	49	Alkylation	764*
4-methoxy		HCl 252-253	52	Alkylation	764*
1-nitro		HCl 235-236	52	Alkylation	445*
10-(3-Dimethylamino)-propyl		HCl 241-242	54	Alkylation	433
1-nitro	239-240	EtOH		Alkylation	444
10-(1-Dimethylamino)-isopropyl	115-116	PhH-petrol, HCl 249-250 (dec)	75	Alkylation	446
10-Ethyl	159	EtOH	45-75	Alkylation	5,434,447
				Hydrolysis of 9-NH$_2$	452
				Redn of 10-vinyl	449
3-chloro-7-methoxy	225			POCl$_3$	448
3,6-diethoxy	180-181			Alkylation	438
3,6-dimethoxy	159-160			Alkylation	438
10-(p-Ethylphenyl)	218-220		44	Dehydrogenation	374
10-Hexyl				Alkylation	435
10-Methoxy	153	Petrol			307
3-chloro-6-nitro	241	Acetone-H$_2$O			130

(Table Continued)

277

Table III. *(Continued)*

10-Substituent; other substituents	mp (°C)	Recrystallizing solvents and derivatives, mp (°C)	Yield (%)	Cyclizing reagent or conditions	Ref.
10-Methyl	(a) 203-204	EtOH	76	H_2SO_4	273
		Thiosemicarbazone 190 (dec) [527]	90	Alkylation	434,435,437
					904
			100	Rearrangement of 9-OMe	414
			80-90	O_2 on 9-CN	903
			22	Dehydrogenation	428
			39	Oxidn	765
	(b)	Chloride 73		$POCl_3$	453
		Borofluoride 208-212		$Me_3O^+ BF_4^-$	
3-benzyloxy-1,2-dimethoxy	175	EtOAc		Alkylation	533*
2-bromo	200				
7-bromo-1-hydroxy-2-methoxy-3,4-methylenedioxy	266-267		43	Alkylation	515
				Br_2-MeOH, pyrolysis	600
7-bromo-1,2,3,4-tetramethoxy	147	MeOH		Bromination	515*

(a) Indicates 9-acridanone.
(b) Indicates 9-chloracridine.

Table III. (Continued)

10-Substituent; other substituents	mp (°C)	Recrystallizing solvents and derivatives, mp (°C)	Yield (%)	Cyclizing reagent or conditions	Ref.
10-Methyl					
2-carbomethoxy-1,3-dimethoxy	213-214	PhH-EtOAc	100	Rearrangement of 9-OMe	455
4-carboxy-1,3-dimethoxy	212-214	MeOH; Me ester 192-193	62	Degradn of acronycin	597
3-chloro	172-173	Toluene		Oxidation	455
6-chloro-2-methoxy	245-246			Alkylation	493
2,6-diamino	308			Alkylation	904
2,7-dibromo	289-290	EtOH	86	Hydrolysis of 9-CN	424
			8	Bromination in MeOH	515,516*
2,6-di(dimethylamino)	275-276			Bromination in CHCl₃	515,516*
2,3-diethoxy-1-hydroxy	173			Hydrolysis of 9-CN	424
				Degradn of evoxanthine	601
3,6-diethoxy	191-192			Alkylation	438
1,3-dihydroxy	288-292 (dec)	EtOH, aq HCl		Degradn of acronycin	408,597
				1,3-diacetyl, hydrolysis	899
1,2-dihydroxy-3,4-dimethoxy	242-243	EtOH; monoacetyl 158-159, diacetyl 204		Degradn	602*

(Table Continued)

279

Table III. (Continued)

10-Substituent; other substituents	mp (°C)	Recrystallizing solvents and derivatives, mp (°C)	Yield (%)	Cyclizing reagent or conditions	Ref.
10-Methyl					
1,4-dihydroxy-2,3-dimethoxy	197-198	Monoacetate 147-148 Diacetate 176-177		Degradn	602*
1,2-dihydroxy-3-ethoxy	259	Monoacetate 255-256 Diacetate 229-230		Degradn	601
1,2-dihydroxy-3-methoxy (nor-arborinine)	242-243	CHCl₃-EtOH Acetyl 255-257, Benzoyl 246-247	64	Hydrolysis	598,601
1,3-dihydroxy-2-methoxy	242-244	MeCN		Hydrolysis	533*
1,3-dimethoxy	163-165	aq MeOH, PhH-petrol HCl 135-136 (dec)[767] Picrate 203-4[731]	62 95	Alkylation Oxidation Extraction Degradn of acronycin	731 408,899,896 767,768, 597
1,4-dimethoxy	88-89	Picrate 229-230	76	Alkylation Oxidation	731 414
2,3-dimethoxy	194-195		88-89	Alkylation	731
2,4-dimethoxy	167-168		100	Oxidation	408
3,4-dimethoxy	115-116		77	Alkylation	731
3,6-dimethoxy	179	PhH	75	Alkylation	438

280

Table III. *(Continued)*

10-Substituent; other substituents	mp (°C)	Recrystallizing solvents and derivatives, mp (°C)	Yield (%)	Cyclizing reagent or conditions	Ref.
10-Methyl					
1,2-dimethoxy-3-ethoxy	141-142			Degradn of evoxanthine	601
2,3-dimethoxy-3-ethoxy	110	Petrol		POCl$_3$	600
3,4-dimethoxy-2-ethoxy-1-hydroxy	70-72	MeOH	36	Alkylation	606
1,3-dimethoxy-4-formyl-1,3-dimethoxy	218-219	MeOH	97	Alkylation Degradn of acronycin	455* 597*
1,3-dimethoxy-2-hydroxy	226-227	Toluene; acetyl 170-1		Degradn of evoxanthine	601
2,3-dimethoxy-1-hydroxy (arborinine)	177-178	MeOH, EtOH; Acetyl 215, 216 Benzoyl 257-258		Hydrolysis Extraction	408,533,601 534,598,734 768,769-771
2,3-dimethoxy-4-hydroxy	176-177			POCl$_3$	218
2,4-dimethoxy-3-hydroxy	176	MeOH			600*
1,2-dimethoxy-3,4-methylenedioxy (melicopine)	181-182	EtOH	71	Alkylation Extraction	220,221 572,772
1,4-dimethoxy-2,3-methylenedioxy (melicopidine)	121-122	MeOH, EtOH		Alkylation Extraction	220,221,770 592,772

(Table Continued)

281

Table III. *(Continued)*

10-Substituent; other substituents	mp (°C)	Recrystallizing solvents and derivatives, mp (°C)	Yield (%)	Cyclizing reagent or conditions	Ref.
10-Methyl					
1,2-dimethoxy-3-prenyloxy	126	EtOAc		Alkylation	533*
1,3-dimethoxy-2-prenyloxy	106	Cyclohexane		Alkylation	533
2,7-dimethyl	190	EtOH		Oxidation	415
3-α,α-dimethylallyloxy-1-hydroxy	263-264	EtOAc-petrol	85	Redn	899
3-α,α-dimethylprop-argyloxy-1-hydroxy	208-209	EtOAc-petrol	67	Alkylation	899
3-ethoxy-1-hydroxy-2-methoxy	194	Acetyl 211		Degradn of evoxanthine	601
3-ethoxy-2-hydroxy-1-methoxy	199-201	Acetyl 209-210		Degradn of evoxanthine	601
3-ethoxy-1,2,4-trihydroxy	239-240	Diacetyl 188-189		Degradn	543
3-ethoxy-1,2,4-trimethoxy	Non crystalline			Degradn	543
1-hydroxy	190			Oxidation	414
1-hydroxy-3-methoxy	175-176	aq EtOH, EtOH-CHCl$_3$		Hydrolysis	408,767
1-hydroxy-3-methoxy (2,4,11-^{14}C)	176	EtOH	49	Hydrolysis	896

282

Table III. (Continued)

10-Substituent; other substituents	mp (°C)	Recrystallizing solvents and derivatives, mp (°C)	Yield (%)	Cyclizing reagent or conditions	Ref.
10-Methyl					
1-hydroxy-4-methoxy	136	aq MeOH, aqDMF; 3-acetyl 200	40	Hydrolysis	414
3-hydroxy-1-methoxy	305			Alkylation and hydrolysis	899
	174-175	aq MeOH		acronycin	597
1-hydroxy-2-methoxy-3-prenyloxy (evoprenine)	143	EtOH		Extraction	533*
1-hydroxy-2,3-methylenedioxy (norevoxanthine)	274-275	Acetyl 230-232 Pyridine, dioxan		Hydrolysis	601, 770
1-hydroxy-2,3,4-trimethoxy (normelicopicine)	129-130	Acetyl 113-115,^{600} tosyl 185-186	92	Hydrolysis	602*
2-hydroxy-1,3,4-trimethoxy	165-166		56	Degradn	602*
3-hydroxy-1,2,4-trimethoxy	177-178	Acetyl 130-131		Degradn	543
4-hydroxy-1,2,3-trimethoxy	191-192			Degradn	602*
1-methoxy			100	Rearrangement of 9-OMe	414

(Table Continued)

283

Table III. (Continued)

10-Substituent; other substituents	mp (°C)	Recrystallizing solvents and derivatives, mp (°C)	Yield (%)	Cyclizing reagent or conditions	Ref.
10-Methyl					
2-methoxy			100	Rearrangement of 9-OMe	414
3-methoxy			100	Rearrangement of 9-OMe	414
4-methoxy	91		100	Rearrangement of 9-OMe	414
1-methoxy-2,3-methylenedioxy (evoxanthine)	225	CHCl$_3$, EtOH, ether PhH; HCl 202, picrate 200, sulfate 215, [773] picrolonate 178 [773]		Extraction Alkylation POCl$_3$	601,768,770 120 220,734
			38		
2-methoxy-3,4-methylenedioxy	203-204	MeOH	15		600*,605*
4-methoxy-1,2-methylenedioxy	164			Rearrangement of 9-OMe	414
1-methoxy-6-nitro	321-322		25	Alkylation	440
2-methoxy-7-nitro	267		70	Alkylation	439
3-methoxy-6-nitro	260-261		70	Alkylation	440
4-methoxy-6-nitro	223-224			Alkylation	439
4-methoxy-7-nitro	235		70	Alkylation	439
2-methoxy-1,3,4-trihydroxy	176-178	Triacetyl 209-210, diacetyl 218-219		Degradn	543

284

Table III. (Continued)

10-Substituent; other substituents	mp (°C)	Recrystallizing solvents and derivatives, mp (°C)	Yield (%)	Cyclizing reagent or conditions	Ref.
10-Methyl					
3-methoxy-1,2,3-trihydroxy	276-277	MeOH; triacetyl 211-213, diacetyl 214-215		Degradn	604
4-methoxy-1,2,3-trihydroxy	227-228	Triacetyl 168-169 diacetyl 193-194		Degradn	543
2-methyl	150-151			Alkylation	904
2,3-methylenedioxy	265-266	EtOH	42	Alkylation	219
2-methylthio				Alkylation	696
2-nitro	285-287	aq DMF	82	Sternbach	674
4-nitro	176-177	Xylene	32	H_2SO_4	135
1,2,3,4-tetrahydroxy	304-306	Triacetyl 193-194		Degradn	604
1,2,3,4-tetramethoxy (melicopicine)	133-134	MeOH		Extraction	592,772
				$POCl_3$	218
2-Trifluoromethyl	180-181				774
3-Trifluoromethyl				Alkylation	696
2-Trifluoromethyl-sulfonyl				Alkylation	82,696
1,2,3-trihydroxy	290-293 (dec)	EtOH		Hydrolysis	897
		MeOH		Bromination	607
1,2,3-trimethoxy	169-170	PhH-petrol	95	Oxidation	408
	178-179			Degradn	601
				$POCl_3$	218

(Table Continued)

285

Table III. (Continued)

10-Substituent; other substituents	mp (°C)	Recrystallizing solvents and derivatives, mp(°C)	Yield (%)	Cyclizing reagent or conditions	Ref.
10-Methyl				Extraction	768
1,2,3,-trimethoxy				Alkylation	534,734,770, 771, 775
1,2,4-trimethoxy	141-142			POCl$_3$	218
1,3,4-trimethoxy	113-115			POCl$_3$	218
2,3,4-trimethoxy	116-118	MeOH	ca. 100	Oxidn	408
				POCl$_3$,	218
				Alkylation	600
10-Phenyl	276	Ether-CHCl$_3$,	100	P$_2$O$_5$-xylene	101
		Toluene	92	POCl$_3$	13,231
				Benzyne	450
2-chloro	229-230	HOAc		PCl$_5$-AlCl$_3$	137
3-chloro	288-289	Toluene		PCl$_5$-AlCl$_3$	137
3-phenyl	218		100	P$_2$O$_5$-xylene	101
10-(n-propyl)	130-131	aq EtOH	82	Alkylation	434,436
10-isopropyl	170-80		4	Alkylation	434
10-(p-tolyl)	277-278	Cyclohexanone		Dehydrogenation	430
2,7-dimethyl	217-219			H$_2$SO$_4$	232
10-(p-Tosyl)					
2-chloro	171-172		80	Sternbach	674
10-Vinyl	181-182	EtOH	92	Acetylene on 9-acridanone	449
3-chloro-7-methoxy	232	EtOH	95	Acetylene on 9-acridanone	448

286

TABLE IV. Diphenylamine-2-carboxylic Acids

Note: All acids containing one or two *types* of substituent, additional *N*-substituted compounds, and further-substituted compounds that direct the reader to references not mentioned elsewhere in the text, are included in this table. Acids with 6- or 2'- and 6'-substituents, which cannot be cyclized to 9-acridanones, are excluded.

Substituents	mp (°)	Recrystallizing solvents and derivatives, mp (°C)	Catalyst	Yield (%)	Reaction Type	Ref.
Unsubstituted	185-187	PhH, MeCN, EtOH, aq EtOH; amide 127-128,[91] Anhydride 161-162,[58] acid chloride 47-50,[58] Et ester b_{15} 203,[886] hydrazide 119-120[929]	None	40-50	Type 1 (Cl)	776
			Cu	98		675, 536
			CuO	82-93		165
			Me salicylate	97		178
			CuO	74	Type 1 (Br)	91, 242
			Cu		Type 2 (Br)	175
				39-57	Chapman	138, 242
				80-89	Oxidn of *N*-phenyl heterocycles	277, 279
3-Acetamido	160		Cu	32	Type 1 (Cl)	147
4-Acetamido	236-238		Cu	40	Type 1 (Cl)	147
5-Acetamido	248-250		Cu	39	Type 1 (Cl)	147

* The cited reference contains details of further compounds of a similar nature.
Type 1: Reaction of halobenzoic acid (halogen in parentheses) and aromatic amine.
Type 2: Reaction of anthranilic acid and halobenzene.

(Table Continued)

287

Table IV. (Continued)

Substituents	mp (°C)	Recrystallizing solvents and derivatives, mp (°C)	Catalyst	Yield (%)	Reaction Type	Ref.
4'-Acetamido	240		Cu	92	Type 1 (Br)	589
	175-176		Cu	65	Type 1 (Cl)	215, 777
4-Acetamido-3'-amino	152-154		Cu	72	Type 1 (Cl)	147
5-Acetamido-3'-amino	182-186		Cu	41	Type 1 (Cl)	147
5-Acetamido-4'-amino	214		Cu	38	Type 1 (Cl)	147
4'-Acetamido-5-chloro	287-288		Cu		Type 1 (Cl)	244
4-Acetamido-3'-nitro	250		Cu	31	Type 1 (Cl)	147
5-Acetamido-3'-nitro	246-248		Cu	44	Type 1 (Cl)	147
3'-Acetamido-4-nitro						77
4'-Acetamido-6-nitro	218	aq EtOH	NaOAc		Type 1 (Br)	158
4'-Acetamido-6-nitro-4-sulfon-amido	250-251		NaOAc	68	Type 1 (Br)	109
4-(4-Acetamidoben-zenesulfonamido)-4'-methoxy	218-220		Cu		Type 1 (Cl)	679*

288

Table IV. (Continued)

Substituents	mp (°C)	Recrystallizing solvents and derivatives, mp (°C)	Catalyst	Yield (%)	Reaction Type	Ref.
3-Acetyl	166	aq EtOH	Cu		Type 1(Cl)	757
2'-Acetyl	207	EtOH	Cu	16	Type 1 (Br)	930
4'-Acetyl	174	EtOH, MeOH; Me ester 80	Cu	60	Type 1 (Br)	930
			Cu-CuI	66	Type 2 (Br)	931
N-Acetyl-5-chloro-4'-methoxy	196-198	aq MeOH		53	Acylation	110
4'-Acetyl-3'-methyl	153	EtOH	Cu-CuI		Type 1 (Cl)	931
4'-Acetyl-2',5'-dimethyl	195	EtOH-actone	Cu-CuI		Type 1 (Cl)	931
4'-Acetyl-3',5'-dimethyl	175	PhH-petrol	Cu-CuI		Type 1 (Cl)	931
4'-Acryloylamino-4-nitro		Me ester 205-207				779*
5-Amino					Redn	883
2'-Amino	206-211	Me ester 71-72, acetyl Me ester 174-176		79-80	Redn	136,236,780-782
3'-Amino	166 (dec)	Xylene; Me ester 102-103	Cu	40	Type 1 (Cl)	124
				75	Redn	124
4'-Amino	205	PhH	Cu	75	Type 1 (Cl)	76, 536
2'-Amino-4-chloro	197-198	Me ester 124-125			Redn	229
2'-Amino-5-chloro	175-177				Redn	229
2'-Amino-4'-chloro	200-205				Redn	229

(Table Continued)

289

Table IV. (Continued)

Substituents	mp (°C)	Recrystallizing solvents and derivatives, mp (°C)	Catalyst	Yield (%)	Reaction Type	Ref.
2'-Amino-5'-chloro	205-208 (dec)			100	Redn	229
2'-Amino-4'-chloro-N-methyl	155				Redn	229
2'-Amino-5'-chloro-N-methyl	155				Redn	229
4-Amino-2',3'-dimethyl		HCl 260-262			Redn	783
2'-Amino-4,N-dimethyl	144-146				Redn	229
5'-Amino-2',3'-dimethyl	204-205	Acetone; 5',N-diacetyl	CuBr₂		Type 1 (Br)-amide	783
4-Amino-4'-dimethylamino		Me ester 111			Esterification	784
2'-Amino-N-(2-dimethylamino)-ethyl		Me ester HCl 205-206			Alkylation	236*
4'-Amino-4,6-dinitro	274				Type 1 (Cl)	785
2'-Amino-4'-ethylsulfonyl		Me ester 128-130			Redn	932
4-Amino-4'-hydroxy	213	Me ester 164			Redn	785
4-Amino-4'-methoxy					Redn	756

290

Table IV. (Continued)

Substituents	mp (°C)	Recrystallizing solvents and derivatives, mp (°C)	Catalyst	Yield (%)	Reaction Type	Ref.
2'-Amino-5-methoxy	182–183				Redn	229
2'-Amino-4'-methoxy	200				Redn	229
2'-Amino-5'-methoxy	278–279				Redn	229
2'-Amino-4'-methoxy-N-methyl	132–134				Redn	229
2'-Amino-4'-methyl	213–215				Redn	229
2'-Amino-N-methyl		Acetyl 145–146			Redn	135
2'-Amino-N-methyl-4'-trifluoromethyl	160				Redn	229
2'-Amino-5'-methylthio	170–172				Redn	229
6-Amino-4-nitro	221				Redn	786
3'-Amino-4-nitro	300		Cu	57	Type 1 (Cl)	77, 147
3'-Amino-5-nitro	300		Cu	70	Type 1 (Cl)	147
3'-Amino-6-nitro	240–280 (slow dec)		Cu	59	Type 1 (Cl)	148
4'-Amino-4-nitro	238–239 (dec)	aq HOAc; acetyl 285 (dec)	Cu		Type 1 (Cl)	883
4'-Amino-5-nitro	282–284 (dec)		Cu	59	Type 1 (Cl)	147
4'-Amino-6-nitro	234–235	aq HOAc; HCl	None Cu(OAc)₂	Poor 49	Type 1 (Br)	158, 214

(Table Continued)

Table IV. (Continued)

Substituents	mp (°C)	Recrystallizing solvents and derivatives, mp (°C)	Catalyst	Yield (%)	Reaction Type	Ref.
4'-Amino-6-nitro-4-sulfonamido	268-270	HCl, Me ester 194-197		68	Hydrolysis	109
2'-Amino-4'-trifluoromethyl	214-215				Redn	229
2'-Aminoacetic acid	212			86	CN⁻,HCHO on 2'-NH₂	136
4'-Aminomethyl	ca. 200 (dec)				Redn of 4'-CN	680
4'-(p-Aminophenyl)	246 (dec)		Cu		Type 1 (Cl)	787
3'-Tert-Amyl			Cu		Type 1 (Cl)	788
4'-Anilino	199	aq EtOH	Cu-CuCl	60	Type 1 (Cl)	913
5'-Anilino-2'-methyl	193 (dec)		Cu	49	Type 1 (Cl)	105
5-Anilino-4-nitro	240 (dec)		None	87	Type 1 (F)	203
p-(bis-N,N'-Anthranyl)-benzene	290-292		Cu-CuCl	87	Type 2 (Br)	226
N-Benzoyl	190-191	Me ester 127-131			Chapman	58, 789
4'-Benzoyl	202	EtOH	Cu-CuI	37	Type 2 (Br)	931
N-Benzoyl-4'-bromo-2'-chloro-6-methyl	186-187	Acetone-PhH; Me ester 138		83	Chapman	257
N-Benzoyl-2'-bromo-6-methyl	198-199	PhH; Me ester 191		80	Chapman	257

292

Table IV. (Continued)

Substituents	mp (°C)	Recrystallizing solvents and derivatives, mp (°C)	Catalyst	Yield (%)	Reaction Type	Ref.
N-Benzoyl-2'-tert-butyl	233-235	EtOH; Me ester 146-147		87	Chapman	151
N-Benzoyl-4'-tert-butyl		Me ester 109-110		87	Chapman	151
N-Benzoyl-2'-carbomethoxymethyl		Me ester 115-119		83	Chapman	790*
N-Benzoyl-2'-chloro		Me ester 123			Chapman	791
N-Benzoyl-4'-chloro		Me ester 139-141			Chapman	789
N-Benzoyl-2'-chloro-4',6-dimethyl	210-211	Me ester 187-188			Chapman	257
N-Benzoyl-2'-chloro-6-methyl	198-199			92	Chapman	711
N-Benzoyl-4'-chloro-4-methoxy	104-106	Acetone-petrol; Me ester 119-121		80 / 30-40	Chapman / Type 1 (C1)	399 / 399
N-Benzoyl-2'-chloro-4'-methoxy-6-methyl	214-215	Acetone-PhH; Me ester 124-125		82	Chapman	257
N-Benzoyl-4,6-dibromo	189-190	PhH; Me ester 134-135		80	Chapman	257
N-Benzoyl-4,4'-di-tert-butyl		EtOH; Me ester 145		89	Chapman	151

(Table Continued)

293

Table IV. (Continued)

Substituents	mp (°C)	Recrystallizing solvents and derivatives, mp (°C)	Catalyst	Yield (%)	Reaction Type	Ref.
N-Benzoyl-2',6'-dichloro	246-248			42	Chapman	242
N-Benzoyl-2',4'-dichloro-6-methyl	201-202	Acetone-PhH; Me ester 131-132		70	Chapman	257
N-Benzoyl-6,4'-dimethyl	158-159	aq EtOH; Me ester 122		68	Chapman	792
N-Benzoyl-2',4'-dimethyl	192-193			66	Chapman	138
N-Benzoyl-4'-fluoro	176-178	Ether; Me ester 114-116		78	Chapman	141,142,789, 793
N-Benzoyl-2'-fluoro-6-methyl	187-188	Acetone-PhH; Me ester 104-106		81	Chapman	257
N-Benzoyl-4'-methoxy	184-187	Acetone-cyclohexane; Me ester 151-153		73	Chapman	789, 793
N-Benzoyl-2'-methyl		Me ester 139-140		85	Chapman	151
N-Benzoyl-4,6,2',4'-tetrabromo		Me ester		80	Chapman	139
3'-Benzyl	153-154		Cu		Type 1 (C1)	794
4'-Benzyloxy	195	Me ester 63-65	Cu		Type 1 (C1)	798
3'-Benzylthio	137-139	K salt 162-164,[795]* Na salt 199-201; PhH	Cu		Type 1 (C1)	795*-797*
2'-Bromo	192-193	PhH, aq EtOH	Cu	32-33	Type 1 (C1)	128,590,685
3'-Bromo	174-175	aq EtOH	Cu	54-60	Type 1 (C1)	590, 799,800

Table IV. (Continued)

Constituents	mp (°C)	Recrystallizing solvents and derivatives, mp (°C)	Catalyst	Yield (%)	Reaction Type	Ref.
4'-Bromo	186	PhH, EtOH	Cu	60	Type 1 (Cl)	246, 536, 800
	210-211	Hydrazide 168-169	CuO	88	Type 1 (Br)	684
					Type 2 (Br)	226
4-Bromo-5,6-dimethyl	224-225				Type 1 (Br)	686
5-Bromo-4'-dimethylamino	222-224	EtOH	Cu	33	Type 1 (Br)	687
2'-Bromo-5'-ethyl	147-149		Cu	37	Type 1 (Br)	125
4-Bromo-4'-fluoro	190		Cu	76	Type 1 (Cl)	208
4'-Bromo-4-iodo	233-234 (dec)		Cu		Type 1 (I)	688
5-Bromo-4-methoxy	198		Cu		Type 1 (Br)	123
5-Bromo-4'-methoxy	205-206				Type 1 (Cl)	690,691
3'-Bromo-5-methoxy	194-195		Cu	71	Type 1 (Cl)	689
3'-Bromo-4'-methoxy	198-199	HOAc	Cu	49	Type 1 (Cl)	123
4-Bromo-4'-nitro	252		Cu	25	Type 2 (Br)	18
2'-Bromo-5'-nitro			Cu	34	Type 1 (Br)	578
4'-Bromo-4-nitro	288-289	HOAc; acid Cl 140-141	None	65-66	Type 1 (Cl)	64,694
			CuO	75	Type 1 (Cl)	683
4'-Bromo-5-nitro	258	HOAc, EtOH	Cu	17-51	Type 1 (Cl)	18,693
5'-Bromo-2'-nitro	240-242		None		Type 2 (Br)	217,692
2'-Bromo-3'-trifluoromethyl	219-220	aq EtOH, PhH	CuBr		Type 1 (Br)	180

(Table Continued)

295

Table IV. (Continued)

Constituents	mp (°C)	Recrystallizing solvents and derivatives, mp (°C)	Catalyst	Yield (%)	Reaction Type	Ref.
4'-n-Butoxy-4-nitro	197-198	aq EtOH	Cu	85	Type 1 (Cl)	14
3'-n-Butyl	116-117		Cu		Type 1 (Br)	801
3'-tert-Butyl	151-153		Cu	59	Type 1 (Cl)	788
					Type 1 (Br)	801
4'-n-Butyl	155-156		Cu	25	Type 1 (Br)	801
	146-149		Cu	33	Type 1 (Cl)	8*
4'-sec-Butyl	179-181	Toluene	Cu	35	Type 1 (Cl)	151
4'-tert-Butyl	205-206		Cu	48	Type 1 (Cl)	151
					Type 1 (Br)	801
5-tert-Butyl-2'-nitro						782
4'-n-Butyl-5-nitro	198	aq EtOH	Cu	65	Type 1 (Cl)	14
4'-sec-Butyl-5-nitro	184-185	Toluene	Cu	58	Type 1 (Cl)	151
3'-n-Butylthio	77-79	N-nitroso [933]	Cu		Type 1 (Cl)	796*,802*,803*
4'-Carbo-n-butoxy	118-119		Cu		Type 1 (Br)	801
4'-Carbo-t-butoxy	165-166		Cu		Type 1 (Br)	801
4'-Carboethoxy	176-177		Cu		Type 1 (Br)	801
			Cu(OAc)$_2$	27	Type 1 (Br)	804*
N-Carboethoxy-2',3'-dimethyl	148-149	PhH-cyclohexane			Isatoic anhydride, hydrolysis	262
4'-Carbomethoxy	180-181				Type 1 (Br)	801
4'-Carbomethoxy-3',5'-dimethoxy	204-205	aq MeOH; Me ester 125-126		38	Type 1 (Cl)	455

Table IV. (Continued)

Substituents	mp (°C)	Recrystallizing solvents and derivatives, mp (°C)	Catalyst	Yield (%)	Reaction Type	Ref.
4'-Carbo-n-propoxy	144-145				Type 1 (Br)	801
2'-Carboxy	301-302	PhH, aq EtOH, aq DMF, EtOH	Cu		Type 1 (Cl)	175,536,805
			Cu	76-88	Type 2 (Cl)	21,188,806
			Cu	84	CO_2 on Li cpd.	275
3'-Carboxy	296 (dec)	EtOH	Cu		Type 1 (Cl)	175
4'-Carboxy	290 (dec)	EtOH	Cu	50	Type 1 (Cl)	175, 188
6-Carboxy-2',3'-dimethyl	233-234 (dec)	aq EtOH	Cu(OAc)$_2$		Type 1 (Br)	801
4'-Carboxy-4,5-dimethyl	273-274 (dec)				Type 1 (Br)	195
2'-Carboxy-4,6-dinitro	153-159	HOAc, aq EtOH;	Cu	64	Type 1 (Br)	200
					Type 1 (Cl)	807
3'-Carboxy-4,6-dinitro	251-252 273	Acetyl 254-255[738] HOAc			Type 2 (Cl)	738
					Type 1 (Cl)	807
4'-Carboxy-4,6-dinitro	264-265	HOAc			Type 1 (Cl)	807
4-Carboxy-4'-methyl	286	EtCOMe		66	Type 1 (Cl)	200
6-Carboxy-4'-methyl	214-215 (dec)	aq EtOH	Cu	74	Type 1 (Br)	200
				81	Type 1 (Br)	200

(Table Continued)

Table IV. (Continued)

Substituents	mp (°C)	Recrystallizing solvents and derivatives, mp (°C)	Catalyst	Yield (%)	Reaction Type	Ref.[*]
2'-Carboxy-5-nitro	324-325		Cu	51-84	Type 2 (Cl)	75,150
N,N'-bis-(2-Carboxy-6-nitrophenyl)-p-phenylenediamine	280		Cu(OAc)$_2$		Type 1 (Br)	214
2'-Carboxymethyl	181-183 (dec)			40	Chapman	790
4'-Carboxymethyl	210 (dec)	BuOAc	Cu	65	Type 1 (Cl)	808
N,N'-bis (2-carboxyphenyl) m-phenylenediamine	262		Cu		Type 1 (Cl)	212
4-Chloro	208	HOAc	Cu		Type 1 (Cl)	809
	177-178				Type 2	137
				80	Chapman	138
5-Chloro	197-198	aq EtOH; anhydride 124-125,[58] acid Cl 102-103[58]	Cu	74-100	Type 1 (Cl)	596,676,698, 733,810,811 934
2'-Chloro	195-196	PhH, MeOH; amide 124[91]	Cu	65-77	Type 1 (Cl)	246,536,800 812
3'-Chloro	170-172	PhH; Et ester, amide 140-141	CuBr$_2$	48-93	Type 1 (Br)	91
			Cu		Type 1 (Cl)	30,121,192, 238,800,813
			CuBr$_2$		Type 1 (Br)	91

298

Table IV. (Continued)

Substituents	mp (°C)	Recrystallizing solvents and derivatives, mp (°C)	Catalyst	Yield (%)	Reaction Type	Ref.
4'-Chloro	177-178	PhH, aq EtOH; amide 156-157	Cu CuBr$_2$	80	Type 1 (Cl) Type 1 (Br)	246, 536, 800 91
3-Chloro-4'-(p-amino-phenyl)			Cu		Type 1 (Cl)	787
5-Chloro-3'-(p-chloro-phenoxy)	162-163	PhH	Cu	25	Type 1 (Cl)	68
5-Chloro-4'-(p-chloro-phenoxy)			Cu		Type 1 (Cl)	787
5-Chloro-4'-(p-chloro-phenyl)			Cu		Type 1 (Cl)	787
6-Chloro-4'-(p-chloro-phenyl)			Cu		Type 1 (Cl)	787
2'-Chloro-N-(o-chloro-phenyl)	212-213		Cu		Type 2	137
4'-Chloro-N-(p-chloro-phenyl)	220-221	aq HOAc	Cu		Type 2	137
5-Chloro-4'-(p-chloro-phenyl)amino			Cu		Type 1 (Cl)	787
2'-Chloro-3'-cyano	238-239		Cu(OAc)$_2$		Type 1 (Br)	194
5-Chloro-3',4'-dimethoxy-4-nitro	240-242				Type 1 (Cl)	41
5-Chloro-3',4'-dimethyl	208-210	PhH	Cu	68	Type 1 (Cl)	814

(Table Continued)

299

Table IV. (Continued)

Substituents	mp (°C)	Recrystallizing solvents and derivatives, mp (°C)	Catalyst	Yield (%)	Reaction Type	Ref.
5-Chloro-4'-dimethyl-amino	246-248		Cu(OAc)$_2$	51-69	Type 1 (C1)	21, 147
3-Chloro-N-dimethyl-aminopropyl				75-85	Reductive condn of aldehyde	238
5-Chloro-4'-dimethyl-sulfonamido	210-214		CuO		Type 1 (C1)	815
3'-Chloro-4-dimethyl-sulfonamido	184-186		CuO		Type 1 (C1)	815*
5'-Chloro-2',4'-dinitro	276	HOAc	Cu		Type 2 (C1)	252
5-Chloro-4'-ethoxy	224-225		Cu		Type 1 (C1)	690
5-Chloro-4'-ethyl	181				Type 1 (C1)	700
5-Chloro-4'-fluoro	212-213	Hydrazide 187-188[929]	Cu	90-94	Type 1 (C1)	701
3'-Chloro-2'-fluoro	210-211	Ph-EtOH	CuBr$_2$		Type 1 (Br)	196
4-Chloro-4-iodo	226-228 (dec)	HOAc	Cu	40-45	Type 1 (I)	688
3-Chloro-4'-methoxy	140-141			28	Chapman	713
4-Chloro-2'-methoxy	200-201		Cu	80	Type 1 (C1)	703
4-Chloro-3'-methoxy	189-190		Cu		Type 1 (C1)	702
4-Chloro-4'-methoxy	191-192		Cu	60	Type 1 (C1)	750, 706
4-Chloro-4'-methoxy			Cu-CuI	75	Type 1 (C1)	704
					Chapman	138

300

Table IV. (Continued)

Substituents	mp (°C)	Recrystallizing solvents and derivatives, mp (°C)	Catalyst	Yield (%)	Reaction Type	Ref.
5-Chloro-4-methoxy	184-185		Cu		Type 1 (Cl)	707
5-Chloro-2'-methoxy	205-206	Me ester 90[886]	Cu	60	Type 1 (Cl)	67, 706
5-Chloro-3'-methoxy	164-165	PhH, EtOH	Cu	14-59	Type 1 (Cl)	68, 706
5-Chloro-4'-methoxy	214-215	Anhydride 139-141	$Cu(OAc)_2$	40	Type 1 (Cl)	708
		acid Cl 110-11	Cu or CuO	77	Type 1 (Cl)	700, 711, 712
6-Chloro-4'-methoxy	147-148		Cu	11	Type 1 (Cl)	251
2'-Chloro-4-methoxy	189-190		Cu	63-75	Type 1 (Cl)	20
2'-Chloro-5-methoxy	278-279		Cu	86	Type 1 (Br)	706, 752
					Chapman	702
3'-Chloro-4-methoxy	190-191		Cu	20	Type 1 (Cl)	20
3'-Chloro-5-methoxy	167-168			83	Chapman	702
3'-Chloro-4'-methoxy	188		Cu	57	Type 1 (Cl)	733
4'-Chloro-4-methoxy	197-198	EtOH; acid Cl 118, explodes	Cu	73	Type 1 (Br)	703, 704
4'-Chloro-5-methoxy	181-182			77	Chapman	399
			Cu	30-40	Type 1 (Cl)	399
4'-Chloro-2'-methoxy-5-methyl	200-202			51	Chapman	399
			Cu		Type 1 (Cl)	699
5'-Chloro-4-methoxy-2'-methyl	169-170	HOAc	Cu	73	Type 1 (Br)	714

(Table Continued)

Table IV. (Continued)

Substituents	mp (°C)	Recrystallizing solvents and derivatives, mp (°C)	Catalyst	Yield (%)	Reaction Type	Ref.
5-Chloro-4-methoxy-2'-nitro	268-269		Cu	60	Type 1 (Cl)	254
5-Chloro-4'-methoxy-2'-nitro	270-272	BuOH			Nitration	253
5-Chloro-4'-methoxy-4-nitro	214		None	80	Type 1 (Cl)	715,716
5-Chloro-2'-methyl	201		Cu	36	Type 1 (Cl)	676
5-Chloro-3'-methyl	197-199	PhCl	Cu	60	Type 1 (Cl)	559,562,697, 717
5-Chloro-4'-methyl	226-231	Me ester 83[886]			Type 1 (Cl)	174, 700
2'-Chloro-6-methyl	181-182			72	Chapman	711
2'-Chloro-3'-methyl	218-219	EtOH, sublimes	CuBr$_2$		Type 1 (Br)	196
3'-Chloro-2'-methyl	207-208	Cyclohexane-PhH	Cu	68	Type 1 (Cl)	182
		Ca salt 134-136	CuBr$_2$		Type 1 (Br)	196
3'-Chloro-4'-methyl	181-182	aq EtOH; amide 132-133[91]	Cu		Type 1 (Cl)	182
			CuBr$_2$		Type 1 (Br)	91
4'-Chloro-5-methyl	203		Cu		Type 1 (Cl)	700
4'-Chloro-2'-methyl	198-200		Cu		Type 1 (Cl)	182
5'-Chloro-2'-methyl	178-180, 200		Cu	60	Type 1 (Br)	153
	234-235	HOAc	Cu	31	Type 1 (Cl)	182,576,816
5'-Chloro-4-methyl-2'-nitro	240-241	HOAc	None		Type 2 (Cl)	217,692

302

Table IV. (Continued)

Substituents	mp (°C)	Recrystallizing solvents and derivatives, mp (°C)	Catalyst	Yield (%)	Reaction Type	Ref.
5'-Chloro-N-methyl-2'-nitro	160	Me ester 92-93	Cu		Type 2	229
5-Chloro-2'-methyl-thio	212-214	EtOAc	Cu		Type 1 (Cl)	817
5-Chloro-4'-methyl-thio	194-195		Cu	34	Type 1 (Cl)	59,718
2'-Chloro-3'-methyl-thio	206-208		Cu		Type 1 (Cl)	818
2'-Chloro-5'-methyl-thio	163-165		Cu		Type 1 (Cl)	818
3-Chloro-6-nitro	203		Cu	56	Type 1 (Cl)	150
4-Chloro-2'-nitro	281-283	Me ester 174-175[780]	Cu		Type 2 (Br)	780, 782
4-Chloro-2'-nitro	232-235	Me ester 138	Cu		Type 2	229
4-Chloro-4'-nitro	257			43	Smiles	264,401
5-Chloro-4-nitro	228-232 (dec)	aq EtOH, CCl$_4$-AmOH	None		Type 1 (Cl)	819
5-Chloro-2'-nitro	281	Me ester 106-107	Cu	35	Type 1 (Cl)	723
			Cu		Type 2	229,782
5-Chloro-4'-nitro	235		Cu	17	Type 1 (Cl)	719
2'-Chloro-4-nitro	273-275		Cu	25	Type 1 (Cl)	500
2'-Chloro-5-nitro	254-256		Cu	15	Type 1 (Cl)	500,926
2'-Chloro-5'-nitro	260-261		Cu	40	Type 1 (Br)	578

(Table Continued)

303

Table IV. (Continued)

Constituents	mp (°C)	Recrystallizing solvents and derivatives, mp (°C)	Catalyst	Yield (%)	Reaction Type	Ref.
3'-Chloro-4-nitro	276		None	60	Type 1 (Cl)	719
3'-Chloro-5-nitro	221-222		Cu	23	Type 1 (Cl)	130
4'-Chloro-4-nitro	285 (dec)	Acid Cl 135-136	Cu	38	Type 1 (Cl)	130,926
			None	86	Type 1 (Cl)	722
4'-Chloro-5-nitro	242		Cu	21	Type 1 (Cl)	500,720,721
			Cu	46	Type 1 (Cl)	293,926
4'-Chloro-2'-nitro	245-248	Et ester 134-136	Cu		Type 2	229
5'-Chloro-2'-nitro	228	Me ester 157-158, Et ester 127-128	Cu		Type 2	153,229
4-Chloro-4'-phenoxy			Cu	35	Type 1 (Cl)	787
5-Chloro-3'-phenoxy	168-169	aq EtOH	Cu		Type 1 (Cl)	68
5-Chloro-4'-phenoxy	190	EtOH	Cu		Type 1 (Cl)	787
4-Chloro-2'-phenyl			Cu		Type 1 (Cl)	787
4-Chloro-N-phenyl	157-158		Cu		Type 2	137
5-Chloro-2'-phenyl	203		Cu	51	Type 1 (Cl)	22
5-Chloro-4'-phenyl	246	PhCl	Cu	67	Type 1 (Cl)	559,562,697
						787
5-Chloro-N-phenyl	138		Cu		Type 2	137
5-Chloro-4-sulfon-amido	245 (dec)	aq MeOH	None	47	Type 1 (Cl)	820,821
5-Chloro-3'-trifluoromethyl	208-210	PhH, $CHCl_3$	Cu	20-47	Type 1 (Cl)	68,814,822

Table IV. (Continued)

Substituents	mp (°C)	Recrystallizing solvents and derivatives, mp (°C)	Catalyst	Yield (%)	Reaction Type	Ref.
2'-Chloro-5'-trifluoromethyl	185-186	aq MeOH; amide 167-169	CuBr$_2$		Type 1 (Br)	91
N-(o-Chlorobenzoyl)		Anilide 214-214			Type 1 (Cl)-amide	13
N-(p-Chlorobenzoyl)-2',3'-dimethyl	117	Me ester			Chapman	823
N-(p-Chlorobenzoyl)-3'-trifluoromethyl		Me ester			Chapman	823
4'-(p-Chlorophenyl)	237-240	Dioxan	Cu	55-60	Type 1 (Cl)	787,824*
4-Cyano	220-222		Cu	40	Type 1 (Cl)	17
5-Cyano	210-212		Cu	60	Type 1 (Cl)	17
3'-Cyano	205		Raney Cu	15-63	Type 1 (Cl)	17,165
4'-Cyano	226-228		Cu	45	Type 1 (Cl)	23
5-Cyano-4'-ethoxy	191-194		Cu	35	Type 1 (Cl)	725
4-Cyano-4'-methoxy	214		Cu	69	Type 1 (Cl)	17
5-Cyano-4'-methoxy	195-196		Cu	38-47	Type 1 (Cl)	718
3'-Cyano-2'-methyl	234-235		Cu(OAc)$_2$		Type 1 (Br)	194
3,2'-Diacetamido		Me ester 168-169			Type 2	199
5,4'-Diacetamido	244-246		Cu	74	Type 1 (Cl)	147
6,3'-Diacetamido	208-210	aq HOAc			Redn and acylation	114
2',4'-Diacetamido-4,6-dinitro	255				Type 1 (Cl)	785

(Table Continued)

Table IV. (Continued)

Substituents	mp (°C)	Recrystallizing solvents and derivatives, mp (°C)	Catalyst	Yield (%)	Reaction Type	Ref.
4,5-Diamino		Me ester 133			Esterification	784
4,6-Diamino	237-238 (dec)				Redn	786
4,4'-Diamino		Me ester 147, Me ester hydrate 86, acetyl Me ester 230			Redn	883
6,4'-Diamino	245-246	aq EtOH			Redn	158
2',4'-Diamino		Me ester 126-129			Redn	932
4,6-Diamino-4'-hydroxy		Me ester 162			Esterification	785
4,6-Diamino-4'-sulfonamido	(dec)	H_2O			Redn	109*
4',6'-Diamino-2'-sulfonamido	233-234	H_2O; diacetyl 232-233			Redn	109
4,4'-Dibromo	232-234	aq acetone	Cu	52	Type 1 (Cl)	139
2',3'-Dibromo	245-247		CuBr		Type 1 (Br)	196
2',4'-Dibromo?	256-7				Halogenation	887
2',5'-Dibromo	229-230	EtOH, HOAc; acid Cl 118-20	Cu	36	Type 1 (Br)	153
3',5'-Dibromo	247-249		Cu	49	Type 1 (Cl)	63,825
3',5'-Dibromo-4-methoxy	205-207	HOAc; acid Cl 161-163	Cu	99	Type 1 (Br)	63

306

Table IV. (Continued)

Substituents	mp (°C)	Recrystallizing solvents and derivatives, mp (°C)	Catalyst	Yield (%)	Reaction Type	Ref.
4,2'-Dibromo-4'-methyl	247	Me ester 93-94			Hydrolysis of 2-anilino-tropone	826
4,4'-Di-tert-butyl	218-219			91	Hydrolysis	151
4,2'-Dichloro		Acid Cl 142-145, amide 234			Type 1 (Cl)	778
4,4'-Dichloro		Acid Cl 134-135, amide 134-135			Type 1 (Cl)	778
5,2'-Dichloro	228-229	HOAc, aq EtOH acid Cl 100-101	Cu(OAc)$_2$	22	Type 1 (Cl)	727
			Cu	47	Type 1 (Cl)	222
5,3'-Dichloro	199-201	PhCl	Cu	42-53	Type 1 (Cl)	68,733
		Hydrazide, 140-141[929]	CuO	6	Type 1 (Cl)	392
			Cu(OAc)$_2$	53	Type 1 (Cl)	727
5,4'-Dichloro	176-178	Hydrazide 185-187[929]			Type 1 (Cl)	691
6,2'-Dichloro		HOAc; acid Cl 89-90	Cu	73	Type 2 (I)	222
2',3'-Dichloro	256-257	EtOH, bis(2-methoxy-ethyl) ether; Me ester 86-88, Et ester 69-71, amide 167-169, di-ethanolamine salt 105-106	Cu	85	Type 1 (Cl)	814
			CuBr$_2$		Type 1 (Br)	91,196

(Table Continued)

Table IV. (Continued)

Substituents	mp (°C)	Recrystallizing solvents and derivatives, mp (°C)	Catalyst	Yield (%)	Reaction Time	Ref.
2',4'-Dichloro	250-251(dec)	EtOH; amide 165-167[929]	Cu	55	Type 1 (Cl)	246
		hydrazide 192-194	CuBr$_2$		Type 1 (Br)	91
					Chapman	138
2',5'-Dichloro	237	HOAc-PhH	Cu	43	Type 1 (Br)	153
				48	Type 1 (Cl)	137
2',6'-Dichloro	216-217	aq acetone	CuO	48	Type 2 (Br)	242
3',4'-Dichloro	178	PhH	Cu		Type 1 (Cl)	816
3',5'-Dichloro	251-252 (dec)	EtOH, aq HOAc, PhH-	Cu	53-61	Type 1 (Br)	576,577
		EtOH; amide 173-175	CuBr		Type 1 (Br)	91
					Type 1 (Cl)	816
2',5'-Dichloro-N-(2,5-dichloro-phenyl	223-227	HOAc, xylene, PhBr	Cu		Type 1 (Cl)	137
4,5-Dichloro-4'-methoxy	233-234		Cu		Type 1 (Cl)	705
3',4'-Dichloro-4-methoxy	187	HOAc; acid Cl 117	Cu	49	Type 1 (Br)	728
3',5'-Dichloro-4'-methoxy	213	HOAc; acid Cl 195	Cu		Type 1 (Cl)	729
2',6'-Dichloro-N-methyl		Me ester b$_{0.005}$ 120-130	None		Methylation	234
4'-(2-Diethylamino)ethoxy-4-nitro	229-230			34	Type 1 (Cl)	935

Table IV. (Continued)

Substituents	mp (°C)	Recrystallizing solvents and derivatives, mp (°C)	Catalyst	Yield (%)	Reaction Time	Ref.
4'-(2-Diethylamino)-ethoxy-5-nitro	226		Cu		Type 1 (Cl)	730
N-(2-Diethylamino)-ethyl-2'-nitro		Me ester HCl 173-174			N-alkylation	236
4,4'-Difluoro	220 (dec)		Cu	69	Type 1 (Br)	208
4,4'-Diiodo	232-233 (dec)		Cu		Type 1 (I)	688
4,4'-Dimethoxy	173-174		Cu	54	Type 1 (Br)	733
5,6-Dimethoxy	155		Cu	15	Type 2 (Br)	508,731
6,2'-Dimethoxy						98
2',3'-Dimethoxy	163-164		Cu	40	Type 1 (Br)	508
			Cu	55	Type 1 (Cl)	731
2',4'-Dimethoxy	207-208		Cu	73	Type 1 (Cl)	731
2',5'-Dimethoxy	166-167		Cu	45-69	Type 1 (Cl)	98,731,732
3',4'-Dimethoxy	180-181		Cu	78-90	Type 1 (Cl)	3,731
3',5'-Dimethoxy	147		Cu	68	Type 1 (Cl)	597,731
2',4'-Dimethoxy-3'-ethoxy	133-135	Petrol	Cu	43	Type 2 (Br)	600
2',4'-Dimethoxy-4-dimethylsulfonamido	184-186		CuO		Type 1 (Cl)	815*
2',3'-Dimethoxy-4',5'-methylenedioxy	173-174	PhH-petrol	Cu	53	Type 2	

(Table Continued)

Table IV. (Continued)

Substituents	mp (°C)	Recrystallizing solvents and derivatives, mp (°C)	Catalyst	Yield (%)	Reaction Time	Ref.
2',5',-Dimethoxy-3',4'-methylenedioxy	166	EtOH	Cu	39	Type 2 (I)	221
2',5'-Dimethoxy-5-nitro	222-223	Aq EtOH			Type 1 (Cl)	927
3',4'-Dimethoxy-4-nitro	221-223		Cu	45	Type 1 (Cl)	14
3',4'-Dimethoxy-5-nitro	228		Cu	34-45	Type 1 (Cl)	14,243
			$Cu(OAc)_2$	50	Type 1 (Cl)	735
4,4'-Dimethyl	184-186	EtOH	CuO	78	Type 2 (Br)	232
5,4'-Dimethyl	178	aq MeOH	Cu/CuCl	44	Type 2 (Br)	81
					Type 1 (Cl)	700
6,2'-Dimethyl	186-187	PhH; Me ester 67-68	Cu	70-75	Type 2 (Br)	10, 23
2',3'-Dimethyl (mefenamic acid, mephenamic acid, Parkemed or Ponstan)	229-230	MeOH, EtOH, aq MeOH aq EtOH, bis (2-methoxyethyl)ether; Me ester 97-99[828,832,936], Et ester $b_{0.4}$ 152-153[832], amide 150-152[386,828], n-PR ester[832]	Cu	58	Type 1 (Cl)	186,827
			CuO		Type 1 (Cl)	828
			Cu,$Cu(OAc)_2$, CuBr		Type 1 (Br)	829
			$Cu(OAc)_2$		Type 1 (I)	204
					Oxidation	783,810,831
					Hydrolysis	261,783,830, 831

310

Table IV. (Continued)

Substituents	mp (°C)	Recrystallizing solvents and derivatives, mp (°C)	Catalyst	Yield (%)	Reaction Time	Ref.
					Hydrogenation	783,830,831
					Desulfuration	204,258,261
					Chapman (Various)	258-261
					Deamination	783
					Carboxylation	274
					Smiles	265
2',4'-Dimethyl	187		CuCl	70	Type 1 (Cl)	240,931
2',5'-Dimethyl	140-141		Cu	40	Type 1 (Cl)	732
3',4'-Dimethyl	188-189		Cu	83	Type 1 (Cl)	122*
2',3'-Dimethyl-4-dimethylsulfonamido	194-195		CuO		Type 1 (Cl)	815
2',3'-Dimethyl-N-formyl	211-212					259
5,5'-Dimethyl-6'-methoxy	185-186		Cu/CuCl	67	Type 2 (I)	81
4',5'-Dimethyl-2',3'-methylenedioxy	166-167		Cu		Type 2 (I)	221
4,5-Dimethyl-2'-nitro						782

(Table Continued)

311

Table IV. (Continued)

Substituents	mp (°C)	Recrystallizing solvents and derivatives, mp (°C)	Catalyst	Yield (%)	Reaction Time	Ref.
2',3'-Dimethyl-4-nitro	265-266 (dec)	aq EtOH	Cu		Type 1 (Br)	783
4',N-Dimethyl-2'-nitro	140-141		Cu		Type 2	229*
5,3'-Dimethyl-2'-nitro-6,4',5'-trimethoxy	212-213	aq MeOH	Cu/CuCl	28	Type 1 (Cl)	119,498
2',3'-Dimethyl-N-phenyl	161	aq HOAc	Cu		Type 2 (I)	230
2',4'-Dimethyl-N-phenyl	185-188	EtOH, HOAc	Cu		Type 2 (I)	833
4,4'-Dimethyl-N-p-tolyl					Type 2 (Br)	232
5-Dimethylamino	171-172	EtOH	Cu(OAc)₂	47	Type 1 (Cl)	117
2'-Dimethylamino	198-200	PhH	Cu	75	Type 1 (Cl)	117
3'-Dimethylamino	154		Cu	40	Type 1 (Cl)	117
4'-Dimethylamino	224-226	PhH; Me ester⁷⁸⁴ 99-100	Cu	35-67	Type 1 (Cl)	117,147,536
3'-Dimethylamino-2'-methyl	180-182	aq EtOH	Cu(OAc)₂	61	Type 1 (Cl)	21
					Redn and methylation	834
4'-Dimethylamino-5-nitro	216-218		Cu	74	Type 1 (Cl)	147

312

Table IV. (Continued)

Substituents	mp (°C)	Recrystallizing solvents and derivatives, mp (°C)	Catalyst	Yield (%)	Reaction Time	Ref.
N-(2-Dimethylamino)ethyl-2'-nitro		Me ester HCl 290-292			N-Alkylation	236*
4-Dimethylsulfonamido	200-201	Me ester 118-119, amide 215-217, acid Cl 154-155	CuO		Type 1 (Cl)	815, 835
4-Dimethylsulfonamido-2'-hydroxy	209-211	Petrol-ether		73	Hydrolysis	266*
4-Dimethylsulfonamido-4'-methoxy	172-174		CuO		Type 1 (Cl)	815
4-Dimethylsulfonamido-6-methyl	220-223		CuO		Type 1 (Cl)	815
3'-Dimethylsulfonamido-2'-methyl	192-194				Redn	836
4-Dimethylsulfonamido-3'-trifluoromethyl	203-204		CuO		Type 1 (Cl)	815
3,2'-Dinitro	265-268	Me ester 131-133	Cu	34	Type 1 (F)	199
4,5-Dinitro	215, 192[786]	aq, EtOH, acetyl 209-210, benzoyl 120-121, Et ester 140, anilide 160-161	None	63	Type 1 (Cl)	203
4,6-Dinitro			None		Type 1 (Cl)	738,786,837
			None		Type 1 (Cl)	156

(Table Continued)

313

Table IV. (Continued)

Substituents	mp (°C)	Recrystallizing solvents and derivatives, mp (°C)	Catalyst	Yield (%)	Reaction Time	Ref.
4,3'-Dinitro	288–290		Cu	9	Type 1 (F)	202
4,4'-Dinitro	293 (cor)		None	33	Type 1 (Cl)	147
5,2'-Dinitro	231		Cu	9	Type 1 (Cl)	210
5,3'-Dinitro	253		Cu	28	Type 1 (Cl)	25
5,4-Dinitro	209–211		Cu	18	Type 1 (Cl)	239
	252 (cor)			79	Type 1 (Cl)	25
					Smiles	264
6,2'-Dinitro	252–254		Cu	20	Type 1 (Br)	583
			Cu	24	Type 1 (Cl)	142
			Cu	22	Type 2 (Br)	583
6,3'-Dinitro	196–198	aq EtOH, aq HOAc; Me ester 143–146	Cu	50	Type 1 (Br)	114
			Cu		Type 1 (Cl)	148
2',4'-Dinitro	270	EtOH-acetone; Me ester 166–167[932]	None	75	Type 2 (Cl)	737,838
			None	89	Type 2 (F)	223
			KF	91–98	Type 2 (F)	225
			KI	85	Type 2 (Cl)	932
3',5'-Dinitro	263		Cu	22	Type 1 (Cl)	26
4,6-Dinitro-2'-chloro		Me ester 178	NaOAc	60	Type 1 (Cl)	205
4,6-Dinitro-4'-chloro		Me ester 182	NaOAc	85	Type 1 (Cl)	205

314

Table IV. (Continued)

Substituents	mp (°C)	Recrystallizing solvents and derivatives, mp (°C)	Catalyst	Yield (%)	Reaction Time	Ref.
4,6-Dinitro-*N*-ethyl	150–151	EtOH			Type 1 (C1)	207
4,6-Dinitro-2'-hydroxy	312				Type 1 (C1)	786
4,6-Dinitro-4'-hydroxy	103	Et ester 150	None		Type 1 (C1)	156
		Acetyl 97–99, benzoyl 123	None		Type i (C1)	738
2',4'-Dinitro-4-hydroxy	266–268	aq THF, aq dioxan	None		Type 2 (F)	224
4,6-Dinitro-2'-methoxy		Et ester 163	None		Type 1 (C1)	156
4,6-Dinitro-4'-methoxy		Me ester 151	NaOAc	80	Type 1 (C1)	205
4,6-Dinitro-2'-methyl	171–172	EtOH			Type 1 (C1)	807
		Me ester 150	NaOAc	75	Type 1 (C1)	205
		Et ester 119	None		Type 1 (C1)	156
					Type 1 (C1)	807
4,6-Dinitro-3'-methyl	203	Et ester 132	None		Type 1 (C1)	156
4,6-Dinitro-4'-methyl	320				Type 1 (C1)	807
2',4'-Dinitro-*N*-methyl	176–178	Me ester 185	NaOAc	78	Type 1 (C1)	205
		EtOH			Type 2 (C1)	932

(Table Continued)

Table IV. (Continued)

Substituents	mp (°C)	Recrystallizing solvents and derivatives, mp (°C)	Catalyst	Yield (%)	Reaction Time	Ref.
4,6-Dinitro-2'-sulfonamido	220-222		NaOAc	40	Type 1 (Br)	109
4,6-Dinitro-4'-sulfonamido	275-276	aq EtOH	NaOAc	63	Type 1 (Br)	109*
4',6'-Dinitro-2'-sulfonamido	287-288	EtOH;	NaOAc	45	Type 2	109
5,N-Diphenyl	222-223	Me ester 230-232		75	Methylation	109
3',5'-Di-(trifluoromethyl)	195-197	EtOH; Na salt 289-293	Cu	75	Type 2 (I)	101
			Cu		Type 1 (Cl)	839
3'-Ethoxy	139		Cu			800
4'-Ethoxy	211	PhH	Cu	92	Type 1 (Cl)	40,193,536, 800
4'-Ethoxy-4-iodo	208		Cu	80	Type 1 (Br)	741
4'-Ethoxy-4-iodo	208-209		Cu	45	Type 1 (I)	741
4'-Ethoxy-4-methoxy	162-163		Cu	71	Type 1 (Cl)	40
4'-Ethoxy-6-methoxy	174		Cu		Type 1 (Cl)	40
5'-Ethoxy-2'-nitro						782
4'-Ethoxy-4-nitro	213-214		Cu	80	Type 1 (Cl)	4
4'-Ethoxy-5-nitro	233-234		Cu	78-80	Type 1 (Cl)	742,743,937
4'-Ethoxy-N-phenyl	135-138		None	85	Type 1 (Cl)	189
5-Ethyl	154-155		Cu	50	Type 2 (I)	833
			Cu		Type 1 (Br)	125

Table IV. (Continued)

Substituents	mp (°C)	Recrystallizing solvents and derivatives, mp (°C)	Catalyst	Yield (%)	Reaction Time	Ref.
2'-Ethyl	169-171	Acetone	Cu	15-60	Type 1 (Cl)	8,23,125
3'-Ethyl	117-119		Cu-Cu(OAc)$_2$	50	Type 1 (Cl)	125
4'-Ethyl	174-175		Cu	35-53	Type 1 (Cl)	8,125
2'-Ethyl-5'-hydroxymethyl	85-90		Cu		Type 1 (Br)	840
N-Ethyl-2'-mercapto	114-116	Petrol		51	CO$_2$ on Li cpd.	276
3'-Ethyl-2'-methyl	206				Type 1 (Br) Chapman	829 829
4'-Ethyl-5-nitro	135-136	EtOH	Cu		Type 1 (Cl)	746
N-Ethyl-2'-nitro			Cu		Type 2 (Br)	135
3',4'-Ethylene-dioxy-5-nitro	249-251	aq EtOH	Cu	69	Type 1 (Cl)	14,747
4'-Ethylsulfonyl-2'-nitro	244-245	EtOH-acetone; Me ester 134-136	Cu	65	Type 2 (Cl)	932
4'-Ethylsulfonyl-N-methyl-2'-nitro	171-172	iPrOH-PhH	Cu		Type 2 (Cl)	932
3'-Ethylthio	114-116		Cu		Type 1 (Cl)	795*,796,797 841
4-Fluoro	190	PhH	Cu		Type 1 (Br)	207
5-Fluoro	183		Cu		Type 1 (Cl)	187
2'-Fluoro	186		CuCl	52	Type 2 (Br)	187
3'-Fluoro	164		Cu	64	Type 1 (I)	121,187

(Table Continued)

317

Table IV. (Continued)

Substituents	mp (°C)	Recrystallizing solvents and derivatives, mp (°C)	Catalyst	Yield (%)	Reaction Time	Ref.
4'-Fluoro	203 (dec)	EtOH, PhH; amide 120-121; hydrazide 135-137[929]	Cu	60	Type 1 (Cl)	799
			Cu	83	Type 1 (Cl)	536,748,800
					Type 1 (Br)	71
					Type 1 (I)	187
5-Fluoro-4'-methoxy	187-189		Cu	64	Type 1 (Cl)	725
3'-Fluoro-4-methoxy			Cu		Type 1 (Br)	67
3-Fluoro-2'-nitro						782
2'-Fluoro-3-nitro		Et ester 105	None	18	Type 1 (Cl)	155
		Amide 184-185	None	49	Type 1 (Cl)	155
3'-Fluoro-3-nitro		Et ester 119-120	None	59	Type 1 (Cl)	155
		Amide 233-234	None	43	Type 1 (Cl)	155
4'-Fluoro-3-nitro		Et ester 121-122	None	34	Type 1 (Cl)	155
		Amide 231-232	None	82	Type 1 (Cl)	155
4'-Fluoro-3'-nitro	175	HOAc	Cu	46	Type 1 (Cl)	208
3'-Heptafluoro-propyl			Cu		Type 1 (Cl)	842
2'-Hydroxy	190-191	aq MeOH	Cu	38	Type 1 (Cl)	175,193
			CuCl	38	Type 1 (Br)	152
3'-Hydroxy	171-172	i-Pr ether	Cu	62	Type 1 (Cl)	794
5'-Hydroxy-2',4'-dinitro	282		None		Type 2 (NO_2)	252
3-Hydroxy-5-methyl	122-123	Anilide 163-164		80	Hydrolysis	11
2'-Hydroxy-4-nitro	251-253	Acetone, petrol		40	Hydrolysis	266
4'-Hydroxy-4-nitro	210		Cu			785

318

Table IV. (Continued)

Substituents	mp (°C)	Recrystallizing solvents and derivatives, mp (°C)	Catalyst	Yield (%)	Reaction Time	Ref.
3'-Hydroxymethyl	143-145		Cu		Type 1 (Br)	840
3'-Hydroxymethyl-2'-methoxy	180-184		Cu		Type 1 (Br)	840*
3'-Hydroxymethyl-2'-methyl	228-232		Cu		Type 1 (Br)	840
4-Iodo	229-230	HOAc	Cu	14	Type 1 (I)	115
2'-Iodo	199-200	CHCl₃	Cu			685
3'-Iodo	168-170		Cu	80-90	Type 1 (Cl)	796,803,843
4'-Iodo	200-202	HOAc	Cu		Type 1 (Cl)	115,800
4-Iodo-2'-ethoxy			Cu		Type 1 (I)	209
4-Iodo-5-methoxy	202-203	HOAc			Chapman	400
4-Iodo-4'-methoxy	208	HOAc	Cu		Type 1 (I)	116
5-Iodo-4'-methoxy	219-220		Cu?		Type 1 (Cl)	690
4'-Iodo-5-methoxy	202-203			50	Chapman	400
4-Iodo-2'-methyl	202-203	HOAc	Cu	95	Type 1 (I)	209
4-Iodo-4'-methyl	218-219		Cu	19	Type 1 (I)	749
2'-Mercapto-N-phenyl		aq MeOH			From phenothiazine	276
3-Methoxy	111		Cu	60	Type 1 (Cl)	307
			Cu/CuCl	24	Type 2 (I)	71
4-Methoxy	171-172	MeOH, EtOH; Me ester 73-75	Cu	80-100	Type 1 (Br)	179,733
5-Methoxy	178		None	83	Type 1 (Cl)	798
			Cu	80	Type 1 (Cl)	751

(Table Continued)

319

Table IV. (Continued)

Substituents	mp (°C)	Recrystallizing solvents and derivatives, mp (°C)	Catalyst	Yield (%)	Reaction Time	Ref.
6-Methoxy	198		Cu/CuCl	39	Type 2 (I)	71
2'-Methoxy	177-178	PhH, EtOH; Me ester 59[886]	Cu Cu	30-87	Type 1 (Cl) Type 1 (Cl)	97 175,193,536, 750,800,813, 913
3'-Methoxy	136	aq EtOH; Et ester $b_{0.4}$ 154-156[798]	CuCl	79	Type 1 (Br) Type 2	152 280
4'-Methoxy	185-187	Me ester 54-57[798] Et Ester 78[886] Hydrazide 129-130[929]	Cu Cu(OAc)$_2$ Cu	45-53 40 64-67	Type 1 (Cl) Type 1 (Cl) Type 1 (Cl)	30,192,750 708 3,750,800, 931
4-Methoxy-4'-iodo	203-204	HOAc	RaCu	86	Type 1 (Cl)	165
3-Methoxy-5-methyl	193	Et ester 74-76	Cu		Type 1 (Br)	116
2'-Methoxy-3'-methyl					Methylation	11
4'-Methoxy-5-methyl	182	aq MeOH	Cu	66	Type 2 (I)	595
4-Methoxy-N-methyl-2'-nitro	168-172		Cu		Type 1 (Cl) Type 2	700 229
4'-Methoxy-N-methyl-2'-nitro	164-166		Cu		Type 2	229
3'-Methoxy-4',5'-methylenedioxy	182-183	EtOH, PhH-petrol	Cu	43	Type 2	14

Table IV. (Continued)

Substituents	mp (°C)	Recrystallizing solvents derivatives, mp (°C)	Catalyst	Yield (%)	Reaction Time	Ref.
4-Methoxy-5-nitro	224-226	EtOH	Cu	73	Type 1 (Br)	120
4-Methoxy-2'-nitro	240		Cu		Type 1(Cl)	241
4-Methoxy-3'-nitro	225		Cu		Type 2	229
5-Methoxy-2'-nitro	235-237		Cu		Type 1 (Br)	132
2'-Methoxy-2'-nitro			Cu		Type 2	229
2'-Methoxy-3-nitro		Et ester 112-113	None	41	Type 1 (Cl)	155
		Amide 216-126	None	59	Type 1 (Cl)	155
2'-Methoxy-4-nitro	249-250	aq EtOH, acetone, ether; Me ester 253-254[266]	Cu	50	Type 1 (Cl)	211
2'-Methoxy-5-nitro		HOAc	Cu	20	Type 1 (Cl)	753
2'-Methoxy-5'-nitro	241-243		Cu	50	Type 1 (Cl)	732
			CuO		Type 1 (Cl)	50
3'-Methoxy-3-nitro		Et ester 81-82	None	46	Type 1 (Cl)	155
		Amide 198-199	None	55	Type 1 (Cl)	155
3'-Methoxy-4-nitro	253-254	aq EtOH	Cu	74	Type 1 (Cl)	211
3'-Methoxy-5-nitro		HOAc	Cu		Type 1 (Cl)	753
4'-Methoxy-3-nitro		Et ester 120-122	None	56	Type 1 (Cl)	155
		Amide 216-217	None	79	Type 1 (Cl)	155
4'-Methoxy-4-nitro	233	EtOH; amilide 168	Cu	70	Type 1 (Cl)	50,756,844
			None	58-60	Type 1 (Cl)	211
4'-Methoxy-5-nitro	235	HOAc	Cu		Type 1 (Cl)	725,742,753
						754

(Table Continued)

321

Table IV. (Continued)

Substituents	mp (°C)	Recrystallizing solvents and derivatives, mp (°C)	Catalyst	Yield (%)	Reaction Time	Ref.
4'-Methoxy-2'-nitro	228-230	Et ester 104	Cu	70-82	Type 2 (Br)	229,255
4'-Methoxy-3'-nitro	193-196	HOAc	Cu	85	Type 1 (Cl)	241
			Cu	21	Type 2	241
5'-Methoxy-2'-nitro	239-240	Me ester 149	Cu		Type 2	229
4'-Methoxy-4-sulfonamido	242		Cu		Type 1 (Cl)	244
4'-Methoxy-5-sulfonamido			Cu		Type 1 (Cl)	244
4'-Methoxy-4-sulfon-diethylamido	170-171		Cu		Type 1 (Cl)	244*
4'-Methoxy-4-sulfonyl (p-sulfonamido-phenyl)amino	246	aq EtOH	Cu		Type 1 (Cl)	757*
2'-Methoxy-5'-tetra methylbutyl	161		Cu		Type 1 (Cl or Br)	758
3-Methyl	145		Cu		Type 1 (Cl)	97
5-Methyl	193-194		Cu		Type 1 (Br)	97
2'-Methyl	191-192	PhH; Me ester 59[886]	Cu	35-87	Type 1 (Cl)	8,51,118,175, 812,931
3'-Methyl	139	PhH; Me ester bp 205°/14 mm[886]	Cu	35-62	Type 1 (Cl)	8,118,175
4'-Methyl	196	PhH; Me ester 50[886]	Cu	36-90	Type 1 (Cl)	8,118,175, 536,682,931

322

Table IV. (Continued)

Substituents	mp (°C)	Recrystallizing solvents and derivatives, mp (°C)	Catalyst	Yield (%)	Reaction Time	Ref.
N-Methyl	103-104	aq MeOH	Cu		Type 1 (C1)	175
	170				*N*-methylation	175
5-Methyl-4'-methyl-thio				14	Carboxylation	273
N-Methyl-5'-methylthio		Ether-petrol; Me ester 102-103			Type 1 (C1)	59
N-Methyl-5'-methylthio				92	Na-MeSH on C1	229
N-Methyl-5'-methyl-thio-2'-nitro		Me ester 102-103	Cu		Type 2	229
2'-Methyl-5'-methyl-sulfonyl	184-186		Cu		Type 1 (C1)	845
2'-Methyl-3'-methyl-sulfoxy					Periodate on SMe	836
6-Methyl-2'-nitro	188-189		Cu	89	Type 2 (Br)	680,782*
2'-Methyl-5-nitro	234-235	HOAc	Cu	59	Type 1 (C1)	693
2'-Methyl-3'-nitro	217-221 (dec)	*i*-PrOH	CuBr₂		Type 1 (Br)	846
2'-Methyl-5'-nitro	221-222		Cu	23-55	Type 1 (C1)	136,732
4'-Methyl-4-nitro	262-265		None		Type 1 (C1)	575
4'-Methyl-5-nitro	222-223	aq EtOH	Cu	80	Type 1 (C1)	693,746
4'-Methyl-2'-nitro	213-215	Et ester 99-100	Cu		Type 2	229

(Table Continued)

323

Table IV. (Continued)

Substituents	mp (°C)	Recrystallizing solvents and derivatives, mp (°C)	Catalyst	Yield (%)	Reaction Time	Ref.
N-Methyl-2'-nitro	136-137	EtOH; Me ester 77-78	Cu		Type 2 (Br)	135
N-Methyl-2'-nitro-4'-trifluoro-methyl	154-156		Cu		Type 2	229
2'-Methyl-N-phenyl	173-176	aq HOAc, aq MeOH	Cu		Type 2 (I)	833
3'-Methyl-N-phenyl	145-149	aq EtOH	Cu		Type 2 (I)	833
4'-Methyl-N-phenyl	174-176	EtOH	Cu		Type 2 (I)	833
2'-Methyl-5'-isopropyl	123-127		Cu	5	Type 1 C1)	8
2'-Methyl-3'-trifluoromethyl	185-187	aq EtOH, PhH -cyclohexane	CuBr		Type 1 (Br)	180
N-Methyl-3'-trifluoro-methyl		Me ester $b_{0.01}$ 125			N-methyla-tion	234
4-Methyl-5,2',4'-trinitro	298				Type 2	759
4'-Methyl-4,6,2'-trinitro ($2\frac{1}{2}H_2O$)	232	aq EtOH	None		Type 1 (C1)	760
3',4'-Methylene-dioxy	165-166	PhH-petrol	Cu	48	Type 2	14
3',4'-Methylene-dioxy-4-nitro	246-247	MeOH	Cu	60	Type 1 (C1)	14*
3',4'-Methylene-dioxy-5-nitro	246-247	aq MeOH	Cu	49-91	Type 1 (C1)	14,747

324

Table IV. (Continued)

Substituents	mp (°C)	Recrystallizing solvents and derivatives, mp (°C)	Catalyst	Yield (%)	Reaction Time	Ref.
4'-Methylsulfonyl	177-179	EtOAc-ether	Cu	40	Type 1 (Cl)	845
4'-Methylsulfonyl-3-nitro		Et ester 189-190	None	10	Type 1 (Cl)	155
4'-Methylsulfonyl-methyl	195-196	Amide 207-208	None	7	Type 1 (Cl)	155
			Cu		Type 1 (Cl)	845
2'-Methylthio	185-185	PhH	Cu		Type 1 (Cl)	817
3'-Methylthio	139-41		Cu	80-90	Type 1 (Cl)	795*,796,797* 841,847
4'-Methylthio	191-192		Cu	35	Type 1 (Cl)	190
				24	Type 2 (Br)	190
5'-Methylthio-2'-nitro		Et ester 187-188	Cu		Type 2	229
4-Morpholinosulfon-amido	221-222		CuO		Type 1 (Cl)	815
3-Nitro	174		Cu	15	Type 1 (Cl)	150
		Et ester 111-112	None	25	Type 1 (Cl)	155
		Amide 184-185	None		Type 1 (Cl)	155
4-Nitro	256-258		None	95	Type 1 (Cl)	5
					Type 1 (Br)	848
5-Nitro	232-234	Me ester 101-102[883]	Cu	78-81	Type 1 (Cl)	147
6-Nitro	195-196		Cu	86	Type 1 (Cl)	147
2'-Nitro	219	PhH, HOAc, xylene;	Cu		Type 1 (Cl)	175

(Table Continued)

325

Table IV. (Continued)

Substituents	mp (°C)	Recrystallizing solvents and derivatives, mp (°C)	Catalyst	Yield (%)	Reaction Time	Ref.
		Me ester 159-160	Cu(OAc)$_2$	50	Type 1 (Cl)	240
			Cu	90	Type 2 (Cl)	136
			Cu	92-95	Type 2 (Br)	118,236,780-782
3'-Nitro	220 (dec)	EtOH	CuCl	45	Type 2 (Br)	233
4'-Nitro	211	EtOH	Cu	60-66	Type 1 (Cl)	118,126,175
			Cu	55	Type 1 (Cl)	124
			Cu		Type 2 (Br)	161,175
2'-Nitro-4'-sulfon-amido				50	Smiles	266
5-Nitro-4'-sulfonic acid		Na salt 291-292	Cu		Type 2 (Cl)	90
			Cu	26	Type 1 (Cl)	582
2'-Nitro-4'-trifluoromethyl	225-226	Me ester 147-148	Cu		Type 2	229
2'-(4-Nitrophenyl)	243		Cu	81	Type 1 Cl	22
3,4,5,6,2',3',4',5'-Octachloro	254-255	PhH; acid Cl 180-181		13	Benzyne	104
3'-Phenoxy	131-133				Type 1 (Cl)	841
4'-Phenoxy	198		Cu		Type 1 (Cl)	787
5-Phenyl	220-221		Cu	86	Type 1 (Cl)	101
2'-Phenyl	148		Cu	35	Type 1 (Cl)	23,787

Table IV. (Continued)

Substituents	mp (°C)	Recrystallizing solvents and derivatives, mp (°C)	Catalyst	Yield (%)	Reaction Time	Ref.
3'-Phenyl	156-157		Cu		Type 1 (C1)	794
4'-Phenyl	234, 250-255	EtOH	Cu	45-63	Type 1 (C1)	23,787,824
3'-Propoxy	112-114				Type 1 (C1)	841
3'-Propylthio	99-101				Type 1 (C1)	796*,797*,802*, 803*
3'-isoPropylthio	114-116	Aniline salt, Ba salt	Cu		Type 1 (C1)	796,841
4-Sulfonic acid	(dec)				Type 1 (Br)	849
3'-Sulfonic acid	265 (dec)	di-Na salt, Ba salt	Cu	9	Type 1 (C1)	938
4'-Sulfonic acid		Me ester (dec)	Cu/CuSO$_4$	29	Type 1 (C1)	78,938 850
3,2',3',5'-Tetramethoxy	125	MeOH	Cu/CuCl	15	Type 2 (Br)	71*
3,2',4',5'-Tetramethoxy	141	aq MeOH	Cu/CuCl	58	Type 2 (I)	71*
2',3',4',5'-Tetramethoxy	157			Very small	Methylation of Quinone	271
3,6,2',5'-Tetramethyl	197-199	PhH-petrol; Me ester 127-128		92	Oxidn of isatin	10
4,6,4'-Triamino		Me ester 151			Redn	785
4,6,4'-Tribromo	222	Me ester 117-118[140]		51	Chapman	139

(Table Continued)

Table IV. (Continued)

Substituents	mp (°C)	Recrystallizing solvents and derivatives, mp (°C)	Catalyst	Yield (%)	Reaction Time	Ref.
4,2',4'-Tribromo	294-296	PhH-HOAc; Me ester 112-113		82	Bromination	140
5,2',4'-Trichloro	262-264	Hydrazide 229-230				929
5,2',5'-Trichloro	238	EtOH	Cu		Type 1 (Cl)	576
3',4',5'-Trichloro	124-125		Cu		Type 1 (Cl)	816
3'-Trifluoromethyl (flufenamic acid, Arlef)	Resolidifies and remelts at 134-136	Cyclohexane, petrol; Amide 127-129, Me ester bp 120-130/0.15mm,[936] N-nitroso,[933] Et ester bp 160-170/0.1 mm	CuO	62	Type 1 (Cl)	814, 851
				83	Type 1 (Cl)	828,852
					Type 1 (Br)	91
			Cu	61	Type 1 (I)	121
4-Trifluoromethyl-sulfonyl			Cu		Type 2 (Br)	82
3,2',5'-Trimethoxy	103	MeOH	Cu/CuCl	27	Type 2 (I)	71
4,2',5'-Trimethoxy	127	aq MeOH	Cu/CuCl	13	Type 1 (Cl)	71*
5,2',5'-Trimethoxy	178	MeOH	Cu/CuCl		Type 1 (Cl)	71
6,2',5'-Trimethoxy	155	MeOH	Cu/CuCl	40	Type 2 (I)	71*
4,6,2'-Trinitro		Me ester 178	NaOAc	62	Type 1 (Cl)	205*
4,6,4',-Trinitro		Me ester 189	NaOAc	70	Type 1 (Cl)	205
2',4',6'-Trinitro	271		None	85	Type 2 (Cl)	94

328

TABLE V. 3-Phenylanthranils

Anthranil substituents	Phenyl substituents	mp (°C)	Recrystallizing solvents and derivatives, mp (°C)	Method of preparation, yield (%)	Ref.
Unsubstituted		53	EtOH; HgCl$_2$ 200 picrate, N-oxide 187-188[333]	A 29-64	301,319,320,310
	4-Amino	113	aq EtOH; HCl 228, acetyl 202, benzoyl 224, benzylidine 148-149 o-nitrobenzylidene 155	B	304,324,331
	4-Bromo	155	HgCl$_2$ 222-223	A 15-58	317
	4-Chloro	156-158		A 7	304,331,853
	4-Chloro-3-sulfonamido	245-246	EtOH	A	853
	2,5-Dihydroxy	200	aq EtOH	B 50	303
	2,4-Dimethoxy	94-95	Hexane	A 100	309
	4-Hydroxy	205	MeOH	B	303
	4-Methyl	96	HgCl$_2$	B	322
	Pentafluoro	114	N-oxide 132-133	A 15	311
6-Amino		135-136	HgCl$_2$ 192, benzoyl 260 (dec)	A	298
5-Bromo		116-118	MeOH	C 79	308,355
	4-Bromo	220-222	HgCl$_2$ 224-225	A	317
	4-Chloro	213-215	EtOAc	C 48	308,335

(Table Continued)

329

Table V. (Continued)

Anthranil substituents	Phenyl substituents	mp (°C)	Recrystallizing solvents and derivatives, mp(°C)	Method of preparation, yield (%)	Ref.
5-Bromo	2,5-Dihydroxy	260	MeOH	B 70	303
	4-Hydroxy	243	aq DMF	B 40	303
	4-Methoxy	134-135	MeOH	C 69	308, 335
5-Bromo-7-chloro	4-Hydroxy	241	aq DMF	B 94	303
4-Carboxy		225	aq HOAc	A	314
5-Chloro		115-117	Petrol, EtOH	A 97	318,334,337,854
				C 46	308,335,336
	4-Amino	208	EtOH; HCl 245, Acetyl 222, benzoyl 242, benzylidene 149, o-nitrobenzylidene 219-220	B	304,323,326,331
	4-Benzyloxy	142			328
	4-Chloro	214-215	CHCl₃, EtOH	C 46	308,325,331,335
	2,5-Dihydroxy	254	MeOH	B 60	
	4-Dimethylamino	162-163	Acetone, EtOH; MeI 184, chloro-platinate 200	B	330
	4-Hydroxy	242	EtOH, DMF; acetyl 172-173, benzoyl 231[328]	B 50-100	303,328,329,855
				C	336

Table V. (Continued)

Anthranil substituents	Phenyl substituents	mp (°C)	Recrystallizing solvents and derivatives, mp (°C)	Method of preparation, yield (%)	Ref.
5-Chloro	4-Hydroxy-3-methoxy	182	HOAc	B 32	327
	2-Hydroxy-5-methyl	210	Acetyl 135[329]	B	329,855
	4-Methoxy	144-145	PhH, acetone	Methylation 85	855
				C 49	308,328,335,336
	2-Methoxy-5-methyl	96-98	aq MeOH	Methylation 75	855
	3,4,5-Trimethoxy	126-129	Hexane	C 86	856
6-Chloro		111-113	MeOH	C	312
5,6-Dichloro		157-158		C 68	335
5,7-Dichloro	4-Hydroxy	251	EtOH	B 100	303
5,6-Dimethoxy		126-129 132-133	$HgCl_2$ 195-207(dec)	A94	313
	4-Methoxy	174	$HgCl_2$ 204-210 (dec)	A 79	313,315
5,6-Dimethyl		132-135	MeOH	A 94	313
6-Nitro		174-175	$HgCl_2$ 190 (dec)	B 32	279,298
	4-Bromo	210-212	EtOAc	B 43	292,300
	4-Chloro	215	EtOH	B 14	130
	2,5-Dimethoxy	213-214		B	927
	4-Fluoro				305
	4-Iodo	200	Petrol	B 54	88,580

(Table Continued)

331

Table V. (Continued)

Anthranil substituents	Phenyl substituents	mp (°C)	Recrystallizing solvents and derivatives, mp (°C)	Method of preparation yield (%)	Ref.
6-Nitro	4-Methyl	211			292,300
5-Trifluoro-methoxy		87-89	aq acetone	C 20	857
5-Trifluoro-methylthio		97-98	aq MeOH	C 36	857

A = From substituted benzophenones, benzhydrols, or diphenylmethanes.
B = From nitrobenzaldehydes and substituted benzenes, under acid conditions.
C = From benzylcyanides and nitrobenzenes, under basic conditions.

TABLE VI. 2-Amino-2'-fluorobenzophenones

Substituents	mp (°C)	Recrystallizing solvents and derivatives, mp (°C)	Yield (%)	Method of preparation	Ref.
Unsubstituted	126-128	EtOH	98	Redn of 5-Cl	376,379,385,858-861
N-benzoyl	113-114	Hexane	87	O₃ on indole	377
N-bromacetyl	117-119	MeOH	75	Acetylation	859-864
N-tosyl	129-130	EtOH	87	Tosylation	376,377,858,863
5-Bromo	101-102	MeOH	64	F-C	377,858,863,865
N-bromacetyl	139-140	MeOH	80	Acylation	862-865
N-glycyl	110-111	PhH-hexane; HCl 184-185	87	NH₃ on bromacetyl	377,862-864
N-methyl	112-113	Ether	50	Hydrolysis of tosyl	377,858,862,863
N-methyl-N-tosyl	154-155	EtOH	86	Alkylation	377,858,862,863
N-tosyl	114-115	MeOH	87	Tosylation	376,377,858,863
5-Carboxamido					
N-acetyl	221-222	Acetone		From 5-CO₂H	223,866
5-Carboxy					
N-acetyl	251-252	MeOH		Oxidn of 5-Me	223,866
5-Chloro	94-95	MeOH	77	F-C	377,383,858-860,863,865,867
N-bromacetyl	133	MeOH	97	Acylation	863-865
N-chloroacetyl	141-142	Ether	90	Acylation	862-864
N(3-dimethyl-amino)propyl	170-171	HCl 172-174		Hydrolysis of diazepine	380*,381*

(Table Continued)

333

Table VI. (Continued)

Substituents	mp (°C)	Recrystallizing solvents and derivatives, mp (°C)	Yield (%)	Method of preparation	Ref.
5-Chloro					
N-(2-ethylamino)-ethyl		HCl 205-15		Hydrolysis of diazepine	378*, 381*
N-glycyl	115-116	PhH-hexane	78	NH₃ on bromacetyl	863-865
N-methyl	119-120	MeOH	92	Hydrolysis of tosyl	377,858,862,863
N-methyl-N-tosyl	151-152	Ether	85	Alkylation	377,858,862,863
N-tosyl	119-120 (132-133)	MeOH	92	F-C	376,377,858,862, 863
5-Cyano	128-129				386
N-acetyl	144-145	MeOH			386,866
4,4'-Diamino-3,5,6,2',3',5',6'-heptafluoro			79	NH₄OH on perfluoro triphenylmethanol	387
	181-182		65	NH₄OH on deca-fluoro-benzophenone	387
3,5-Dinitro	350	CH₂Cl₂-MeOH		Hydrolysis of diazepine	859
5,5'-Dinitro	228-230	EtOH	96	Hydrolysis of diazepine	379
5-Methyl	69	PhH-hexane	24	F-C	384,858,862,863, 866
N-acetyl	162-163	PhH		Acylation	386,866
5-Nitro	154-158 (161-163)	Ether		F-C; Hydrolysis of diazepine	382,859-861,868

Table VI. (Continued)

Substituents	mp (°C)	Recrystallizing solvents and derivatives, mp (°C)	Yield (%)	Method of preparation	Ref.
5-Nitro *N*-methyl	186-187	MeOH		Hydrolysis of diazepine	859,868
5-Chloro-2'-fluoro-2-nitrobenzophenone	111-114			Peracid on 2-NH$_2$	316

* The cited reference contains details of further compounds of a similar nature.
F-C: Friedel-Crafts reaction.

TABLE VII. Cyclization of Diphenylamine-2-carboxylic Acids with 3'-Substituents

Diphenylamine-2-carboxylic acid cyclized	% Yield or ratio of products*		Separation	Ref.
	1-substituted	3-substituted		
4-Acetamido-3'-amino	–	84		147
5-Acetamido-3'-amino	–	41		147
4-Acetamido-3'-nitro	93	–		147
5-Acetamido-3'-nitro	92	–		147
3'-Amino	20	80		124
3'-Amino-4-nitro	–	38		147
3'-Amino-5-nitro	–	87		147
3'-Amino-6-nitro	–	86		148
3'-Bromo	–	–	(b) P	879
3'-Bromo-4'-methoxy	–	25	(b) F-BP	123
3'-Chloro	75-82	20-28		30
	40	60		121
3'-Chloro-4-methoxy	24	–		20,67
5-Chloro-3'-methoxy	36	54	(b) F-M	706, 708
3'-Chloro-4-nitro	25	–	(b) F-B	719
	78	22		130
3'-Chloro-5-nitro	68	32	(b)^i	130
5-Chloro-3'-trifluoromethyl	–	40		68,822
3'-Cyano	–	–	(b)	17,879
3',5-Dichloro	26	74	(b) F-M	392, 727
	–	22-31		68,707
3',4'-Dichloro-4-methoxy	–	–	(b) F-T, K	728
3',4'-Dimethyl	–	80	(a) F-aq A	122, 686
			(b) F-C or M	122
3'-Dimethylamino	–	30		117
3',4-Dinitro	73	–		147
3',5-Dinitro	90	10		239
	93	–		147

(Table Continued)

Table VII. (Continued)

Diphenylamine-2-carboxylic acid cyclized	% Yield or ratio of products*		Separation	Ref.
	1-substituted	3-substituted		
3',6-Dinitro	98	–		148
3'-Fluoro	40	60		121,124, 187
3'-Fluoro-4'-nitro	–	–	(b) F-T	208
3'-Methoxy	40	60	(b) F-B, H	23,30,549, 643
	22	78	(c)[ii]	414
	24	53	(b) F-M	750
	36	62		708
	–	–	(b) P	879
3'-Methoxy-4', 5'-methylene-dioxy	65	17		120
3'-Methoxy-5-nitro	15	58	(b) F-B	753
4-Methoxy-3'-nitro	8	–		132
3'-Methyl	80	20	(b) H, F-B	23,30,97, 121,129, 541
			(b) P	879
3'-Nitro	80	20	(a) F-A	30,147,591
	25	20	(b) F-B, E	761
	35	30	(c)[iii]	760
	88	12		24
3',4'- Tetramethyl-ene	–	78		122
3'-Trifluoromethyl	65	35		121,877
3',4'-Trimethylene	–	83		122

Note: Also (no details given): 3'-Chloro-4,4'-dimethoxy,[707] 3'-chloro-4'-methoxy,[707] 3'-chloro-5-methoxy,[702] 4-chloro-3'-methoxy,[702] 3',5-dichloro-4'-methoxy,[707] 3',4'-dimethoxy,[3] 3'-ethyl,[125] 3'-fluoro-4-methoxy.[148]
* = Indicates not isolated or not given.
[i] =Reduced to amino and fractional crystallization; [ii] =as hydrochloride;
[iii] = counter current separation between dilute hydrochloric acid and ethyl acetate.
F = Fractional crystallization; H = differential hydrolysis; P = pyridine complex formation.
(a) 9-Acridanone; (b) 9-chloroacridine; (c) 9-aminoacridine.
A = acetic acid; B = benzene; C = cyclohexane; E = ethyl acetate; K = acetone; M = alcoholic ammonia; P = petrol; and T = toluene.

TABLE VIII. Acridinequinones

Substituents	mp (°C)	Recrystallizing solvents and derivatives, mp (°C)	Yield (%)	Method of preparation	Ref.
1,2-Acridinequinone 4-N-piperidino	186-188 (dec)	CHCl$_3$-hexane	73	O$_2$-Cu(OAc)$_2$-piperidine on hydroxyacridine	610
9-Acridanone-1,2-quinones					
Unsubstituted	>250 (dec)	Phenazine 300	68	Oxidn	595
3,4-dihydroxy-10-methyl				Degradn	604
3,4-dimethoxy-10-methyl	233-235	PhH; phenazine 224-225; 2,4-DNP 260-261	8	Degradn	604
3-ethoxy-10-methyl	261	Phenazine 304-305		Degradn	601
3-ethoxy-4-methoxy-10-methyl	255-256	Phenazine 213-214		Degradn	543
3-hydroxy-4-methoxy-10-methyl	Unstable			Oxidn	543
3-methoxy-10-methyl	279-280 (dec)	Hydrate 118; Phenazine 285-287		Oxidn; Degradn	734,869; 601
Acetal-type Derivatives of 1,2-Quinones					
2,3-Dimethoxy-10-methyl-2,3-methylenedioxy-1-oxo-1,2-dihydro-9-acridanone	252-254	CHCl$_3$-petrol		Degradn	606

338

Table VIII. (*Continued*)

Substituents	mp (°C)	Recrystallizing solvents and derivatives, mp (°C)	Yield (%)	Method of preparation	Ref.
1-Ethoxy-10-methyl-1-oxo-2,3,4-trimethoxy-1,2-dihydro-9-acridanone	142-144	CHCl$_3$-petrol		Degradn	606
2-Methoxy-10-methyl-2,3-methylenedioxy-1-oxo-1,2-dihydro-9-acridanone	241-242	MeOH		Degradn	607
10-Methyl-2,3-methylenedioxy-1,1,2-trimethoxy-1,2-dihydro-9-acridanone	171-173	CHCl$_3$-petrol		Degradn	607*
10-Methyl-1-oxo-2,2,3,4-tetramethoxy-1,2-dihydro-9-acridanone	200-201	Ether-petrol		Degradn	606
9-Acridanone-1,4-quinones					
Unsubstituted	>250 (dec)		65	Oxidn	595
2,8-dihydroxy			8	Oxidn	595
3,5-dihydroxy			3	Oxidn	595
2,3-dimethoxy-10-methyl	201-202	PhH, MeOH 2,4-DNP 255-256(dec)	50	Degradn	604,606

339

(Table Continued)

Table VIII. (Continued)

Substituents	mp (°C)	Recrystallizing solvents and derivatives, mp (°C)	Yield (%)	Method of preparation	Ref.
3-ethoxy-2-hydroxy-10-methyl	197-198	Acetyl 218-219		Oxidn	543
3-ethoxy-2-methoxy--10-methyl	195-196			Degradn	543
2-hydroxy			35	Oxidn	595
6-hydroxy			24	Oxidn	595
2-hydroxy-3-methoxy-10-methyl	231-233 247-248 (dec)	Acetyl 236-237 Phenazine 258-259		Degradn	543,600
3-hydroxy-2-methoxy-10-methyl	130 (dec)	Acetyl 230-232 (dec) Phenazine 281-283		Oxidn	604,606
3-methoxy			27	Oxidn	595
3-methyl			95	Oxidn	595
Acetal-type Derivative of 1,4-Quinone					
2-Methoxy-10-methyl-3,4-methylenedioxy-1-oxo-1,4-dihydro-9-acridanone	207-208			Degradn	515
1,2,3,4-Tetraoxo-9-acridanone 10-methyl		Dihydrate 263-266 (dec)		Oxidn	604

*The cited reference contains details of further compounds of a similar nature.

TABLE IX. Actinomycinol [De(s)peptidoactinomycin] and Derivatives

Substituents	Recrystalizing solvents, mp (°C)
Actinomycinol (2,5-dihydroxy-3,6-dimethyl-9-acridanone-1,4-quinone)	mp >230 (dec),[81] > 300 (dec)[617] ex nitrobenzene Sublimes 180-190/0.001 mm[870] Dihydro >280 (dec) [81,62]
Acyl derivatives:	Diacetate 189-191[119,498,613,616] Diacetate perchlorate 158[616] Triacetate 164,[615,870] 181,[616] 210[617] Dihydrotetraacetate 272-5[81,621] Dihydropentaacetate 237-238[617] 249-251,[615,616,621] 269[870] Dihydrodeoxytetraacetate 235-236[616,621] Tribenzoate 223-225[616,621]
Ethers:	Monomethylether 215-216[81,621] Trimethylether 168[616,621] Dihydrotetramethylether 183-184[81] 192[621]
Other Derivatives:	Phenazine > 240[81,621] Phenazine triacetate 258[81,621]

TABLE X. Dihydroacridinones other than 9-Acridanone (including "carbazons" and "carbazims")

Substituents	mp (°C)	Recrystallizing solvents and derivatives, mp (°C)	Yield (%)	Cyclizing reagent or conditions	Ref.
7-Amino-9,9-diethyl-2-imino				Oxidn	871
7-Amino-9,9-dimethyl-2-imino			16	Oxidn	639,883,917
7-Amino-9,9-dimethyl-2-oxo		Acetyl 235		Oxidn, or hydrolysis	639
5-Amino-9,9-diphenyl-2-imino		HCl, chloroplatinate	<10	Oxidn	633,883,917
7-Amino-9,9-diphenyl-2-imino	240-250	HCl, chloroplatinate; ether		Oxidn	633
5-Amino-9,9-diphenyl-2-oxo		Acetyl 230		Hydrolysis	637
7-Amino-9,9-diphenyl-2-oxo	248-249	Acetyl 245-246		Hydrolysis or oxidn	636-638
9-Carboxy-1,8-dimethyl-6-hydroxy-3-oxo				Dicarboxylic acid–urea-phenol	628
9-Carboxy-6-hydroxy-3-oxo			50	Dicarboxylic acid–urea-phenol	628
9-Carboxy-3-oxo-1,6,8-trihydroxy				Dicarboxylic acid–urea-phenol	628
9-Carboxy-3-oxo-4,5,6-trihydroxy				Dicarboxylic acid–urea-phenol	628
4,7-Diamino-9,9-diphenyl-2-imino		HCl, chloroplatinate		Oxidn	633, 638
5,7-Diamino-9,9-diphenyl-2-oxo	277-278	Acetyl 236-237		Oxidn, or hydrolysis	636, 638
9,9-Diethyl-7-dimethylamino-2-imino				Oxidn	871

342

Table X. *(Continued)*

Substituents	mp (°C)	Recrystallizing solvents and derivatives, mp (°C)	Yield (%)	Cyclizing reagent or conditions	Ref.
9,9-Diethyl-2-imino				Oxidn	871
7-Diethylamino-9,9-diphenyl-2-imino	325			Oxidn	635
2-Diethylimonium-9,9-diphenyl perchlorate	114 (dec)			Oxidn	635
9,9-Dimethyl-7-hydroxy-2-oxo	200 (dec)			Hydrolysis	639
1,8-Dimethyl-6-hydroxy-3-oxo-9β-propionic acid				Dicarboxylic acid-urea-phenol	628
9,9-Dimethyl-2-imino				Oxidn	639
7-Dimethylamino-9,9-dimethyl-2-imino				Oxidn	639
7-Dimethylamino-9,9-diphenyl-2-imino				Oxidn	634
2-Dimethylimonium-9,9-diphenyl chloride				Oxidn	634
9,9-Diphenyl-7-hydroxy-2-oxo	232-233	Acetyl 208-209		Hydrolysis	636
9,9-Diphenyl-2-imino	160 (dec)	HCl, HClO₄; diacetyl 268		Oxidn	633
9,9-Diphenyl-2-imino-4,5,7-triamino		Ether		Oxidn	633
9,9-Diphenyl-2-oxo-4,5,7-triamino	190-230	Triacetyl 310 (dec)		Hydrolysis	637

(Table Continued)

343

Table X. *(Continued)*

Substituents	mp (°C)	Recrystallizing solvents and derivatives, mp (°C)	Yield (%)	Cyclizing reagent or conditions	Ref.
6-Hydroxy-3-oxo-9β-propionic acid			50	Dicarboxylic acid-urea-phenol	628
10-Methyl-1-oxo	280 (dec)	Sinters 145		Base on HCl	629
10-Methyl-2-oxo	188 (dec)	Sinters 133		Base on HCl	629
10-Methyl-3-oxo	157			Base on HCl	629
10-Methyl-4-oxo	(dec) no mp			Base on HCl	629
10-Methyl-3-oxo-9-phenyl	231			Heat on 3-OH.HCl	626
1-Oxo	250	Darkens 230		Demethylation	629
2-Oxo	282–284			Demethylation	629
3-Oxo	292			Demethylation	629
4-Oxo	116–117			Demethylation	629
Spiro-(cyclohexa-2, 5-dien-4-one)-1,9-(7-hydroxyacridan-2-one)	248 (dec)	H₂O		Phenol, acid, nitrite	917*
Spiro-(7-hydroxyacridan-2-one)-9,1-(4-hydroxycyclohexane)	225	H₂O		Reduction	917

*The cited reference contains details of further compounds of a similar nature.

344

TABLE XI. 9-Acridanthiones, 9-Acridanselenones and Derivatives

9-Acridanthione, or 9-substituted acridine	mp (°C)	Recrystallizing solvents and derivatives, mp (°C)	Yield (%)	Reagent or conditions	Ref.
9-Acridanselenone	238	aq acetone	80	NaSeH	646
9-Acridanthione	275	Xylene, MeOH	79	Acridine and S NaHS	346,644, 647
			93	P_2S_5 on 9-acridanone	642
			58	Ac_2S on 9-acridanone 10-oxide	921
10-oxide	147 (dec)	aq EtOH	86	Na_2S on 9-Cl-10-oxide	669
9-Benzoylthio	209	EtOH		Acylation	662*
9-Benzoylthio-2-chloro	179			Alkylation	662
9-Benzylthio	109			Aklylation	662,663
9-Benzylthio-2-chloro	103			Aklylation	662
9-Benzylselenyl	110	aq EtOH			646
3,6-Bis(dimethylamino)	360-361	Pyridine, DMF	85	Acridine and S	645
3,6-Bis(dimethylamino)-9-methylthio	178-179	aq EtOH	81	Alkylation	645*
2-Bromo	256	aq pyridine, HOAc		Na_2S on 9-Cl	206
9-Carboxymethylthio-6-chloro-2-methoxy	222 (dec)	HOAc	57	RSH on atebrin	651
2-Chloro	272-274	aq pyridine, HOAc		Na_2S on 9-Cl	649
3-Chloro	220-221		97	$EtOCS_2K$ on 9-Cl	647
2-Chloro-9-ethylthio	80-81			Alkylation	662

(Table Continued)

345

Table XI. (Continued)

9-Acridanthione, or 9-substituted acridine	mp (°C)	Recrystallizing solvents and derivatives, mp (°C)	Yield (%)	Reagent or conditions	Ref.
6-Chloro-9-ethylthio-2-methoxy	125	Petrol	94	RSH on atebrin	283,651
6-Chloro-2-methoxy	245-248	(two crystal forms, one + H_2O)		Ca polysulfide on 9-Cl	283
	266	MeOH; Au 247, Ag 290			
			55	H_2S on atebrin	651
			90	$EtOCS_2K$ on 9-Cl	647,648
					650
2-Chloro-9-methylthio	115			Alkylation	662
6-Chloro-2-methoxy-9-phenylthio	154-155	Petrol, EtOH	70	RSH on atebrin	651
9-(o-Chlorophenyl)thio	235-237	$CHCl_3$-PhH; picrate 252-253	90	RSH on 9-Cl	668*
9-(p-Chlorophenyl)thio	225-226	$CHCl_3$-PhH; picrate 207-208	90	RSH on 9-Cl	668*
2,7-Dibromo	311	Pyridine, PhH-$PhNO_2$		Na_2S on 9-Cl	649
2,7-Dibromo-4,5-dichloro	244	Pyridine		Na_2S on 9-Cl	206
4,5-Dibromo-2,7-dichloro	265	aq pyridine		Na_2S on 9-Cl	206
2,4-Dibromo-9-methylthio	215			Alkylation	662*
2,7-Dichloro	314-316	aq pyridine		Na_2S on 9-Cl	649
2,4-Dichloro-9-methylthio	185-187			Alkylation	662*
1,N-Dimethyl	130	EtOH			541
2,N-Dimethyl	220	Xylene			541
3,N-Dimethyl	240	EtOH			541
4,N-Dimethyl	158	EtOH			541

346

Table XI. (Continued)

9-Acridanthione, or 9-substituted acridine	mp (°C)	Recrystallizing solvents and derivatives, mp (°C)	Yield (%)	Reagent or conditions	Ref.
10-(2-Dimethylaminoethyl)-1-methoxy		HCl 248-250	63	Alkylation	764*
10-(2-Dimethylaminoethyl)-2-methoxy		HCl 234-235	52	Alkylation	764*
10-(2-Dimethylaminoethyl)-3-methoxy		HCl 254-255	59	Alkylation	764*
10-(2-Dimethylaminoethyl)-4-methoxy		HCl 222-224	64	Alkylation	764*
10-(2-Dimethylaminoethyl)-1-nitro		HCl 228-229	52	Alkylation	445*
2-Ethoxy-6-nitro	285-286		100	$KSPS(OPh)_2$ on 9-Cl	652
10-Ethyl-9-acridanselenone	242-243	Xylene			550
10-Ethyl	218	Xylene			550
10-Ethylthio	38,65	Petrol		Alkylation	646,662
9-(p-Hydroxyphenylthio)	275-276			RSH on 9-Cl	668
1-Methoxy	240-241	EtOH (dry)	70	Na_2S on 9-Cl	764
2-Methoxy	232-233	EtOH	59	Na_2S on 9-Cl	650,764
		Au 219-220, Ag 261	90	$EtOCS_2K$ on 9-Cl	647
3-Methoxy	220-222		71	Na_2S on 9-Cl	764
4-Methoxy	200-201		73	Na_2S on 9-Cl	764
1-Methoxy-10-methyl	122	EtOAc-petrol, ether			541

(Table Continued)

347

Table XI. (Continued)

9-Acridanthione, or 9-substituted acridine	mp (°C)	Recrystallizing solvents and derivatives, mp (°C)	Yield (%)	Reagent or conditions	Ref.
2-Methoxy-10-methyl	186	Xylene			541
3-Methoxy-10-methyl	227	Xylene			541
4-Methoxy-10-methyl	114	EtOH			541
10-Methyl-9-acridanselenone	259	Xylene			550
2-Methyl	263		90	NaHS on 9-Cl	469,647
10-Methyl	267	Xylene	90	EtOCS$_2$K on 9-gem diCl	266,541
			75	P$_2$S$_5$ on 9-acridanone	643
10-Methyl-9-methylthio	113–114	Iodide 260			643
9-Methylthio	168	Petrol, ether, aq acetone			646,662
10-oxide		aq acetone		Alkylation	669
9-Methylselenyl	108	aq EtOH			646
9-Methylsulfinyl	177–179	aq acetone		Peroxide on 9-MeS	669,677
4-Nitro	236				136
10-Phenyl-9-acridanselenone		Could not be purified			550
10-Phenyl	229	HOac	92	EtOCS$_2$K on 9-gem diCl	672,673
2,4,5,7-Tetrabromo	302–303	Pyridine, toluene		Na$_2$S on 9-Cl	649
2,4,5,7-Tetrabromo-9-methylthio	267			Alkylation	662*
2,4,5,7-Tetrachloro	235	Pyridine		Na$_2$S on 9-Cl	649
2,4,5,7-Tetrachloro-9-methylthio	233			Alkylation	662*

*The cited reference contains details of further compounds of a similar nature.

References*

1. V. Zanker and A. Wittwer, *Z. Phys. Chem. (Frankfurt am Main)*, **24**, 183 (1960); *Chem. Abstr.*, **54**, 16172 (1960).
2. T. W. Stanley and W. C. Elbert, *Talanta*, **14**, 431 (1967).
3. W. Borsche, F. Runge, and W. Trautner, *Chem. Ber.*, **66**, 1315 (1933); *Chem. Abstr.* **27**, 5744 (1933).
4. W. Lesnianski and K. Dziewonski, *Przemysl. Chem.*, **13**, 401 (1929); *Bull. Intern. Acad. Polon. Sci.*, *1929 A*, 81; *Chem. Abstr.* **24**, 120, 1112 (1930); *Z* 100, I, 3105 (1929).
5. C. Graebe and K. Lagodzinski, *Justus Liebigs Ann. Chem.*, **276**, 35 (1893); *Chem. Soc. Abstr.* **64**, 649.
6. A. Albert and B. Ritchie, *Org. Syntheses*, **22**, 5 (1942).
7. N. S. Drozdov, *J. Gen. Chem. USSR*, **4**, 1 (1934); *Chem. Abstr.*, **28**, 5486 (1934).
8. R. M. Acheson and R. G. Bolton, Private communication.
9. N. S. Drozdov, *Zh. Obshch. Khim.*, **4**, 117 (1934); *Chem. Abstr.*, **28**, 5456 (1934).
10. M. S. Newman and W. H. Powell, *J. Org. Chem.*, **26**, 812 (1961).
11. I. A. Zaitsev, M. M. Shestaeva, and V. A. Zagorevskii, *Zh. Org. Khim.*, **2**, 1769 (1966); *Chem. Abstr.*, **66**, 55194 (1967).
12. K. Gleu, S. Nitzsche, and A. Schubert, *Chem. Ber.* **72B**, 1093 (1939); *Chem. Abstr.* **33**, 5854 (1939).
13. P. A. Petyunin, M. E. Konshin, and N. G. Panferova, *Khim. Geterotsikl. Soedin.*, *Akad. Nauk Latv. SSR*, **1965**, 257; *Chem. Abstr.*, **63**, 11496 (1965).
14. E. A. Steck J. S. Buck, and L. T. Fletcher, *J. Amer. Chem. Soc.*, **79**, 4414 (1957).
15. K. Gleu and A. Schubert, *Chem. Ber.*, **73**, 805 (1940); *Chem. Abstr.*, **35**, 3258 (1941).
16. N. S. Drozdov, *J. Gen. Chem. USSR*, **16**, 455 (1946); *Chem. Abstr.*, **41**, 966 (1947).
17. A. A. Goldberg and W. Kelly, *J. Chem. Soc.*, 637 (1947).
18. A. Kliegl and L. Schaible, *Chem. Ber.*, **90**, 60 (1957); *Chem. Abstr.*, **51**, 10525 (1957).
19. A. Albert and R. Royer, *J. Chem. Soc.*, 1148 (1949).
20. W. G. Dauben, *J. Amer. Chem. Soc.*, **70**, 2420 (1948).
21. N. S. Drozdov and A. F. Bekhli, *J. Gen. Chem. USSR*, **8**, 1505 (1938); *Chem. Abstr.*, **33**, 4595 (1939).
22. J. H. Wilkinson, *J. Chem. Soc.*, 464 (1950).
23. A. Albert and W. Gledhill, *J. Soc. Chem. Ind.*, **64**, 169 (1945).
24. A. Hampton and D. Magrath, *J. Chem. Soc.*, 1008 (1949).
25. A. Albert and W. H. Linnell, *J. Chem. Soc.*, 1614 (1936).
26. A. Albert and W. H. Linnell, *J. Chem. Soc.*, 22, 1938.
27. K. Matsumura, *J. Amer. Chem. Soc.*, **57**, 1533 (1935).
28. A. S. Samarin and A. G. Lebekhov, *Tr. Tomskogo Gos. Univ. Ser. Khim.*, **154**, 253 (1962); *Chem. Abstr.*, **60**, 4110 (1964).
29. O. Yu. Magidson and A. M. Grigorovski, *Chem. Ber.*, **66**, 866 (1933); *Chem. Abstr.*, **27**, 3713 (1936).
30. K. Lehmstedt and K. Schrader, *Chem. Ber.*, **70B**, 838 (1937); *Chem. Abstr.*, **31**, 4670 (1937).
31. G. Illuminati, G. Marino, and O. Piovesana, *Ric. Sci. Rend. Sez. A.*, **4**, 437 (1964); *Chem. Abstr.*, **61**, 14504 (1964).

*Chemical Abstracts (*Chem. Abstr.*) and *British Chemical Society Abstracts* (*Chem. Soc. Abstr.*) references are included, except after references to American, British, and International journals.

32.　N. B. Chapman and D. Q. Russell-Hill, *J. Chem. Soc.*, 1563 (1956).
33.　S. Hünig and H. Herrmann, *Justus Liebigs Ann. Chem.*, **636**, 21 (1960); *Chem. Abstr.*, **55**, 12857 (1961).
34.　O. Piovesana, *Univ. Studi. Trieste, Fac. Sci. Inst. Chim.*, **14**, 65 (1961); *Chem. Abstr.*, **59**, 2608 (1963).
35.　Z. V. Pushkareva and L. V. Varyukhina, *Dokl. Akad. Nauk SSSR*, **103**, 257 (1955); *Chem. Abstr.*, **50**, 5667 (1956).
36.　E. F. Elslager and F. H. Tendick, *J. Med. Pharm. Chem.*, **5**, 1149 (1962).
37.　J. Peryt, E. Zylkiewicz, and A. Ledochowski, *Rocz. Chem.*, **43**, 623 (1969); *Chem. Abstr.*, **71**, 12990 (1969).
38.　V. Zanker and W. Fluegel, *Z. Naturforsch.*, **19b**, 376 (1964); *Chem. Abstr.*, **61**, 10559 (1964).
39.　W. W. Carlson and L. H. Cretcher, *J. Amer. Chem. Soc.*, **70**, 597 (1948).
40.　W. H. Linnell and R. E. Stuckey, *Quart. J. Pharm. Pharmacol.*, **13**, 162 (1940).
41.　A. M. Grigorovskii and T. A. Veselitskaya, *Zh. Obshch. Khim.*, **26**, 466 (1956); *Chem. Abstr.* **50**, 13928 (1956).
42.　I. L. Knunyantz, G. V. Chelintsev, Z. V. Benevolenska, E. D. Osetrova, and A. I. Kursanova, *Bull. Acad. Sci. USSR*, **1934**, 165; *Chem. Abstr.*, **28**, 4837 (1934).
43.　N. S. Drozdov and O. M. Cherntzov, *Zh. Obshch. Khim.*, **5**, 1736 (1935); *Chem. Abstr.*, **30**, 3432 (1936).
44.　K. Kitani, *J. Chem. Soc. Jap.*, **75**, 396 (1954); *Chem. Abstr.*, **49**, 10296 (1955).
45.　J. T. Stewart, T. D. Shaw, and A. B. Ray, *Anal. Chem.*, **41**, 360 (1969).
46.　J. T. Stewart, A. B. Ray, and W. B. Fackler, *J. Pharm. Sci.*, **58**, 1261 (1969).
47.　A. Ledochowski, *Rocz. Chem.*, **40**, 2015 (1966); *Chem. Abstr.*, **67**, 43066 (1967).
48.　A. Ledochowski, *Rocz. Chem.*, **41**, 717 (1967); *Chem. Abstr.*, **67**, 53204 (1967).
49.　A. Ledochowski, *Rocz. Chem.*, **41**, 1255 (1967); *Chem. Abstr.*, **68**, 104289 (1968).
50.　N. S. Drozdov and N. S. Leznova, *J. Gen. Chem. USSR*, **5**, 690 (1935); *Chem.Abstr.*, **29**, 7334 (1935).
51.　K. Lehmstedt, W. Bruns, and H. Klee, *Chem. Ber.*, **69**, 2399 (1936); *Chem. Abstr.*, **31**, 407 (1937).
52.　A. Ledochowski and E. Zylkiewicz, *Rocz. Chem.*, **42**, 595 (1968); *Chem. Abstr.*, **71**, 38028. (1969).
53.　A. Ledochowski, E. Zylkiewicz, and F. Muzalewski, *Rocz. Chem.*, **42**, 445 (1968); *Chem. Abstr.*, **69**, 85854 (1968).
54.　A. Ledochowski and E. Zylkiewicz, *Rocz. Chem.*, **43**, 291 (1969); *Chem. Abstr.*, **71**, 12218 (1969).
55.　A. Ledochowski, *Rocz. Chem.*, **41**, 1561 (1967); *Chem. Abstr.*, **69**, 2208 (1968).
56.　K. Noda, *J. Pharm. Soc. Jap.*, **64B**, 79 (1954); *Chem. Abstr.*, **48**, 10026 (1956).
57.　A. M. Grigorovskii and N. S. Milovanov, *Zh. Priklad. Khim.*, **23**, 192 (1950); English Transl. p. 197; *Chem. Abstr.*, **45**, 1596 (1951).
58.　A. M. Grigorovskii, *Zh. Obshch. Kihm.*, **19**, 1744 (1949); *Chem. Abstr.*, **44**, 2953 (1950); *J. Gen. Chem. USSR*, **19**, 185 (1949).
59.　F. Mietzsch, H. Mauss, and I. G. Farbenindustrie, German Patent 630,842; *Chem. Abstr.*, **31**, 112 (1937); British Patent, 437,953; *Chem. Abstr.*, **30**, 2578 (1936).
60.　I. G. Farbenindustrie, British Patent 450,254; *Chem. Abstr.*, **30**, 8244 (1936).
61.　R. O. Goodall and W. O. Kermack, *J. Chem. Soc.*, 1546 (1936).
62.　N. S. Drozdov, *J. Gen. Chem. USSR*, **8**, 1192 (1938); *Chem. Abstr.*, **33**, 4251 (1939).
63.　G. S. Patel, S. R. Patel, and K. S. Nargund, *J. Indian Chem. Soc.*, **34**, 477 (1957); *Chem. Abstr.* **52**, 10086 (1958).
64.　K. Singh and G. Singh, *Indian J. Pharm.*, **14**, 47 (1952); *Chem. Abstr.*, **47**, 6947 (1953).

65. F. H. Tendick, P. E. Thompson, and E. F. Elslager, U.S. Patent 3,012,036; British Patent 871,275; *Chem. Abstr.*, **55**, 25994 (1961).
66. G. Singh, T. Singh, and M. Singh, *J. Indian Chem. Soc.*, **23**, 466 (1946); *Chem. Abstr.*, **42**, 1278 (1948).
67. E. A. H. Friedheim and E. D. Bergmann, Biritish Patent 587,050; *Chem. Abstr.*, **42**, 621 (1948).
68. N. B. Ackerman, D. K. Haldorsen, F. H. Tendick, and E. F. Elslager, *J. Med. Chem.*, **11**, 315 (1968).
69. F. Jourdan, *Chem. Ber.*, **18**, 1444 (1885); *Chem. Soc. Abstr.*, **48**, 987.
70. R. M. Acheson, M. L. Burstall, and M. J. T. Robinson, unpublished results.
71. H. Brockmann, H. Muxfeldt, and G. Haese, *Chem. Ber.*, **89**, 2174 (1956); *Chem. Abstr.*, **51**, 7374 (1957).
72. C. K. Bradsher, *Chem. Rev.*, **38**, 447 (1946).
73. M. Polaczek, *Rocz. Chem.*, **15**, 565 (1935); *Chem. Abstr.*, **30**, 4499 (1936).
74. A. Albert, *The Acridines*, 2nd ed., Arnold, London, 1966, p. 30.
75. K. Matsumura, *J. Amer. Chem. Soc.*, **60**, 591 (1938).
76. F. Ullmann and W. Bader, *Justus Liebigs Ann. Chem.*, **355**, 339 (1907); *Chem. Abstr.*, **2**, 87 (1908).
77. A. A. Goldberg and W. Kelly, U.S. Patent 2,493,191; *Chem. Abstr.*, **44**, 3038 (1950).
78. A. A. Goldberg and W. Kelly, British Patent 602,334; *Chem. Abstr.*, **42**, 8828 (1946).
79. L. Villemey, *Ann. Chim.*, **5**, 570 (1950); *Chem. Abstr.*, **45**, 2947 (1951).
80. T. Takahashi, S. Shimada, and K. Niwa, *J. Pharm. Soc. Jap.*, **66**, 31 (1946); *Chem. Abstr.*, **45**, 8533 (1951).
81. H. Brockmann and H. Muxfeldt, *Chem. Ber.*, **89**, 1379 (1956); *Chem. Abstr.*, **51**, 5079 (1957).
82. Smith, Kline, and French Laboratories, Dutch Patent 299,129; *Chem. Abstr.*, **64**, 2072 (1966).
83. J. R. Beck, R. N. Booher, A. C. Brown, R. Kwok, and A. Pohland, *J. Amer. Chem. Soc.*, **89**, 3934 (1967).
84. W. D. Smart, U.S. Patent 3,021,334; *Chem. Abstr.*, **57**, 788 (1962).
85. R. R. Goodall and W. O. Kermack, *J. Chem. Soc.*, 1163 (1936).
86. F. Brody (American Cyanamid Co.), U.S. Patent 2,725,375; *Chem. Abstr.*, **50**, 6801 (1956).
87. F. Brody (American Cyanamid Co.), U.S. Patent 2,694,713; *Chem. Abstr.*, **49**, 7255 (1955).
88. V. A. Sklyarov, *Zh. Obshch. Khim.*, **30**, 3743 (1960); *Chem. Abstr.*, **55**, 21115
89. N. N. Vorozhtsov and N. D. Genkin, *USSR Patent* 57,680; *Chem. Abstr.*, **38**, 5228 (1944).
90. C. Zickendraht, *German Patent* 1,802,562; *Chem. Abstr.*, **71**, 125993 (1969).
91. P. F. Juby, T. W. Hudyma, and M. Brown, *J. Med. Chem.*, **11**, 111 (1968); R. A. Scherrer, C. V. Winder, and F. W. Short, *Abstr. 9th Nat. Med. Chem. Symp.*, *American Chemical Society*, Minneapolis, Minn., June 1966.
92. W. Dirscherl and H. Thron, *Justus Leibigs Ann. Chem.*, **504**, 297 (1933); *Chem. Abstr.*, **27**, 4801 (1933).
93. N. S. Drozdov, *J. Gen. Chem. USSR*, **8**, 937 (1938); *Chem. Abstr.*, **33**, 1330 (1936).
94. G. Schroeter and O. Eisleb, *Justus Leibigs Ann. Chem.*, **367**, 114 (1909); *Chem. Soc. Abstr.*, **96**, 575.
95. F. Ullmann, W. Bader, and H. Labhardt, *Chem. Ber.*, **40**, 4795 (1907); *Chem. Soc. Abstr.*, **94**, 52.

96. F. Ullmann and I. C. Das-Gupta, *Chem. Ber.*, **47**, 553 (1914); *Chem. Soc. Abstr.*, **106**, 413.

97. K. Gleu and S. Nitzsche, *J. Prakt. Chem.*, **153**, 200 (1939); *Chem. Abstr.*, **33**, 6854 (1939).

98. S. M. Sherlin, G. I. Braz, A. Ya. Yakubovich, E. I. Vorob'eva, and F. E. Rabinovich, *J. Gen. Chem. USSR*, **8**, 884 (1938); English Transl., p. 898; *Chem. Abstr.*, **33**, 1330 (1939).

99. F. Ullmann and H. W. Ernst, *Chem. Ber.*, **39**, 298 (1906); *Chem. Soc. Abstr.*, **90**, 205.

100. W. Knapp, *Monatsh. Chem.*, **70**, 251 (1937); *Chem. Abstr.*, **31**, 5348 (1937).

101. G. M. Kosolapoff and C. S. Schoepfle, *J. Amer. Chem. Soc.*, **76**, 1276 (1954).

102. M. Schöpff, *Chem. Ber.*, **26**, 2589 (1893); *Chem. Soc. Abstr.*, **66**, 41.

103. H. Liebermann, H. Kirchhoff, W. Gliksman, L. Loewy, A. Gruhn, T. Hammerich, N. Anitschkoff, and B. Schulze, *Justus Liebigs Ann. Chem.*, **518**, 245 (1935); *Chem. Abstr.*, **29**, 5844 (1935).

104. R. Howe, *J. Chem. Soc. C*, 478 (1966).

105. R. Weiss and J. L. Katz, *Monatsh. Chem.*, **50**, 225 (1928); *Sitzungsber.*, **137**, 701 (1928); *Chem. Abstr.*, **23**, 1131 (1929).

106. Farbwerke vorm. Meister, Lucius, and Brünning, German Patent 243,586; French Patent 422,956; *Chem. Abstr.*, **6**, 2001, 2325 (1912).

107. I. G. Farbenindustrie, German Patent 531,013; British Patent 334,240; *Chem. Abstr.*, **25**, 1093, 5574 (1931).

108. S. Tada, Japanese Patent 69 14,706; *Chem. Abstr.*, **71**, 125995 (1969).

109. R. B. Herbert and F. G. Holliman, *Tetrahedron*, **21**, 663 (1965).

110. L. J. Sargent, *J. Org. Chem.*, **14**, 285 (1949).

111. A. M. Grigorovskii and T. A. Veselitskaya, *Zh. Obshch. Khim.*, **18**, 1795 (1948); *Chem. Abstr.* **43**, 3826 (1949).

112. M. Slade, Ph.D., Dissertation, University of California, San Francisco, 1966; *Diss. Abst.*, (*B*), **28**, 1439 (1967).

113. B.I.O.S. Final Rep. No. 1088.

114. D. J. H. Brock and F. G. Holliman, *Tetrahedron*, **19**, 1903 (1963).

115. S. Singh and M. Singh, *J. Sci. Ind. Res.* (*India*), **10B**, 298 (1951); *Chem. Abstr.*, **48**, 8785 (1954).

116. S. Singh and M. Singh, *J. Sci. Ind. Res.* (*India*), **10B**, 180 (1951); *Chem. Abstr.*, **48**, 4547 (1954).

117. A. Ledochowski and B. Kozinska, *Rocz. Chem.*, **39**, 357 (1965); *Chem. Abstr.*, **63**, 16302 (1965).

118. F. Ullmann, *Justus Liebigs Ann. Chem.*, **355**, 312 (1907); *Chem. Abstr.*, **2**, 87 (1908); *Chem. Soc. Abstr.*, **92**, 842.

119. H. Brockmann and H. Muxfeldt, *Angew. Chem.*, **67**, 618 (1955); *Chem. Abstr.*, **50**, 7803 (1956).

120. T. R. Govindachari, R. P. Pai, P. S. Subramaniam, and V. Subramanyam, *Tetrahedron*, **23**, 1827 (1967).

121. J. H. Wilkinson and I. L. Finar, *J. Chem. Soc.*, 32 (1948).

122. P. Kränzlein, *Chem. Ber.*, **70**, 1776 (1937); *Chem. Abstr.*, **31**, 7432 (1937).

123. S. R. Patel and K. S. Nargund, *J. Indian Chem. Soc.*, **32**, 187 (1955); *Chem. Abstr.*, **50**, 4954 (1956).

124. A. Albert and B. Ritchie, *J. Soc. Chem. Ind.*, **60**, 120 (1941).

125. L. J. Sargent, *J. Org. Chem.*, **19**, 599 (1954).

126. K. Lehmstedt, *Chem. Ber.*, **71**, 1609 (1938); *Chem. Abstr.*, **32**, 8425 (1938).

127. K. Matsumura, *J. Amer. Chem. Soc.*, **61**, 2247 (1939).
128. G. S. Chandler, R. A. Jones, and W. H. F. Sasse, *Aust. J. Chem.*, **18**, 108 (1965); *Chem. Abstr.*, **62**, 6459 (1965).
129. A. Ledochowski, B. Stefanska, and B. Kozinska, *Rocz. Chem.*, **38**, 421 (1964); *Chem. Abstr.*, **61**, 1830 (1964).
130. F. R. Bradbury and W. H. Linnell, *J. Chem. Soc.*, 377 (1942).
131. B. Cairns and W. O. Kermack, *J. Chem. Soc.*, 1322 (1950).
132. K. G. Yekundi and S. R. Patel, *Chem. Ber.*, **90**, 2448 (1957); *Chem. Abstr.*, **52**, 14617 (1958).
133. H. Gilman and S. M. Spatz, *J. Org. Chem.*, **17**, 860 (1951).
134. R. Weiss and J. L. Katz, *Monatsh. Chem.*, **50**, 109 (1928); *Chem. Abstr.*, **23**, 839 (1929).
135. H. Burton and C. S. Gibson, *J. Chem. Soc.*, 2501 (1924).
136. G. R. Clemo, W. R. Perkin, and R. Robinson, *J. Chem. Soc.*, 1751 (1924).
137. M. Gomberg and D. L. Tabern, *J. Amer. Chem. Soc.*, **48**, 1345 (1926).
138. M. M. Jamison and E. E. Turner, *J. Chem. Soc.*, 1954 (1937).
139. R. M. Acheson and M. J. T. Robinson, *J. Chem. Soc.*, 232 (1953).
140. R. M. Acheson and M. J. T. Robinson, *J. Chem. Soc.*, 484 (1956).
141. H. M. Blatter, U.S. Patent 3,403,152; *Chem. Abstr.*, **70**, 37833 (1969).
142. H. M. Blatter, R. W. J. Carney, and G. de Stevens, U.S. Patent 3,385,856; *Chem. Abstr.*, **69**, 36168 (1968).
143. W. H. Linnell and W. H. Perkin, *J. Chem. Soc.*, 2451 (1924).
144. M. Freund and A. Schwarz, *Chem. Ber.*, **56B**, 1828 (1923); *Chem. Abstr.*, **18**, 231 (1924).
145. A. Albert and B. Ritchie, *J. Proc. Roy. Soc. New South Wales*, **74**, 77,373 (1940); *Chem. Abstr.*, **34**, 7286 (1940); **35**, 4748 (1941).
146. F. Bradbury and W. Linnell, *Quart. J. Pharm. Pharmacol.*, **11**, 240 (1930).
147. A. A. Goldberg and W. Kelly, *J. Chem. Soc.*, 102 (1946).
148. A. A. Goldberg and W. Kelly, *J. Chem. Soc.*, 595 (1947).
149. K. Lehmstedt and F. Dostal, *Chem. Ber.*, **72B**, 804 (1939); *Chem. Abstr.*, **33**, 5403 (1939).
150. K. Lehmstedt and K. Schrader, *Chem. Ber.*, **70**, 1526 (1937); *Chem. Abstr.*, **31**, 6659 (1937).
151. C. W. C. Harvey, D. Phil. Dissertation, University of Oxford, England, 1970.
152. H. Gurien, D. H. Malarek, and A. I. Rachlin, *J. Heterocycl. Chem.*, **3**, 527 (1966).
153. H. B. Nisbet, *J. Chem. Soc.*, 1372 (1933).
154. G. J. Marriott and R. Robinson, *J. Chem. Soc.*, 134 (1939).
155. M. Day and A. T. Peters, *J. Soc. Dyers Colour.*, **83**, 137 (1967).
156. A. B. Sen, *J. Indian Chem. Soc.*, **23**, 53 (1946); *Chem. Abstr.*, **40**, 5412 (1946).
157. C. Graebe and K. Lagodzinski, *Chem. Ber.*, **25**, 1733 (1892); *Chem. Soc. Abstr.*, **62**, 1086.
158. F. G. Hollimann, B. A. Jeffery, and D. J. H. Brock, *Tetrahedron*, **19**, 1841 (1963).
159. F. Ullmann, *Chem. Ber.*, **36**, 2382 (1903); *Chem. Soc. Abstr.*, **84**, 692.
160. I. Goldberg and F. Ullmann, German Patent 173,523; *Chem. Soc. Abstr.*, **90**, 953; *Friedl.*, **8**, 161 (1908).
161. I. Goldberg, *Chem. Ber.*, **39**, 1691 (1906); *Chem. Soc. Abstr.*, **90**, 426.
162. L. A. Elson and C. S. Gibson, *J. Chem. Soc.*, 2381 (1931).
163. R. Q. Brewster and T. Groening, *Org. Syntheses*, **14**, 66 (1934).
164. R. G. Bolton, Private communication.
165. K. Bauer, *Chem. Ber.*, **83**, 10 (1950); *Chem. Abstr.*, **44**, 5362 (1950).

166. L. Faucounau, *Bull. Soc. Chim. Fr.*, **4**, 58 (1937); *Chem. Abstr.*, **31**, 3217 (1937).
167. P. W. Reynolds and I. C. I., British Patent 621,749; *Chem. Abstr*, **43**, 6762 (1949).
168. F. Giral and L. Calderon, *Ciencia* (Mexico), **6**, 369 (1945); *Chem. Abstr.*, **40**, 7181 (1946).
169. P. E. Weston and H. Adkins, *J. Amer. Chem. Soc.*, **50**, 859 (1928).
170. J. F. Bunnett and R. E. Zahler, *Chem. Rev.*, **49**, 273 (1951).
171. N. Tuttle, *J. Amer. Chem. Soc.*, **45**, 1906 (1923).
172. W. A. Waters, *Chemistry of Free Radicals*, Oxford University Press, London, 1946, p. 171.
173. K. W. Rosenmund, K. Luxat, and W. Tiedemann, *Chem. Ber.*, **56**, 1950 (1923); *Chem. Abstr.*, **17**, 3837 (1923).
174. O. Yu. Magidson and A. M. Grigorovski, *Chem. Ber.*, **69B**, 396 (1936); *Chem. Abstr.*, **30**, 3822 (1936).
175. P. Grammaticakis, *Bull. Soc. Chim. Fr.*, **1954**, 92; *Chem. Abstr.*, **48**, 9198 (1954).
176. W. G. Dauben, *J. Amer. Chem. Soc.*, **70**, 2420 (1948).
177. A. A. Goldberg, *J. Chem. Soc.*, 4368 (1952).
178. I.C.I. British Patent 850,870; *Chem. Abstr.*, **55**, 22237 (1961).
179. F. Ullmann and H. Kipper, *Chem. Ber.*, **38**, 2120 (1905); *Chem. Soc. Abstr.*, **88**, 596.
180. Parke, Davis, and Co., British Patent 1,027,030; *Chem. Abstr.*, **65**, 7108 (1966).
181. F. Ullmann and R. Dahmen, *Chem. Ber.*, **41**, 3744 (1908); *Chem. Soc. Abstr.*, **94**, 975.
182. Aktieselskabet Gea., Dutch Patent 6,600,251; *Chem. Abstr.* **66**, 2377 (1967).
183. A. Albert and W. Gledhill, *J. Soc. Chem. Ind.*, **64**, 169 (1945).
184. A. Albert, *The Acridines*, 2nd ed., Arnold, London, 1966, p. 60.
185. C. C. Price and R. M. Roberts, *J. Org. Chem.*, **11**, 463 (1946).
186. J. Wrotek, Z. Lukasiewicz, and S. Sabiniewicz, Polish Patent 53,136; *Chem. Abstr.* **68**, 29430 (1968).
187. J. H. Wilkinson and I. L. Finar, *J. Chem. Soc.*, 759 (1947).
188. F. Ullmann and H. Hoz, *Justus Liebigs Ann. Chem.*, **355**, 352 (1907); *Chem. Abstr.*, **2**, 87 (1908).
189. L. O. Rusetskii, *Med. Prom. SSSR*, **17**, 26 (1963); *Chem. Abstr.*, **59**, 9866 (1963).
190. P. G. Sergeev, *J. Gen. Chem. USSR*, **1**, 279 (1931); *Chem. Abstr.*, **26**, 2184 (1932).
191. R. C. Elderfield, W. J. Gensler, and O. Birstein, *J. Org. Chem.*, **11**, 812 (1946).
192. Egyesult Gyogyszer es Tapszergyar, Hungarian Patent 151,163; *Chem. Abstr.*, **60**, 13191 (1964).
193. F. Ullmann and H. Kipper, *Justus Liebigs Ann. Chem.*, **355**, 342 (1907); *Chem. Abstr.*, **2**, 87 (1908).
194. Parke, Davis, and Co., British Patent 1,075,494; *Chem. Abstr.*, **68**, 2712 (1968).
195. Parke, Davis, and Co., German Patent 1,188,609; *Chem. Abstr.*, **63**, 1741 (1965).
196. R. A. Scherrer and F. W. Short, U.S. Patent 3,313,848; *Chem. Abstr.*, **67**, 73368 (1967).
197. P. F. Juby, U.S. Patent 3,294,813; *Chem. Abstr.*, **66**, 65475 (1967).
198. F. Ullmann and P. Dootson, *Chem. Ber.*, **51**, 9 (1918); *Chem. Soc. Abstr.*, **114**, 189.
199. D. J. H. Brock and F. G. Holliman, *Tetrahedron*, **19**, 1911 (1963).
200. G. Stapleton and A. I. White, *J. Amer. Pharm. Assn.*, **43**, 193 (1954).
201. A. Albert, *The Acridines*, 1st ed., Arnold, London, 1951, p. 48.
202. H. Goldstein and A. Giddey, *Helv. Chim. Acta*, **37**, 1121 (1954); *Chem. Abstr.*, **49**, 10231 (1955).
203. H. Goldstein and M. Urvater, *Helv. Chim. Acta*, **34**, 1350 (1951); *Chem. Abstr.*, **46**, 5553 (1952).
204. Parke, Davis, and Co., Dutch Patent 6,507,783; *Chem. Abstr.*, **64**, 19501 (1966).

205. T. C. Mathur and C. Lal, *J. Indian Chem. Soc.*, **40**, 975 (1963); *Chem. Abstr.*, **60**, 5455 (1964).

206. M. Ionescu and D. Postescu, *Studia Univ. Babes-Bolyai, Ser. Chem.*, **11**, 73 (1966); *Chem. Abstr.*, **65**, 20096.

207. N. N. Quang, B. K. Diep, and N. P. Buu-Hoi, *Rec. Trav. Chim. Pays-Bas*, **83**, 1142 (1964); *Chem. Abstr.*, **62**, 1590 (1965).

208. K. C. Joshi and S. C. Bahel, *J. Indian Chem. Soc.*, **38**, 877 (1961); *Chem. Abstr.*, **56**, 15483 (1962).

209. S. Singh and M. Singh, *J. Sci. Ind. Res.* **13B**, 405, (1954); *Chem. Abstr.*, **49**, 12476 (1955).

210. M. T. Bogert, A. D. Hirschfelder, and P. G. I. Lauffer, *Collect. Czech. Chem. Commun.*, **2**, 383 (1930); *Chem. Abstr.*, **24**, 4516 (1930).

211. V. I. Kikhteva and N. N. Dykhanov, *Metody Polucheniya Khim. Reaktivov Preparatov*, *Gos. Kom. Sov. Min. SSSR Khim.*, **1964**, 74; *Chem. Abstr.*, **65**, 7087 (1966).

212. G. Leandri and C. Angelini, *Boll. Sci. Fac. Chim. Ind. Bologna*, **9**, 32 (1951); *Chem. Abstr.*, **45**, 10245 (1951).

213. F. Ullmann and R. Maag, *Chem. Ber.*, **39**, 1693 (1906); *Chem. Soc. Abstr.*, **90**, 459.

214. W. S. Struve, U.S. Patent 2,830,990; *Chem. Abstr.*, **52**, 14707 (1958).

215. F. Montanari, *Boll. Sci. Fac. Chim. Ind. Bologna*, **17**, 33 (1959); *Chem. Abstr.*, **54**, 3291 (1960).

216. A. S. Samarin and N. A. Samarina, USSR Patent 234,418; *Chem. Abstr.*, **70**, 106208 (1969).

217. J. R. Geigy A-G., French Patent 1,509,386; *Chem. Abstr.*, **70**, 88815 (1969).

218. G. K. Hughes, K. G. Neill, and E. Ritchie, *Aust. J. Sci. Res.*, **3A**, 497 (1950); *Chem. Abstr.*, **46**, 4544 (1950).

219. F. Dallacker and G. Adolphen, *Justus Liebigs Ann. Chem.*, **691**, 134 (1966); *Chem. Abstr.*, **64**, 19699 (1966).

220. F. Dallacker and G. Adolphen, *Justus Liebigs Ann. Chem.*, **691**, 138 (1966); *Chem. Abstr.*, **64**, 19700 (1966).

221. F. Dallacker and G. Adolphen, *Tetrahedron Lett.*, 2023 (1965).

222. G. S. Patel, S. R. Patel, and K. S. Nargund, *J. Indian Chem. Soc.*, **34**, 371 (1957); *Chem. Abstr.*, **52**, 3818.

223. H. Brauninger and K. Spangenberg, *Pharmazie*, **12**, 335 (1957); *Chem. Abstr.*, **54**, 22668 (1960).

224. H. J. Zeitler, *Z. Physiol. Chem.*, **340**, 73 (1965); *Chem. Abstr.*, **63**, 5553 (1965).

225. N. N. Vorozhtsov and G. G. Yakobson, *Zh. Obshch. Khim.*, **28**, 40 (1958); *Chem. Abstr.*, **52**, 12784 (1958).

226. G. M. Badger and R. Pettit, *J. Chem. Soc.*, 1874 (1952).

227. K. Lehmstedt and H. Hundertmark, *Chem. Ber.*, **64**, 2386 (1931); *Chem. Abstr.*, **26**, 462 (1932).

228. J. Houben and T. Arendt, *Chem. Ber.*, **43**, 3533 (1910); *Chem. Abstr.*, **5**, 1104 (1911).

229. F. Hunziker, H. Lauener, and J. Schmutz, *Arzneim-Forsch.*, **13**, 324 (1963); *Chem. Abstr.*, **59**, 8753 (1963).

230. J. Riviere, French Patent M4332; *Chem. Abstr.*, **67**, 111431 (1967).

231. I. Goldberg and M. Nimerovsky, *Chem. Ber.*, **40**, 2448 (1907); *Chem. Soc. Abstr.*, **92**, 621.

232. H. Okawa and T. Yoshino, *Bull. Chem. Soc. Jap.*, **42**, 1934 (1969); *Chem. Abstr.*, **71**, 91256 (1969).

233. D. H. Hey and R. D. Mulley, *J. Chem. Soc.*, 2276 (1952).

234. A. Sallmann and R. Pfister, South African Patents 67 05,988; 05,992; *Chem. Abstr.*, **70**, 67937, 106221 (1969).
235. Park Davis, and Co., French Patent, M2935; *Chem. Abstr.*, **62**, 13091 (1965).
236. A. R. Hanze, R. E. Strube, and M. E. Greig, *J. Med. Chem.*, **6**, 767 (1963).
237. W. E. Coyne and J. W. Cusic, *J. Med. Chem.*, **10**, 541 (1967).
238. Panorganic South America, Mexican Patent 58,887; *Chem. Abstr.*, **54**, 24820 (1960).
239. A. Albert and W. H. Linnell, *J. Chem. Soc.*, 88 (1936).
240. F. Ullmann and P. Dieterle, *Justus Liebigs Ann. Chem.*, **355**, 323 (1907); *Chem. Abstr.*, **2**, 87 (1908).
241. A. T. Troshchenko, *Zh. Obshch. Khim.*, **28**, 2207 (1958); *Chem. Abstr.*, **53**, 2230 (1959).
242. L. Levai, M. Ritvay, A. Vedres, L. Gyongyossy, and G. Balogh, Hungarian Patent, 154,230; *Chem. Abstr.*, **69**, 43635 (1968).
243. C. S. Miller and C. A. Wagner, *J. Org. Chem.*, **13**, 891 (1948).
244. U. P. Basu and S. J. Das-Gupta, *J. Indian Chem. Soc.*, **16**, 100 (1939); *Chem. Abstr.*, **33**, 5852 (1939).
245. D. G. I. Felton, private communication.
246. F. Ullmann and E. Tedesco, *Justus Liebigs Ann. Chem.*, **355**, 339 (1907); *Chem. Soc. Abstr.*, **92**, 844.
247. H. Goldstein and W. Rodel, *Helv. Chim. Acta*, **9**, 765 (1926); *Chem. Abstr.*, **21**, 232 (1927).
248. H. Goldstein and M. de Simo, *Helv. Chim. Acta*, **10**, 603 (1927); *Chem. Abstr.*, **22**, 1581 (1928).
249. D. P. Spalding, E. C. Chapin, and H. S. Mosher, *J. Org. Chem.*, **19**, 357 (1954).
250. G. B. Bachman and J. W. Wetzel, *J. Org. Chem.*, **11**, 454 (1946).
251. A. M. Grigorovski and E. M. Terentieva, *J. Gen. Chem. USSR*, **17**, 517 (1947); *Chem. Abstr.*, **42**, 910 (1948).
252. G. Leandri and C. Angelini, *Ann. Chim.*, **40**, 682 (1950); *Chem. Abstr.*, **46**, 507 (1952).
253. D. Shapiro, *J. Amer. Chem. Soc.*, **72**, 2786 (1950).
254. I. L. Knunyantz and Z. V. Benevolenskaya, *Zh. Obshch. Khim.*, **10**, 1415 (1940); *Chem. Abstr.*, **35**, 3642 (1941).
255. B. V. Samant, *Chem. Ber.*, **75**, 1008 (1942); *Chem. Abstr.*, **37**, 4400 (1943).
256. A. W. Chapman, *J. Chem. Soc.*, 1743 (1927).
257. M. M. Harris, W. G. Potter, and E. E. Turner, *J. Chem. Soc.*, 145 (1955).
258. Parke, Davis, and Co., Dutch Patent 292,085; *Chem. Abstr.*, **64**, 19506 (1966).
259. Parke, Davis, and Co., German Patent 1,186,870; *Chem. Abstr.*, **63**, 543 (1965).
260. Parke, Davis, and Co., German Patent 1,186,871; *Chem. Abstr.*, **63**, 544 (1965).
261. Parke, Davis, and Co., German Patent 1,190,951; *Chem. Abstr.*, **63**, 4209 (1965).
262. Parke, Davis, and Co., Belgian Patent 637,515; *Chem. Abstr.*, **63**, 4307 (1965).
263. M. Ionescu, I. Goia, and I. Felmeri, *Acad. Rep. Populare Romine, Filiala Cluj, Stud. Cercet. Chim.*, **8**, 351 (1957); *Chem. Abstr.*, **54**, 4587 (1960).
264. I. A. Solov'eva and A. G. Guseva, *Zh. Org. Khim.*, **4**, 1973 (1968); *English Transl.* p. 1905; *Chem. Abstr.*, **70**, 28585 (1969).
265. Parke, Davis, and Co., German Patent 1,186,073; *Chem. Abstr.*, **62**, 11741 (1965).
266. F. Kuenzle and J. Schmutz, *Helv. Chim. Acta*, **52**, 622 (1969); *Chem. Abstr.*, **70**, 106484 (1969).
267. J. Ville and C. Astre, *Bull. Soc. Chim. Fr.*, **13**, 746 (1895); *Chem. Soc. Abstr.*, **72**, 525.
268. O. Suchanek, *J. Prakt. Chem.* **90**, 472 (1914); *Chem. Soc. Abstr.*, **108**, 269.
269. J. Ville and C. Astre, *C. R. Acad. Sci., Paris*, **120**, 684 (1895); *Chem. Soc. Abstr.*, **68**, 465.

270. C. Astre, *Bull. Soc. Chem. Fr.*, **15**, 1025 (1896).
271. R. M. Acheson and B. F. Sansom, *J. Chem. Soc.*, 4440 (1955).
272. W. G. Hanger, W. C. Howell, and A. W. Johnson, *J. Chem. Soc.*, 496 (1958).
273. H. Gilman and S. M. Spatz, *J. Org. Chem.*, **17**, 860 (1951).
274. Parke, Davis, and Co., German Patent 1,186,074; *Chem. Abstr.*, **62**, 11741 (1965).
275. E. R. H. Jones and F. G. Mann, *J. Chem. Soc.*, 786 (1956).
276. H. Gilman and J. J. Dietrich, *J. Amer. Chem. Soc.*, **80**, 380 (1958).
277. E. Ziegler and T. Knappe, *Monatsh. Chem.*, **94**, 736 (1963); *Chem Abstr.*, **59**, 13948 (1963).
278. A. I. Kizber, *Zh. Obshch. Khim.*, **24**, 2195 (1954); *Chem. Abstr.*, **50**, 207 (1956).
279. M. I. Rogovik, *Nauk. Zap. Chernivets'k. Univ.*, **53**, 74 (1961); *Chem. Abstr.*, **61**, 4518 (1964).
280. Yu. S. Rozum, *Ukr. Khim. Zh.*, **21**, 361 (1955); *Chem. Abstr.*, **49**, 14772 (1955).
281. A. Albert and B. Ritchie, *J. Chem. Soc.*, 458 (1943).
282. H. J. Barber, J. H. Wilkinson, and W. G. H. Edwards, *J. Soc. Chem. Ind.*, **66**, 411 (1947).
283. K. Kitani, *J. Chem. Soc. Jap.*, **75**, 482 (1954); *Chem. Abstr.*, **49**, 10296 (1955).
284. R. M. Acheson and C. W. Jefford, *J. Chem. Soc.*, 2676 (1956).
285. D. L. Hammick, S. F. Mason, and G. W. Meacock, *J. Chem. Soc.*, 4745 (1952).
286. E. P. Taylor, *J. Chem. Soc.*, 5048 (1952).
287. I. S. Ioffe and N. I. Devyatova, *Zh. Obshch. Khim.*, **30**, 884 (1960); *Chem. Abstr.*, **55**, 546 (1961).
288. K. Lehmstedt, *Chem. Ber.*, **65**, 834 (1932); *Chem. Abstr.*, **26**, 4051 (1932).
289. I. Tanasescu, *Bull. Soc. Chim. Fr.*, **41**, 528 (1927); *Chem. Abstr.*, **21**, 3905 (1927).
290. K. Drechsler, *Monatsh. Chem.*, **35**, 533 (1914); *Chem. Soc. Abstr.*, **106**, 991; *Sitzungsber.*, **73**, 51.
291. I. Tanasescu, L. Almasi, and A. Hantz, *Acad. Rep. Populare Romine, Filiala Cluj, Stud. Cercet. Chim.*, **11**, 115 (1960); *Chem. Abstr.*, **55**, 11415.
292. I. Tanasescu, M. Ionescu, I. Goia, and H. Mantsch, *Bull. Soc. Chim. Fr.*, 698 (1960); *Chem. Abstr.*, **55**, 23536.
293. K. Lehmstedt, *Chem. Ber.*, **65**, 999 (1932); *Chem. Abstr.*, **26**, 4817 (1932).
294. E. Bamberger, *Chem. Ber.*, **42**, 1707 (1909); *Chem. Abstr.*, **3**, 2303 (1909).
295. K.-H. Wünsch and A. J. Boulton, *Adv. Heterocycl. Chem.*, **8**, 277 (1967).
296. A. Silberg and Z. Frenkel, *Stud. Univ. Babes-Bolyai, Ser. Chem.*, **7**, 53 (1962); *Chem. Abstr.*, **62**, 523 (1965).
297. I. Tanasescu and M. Macarovici, *Bull. Soc. Chim. Fr.*, **53**, 372 (1933); *Chem. Abstr.*, **27**, 4802 (1933).
298. I. Tanasescu and E. Ramontianu, *Bull. Soc. Chim. Fr.*, **53**, 918 (1933); *Chem. Abstr.*, **28**, 770 (1934).
299. I. Tanasescu and M. Macarovici, *Bull. Soc. Chim. Fr.*, **49**, 1295 (1931); *Chem. Abstr.*, **26**, 1285 (1932).
300. I. Tanasescu and Z. Frenkel, *Bull. Soc. Chim. Fr.*, 693 (1960); *Chem. Abstr.*, **55**, 23536 (1961).
301. A. Silberg and Z. Frenkel, *Rev. Roum. Chim.*, **10**, 1035 (1965); *Stud. Cercet. Chim.*, **13**, 1071 (1965); *Chem. Abstr.*, **64**, 12641 (1966).
302. A. Kliegl, *Chem. Ber.*, **42**, 591 (1909); *Chem. Soc. Abstr.*, **96**, 255.
303. J. D. Loudon and G. Tennant, *J. Chem. Soc.*, 3092 (1962).
304. L. Tanasescu and A. Silberg, *Bull. Soc. Chim. Fr.*, **3**, 2383 (1936); *Chem. Abstr.*, **32**, 568 (1938).

305. I. Tanasescu, L. Almasi, and A. Hantz, *Acad. Rep. Populare Romine, Filiala Cluj, Stud. Cercet. Chim.*, **11**, 105 (1960); *Chem. Abstr.*, **55**, 11415 (1961).

306. I. Tanasescu, *Bull. Soc. Chim. Fr.*, **53**, 381 (1933); *Chem. Abstr.*, **27**, 4802 (1933).

307. A. Kliegl and A. Fehrle, *Chem. Ber.*, **47**, 1629 (1914); *Chem. Abstr.*, **8**, 2726 (1914).

308. R. B. Davis and L. C. Pizzini, *J. Org. Chem.*, **25**, 1884 (1960).

309. R. Kwok and P. Pranc, *J. Org. Chem.*, **33**, 2880 (1968).

310. E. Bamberger and S. Lindberg, *Chem. Ber.*, **42**, 1723 (1909); *Chem. Abstr.*, **3**, 2302.

311. P. L. Coe, A. E. Jukes and J. C. Tatlow, *J. Chem. Soc. C*, 2020 (1966).

312. M. Ogata, H. Matsumoto, and H. Kano, *Tetrahedron*, **25**, 5205 (1969).

313. F. Korte and O. Behner, *Justus Liebigs Ann. Chem.*, **621**, 51 (1959); *Chem. Abstr.*, **53**, 19992 (1959).

314. J. Tirouflet, *Bull. Soc. Sci. Bretagne, Spec. No.* **26**, 69 (1951); *Chem. Abstr.*, **47**, 8694 (1953).

315. A. F. Aboulezz and R. Quelet, *J. Chem. UAR*, **5**, 137 (1962); *Chem. Abstr.*, **63**, 11411 (1965).

316. A. I. Rachlin, U.S. Patent 3,261,870; *Chem. Abstr.*, **65**, 15277 (1966).

317. N. Campbell and H. F. Andrew, *Proc. Roy. Soc. Edinburgh*, *A66*, 252 (1963–1964); *Chem. Abstr.*, **62**, 16157.

318. W. B. Dickinson, *J. Amer. Chem. Soc.*, **86**, 3580 (1964).

319. P. A. S. Smith, B. B. Brown, R. K. Putney, and R. F. Reinisch, *J. Amer. Chem. Soc.*, **75**, 6335 (1953).

320. J. Meisenheimer, O. Senn, and P. Zimmerman, *Chem. Ber.*, **60**, 1736 (1927); *Chem. Abstr.*, **22**, 235 (1928).

321. D. R. Eckroth, *Diss. Abstr.*, *B*, **27**, 102 (1966).

322. A. Kliegl, *Chem. Ber.*, **41**, 1845 (1908); *Chem. Abstr.*, **2**, 2696 (1908).

323. I. Tanasescu and M. Suciu, *Bull. Soc. Chim. Fr.*, **3**, 1753 (1936); *Chem. Abstr.*, **31**, 7861 (1937).

324. I. Tanasescu and A. Silberg, *Bull. Soc. Chim. Fr.*, **51**, 1357 (1932); *Chem. Abstr.*, **27**, 2439 (1933).

325. I. Tanasescu and M. Suciu, *Bull. Soc. Chim. Fr.*, **4**, 245 (1937); *Chem. Abstr.*, **32**, 531 (1938).

326. I. Tanasescu, C. Anghel, and A. Popescu, *Studia Univ, Babes-Bolyai, Ser. Chem.*, **9**, 89 (1964); *Chem. Abstr.*, **61**, 16004 (1964).

327. I. S. Ioffe and B. G. Belen'kii, *Zh. Obshch. Khim.*, **23**, 1525 (1953); *Chem. Abstr.*, **48**, 1685 (1954).

328. A. Guyot and A. Haller, *Bull. Soc. Chim. Fr.*, **31**, 530 (1904); *Chem. Soc. Abstr.*, **86**, 530.

329. T. Zincke and K. Siebert, *Chem. Ber.*, **39**, 1930 (1906); *Chem. Soc. Abstr.*, **90**, 515.

330. T. Zincke and W. Prenntzell, *Chem. Ber.*, **38**, 4116 (1905); *Chem. Soc. Abstr.*, **90**, 110.

331. I. Tanasescu, C. Anghel and A. Popescu, *Studia Univ. Babes-Bolyai, Ser. Chem.*, **8**, 141 (1963); *Chem. Abstr.*, **61**, 13279 (1964).

332. O. E. Schultz and L. Geller, *Arch. Pharm.*, **288**, 234 (1955); *Chem. Abstr.*, **50**, 4142 (1956).

333. H. H. Szmant and C. M. Harmuth, *J. Amer. Chem. Soc.*, **81**, 962 (1959).

334. H. J. Heidrich, S. Henker, and H. Roehnert, East German Patent, 61,265; *Chem. Abstr.*, **70**, 37806 (1969).

335. R. B. Davis, U.S. Patent 3,156,704; *Chem. Abstr.*, **62**, 2743 (1965).

336. G. N. Walker, *J. Org. Chem.*, **27**, 1929 (1962).

337. Romania, Ministry of Petroleum and Chem. Ind., *Belgian Patent* 670,674; *Chem. Abstr.*, **65**, 13715 (1966).

338. R. B. Davis, L. C. Pizzini, and J. D. Benigni, *J. Amer. Chem. Soc.*, **82**, 2913 (1960).
339. R. B. Davis, L. C. Pizzini, and E. J. Bara, *J. Org. Chem.*, **26**, 4270 (1961).
340. H. Neresheimer and W. Ruppel, German Patent 603,622; *Chem. Abstr.*, **29**, 817; U.S. Patent 2,080,057; *Chem. Abstr.*, **31**, 4830 (1937).
341. A. H. Riebel, R. E. Erickson, S. J. Abshire, and P. S. Bailey, *J. Amer. Chem. Soc.*, **82**, 1801 (1960).
342. A. Pictet and E. Patry, *Chem. Ber.*, **26**, 1962 (1893); *Chem. Soc. Abstr.*, **64**, 722.
343. K. Lehmstedt, *Chem. Ber.*, **64**, 1232 (1931); *Chem. Abstr.*, **25**, 4270 (1931).
344. G. Graebe and H. Caro, *Chem. Ber.*, **13**, 99 (1880); *Chem. Soc. Abstr.*, **38**, 398.
345. H. Schmidt and W. E. Leutenegger, *Helv. Chim. Acta*, **30**, 1965 (1947); *Chem. Abstr.*, **42**, 3758 (1948).
346. A. P. Wolf and R. C. Anderson, *J. Amer. Chem. Soc.*, **77**, 1608 (1955).
347. E. J. Moriconi and F. A. Spano, *J. Amer. Chem. Soc.*, **86**, 38 (1964).
348. R. M. Acheson and M. L. Burstall, *J. Chem. Soc.*, 3240 (1954).
349. K. Lehmstedt, *Chem. Ber.*, **60**, 1370 (1927); *Chem. Abstr.*, **21**, 2903 (1927).
350. K. Lehmstedt, *Chem. Ber.*, **71**, 808 (1938); *Chem. Abstr.*, **32**, 4589 (1938).
351. V. S. Fedorov, *Zh. Obshch. Khim.*, **26**, 591 (1956); *Chem. Abstr.*, **50**, 13929 (1956).
352. H. Gilman, J. Eisch, and T. Soddy, *J. Amer. Chem. Soc.*, **79**, 1245 (1957).
353. E. Ochiai and I. Iwai, *J. Pharm. Soc. Jap.*, **69**, 413 (1949); *Chem. Abstr.*, **44**, 1988 (1950).
354. O. Tsuge and M. Nishinohara, *Bull. Chem. Soc. Jap.*, **38**, 2034 (1965); *Chem. Abstr.*, **64**, 12641 (1966).
355. O. Tsuge, M. Nishinohara, and K. Sadano, *Bull. Chem. Soc. Jap.*, **38**, 2037 (1965); *Chem. Abstr.*, **64**, 12641 (1966).
356. K. Soda and K. Nishiide, Japanese Patent 69 27,388; *Chem. Abstr.*, **72**, 31645 (1970).
357. I. Tanasescu and E. Ramontianu, *Bull. Soc., Chim. Fr.*, **1**, 547 (1934); *Chem. Abstr.*, **28**, 6149 (1934).
358. S. Oae, S. Kozuka, Y. Sakaguchi, and K. Hiramatsu, *Tetrahedron*, **22**, 3143 (1966).
359. J. H. Markgraf and C. C. Carson, *J. Org. Chem.*, **29**, 2806 (1964).
360. J. H. Markgraf and M-K. Ahn, *J. Amer. Chem. Soc.*, **86**, 2699 (1964).
361. S. Oae and S. Kozuka, *Tetrahedron*, **21**, 1971 (1965).
362. H. Otaka, *Nichidai Igaku Zasshi*, **13**, 2330 (1954); *Chem. Abstr.*, **52**, 10390 (1958).
363. M. Kumasaka, *Nichidai Igaku Zasshi*, **19**, 3726 (1960); *Chem. Abstr.*, **61**, 7467 (1964).
364. (a) Y. Hashimoto and A. Ogino, *Nippon Univ. J. Med.*, **2**, 247 (1960). (b) R. Ito, M. Kumasaka, and Y. Hashimoto, *Nippon Univ. J. Med.*, **2**, 275 (1960); *Chem. Abstr.*, **56**, 16084 (1962).
365. H. Otaka, I. Ikeda, and R. Ito, *Nippon Univ. J. Med.*, **3**, 235 (1961); *Chem. Abstr.*, **59**, 10652 (1963).
366. H. Otaka and Y. Hashimoto, *Nippon Univ. J. Med.*, **2**, 1 (1960); *Chem. Abstr.*, **56**, 16084 (1962).
367. Y. Hashimoto and Y. Oyamada, *Nippon Univ. J. Med.*, **1**, 339 (1959); *Chem. Abstr.*, **56**, 10565 (1962).
368. H. Otaka, K. Kono, and R. Ito, *Nippon Univ. J. Med.*, **3**, 235 (1961); *Chem. Abstr.*, **59**, 10652 (1963).
369. R. Ito, H. Otaka, A. Sasaki, and Y. Hashimoto, *Nippon Univ. J. Med.*, **3**, 93 (1961); *Chem. Abstr.*, **58**, 12952 (1963).
370. R. A. Reed, *J, Chem. Soc.*, 186 (1945).
371. A. Braun, *Rocz. Chem.*, **36**, 151 (1962); *Chem. Abstr.*, **57**, 15071 (1962).
372. R. A. Reed, *J. Chem. Soc.*, 425 (1944).
373. H. Tiedtke, *Chem. Ber.*, **42**, 621 (1909); *Chem. Abstr.*, **3**, 1153 (1909).

374. A. Braun and W. Kirkor, *Lodz. Towarz. Nauk. Wydzial III Acta Chim.*, **4**, 137 (1960); *Chem Abstr.*, **59**, 5136 (1963).
375. J. von Braun, A. Heymons, and G. Mauz, *Chem. Ber.*, **64**, 227 (1931); *Chem. Abstr.*, **25**, 3345 (1931).
376. F. Hoffmann-La Roche and Co., A-G., French Patent 1,375,300; *Chem. Abstr.*, **62**, 7694 (1965).
377. L. H. Sternbach, R. I. Fryer, W. Metlesics, G. Sach, and A. Stempel, *J. Org. Chem.*, **27**, 3781 (1962).
378. R. J. Fryer, E. Reeder, and L. H. Sternbach, French Patent 1,508,140; *Chem. Abstr.*, **70**, 57431 (1969).
379. R. I. Fryer, B. Brust, and L. H. Sternbach, *J. Chem. Soc.*, 4977 (1963).
380. R. I. Fryer, E. Reeder, and L. H. Sternbach, French Patent 1,500,341; *Chem. Abstr.*, **69**, 106259 (1968).
381. E. Reeder, L. H. Sternbach, and R. I. Fryer, French Patent 1,493,255; *Chem. Abstr.*, **69**, 67111 (1968).
382. F. Hoffmann-La Roche and Co. A-G., Dutch Patent 6,405,644; *Chem. Abstr.*, **62**, 16137 (1965).
383. F. Hoffmann-La Roche and Co. A-G., Belgian Patent 632,685; *Chem. Abstr.*, **62**, 577 (1965).
384. L. H. Sternbach, G. Saucy, F. A. Smith, M. Mueller, and J. Lee, *Helv. Chim. Acta*, **46**, 1720 (1963); *Chem. Abstr.*, **59**, 11495 (1963).
385. F. Hoffmann-La Roche and Co. A-G., Belgian Patent 622,079; *Chem. Abstr.*, **59**, 12711 (1963).
386. G. Saucy, F. A. Smith, and L. H. Sternbach, U.S. Patent 3,329,701; *Chem. Abstr.*, **68**, 69057 (1968).
387. T. N. Gerasimova, E. G. Lokshina, V. A. Barkhash, and N. N. Vorozhtsov, *Zh. Obshch. Khim.*, **37**, 1300 (1967); *Chem. Abstr.*, **68**, 49250 (1968).
388. F. Ullmann and W. Denzler, *Chem Ber.*, **39**, 4332 (1906); *Chem. Soc. Abstr.*, **92**, 142.
389. C. W. Pohlmann, *Rec. Trav. Chim. Pays-Bas*, **55**, 737 (1936); *Chem. Abstr.*, **30**, 7110 (1936).
390. W. Staedel, *Chem. Ber.*, **27**, 3362 (1894); *Chem. Soc. Abstr.*, **68**, 147.
391. R. T. Parfitt, *J. Chem. Soc.*, *C*, 87 (1966).
392. D. P. Spalding, G. W. Moersch, H. S. Mosher, and F. C. Whitmore, *J. Amer. Chem. Soc.*, **68**, 1596 (1946).
393. K. Matsumura, *J. Amer. Chem. Soc.*, **51**, 816 (1929).
394. M. Schöpff, *Chem. Ber.*, **27**, 2316 (1894); *Chem. Soc. Abstr.*, **66**, 598.
395. C. Graebe and F. Ullmann, *Chem. Ber.*, **27**, 3483 (1894); *Chem. Soc. Abstr.*, **68**, 147.
396. R. M. Acheson and B. Adcock, *J. Chem. Soc. C*, 1045 (1968).
397. D. H. Hey, C. W. Rees, and A. R. Todd, *Chem. Ind.*, 1332 (1962).
398. G. Ege, *Chem. Ber.*, **101**, 3079 (1968); *Chem. Abstr.*, **69**, 95678 (1968).
399. G. Singh, S. Singh, A. Singh, and M. Singh, *J. Indian Chem. Soc.*, **28**, 459 (1951); *Chem. Abstr.*, **46**, 11205 (1952).
400. S. Singh, *J. Sci. Ind. Res.*, **10B**, 82 (1951); *Chem. Abstr.*, **46**, 2547 (1952).
401. M. Ionescu and I. Goia, *Rev. Chim. Acad. Rep. Populaire Roumaine*, **5**, 85 (1960); *Chem. Abstr.*, **55**, 9402 (1961).
402. A. Senier and F. G. Shepheard, *J. Chem. Soc.*, **95**, 441 (1909).
403. A. Pictet and A. Hubert, *Chem. Ber.*, **29**, 1189 (1896); *Chem. Soc. Abstr.*, **70**, 503.
404. G. I. Braz and S. A. Kore, *Zh. Obshch. Khim.*, **23**, 868 (1953); *Chem. Abstr.*, **48**, 3979 (1954).
405. G. Koller and E. Krakauer, *Monatsh. Chem.*, **50**, 51 (1928); *Chem. Abstr.*, **22**, 3663 (1928).

406. D. Groeger and S. Johne, *Z. Naturforsch.*, **23B**, 1072 (1968); *Chem. Abstr.*, **69**, 74536 (1968).
407. D. P. Chakraborty, *J. Indian Chem. Soc.*, **46**, 177 (1969); *Chem. Abstr.*, **71**, 57519 (1969).
408. G. K. Hughes and E. Ritchie, *Aust. J. Sci. Res.*, **A4**, 423 (1951); *Chem. Abstr.*, **46**, 2548 (1952).
409. E. Leete, *Biogenesis of Natural Compounds*, P. Bernfeld, ed., Pergamon Press, Elmsford, N.Y., 1963, p. 780; *Ann. Rev. Plant Physiol.*, **18**, 179 (1967).
410. S. Niementowski, *Chem. Ber.*, **29**, 76 (1896); *Chem. Soc. Abstr.*, **70**, 261.
411. W. L. Baczynski and S. Niementowski, *Chem. Ber.*, **52**, 461 (1919); *Chem. Abstr.*, **13**, 2518 (1919).
412. J. R. Beck, et al., *J. Amer. Chem. Soc.*, **90**, 4706 (1968).
413. H. Decker and W. Petsch, *J. Prakt. Chem.*, **143**, 211 (1935); *Chem. Abstr.*, **30**, 462 (1936).
414. G. K. Hughes, N. K. Matheson, A. T. Norman, and E. Ritchie, *Aust. J. Sci. Res.*, **A5**, 206 (1952); *Chem. Abstr.*, **47**, 2176 (1953).
415. O. Dimroth and R. Criegee, *Chem. Ber.*, **90**, 2207 (1957); *Chem. Abstr.*, **52**, 10084 (1958).
416. F. Kröhnke and H. L. Honig, *Chem. Ber.*, **90**, 2215 (1957); *Chem. Abstr.*, **52**, 10086 (1958).
417. H. Decker and G. Dunant, *Chem. Ber.*, **42**, 1176 (1909); *Chem. Soc. Abstr.*, **96**, 433.
418. H. Decker, *J. Prakt. Chem.*, **45**, 161 (1892); *Chem. Soc. Abstr.*, **72**, 879.
419. A. Pictet and E. Patry, *Chem. Ber.*, **35**, 2534 (1902); *Chem. Soc. Abstr.*, **82**, 644.
420. H. Decker and T. Hock, *Chem. Ber.*, **37**, 1564 (1904); *Chem. Soc. Abstr.*, **86**, 620.
421. A. Albert and G. Catterall, *J. Chem. Soc.*, 4657 (1965).
422. V. Zanker and H. Cnobloch, *Z. Naturforsch.*, **176**, 819 (1962); *Chem. Abstr.*, **58**, 6366 (1963).
423. A. Kaufmann and A. Albertini, *Chem. Ber.*, **42**, 1999 (1909); *Chem. Soc. Abstr.*, **96**, 606.
424. P. Ehrlich and L. Benda, *Chem. Ber.*, **46**, 1931 (1913); *Chem. Abstr.*, **7**, 2944 (1913).
425. F. McCapra and R. A. Hann, *Chem. Commun.*, 442 (1969).
426. E. G. Janzen, J. B. Pickett, J. W. Happ, and W. DeAngelis, *J. Org. Chem.*, **35**, 88 (1970).
427. I. G. Farbenindustrie, French Patent 773,918; *Chem. Abstr.*, **29**, 2177 (1935).
428. E. Bergmann and O. Blum-Bergmann, *Chem. Ber.*, **63**, 757 (1930); *Chem. Abstr.*, **24**, 3511 (1930).
429. F. Ullmann and R. Maag, *Chem. Ber.*, **40**, 2515 (1907); *Chem. Abstr.*, **1**, 2600 (1907).
430. A. Chrzaszczewska, A. Braun, and M. Nowaczyk, *Lodz Towarz. Nauk.*, *Wydzial III*, *Acta Chim.*, **3**, 93 (1958); *Chem. Abstr.*, **53**, 13148 (1959).
431. A. Braun, *Lodz. Towarz. Nauk.*, *Wydzial III*, *Acta Chim.*, **12**, 101 (1967); *Chem. Abstr.*, **71**, 124189 (1969).
432. Cassella Farbwerke Mainkur A.-G., *British Patent* 793,088; *Chem. Abstr.*, **52**, 20203 (1958).
433. N. N. Dykhanov and G. A. Gorlach, *Med. Prom. SSSR*, **15**, 8 (1961); *Chem. Abstr.*, **56**, 4728 (1962).
434. K. Kormendy, *Acta. Chim. Acad. Sci. Hung.*, **21**, 83 (1959); *Chem. Abstr.*, **54**, 18524 (1960).
435. D. L. Tabern (Abbott Laboratories), U.S. Patent 2,645,594; *Chem. Abstr.*, **48**, 9409 (1954).
436. A. Chrzaszczewska, W. Kirkor, J. Bajan, and M. Nowaczyk, *Lodz. Towarz. Nauk.*, *Wydzial III*, *Acta Chim.*, **6**, 49 (1960); *Chem. Abstr.*, **55**, 18729 (1955).

437. A. Eckert and K. Steiner, *Monatsh. Chem.*, **36**, 175 (1915); *Sitzungsber.*, *IIB*, **123**, 1141 (1914); *Chem. Soc. Abstr.*, **108**, 564.

438. K. Ishihara, *J. Chem. Soc. Jap.*, **55**, 458 (1934); *Chem. Abstr.*, **28**, 5455 (1934).

439. V. I. Kikhteva and N. N. Dykhanov, *Metody Polucheniya Khim. Reaktivov Preparatov, Gos. Kom. Sov. Min. SSSR Khim.*, 70 (1964); *Chem. Abstr.*, **65**, 16940 (1966).

440. V. P. Maksimets and A. K. Sukhomlinov, *Khim. Geterotsikl. Soedin.*, **1966**, 739; *Chem. Abstr.*, **66**, 54868 (1967).

441. J. R. Cannon, G. K. Hughes, K. G. Neill, and E. Ritchie, *Aust. J. Sci. Res.*, **A5**, 406 (1952); *Chem. Abstr.*, **47**, 3858 (1953).

442. O. Eisleb, *Chem. Ber.*, **74**, 1433 (1941); *Chem. Abstr.*, **36**, 5465 (1942).

443. Z. Ledochowski and B. Wysocka-Skrzela, *Rocz. Chem.*, **38**, 225 (1964); *Chem. Abstr.*, **60**, 14471 (1964).

444. A. Ledochowski, B. Wysocka-Skrzela, and C. Radzikowski, Polish Patent 56,606; *Chem. Abstr.*, **70**, 106402 (1969).

445. B. Wysocka-Skrzela, Z. Ledochowski, and A. Ledochowski, *Rocz. Chem.*, **43**, 1279 (1969); *Chem. Abstr.*, **71**, 124187 (1969).

446. R. W. Stoughton (Mallinckrodt Chem. Works), U.S. Patent 2,709,171; *Chem. Abstr.*, **50**, 5775 ((1956).

447. R. R. Schumaker, French Patent 1,527,778; *Chem. Abstr.*, **71**, 81175 (1969).

448. B. I. Mikhant'ev and V. A. Sklyarov, *Zh. Obshch. Khim.*, **27**, 1697 (1957); *Chem. Abstr.*, **52**, 3818 (1958).

449. B. I. Mikhant'ev and V. A. Sklyarov, *Zh. Obshch. Khim.*, **26**, 784 (1956); *Chem. Abstr.*, **50**, 14760 (1956).

450. G. Wittig and K. Niethammer, *Chem. Ber.*, **93**, 944 (1960); *Chem. Abstr.*, **54**, 18516 (1956).

451. N. S. Drozdov, *J. Gen. Chem. USSR*, **7**, 2292 (1937); *Chem. Abstr.*, **32**, 568 (1938).

452. I. S. Ioffe and N. A. Selezneva, *Zh. Obshch. Khim.*, **31**, 50 (1961); *Chem. Abstr.*, **55**, 24751 (1961).

453. H. Quast and S. Huenig, *Chem. Ber.*, **101**, 435 (1968); *Chem. Abstr.*, **68**, 88180 (1968).

454. K. Lehmstedt, *Chem. Ber.*, **68**, 1455 (1935); *Chem. Abstr.*, **29**, 6893 (1935).

455. P. L. Macdonald and A. V. Robertson, *Aust. J. Chem.*, **19**, 275 (1966); *Chem. Abstr.*, **64**, 12745 (1966).

456. L. Knorr and E. Fertig, *Chem. Ber.*, **30**, 937 (1897); *Chem. Soc. Abstr.*, **72**, 371.

457. H. H. Perkampus and T. Rossel, *Z. Elektrochem.*, **62**, 94 (1958); *Chem. Abstr.*, **52**, 9760 (1958).

458. R. Itoh, *J. Phys. Sov. Jap.*, **14**, 1224 (1959); *Chem. Abstr.*, **54**, 9490 (1960).

459. H. Kokubun, *Naturwissenschaften.*, **44**, 233 (1957); *Chem. Abstr.*, **51**, 14427 (1957).

460. A. Albert and L. N. Short, *J. Chem. Soc.*, 760 (1945).

461. E. Sawicki, T. W. Stanley, W. C. Elbert, and M. Morgan, *Talanta*, **12**, 605 (1965).

462. S. Caroli and M. Lederer, *J. Chromatogr.*, **37**, 333 (1968).

463. M. Lederer and G. Roch, *J. Chromatogr.*, **31**, 618 (1967).

464. A. Albert and R. Goldacre, *J. Chem. Soc.*, 454 (1943).

465. H. Gilman, C. G. Stuckwisch, and A. R. Kendall, *J. Amer. Chem. Soc.*, **63**, 1758 (1941).

466. A. Albert and J. N. Phillips, *J. Chem. Soc.*, 1294 (1956).

467. A. K. Sukhomlinov and V. P. Maksimets, *Khim. Geterotsikl. Soedin.*, *Akad. Nauk Latv. SSR*, 416 (1963); *Chem. Abstr.*, **65**, 10464 (1966).

468. D. Cook, *Can. J. Chem.*, **41**, 2575 (1963); *Chem. Abstr.*, **59**, 11216 (1963).

469. R. M. Acheson, M. L. Burstall, C. W. Jefford, and B. F. Sansom, *J. Chem. Soc.*, 3742 (1954).

470. J. P. Kokko and J. H. Goldstein, *Spectrochim. Acta*, **19**, 1119 (1963).
471. A. Marzin, *J. Prakt. Chem.*, **138**, 99 (1933); *Chem. Abstr.*, **27**, 5332 (1933).
472. I. Tanasescu and E. Ramontianu, *Bull. Soc. Chim. Fr.*, **2**, 1485 (1935); *Chem. Abstr.*, **29**, 7335 (1935).
473. A. I. Gurevich and Yu. N. Sheinker, *Zh. Fiz. Khim.*, **33**, 883 (1959); *Chem. Abstr.*, **54**, 8285 (1960).
474. A. I. Gurevich and Yu. N. Sheinker, *Zh. Vsesoyuz. Khim. Obschest. D. I. Mendeleeva*, **6**, 116 (1961); *Chem. Abstr.*, **55**, 14460 (1961).
475. S. F. Mason, *J. Chem. Soc.*, 1281 (1959).
476. L. N. Short, *J. Chem. Soc.*, 4584 (1952).
477. A. K. Sukhomlinov, *Zh. Obshch. Khim.*, **28**, 1038 (1958); *Chem. Abstr.*, **52**, 16350 (1958).
478. S. J. Angyal and C. L. Angyal, *J. Chem. Soc.*, 1461, (1952).
479. Z. V. Pushkareva and Z. Yu. Kokoshko, *Doklady Akad. SSSR*, **93**, 77 (1953); *Chem. Abstr.*, **49**, 3194 (1955).
480. A. N. Aleksandrov, A. V. Karyakin, and N. G. Yaroslavskii, *Vestnik Leningrad University*, **5**, 121 (1950); *Chem. Abstr.*, **49**, 5476 (1955).
481. A. V. Karyakin and A. V. Shablya, *Doklady Akad. Nauk SSSR*, **116**, 969 (1957); *Chem. Abstr.*, **52**, 4324 (1958).
482. A. V. Karyakin, A. M. Grigorovskii, and N. G. Yaroslavskii, *Doklady Akad. Nauk SSSR*, **67**, 679 (1949); *Chem. Abstr.*, **44**, 3999 (1950).
483. A. Albert and J. B. Willis, *J. Soc. Chem. Ind.*, **65**, 26 (1946).
484. A. I. Gurevich and Yu. N. Sheinker, *Zh. Fiz. Khim.*, **36**, 734 (1962); *Chem. Abstr.*, **57**, 7231 (1962).
485. A. K. Sukhomlinov and V. I. Bliznyukov, *Zh. Obshch. Khim.*, **29**, 1316 (1959); *Chem. Abstr.*, **54**, 9928 (1960).
486. L. Hunter, *J. Chem. Soc.*, 806, (1945).
487. A. G. Cairns-Smith, *J. Chem. Soc.*, 182 (1961).
488. H. Kokubun, *Z. Phys. Chem.* (Frankfurt Am Main), **17**, 281 (1958); *Chem. Abstr.*, **55**, 2272 (1961).
489. D. N. Shigorin, A. K. Piskunov, G. A. Ozerova, N. A. Shchaglova, and N. V. Verein, *Doklady Akad. Nauk SSSR*, **158**, 432 (1965); *Chem. Abstr.*, **62**, 153 (1965).
490. Uu. N. Sheinker and Yu. I. Pomerantsev, *Zh. Fiz. Khim.*, **30**, 79 (1956); *Chem. Abstr.*, **50**, 14780 (1956).
491. S. F. Mason, *J. Chem. Soc.*, 4874 (1957).
492. J. T. Braunholtz, E. A. V. Ebsworth, F. G. Mann, and N. Sheppard, *J. Chem. Soc.*, 2780 (1958).
493. S. Nitzsche, *Chem. Ber.*, **77**, 337 (1944); *Chem. Abstr.*, **40**, 5053 (1946).
494. A. Weizmann, *Trans. Faraday Soc.*, **36**, 978 (1940).
495. J. Locher, *Justus Liebigs Ann. Chem.*, **279**, 275 (1894); *Chem. Soc. Abstr.*, **66**, 530.
496. S. Kahn, *Justus Liebigs Ann. Chem.*, **279**, 270 (1894); *Chem. Soc. Abstr.*, **66**, 529.
497. R. A. Reed, *J. Chem. Soc.*, 679 (1944).
498. H. Brockmann and H. Muxfeldt, *Chem. Ber.*, **89**, 1397 (1956); *Chem. Abstr.*, **51**, 5080 (1957).
499. N. S. Drozdov and O. M. Cherntsov, *Zh. Obshch. Khim.*, **21**, 1710 (1951); *Chem. Abstr.*, **46**, 4009 (1952).
500. F. Bradbury and W. Linnell, *Quart. J. Pharm. Pharmacol.*, **15**, 31 (1942).
501. A. Eckert and R. Pollack, *Monatsh. Chem.*, **38**, 11 (1917); *Chem. Abstr.*, **11**, 2772 (1917).

502. Von Schuckmann, *Frames* 3306-8, *Publishing Board's Microfilm PB 82023 on I. G. Farbenindustrie*, U.S. Dept. of Commerce, 1942.

503. W. J. Wechter, *J. Org. Chem.*, **28**, 2935 (1963).

504. E. R. Blout and R. S. Corley, *J. Amer. Chem. Soc.*, **69**, 763 (1947).

505. K. Lehmstedt, *Chem. Ber.*, **64**, 2381 (1931); *Chem. Abstr.*, **26**, 461 (1932).

506. F. Ullmann, *Chem. Ber.*, **36**, 1017 (1903); *Chem. Soc. Abstr.*, **84**, 519.

507. V. Kaufmann, *Justus Leibigs Ann. Chem.*, **279**, 281 (1894); *Chem. Soc. Abstr.*, **66**, 531.

508. J. N. Graves, G. K. Hughes, and F. Lions, *J. Proc. Roy. Soc. New South Wales*, **71**, 255 (1938); *Chem. Abstr.*, **32**, 7461 (1938).

509. A. M. Grigorovskii and V. S. Federov, *Zh. Prikl. Khim.*, **21**, 529 (1948); *Chem.Abstr.* **43**, 646 (1949).

510. A. L. Tarnoky, *Biochem. J.*, **46**, 297 (1950).

511. R. M. Acheson, T. G. Hoult, and K. A. Barnard, *J. Chem. Soc.*, 4142 (1954).

512. R. L. Bowman and N. Alexander, *Science.* **154**, 1454 (1966).

513. I. Tanasescu and E. Ramontianu, *Bull. Soc. Chim. Fr.*, **6**, 486 (1939); *Chem. Abstr.*, **33**, 5853 (1939).

514. M. Ionescu and I. Goia, *Acad. Rep. Populare Romine, Filiala Cluj, Stud. Cercet. Chim.*, **10**, 335 (1959); *Chem. Abstr.*, **55**, 533 (1961).

515. R. H. Prager and H. M. Thredgold, *Aust. J. Chem.*, **22**, 1503 (1969); *Chem. Abstr.*, **71**, 49742 (1969).

516. R. H. Prager and H. M. Thredgold, *Aust. J. Chem.*, **22**, 1511 (1969); *Chem. Abstr.*, **71**, 48939 (1969).

517. S. Kruger, *J. Chem. Soc.*, 3648 (1952).

518. S. M. Sherlin, G. I. Bras, A. I. Jakubovich, E. I. Vorob'ova, and A. P. Sergeev, *Justus Liebigs Ann. Chem.*, **516**, 218 (1935); *Chem. Abstr.*, **29**, 3677 (1935).

519. K. Toei, *Nippon Kagaku Zasshi*, **76**, 1085 (1955); *Chem. Abstr.*, **51**, 11913 (1957).

520. R. M. Acheson, B. Adcock, G. M. Glover, and L. E. Sutton, *J. Chem. Soc.*, 3367 (1960).

521. T. D. Perrine and L. J. Sargent, *J. Org. Chem.*, **14**, 583 (1949).

522. W. L. Semon and D. Craig *J. Amer. Chem. Soc.*, **58**, 1278 (1936).

523. P. N. Craig, U.S. Patent 3,043,842; *Chem. Abstr.*, **58**, 4527 (1963).

524. N. S. Drozdov and O. M. Cherntzov, *Zh. Obshch. Khim.*, **14**, 181 (1944); *Chem Abstr.*, **39**, 2290 (1945).

525. N. S. Drozdov and O. M. Cherntzov, *Zh. Obshch. Khim.*, **5**, 1576 (1935); *Chem. Abstr.*, **30**, 2195 (1936).

526. A. Albert, *J. Chem. Soc.*, 4653 (1965).

527. I. S. Ioffe and A. B. Tomchin, *Zh. Obshch. Khim.*, **39**, 1156 (1969); *Chem. Abstr.*, **71**, 70475 (1969).

528. A. K. Sheinkman, S. G. Potashnikova, and S. N. Baranov, *Khim. Geterotsikl. Soedin.*, 563 (1969); *Chem. Abstr.*, **71**, 124188 (1969).

529. O. N. Chupakhin, V. A. Trofimov, and Z. V. Pushkareva, *Khim. Geterotsikl. Soedin.*, 954 (1969); *Chem. Abstr.*, **72**, 111270 (1970).

530. H. Gilman and D. A. Shirley, *J. Amer. Chem. Soc.*, **72**, 2181 (1950).

531. V. K. Potapov, A. D. Filyugina, D. N. Shigorin, and G. A. Ozerova, *Doklady Akad. Nauk SSSR*, **180**, 398 (1968); *Chem. Abstr.*, **69**, 76408 (1968).

532. J. H. Bowie, R. G. Cooks, R. H. Prager, and H. M. Thredgold, *Aust. J. Chem.*, **20**, 1179 (1967); *Chem. Abstr.*, **67**, 37470 (1967).

533. J. A. Diment, E. Ritchie, and W. C. Taylor, *Aust. J. Chem.*, **20**, 1719 (1967).

534. S. C. Pakrashi, S. K. Roy, L. F. Johnson, T. George, and C. Djerassi, *Chem. Ind.* (London), 464 (1961).

535. L. Villemey, *C. R.*, *Acad. Sci. Paris*, **230**, 303 (1950); *Chem. Abstr.*, **44**, 5218 (1950).

536. L. Villemey, *Ann. Chim.*, **5**, 570, 642, 779 (1950); *Chem. Abstr.*, **45**, 2947, 3719 (1951).

537. F. Kröhnke and H. L. Honig, *Chem. Ber.*, **90**, 2226 (1957); *Chem. Abstr.*, **52**, 10088 (1958).

538. K. Lehmstedt and H. Hundertmark, *Chem. Ber.*, **62**, 1065 (1929); *Chem. Abstr.*, **23**, 4219 (1929).

539. H. Decker and G. Dunant, *Chem. Ber.*, **39**, 2720 (1906); *Chem. Soc.*, *Abstr.*, **90**, 901.

540. K. Gleu and R. Schaarschmidt, *Chem. Ber.*, **73**, 909 (1940); *Chem. Abstr.*, **35**, 3259 (1941).

541. K. Gleu and S. Nitzsche, *J. Prakt. Chem.*, **153**, 225 (1939); *Chem. Abstr.*, **33**, 6855 (1939).

542. K. Gleu and S. Nitzsche, *J. Prakt. Chem.*, **153**, 233 (1939); *Chem. Abstr.*, **33**, 6854 (1939).

543. W. D. Crow and J. R. Price, *Aust. J. Sci. Res.*, **2A**, 282 (1949); *Chem. Abstr.*, **46**, 4041 (1952).

544. K. Ishihara, *J. Chem. Soc. Jap.*, **55**, 557 (1934).

545. N. S. Drozdov and V. A. Sklyarov, *J. Gen. Chem. USSR*, **14**, 945 (1944); *Chem. Abstr.*, **39**, 4613 (1945).

546. N. S. Drozdov, *Trudy Kafedry Biokhim. Moskov. Zootekh. Inst. Konevodstva*, **1944**, 33 (1945); *Chem. Abstr.*, **41**, 764 (1947).

547. K. Junghans and A. Schoenberg, German Patent 1,295,548; *Chem. Abstr.*, **71**, 38826 (1969).

548. A. Schoenberg and K. Junghans, *Chem. Ber.*, **99**, 1015 (1966); *Chem. Abstr.*, **64**, 15837 (1966).

549. A. Chrzaszczewska and A. Braun, *Lodz. Towarz. Nauk.*, *Wydzial III*, *Acta Chim.*, **9**, 189 (1964); *Chem. Abstr.*, **62**, 9103 (1965).

550. K. Gleu and R. Schaarschmidt, *Chem. Ber.*, **72**, 1246 (1939); *Chem. Abstr.*, **33**, 7301 (1939).

551. (a) N. S. Drozdov, *Trudy Kafedry Biokhim. Moskov. Zootekh. Inst. Konevodstva*, **1944**, 48 (1945); *Chem. Abstr.*, **41**, 764; (b) *J. Gen. Chem. USSR*, **16**, 243 (1946); *Chem. Abstr.*, **41**, 130 (1947).

552. H. Bünzly and H. Decker, *Chem. Ber.*, **37**, 575 (1904); *Chem. Soc. Abstr.*, **86**, 344.

553. H. Decker and R. Pschorr, *Chem. Ber.*, **37**, 3396 (1904); *Chem. Soc. Abstr.*, **86**, 926.

554. C. Kaiser and C. L. Zirkle, *U.S. Patent* 3,391,143; *Chem. Abstr.*, **69**, 96501 (1968).

555. C. L. Zirkle, *U.S.* Patent 3,131,190; *Chem. Abstr.*, **61**, 4325 (1964).

556. M. Gomberg and L. H. Cone, *Justus Liebigs*, *Ann. Chem.*, **370**, 203 (142) (1909); *Chem. Soc. Abstr.*, **98**, 55.

557. I. Tanasescu and E. Ramontianu, *Bull. Soc.*, *Chim. Fr.*, **3**, 2009 (1936); *Chem. Abstr.*, **31**, 8531 (1937).

558. A. Kliegl and A. Brösamle, *Chem. Ber.*, **69**, 197 (1936); *Chem. Abstr.*, **30**, 3433 (1936).

559. E. F. Elslager, R. E. Bowman, F. H. Tendick, D. J. Tivey, and D. F. Worth, *J. Med. Pharm. Chem.*, **5**, 1159 (1962).

560. N. M. Voronina, Z. V. Pushkareva, L. B. Radina, and N. W. Babikova, *Zh. Obshch. Khim.*, **30**, 3476 (1960); *Chem. Abstr.*, **55**, 19921 (1961).

561. A. Ledochowski, *Rocz. Chem.*, **42**, 1973 (1968); *Chem. Abstr.*, **70**, 106357 (1969).

562. E. F. Elslager and F. H. Tendick, U.S. Patent 2,880,210; *Chem. Abstr.*, **53**, 15100 (1959).

563. K. Lehmstedt and H. Klee, *Chem. Ber.*, **69**, 1514 (1936); *Chem. Abstr.*, **30**, 5991 (1936).

564. M. Ionescu, H. Mantsch, and I. Goia, *Chem. Ber.*, **96**, 1726 (1963); *Chem. Abstr.*, **60**, 390 (1964).

565. I. Tanasescu and Z. Frenkel, *Acad. Rep. Populare Romine, Stud. Cercet. Chim.*, **4**, 227 (1956); *Chem. Abstr.*, **51**, 10527 (1957).

566. K. Lehmstedt and F. Dostal, *Chem. Ber.*, **72**, 1071 (1939); *Chem. Abstr.*, **33**, 5853 (1939).

567. H. Decker, T. Hock and C. Djiwonsky, *Chem. Ber.*, **35**, 3068 (1902); *Chem. Soc. Abstr.*, **82**, 830.

568. I. Tanasescu and M. Macarovici, *Bull. Soc. Chim. Fr.*, **4**, 240 (1937); *Chem. Abstr.*, **32**, 568 (1938).

569. I. Tanasescu and E. Ramontianu, *Chem. Ber.*, **69**, 1825 (1936); *Chem. Abstr.*, **31**, 8532 (1937).

570. K. Lehmstedt and H. Klee, *Chem. Ber.*, **69**, 1155 (1936); *Chem. Abstr.*, **30**, 5224 (1936).

571. K. Lehmstedt and F. Dostal, *Chem. Ber.*, **71**, 2432 (1938); *Chem. Abstr.*, **33**, 987 (1939).

572. G. A. Dima and P. Poganceanu, *Bull. Sect. Sci. Acad. Roumaine*, **22**, 19 (1939); *Chem Abstr.*, **34**, 5754 (1940).

573. M. Ionescu, A. R. Katritzky, and N. Ternai, *Tetrahedron*, **22**, 3227 (1966).

574. I. Tanasescu, V. Farcasan, and C. Toma, *Bull. Soc. Chim. Fr.*, 691 (1960); *Chem. Abstr.*, **55**, 23535.

575. N. S. Drozdov, *J. Gen. Chem. USSR*, **7**, 219 (1937); *Chem. Abstr.*, **31**, 4320 (1937).

576. A. Archer (Sterling Drug, Inc.), U.S. Patent 2,647,901; *Chem. Abstr.*, **48**, 13729 (1954).

577. S. Archer, L. B. Rochester, and M. Jackman, *J. Amer. Chem. Soc.*, **76**, 588 (1954).

578. H. B. Nisbet and A. B. Goodlet, *J. Chem. Soc.*, 2772 (1932).

579. I. Tanasescu and Z. Frenkel, *Bull. Soc. Chim. Fr.*, 696 (1960); *Chem. Abstr.*, **55**, 23536 (1961).

580. I. Tanasescu and Z. Frenkel, *Studia Univ. Babes-Bolyai*, **2**, 145 (1959); *Chem. Abstr.*, **55**, 5496 (1961).

581. W. Gruszecki and E. Borowski, *Rocz. Chem.*, **42**, 733 (1968); *Chem. Abstr.*, **71**, 38774 (1969).

582. K. Matsumura, *J. Amer. Chem. Soc.*, **60**, 593 (1938).

583. E. R. Klein and F. N. Lahey, *J. Chem. Soc.*, 1418 (1947).

584. H. Medenwald, *Med. Chem. Abhandl. Med.-Chem. Forschungsstatten Farbenfabrik Bayer*, **5**, 206 (1956); *Chem. Abstr.*, **55**, 8403 (1961).

585. E. Koft and F. H. Case, *J. Org. Chem.*, **27**, 865 (1962).

586. E. Aarons and A. Albert, *J. Chem. Soc.*, 183 (1942).

587. A. Albert, *The Acridines*, 2nd ed., Arnold, London, 1966, pp. 5, 377.

588. B. S. Joshi, N. Parkash, and K. Venkataraman, *J. Sci. Ind. Res.*, **14B**, 325 (1955); *Chem. Abstr.*, **50**, 11341.

589. M. M. Barnett, A. H. C. Gillieson and W. O. Kermack, *J. Chem. Soc.*, 433, (1934).

590. I. Tanasescu, V. Farcasan, and O. Piringer, *Acad. Rep. Populare Romine, Filiala Cluj, Stud. Cercert. Chim.*, **12**, 285 (1961); *Chem. Abstr.*, **59**, 573 (1963).

591. I. Tanasescu and V. Farcasan, *Analele Acad. Rep. Populare Romane, Sect. Stiinte Geol., Geograf. Biol., Ser. A2, Mem, No. 16*, 1 (1949); *Chem. Abstr.*, **46**, 8118 (1952).

592. J. R. Price, *Aust. J. Sci. Res.*, **2A**, 249 (1949); *Chem. Abstr.*, **46**, 4010 (1952).

593. J. R. Price, *Aust. J. Sci. Res.*, **2A**, 272 (1949); *Chem. Abstr.*, **46**, 4013 (1952).

594. J. R. A. Anderson, K. G. O'Brien, and F. H. Reuter, *Anal. Chim. Acta*, **7**, 226 (1952).

595. H. Brockmann, H. Muxfeldt, and G. Haese, *Chem. Ber.*, **90**, 44 (1957); *Chem. Abstr.*, **51**, 10526 (1957).

596. T. R. Govindachari, B. R. Pai, and V. N. Ramachandran, *Indian J. Chem.*, 179 (1968); *Chem. Abstr.*, **69**, 106585.

597. L. J. Drummond and F. N. Lahey, *Aust. J. Sci. Res.*, **A2**. 630 (1949); *Chem. Abstr.*, **47**, 3862 (1953).

598. S. K. Banerjee, D. Chakravarti, R. N. Chakravarti, H. M. Fales, and D. L. Klayman, *Tetrahedron*, **16**, 251 (1961); *Bull. Calcutta School Trop. Med.*, **9**, 116 (1961); *Chem. Abstr.*, **57**, 12566 (1962).

599. F. M. Dean, J. Goodchild, L. E. Houghton, J. A. Martin, R. B. Morton, B. Parton, A. W. Price, and N. Somvichien, *Tetrahedron Lett.*, 4153 (1966).

600. R. H. Prager and H. M. Thredgold, *Aust. J. Chem.*, **21**, 229 (1968); *Chem. Abstr.*, **68**, 87426 (1968).

601. G. K. Hughes and K. G. Neill, *Aust. J. Sci. Res.*, **2A**, 429 (1949); *Chem. Abstr.*, **46**, 117 (1952).

602. W. D. Crow and J. R. Price, *Aust. J. Sci. Res.*, **2A**, 255 (1949); *Chem. Abstr.*, **46**, 4011 (1952).

603. N. Steiger (Hoffmann-La Roche, Inc.), U.S. Patent 2,732,373; *Chem. Abstr.*, **50**, 10795 (1956).

604. W. D. Crow, *Aust. J. Sci. Res.*, **2A**, 264 (1949); *Chem. Abstr.*, **46**, 4011 (1952).

605. R. H. Prager and H. M. Thredgold, *Tetrahedron Lett.*, 4909 (1966).

606. R. H. Prager and H. M. Thredgold, *Aust. J. Chem.*, **22**, 1477 (1969); *Chem. Abstr.*, **71**, 50287 (1969).

607. R. H. Prager and H. M. Thredgold, *Aust. J. Chem.*, **22**, 1493 (1969); *Chem. Abstr.*, **71**, 50288 (1969).

608. G. K. Rollefson and R. W. Stoughton, *J. Amer. Chem. Soc.*, **63**, 1517 (1941).

609. Yu. S. Tsizin and M. V. Rubtsov, USSR Patent 188,495; *Chem. Abstr.*, **67**, 90696 (1967).

610. Yu. S. Tsizin and M. V. Rubtsov, *Khim. Geterotsikl. Soedin., Sb. 1; Azotsoderzhashchie Geterotsikly*, 285 (1967); *Chem. Abstr.*, **70**, 87520 (1969).

611. A. W. Johnson, *Ciba Foundation Symposium, Aminoacids Peptides Antimetabolic Activity*, 123, (1958).

612. (a) A. W. Johnson, in *Actinomycin*, S. A. Waksman, ed., Interscience, New York, 1968, p. 33; (b) H. Brockmann, *Progr. Chem. Org. Nat. Prod.*, **18**, 1 (1960); *Pure Appl. Chem.*, **2**, 405 (1961); *Angew. Chem.*, **66**, 1 (1964).

613. H. Brockmann, G. Pampus, and J. H. Manegold, *Chem. Ber.*, **92**, 1294 (1959); *Chem. Abstr.*, **54**, 545 (1960).

614. H. Brockmann and R. Vohwinkel, *Naturwissenschaften*, **41**, 257 (1954); *Chem. Abstr.*, **49**, 5571 (1955).

615. H. Brockmann and N. Grubhofer, *Naturwissenchaften*, **37**, 494 (1950); *Chem. Abstr.*, **45**, 7997 (1951).

616. H. Brockmann and K. Vohwinkel, Chem. *Ber.*, **89**, 1373 (1956); *Chem. Abstr.*, **51**, 5078 (1957).

617. A. W. Johnson, A. R. Todd, and L. C. Vining, *J. Chem. Soc.*, 2672 (1952).

618. S. J. Angyal, E. Bullock, W. G. Hanger, and A. W. Johnson, *Chem. Ind. (London)*, 1295 (1955).

619. H. Brockmann and H. Muxfeldt, *Angew. Chem.*, **67**, 617 (1955); *Chem. Abstr.*, **50**, 1847 (1956).

620. E. Bullock and A. W. Johnson, *J. Chem. Soc.*, 1602 (1957).

621. H. Brockmann and H. Muxfeldt, *Naturwissenschaften*, **41**, 500 (1954); *Chem. Abstr.*, **49**, 10963 (1955).

622. W. H. Linnell and L. K. Sharp, *Quart. J. Pharm. Pharmacol.*, **21**, 58 (1948).

623. S. Nitzsche, *Angew. Chem.*, **52,** 517 (1939).

624. S. F. Mason, *J. Chem. Soc.*, 674 (1958).

625. S. F. Mason, *J. Chem. Soc.*, 5010 (1957).

626. F. Kehrmann and Zd. Matusinsky, *Chem. Ber.*, **45,** 3498 (1912); *Chem. Abstr.*, **7,** 2394 (1913).

627. N. Campbell and A. G. Cairns-Smith, *J. Chem. Soc.*, 1191 (1961).

628. H. Baumann, German Patent 1,232,295; *Chem. Abstr.*, **66,** 76933 (1967).

629. S. Nitzsche, *Chem. Ber.*, **76,** 1187 (1943); *Chem. Abstr.*, **39,** 1170 (1945).

630. F. Kehrmann and A. Stepanoff, *Chem. Ber.*, **41,** 4133 (1908); *Chem. Abstr.*, **3,** 539 (1909).

631. F. Kehrmann, H. Goldstein, and P. Tschudi, *Helv. Chim. Acta*, **2,** 315 (1919); *Chem. Abstr.*, **13,** 3177 (1919).

632. F. Kehrmann and J. Tschui, *Helv. Chim. Acta*, **8,** 27 (1925); *Chem. Abstr.*, **19,** 1280 (1925).

633. F. Kehrmann, H. Goldstein, and P. Tschudi, *Helv. Chim. Acta*, **2,** 379 (1919); *Chem. Abstr.*, **13,** 3178 (1919).

634. H. Goldstein and M. Piolino, *Helv. Chim. Acta*, **10,** 334 (1927); *Chem. Abstr.*, **21,** 2268 (1927).

635. H. Goldstein and M. de Simo, *Helv. Chim. Acta*, **10,** 596 (1927); *Chem. Abstr.*, **22,** 1590 (1928).

636. F. Kehrmann and J. Tschui, *Helv. Chim. Acta*, **8,** 23 (1925); *Chem. Abstr.*, **19,** 1280 (1925).

637. F. Kehrmann and F. Brunner, *Helv. Chim. Acta*, **9,** 216 (1926); *Chem. Abstr.*, **20,** 1801 (1926).

638. G. Goldstein and J. Vaymatchar, *Helv. Chim. Acta*, **11,** 245 (1928); *Chem. Abstr.*, **22,** 3129 (1928).

639. H. Goldstein and W. Kopp, *Helv. Chim. Acta*, **11,** 478 (1928); *Chem. Abstr.*, **22,** 2944 (1928).

640. F. Kehrmann, H. Goldstein and F. Brunner, *Helv. Chim. Acta*, **9,** 222 (1926); *Chem. Abstr.*, **20,** 1802 (1926).

641. A. Edinger, *Chem. Ber.*, **33,** 3769 (1900); *Chem. Soc. Abstr.*, **80,** 166.

642. V. Farcasan and I. Balazs, *Studia Univ. Babes-Bolyai, Ser. Chem.*, **14,** 43 (1969); *Chem. Abstr.*, **72,** 43555.

643. Gevaert PhotoProducten N.V., German Patent 1,146,751; *Chem. Abstr.*, **59,** 15421 (1963).

644. A. Edinger and W. Arnold, *J. Prakt. Chem.*, **64,** 182 (1901); *Chem. Soc. Abstr.*, **80,** 753.

645. E. F. Elslager, *J. Org. Chem.*, **27,** 4346 (1962).

646. A. Edinger and I. C. Ritsema, *J. Prakt. Chem.*, **68,** 72 (1903); *Chem. Soc. Abstr.*, **84,** 719.

647. O. M. Cherntzov, *J. Gen. Chem. USSR*, **14,** 186 (1944); *Chem. Abstr.*, **39,** 2291 (1945).

648. R. S. Asquith, D. L. Hammick, and P. L. Williams, *J. Chem. Soc.*, 1181 (1948).

649. M. Ionescu, H. Mantsch, and I. Goia, *Acad. Rep. Populare Romine, Filiala Cluj, Stud. Cercet. Chim.*, **12,** 135 (1961); *Chem. Abstr.*, **57,** 12432 (1962).

650. S. J. Das-Gupta, *J. Indian Chem. Soc.*, **17,** 244 (1940); *Chem. Abstr.*, **34,** 6282 (1940).

651. F. Wild and J. M. Young, *J. Chem. Soc.*, 7261 (1965).

652. V. G. Pesin, I. G. Vitenberg, and A. M. Khaletskii, *Zh. Obshch. Khim.*, **34,** 2769 (1964); *Chem. Abstr.*, **61,** 14663 (1964).

653. V. G. Pesin, I. G. Vitenberg, and A. M. Khaletskii, *Zh. Obshch. Khim.*, **34,** 1276 (1964); *Chem. Abstr.*, **61,** 1853 (1964).

654. V. G. Pesin, A. M. Khaletskii, and I. G. Vitenberg, *Zh. Obshch. Khim.*, **31**, 2522 (1961); *Chem. Abstr.*, **56**, 14046 (1962).
655. G. Hopkins and L. Hunter, *J. Chem. Soc.*, 638, (1942).
656. J. I. Mueller, K. E. Daugherty, and R. J. Robinson, *Anal. Chem.*, **36**, 2195 (1964).
657. M. W. Goheen, K. E. Daugherty, and R. J. Robinson, *Anal. Chim. Acta*, **32**, 81 (1965).
658. S. J. Das-Gupta, *J. Indian Chem. Soc.*, **18**, 43 (1941); *Chem. Abstr.*, **36**, 56 (1942).
659. K. E. Daugherty, R. J. Robinson, and J. I. Mueller, *Anal. Chem.*, **36**, 1098 (1964).
660. K. Kitani, *J. Chem. Soc. Jap.*, **75**, 480 (1954); *Chem. Abstr.*, **49**, 10296 (1955).
661. Mitsui Chem. Ind. Co., Japanese Patent 163,130; *Chem. Abstr.*, **42**, 6382 (1948).
662. M. Ionescu, I. Goia, and H. Mantsch, *Acad. Rep. Populare Romine, Filiala Cluj, Stud. Cercet. Chim.*, **13**, 95 (1962); *Chem. Abstr.*, **60**, 5456 (1964).
663. A. Edinger and W. Arnold, *J. Prakt. Chem.*, **64**, 471 (1901); *Chem. Soc. Abstr.*, **82**, 181.
664. S. J. Das-Gupta, *J. Indian Chem. Soc.*, **20**, 137 (1943); *Chem. Abstr.*, **37**, 6665 (1943).
665. R. O. Clinton and C. M. Suter, *J. Amer. Chem. Soc.*, **70**, 491 (1948).
666. L. Monti and G. Franchi, *Gazz. Chim. Ital.*, **81**, 764 (1951); *Chem. Abstr.*, **48**, 3976 (1954).
667. A. Albert, *The Acridines*, 2nd ed., Arnold, London, 1966, pp. 28, 386.
668. S. P. Acharya and K. S. Nargund, *J. Karnatak University*, **6**, 37 (1961); *Chem. Abstr.*, **59**, 7490 (1963).
669. I. Goia, H. Mantsch and M. Ionescu, *Rev. Roum. Chim.*, **13**, 1511 (1968); *Chem. Abstr.*, **71**, 49744 (1969).
670. G. M. Badger, N. Kowanko, and W. H. F. Sasse, *J. Chem. Soc.*, 440 (1959).
671. K. Lehmstedt and H. Hundertmark, *Chem. Ber.*, **63**, 1229 (1930); *Chem. Abstr.*, **24**, 4040 (1930).
672. A. Schönberg, O. Schütz, S. Nickel, H. Krüll, W. Marschner, and F. Kaplan, *Chem. Ber.*, **61**, 1375 (1928); *Chem. Abstr.*, **22**, 4510 (1928).
673. A. Schoenberg and E. Frese, *Chem. Ber.*, **101**, 701 (1968); *Chem. Abstr.*, **68**, 68597 (1968).
674. R. I. Fryer, J. Earley, and L. H. Sternbach, *J. Chem. Soc.*, 4979 (1963).
675. C. F. Allen and G. H. W. McKee, *Org. Syntheses*, **19**, 6 (1939).
676. C. D. Hurd and O. E. Fancher, *J. Amer. Chem. Soc.*, **69**, 716 (1947).
677. M. Ionescu, I. Goia, and H. Mantsch, *Rev. Roum. Chim.*, **11**, 243 (1966); *Chem. Abstr.*, **65**, 3713 (1966).
678. Z. V. Pushkareva, L. V. Varyukhina, and Z. Yu. Kokoshko, *Doklady Akad. Nauk SSSR*, **108**, 1098 (1956); *Chem. Abstr.*, **51**, 21 (1957).
679. S. J. Das-Gupta, *J. Indian Chem. Soc.*, **18**, 25 (1941); *Chem. Abstr.*, **36**, 91 (1942).
680. A. Albert and R. Goldacre, *J. Chem. Soc.*, 706 (1946).
681. W. Gruszecki and E. Borowski, *Rocz. Chem.*, **41**, 1611 (1967); *Chem. Abstr.*, **68**, 95656 (1968).
682. N. S. Drozdov, *J. Gen. Chem. USSR*, **6**, 1641 (1936); *Chem. Abstr.*, **31**, 2610 (1937).
683. M. Polaczek, *Rocz. Chem.*, **16**, 76 (1936); *Chem. Abstr.*, **31**, 2218 (1937).
684. Z. Ledochowski, et al., *Rocz. Chem.*, **34**, 63 (1960); *Chem. Abstr.*, **54**, 16452 (1960).
685. A. Ledochowski and B. Kozinska, *Rocz. Chem.*, **40**, 127 (1966); *Chem. Abstr.*, **65**, 16940 (1966).
686. Z. Ledochowski, B. Stefanska, and A. Ledochowski, *Rocz. Chem.*, **40**, 291 (1966); *Chem. Abstr.*, **65**, 2218 (1966).
687. A. Ledochowski, Z. Ledochowski, and R. Dobrzeniecka, *Rocz. Chem.*, **36**, 1101 (1962); *Chem. Abstr.*, **58**, 12511 (1963).

688. S. Singh and M. Singh, *J. Sci. Ind. Res.*, **11B**, 102 (1952); *Chem. Abstr.*, **48**, 8786 (1954).
689. Z. Ledochowski, et al., *Rocz. Chem.*, **34**, 53 (1960); *Chem. Abstr.*, **54**, 16452 (1960).
690. H. Mauss and I. G. Farbenindustrie, German Patent 565,411, U.S. Patent 855,302; *Chem Abstr.*, **27**, 999 (1933).
691. O. Yu. Magidson, A. M. Grigorovskii, and E. P. Gal'perin, *Zh. Obshch. Khim.*, **8**, 56 (1938); *Chem. Abstr.*, **32**, 5405 (1938).
692. J. R. Geigy A.-G., British Patent 1,127,721; *Chem. Abstr.*, **70**, 79162.
693. V. P. Maksimets, A. K. Sukhomlinov, and N. N. Shtefan, *Khim. Geterotsikl. Soedin.*, **1969**, 947; *Chem. Abstr.*, **72**, 121338 (1970).
694. K. Singh and G. Singh, *Indian J. Pharm.*, **14**, 47 (1952); *Chem. Abstr.*, **47**, 6947 (1953).
695. G. B. Bachman and F. M. Cowen, *J. Org. Chem.*, **13**, 89 (1948).
696. Smith, Kline, and French Laboratories, French Patent M3550; *Chem. Abstr.*, **64**, 8157 (1966).
697. E. F. Elslager and F. H. Tendick (Parke, Davis, and Co.), British Patent 829,728; *Chem. Abstr.*, **54**, 18559 (1960).
698. A. Albert, F. J. Dyer, and W. H. Linnell, *Quart. J. Pharm. Pharmacol.*, **10**, 649 (1937).
699. N. Steiger (Hoffman-La Roche, Inc.), U.S. Patent 2,732,374; *Chem. Abstr.*, **50**, 10795 (1956).
700. H. Mauss, F. Mietzsch, and I. G. Farbenindustrie, German Patent 571,499; *Chem. Abstr.*, **27**, 3036 (1933).
701. H. L. Bradlow and C. A. Van der Werf, *J. Amer. Chem. Soc.*, **70**, 654 (1948).
702. G. Singh, A. Singh, S. Singh, and M. Singh, *J. Indian Chem. Soc.*, **28**, 698 (1951); *Chem. Abstr.*, **47**, 12391.
703. F. Singh, T. Singh and G. Singh, *J. Indian Chem. Soc.*, **24**, 51 (1947); *Chem. Abstr.*, **42**, 1278 (1948).
704. G. Singh, G. Singh, and M. Singh, *J. Indian Chem. Soc.*, **24**, 79 (1947); *Chem. Abstr.*, **42**, 1279 (1948).
705. I. Kh. Fel'dman and E. L. Kopeliovich, *Arch. Pharm.*, **273**, 488 (1935); *Chem. Abstr.*, **30**, 1378 (1936).
706. A. Ledochowski, et al., *Rocz. Chem.*, **36**, 827 (1962); *Chem. Abstr.*, **59**, 570 (1963).
707. K. C. Kshatriya, K. S. Nargund, and S. R. Patel, *J. Univ. Bombay*, **17A**, 13 (1948); *Chem Abstr.*, **43**, 6632 (1949).
708. A. F. Bekhli, *Sbornik Statei Obshch. Khim.*, **2**, 1130 (1953); *Chem. Abstr.* **49**, 5479 (1955).
709. G. Singh, S. Singh, A. Singh, and M. Singh, *J. Indian Chem. Soc.*, **29**, 783 (1952); *Chem. Abstr.*, **47**, 12390 (1953).
710. R. E. Bowman and D. J. Tivey, British Patent 799,366; *Chem. Abstr.*, **53**, 5293 (1959).
711. D. M. Hall and E. E. Turner, *J. Chem. Soc.*, 694 (1945).
712. L. W. Greene, *Amer. J. Pharm.*, **120**, 39 (1948).
713. W. G. Dauben and R. L. Hodgson, *J. Amer. Chem. Soc.*, **72**, 3479 (1950).
714. V. K. Mehta and S. R. Patel, J. Indian Chem. Soc., **43**, 235 (1966); *Chem. Abstr.*, **65**, 5439 (1966).
715. N. N. Dykhanov, G. A. Gorlach, and V. P. Sergovskaya, *Med. Prom. SSSR*, **14**, 22 (1960); *Chem. Abstr.*, **55**, 4505 (1961).
716. N. N. Dykhanov, USSR Patent 126,497; *Chem. Abstr.*, **54**, 18559 (1960).
717. R. E. Bowman and D. J. Tivey, *British Patent* 799,631; *Chem. Abstr.*, **53**, 5293 (1959).
718. O. Yu. Magidson and A. I. Travin, *Chem. Ber.*, **69**, 537 (1936); *Chem, Abstr.*, **30**, 4498 (1936).

719. M. L. Aggarwal, I. Sen-Gupta and B. Ahmad, *J. Indian Chem. Soc.*, **22**, 41 (1945); *Chem. Abstr.*, **40**, 576 (1946).

720. K. Singh and B. Ahmad, *J. Indian Chem. Soc.*, **26**, 175 (1949); *Chem. Abstr.*, **44**, 3998 (1950).

721. Y. Ahmad, *Pakistan J. Sci. Res.*, **1**, 36 (1949); *Chem. Abstr.*, **46**, 4545 (1950).

722. G. Singh and M. Singh, *J. Indian Chem. Soc.*, **25**, 227 (1948); *Chem. Abstr.*, **43**, 7937 (1949).

723. N. J. Leonard and L. C. Smith, *J. Amer. Chem. Soc.*, **69**, 3147 (1947).

724. K. Lehmstedt, *Chem. Ber.*, **67**, 336 (1934); *Chem. Abstr.*, **28**, 2342 (1934).

725. O. Yu. Magidson and A. I. Travin, *J. Gen. Chem. USSR*, **11**, 243 (1941); *Chem. Abstr.*, **35**, 7965 (1941).

726. A. A. Goldberg (Ward, Blenkinsop, and Co.), British Patent 602,331; *Chem. Abstr.*, **42**, 8827 (1948).

727. A. F. Behkli, *Zh. Obshch. Khim.*, **23**, 329 (1953); *Chem. Abstr.*, **48**, 2713 (1954).

728. A. G. Munshi, P. M. Dholkia, and K. S. Nargund, *J. Indian Chem. Soc.*, **35**, 130 (1958); *Chem. Abstr.*, **53**, 13150 (1959).

729. A. G. Munshi and K. S. Nargund, *J. Indian Chem. Soc.*, **36**, 115 (1959); *Chem. Abstr.*, **53**, 21965 (1959).

730. O. Eisleb and H. Jensch, German Patent 488,680; British Patent 283,510; *Chem. Abstr.*, **24**, 2149 (1930).

731. M. Ionescu and I. Mester, *Rev. Roum. Chem.*, **14**, 789 (1969); *Chem. Abstr.*, **72**, 21587 (1970).

732. B. Horowska and A. Ledochowski, *Rocz. Chem.*, **42**, 1351 (1968); *Chem. Abstr.*, **70**, 28802 (1969).

733. K. C. Kshatriya, S. R. Patel, and K. S. Nargund, *J. Univ. Bombay*, **15A**, 42 (1946); *Chem. Abstr.*, **41**, 6214 (1947).

734. G. K. Hughes, K. G. Neill, and E. Ritchie, *Aust. J. Sci. Res.*, **5A**, 401 (1952); *Chem. Abstr.*, **47**, 3857 (1953).

735. A. K. Sukhomlinov, V. A. Ruzhnikov, and V. P. Maksimets, *Izv. Vyssh. Ucheb. Zaved. Khim. Khim. Teknol.*, **9**, 246 (1966); *Chem. Abstr.*, **65**, 10563 (1966).

736. H. Mauss and F. Mietzsch, (I. G. Farbenindustrie), German Patent 632,224; British Patent 441,132; U.S. Patent, 2,121,207; *Chem. Abstr.*, **30**, 6761 (1936).

737. N. S. Drozdov and S. S. Drozdov, *J. Gen. Chem. USSR*, **4**, 1 (1934); *Chem. Abstr.*, **28**, 5456 (1934).

738. P. Cohn, *Monatsh. Chem.*, **22**, 385 (1901); *Chem. Soc. Abstr.*, **80**, 642.

739. K. Matsumura, *J. Amer. Chem. Soc.*, **49**, 810 (1927).

740. G. N. Kosheleva, *Zh. Obshch. Khim.*, **26**, 2567 (1956); *Chem. Abstr.*, **51**, 5081 (1957).

741. S. Singh, *J. Sci. Ind. Res.*, **9B**, 226 (1950); *Chem. Abstr.*, **45**, 7573 (1957).

742. Meister, Lucius, and Brünning, (Farbwerke) German Patents 364,031–364,037; Swiss Patents, 93,439; 93,752; 93,753; 94,625; 94,626; 94,982; 96,608; 96,609; British Patent 176,038; U.S. Patent 1,629,873; *Chem. Abstr.*, **18**, 1131 (1924).

743. A. Albert and W. Gledhill, *J. Soc. Chem. Ind.*, **61**, 159 (1942).

744. Z. Yu. Kokoshko and Z. V. Pushkareva, *J. Gen. Chem. USSR*, **24**, 875 (1954); *Chem. Abstr.*, **49**, 8284 (1955).

745. A. M. Khaletskii, V. G. Pesin, and Tsin Chou, *Zh. Obshch. Khim.*, **28**, 2821 (1958); *Chem. Abstr.*, **53**, 9224 (1959).

746. F. Mietzsch, H. Mauss, J. Klarer, and I. G. Farbenindustrie, German Patent 547,983; *Chem. Abstr.*, **26**, 3522 (1932); British Patent, 367,024; *Chem. Abstr.*, **27**, 1892 (1933).

747. E. A. Steck, U.S. Patent 2,762,806; *Chem. Abstr.*, **51**, 4444 (1957).

748. J. Lichtenberger and R. Thermet, *Bull. Soc. Chim. Fr.*, 318 (1951); *Chem. Abstr.*, **46,** 3509 (1952).
749. S. Singh and M. Singh, *J. Sci. Ind. Res.*, **9B,** 27 (1950); *Chem. Abstr.*, **44,** 6863 (1950).
750. A. Ledochowski, B. Kozinska, and B. Stefanska, *Rocz. Chem.*, **38,** 219 (1964); *Chem. Abstr.*, **60,** 14471 (1964).
751. F. Ullmann and C. Wagner, *Justus Liebigs Ann. Chem.*, **355,** 359 (1907); *Chem. Abstr.*, **2,** 87 (1908).
752. G. Singh and M. Singh, *J. Indian Chem. Soc.*, **23,** 224 (1946); *Chem. Abstr.*, **41,** 2419 (1947).
753. A. K. Sukhomlinov and V. P. Maksimets, *Khim. Geterotsikl. Soedin.*, *Akad. Nauk Latv. SSR*, 99 (1965); *Chem. Abstr.*, **63,** 5602 (1965).
754. F. Mietzsch, H. Mauss, and I. G. Farbenindustrie, German Patent 553,072; British Patent 363,392; U.S. Patent 2,113,357; *Chem. Abstr.*, **26,** 4683 (1932).
755. V. I. Kikhteva and N. N. Dykhanov, *Metody Polucheniya Khim. Reaktivov Preparatov, Gos. Kom. Sov. Min. SSSR Khim.*, 78 (1964); *Chem. Abstr.*, **64,** 15837 (1966).
756. I. G. Farbenindustrie, German Patent 642,758; *Friedl.*, **23,** 541 (1936).
757. S. J. Das-Gupta, *J. Indian Chem. Soc.*, **16,** 364 (1939); *Chem. Abstr.*, **34,** 2379 (1940).
758. J. B. Niederl and M. B. Hundert, *J. Amer. Chem. Soc.*, **72,** 4071 (1950).
759. G. Errera and R. Maltese, *Gazz. Chim. Ital.*, **35,** 370 (1905); *Chem. Soc. Abstr.*, **90,** 84.
760. S. Cuttitta, *Gazz. Chem. Ital.*, **36,** 325 (1906); *Chem. Soc. Abstr.*, **90,** 697.
761. Starogardzkie Zaklady Farmaceutyczne "Polfa," French Patent 1,458,183; *Chem. Abstr.* **68,** 39493 (1968).
762. A. Ledochowski and B. Stefanska, *Rocz. Chem.*, **40,** 301 (1966); *Chem. Abstr.*, **65,** 2218 (1966).
763. M. Schöpff, *Chem. Ber.*, **26,** 1980 (1892); *Chem. Soc. Abstr.*, **62,** 1223.
764. B. Wysocka-Skrzela and Z. Ledochowski, *Rocz. Chem.*, **42,** 1755 (1968); *Chem. Abstr.*, **70,** 47270 (1969).
765. O. Fischer and K. Demeler, *Chem. Ber.*, **32,** 1307 (1899), *Chem. Soc. Abstr.*, **76,** 635.
766. K. Gleu and A. Schubert, *Chem. Ber.*, **73,** 757 (1940); *Chem. Abstr.*, **35,** 2519 (1941).
767. J. A. Lamberton and J. R. Price, *Aust. J. Chem.*, **6,** 66 (1953); *Chem. Abstr.*, **47,** 11210 (1953).
768. A. K. Ganguly, T. R. Govindachari, A. Manmade, and P. A. Mohamed, *Indian J. Chem.*, **4,** 334 (1966); *Chem. Abstr.*, **65,** 18990 (1966).
769. K. C. Das and P. K. Bose, *Trans. Bose Res. Inst.*, **26,** 129 (1963); *Chem. Abstr.* **64,** 9777 (1966).
770. R. J. Gell, G. K. Hughes, and E. Ritchie, *Aust. J. Chem.*, **8,** 114 (1955); *Chem. Abstr.*, **50,** 1050 (1956).
771. F. N. Lahey, M. McCamish, and T. McEwan, *Aust. J. Chem.*, **22,** 447 (1969); *Chem. Abstr.*, **70,** 97020 (1969).
772. F. N. Lahey and W. C. Thomas, *Aust. J. Sci. Res.*, **2A,** 423 (1949); *Chem. Abstr.*, **45,** 5696 (1951).
773. R. R. Paris and A. Stambouli, *Acad. Sci., Paris*, **247,** 2421 (1958); *Chem. Abstr.*, **53,** 13510 (1959).
774. C. L. Zirkle, U.S. Patent 3,449,334; *Chem. Abstr.*, **71,** 49800 (1969).
775. L. Fonzes and F. Winternitz, *C. R. Acad. Sci., Paris, Ser. C*, **266,** 930 (1968); *Chem. Abstr.*, **69,** 36303 (1968).
776. T. A. Avrorova, *Zavodskaya Laboratory*, **16,** 232 (1950); *Chem. Abstr.*, **44,** 7275 (1950).
777. A. Albert, *The Acridines*, 1st ed., Arnold, London, 1951, p. 55.

778. S. R. Patel and K. S. Nargund, *J. Indian* Chem. *Soc.*, **32**, 770 (1955); *Chem. Abstr.*, **50**, 11341 (1956).
779. Kalle A.-G., German Patent 1,114,704; *Chem. Abstr.*, **57**, 1792 (1962).
780. R. E. Strube, U.S. Patent 3,291,790; *Chem. Abstr.*, **66**, 55529 (1967).
781. A. M. Monro, R. M. Quinton, and T. I. Wrigley, *J. Med. Chem.*, **6**, 255 (1963).
782. Upjohn Co., Dutch Patent 297,030; *Chem. Abstr.*, **64**, 8218 (1966).
783. Parke, Davis, and Co., German Patent 1,185,622; *Chem. Abstr.*, **62**, 16138 (1965).
784. H. Goldstein and M. Piolino, *Helv. Chim. Acta*, 10, 334 (1927); *Chem. Abstr.*, **21**, 2259 (1927).
785. H. Goldstein and J. Vaymatchar, *Helv. Chim. Acta*, **11**, 239 (1928); *Chem. Abstr.*, **22**, 3129 (1928).
786. F. Ullmann, G. Engi, N. Wosnessensky, E. Kuhn, and E. Herre, *Justus Liebigs Ann. Chem.*, **366**, 79 (1909); *Chem. Abstr.* 3, 2298 (1909).
787. Ward, Blenkinsop, and Co., and A. A. Goldberg, British Patent 649,147; U.S. Patent 2,553,914; *Chem. Abstr.*, **46**, 3081 (1952).
788. H. L. Yale and F. A. Sowinski, U.S. Patent 2,931,810; *Chem. Abstr.*, **54**, 19724 (1960).
789. H. M. Blatter, H. Lukaszewski, and G. de Stevens, *J. Org. Chem.*, **30**, 1020 (1965).
790. J. W. Schulenburg and S. Archer, *J. Amer. Chem. Soc.*, **83**, 3091 (1961).
791. C. Buchanan and S. H. Graham, *J. Chem. Soc.*, 500 (1950).
792. J. W. Brooks, M. M. Harris, and K. E. Howlett, *J. Chem. Soc.*, 2380 (1957).
793. H. M. Blatter, R. W. J. Carney, and G. de Stevens, U.S. Patent 3,403,153; *Chem. Abstr.*, **70**, 4142 (1969).
794. K. Soda, K. Nishiide, T. Atoka, K. Noga, K. Sakai, and E. Fujihira, Japanese Patent 68 09,051; *Chem. Abstr.*, **70**, 28654 (1969).
795. J. Renz, J. P. Bourquin, G. Gamboni, and G. Schwarb, U.S. Patent 3,358,019; *Chem. Abstr.*, **69**, 10267 (1968).
796. J. P. Bourquin, et al., Helv. Chim. Acta, **41**, 1061 (1958); *Chem. Abstr.*, **52**, 18423 (1958).
797. Sandoz, Ltd., Swiss Patent 364,793; *Chem. Abstr.*, **59**, 13887 (1963).
798. I. Molnar and T. Wagner-Jauregg, *Helv. Chim. Acta*, **52**, 401 (1969); *Chem. Abstr.*, **70**, 87542 (1969).
799. S. Kurzepa and J. Cieslak, *Rocz. Chem.*, **34**, 111 (1960); *Chem. Abstr.*, **54**, 19689 (1960).
800. R. Gryglewski and T. A. Gryglewska, *Biochem. Pharmacol.*, **15**, 1171 (1966).
801. G. Picciola, R. Gaggi, and W. Caliari, *Farmaco, Ed. Sci.*, **23**, 502 (1968); *Chem. Abstr.*, **69**, 26946 (1968).
802. Sandoz, Ltd., Swiss Patent 355,145; *Chem. Abstr.*, **57**, 15077 (1962).
803. Westminster Bank, Ltd., British Patent 890,732; *Chem. Abstr.*, **57**, 16574 (1962).
804. Italfarmaco S.P.A., British Patent 1,158,954; *Chem. Abstr.*, **71**, 91107 (1969).
805. A. Young and T. R. Sweet, *J. Amer. Chem. Soc.*, **80**, 800 (1958).
806. S. Somasekhara, G. M. Shah, and S. L. Mukherjee, *Current Sci. India*, **33**, 521 (1964); *Chem. Abstr.*, **61**, 13307 (1964).
807. A. Purgotti and C. Lunini, *Gazz. Chim. Ital.*, **33**, 324 (1903); *Chem. Soc. Abstr.*, **86**, 315.
808. Rhone-Poulenc S. A., Dutch Patent 6,515,071; *Chem. Abstr.*, **66**, 2379 (1967).
809. F. Ullmann and C. Wagner, *Justus Liebigs Ann. Chem.*, **371**, 388 (1909); *Chem. Abstr.*, **4**, 2122 (1910).
810. Yoshitomi Drug Manufacturing Co., Japanese Patent 5570 (59); *Chem. Abstr.*, **54**, 14277 (1960).

811. N. N. Dykhanov and M. I. Siderova, *Med. Prom. SSSR*, **10**, 11 (1956); *Chem. Abstr.*, **52**, 7208 (1958).

812. S. P. Massie and P. K. Kadaba, *J. Org. Chem.*, **21**, 347 (1956).

813. G. Csepreghy, G. Gelegonya, and K. Horvath, Hungarian Patent, 147,993; *Chem. Abstr.*, **58**, 5577 (1963).

814. R. B. Moffett and B. D. Aspergren, *J. Amer. Chem. Soc.*, **82**, 1600 (1960).

815. CIBA, Ltd., Belgian Patent 629,369; *Chem. Abstr.*, **61**, 617 (1964).

816. L. A. Elson and C. S. Gibson, *J. Chem. Soc.*, 294 (1931).

817. M. H. Sherlock, N. Sperber, and D. Papa, U.S. Patent 2,889,328; *Chem. Abstr.*, **54**, 411 (1960).

818. Sandoz, Ltd., Dutch Patent 6,605,691; *Chem. Abstr.*, **67**, 90541 (1967).

819. H. Goldstein and E. Schaaf, *Helv. Chim. Acta*, **40**, 369 (1957); *Chem. Abstr.*, **51**, 11289 (1957).

820. Farbwerke Hoechst A.-G., German Patent 1,122,541; *Chem. Abstr.*, **56**, 14032 (1962).

821. K. Sturm, W. Siedel, R. Weyer, and H. Ruschig, *Chem. Ber.*, **99**, 328 (1966); *Chem. Abstr.*, **64**, 8112 (1961).

822. G. B. Bachman and G. M. Picha, *J. Amer. Chem. Soc.*, **68**, 2112 (1946).

823. J. Riviere, French Patent M4333; *Chem. Abstr.*, **68**, 68754 (1968).

824. I. Ya. Postovskii, R. O. Matevosyan, and A. K. Chirkov, *Zh. Obshch. Khim.*, **29**, 3106 (1959); *Chem. Abstr.*, **54**, 13058 (1960).

825. J. Hebky, O. Radek, and J. Kejha, *Coll. Czech. Chem. Commun.*, **24**, 3988 (1959); *Chem. Abstr.*, **54**, 6730 (1960).

826. S. Iseda, *Bull. Chem. Soc. Jap.*, **30**, 694 (1957); *Chem. Abstr.*, **52**, 9058 (1958).

827. P. Axerio, *Farm. Nueva*, **30**, 1 (1965); *Chem. Abstr.*, **62**, 14539 (1965).

828. Bristol-Meyers Co., Dutch Patent 6,604,860; *Chem. Abstr.*, **66**, 55513 (1967).

829. Parke, Davis, and Co., French Patent 1,315,030; *Chem. Abstr.*, **59**, 1538 (1963).

830. Parke, Davis, and Co., Dutch Patents 292,084; 292,122; 292,123; *Chem. Abstr.*, **65**, 657 (1966); 292,081; *Chem. Abstr.*, **65**, 658 (1966); 292,078; *Chem. Abstr.*, **65**, 661 (1966).

831. Parke, Davis, and Co., German Patent 1,190,952; *Chem. Abstr.*, **63**, 6921 (1965).

832. Parke, Davis, and Co., French Patent 1,374,577; *Chem. Abstr.*, **62**, 9070 (1965).

833. C. P. Krimmel, U.S. Patent 2,596,156; *Chem. Abstr.*, **47**, 2210 (1953).

834. Parke, Davis, and Co., British Patent 1,092,921; *Chem. Abstr.*, **69**, 18818 (1968).

835. CIBA, Ltd., French Patent M2599; *Chem. Abstr.*, **62**, 1602 (1965).

836. Parke, Davis, and Co., British Patent 1,027,060; *Chem. Abstr.*, **65**, 16904 (1966).

837. A. Purgotti and A. Contardi, *Gazz. Chim. Ital.*, **32**, 573 (1902); *Chem. Soc. Abstr.*, **82**, 778.

838. M. W. Whitehouse, *Biochem. Pharmacol.*, **16**, 753 (1967).

839. A. Sallmann and R. Pfister, German Patent 1,815,804; *Chem. Abstr.*, **71**, 112637 (1969).

840. Sandoz, Ltd., Dutch Patent 6.505,542; *Chem. Abstr.*, **64**, 12606 (1966).

841. Sandoz, Ltd., British Patent 808,112; *Chem. Abstr.*, **53**, 16164 (1959).

842. Smith, Kline, and French Laboratories, British Patent 829,246; *Chem. Abstr.*, **54**, 17428 (1960).

843. Sandoz, Ltd., Swiss Patent 358,080; *Chem. Abstr.*, **59**, 8653 (1963).

844. M. L. Dhar, K. S. Narang, and J. N. Ray, *J. Chem. Soc.*, 304 (1938).

845. H. Hirano and K. Masuda, Japanese Patent 67 19,583; *Chem. Abstr.*, **69**, 18858 (1968).

846. Parke, Davis, and Co., Belgian Patent 612,423; *Chem. Abstr.*, **57**, 13690 (1962).

847. A. Stoll, J. P. Bourquin, and J. Renz, U.S. Patent 2,902,491; *Chem. Abstr.*, **54**, 2368 (1960).

848. M. Schöpff, *Chem. Ber.*, **23**, 3435 (1890); *Chem. Soc. Abstr.*, **60**, 295.

849. P. Fischer, *Chem. Ber.*, **24,** 3785 (1891); *Chem. Soc. Abstr.*, **62,** 331.

850. V. M. Cherkasov, *Zh. Obshch. Khim.*, **23,** 197 (1953); *English Transl.*, p. 201; *Chem. Abstr.*, **48,** 2655 (1954); **49,** 4558 (1955).

851. Parke, Davis, and Co., French Patent M1341; *Chem. Abstr.*, **58,** 10130 (1963).

852. Parke, Davis, and Co., French Patent M2948; *Chem. Abstr.*, **63,** 8269 (1965).

853. W. Graf, E. Girod, E. Schmid, and W. G. Stoll, *Helv. Chim. Acta*, **42,** 1085 (1959); *Chem. Abstr.*, **54,** 402 (1960).

854. F. Hoffmann-La Roche and Co., Dutch Patent 6,407,011; *Chem. Abstr.*, **63,** 583 (1965).

855. J. C. E. Simpson and O. Stephenson, *J. Chem. Soc.*, 353 (1942).

856. J. Iwao, Japanese Patent 69 26,663; *Chem. Abstr.*, **72,** 12708 (1970).

857. F. J. McEvoy, E. N. Greenblatt, A. C. Osterberg, and G. R. Allen, *J. Med. Chem.*, **11,** 1248 (1968).

858. E. Reeder and L. H. Sternbach, U.S. Patent 3,243,427; 3,239 564; *Chem. Abstr.* **64,** 19647, 19498 (1966).

859. O. Keller N. Steiger and L. H. Sternbach U.S. Patent 3,203,990; *Chem. Abstr.*, **64,** 3576 (1966).

860. J. Kariss and H. L. Newmark, U.S. Patent 3,123,529; *Chem. Abstr.*, **60,** 12035 (1964)

861. J. Kariss and H. L. Newmark, U.S. Patent 3,116,203; *Chem. Abstr.*, **60,** 5529 (1964).

862. E. Reeder and L. H. Sternbach, U.S. Patent 3,402,171; *Chem. Abstr.*, **70,** 37805 (1969).

863. E. Reeder and L. H. Sternbach, U.S. Patent 3,136,815; *Chem. Abstr.*, **61,** 9515 (1964).

864. L. H. Sternbach et al., *J. Org. Chem.*, **27,** 3788 (1962).

865. E. Reeder, L. H. Sternbach, O. Keller, N. Steiger, and A. Stempel, Swiss Patent 414,652; *Chem. Abstr.*, **68,** 69054 (1968).

866. F. Hoffmann-La Roche and Co., Belgian Patent 615,194; *Chem. Abstr.*, **59,** 12827 (1963).

867. (a) F. Hoffmann-La Roche and Co., German Patent 1,145,626; Chem. Abstr., **60,** 12033 (1964). (b) O. Keller, N. Steiger, and L. H. Sternbach, U.S. Patent 3,121,075; *Chem. Abstr.*, **61,** 5672 (1964). (c) E. Reeder and L. H. Sternbach, U.S. Patent 3,109,843; 3,141,890; *Chem. Abstr.*, **60,** 2994 (1964); **61,** 9514 (1964).

868. L. H. Sternbach, R. I. Fryer, O. Keller, W. Metlesics, G. Sach, and N. Steiger, *J. Med. Chem.*, **6,** 261 (1963).

869. S. K. Banerjee and R. N. Chakravarti, *Bull. Calcutta School Trop. Med.*, **13,** 60 (1965); *Chem. Abstr.*, **64,** 3475 (1966).

870. H. Brockmann and N. Grubhofer, *Chem. Ber.*, **86,** 1407 (1953); *Chem. Abstr.*, **49,** 1769 (1955).

871. H. Goldstein and W. Kopp, *Helv. Chim. Acta*, **11,** 486 (1928); *Chem. Abstr.*, **22,** 2944 (1928).

872. A. D. Samarin and I. G. Shchurova, *Nauch. Tr. Perm. Politekh. Inst.*, **44,** 74 (1968); *Ref. Zh. Khim.*, *Abstr. No.* 15Zh334 (1969); *Chem. Abstr.*, **73,** 120475 (1970).

873. D. Mayer, G. Hermann, and K. Sasse, German Patent 1,811,409; *Chem. Abstr.*, **73,** 55983 (1970).

874. A. Ledochowski, M. Bogucka, B. Stefanska, B. Horowska, J. Zielinski, and C. Radzilowski, Polish Patent 60,640; *Chem. Abstr.*, **74,** 125479 (1971).

875. A. S. Samarin and A. F. Vyatkin, *Nauch. Tr. Perm. Politekh. Inst.*, **52,** 53 (1969); *Ref. Zh. Khim.*, *Abstr. No.* 24Zh337 (1969); *Chem. Abstr.*, **73,** 35201 (1971).

876. A. Ledochowski and S. Wojenski, *Rocz. Chem.*, **44,** 43 (1970); *Chem. Abstr.*, **73,** 24571 (1970).

877. H-D. Dell and R. Kamp, *Arch. Pharm.* (*Weinheim*), **303**, 785 (1970); *Chem. Abstr.*, **74**, 34662 (1971).
878. G. Devaux, P. Mesnard, and A. M. Brisson, *Ann. Pharm. Fr.*, **27**, 239 (1969); *Chem. Abstr.*, **71**, 94809 (1969).
879. A. Ledochowski, W. Gruszecki, B. Stefanska, and B. Horowska, Polish Patent 60,794; *Chem. Abstr.*, **74**, 111930 (1971).
880. Instituto Luso Farmaco d'Italia S.R.L., British Patent 1,199,386; *Chem. Abstr.*, **73**, 87653 (1970).
881. K. H. Bolze, O. Brendler, and D. Lorenz, German Patent 1,939,111-2; *Chem. Abstr.*, **74**, 76185-6 (1971).
882. E. Fujihira, S. Otomo, K. Sota, and M. Nakazawa, *Yakugaku Zasshi*, **91**, 143 (1971); *Chem. Abstr.*, **74**, 139208 (1971).
883. F. R. Bradbury and W. H. Linnell, *Quart. J. Pharm. Pharmacol.*, **11**, 240 (1938).
884. B. Linke, *J. Prakt. Chem.*, **101**, 265 (1920); *Chem. Abstr.*, **15**, 1293 (1921).
885. E. Wolthuis, B. Bouma, J. Modderman, and L. Sytsma, *Tetrahedron Lett.*, 407 (1970).
886. L. Legrand and N. Lozac'h, *Bull. Soc. Chim. Fr.*, 1173 (1969); *Chem. Abstr.*, **71**, 70573 (1969).
887. M. S. Rovinskii and E. Z. Sterina, *Khim. Tekhnol.*, **13**, 63 (1968); *Chem. Abstr.*, **71**, 80873 (1969).
888. J. I. G. Cadogan, *Synthesis*, **1**, 11 (1969).
889. J. I. G. Cadogan, R. K. Mackie, and M. J. Todd, *Chem. Commun.*, 491 (1966).
890. J. I. G. Cadogan, R. Marshall, D. M. Smith, and M. J. Todd, *J. Chem. Soc. C*, 2441 (1970).
891. H. Führer, *Arch. Exptl. Path. Pharmacol.*, **51**, 391 (1904); *Zentr. 1904 II*, 720.
892. W. H. Perkin and W. G. Sedgwick, *J. Chem. Soc.*, **125**, 2437 (1924).
893. I. H. Bowen, P. Gupta, and J. R. Lewis, *J. Chem. Soc., D* (*Chem. Commun.*), 1625 (1970).
894. K. Isagawa, T. Ishiwaka, M. Kawai, and Y. Fushizaki, *Bull. Chem. Soc. Jap.*, **42**, 2066 (1969); *Chem. Abstr.*, **71**, 70578 (1969).
895. R. Robinson, *The Structural Relations of Natural Products*, Oxford University Press, London, 1955, p. 94.
896. S. Johne, H. Bernasch, and D. Groeger, *Pharmazie*, **25**, 777 (1970); *Chem. Abstr.*, **74**, 72796 (1971).
897. R. H. Prager and H. M. Thredgold, *Aust. J. Chem.*, **22**, 2627 (1969); *Chem. Abstr.*, **72**, 43936 (1970).
898. J. Eliasberg and P. Friedländer, *Chem. Ber.*, **25**, 1752 (1892).
899. J. R. Hlubucek, E. Ritchie, and W. C. Taylor, *Aust. J. Chem.*, **23**, 1881 (1970); *Chem. Abstr.*, **73**, 99073 (1970).
900. D. L. Dreyer, *Tetrahedron*, **22**, 2923 (1966).
901. E. Wenkert, *Experientia*, **15**, 165 (1959).
902. G. E. Ivanov and V. A. Izmail'skii, *Khim. Geterotsikl. Soedin.*, 1119 (1970); *Chem. Abstr.*, **74**, 76294 (1971).
903. J. W. Happ, E. G. Janzen, and B. C. Rudy, *J. Org. Chem.*, **35**, 3382 (1970).
904. N. S. Drozdov, *J. Gen. Chem. USSR*, **9**, 1456 (1939); *Chem. Abstr.*, **34**, 2847 (1940).
905. S. Steenken, *Tetrahedron Lett.*, 4791 (1970).
906. Smith, Kline, and French Laboratories, French Patent 1,530,413; *Chem. Abstr.*, **72**, 3397 (1970).
907. M. Ionescu, I. Goia, and M. Vlassa, *Rev. Roum. Chim.*, **15**, 1785 (1970); *Chem. Abstr.*, **74**, 87793 (1971).

908. I. Goia and M. Ionescu, *Rev. Roum Chim.*, **15**, 1233 (1970); *Chem. Abstr.*, **74**, 64192 (1971).

909. V. A. Trofimov, O. N. Chupakhin, and Z. V. Pushkareva, USSR Patent 271, 697; *Chem. Abstr.*, **74**, 4699 (1971).

910. O. N. Chupakhin, V. A. Trofimov, and Z. V. Pushkareva, *Khim. Geterotsikl. Soedin.*, 1674 (1970); *Chem. Abstr.*, **74**, 87189 (1971).

911. K. Lehmstedt, *Chem. Ber.*, **70B**, 172 (1937); *Chem. Abstr.*, **31**, 8532 (1937).

912. S. Tada, Japanese Patent 70 20,182; *Chem. Abstr.*, **73**, 121549 (1970).

913. L. Kalb, *Chem. Ber.*, **43**, 2209 (1910); *Chem. Abstr.*, **4**, 2929 (1910).

914. J. Hlubucek, E. Ritchie, and W. C. Taylor, *Chem. Ind.*, (London) 1809 (1969).

915. W. M. Bandaranayake, L. Crombie, and D. A. Whiting, *Chem. Commun.*, 970 (1969).

916. W. Baczynski and S. Niementowski, *Chem. Ber.*, **38**, 3009 (1905); *Chem. Soc. Abstr.*, **88**, 927.

917. G. R. Bedford, R. Hill, and B. R. Webster, *J. Chem. Soc.*, C, 2462 (1970).

918. C. Liebermann, *Chem. Ber.*, **7**, 1098 (1874); *Chem. Soc. Abstr.*, **28**, 167.

919. C. Kraemer, *Chem. Ber.*, **17**, 1875 (1884); *Chem. Soc. Abstr.*, **46**, 1340.

920. A. Baeyer and H. Caro, *Chem. Ber.*, **7**, 963 (1874); *Chem. Soc. Abstr.*, **28**, 83.

921. J. H. Markgraf, M-K. Ahn, C. G. Carson, and G. A. Lee, *J. Org. Chem.*, **35**, 3983 (1970).

922. V. G. Pesin and I. G. Vitenberg, *Zh. Obshchei Khim.*, **35**, 930 (1965); *Chem. Abstr.*, **63**, 6897 (1965).

923. V. P. Maksimets and O. N. Popilin, *Khim. Geterotsikl. Soedin*, 191 (1970); *Chem. Abstr.*, **73**, 13681 (1970).

924. D. N. Kravtsov, E. M. Rokhlina, and A. N. Nesmeyanov, *Izv. Akad. Nauk SSSR, Ser. Khim.*, 1035 (1968); *Chem. Abstr.*, **70**, 4254 (1969).

925. K. Lehmstedt, German Patent 581,328; *Chem. Abstr.*, **27**, 5083 (1933).

926. V. I. Bliznyukov and A. A. Martinovskii, *Farm. Zh. (Kiev)*, **24**, 30 (1969); *Chem. Abstr.*, **72**, 31577 (1970).

927. M. Ionescu and I. Hopartean, *Studia Univ. Babes-Bolyai, Ser. Chem.*, **15**, 77 (1970); *Chem. Abstr.*, **73**, 130863 (1970).

928. L. Erdey, E. Banyai, E. Zalay, and M. Tesy, *Acta Chim. Acad. Sci. Hung.*, **15**, 65 (1958) *Chem. Abstr.*, **53**, 6119 (1959).

929. N. H. Berner, R. S. Varma, and D. W. Boykin, *J. Med. Chem.*, **13**, 552 (1970).

930. S. G. P. Plant and C. R. Worthing, *J. Chem. Soc.*, 1278 (1955).

931. J. Itier and A. Casadevall, *Bull. Soc. Chim. Fr*, 2342 (1969); *Chem. Abstr.*, **71**, 80862 (1969).

932. M. Takeda, M. Matsubara, and H. Kugita, *Yakugaku Zasshi*, **89**, 158 (1969); *Chem. Abstr.*, **71**, 3368 (1969).

933. G. Zoni, M. L. Molinari, and S. Banfi, *Farmaco, Ed. Sci.*, **26**, 191 (1971); *Chem. Abstr.*, **74**, 110121 (1971).`

934. E. D. Mech, *Gigiena Truda Professional. Zabolevaniya*, **1**, 49 (1957); *Chem. Abstr.*, **52**, 2020 (1958).

935. F. L. Bach, J. C. Barclay, and E. Cohen, *J. Med. Chem.*, **10**, 802 (1967).

936. B. Unterhalt, *Arch. Pharm. (Weinheim)*, **303**, 445 (1970); *Chem. Abstr.*, **73**, 35267 (1971).

937. I. G. Farbenindustrie, Swiss Patents 137,134-137,141; *Zentr. 1930 II*, 625.

938. Meister, Lucius and Bruning, German Patents 145,189; 146,102; *Zentr. 1903 II*, 1097; 1152.

CHAPTER IV

The Acridine Alkaloids

J. E. SAXTON

*The University of Leeds,
Leeds, England*

1. Introduction

The first naturally occurring derivatives of acridine to be discovered were the principal alkaloids of *Melicope fareana* F. Muell., *Evodia xanthoxyloides* F. Muell., and *Acronychia baueri* Schott, which were isolated and character-ized by Hughes and his collaborators[1] in 1948 during a survey of the alkaloidal constituents of certain Australian flora. To date, 24 acridine alkaloids of established structure have been isolated, and one other of undetermined structure has been reported; all of them occur in various members of the family Rutaceae. With the exception of 10-methyl-9-acridanone, all the alka-loids are highly crystalline, yellow derivatives of the mono- or poly-hydroxy-9-acridanones. Most of the major alkaloids have now been synthesized, including acronycine, which exhibits a broad antitumor spectrum and has been shown to have significant activity against experimental neoplasms in laboratory animals.[2a,b] The accompanying table lists all the acridine alkaloids known up to the end of 1970.

The acridine alkaloids commonly occur in association with furanoquino-line alkaloids, e.g., with skimmianine (**1**) and acronycidine (**2**) in *M. fareana*,[3] with kokusaginine (**3**), evodine (**4**), evoxine (**5**), and evoxoidine (**6**) in *E. xanthoxyloides*,[6, 14] and with acronidine (**7**) in *A. baueri*.[4] The two groups of alkaloids would thus appear to have a common biosynthetic origin and follow

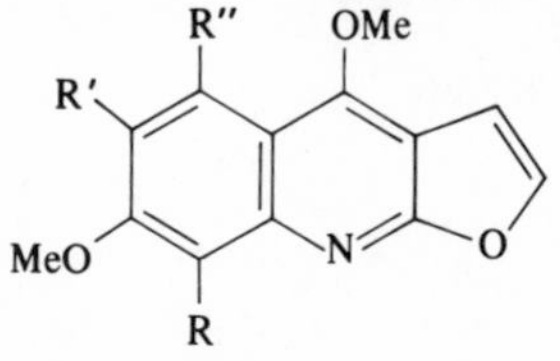

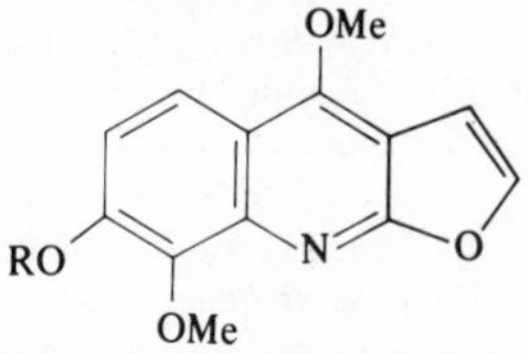

1; R = OMe, R' = R'' = H

2; R = R'' = OMe, R' = H

3; R' = OMe, R = R'' = H

4; R = Me C CH CH₂– (with CH₂ double bond and OH)

5; R = HO C Me₂ CH CH₂– (with OH)

6; R = Me₂CH CO CH₂–

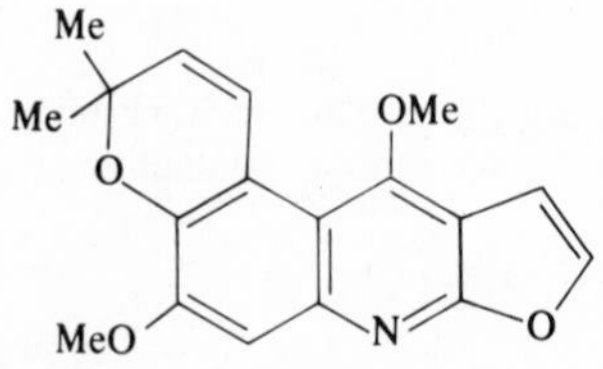

7

TABLE I. The Acridine Alkaloids

Alkaloid	Molecular formula	mp (°C)	Structure	Occurrence	Ref.
Melicopicine	$C_{18}H_{19}NO_5$	133-134	9	a, c	1, 3, 4, 5
Melicopidine	$C_{17}H_{15}NO_5$	121-122	22	a, b, c, e	1, 3, 4, 5, 6, 7, 8
Melicopine	$C_{17}H_{15}NO_5$	178.5-9.5	23	a, c, j	1, 3, 4, 5, 9
Normelicopicine	$C_{17}H_{17}NO_5$	129-129.5	13	c	10, 11
Normelicopidine	$C_{16}H_{13}NO_5$	211-212		c, e	10, 11, 8
Normelicopine	$C_{16}H_{13}NO_5$	235.5-6.5		c	10, 11
Evoxanthine	$C_{16}H_{13}NO_4$	217-218	83	b, e, f, g, l, q, s	1, 6, 7 8, 12, 13 14, 15, 16, 17, 18
Evoxanthidine	$C_{15}H_{11}NO_4$	312-313	84	b	14
Norevoxanthine	$C_{15}H_{11}NO_4$	274-275	85	f	19
Xanthevodine	$C_{16}H_{13}NO_5$	213-214	82	b, c	6, 20
Xanthoxoline	$C_{15}H_{13}NO_4$	256-257	103	b	14
1-Hydroxy-2,3-dimethoxy-10-methyl-9-acridanone (Arborinine)	$C_{16}H_{15}NO_4$	176-177	104	b, e, h, k, l, m, n, p, q, r	7, 8, 14, 17, 18, 21 22-28
1,2,3-Trimethoxy-10-methyl-9-acridanone	$C_{17}H_{17}NO_4$	169-170	105	e, l	7, 8, 17
1,3-Dimethoxy-10-methyl-9-acridanone	$C_{16}H_{15}NO_3$	161-163	115	c, l, s	4, 13, 17
Evoprenine	$C_{20}H_{21}NO_4$	143	109	e	8
Acronycine	$C_{20}H_{19}NO_3$	175-176	116	c	1, 5

(Table Continued)

381

Table I. (Continued)

Alkaloid	Molecular formula	mp (°C)	Structure	Occurrence	Ref.
Noracronycine	$C_{19}H_{17}NO_3$	198-200	**117**	h	29
Des-10-methyl-acronycine	$C_{19}H_{17}NO_3$	268-270		h	29
Des-10-methyl-noracronycine	$C_{18}H_{15}NO_3$	245-246	**161b**	h	29
1-Hydroxy-10-methyl-9-acridanone	$C_{14}H_{11}NO_2$	192-194		m	30
Rutacridone	$C_{19}H_{17}NO_3$	161-162	**166**	m	31
10-Methyl-9-acridanone	$C_{14}H_{11}NO$	199-201		o	32
Tecleanthine	$C_{17}H_{15}NO_5$	158	**167**	q	18
Atalaphylline	$C_{23}H_{25}NO_4$	246	**171**	t	33
N-Methylatalaphylline	$C_{24}H_{27}NO_4$	192-193	**176**	t	33

Key: *a, Melicope fareana* F. Muell.; *b, Evodia xanthoxyloides* F. Muell.; *Acronychia baueri* Schott.; *d, Haplophyllum dubium (Aplophyllum dubium* Korovin); *e, Evodia alata* F. Muell.; *f, Teclea grandifolia* Engl.; *g, Balfourodendron riedelianum* Engl.; *h, Glycosmis pentaphylla* (Retz.) Correa; *j, Acronychia acidula* F. Muell.; *k, Fagara leprieurii* Engl.; *l, Vepris bilocularis* Engl.; *m, Ruta graveolens* L.; *n, Acronychia haplophylla* (F. Muell.) Engl.; *o, Thamnosma montana* Torr. et Frem.; *p, Glycosmis arborea* Correa; *q, Teclea natalensis* (Sond.) Engl.; *r, Ravenia spectabilis* Engl.; *s, Vepris ampody* H. Perr.; *t, Atalantia monophylla* Correa

382

divergent pathways only in the later stages of the biosynthesis. A point of interest and possible biosynthetic significance is that the furanoquinoline alkaloids in this group are all 4-methoxyquinoline derivatives; in contrast, all the acridine alkaloids are 9-acridanone derivatives, and none of them possesses a methoxyl group at the 9-position. An exception to this generalization is provided by *Balfourodendron riedelianum*, which contains evoxanthine as the sole representative of the acridine alkaloids, together with two furanoquinoline alkaloids (maculosidine and flindersiamine) and three other alkaloids that are all derivatives of *N*-methyl-4-quinolone.[16]

Evolidine, the remaining "alkaloid" of *E. xanthoxyloides*,[14] is now known to be a cyclic heptapeptide with the following structure,[34, 35] and it has been synthesized.[36]

cyclo[-L-Ser-L-Phe--L-Leu-L-Pro-L-Val-L-Asp(β-NH$_2$)-L-Leu-]

2. Melicopicine, Melicopidine, and Melicopine

A. Isolation and General Properties

The bark and leaves of *M. fareana*, on percolation or Soxhlet extraction with methanol, were found by Price to yield five alkaloids, of which melicopicine, melicopidine and melicopine are 9-acridanone derivatives (the others are acronycidine and skimmianine).[3] The separation of the alkaloids depends on the insolubility of melicopine in ether, and on the different basicities of the other alkaloids. Dilution of a cold solution of melicopicine and melicopidine in 10% aqueous hydrochloric acid, until the acid concentration has fallen to 5%, results in the precipitation of melicopicine; further dilution to 2% of acid then causes the precipitation of melicopidine hydrochloride. Acronycidine and skimmianine can be isolated from the mother liquors. Since their original isolation, melicopine, melicopicine, and melicopidine have also been shown to occur in other plants, notably in *A. baueri*, which is also reported to contain the *nor* alkaloids (see Table I).

Melicopine and melicopidine each contain two methoxyl groups and a methylenedioxy group, whereas melicopicine contains four methoxyl groups but no methylenedioxy group; all three alkaloids contain one methylimino group.[3] A subsequent examination by Crow and Price showed that all three alkaloids can be recovered unchanged after attempted acetylation, and they do not react with potassium permanganate in acetone.[37] They do not react with 2,4-dinitrophenylhydrazine, nitrous acid, or alkyl halides and are insoluble in and stable to hot aqueous alkali. The color of the alcoholic solutions of the alkaloids is not altered by alkali. Hence the methylimino group and the oxygen atom not involved in the methoxy and methylenedioxy groups must be in unreactive positions in each compound.

B. Structure and Synthesis of Melicopicine

Price[38] found that the vigorous oxidation of melicopine, melicopidine, and melicopicine gave the same acid, $C_{11}H_9NO_3$, which was identified as 1-methyl-4-quinolone-3-carboxylic acid (8) by comparison with a synthetic specimen.[38, 39] In the case of melicopicine, this result coupled with the analytical data leads to a unique structure (9), which was later confirmed by synthesis. Melicopidine and melicopine must necessarily have similar structures in which two adjacent methoxyl groups in (9) are replaced by a methylenedioxy group.[40]

The synthesis of melicopicine was accomplished from 2',3',4',5'-tetra-methoxydiphenylamine-2-carboxylic acid (10), prepared from anthranilic acid and the appropriate iodobenzene (o-chlorobenzoic acid and the required aniline gave only intractable tars).[41] This acid (10), when heated under reflux

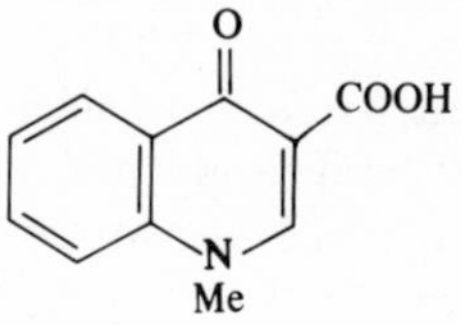

8

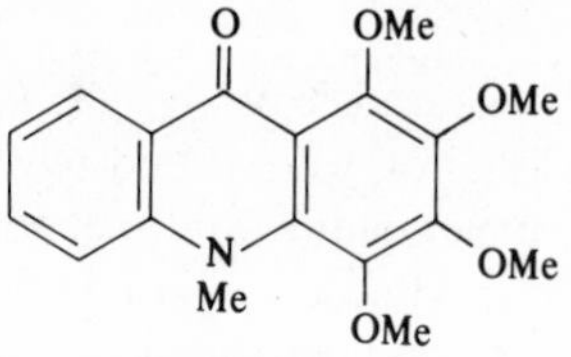

9, Melicopicine

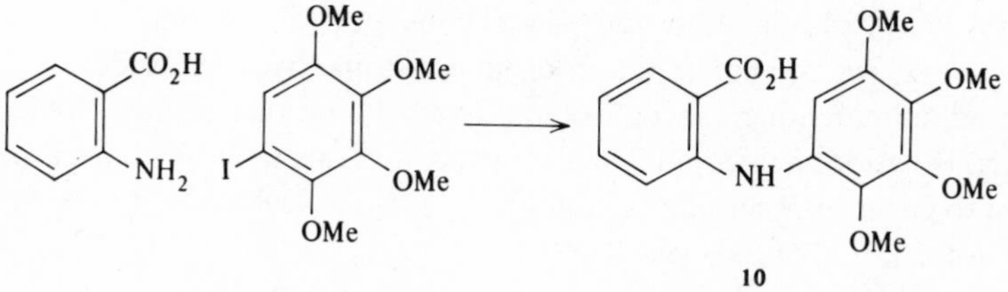

10

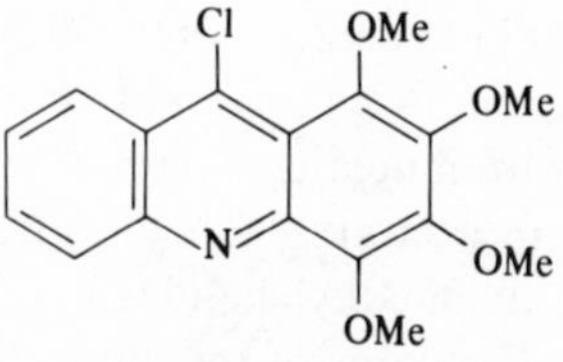

11

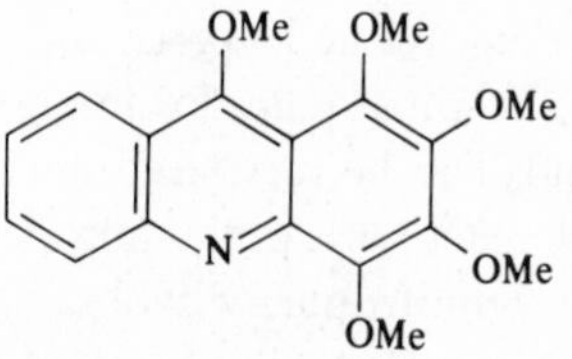

12

with phosphorus oxychloride, cyclized to the 9-chloroacridine derivative (**11**), which with sodium methoxide gave the 9-methoxyacridine (**12**) as a dark oil that could not be easily purified. However, when heated with methyl iodide, it gave melicopicine (**9**), identical with the naturally occurring compound, in 15% overall yield.

The 9-acridanone structure of the alkaloids was confirmed by their uv absorption spectra,[42] and by their reduction[40] to $\Delta^{9,9'}$-biacridans with zinc and hydrochloric acid in acetic acid solution. However, only the product from melicopine was isolated.

Melicopidine is a much stronger base than melicopine, which is appreciably more basic than melicopicine. In view of the very close structural similarity of the alkaloids, it is difficult to account for this; however, it should be noted that pK_a values have not been determined. Melicopidine hydrochloride, perhaps surprisingly, is precipitated when a solution of the alkaloid in hydrochloric acid is diluted. This has been tentatively ascribed to oxonium salt formation.[40]

C. The *nor* Alkaloids

During the isolation of the acridine alkaloids, Price noticed that all were slowly attacked by dilute mineral acid.[3] In solution the alkaloid hydrochlorides slowly decomposed to products more deeply colored than the original materials. Crow and Price found that refluxing the alkaloids for 1 hr with ethanolic hydrogen chloride caused almost quantitative decomposition into these highly colored decomposition products,[37] the *nor* alkaloids, which analyzed in each case for the original alkaloids less CH_2. When dry melicopidine hydrochloride was heated, decomposition also occurred.

The sensitivity of melicopicine, melicopine, and melicopidine to mineral acid raises the question of the status of the *nor* compounds as true alkaloids, since in most cases hydrochloric acid has been used at one stage or another in the isolation of the alkaloids from the plant materials; thus the *nor* compounds may well be artifacts. Both Svoboda[10] and Diment, Ritchie, and Taylor[8] regard them as artifacts, since the *nor* compounds have never been isolated following extraction procedures that avoid the use of acidic reagents. However, it is relevant to note that evoprenine and 1-hydroxy-2,3-dimethoxy-10-methyl-9-acridanone were isolated,[8, 22] even when the use of acids was avoided, so that the plants do contain some cryptophenolic constituents.

The *nor* compounds from melicopicine, melicopine, and melicopidine resemble the parent alkaloids in solubility but are more deeply colored (orange → red), very weakly basic, and insoluble in alkali. The colors of their ethanolic solutions are enhanced by the addition of sodium hydroxide. However, all the *nor* compounds give yellow acetyl derivatives with acetic anhydride and sodium acetate, and methylation with dimethyl sulfate and potassium carbonate in acetone regenerates the original alkaloid in each case. This suggests that the *nor* compounds are unreactive phenols but leaves the position of demethylation uncertain. The oxidation of normelicopicine (13) with nitric or nitrous acid to a mixture of dimethoxyquinones (14 and 15) shows that demethylation must have occurred at either position 1 or 4; the fact that quinones are produced by the loss of only one methoxyl group shows that the unreactive OH in normelicopicine must be phenolic and convertible into a CO group. It was not possible at that time to allocate the position of demethylation by analogy, as similar demethylations of appropriate model compounds had not been carried out, but later work showed that of the monomethoxy-10-methyl-9-acridanones, the 1-methoxy compound was by far the most easily demethylated.

The fact that the *nor* alkaloids are unreactive suggests that the phenolic group is involved in hydrogen bonding. This is only possible if the hydroxyl group is in position 1, where it can form a hydrogen bond with the peri-situated 9-acridanone oxygen (13a). Other evidence given below also supports the conclusion that the hydroxyl group is in this position.

13a   13b

In contrast to normelicopicine, the isomeric 2-, 3-, and 4-hydroxytri-methoxy-10-methyl-9-acridanones methylate easily and have normal phenolic properties; they are also less highly colored, more soluble in polar solvents and less soluble in nonpolar solvents than normelicopicine. Hydrogen bonding in normelicopicine is also responsible for the small depression of melting point observed with wet normelicopicine (7–10°), compared with the much larger depression (45°) observed with 2-hydroxy-1,3,4-trimethoxy-10-methyl-9-acridanone, where intramolecular hydrogen bonding cannot take place. The ir absorption spectra of the hydroxytrimethoxy-10-methyl-9-acrida-nones, examined in the solid state, also lead to the same conclusion.[40] The 2-, 3-, and 4-hydroxy isomers show bands at 3257, 3278, and 3278 cm^{-1}, respectively, which are attributable to "free" OH groups. In contrast, this band is not present in the spectrum of normelicopicine; neither is it present in the spectra of 1-hydroxyanthraquinone and its derivatives in which the hydroxyl group is also situated *peri* to the carbonyl group. The stability of such a hydrogen bond will be greater the larger the contribution of ionic structures such as (13b); the latter should also result in a lower basicity of the *nor* alkaloids compared with the parent alkaloids, in agreement with the facts. The ease of demethylation is presumably also a result of the high degree of stability of the products owing to hydrogen bonding.

Both normelicopicine and melicopicine on nitric acid oxidation[43] give the same mixture of two isomeric quinones of molecular formula $C_{16}H_{13}NO_5$. One of these compounds (15) is bright red, has mp 200–200.5°, and is insoluble in saturated aqueous sodium bisulfite, while the other (14) is dark red, has mp 233–235°, and is soluble in this reagent. Both quinones contain two methoxyl groups and give 2,4-dinitrophenylhydrazones; phenylhydrazine itself merely causes reduction to the corresponding quinols. Sulfur dioxide also reduces both quinones to the orange quinols (16 and 17) which with acetic anhydride give mono- and diacetates, respectively, in the presence of pyridine and sodium acetate. The monoacetates can also be prepared by the reductive acetylation of the quinones. Nitric acid oxidizes the quinols back to the quinones. The bright red quinone (15) does not react with *o*-phenylene-diamine at room temperature, but its dark red isomer gives a yellow phenazine and is therefore the orthoquinone (14).[43]

Demethylation of both quinones with hot aqueous sodium carbonate (but not with sodium hydroxide) gives the same red compound, 2-hydroxy-3-methoxy-10-methyl-9-acridanone-1,4-quinone (18). On reductive acetylation this gives a diacetyl derivative, which can also be prepared by reducing the quinone to the quinol (19) followed by acetylation. In alkaline solution this quinol is oxidized back to the quinone (18) by air. This quinone [as also does 2,3-dimethoxy-10-methyl-9-acridanone-1,4-quinone (15) on prolonged reaction] forms a phenazine with *o*-phenylenediamine, which gives a mono-acetyl derivative with acetic anhydride and pyridine, suggesting that the acylable center is not that present in normelicopicine and is therefore in position 4. Boiling 46% hydrobromic acid demethylates the monomethoxy-quinone (18) to 2,3-dihydroxy-10-methyl-9-acridanone-1,4-quinone (20), a red microcrystalline compound which with sulfur dioxide gives the quinol (21).

Reductive methylation of 14, 15, or 20, and methylation of 16, 17, 19, or 21 in one or two stages gives back melicopicine, showing that no fundamental change in the molecule takes place during the degradations. Oxidation of the dihydroxyquinone (20) with hydrogen peroxide in 5% aqueous sodium hydroxide gives 1-methyl-4-quinolone-3-carboxylic acid (8), the product of vigorous nitric acid oxidation of the alkaloids.

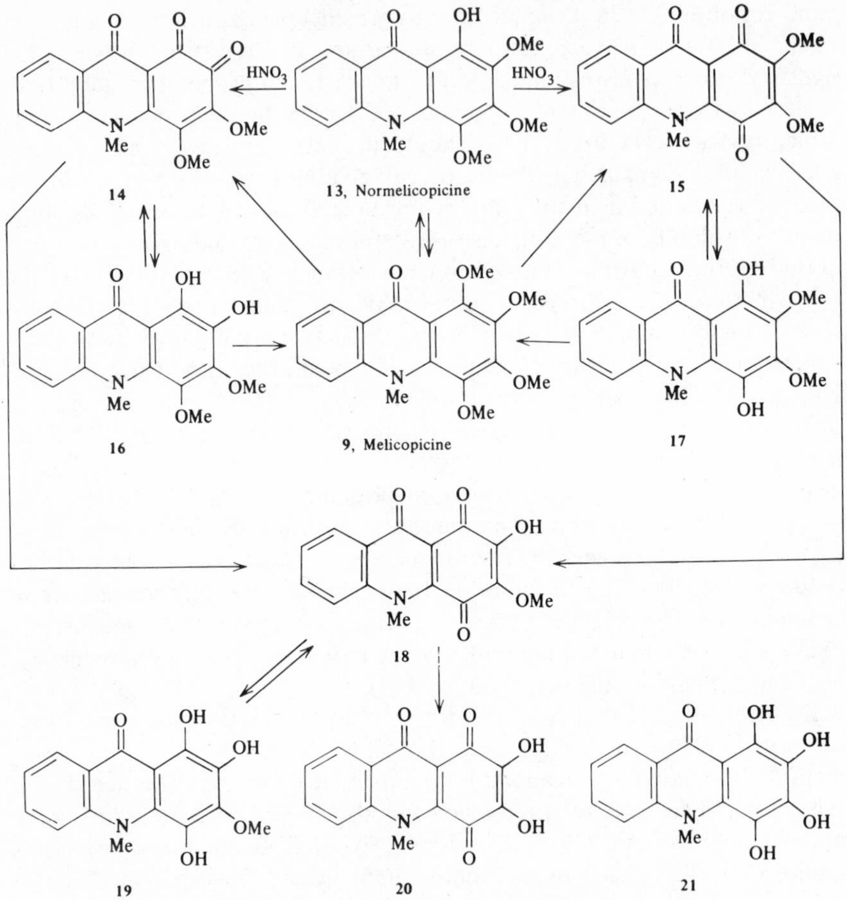

14   **13, Normelicopicine**   **15**

16   **9, Melicopicine**   **17**

18

19   **20**   **21**

D. Action of Alkali Metal Alkoxides on the Alkaloids

(1) *Structure of Melicopidine*

Although unattacked by aqueous alkali, melicopine and melicopidine were found to react slowly with potassium hydroxide in alcoholic solution to give isomeric phenols.[37] The reaction is slow in methanol and ethanol but fast in isopropyl alcohol. The reaction products are soluble in alkali, give no methylenedioxy color reaction, and contain three alkoxyl groups, one of which is derived from the solvent. It is clear that the methylenedioxy ring in the alkaloids has been opened by the alkali, a known reaction of methylenedioxy compounds:[44]

$$\text{(benzodioxole-}CH_2) + ROH \longrightarrow \text{(catechol, OH/OR)} + CH_2O$$

The two trimethoxy compounds obtained in methanol solution are isomeric with normelicopicine. The product from melicopine, 4-hydroxy-1,2,3-tri-methoxy-10-methyl-9-acridanone (24; R = Me) on nitric acid oxidation gives the same 1,4-quinone (15) as is obtained from the oxidation of normelicopicine. Demethylation of 24 (R = Me) with methanolic hydrogen chloride gives the quinol (17), corresponding to this quinone. This shows that the hydroxyl group formed in the demethylation of melicopicine to normelicopicine, and that formed during the opening of the methylenedioxy ring in melicopine, must be in *para* positions to each other in the benzenoid ring; therefore, in melicopine (23), the methylenedioxy group must be at positions 3 and 4.

The phenol from melicopidine, 2-hydroxy-1,3,4-trimethoxy-10-methyl-9-acridanone (25), on nitric acid oxidation gives an *o*-quinone (14) identical with that obtained from the oxidation of normelicopicine, and on demethylation with methanolic hydrogen chloride gives the corresponding quinol (16). On the assumption that in the formation of normelicopidine demethylation takes place in the same position as in melicopicine, melicopidine must be 1,4-dimethoxy-2,3-methylenedioxy-10-methyl-9-acridanone (22). A complete proof of this, and hence that demethylation did in fact take place in the same positions in both alkaloids, was obtained from a study of the dimethoxy-ethoxyphenols (24; R = Et, and 26), prepared by the reaction of ethanolic potash with melicopine and melicopidine, respectively.[40]

These phenols on oxidation gave, respectively, an ethoxymethoxy-*p*-quinone (27) and an ethoxymethoxy-*o*-quinone (28). Both these quinones are insoluble in cold aqueous sodium carbonate but when heated they dissolve and hydrolyze to the same ethoxyhydroxyquinone (29). This shows conclusively that one oxygen atom of the methylenedioxy group in both melicopidine and melicopine must be attached to the same position, i.e., position 3 in the benzene ring, and therefore melicopidine has the structure 22.

The alkaline hydrolysis of alkoxyl groups in different positions in the two quinones might be expected on electronic grounds. In both quinones the effect of the carbonyl group at position 1 on the reactivity of position 3 toward nucleophilic attack will be considerably diminished by resonance stabilization involving limiting structures of types 27a and 27b. At the same time the influence of the carbonyl groups at position 4 (in 27) and 2 (in 28) on their respective β positions (positions 2 and 4, respectively), will hardly be affected. This suggests that nucleophilic reagents should attack position 2 in

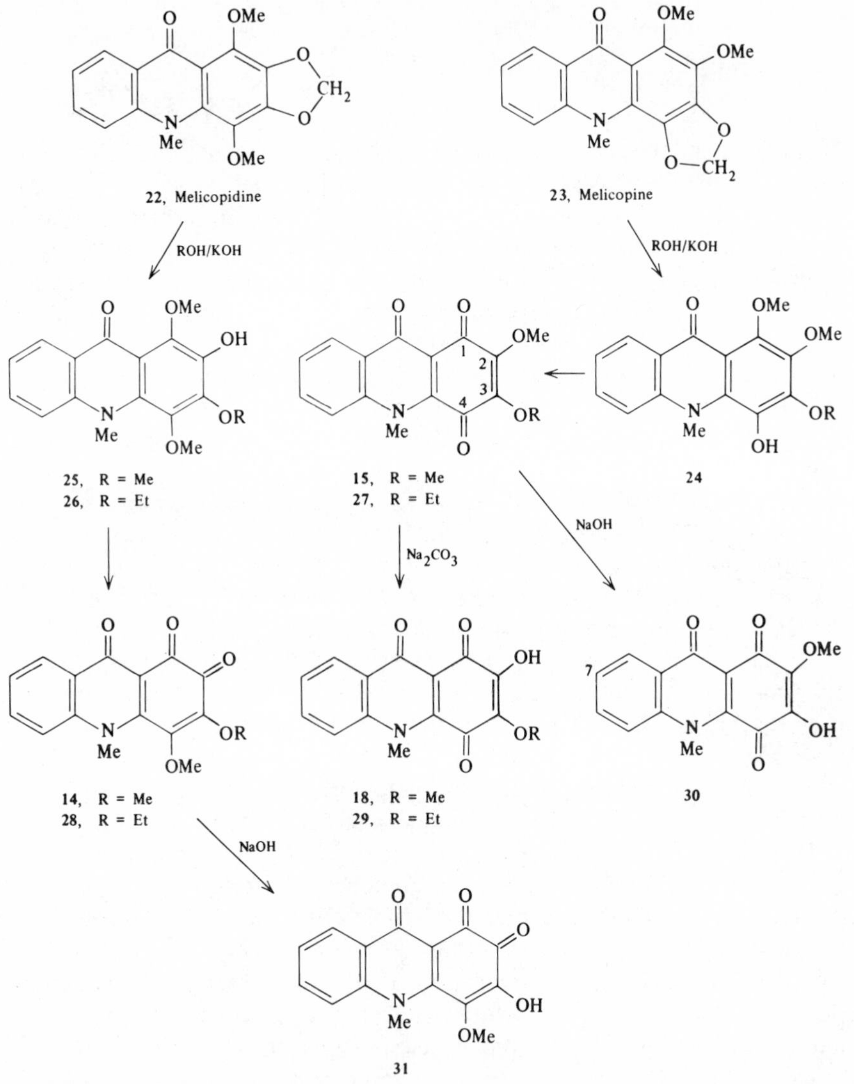
22, Melicopidine
23, Melicopine
ROH/KOH
ROH/KOH
25, R = Me
26, R = Et
15, R = Me
27, R = Et
24
Na2CO3
NaOH
14, R = Me
28, R = Et
18, R = Me
29, R = Et
30
NaOH
31

27 and position 4 in **28**, which agrees with the results of the sodium carbonate hydrolyses.[40] However, the differences in energy requirements for attack at position 3 and at position 2 (in **27**) and 4 (in **28**) must be small, since hydrolysis of these quinones with cold aqueous sodium hydroxide results in attack at position 3, with formation of the isomeric monomethoxyquinones (**30**) and (**31**), respectively.[40] The fact that analogous results are obtained with the corresponding dimethoxyquinones **14** and **15** disposes of the possibility that

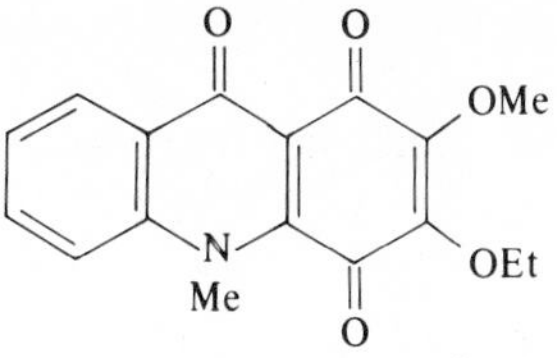

27

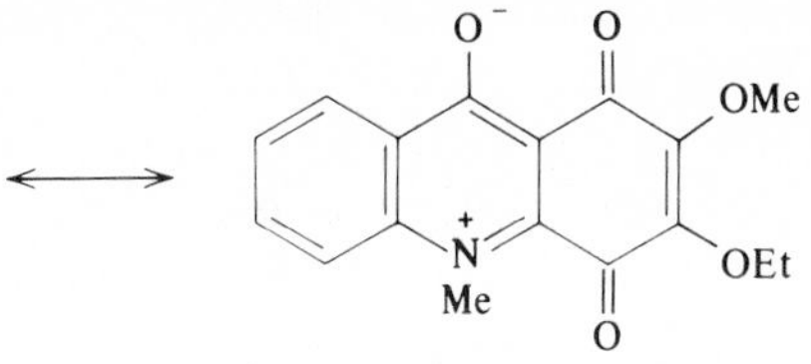

27a

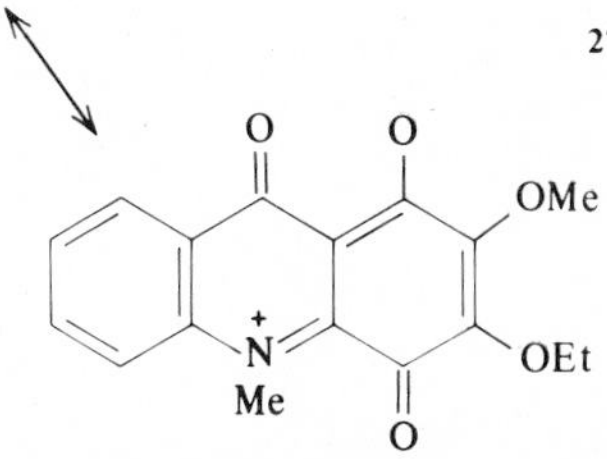

27b

the ethoxyl groups in 27 and 28 are preferentially attacked by this reagent. No satisfactory explanation of this unexpected result has yet been proposed.

The quinones 30 and 31 can also be obtained by nitrous acid oxidation of normelicopine and normelicopidine, respectively. The *o*-quinone (31) is unstable and could not be purified; hence it was characterized as its diacetyl-dihydro derivative. The *p*-quinone (30), unlike the typical quinones in this series (e.g., 15 and 18) which are stable, orange-red substances, is unstable and greenish yellow; however, all three quinones give bright red acetyl derivatives and can be reduced to orange-red quinols. Hence, in the case of the quinone (30) reduction is accompanied by an apparent bathochromic effect, instead of the usually observed hypsochromic effect. Another anomalous property of the *p*-quinone (30) is its behavior toward boiling 10% aqueous sodium hydroxide. Whereas the quinone (18) is stable under these conditions, the *p*-quinone (30) is completely decomposed to a product, $C_{13}H_{13}NO_3$, which contains one methoxyl group. This is presumably 3-methoxyacetyl-1-methyl-4-quinolone, since it gives a 2,4-dinitrophenylhydrazone and is oxidized by acid permanganate to 1-methyl-4-quinolone-3-carboxylic acid (8).[40]

(2) *Structure of Melicopine*

Although single structures for melicopidine and melicopicine can be written without definite knowledge as to which (1- or 4-) methoxyl group suffers demethylation in the formation of the *nor* alkaloids, this is not so for melicopine. Two structures, 23 and 23a, are possible. On the assumption, however, that the *nor* alkaloids are all 1-hydroxy-10-methyl-9-acridanones (these structures being the only ones consistent with their chemical and physical properties), melicopine must have structure **23**. The other structure would require the opening of the methylenedioxy ring in order to form a 1-hydroxy-9-acridanone. This is in conflict with the undisputed presence of an intact methylenedioxy ring in normelicopine.

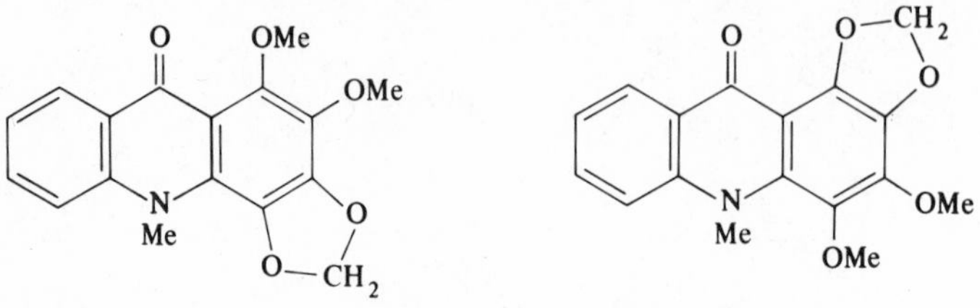

The position of the methylenedioxy ring in melicopine was originally deduced from a consideration of the mechanism of its reaction with alkali metal alkoxides.[40] The opening of a methylenedioxy ring by this reagent to give a phenolic ether and formaldehyde was first reported by Robinson and Robinson,[44] who observed that the reaction of 1,2-methylenedioxy-4-nitro-benzene (**32**) with methanolic potash gave rise to 2-hydroxy-4-nitroanisole (**34**). In modern terms it is clear that the nitro group facilities attack at the *para* position by the nucleophilic methoxide ion by lowering the energy of the transition state leading to the intermediate **33**; the loss of formaldehyde from the latter then leads to the product, 2-hydroxy-4-nitroanisole:

For exactly similar reasons, the carbonyl group at position 9 in melicopidine activates positions 1 and 3 toward nucleophilic attack. In this particular case attack must take place at position 3, since position 1 cannot be involved in the methylenedioxy ring. Since it has been shown that ethanolic potash attacks both melicopidine and melicopine at the same position, only one structure (**23**) is possible for melicopine. This conclusion was supported by three other observations:

1. The *nor* alkaloids showed the expected properties of 1-hydroxy-10-methyl-9-acridanone derivatives.

2. The reaction of melicopicine with ethanolic potash, which gave a mixture of phenolic and nonphenolic products, was also shown to involve nucleophilic attack at position 3. The phenol, $C_{17}H_{15}NO_5$, gave a monoacetate and was methylated to melicopicine. It was not identical with any of the three previously prepared monohydroxytrimethoxy-10-methyl-9-acridanones and is therefore the 3-hydroxy isomer **35**.

The nonphenolic and cryptophenolic fraction was separated into two compounds. The less basic material, $C_{18}H_{19}NO_5$, was isomeric with melicopicine but possessed the properties of normelicopicine and was not demethylated by alcoholic hydrogen chloride. On nitric acid oxidation, it gave 3-ethoxy-2-methoxy-10-methyl-9-acridanone-1,4-quinone. Since the compound is not 1,2-dimethoxy-3-ethoxy-4-hydroxy-10-methyl-9-acridanone **(36)**, it must therefore be 2,4-dimethoxy-3-ethoxy-1-hydroxy-10-methyl-9-acridanone **(37)** and is formed by demethylation at position 1 and ether exchange at position 3.

35

36

37

The more basic, noncrystalline fraction gave a crystalline hydrochloride, which with ethanolic hydrogen chloride was converted into 2,4-dimethoxy-3-ethoxy-1-hydroxy-10-methyl-9-acridanone **(37)**. The noncrystalline material is therefore the methyl or ethyl ether of **(37)**, showing that in addition to some demethylation at position 1, ether exchange also occurred at position 3.

3. Some evidence for activation at position 1 in melicopine was found in the isolation of a small quantity of nonphenolic by-product from the reaction of melicopine with ethanolic potash. This by-product was a yellow crystalline base resembling melicopine but melted ca. 30° lower. It could not be separated into pure components by fractional crystallization or by differences in basic

strength; however, with methanolic hydrogen chloride, it gave 89% of its weight of normelicopine. It is, therefore, almost certainly a mixture of melicopine with 1-ethoxy-2-methoxy-3,4-methylenedioxy-10-methyl-9-acridanone formed by ether exchange.[40]

E. Acidic Degradation Products

A colorless dibasic acid (38), $C_{13}H_9NO_6$, was obtained when melicopidine reacted with nitrous acid, when the 3-alkoxy-1,2-dihydroxy-4-methoxy-10-methyl-9-acridanones 16 and 39 were oxidized by air in alkaline solution, and as a by-product in a number of other reactions.[40] It gave 1-methyl-4-quinolone when heated with zinc dust and 1-methyl-4-quinolone-3-carboxylic acid (8) on oxidation with 68% nitric acid or potassium permanganate. Decarboxylation in butyl phthalate with copper bronze also gave 8 (in small yield) and a base $C_{11}H_9NO_2$. This base gave a 2,4-dinitrophenylhydrazone and was easily oxidized to 1-methyl-4-quinolone-3-carboxylic acid. Therefore, it is 1-methyl-4-quinolone-3-carboxaldehyde (40). As the second carboxyl group must be in position 2, the colorless dibasic acid has structure 38.

A by-product from the aerial oxidation of 16 and 39, but not of the corresponding o-quinones, which were stable under the conditions used, was the colorless acid $C_{13}H_9NO_5$ (41), also obtained in good yield from the oxidation of 3-hydroxy-2-methoxy-10-methyl-9-acridanone-1,4-quinone (30) with alkaline permanganate. This acid was also formed from 38 by reduction with zinc dust and sodium hydroxide. Its structure is represented by 41. This was confirmed by oxidation to the acid 8 and by decarboxylation to a basic lactone $C_{12}H_9NO_3$ (42).

An orange-red solid (43), $C_{13}H_9NO_4$, whose weakly acidic aqueous solutions are yellow with a green fluorescence, was formed as a by-product of several reactions. It may be prepared in quantity by the aerial oxidation of alkaline solutions of the dialkoxy-p-quinols (17 and 44), and of 24 (R = Me or Et), or by the action of an excess of nitrous acid on normelicopine. The p-quinones are not intermediates in this reaction, since they are not oxidized under the same conditions. The orange-red compound did not acetylate or react with 2,4-dinitrophenylhydrazine, but oxidation with 68% nitric acid gave 1-methyl-4-quinolone-3-carboxylic acid; with cold alkaline potassium permanganate 38 was obtained. The only structure that accounts for these properties is 43; the failure to acetylate or react with ketonic reagents[40] is also characteristic of croconic acid (44a).

2,3-Dihydroxy-10-methyl-9-acridanone-1,4-quinone (20) reacts with cold concentrated nitric acid to give a compound of empirical formula $C_{14}H_{11}NO_7$, which is tentatively considered to be 45, although other formulations are not excluded by the evidence available.[43] This structure is in accord with its mode of formation and its properties but does not account for the complete failure of all attempts to reduce it to the corresponding tetrahydroxy compound. When its emerald green solution in aqueous sodium hydroxide is shaken in air degradation to the orange-red acid (43) occurs. This is considered analogous to the formation of croconic acid from hexaketocyclohexane, which has been formulated as a benzilic acid rearrangement.

In contrast to the above oxidation reactions is the behavior of the ethyl ether (26) toward alkaline hydrogen peroxide. The product is a dibasic acid, $C_{17}H_{17}NO_7$, which contains two alkoxyl groups. Since only one carbon atom has been lost during the oxidation, the ring is presumed to have been severed between positions 1 and 2. If this assumption is correct, this acid,[40] which readily decarboxylates to another acid $C_{16}H_{17}NO_5$, must have the constitution 46.

F. Reaction of *Melicope* Alkaloids with Bromine

In an attempt to prepare 7-bromomelicopine, Prager and Thredgold[45] investigated the reaction of melicopine with bromine in methanol. The product, however, was not the desired 7-bromo derivative but an unstable aryl hypobromite salt to which the structure 47 was ultimately assigned.[46] When the reaction was performed in ethanol solution, the corresponding ethoxy isomer 48 was obtained. The reaction of 47 with potassium iodide affords the product 49; this is also the product of reaction of 47 with dilute alkali.[45] It can also be obtained by reaction of the tetraphenylborate salt corresponding to 47 with iodide ion, a reaction that proceeds with liberation of only 1 mole of iodine. The reverse reaction can be achieved by treatment of 49 with bromine in methanol,[46] which gives as sole product the hypobromite salt (47).

That demethylation of melicopine at position 1 occurs in the formation of 47 is established by the fact that 47 is also obtained from the reaction of normelicopine with bromine.

Hydrogenation of 49 in the presence of Adams's catalyst gives the enol (50), which loses methanol on brief treatment with sulfuric acid with the formation of normelicopine. Normelicopine can also be produced directly (and quantitatively) by hydrogenation of 47 itself.[45] Borohydride reduction of 49 at 0° also affords the enol 50; at 50°, however, 50 is obtained, together with three other products to which the structures 51, 52, and 53 were assigned. The structure of demethoxymelicopine (52) was subsequently established by its preparation from normelicopine. In an exactly analogous manner, the structure of the methyl ether 54 was confirmed by its preparation from normelicopicine. Finally, the position of the phenolic hydroxyl group in 53 was proved by unambiguous synthesis of its ethyl ether (55). These 2,3,4-trisubstituted derivatives of 10-methyl-9-acridanone are presumably formed from 49 via its tetrahydro derivative 56; the production of the phenolic compound 53 provides clear evidence for the presence of a methoxyl group[45] at position 4 in 49, and therefore also in 47.

When a solution of melicopine and bromine in methanol was refluxed for 9 hr, the reaction proceeded further and gave a compound, $C_{19}H_{21}NO_8$,

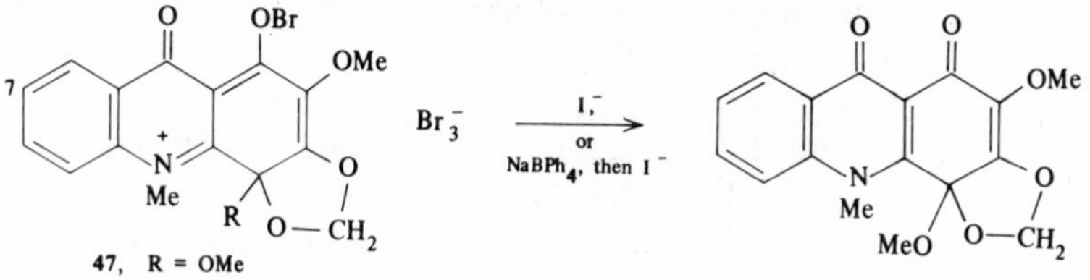

47, R = OMe
48, R = OEt
61, R = Br

49

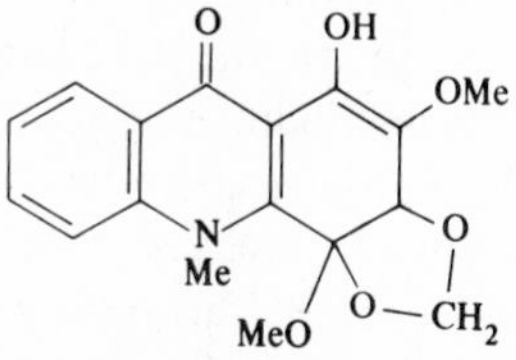

50

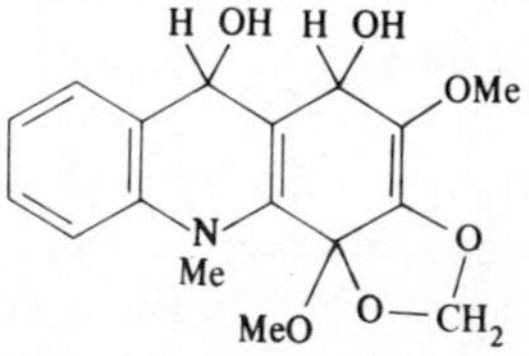

51

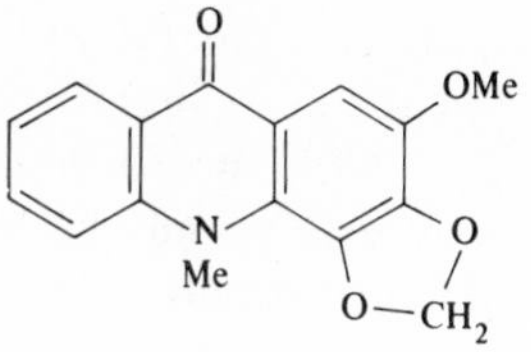

52

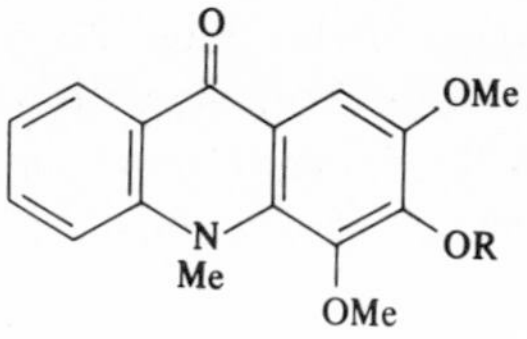

53, R = H
54, R = Me
55, R = Et

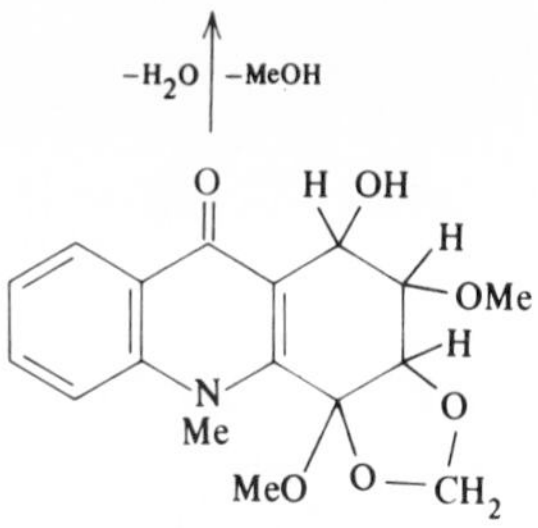

−H₂O −MeOH

56

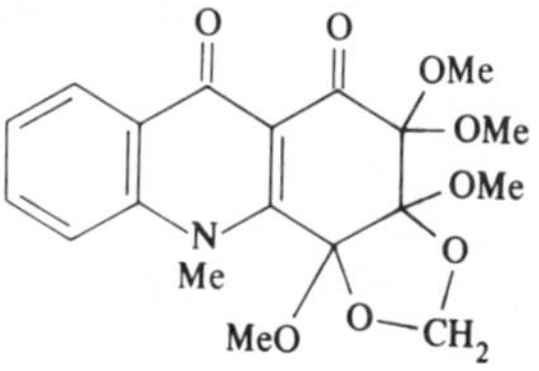

57

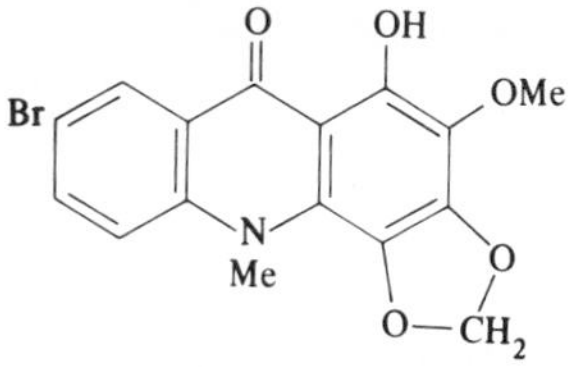

58

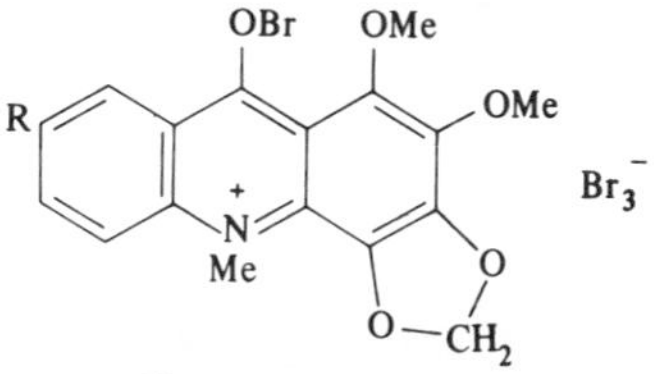

59, R = H
60, R = Br

397

which appeared to be a derivative of 1-methyl-4-quinolone.[45] On the assumption that **47** is an intermediate in this reaction, this product is formulated as the complex ketal **57**. The presence of a carbonyl group in this compound was shown by its reduction by sodium borohydride in refluxing ethanol solution, but the best evidence for its constitution was derived from its reduction by zinc and acetic acid, which afforded normelicopicine (**13**) in good yield.[47]

When the hypobromite (**47**) was allowed to stand for several weeks, a deep red-purple compound of unknown constitution was obtained. In an attempt to accelerate this transformation, **47** was heated to its melting point. The product, surprisingly, proved to be 7-bromonormelicopine (**58**) and was identified by comparison with authentic material prepared independently.

The bromination of melicopine in other solvents also proves to be a complex reaction. Thus, in acetic acid the products include 7-bromonormelicopine, normelicopine, and the quinone (**30**), together with its 7-bromo derivative. Bromination in chloroform resulted in the formation of a pink-red compound containing four bromine atoms, which was formulated as the hypobromite perbromide **59**. The treatment of this compound with alkali regenerated melicopine; with aqueous potassium iodide, 2 moles of iodine were produced, in addition to melicopine. When kept in chloroform solution, the salt **59** was slowly converted into its 7-bromo derivative (**60**), from which 7-bromomelicopine could be isolated after treatment with alkali.[46]

A similar series of bromination reactions was also carried out on melicopicine.[47] In methanol solution the addition of bromine caused the rapid precipitation of an orange compound, $C_{20}H_{25}Br_4NO_7$, for which the structure (**62**) is proposed. In accordance with this constitution the reaction of **62** with iodide ion liberated 2 moles of iodine; the product from this reaction, or from brief treatment of **62** with alkali, was the ketal (**63**). The salt (**62**) could readily be reformed from the ketal (**63**) by reaction with bromine in methanol. Hydrolysis of **63** with dilute acid gave a mixture of the quinones **15** and **18**, the latter presumably being formed by the known selective demethylation of **15**. From the bromination of melicopicine in ethanol, compounds analogous to **62** and **63** were formed. Acid hydrolysis of the ketal, corresponding to **63**, gave three compounds only, identified as **15**, **18**, and **64**. The formation of **64** proves conclusively that the bromination reactions result in the introduction of an alkoxyl group at position 2; the other alkoxyl group is then necessarily situated at position 1, since any other alternatives give a structure for the ketal **63** which should have a carbonyl group reducible by sodium borohydride, a reagent which is without effect on **63**. Attempts to confirm the presence of an additional methoxyl group at position 1 in **63** by partial hydrolysis to compound **65**, obtained from normelicopicine, were unsuccessful. Even after only 1 min with cold water, the ketal (**63**) had been completely hydrolyzed to the quinone (**15**) and another, unidentified, compound.[47]

The bromination of melicopicine in chloroform solution appears to give a mixture of the hypobromite perbromide salts, **66** and **67**. However, these compounds are noncrystalline, unstable, and extremely difficult to isolate. The removal of the chloroform below room temperature results in demethylation at C-1 and bromination at C-7. Treatment of the reaction mixture with sodium thiosulfate gives mainly melicopicine with a small amount of 7-bromomelicopicine.[46]

The bromination of normelicopicine in methanol gives a salt (**68**), which gives the ketone (**65**) on mild treatment with alkali. The presence of a carbonyl function in **65** was confirmed by its reduction with zinc and acetic acid or by hydrogenation, which regenerated normelicopicine. The treatment of **65** with dilute acid gives a mixture of the quinones **15** and **18**. Similarly, bromination of normelicopicine in ethanol yields the salt (**69**) which, with alkali, affords the ketone **70**. Acid hydrolysis of **70** gives a mixture of the quinones **15**, **18**, and **64**, thereby unambiguously locating the position of introduction of the alkoxyl group.[47]

In chloroform solution the bromination of normelicopicine gives a product that possibly has the structure **71**, since reaction with aqueous alkali affords a neutral product,[46] probably the keto-ester **72**, which, in the presence of dilute alkali, decomposes to **43**.

The formation of an aryl hypobromite perbromide, and the addition of alkoxyl groups to positions 1 and 2, are also characteristic of the reaction of melicopidine with bromine in methanol, which affords[47] the orange salt **73**. Like the analogous compound (**62**) derived from melicopicine, the salt **73** was converted by dilute alkali into the corresponding derivative (**74**) of 1-methyl-4-quinolone, which could be reconverted into **73** by reaction with bromine in methanol. The salt **73** also liberated 2 moles of iodine when titrated with aqueous potassium iodide. Unlike **63**, however, the melicopidine-derived ketal, **74**, could be selectively hydrolyzed by dilute hydrochloric acid, the product being the ketone **75**.

The uv spectrum of the ketal **74** was almost identical with that of the corresponding compound **63**, derived from melicopicine, suggesting that the two introduced methoxyl groups were situated at C-1 and C-2. However, all attempts to confirm the placing of one methoxyl group at C-1 by reconversion of **74** into melicopidine proved unsuccessful. Thus reduction with zinc and acetic acid afforded normalicopidine, owing to preferential acid hydrolysis of the ketal function at position 1. Boron trifluoride also gave normelicopidine, while the reduction of **74** with sodium in liquid ammonia yielded a

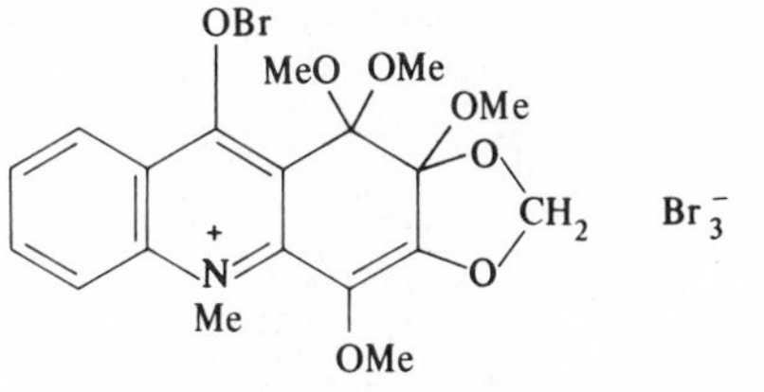

73

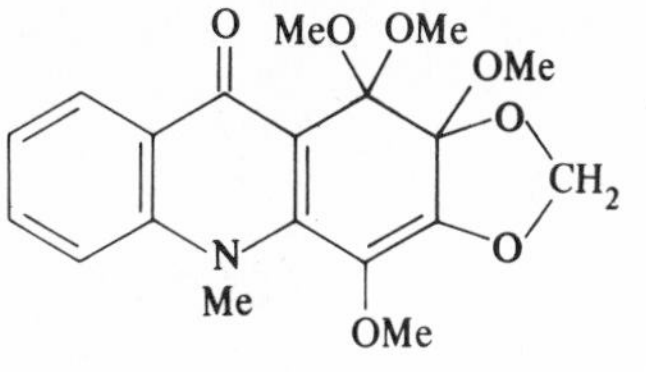

74

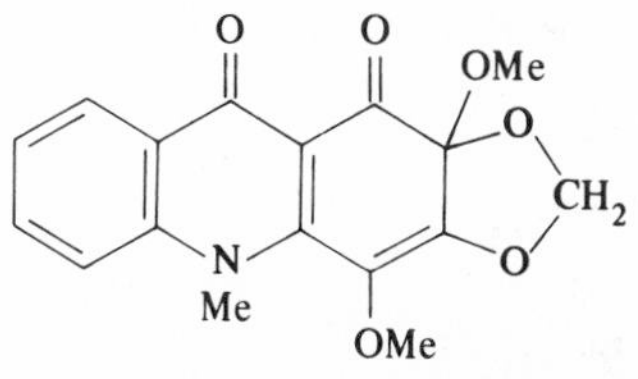

75

mixture of products, none of which was melicopidine, and the compound was inert to hydrogenation.[47]

It was expected that bromination of normelicopidine would give products similar to those from either normelicopicine or normelicopine, in that a methoxyl group would be introduced into position 2 or 4. In fact, the bromination of normelicopidine in methanol with 2 moles of bromine gave rise to a mixture of two products, readily separable by fractional crystallization. When 3 moles of bromine were used, only the lower melting product was obtained. The structure of the higher melting compound was shown to be 76, since treatment with dilute alkali gave the known ketone, 75. When the latter was treated with bromine in chloroform, the salt 76 was regenerated, but in methanol the product was the second salt, 77. This salt contains two more methoxyl groups than normelicopidine (nmr spectrum) and is converted by alkali into a colorless compound (78), whose uv spectrum is identical with that of the compound 57 derived from melicopine.

The bromoketone 78 was only slowly reduced by sodium borohydride; the product was an ill-defined, noncrystalline substance that was not characterized. However, reduction with zinc and acetic acid gave normelicopidine, and the bromine atom was readily replaced by a methoxyl group on reaction with methanolic silver nitrate. Two structures, 78 and 79, are thus possible for the bromoketone, and a differentiation was achieved using tritium-labeled methanol in its preparation from normelicopidine. When reduced with zinc and acetic acid, the tritium-labeled bromoketone afforded radioactive normelicopidine, showing that the second methoxyl group had been introduced at position 4, as required by structure 78.

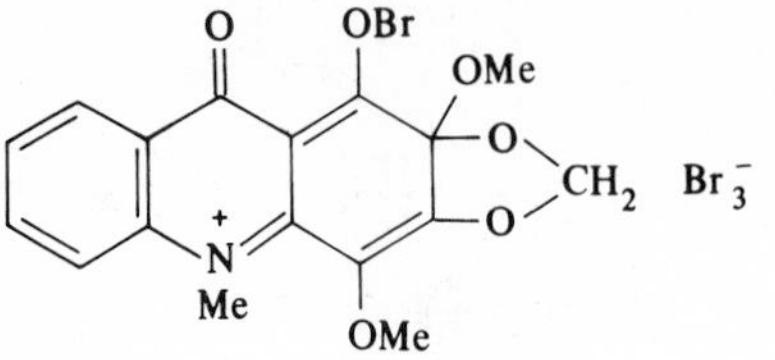

76 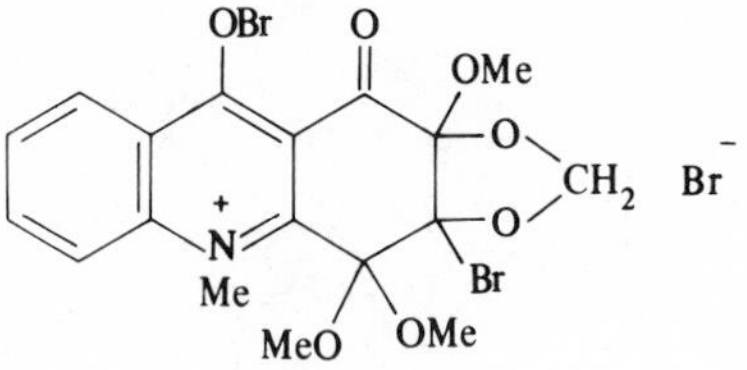 77

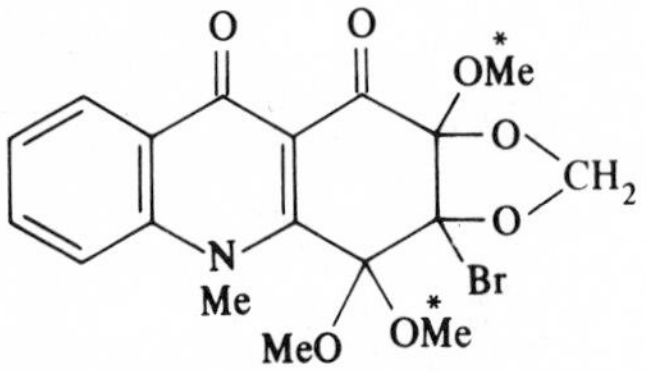

78 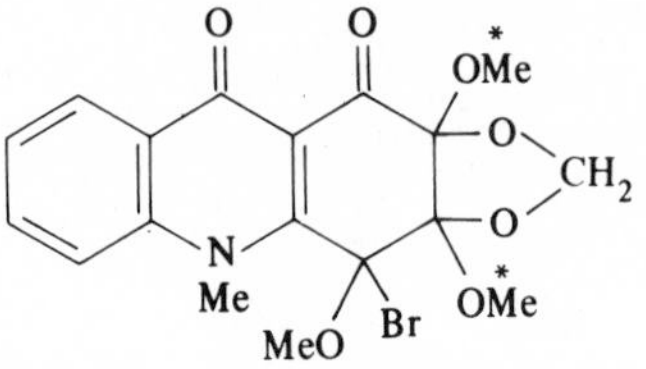 79

G. Synthesis of Melicopidine and Melicopine

Although the structures of melicopidine and melicopine were elucidated in 1949, the synthesis of these alkaloids was not achieved until 1965.[48] 2,5-Dimethoxy-3,4-methylenedioxyiodobenzene was condensed with anthranilic acid in the presence of copper, and the diphenylamine derivative (80) so obtained was cyclized by means of phosphorus oxychloride to give the 9-chloroacridine derivative (81). Hydrolysis of this product by hot dilute hydrochloric acid gave the corresponding 9-acridanone (82). which was identical with the alkaloid xanthevodine. Finally, N-methylation of xanthevodine with methyl iodide and potassium hydroxide in acetone gave melicopidine.

An exactly analogous synthesis, starting from 4,5-dimethoxy-2,3-methylenedioxyiodobenzene, afforded melicopine.[48]

3. Xanthevodine

Xanthevodine, $C_{16}H_{13}NO_5$, is a pale yellow, crystalline alkaloid, mp 213–214°, which occurs in the leaves of E. xanthoxyloides[14]; apparently it is not present in the bark.[6] Aside from melicopine, melicopicine, and melicopidine, xanthevodine is the only other derivative of 1,2,3,4-tetrahydroxy-9-acridanone encountered so far among the acridine alkaloids. It contains two methoxyl groups and no methylimino group, and gives 4-quinolone-3-

carboxylic acid on oxidation. *N*-Methylation converts it into melicopidine (22), so that it can only be 1,4-dimethoxy-2,3-methylenedioxy-9-acridanone (82).[49] This structure is amply confirmed by its total synthesis.[47]

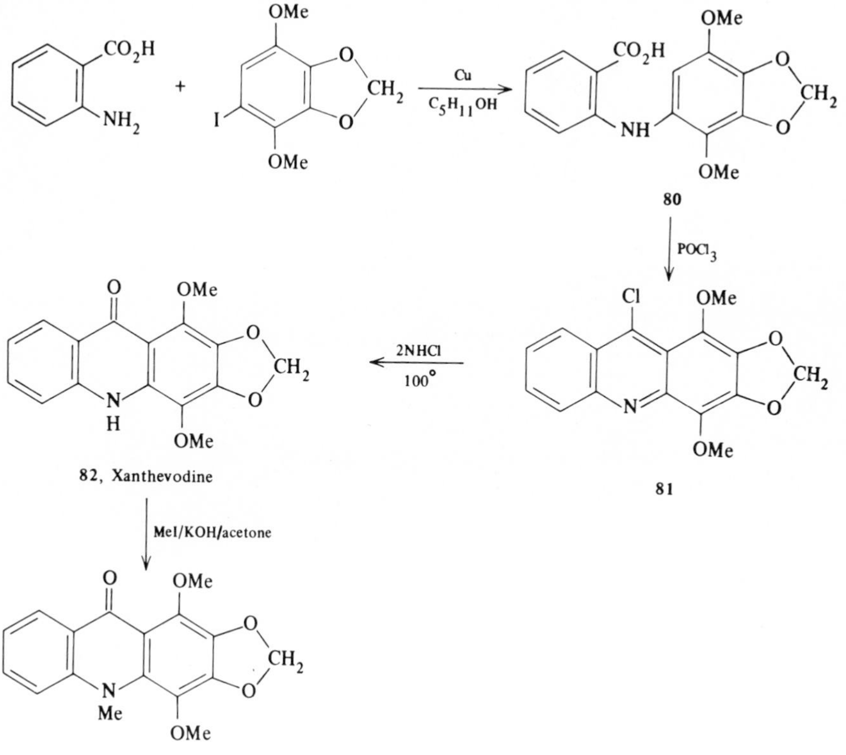

82, Xanthevodine

Melicopidine, 22

4. Derivatives of 1,2,3-Trihydroxy-9-acridanone

A. Evoxanthine, Evoxanthidine, and Norevoxanthine

Among the acridine alkaloids are seven derivatives of 1,2,3-trihydroxy-9-acridanone. Three of these (evoxanthine, evoxanthidine, and norevoxanthine) bear a methylenedioxy group at positions 2 and 3, whereas the remaining four bear alkoxyl groups at these positions.

The first of these alkaloids to be encountered was evoxanthine (83), which was isolated, together with melicopidine (22) and kokusaginine (3), from the bright yellow bark of *E. xanthoxyloides*.[6] Subsequently, evoxanthidine (84) was extracted, together with evoxanthine and seven other alkaloids, from the leaves of the same species.[14] Evoxanthine appears to occur widely in the Rutaceae (see Table I); in contrast, evoxanthidine has so far only been obtained from *E. xanthoxyloides*. Similarly, norevoxanthine (85) has so far only been reported to occur in *Teclea grandifolia* Engl.[19]; its isolation from *E. xanthoxyloides*[14] may have been the result of demethylation of evoxanthine during the extraction procedure.

Evoxanthine, mp 217–218°, crystallizes as yellow cubes from its yellow solution (strong blue-violet fluorescence) in ethanol, or as needles from benzene or ethyl acetate.[6] Its empirical formula is $C_{16}H_{13}NO_4$; it contains one methoxyl, one *N*-methyl, and one methylenedioxy group. The fourth oxygen atom shows no phenolic or ketonic properties.

Nitric acid oxidation of evoxanthine (83) gives 1-methyl-4-quinolone-3-carboxylic acid (8), showing that it is structurally related to the other acridine alkaloids and is probably a derivative of 10-methyl-9-acridanone. This is consistent with its uv absorption spectrum.[42] When evoxanthine is heated with methanolic hydrogen chloride, demethylation occurs (cf. melicopine and its congener) with formation of norevoxanthine (85). This compound contains no methoxyl groups and can be remethylated to evoxanthine. Norevoxanthine, mp 274–275°, crystallizes from pyridine or dioxan as orange needles and forms a yellow acetyl derivative, mp 240–242°, only with difficulty. It is a very weak base, soluble only in concentrated hydrochloric acid, and is insoluble in aqueous sodium hydroxide. This contrasts with evoxanthine which, in spite of being a very weak base (pK_a 2.6), forms a crystalline hydrochloride which separates when a warm solution of the alkaloid in 2% aqueous hydrochloric acid is cooled.[6]

The methylenedioxy ring in evoxanthine is opened by hot methanolic or ethanolic potash with the formation of the phenols 86 (R = Me or Et), neither of which gives a positive test for a methylenedioxy group. Both compounds are easily demethylated by methanolic hydrogen chloride to 87 (R = Me or Et). Methylation of 87 (R = Me or Et) gives the 3-alkoxy-1-hydroxy-2-methoxy-10-methyl-9-acridanones 88 (R = Me or Et), which resist further methylation. The cryptophenols 88 (R = Me or Et) have also been prepared from the phenols 86 (R = Me or Et) by methylation to 89, followed by demethylation at position 1 with methanolic hydrogen chloride.

On oxidation with concentrated nitric acid, the quinols (87) and the phenols (88) give the corresponding deep red quinones (90; R = Me or Et), which can be reduced back to the quinols by sulfur dioxide. Both quinones give quinoxaline derivatives (91; R = Me or Et) with *o*-phenylenediamine. The alkoxyl groups, introduced by the action of alcoholic potash on the alkaloid, are retained in these quinoxaline derivatives. This shows that the *o*-quinones cannot arise from the methylenedioxy bridge alone, and that the

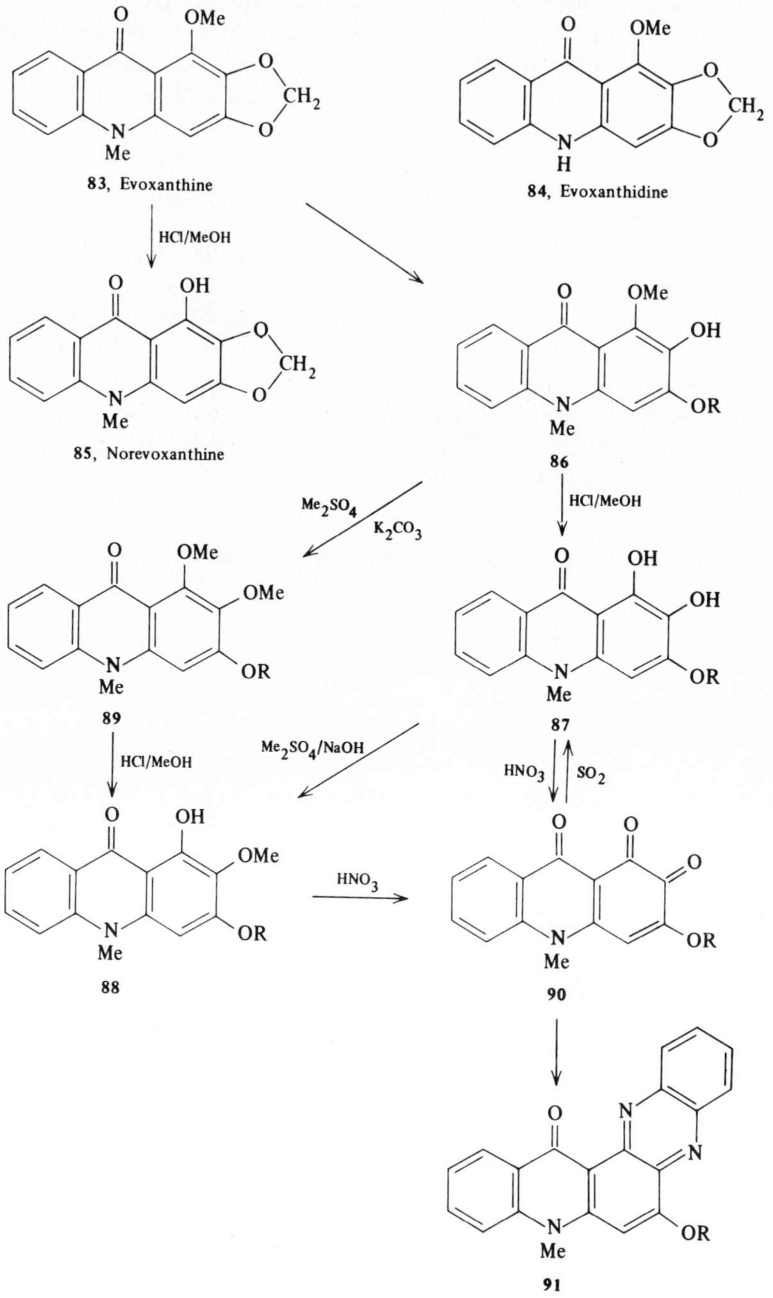

83, Evoxanthine

84, Evoxanthidine

HCl/MeOH

85, Norevoxanthine

86

Me₂SO₄ K₂CO₃

HCl/MeOH

89

87

HCl/MeOH

Me₂SO₄/NaOH

HNO₃ SO₂

88

HNO₃

90

91

methoxyl group must be adjacent to one end of this bridge. This methoxyl group in evoxanthine is considered to be in position 1. Its demethylation to norevoxanthine (85) with alcoholic hydrochloric acid is then analogous to the similar effect of this reagent on the other 9-acridanone alkaloids. The properties of norevoxanthine are in accord with this, but no other, formulation. It is similar to the other *nor* alkaloids in color, spectral properties, and the unreactivity of its hydroxyl group. Evoxanthine, therefore, is 1-methoxy-2,3-methylenedioxy-10-methyl-9-acridanone (83).[6]

Evoxanthidine, $C_{15}H_{11}NO_4$, separates from its yellow solution (strong blue fluorescence) in alcohol as pale yellow needles, mp. 312–313°. It contains one methoxyl but no methylimino group, and gives a positive color reaction for a methylenedioxy group. Nitric acid oxidation gives 4-quinolone-3-carboxylic acid, which suggests that the alkaloid is a methoxymethylenedioxy-9-acridanone. Methylation with methyl iodide and potassium carbonate converted it into evoxanthine; therefore, evoxanthidine must be 1-methoxy-2,3-methylenedioxy-9-acridanone (84).[49]

B. Bromination of Evoxanthine and Norevoxanthine

The bromination of evoxanthine in methanol[50] is similar to that of melicopicine and melicopidine, in that addition occurs at positions 1 and 2. The product is a hypobromite-perbromide salt (92), which gives the highly unstable bromoether (93) with dilute alkali; the reverse stage may be accomplished by means of bromine in methanol. The uv spectrum of 93 is very similar to that of the compounds 63 and 74 previously obtained from melicopicine and melicopidine, and their nmr spectra are also similar. The bromoether 93 decomposes on standing to a mixture of evoxanthine, norevoxanthine, and 4-bromonorevoxanthine; evoxanthine is also the product of the reaction of 93 with tetraethylammonium iodide (which also results in the liberation of 1 mole of iodine), dilute acid, or dilute alkali. Methanolic or ethanolic silver nitrate reacts with 93 to give the appropriate ketal (94; R = Me or Et), which with dilute acid gives the ketone 95. The fact that both ketals give the same ketone 95 indicates that the alkoxyl group introduced in the reaction with alcoholic silver nitrate, and hence the original bromine atom in the bromoether 93, must be situated at position 1; the methoxyl group[50] introduced in the formation of 92 and 93 must, therefore, be located at position 2.

The ketone 95 is readily reduced to norevoxanthine by zinc and acetic acid, and gives 1,2,3-trihydroxy-10-methyl-9-acridanone (96) when warmed with dilute alkali. The formation of this trihydroxy compound requires a reduction from the quinone stage of oxidation of the starting material (95); this could be

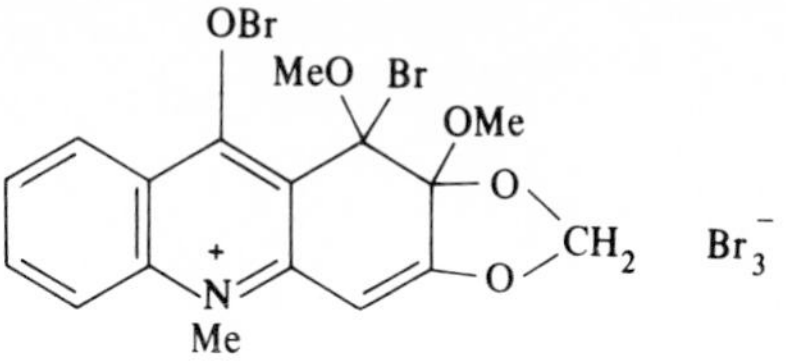

92

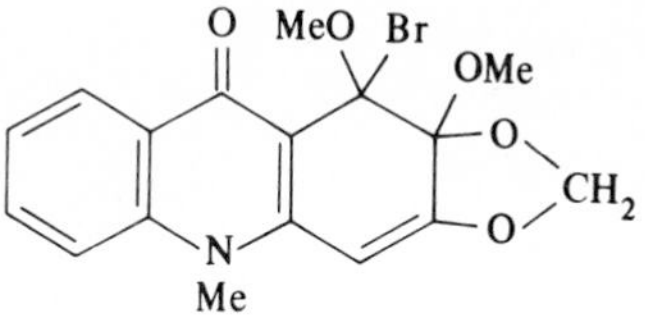

93

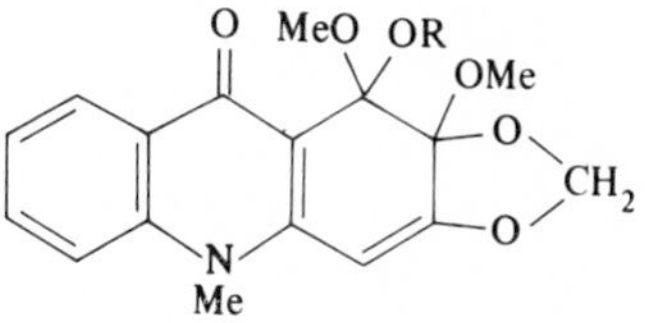

94

achieved by the formaldehyde released in the hydrolysis of the acetal-ketal function of **95**, or it could proceed via an intramolecular redox reaction involving the methoxyl group [50] at position 2:

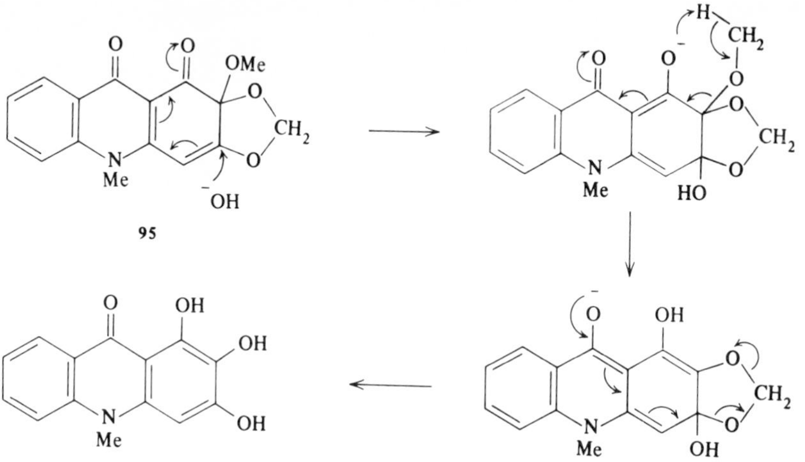

95

96

Norevoxanthine, unlike the alkaloids already discussed, contains a very reactive, unsubstituted aromatic position at C-4 and therefore might be expected on bromination to yield a 4-bromo derivative directly. However, this was not observed and, as with the other alkaloids, bromination in methanol resulted in addition to the oxygenated aromatic ring; the product was the complex salt **97**, which with dilute alkali gave rise to the ketone **98**. When this

ketone was briefly warmed with dilute sodium hydroxide 4-bromonor-
evoxanthine (99) was obtained, a reaction that can also be explained mecha-
nistically by an intramolecular redox reaction involving the methoxyl group
(cf. arrows in Structure 100).[50]

One further interesting reaction of the compound (98) is its reduction by
zinc and acetic acid to norevoxanthine. Presumably, this reduction proceeds
via the unbrominated analogue (95), since no significant reduction of 4-bro-
monorevoxanthine occurs under these conditions.[50]

C. Synthesis of Evoxanthine and Evoxanthidine

Two independent syntheses of evoxanthine and evoxanthidine, by essentially the same route, have been reported.[51, 52] 3-Methoxy-4,5-methylene-
dioxydiphenylamine-2'-carboxylic acid (101) was prepared by the Ullmann
reaction, either from 3-methoxy-4,5-methylenedioxyiodobenzene and anthranilic acid,[51] or from 3-methoxy-4,5-methylenedioxyaniline and o-bromobenzoic acid,[52] and then cyclized by means of phosphorus oxychloride to the
9-chloroacridine derivative (102), together with some of the alternative
cyclization product. The hydrolysis of 102 by means of dilute hydrochloric
acid then afforded evoxanthidine (84), which on methylation in acetone with
methyl iodide and potassium carbonate or hydroxide gave evoxanthine.

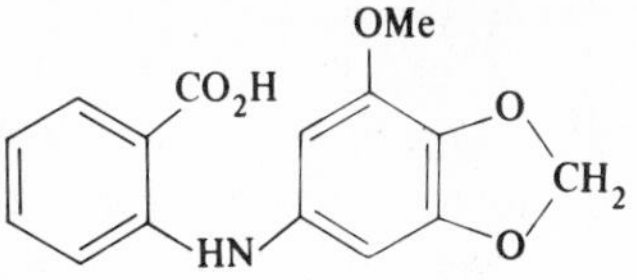

101 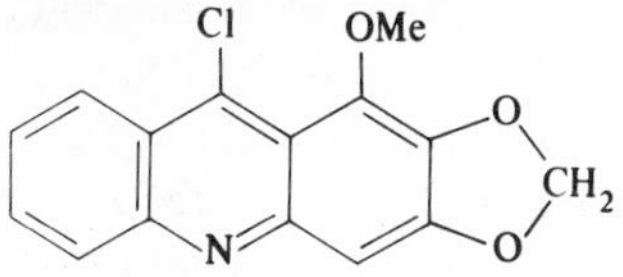   102

D. Xanthoxoline, 1-Hydroxy-2,3-dimethoxy- and 1,2,3-Trimethoxy-10-methyl-9-acridanone

Xanthoxoline (**103**) has so far only been isolated from the leaves of *Evodia xanthoxyloides*.[14] Its *N*-methyl derivative, arborinine (**104**), in contrast, occurs more widely (see Table I), but in the earlier extractions it was not claimed to be a genuine alkaloid,[7, 14] since it occurred in association with 1,2,3-trimethoxy-10-methyl-9-acridanone (**105**), which, it was suspected, might have suffered demethylation during the acid extraction procedure. However, in extractions of *Evodia alata*[8] and *Acronychia haplophylla*,[22] which were carried out much more recently, the use of acid was carefully avoided; hence it seems that 1-hydroxy-2,3-dimethoxy-10-methyl-9-acridanone is an actual constituent of the intact plant. It is also of interest that **104** has been isolated from *Fagara leprieurii*,[21] but no trace of its *O*-methyl derivative (**105**) could be found.

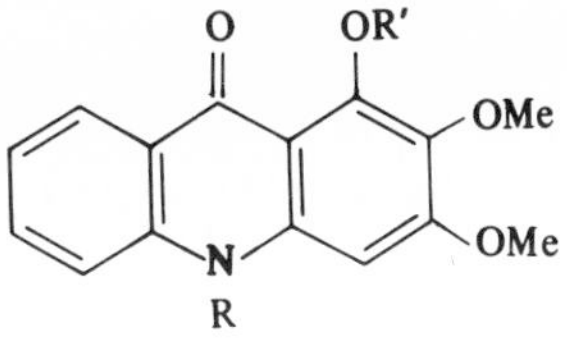

106

103, R = R' = H
104, R = Me, R' = H
105, R = R' = Me

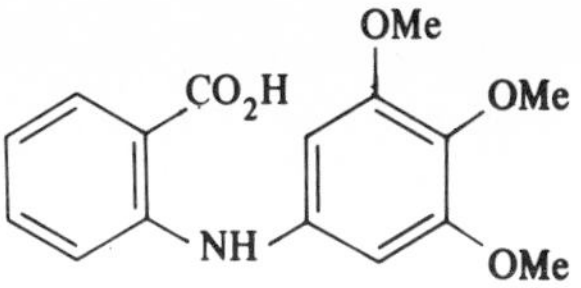

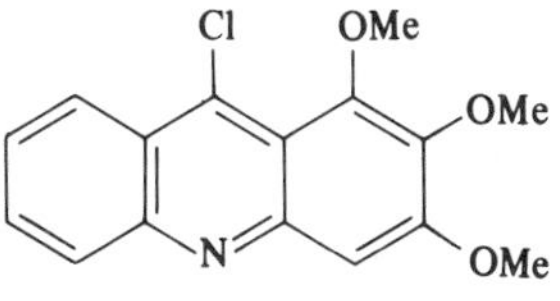

107

Xanthoxoline, $C_{15}H_{13}NO_4$, crystallizes as golden yellow needles, mp 265–268° (dec), from ethanol, but its solution shows no fluorescence in contrast to those of the other acridanone alkaloids. It contains two methoxyl groups, but no methylimino or methylenedioxy groups, and is insoluble in 10% aqueous sodium hydroxide. It is a much weaker base than evoxanthidine or xanthevo-

dine but on vigorous acetylation gives a monoacetyl derivative, that forms
fluorescent solutions. Xanthoxoline contains a cryptophenolic hydroxyl
group, therefore, and like the *nor* alkaloids gives a green ferric chloride color
in ethanol. Methyl iodide and potassium carbonate in acetone convert it into a
mixture of 1,2,3-trimethoxy-10-methyl-9-acridanone (**105**) and 1-hydroxy-
2,3-dimethoxy-10-methyl-9-acridanone (**104**), easily separated as only the
former is sufficiently basic to dissolve in 3*N* acetic acid.

Xanthoxoline is **103**, and its structure, together with the structure of its
N-methyl (**104**) and *N,O*-dimethyl (**105**) derivatives, have been amply confirmed by synthesis. 3′,4′,5′-Trimethoxydiphenylamine-2-carboxylic acid
(**106**), prepared from 3,4,5-trimethoxyiodobenzene and anthranilic acid, was
cyclized by means of phosphorus oxychloride to the 9-chloroacridine derivative **107**. The reaction of **107** with sodium methoxide afforded 1,2,3,9-
tetramethoxyacridine, which, when heated with methyl iodide in a sealed
tube,[41] gave 1,2,3-trimethoxy-10-methyl-9-acridanone (**105**). Alternatively,
the hydrolysis of **107** gave 1,2,3-trimethoxy-9-acridanone, the hydrochloride
of which decomposed at 150–160° to xanthoxoline.[49]

1,2,3-Trimethoxy-10-methyl-9-acridanone (**105**) has also been synthesized
by condensation of *o*-aminobenzaldehyde with 1,2,3,5-tetrahydroxybenzene,
which gave 1,2,3-trihydroxyacridine.[53] Methylation of this product with
diazomethane afforded 1,2,3-trimethoxyacridine whose methosulfate, when
oxidized in alkaline solution with potassium ferricyanide, yielded the compound **105**. Demethylation of **105** with hydrochloric acid in ethanol[53] then
gave 1-hydroxy-2,3-dimethoxy-10-methyl-9-acridanone (**104**).

E. Evoprenine

Evoprenine, $C_{20}H_{21}NO_4$, mp 143°, a minor constituent of the bark of
E. alata, is a yellow, weakly basic alkaloid, which exhibits the spectrographic
and chemical properties of a cryptophenolic member of the *nor* series.[8] The
skeleton of evoprenine and the functional groups present were clearly
revealed by the nmr spectrum. Thus a one-proton signal centered at τ -4.65,
which disappeared on deuterium exchange, was assigned to the proton of the
cryptophenolic hydroxyl group, strongly bonded to the carbonyl group.
Aside from a 3-proton multiplet assigned to protons at C-5, 6, and 7, the
aromatic region exhibited a broadened doublet centered at τ 1.8, owing to
hydrogen at C-8 (deshielded by the carbonyl group) and a singlet at τ 3.95,
owing to a hydrogen atom at C-4. Two 3-proton singlets at τ 6.45 and 6.15
were assigned to methylimino and methoxyl protons, respectively. Of greater
interest were the remaining signals, namely, a broadened one-proton triplet
at τ 4.45, a two-proton doublet at τ 5.35, and a broad six-proton singlet at
τ 8.2, which obviously arose from a γ,γ-dimethylallyl (prenyl) ether function.[8]

Evoprenine is thus a derivative of 1-hydroxy-10-methyl-9-acridanone, and if the normal oxygenation pattern in the trihydroxy-9-acridanone series of alkaloids is observed, evoprenine can only be **108** or **109**. Since the amount of alkaloid available for degradation was small, synthetic experiments offered the best means of establishing its structure. The treatment of evoxanthine (**83**) with methanolic potassium hydroxide gave the phenol **86** (R = Me), which was then demethylated by alcoholic hydrochloric acid to **87** (R = Me). Partial alkylation of this product by means of γ,γ-dimethylallyl bromide and potassium carbonate in acetone readily afforded the ether **108**, but this proved not to be identical with evoprenine.

In the first attempt to synthesize the second isomer (**109**), evoxanthine was treated with γ,γ-dimethylallyl alcohol and potassium hydroxide to yield the phenolic ether (110), which was methylated to 111. Although acid-catalyzed cleavage of 1-methoxyl groups occurs with facility in the 9-acridanone series, conditions for effecting this reaction selectively on 111 could not be found; fission of the sensitive allyl ether function also occurred and 112 was the only isolable product. This substance, which was also obtained by the action of acid on evoprenine, was more conveniently prepared as follows. Evoxanthine was heated with benzyl alcohol and potassium hydroxide to yield the crude benzyl ether (113), which was methylated to 114. This last product was readily demethylated by acid to (88; $R = CH_2C_6H_5$), but under more vigorous conditions afforded 112. Partial alkylation of 112 with γ,γ-dimethylallyl bromide then furnished 109, which was identical with evoprenine. Since evoxanthine has been synthesized, this sequence also constitutes a formal synthesis of evoprenine.[8]

5. Derivatives of 1,3-Dihydroxy-9-acridanone

A. Acronycine and Its Derivatives

Among the acridine alkaloids are five derivatives of 1,3-dihydroxy-9-acridanone, of which the simplest is 1,3-dimethoxy-10-methyl-9-acridanone (115), a constituent of *A. baueri*[4] and *Vepris bilocularis*.[17] The other alkaloids of this group are the important tumor-inhibitor acronycine and three of its demethyl derivatives. So far acronycine has been obtained only from the bright yellow bark of *A. baueri*,[1, 5] the Australian scrub ash or scrub yellow-wood, and its demethyl derivatives, noracronycine, des-10-methylacronycine, and des-10-methylnoracronycine, from the root bark of *Glycosmis penta-phylla*.[29]

Acronycine, $C_{20}H_{19}NO_3$, mp 175–176°, is a weak base readily soluble in most organic solvents. It crystallizes best from ethyl acetate, in which it gives a yellow solution with a bright green fluorescence. Acronycine forms a red hydrochloride, mp 120–130° (dec), sulfate, mp 158–159° (dec) and an orange picrate,[5] mp 150–154°. It contains one *N*-methyl and one methoxyl group, and gives a negative methylenedioxy color reaction. It does not react with acetic anhydride, benzoyl chloride, or ketonic reagents.

Acronycine (116) is attacked by mineral acids in the same way as the other acridanone alkaloids but rather less readily. Hot alcoholic hydrogen chloride gives a polymeric product, but demethylation to noracronycine, $C_{19}H_{17}NO_3$, occurs when dry acronycine hydrochloride is heated.[54] Noracronycine (117) is very similar in properties to the other *nor* alkaloids. It is deeper in color than the parent alkaloid, forms no stable salts, and is not methylated by diazomethane. Methylation is effected by methyl sulfate and potassium carbonate in acetone, which regenerates acronycine, and a monoacetyl derivative is formed with acetic anhydride.

Acronycine contains one reactive double bond that can be reduced by hydrogen in the presence of Raney nickel to give dihydroacronycine (118), $C_{20}H_{21}NO_3$. This forms a hydrochloride that is demethylated to dihydronoracronycine (119), when heated alone or in alcoholic hydrogen chloride solution.[54]

Both noracronycine and dihydronoracronycine on oxidation with warm concentrated nitric acid give 1-methyl-4-quinolone-3-carboxylic acid (8).[39] Acronycine itself under similar conditions yields an insoluble orange trinitro compound. This material on prolonged heating with an excess of concentrated nitric acid slowly dissolves to give a cream-colored crystalline solid, $C_{11}H_8N_3O_5$, mp. 262–263°, which was shown to be 1-methyl-6-nitro-4-quinolone-3-carboxylic acid (120) by comparison with a synthetic specimen.[39]

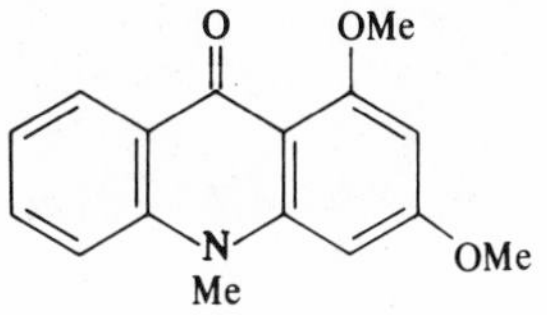

115

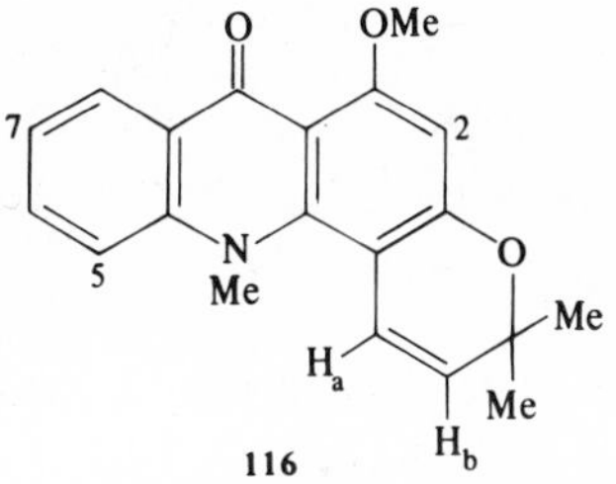

116
Acronycine

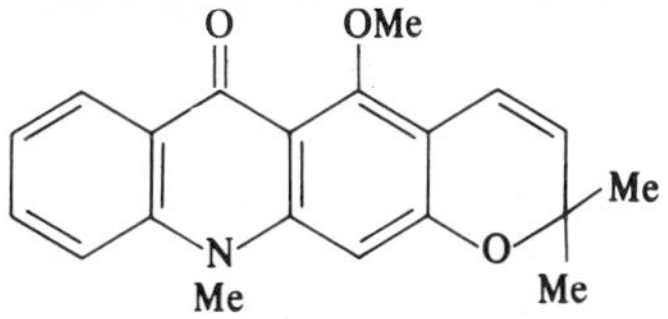

116a

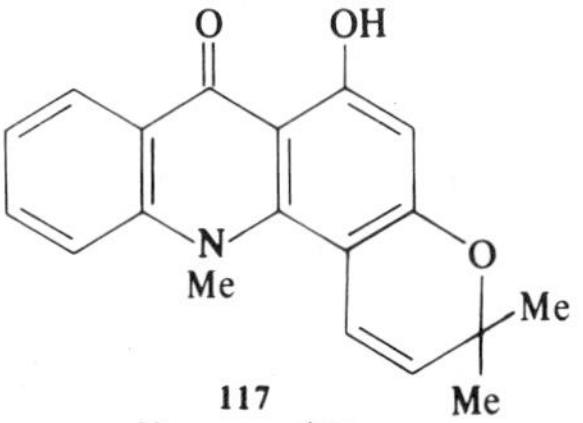

117
Noracronycine

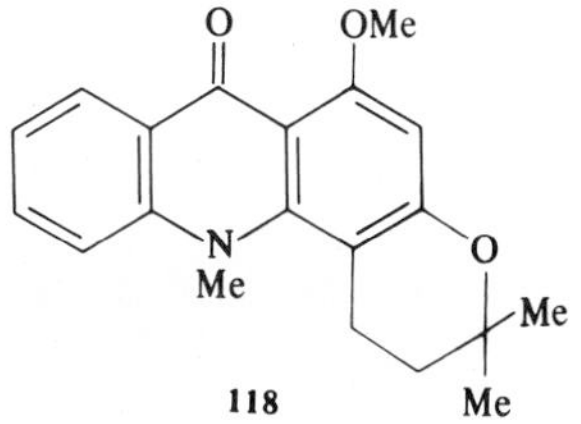

118

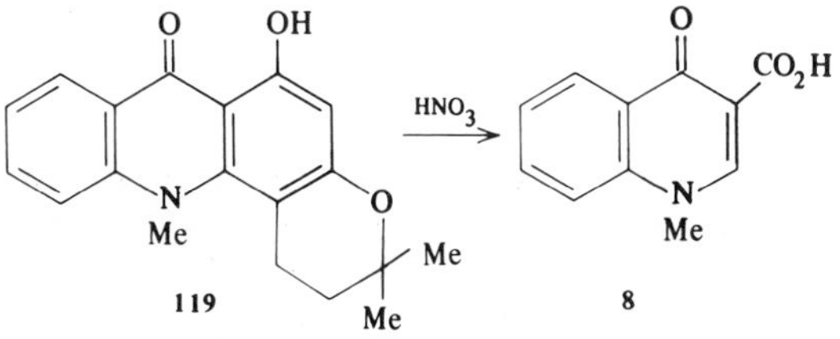

119

8

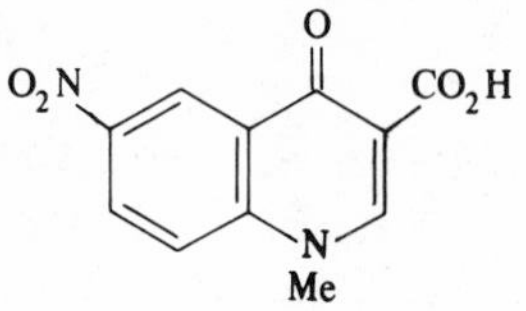

120

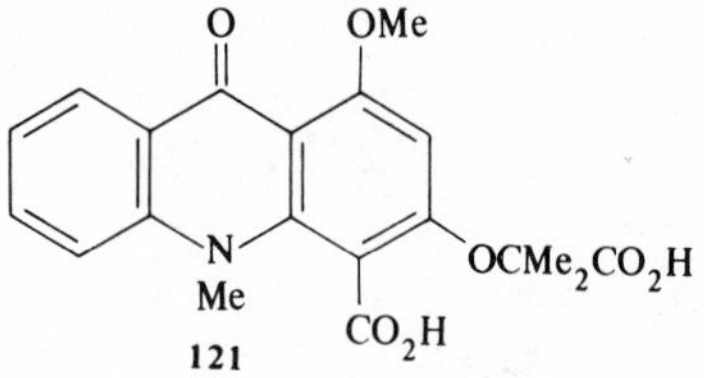

121

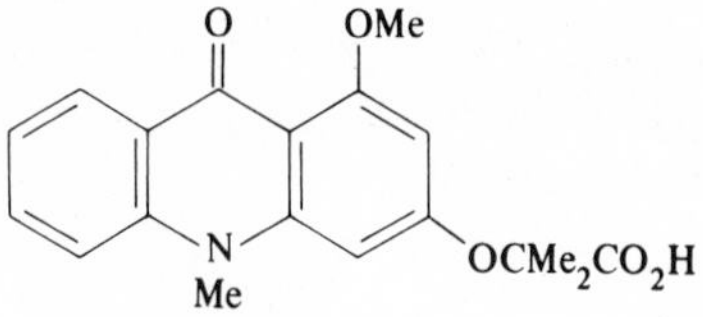

122

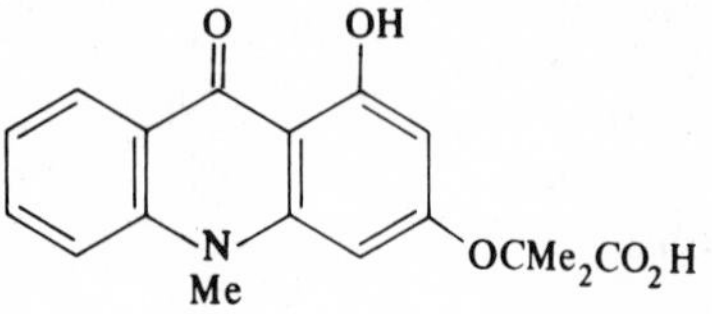

123

HO CMe$_2$CO$_2$H

124

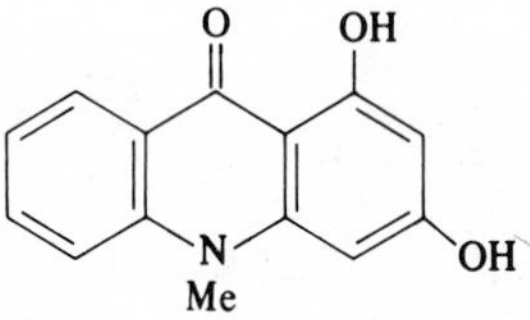

125

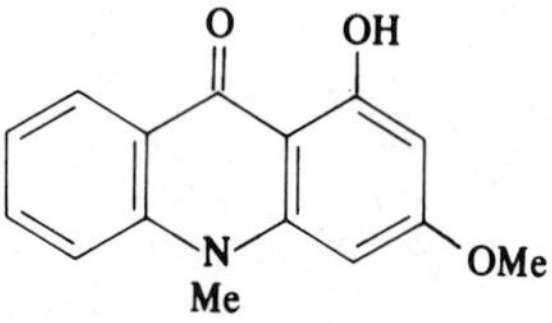

126

The oxidation of acronycine with potassium permanganate in acetone gives a dibasic acid (121), $C_{20}H_{19}NO_7$, without the loss of carbon. The reactive double bond in the alkaloid is therefore contained in a ring. This dibasic acid decarboxylates when crystallized from 10% aqueous hydrochloric acid with the formation of acronycinic acid (122), $C_{19}H_{19}NO_5$, mp 227° (dec). Similar treatment of noracronycine affords noracronycinic acid (123), which, when heated above its melting point, decomposes to α-hydroxyisobutyric acid (124) and a phenol, $C_{14}H_{11}NO_3$ (125). The thermal decomposition of acronycinic acid (122) similarly gives α-hydroxyisobutyric acid, the same phenol $C_{14}H_{11}NO_3$, and its methyl ether, $C_{15}H_{13}NO_3$ (126), which was also prepared from the phenol (125) with diazomethane. The methyl ether contains an unreactive hydroxyl group that is, however, methylated by methyl sulfate.

These reactions can only be explained if acronycine contains a dimethyl-pyran ring. However, no trace of acetone could be detected in the reaction of acronycine (116) with concentrated alkali, as normally happens with di-methylpyran derivatives under these conditions.[55a, b, c] The products of reaction of acronycine with concentrated alkali were two phenols (127 and 128), both of which could be reconverted into acronycine by warm hydro-chloric acid. One of the phenols, $C_{20}H_{19}NO_3$ (127), is isomeric with acro-nycine, while the other, 128, has the composition $C_{20}H_{19}NO_3 \cdot 2H_2O$. The uv absorption spectrum of the former phenol indicates additional conjugation, which is consistent with its formulation.[54]

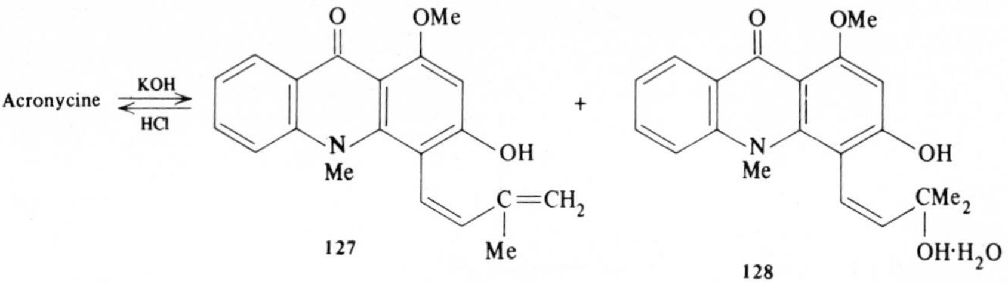

Acronycine $\rightleftarrows^{KOH}_{HCl}$

127   +   128

Acronycine reacts rapidly with bromine in chloroform to give a deep red precipitate of monobromoacronycine hydrobromide.[54] Further reaction of monobromoacronycine (129) with bromine results in demethylation with the formation of the dimorphic dibromonoracronycine, which is also formed when noracronycine is brominated directly.[54] The bromine atom in mono-bromoacronycine must be present in the pyran ring, since oxidation gives acronycinic acid (122). It must also be present on the carbon atom adjacent to the benzenoid ring because on ozonolysis a phenolic acid (130) is formed; in contrast, the ozonolysis of acronycine[39] yields a phenolic aldehyde (131).

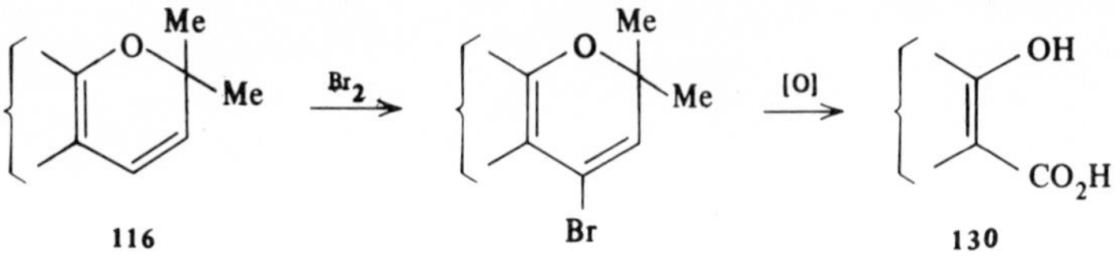

116
partial structure

129

130

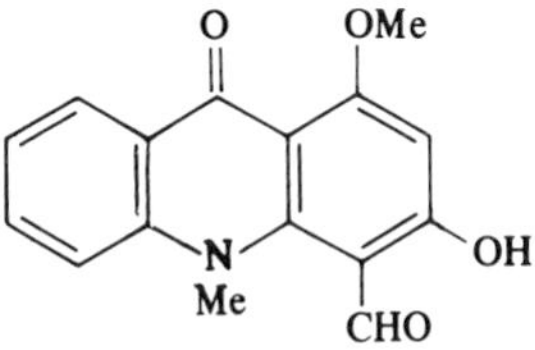

131

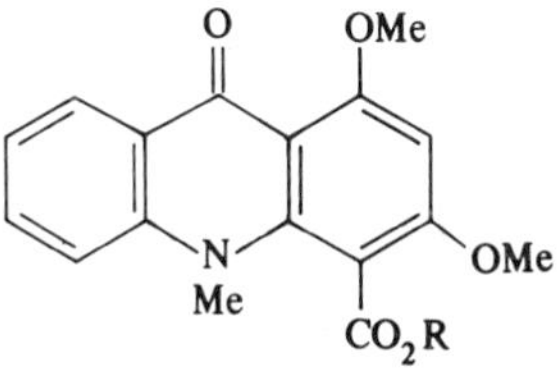

132, R = H
134, R = Me

The nitration of acronycine with concentrated nitric acid in ethanol (1:5) gives a mononitroacronycine in which the nitro group is in the pyran ring, since oxidation affords acronycinic acid. The position of the nitro group is probably the same as that of the bromine in monobromoacronycine.[39]

The degradation of acronycine and noracronycine to 1-methyl-4-quinolone-3-carboxylic acid and its 6-nitro derivative suggests that the phenol $C_{14}H_{11}NO_3$ (125) obtained by degradation of the dimethylpyran ring is almost certainly a dihydroxy-10-methyl-9-acridanone, in which both hydroxyl groups are present in the same benzenoid ring. One of the hydroxyl groups must be in position 1, since the behavior of acronycine and noracronycine indicate that the methoxyl group in the alkaloid must be in this position. This is in agreement with the low reactivity of one of the hydroxyl groups in the dihydric phenol. Since a quinone could not be obtained from this phenol, it was considered to be 1,3-dihydroxy-10-methyl-9-acridanone (125), a proposal that was confirmed by the synthesis of 125 from 3′,5′-dimethoxydiphenylamine-2-carboxylic acid:[39]

Additional confirmation of the positions of the oxygen atoms in acronycine was provided by ozonolysis, which afforded a phenolic aldehyde (131). Methylation of (131), followed by permanganate oxidation, gave the corresponding dimethoxy-10-methyl-9-acridanonecarboxylic acid, mp 195–196° (132), which, when heated to 150° in butyl phthalate, decarboxylated to 1,3-dimethoxy-10-methyl-9-acridanone, identical with a synthetic specimen.[39]

The properties of acronycine and noracronycine clearly indicate that the methoxyl group in the alkaloid is situated in position 1. Consequently, the hydroxyl group in position 3 in the phenol 125 must be derived from the dimethylpyran ring. This requires that acronycine has one of two structures, 116 or 116a; all the evidence cited above is consistent with either of these formulations. The only apparent inconsistency is found in the isolation of the phenol ether (126), but this is presumably formed during the degradation by remethylation of the 1,3-dihydroxy compound 125, or by intermolecular *trans*-methylation of the initially formed 1-methoxy-3-hydroxy-10-methyl-9-acridanone. An alternative, less likely, explanation is that after the elimination of α-hydroxyisobutyric acid, the acridanone ring opens and then recyclizes, yielding the more stable arrangement with the hydroxyl group situated in the *peri* position with respect to the carbonyl group.[39]

The first, tentative, proposal concerning the structure of acronycine, based on unpublished material, favored the linear isomer 116a, and the uv spectra of acronycine and its derivatives were discussed on this basis.[42]

The first chemical proof of the structure of acronycine was provided by Macdonald and Robertson,[56] who synthesized 1,3-dimethoxy-2-methoxy-carbonyl-10-methyl-9-acridanone (133) by the following route, starting from methyl 4-amino-2,6-dimethoxybenzoate:

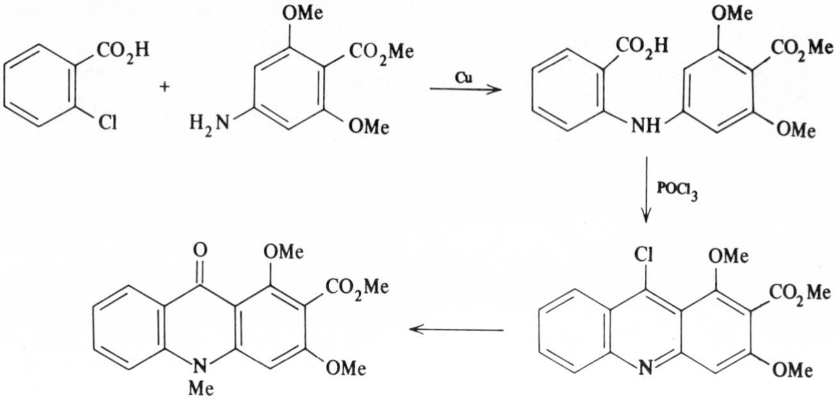

If acronycine had the linear structure 116a, the product of ozonolysis, methylation, oxidation, and esterification would have the structure 133. It was shown, however, that the ester 133 was different from the degradation product of acronycine, which, it must therefore be concluded, has the structure 134; acronycine[56] must then have the structure 116.

Subsequently, Govindachari and his collaborators furnished independent proof of the structure 116 for acronycine.[29] Among the alkaloids of *Glycosmis pentaphylla* are three acridine alkaloids that were shown to be noracronycine (117), des-*N*-methylacronycine, and des-*N*-methylnoracronycine. Desulfurization of the toluene-*p*-sulfonate of noracronycine with Raney nickel in a current of hydrogen afforded a colorless deoxy compound. This, surprisingly, was not the expected product ($C_{19}H_{19}NO_2$) of deoxygenation and saturation of the dimethylpyran double bond but a compound of molecular formula, $C_{19}H_{23}NO_2$, which was clearly obtained from the expected product (**135**, or linear isomer) by saturation of one of the aromatic rings. This was strongly supported by the uv absorption spectrum of the product, no longer characteristic of a 9-acridanone but typical of a 4-quinolone. Which of the two aromatic rings (A or B) had suffered hydrogenation was readily determined from the nmr spectrum of the product, which exhibited only two aromatic protons. Thus there remain only two possible structures for this product, **136** and **137**, depending on the structure (**116** or **116a**) of acronycine. Since the nmr signals resulting from the aromatic protons in the desulfurization product comprised an *ortho*-coupled AB system with $J_{A,B} = 9$ Hz, the two protons must be attached to adjacent carbon atoms, and the only acceptable structure is (**136**). Hence acronycine is (**116**).[29]

Finally, a direct proof of the structure of acronycine,[57] by means of a three-dimensional single crystal X-ray diffraction analysis of bromodihydroacronycine (**138**), has been carried out by Gougoutas and Kaski.*

*It is of some interest to note that a choice between the two possible structures (**116** and **116a**) for acronycine may very simply be made on the basis of the Nuclear Overhauser Effect. Thus irradiation of the *N*-methyl signal at τ 6.3 in the nmr spectrum of acronycine results in a 17% increase in the integrated intensity of the doublet at τ 3.54 (*J* 9.5 Hz) owing to H_a (see **116**). Also, irradiation of the singlet at τ 6.06 (OMe group) results in a 48% selective increase in the integrated intensity of the signal at τ 3.72 owing to the proton at C-2. These observations confirm the proximity of H_a to the *N*-methyl group, and of the methoxyl group to the lone aromatic proton,[58] as required by **116** but not by **116a**. Confirmatory evidence was obtained from an analogous study on noracronycine and a nonlinear model benzacridanone derivative.

The structure (116) of acronycine has been established by three interrelated syntheses, which have been completed in the Lilly Research Laboratories.[2a,b] Cyclization of the β-propionyl derivative (139) of 3,5-dimethoxyaniline by means of zinc chloride and sodium chloride at 155° gave the 3,4-dihydro-carbostyril derivative (140), which was converted into the N-o-carboxyphenyl derivative (141) by reaction with o-iodobenzoic acid in the presence of cuprous iodide. Polyphosphoric acid cyclization of 141 gave a mixture of the tetracyclic 9-acridanone lactam (142) and the corresponding 9-acridanone-carboxylic acid (143), both of which with methanolic hydrogen chloride gave the ester 144. An attempted conversion of this ester into the appropriate tertiary alcohol by reaction with methyl magnesium iodide simply resulted in demethylation of the methoxyl group at position 1; the same product (145) could be obtained by demethylation of 144 with boron trichloride. Fortunately, the ester function in 145 reacted with methyl lithium in tetrahydrofuran at −18° to give the desired tertiary alcohol (146).

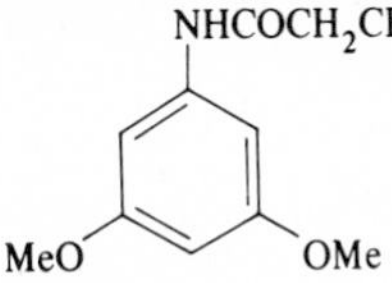

139

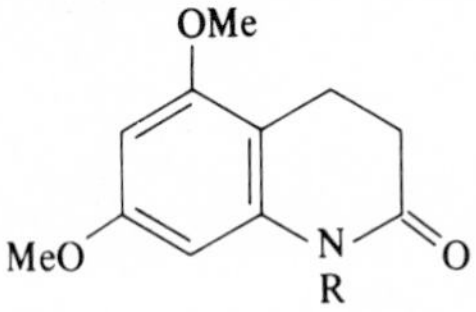

140, R = H

141, R =
HO₂C

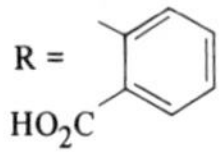

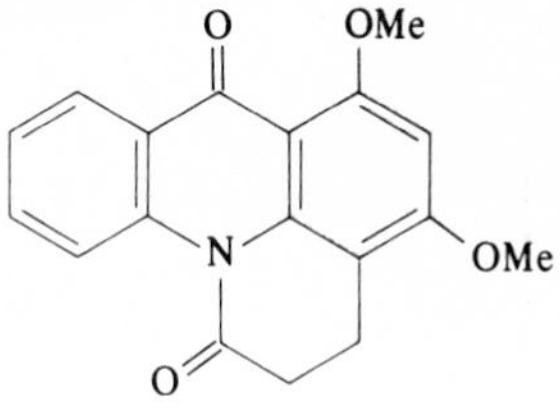

142

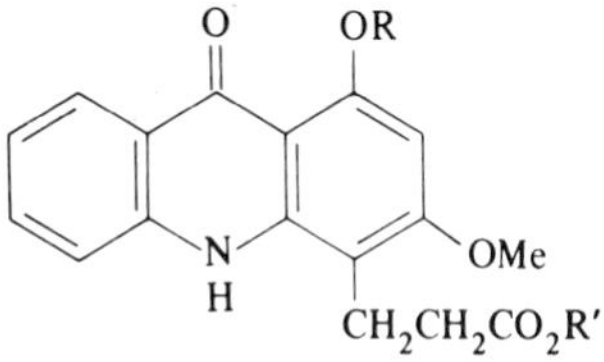

143, R = Me, R′ = H
144, R = R′ = Me
145, R = H, R′ = Me

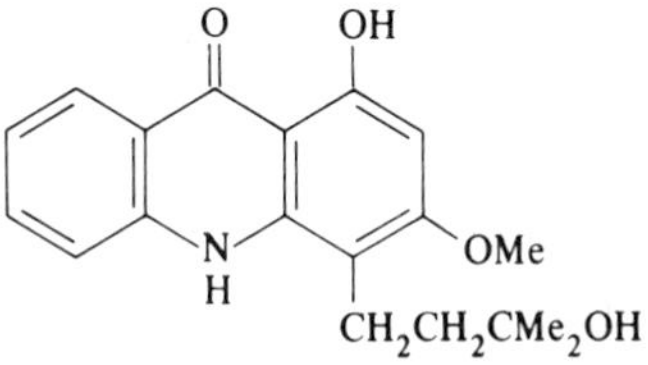

146

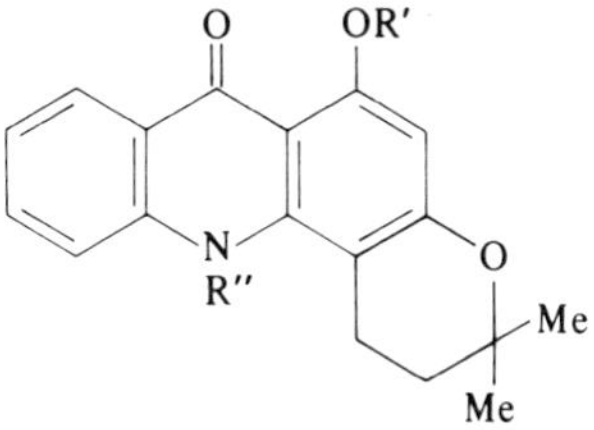

147, R′ = R″ = H
148, R′ = H, R″ = Me
157, R′ = Me, R″ = H
158, R′ = R″ = Me

Fusion of 146 with pyridine hydrochloride at 200° resulted in demethylation and cyclization, with formation of the tetracyclic product (147). Because of its insolubility, 147 was not normally isolated but methylated with methyl iodide and potassium carbonate to give its N-methyl derivative (148), which was identified as dihydronoracronycine. Dehydrogenation of 148 was accomplished by utilizing 2,3-dichloro-5,6-dicyanobenzoquinone in refluxing toluene or dioxan; the noracronycine so produced was methylated to acronycine, identical with authentic material.[2a, b]

An independent synthesis of 147 afforded a second route to acronycine. The condensation of anthranilic acid with 1,3,5-trihydroxybenzene in butanol in the presence of zinc chloride gave 1,3-dihydroxy-9-acridanone (149), which, when alkylated with 1-chloro-3-methyl-2-butene in trifluoroacetic acid with zinc chloride as catalyst, gave the acronycine derivative 147, together with the bischromane derivative 150.[2]

149   150

The third synthesis of acronycine started with the known 7-hydroxy-2,2-dimethyl-4-chromanone (151), which was methylated to 152 and hydrogenated to the chroman derivative 153. Alternatively, 153 could be prepared by the hydrogenation of 7-methoxy-2,2-dimethylchromene (154), obtained by reduction of 152 with lithium aluminum hydride followed by dehydration with phosphorus oxychloride in pyridine. Bromination of 153 gave the 6-bromo derivative 155 (identified by its nmr spectrum); this was converted into the 5-amino derivative 156 (also identified by its nmr spectrum) by treatment with an excess of sodamide in liquid ammonia. Ullmann condensation of 156 with o-bromobenzoic acid gave the corresponding diphenylamine, which was cyclized without isolation to N-desmethyldihydroacronycine (157) by treatment with polyphosphoric acid at 90°. The methylation of 157 by means of methyl iodide and potassium carbonate gave dihydroacronycine (158), which was finally dehydrogenated to acronycine.[2]

A fourth synthesis of acronycine[58] utilizes a method recently developed for the synthesis of 2,2-dimethylchromene derivatives. The alkylation of 1,3-dihydroxy-10-methyl-9-acridanone with 3-chloro-3-methylbut-1-yne in di-

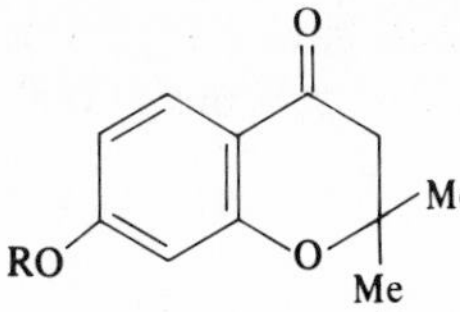

151, R = H
152, R = Me

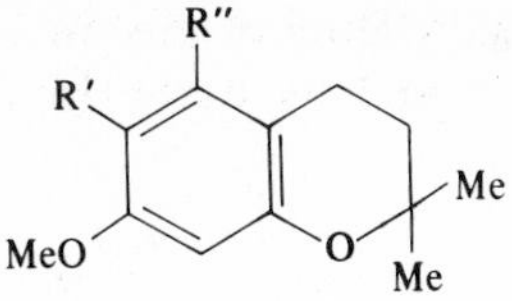

153, R' = R'' = H
155, R' = Br, R'' = H
156, R' = H, R'' = NH$_2$

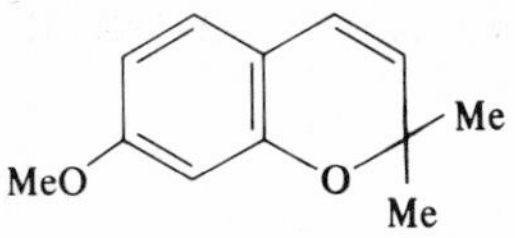

154

methylformamide at 52° in the presence of potassium carbonate gave the acetylenic ether (159), which when heated in N,N-diethylaniline under reflux gave noracronycine (117) in almost quantitative yield. Acronycine was then obtained by the usual methylation procedure.

When the preparation of the ether (159) was carried out at a somewhat higher temperature (70°), the product was found to consist of the ether (159), together with substantial amounts of noracronycine. Hence it would appear that the Claisen rearrangement of the ether (159) occurs quite rapidly, even at 70°.

Hydrogenation of the acetylenic ether (159) in the presence of the Lindlar catalyst yielded the α,α-dimethylallyl ether (160), which rearranged in boiling diethylaniline to give the normal Claisen product (161). Noracronycine (117) was also obtained from this product by cyclodehydrogenation with chloranil in boiling dioxan.[58]

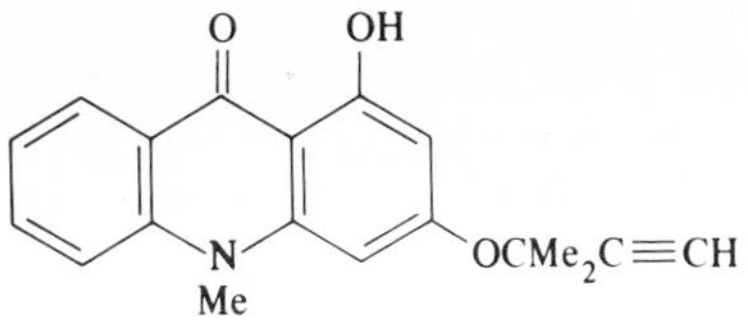

159

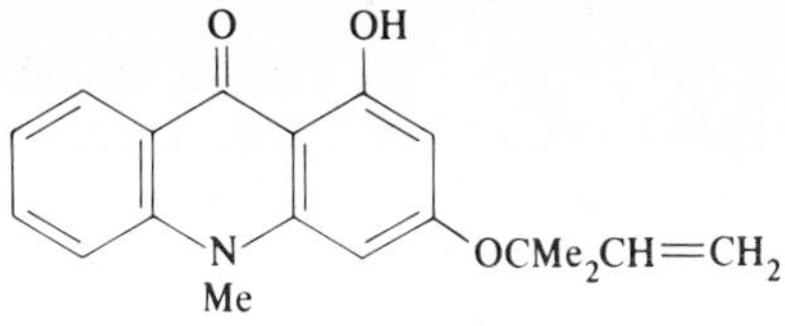

160

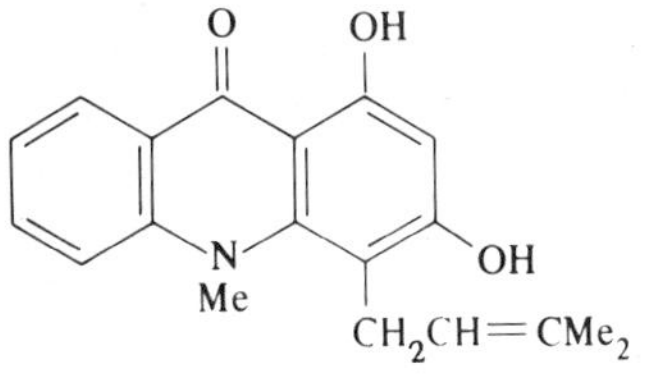

161

Almost simultaneously, still another direct synthesis of acronycine was reported.[59a] The condensation of 1,3-dihydroxy-9-acridanone (149) with 3-hydroxyisovaleraldehyde dimethyl acetal (161a) in the presence of pyridine at 150° afforded a mixture of des-10-methylnoracronycine (161b) and its linear isomer (**161c**). The methylation of this mixture then gave a separable mixture of acronycine (**116**) and its linear isomer isoacronycine (**116a**).

149   161a

Pyridine/150°/8 hrs.

161b   161c

Isoacronycine has also been synthesized[59b] from 1,3-dihydroxy-10-methyl-9-acridanone by a Pechmann condensation with malic acid to give the linear 5-hydroxy-11-methyl-2*H*-pyrano[2,3-*b*]acridan-2,6-dione, which with excess methyl lithium, followed by methylation, gave isoacronycine (**116a**).

Finally, mention should be made of the synthesis of hexahydrodesoxynoracronycine (**162**), the product of reduction of noracronycine tosylate by hydrogen in the presence of Raney nickel.[60] 3-Methoxy-9-acridanone was demethylated with hydriodic acid and acetic anhydride to the hitherto unknown 3-hydroxy-9-acridanone (**163**), which was condensed with 3,3-dimethylallyl diphenyl phosphate at 130–140°. Although a mixture of products was obtained, only one compound could be isolated, which was shown by analysis of its nmr spectrum to be the angular chroman derivative **164**. This conclusion was confirmed by *N*-methylation of **164**, which afforded desoxydihydronoracronycine (**165**). The hydrogenation of **165** then gave hexahydrodesoxynoracronycine (**162**), identical with the degradation product of noracronycine.[60]

In view of the reported antitumor properties of acronycine (acronine), its metabolism in mammals is of considerable interest. Preliminary experiments[61] have shown that in five species (man, dog, guinea pig, mouse, and rat), the major oxidation route appears to involve hydroxylation at positions 5 and 7 of the acridine ring, indicating a similarity of the enzyme systems involved in these animals. In four of the above species (the exception being the guinea pig), hydroxylation of one of the methyl groups in the dimethylpyran ring was also observed. In general, demethylation appears not to be an important route of biotransformation, although it was observed in the guinea pig and, to a lesser extent, in the mouse.

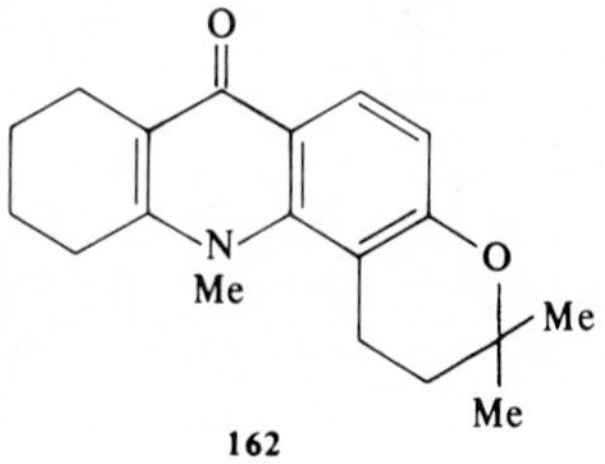

162

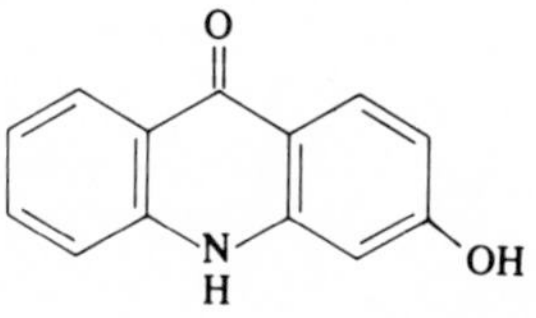

163

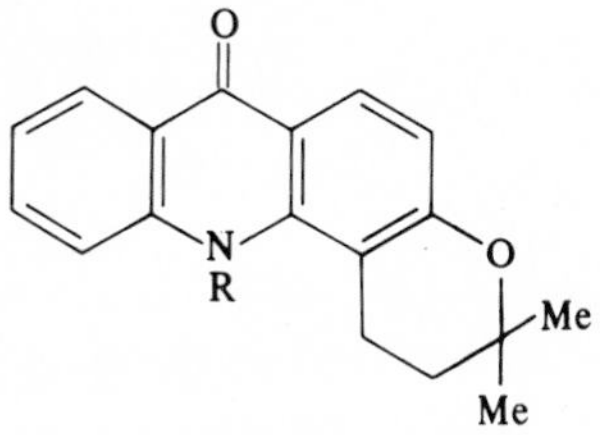

164, R = H
165, R = Me

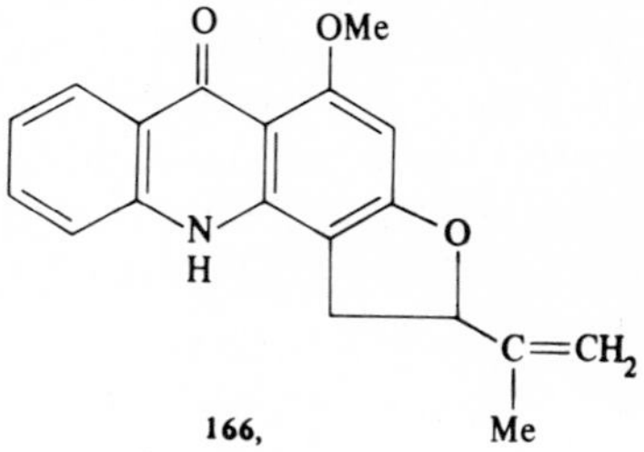

166,
Rutacridone ?

B. Rutacridone

Rutacridone, $C_{19}H_{17}NO_3$, mp 161–162, $[\alpha]_D^{21°} - 43°$, a constituent of *Ruta graveolens*, is the first presumed acridine alkaloid to exhibit optical activity. Its structure is not yet known with certainty, and it has not even been rigorously proved to belong to the acridine series. On the basis of its ir and nmr spectra, however, the structure **166** is tentatively proposed for rutacridone, although the linear isomer cannot at present be excluded from consideration.[31]

6. Derivatives of 1,2,3,5-Tetrahydroxy-9-acridanone, Tecleanthine

The only naturally occurring derivative of 1,2,3,5-tetrahydroxy-9-acridanone hitherto encountered is tecleanthine (**167**), $C_{17}H_{15}NO_5$, mp 158°, which occurs in the bark of *Teclea natalensis* (Song.) Engl.[18] Tecleanthine is a weak base that yields a picrate and a picrolonate, and gives a positive test for a methylenedioxy group. The uv spectrum and mass spectral fragmentation pattern resemble those of evoxanthine, suggesting that tecleanthine is also a derivative of 9-acridanone.

The nmr spectrum of tecleanthine discloses the presence of two methoxyl groups (3H singlets at τ 5.88 and 6.20), an N-methyl group (3H singlet at τ 6.10) and a methylenedioxy group (2H singlet at τ 4.01). Of the four aromatic protons, three are present in an ABX system, in which the proton at lowest field (C-8H) gives rise to a quartet at τ 2.00 (J 6.6 and 3.3 Hz), its downfield position being the consequence of deshielding by the carbonyl group; the other two protons give rise to a multiplet at τ 2.81–2.94. The fourth proton is responsible for a singlet at τ 3.40 and is presumably an isolated proton in a highly substituted aromatic ring.

The ABX pattern of the protons in ring A indicates that the methoxyl group in this ring is situated at C-5. The position of the second methoxyl group becomes apparent from the ready hydrolysis of tecleanthine with ethanolic hydrochloric acid, which affords nortecleanthine (168), a crypto-phenolic substance insoluble in potassium hydroxide solution. Hence the second methoxyl group is situated at position 1, and the methylenedioxy group must occupy positions 2 and 3, or 3 and 4. Unfortunately, the position of this group cannot be deduced from the chemical shift of the signal from the lone hydrogen in ring B. The assignment of the methylenedioxy group was finally made after comparison of the uv spectra of tecleanthine, evoxanthine (83), melicopidine (22), and melicopine (23). The first three alkaloids exhibit a bathochromic shift of 60–73 nm in the 300 nm region in acid solution; whereas melicopine, containing a 3,4-methylenedioxy group, shows no shift

167, Tecleanthine

168

169a, R = OH, Balfouridine
169b, R = H, Lunacrine

170

under these conditions. Further, melicopine is the only one of these four alkaloids not to give a deep green color with alcoholic ferric chloride. Therefore, it is concluded that tecleanthine is 1,5-dimethoxy-2,3-methylenedioxy-10-methyl-9-acridanone (167), and it is the first naturally occurring acridine derivative oxygenated in ring A to be encountered.[18] Several quinolone alkaloids containing the same methoxy-*N*-methylquinolone part structure as tecleanthine are known, and it is of interest to note that they exhibit nmr spectra very similar to that of tecleanthine in the aromatic proton region[18]; these include balfouridine (169a), lunacrine (169b), and the pyranoquinolone (170).

7. Derivatives of 1,3,5-Trihydroxy-9-acridanone Atalaphylline and *N*-Methylatalaphylline

Two further derivatives of 9-acridanone oxygenated at position 5 have recently been isolated from *Atalantia monophylla* Correa (fam. Rutaceae), a thorny tree that has been claimed to be useful in the treatment of snakebite and rheumatism. From the root bark of plants collected in Madras, two new alkaloids, atalaphylline and its *N*-methyl derivative, have been isolated.[33]

Atalaphylline (171), $C_{23}H_{25}NO_4$, is a 9-acridanone derivative (uv and ir spectra) that contains at least one free phenolic hydroxyl group (ferric chloride reaction). It is devoid of methoxyl, methylenedioxy, or *N*-methyl groups (nmr spectrum) but contains two dimethylallyl (prenyl) units and three aromatic protons that give rise to an ABX pattern of signals very similar to those exhibited by lunacrine (169b) and tecleanthine (167). The presence of the prenyl units is confirmed by the mass spectrum, which contains ions due to the loss of one and two prenyl units, and by the nmr spectrum of tetrahydroatalaphylline which discloses the presence of four secondary C-methyl groups.

The methylation of atalaphylline by means of diazomethane affords a dimethyl ether (172), which still contains one phenolic hydroxyl group (ferric chloride reaction). Methylation with methyl iodide and potassium carbonate under forcing conditions yields a tetramethyl derivative (173), methylation of the three hydroxyl groups being accompanied by *N*-methylation.

The reaction of atalaphylline with formic acid results in cyclization of both prenyl groups with adjacent hydroxyl groups to give bicycloatalaphylline (174), whose nmr spectrum confirms the presence of two dimethylchroman rings in the molecule. One of the hydroxyl groups involved in this double cyclization must be the cryptophenolic group at position 1, since bicyclo-atalaphylline is readily methylated by diazomethane to an *O*-methyl ether

(175), whose nmr spectrum is completely in accord with its formulation as a 5-methoxy-1,2,3,4-tetrasubstituted-9-acridanone derivative.[33]

It is clear from the above evidence that atalaphylline contains a hydroxyl group at C-5 as in tecleanthine, and that one of the remaining hydroxyl groups must be at position 1 to account for the cryptophenolic properties. The cyclization to bicycloatalaphylline necessarily means that one prenyl group is situated at position 2, so that the remaining hydroxyl and prenyl groups must be situated at positions 3 and 4, not necessarily respectively. Thus there remain two possible structures for atalaphylline, but since all known acridanone alkaloids are oxygenated at positions 1 and 3, in consequence of their presumed biosynthesis from acetate via phloroglucinol or its biochemical equivalent, atalaphylline almost certainly has the structure (171).

The minor alkaloid from *Atalantia monophylla* exhibits spectra closely similar to those of atalaphylline, except that its nmr spectrum contains an additional 3-proton singlet, resulting from an *N*-methyl group. Its identity as *N*-methylatalaphylline (176) was established by vigorous methylation, which yielded a trimethyl ether identical with the exhaustive methylation product (173) derived from atalaphylline.[33]

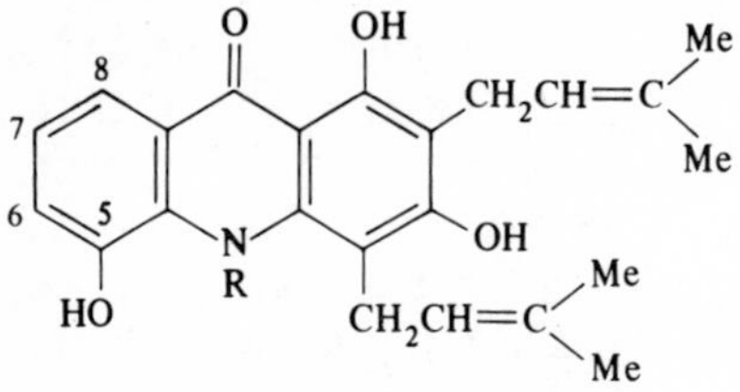

171, R = H; Atalaphylline
176, R = Me; N−Methylatalaphylline

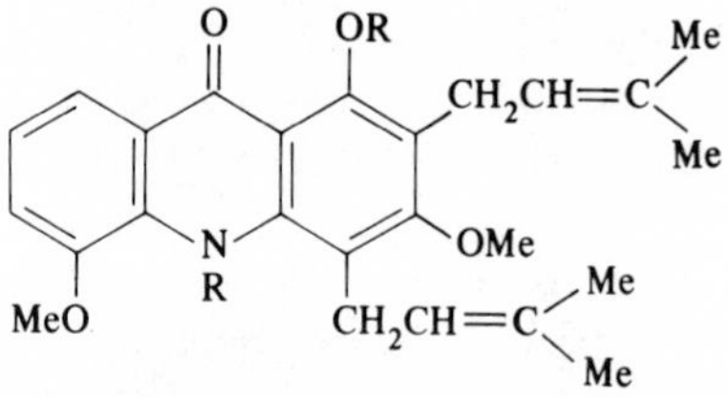

172,   R = H
173,   R = Me

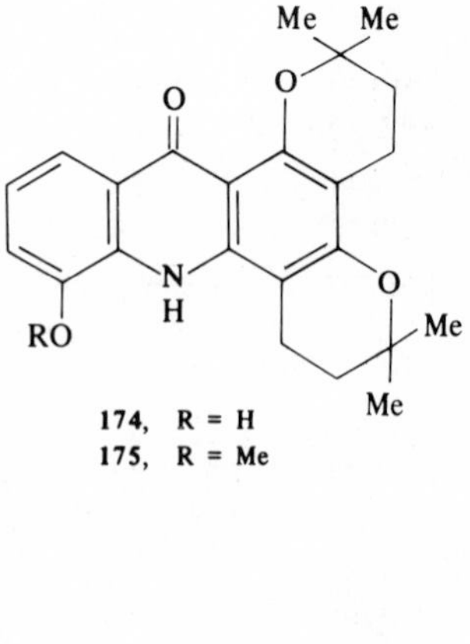

174,   R = H
175,   R = Me

8. Biosynthesis of the Alkaloids

Various authors have speculated on the possible mode of biosynthesis of the acridine alkaloids. Hughes and Ritchie pointed out that all the acridine alkaloids hitherto isolated contained one unsubstituted benzene ring and at least two substituents invariably present in positions 1 and 3 of the other benzene ring.[53] This prompted them to suggest that the alkaloids may be formed by the reaction of *o*-aminobenzaldehyde or its biochemical equivalent with phloroglucinol (or a derivative) to give 1,3-dihydroxyacridine (**177**), which is then oxidized to the appropriate polyhydroxy-9-acridanone and methylated or methylenated to give the alkaloids:

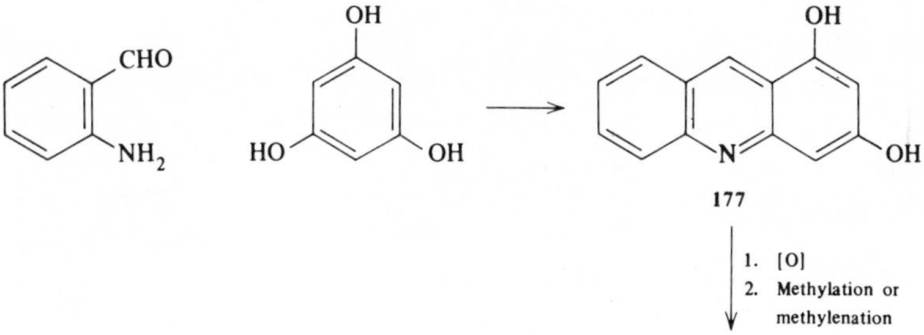

Although this route has its laboratory analogy, e.g., the synthesis of 1,3-dimethoxy-10-methyl-9-acridanone (**115**) via 1,3-dihydroxyacridine (**177**), there is no evidence that a similar route operates in the plant.[53]

Robinson postulated the condensation of anthranilic acid with 3,5-dioxo-hexanoic acid (from acetate) to form 1,3-dihydroxy-9-acridanone (**149**), which is then suitably modified.[62] An alternative possibility is the intervention of the keto acid (**178**), a known oxidation product of tryptophan,[62] which could then condense with acetate to form (**149**). The subsequent stages leading to the alkaloids are then unexceptional.

Leete's suggestion is based on the condensation of anthranilic acid with successive molecules of acetic acid, and it is naturally assumed that the isopentenyl unit in such molecules as acronycine (**116**) is derived from mevalonic acid.[63]

All these proposals require nuclear oxidation of the intermediate acridine or acridanone at some convenient point in the biosynthesis of the derivatives of tri- and tetra-hydroxy-9-acridanone (i.e., the evoxanthine and melicopine series).

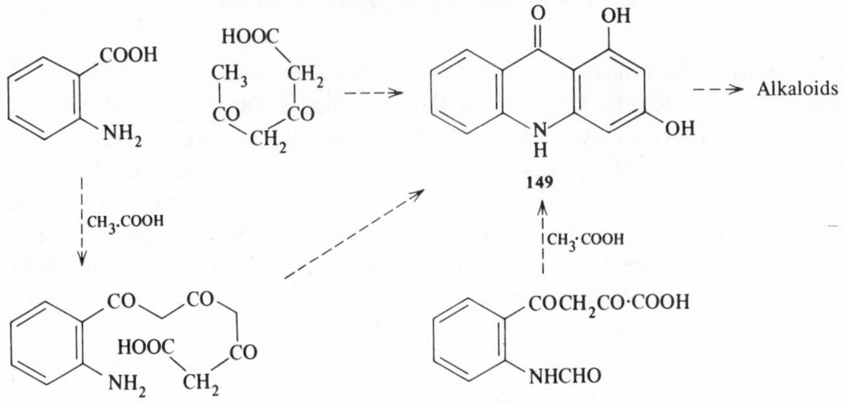

149

178

The first experimental evidence relating to the biosynthesis of the acridine alkaloids was provided by Gröger and Johne,[64] who administered non-specifically tritiated anthranilic acid to *Glycosmis arborea* and isolated radioactive arborinine (**104**). Oxidation of the arborinine with nitric acid gave 1-methyl-4-quinolone-3-carboxylic acid (**8**), which had the same specific radioactivity as the parent alkaloid. Hence it may be concluded that anthranilic acid is a precursor of ring A of arborinine,[64] and presumably also of the other acridine alkaloids.

This conclusion has been confirmed by Prager and Thredgold,[65] who examined the rôle of 5T-anthranilic acid (COOH = 1) in the biosynthesis of alkaloids in *Acronychia baueri*. Extraction of the plant 8 days after initial administration of 5T-anthranilic acid yielded radioactive melicopine, melicopidine, and melicopicine. Bromination of the tritiated melicopicine afforded inactive 7-bromonormelicopicine (**179**), showing that ring A and not ring B was derived from anthranilic acid. In connection with Robinson's proposal that tryptophan might be implicated in acridine alkaloid biosynthesis, via the oxidation product (**178**), it is of interest to note that 3-^{14}C-tryptophan was not incorporated into the alkaloids of *A. baueri* under the conditions studied.[65] Hence, if tryptophan is a precursor, it must first be degraded (to anthranilic acid?), rather than incorporated via the acid (**178**).

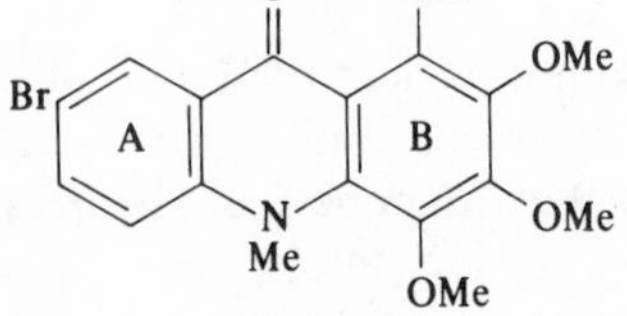

179

In further experiments the incorporation of 4-hydroxy-2-quinolone and its *N*-methyl derivative was observed; the greater incorporation of the *N*-methyl compound suggests that it is closer to the end of the biosynthetic pathway.

In an independent experiment to ascertain at which stage *N*-methylation was occurring, 5-T-*N*-methylanthranilic acid was administered to *E. xanth-oxyloides* and radioactive evoxanthine was isolated; the incorporation observed was comparable to that observed with other precursors in *A. baueri*. Unfortunately, the incorporation of *N*-methylanthranilic acid in *A. baueri* has not yet been investigated; however, there is no reason to believe that the biosynthetic pathways to the acridine alkaloids differ appreciably in these two plants. In connection with alkaloid biosynthesis in *A. baueri*, it is relevant to note that tritiated *N*-methylanthranilic acid was isolated after administration of labeled anthranilic acid. It is suggested, therefore, that the main pathway for acridine alkaloid biosynthesis is via *N*-methylanthranilic acid, with a minor pathway involving *N*-methylation at a later stage:

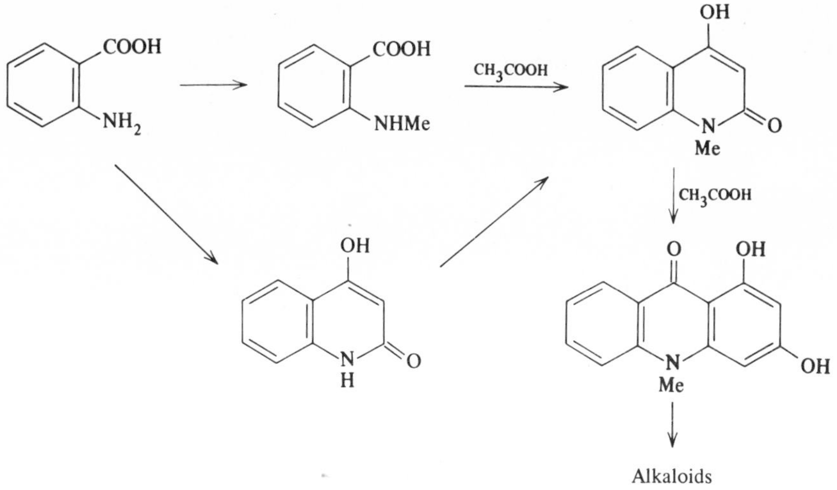

The participation of anthranilic acid in the biosynthesis of the acridine alkaloids thus seems well established, although there is still room for speculation concerning the detailed pathway by which it is incorporated into the alkaloids. Very recently, Lewis and his collaborators,[66] noting the similarity in the substitution pattern between the naturally occurring xanthones and the acridine alkaloids, have suggested that analogous biogenetic pathways may be involved. This receives vicarious support from the *in vitro* conversion of hydroxylated aminobenzophenones into hydroxy-9-acridanones. Thus oxidative cyclization of 2-amino-3'-hydroxybenzophenone (**180**) by manganese triacetate or potassium persulfate gave low yields of the hydroxyacridanones **181** and **182**, while the tin and hydrochloric acid reduction of 2-nitro-2′,4′,6′-trihydroxybenzophenone (**183**) gave a quantitative yield of the expected dihydroxyacridanone **149**, presumably via the related aminotrihydroxybenzophenone, **184**.

It is therefore suggested that the condensation of anthranilic acid with polyacetate to give a dihydroxyacridanone (e.g., **149**) proceeds via an aminotrihydroxybenzophenone (**184**); alternatively, condensation of anthranilic acid with trihydroxybenzoic acid (both derived from shikimic acid) may give rise to an isomeric aminotrihydroxybenzophenone (**185**), which is then cyclized and suitably modified to give the alkaloids. The co-occurrence of 1,3-dioxygenated and 1,2,3,4-tetraoxygenated 9-acridanones in *Acronychia* species suggests that the former route, with or without hydroxylation stages, is the more probable.[1, 5, 67]

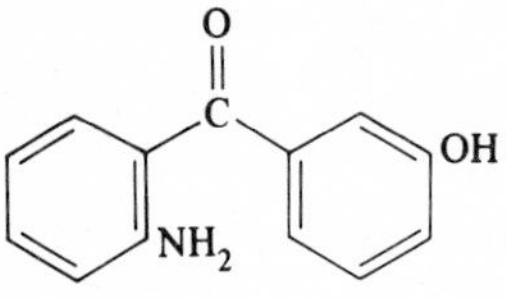

180

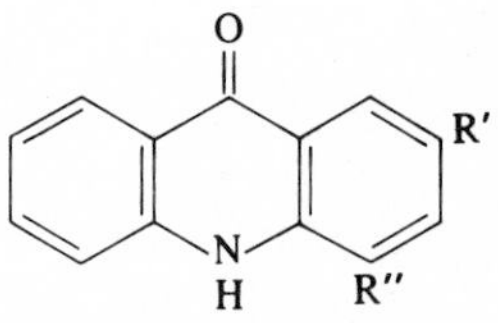

181, R′ = OH, R″ = H
182, R′ = H, R″ = OH

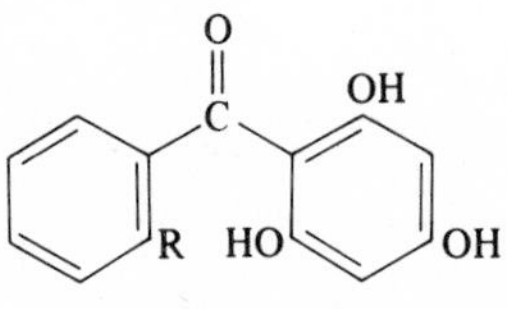

183, R = NO$_2$
184, R = NH$_2$

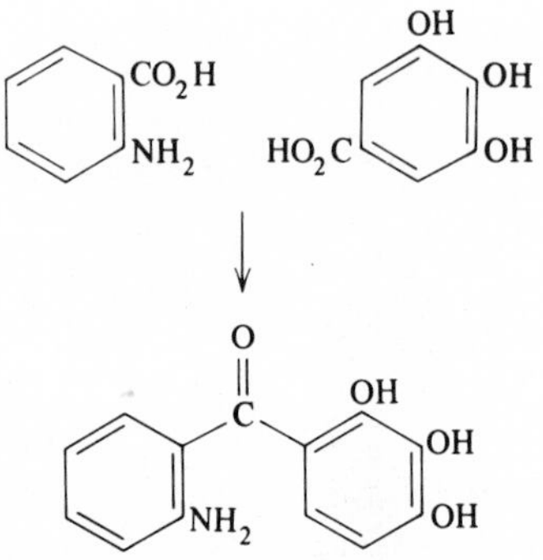

185

All the routes proposed involve acetic acid, so that it is not surprising to find that sodium 1-^{14}C-acetate is incorporated into melicopicine in *A. baueri*. However, the mode of incorporation has not yet been rigorously established, since degradation of the melicopicine by published methods proved not to be practicable on a very small scale.[65]

Finally, it has been shown that tritiated 1,2,3-trihydroxy-10-methyl-9-acridanone is not incorporated into *E. xanthoxyloides* alkaloids,[65] either by wick feeding of the seedlings or by absorption through the roots. However, the implications of this observation are not clear at present.

References

1. G. K. Hughes, F. N. Lahey, J. R. Price, and L. J. Webb, *Nature*, **162**, 223 (1948).
2. (a) J. R. Beck, R. N. Booher, A. C. Brown, R. Kwok, and A. Pohland, *J. Amer. Chem. Soc.*, **89**, 3934 (1967); (b) J. R. Beck et al. *J. Amer. Chem. Soc.*, **90**, 4706 (1968).
3. J. R. Price, *Aust. J. Sci. Res.*, A2, 249 (1949).
4. J. A. Lamberton and J. R. Price, *Aust. J. Chem.*, **6**, 66 (1953).
5. F. N. Lahey and W. C. Thomas, *Aust. J. Sci. Res.*, A2, 423 (1949).
6. G. K. Hughes and K. G. Neill, *Aust. J. Sci. Res.*, A2, 429 (1949).
7. R. Gell, G. K. Hughes, and E. Ritchie, *Aust. J. Chem.*, **8**, 114 (1955).
8. J. A. Diment, E. Ritchie, and W. C. Taylor, *Aust. J. Chem.*, **20**, 1719 (1967).
9. F. N. Lahey and J. A. Lamberton, cited by J. R. Price, *The Alkaloids*, Vol. 2, Academic Press, New York, 1952, p. 355.
10. G. H. Svoboda, *Lloydia*, **29**, 206 (1966).
11. G. H. Svoboda, G. A. Poore, P. J. Simpson, and G. B. Boder, *J. Pharm. Sci.*, **55**, 758 (1966).
12. S. R. Johns and J. A. Lamberton, *Aust. J. Chem.*, **19**, 895 (1966).
13. C. Kan-Fan, B. C. Das, P. Boiteau, and P. Potier, *Phytochemistry*, **9**, 1283 (1970).
14. G. K. Hughes, K. G. Neill, and E. Ritchie, *Aust. J. Sci. Res.*, A5, 401 (1952).
15. R. R. Paris and A. S. Stambouli, *C. R. Acad. Sci., Paris*, **247**, 2421 (1958).
16. H. Rapoport and K. G. Holden, *J. Amer. Chem. Soc.*, **82**, 4395 (1960).
17. A. K. Ganguly, T. R. Govindachari, A. Manmade, and P. A. Mohamed, *Indian J. Chem.*, **4**, 334 (1966).
18. K. H. Pegel and W. G. Wright, *J. Chem. Soc., C*, 2327 (1969).
19. F. D. Popp and D. P. Chakraborty, *J. Pharm. Sci.*, **53**, 968 (1964).
20. J. A. Lamberton, *Aust. J. Chem.*, **19**, 1995 (1966).
21. L. Fonzes and F. Winternitz, *C. R. Acad. Sci., Paris, Ser. C*, **266**, 930 (1968).
22. F. N. Lahey, M. McCamish, and T. McEwan, *Aust. J. Chem.*, **22**, 447 (1969).
23. D. Chakravarti, R. N. Chakravarti, and S C Chakravarti, *J. Chem. Soc.*, 3337 (1953).
24. S. C. Pakrashi, S. K. Roy, L. F. Johnson, T. George, and C. Djerassi, *Chem. Ind.* (London), 464 (1961).
25. S. K. Banerjee, D. Chakravarti, R. N. Chakravarti, H. M. Fales, and D. L. Klayman, *Tetrahedron*, **16**, 251 (1961).
26. T. R. Govindachari et al., *Indian J. Chem.*, **7**, 308 (1969).
27. K. C. Das and P. K. Bose, *Trans. Bose Res. Inst.*, **26**, 129 (1963), quoted by B. D. Paul and P. K. Bose, *Indian J. Chem.*, **7**, 678 (1969); *Chem. Abstr.*, **64**, 9777 (1966).
28. I. Novak, G. Buzas, E. Minker, M. Koltai, and K. Szendrei, *Pharmazie*, **20**, 655 (1965); *Acta Pharm. Hung.*, **37**, 131 (1967); *Chem. Abstr.*, **67**, 25374 (1967).
29. T. R. Govindachari, B. R. Pai, and P. S. Subramanian, *Tetrahedron*, **22**, 3245 (1966).
30. J. Reisch, K. Szendrei, I. Novák, E. Minker, and Z. Rózsa, *Experientia*, **27**, 1005 (1971).
31. J. Reisch, K. Szendrei, E. Minker, and I. Novak, *Acta Pharm. Suecica*, **4**, 265 (1967); *Chem. Abstr.*, **68**, 39861 (1968).
32. D. L. Dreyer, *Tetrahedron*, **22**, 2923 (1966).

33. T. R. Govindachari, N. Viswanathan, B. R. Pai, V. N. Ramachandran, and P. S. Subramanian, *Tetrahedron*, **26**, 2905 (1970).

34. F. W. Eastwood, G. K. Hughes, E. Ritchie, and R. M. Curtis, *Aust. J. Chem.*, **8**, 552 (1955).

35. (a) H. D. Law, I. T. Millar, H. D. Springall, and A. J. Birch, *Proc. Chem. Soc.*, 198 (1958); (b) H. D. Law, I. T. Millar, and H. D. Springall, *J. Chem. Soc.*, 279 (1961).

36. R. O. Studer and W. Lergier, *Helv. Chim. Acta*, **48**, 460 (1965).

37. W. D. Crow and J. R. Price, *Aust. J. Sci. Res.*, **A2**, 255 (1949).

38. J. R. Price, *Aust. J. Sci. Res.*, **A2**, 272 (1949).

39. L. J. Drummond and F. N. Lahey, *Aust. J. Sci. Res.*, **A2**, 630 (1949).

40. W. D. Crow and J. R. Price, *Aust. J. Sci. Res.*, **A2**, 282 (1949).

41. G. K. Hughes, K. G. Neill, and E. Ritchie, *Aust. J. Sci. Res.*, **A3**, 497 (1950).

42. R. D. Brown and F. N. Lahey, *Aust. J. Sci. Res.*, **A3**, 593 (1950).

43. W. D. Crow, *Aust. J. Sci. Res.*, **A2**, 264 (1949).

44. R. Robinson and G. M. Robinson, *J. Chem. Soc.*, 929 (1917).

45. R. H. Prager and H. M. Thredgold, *Tetrahedron Lett.*, 4909 (1966); *Aust. J. Chem.*, **21**, 229 (1968).

46. R. H. Prager and H. M. Thredgold, *Aust. J. Chem.*, **22**, 1503 (1969).

47. R. H. Prager and H. M. Thredgold, *Aust. J. Chem.*, **22**, 1477 (1969).

48. F. Dallacker and G. Adolphen, *Tetrahedron Lett.*, 2023 (1965); *Justus Liebigs Ann. Chem.*, **691**, 138 (1966).

49. J. R. Cannon, G. K. Hughes, K. G. Neill, and E. Ritchie, *Aust. J. Sci. Res.*, **A5**, 406 (1952).

50. R. H. Prager and H. M. Thredgold, *Aust. J. Chem.*, **22**, 1493 (1969).

51. F. Dallacker and G. Adolphen, *Justus Liebigs Ann. Chem.*, **691**, 134 (1966).

52. T. R. Govindachari, B. R. Pai, P. S. Subramanian, and V. Subramanyam, *Tetrahedron*, **23**, 1827 (1967).

53. G. K. Hughes and E. Ritchie, *Aust. J. Sci. Res.*, **A4**, 423 (1951).

54. R. D. Brown, L. J. Drummond, F. N. Lahey, and W. C. Thomas, *Aust. J. Sci. Res.*, **A2**, 622 (1949).

55. (a) W. Bridge, R. G. Heyes, and A. Robertson, *J. Chem. Soc.*, 279 (1937); (b) A. McGookin, F. P. Reed, and A. Robertson, *J. Chem. Soc.*, 748 (1937); (c) S. W. George and A. Robertson, *J. Chem. Soc.*, 1535 (1937).

56. P. L. Macdonald and A. V. Robertson, *Aust. J. Chem.*, **19**, 275 (1966).

57. J. Z. Gougoutas and B. A. Kaski, *Acta Crystallogr.*, **26B**, 853 (1970).

58. J. Hlubucek, E. Ritchie, and W. C. Taylor, *Chem. Ind.* (London), 1809 (1969); *Aust. J. Chem.*, **23**, 1881 (1970).

59. (a) W. M. Bandaranayake, L. Crombie, and D. A. Whiting, *Chem. Commun.*, 970 (1969); (b) S. O. Chan and C. V. Greco, *J. Heterocycl. Chem.*, **7**, 261 (1970).

60. T. R. Govindachari, B. R. Pai, and V. N. Ramachandran, *Indian J. Chem.*, **6**, 179 (1968).

61. H. R. Sullivan, R. E. Billings, J. L. Occolowitz, H. E. Boaz, F. J. Marshall, and R. E. McMahon, *J. Med. Chem.*, **13**, 904 (1970).

62. Sir R. Robinson, *The Structural Relations of Natural Products*, Clarendon Press, Oxford, England, 1955, p. 94.

63. E. Leete, in Biogenesis of Natural Compounds, P. Bernfeld, ed., Pergamon Press, Elmsford, N.Y., 1963, p. 780.

64. D. Gröger and S. Johne, *Z. Naturforsch.*, *B*, **23**, 1072 (1968).

65. R. H. Prager and H. M. Thredgold, *Aust. J. Chem.*, **22**, 2627 (1969).

66. I. H. Bowen, P. Gupta, and J. R. Lewis, *Chem. Commun.*, 1625 (1970).

67. G. H. Svoboda and R. W. Kattau, *Lloydia*, **30**, 364 (1967).

CHAPTER V

Acridinium Salts and Reduced Acridines

I. A. SELBY

Pharmaceutical Division, Reckitt and Colman,
Hull, England

1. Acridinium Salts

A. Methods of Preparation

Acridinium salts are generally prepared by direct alkylation of acridines, by reduction of 10-alkyl-9-acridanones and oxidation of the resulting acridans, or by the reaction of 10-alkyl-9-acridanones with phosphoryl chloride, and subsequent nucleophilic displacement of the 9-chloro group, if desired, to give 9-substituted acridinium salts.

(1) *Alkylation of Acridines*

The 10-position of acridine (1) is less nucleophilic than the 1-position of either pyridine or quinoline. This may be partly caused by the steric hindrance offered by the 4- and 5-hydrogen atoms of acridine to the approaching electrophile, since the nitrogen electron density in acridine is higher than in the other two cases.[1] The quaternization reaction is favored by the use of polar solvents: acridine did not react with methyl iodide at 60°,[2] although it did at 100°,[3] but the use of dimethylformamide as the solvent gave the iodide at room temperature in 2 weeks.[4] 9-Benzyl-[5] and 9-phenyl-acridines[6] have been methylated with methyl iodide at 100°, and allyl bromide gave a high yield of 10-allylacridinium bromide.[4] Attempts to quaternize acridine with propyl iodide were successful when refluxing xylene was used as a solvent, but milder conditions failed and dimethylformamide was not effective in this case. All attempts to quaternize acridine with ω-iodo aliphatic long-chain esters gave rise only to acridinium iodide.[7] In general, higher 10-alkyl-acridinium salts are better prepared via the appropriate 9-acridanone. A better general methylation technique used methyl sulfate or p-toluene-sulfonate in nitrobenzene at elevated temperatures.[8, 9, 10a, b] The iodides were formed by treating the products with aqueous potassium iodide. Certain 9-phenylacridines, however, have proved rather unreactive toward methyl sulfate, even at its boiling point.[11] Acridines have been successfully alkylated using the butyl, benzyl and phenyl esters of p-toluenesulfonic acid.[12]

Aminoacridines may be quaternized directly; it is usual to protect the amino group by acetylation,[2, 9, 13] although this is not always necessary. 3,6-Diamino-2,7-dimethylacridine was quaternized exclusively at position 10 by methyl sulfate or p-toluenesulfonate, but methyl or ethyl iodide preferentially attacked the primary amino group.[14] 3,6-Bisdimethylamino-acridine[14] and 8-dimethylamino-9-methylbenz[a]acridine[15] were quaternized by methyl sulfate in boiling toluene without the involvement of the primary amino groups, and 6,9-diamino-2-ethoxyacridine (rivanol) is stated to give the quaternary salts with either methyl iodide or chloride at room temperature.[16] 9-Aminoacridine[2] reacted with methyl iodide exclusively at the 10-position, whereas the isomeric aminoacridines all reacted preferentially at the primary amino groups. Treatment of 1-, 2-, or 3-acetamidoacridines with methyl p-toluenesulfonate gave the acridinium salts, but the 4 isomer failed to react, perhaps for steric reasons.[2]

(2) Reduction and Reoxidation of 9-Acridanones

An alternative technique for the preparation of 10-alkylacridinium salts involves the reduction of the appropriate 10-alkyl-9-acridanone to the acridan followed by reoxidation, generally with ferric chloride, to the acridinium salt. The reduction of the 9-acridanone, which is usually performed with sodium amalgam, sodium in ethanol, or zinc and hydrochloric acid, is discussed in detail in Section 2.A(2). Reoxidation to the quaternary salt is brought about by adding ferric chloride to an acidified suspension or solution of the acridan. The technique has been applied to many amino-acridans[13, 14, 17–19] and other types of acridan.[20, 21, 22] The presence of electron-attracting groups generally makes boiling necessary.

Oxidation of 9-*t*-butyl-3-diethylaminoacridan, however, resulted in the loss of the *t*-butyl group, although a *sec*-butyl group was left intact.[23] The loss parallels the dealkylation that sometimes occurs during the oxidation of dihydropyridines obtained by Hantzsch's synthesis.[24]

Oxidation of the acridan has also been achieved with nitric,[25a, b] nitrous[26a–c] and chromic[27] acids.

If the 9-acridanone reduction is conducted using a Grignard reagent, 9,10-dialkylacridinium salts are obtained on treatment of the resulting 9-hydroxyacridan with acid. 10-Methyl-9-acridanone (2) reacted with phenyl[28] or benzyl[29] magnesium bromides to give 9-hydroxy-10-methyl-9-phenylacridan (3) or 9-benzylidene-10-methylacridan (4); these were converted to the corresponding salts with acid.

10-Phenyl-9-acridanone[30, 31] and 10-(2-diethylaminoethyl)-9-acridanone[32] behaved similarly, and phenyl lithium reacted analogously with 10-methoxy-9-acridanone.[33] All the monochloro-9,10-diphenylacridinium chlorides have been prepared, using aryl magnesium bromides and 10-phenyl-9-acridanones.[34]

9-Substituents may also be inserted by reaction of the 10-alkyl- or 10-aryl-9-acridanone with phosphoryl,[35a, b] thionyl[36] or oxalyl[37] chloride. The 9-chloro groups of the 9-chloroacridinium salts that are obtained are even more labile toward nucleophiles than those of 9-chloroacridine and can be displaced by gaseous ammonia[36] to give 9-aminoacridinium salts, and by amines[36, 38] to give 9-aryl- or 9-alkyl-aminoacridinium salts. Water causes decomposition to the 9-acridanones.

B. General Properties and Reactions

Quaternary acridinium salts form yellow solutions with a green fluorescence and, like many other quaternary salts derived from heterocyclic bases, are usually highly crystalline and comparatively soluble in water. They are not of great value in the acridine series for characterization purposes, as their melting points are usually decomposition points and vary considerably with the rate of heating.

Anion exchange of acridinium salts may be achieved in a variety of ways. If a 9-alkyl or aryl substituent is present, the carbinol obtained by treatment with sodium hydroxide is dissolved in the appropriate acid. If no 9-substituent is present, treatment with a base gives the 9-acridanone and one of the following alternative techniques must be employed. Acridinium bicarbonates are usually sparingly soluble in water; this property was taken advantage of in the conversion of 3,6-diamino-10-methylacridinium chloride to the acetate.[39] The bicarbonate is precipitated by the addition of sodium bicarbonate solution and redissolved in acetic acid. Saturated aqueous solutions of acridinium sulfates or methosulfates gave precipitates of the halides with concentrated sodium or potassium halide solutions, and iodides have been converted to chlorides by shaking with freshly precipitated silver chloride.[40]

The characteristic reactions of acridinium salts occur at the highly electron-deficient 9-position. Nucleophilic attack at this position occurs more readily than at the corresponding position in the pyridine or quinoline analogues; this is partly a reflection of the fact that the loss of aromaticity in the central ring is offset by a gain in one of the benzene rings.

Nu = nucleophile

The range of nucleophiles capable of this type of addition is wide and includes H_2O, OH^-, NH_3, CN^-, SH^-, PhSH, SO_3^-, NH_2OH, aromatic amines, Grignard reagents, Michael reagents (carbanions), and tetramethoxyethylene. Alkyl groups at position 9 of acridinium salts are strongly activated toward diazonium salts, and cyanine-type dyestuffs have been prepared in this way (see Chapter I, p. 87).

The reaction of quaternary acridinium compounds with the hydroxide ion has been the subject of much study.

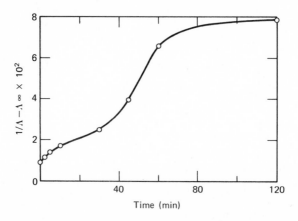

The reaction of 10-methyl-9-phenylacridinium chloride with sodium hydroxide, and of the sulfate with barium hydroxide, was studied conducti-metrically by Hantzsch and Kalb.[41] It was found that after addition of an equivalent of hydroxide ion at 0°, the conductivity slowly diminished and initially nothing could be extracted with ether. The observed drop in conductivity contrasted with the case of the pyridinium and quinolinium analogues, although the latter did show a very slight diminution. It was also observed that the conductivity dropped more rapidly if 10-methylacridinium chloride or 9,10-dimethylacridinium chloride was used.

These results were considered to show that the initially formed hydroxide slowly (or in the last-mentioned cases, more rapidly) isomerized to the pseudobasic acridan. However, it has been shown that the concentration in Hantzsch's experiment of the acridan/acridinium hydroxide would have been higher than its measured solubility[42]; it therefore follows that the solutions were supersaturated and precipitation was at least partly responsible for the drop in conductivity. The shape of the curve is not that of an approach to a simple equilibrium.

The nature of the product was demonstrated by uv spectroscopy. The material precipitated from solutions of 10-methyl-9-phenylacridinium chloride

Fig. 1.   Graph of Hantzsch's data on 10-methyl-9-phenylacridinium hydroxide.[42]

with sodium hydroxide was soluble in ether or chloroform, and the uv spectrum of these solutions was closely similar to that of 9-phenylacridan but different from that of 10-methyl-9-phenylacridinium iodide.[43] The ether and chloroform solutions were colorless and did not fluoresce. The molecular weight determined in benzene was normal.[44] All these facts suggest that the compound was present as the 9-hydroxyacridan.

When dissolved in 50% aqueous methanol, however, the material gave a yellow solution that fluoresced; its uv spectrum was matched by that of a solution containing 75% 9-phenylacridan and 25% 10-methyl-9-phenyl-acridinium iodide.[43, 45] The composition of the equilibrium mixture was altered in the expected ways: the addition of alkali caused the spectrum to revert to the acridan type and the addition of water increased the proportion of ionized material. Similar observations were made with 10-methylacridinium iodide and 9,10-dimethylacridinium iodide in ethanolic solution.[46] Further proof of the ionic nature of the aqueous solutions is provided by the observation that the treatment of a solution of 10-methyl-9-phenylacridinium iodide with sulfuric acid led to no change in the optical properties. The product is the ionic 10-methyl-9-phenylacridinium sulfate.[42] Moreover, the activity coefficients for the hydroxy compound in aqueous potassium and barium chloride solutions are only consistent with its being a strong electrolyte.[47] The solid is colorless, and therefore probably in the acridan form.

Conductivity studies have also been made of the product obtained from aqueous solutions of certain amino-10-methylacridinium bromides and sodium hydroxide,[48] but precipitation occurs, so that attempts to interpret the observed falls in conductivity are open to the same objections as the Hantzsch studies. In addition, some of the compounds were not substituted at position 9, and it is known that disproportionation and oxidation to the 9-acridanone is facile for such compounds.[49]

Addition of alcoholic sodium hydroxide to 9-cyano-10-methylacridinium methosulfate gave the 9-hydroxyacridan (5), which underwent a chemiluminescent oxidation with hydrogen peroxide solution (see Chapter IX, p. 622).[50]

The hydroxyl group of these 9-hydroxyacridans is easily etherified in the presence of alcohols, perhaps via the ionic form. Recrystallization of 9-

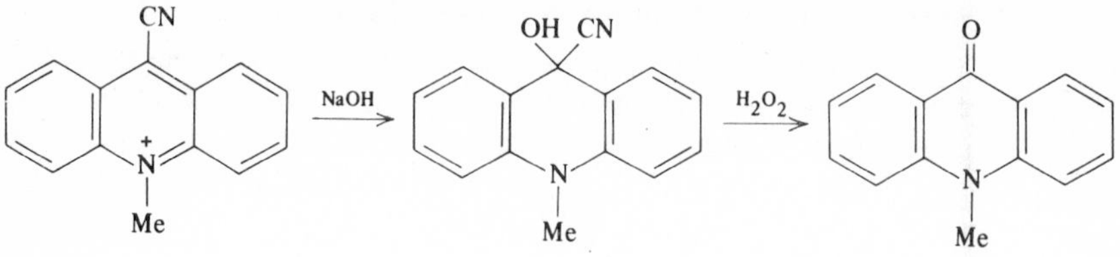

5

hydroxy-10-methyl-9-phenylacridan from ethanol, methanol, or isobutanol gave the 9-alkoxyacridan in each case.[44] The process was reversed with boiling water. The hydroxyl group was also replaced by cyanide on treatment of 10-methyl-9-phenylacridinium hydroxide with aqueous potassium cyanide,[41] and a chloroform or ether solution of the product had an acridan-type uv spectrum.[46] Like the hydroxyacridans, this cyano derivative was partly ionized in the more polar solvent ethanol,[45] but the addition of cyanide ion caused the spectrum to revert to the acridan type. The cyanoacridans are resistant to acid, but hot concentrated sulfuric acid gave the corresponding acridinium sulfates.[8, 51] As with the hydroxyacridans, the absence of a 9-substituent renders these compounds liable to oxidation by air to the 9-acridanone.[8, 51]

Concentrated aqueous ammonia converted 10-methyl-9-phenylacridinium iodide to the 9-amino-10-methyl-9-phenylacridan[52]; recrystallization of this compound from ethanol gave the 9-ethoxyacridan. Heating the 9-hydroxy- or 9-amino-acridans with aniline gave **6**.

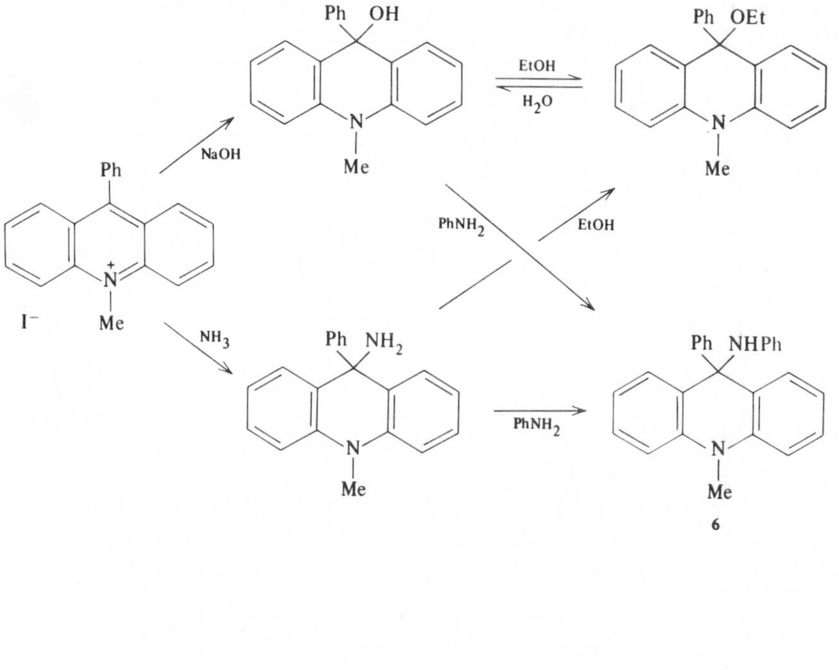

6

Hydroxylamine and *p*-dimethylaminoaniline reacted in the same way as ammonia,[53a, b] but aromatic amines in the presence of sulfur attacked the acridinium compound through the *para* position of the amine. For example, aniline, sulfur, and acridinium methiodide[54] gave a 90% yield of **7**.

7

The hydrosulfide ion attacked quaternary acridinium salts to give 10-alkyl-acridan-9-thiols,[55] and the thiophenate[55] and sulfite[56] ions reacted analogously, However, sodium benzenesulfinate and sodium *p*-toluenesulfinate gave the salts, rather than the acridans, and the products had solubility properties in accord with the ionic structure.[55] The relative affinities of the hydroxyl and thiophenate ion for the 9-carbon atom have been assessed spectrophotometrically, using the 10-methyl-9-phenylacridinium ion.[57]

Most types of carbanions active in the Michael reaction added to the 9-position of acridinium salts in the presence of sodium methoxide,[4, 58, 59] reoxidation of the products with bromine gave the 9-substituted analogues of the starting material. Ethyl magnesium bromide gave the 10-alkyl-9-ethyl-acridans.[3] Tetramethoxyethylene was added in 1 hr at 60° to 10-methyl-acridinium iodide in the presence of acetonitrile[60] to give the ester **9** via the intermediate **8**.

Derivatives in which the 9-carbon atom is part of a cyclic lactone or lactam have been prepared by cyclization of **10** with caustic soda[61] or ammonia or primary amines.[62] Compound **10a** was obtained from 9-(2-carboxyphenyl)-acridine and methyl iodide.[61] Other analogous reactions have also been reported.[63, 64]

The elimination of water from suitably substituted acridinium salts gives rise to the anhydrobases, and the thermochromic benzylidene compound **11**, for example,[65] was prepared by the treatment of 9-benzyl-10-methylacridinium iodide with sodium bicarbonate solution.[5]

Oxidation cleaved the benzylidene double bond, and acids caused reversion to the acridinium salt.

9-Imino-10-methylacridan (**12**, R = H) was obtained by dehydration at 130° of 9-amino-9-hydroxy-10-methylacridan,[2, 66] and treatment of 10-methyl-9-phenylaminoacridinium hydroxide with aqueous ammonia[67] gave the 9-phenylimino analogue **12** (R = Ph). Another preparation of **12** (R = Ph) is given in Section 2.B.

An alternative approach was the amination of 9-chloro-10-methylacridinium chlorosulfites to give 9-alkylimino- or 9-phenylimino-acridans.[36] Other types of iminoacridan have also been obtained. 3-Amino-2,7,10-trimethylacridinium hydroxide (or the acridan) gave a 3-imino-3,10-dihydro-acridine (13) on boiling in nitrobenzene.[68] Compound 14 gave an anhydrobase (15, R = Ac) on treatment with cold aqueous ammonia through elimination of an acetyl group.[69] The removal of the other acetyl group required boiling

35% sulfuric acid, and basification gave 15 (R = H). Other examples of the reaction have been reported.[9, 13, 70]

Structural isomers of 9-acridanone, known in earlier literature as "iso-acridones," were obtained by demethylation of 1-, 2-, 3-, or 4-methoxy-10-methylacridinium chlorides with aluminium chloride in xylene.[27] Boiling hydrobromic acid only demethylated the 2- and 3-methoxy compounds.

Classical structures can be written only for the 1,10- and 3,10-dihydro-acridine-1- (16) and -3-ones (16a), but the similarity of their uv spectra[71] suggests that all four are best represented by mesoionic structures (cf. 17 and 17a). The "isoacridones" do not resemble 9-acridanone and all of them form stable hydrochlorides, whereas 10-methyl-9-acridanone hydrochloride is hydrolyzed by water. The uv spectra of these "isoacridones" and their hydrochlorides closely resemble those of pyocyanin and its hydrochloride. This is expected, since the replacement of CH by nitrogen in such an aromatic system usually causes little change in the uv spectrum.

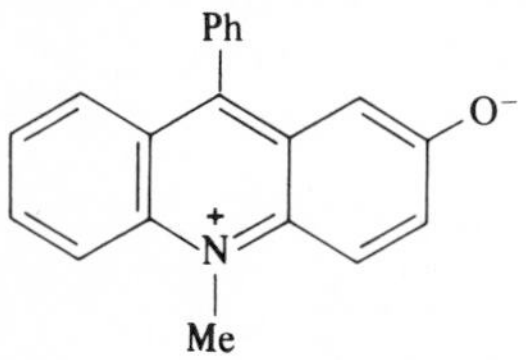

18

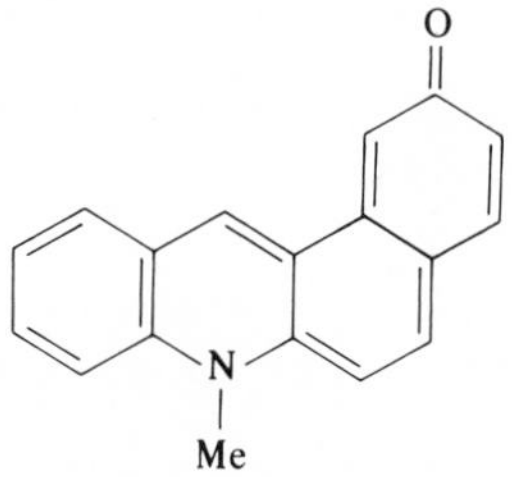

19

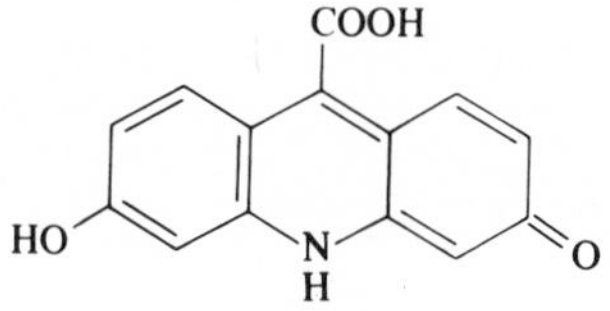

20

The dihydroacridinone 18 was prepared by dehydration of the 2,9-di-hydroxyacridan,[72] and 19 by hydrolysis of the 2-hydroxy iodide with alkali.[73] Compound 20 was prepared in 50% yield by the reaction of resorcinol, formic acid, and urea.[74]

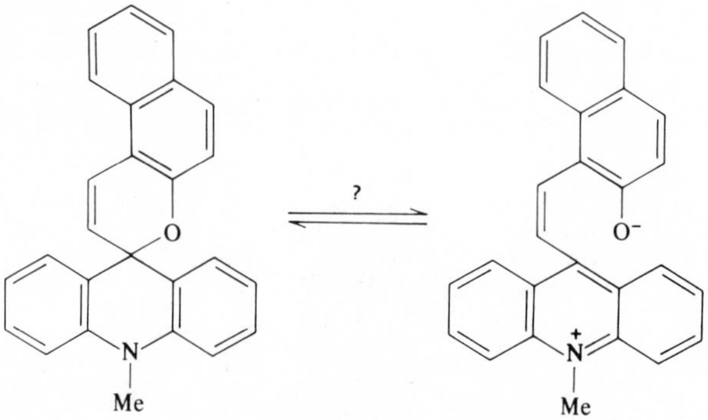

21

22

The compound **21** is a colorless solid that melts to a blue liquid. Its cold solutions are pale blue, but the color intensifies reversibly on heating, possibly because of the formation of the zwitterion **22**, or perhaps free radicals.[64]

10-Bromoacridinium bromides appear to be formed from acridine or 9-phenylacridine and bromine in carbon tetrachloride. They are decomposed by acid to the salts of the original acridines.[75]

2. Acridans

A. Methods of Preparation

Acridans may be prepared either by (1) the reduction of acridines, 9-acridanones, or acridinium salts, or (2) by cyclization of diphenylamines or diphenylmethanes. These two techniques are very general and have been used in the preparation of a wide range of acridans. Other less general reactions leading to acridans are also of preparative value.

(1) *Reduction of Acridines*

Acridine itself has been hydrogenated to acridan using a number of catalysts, including Raney nickel (85% yield),[76] copper chromite (90% yield),[76] and ruthenium oxide (98.5% yield).[77] The most convenient of these procedures is probably the first, since it is conducted at room temperature and is rapid. Raney nickel has been employed successfully in the hydrogenation of all the aminoacridines to the acridans.[2] Palladium on barium sulfate was an effective catalyst in the hydrogenation of methyl acridine-9-carboxylate to the acridan[78]; although the free acid could not be reduced in this way, the acridan was obtained using sodium amalgam.[79] 9-Chloroacridine and 3,9-dichloro-7-methoxyacridine on hydrogenation over Raney nickel in alkaline aqueous methanol were both reduced to the 9-dechlorinated acridans.[80] The other chlorine atom was unaffected. Heating acridine with Raney nickel and triethylammonium formate at 160° also gives rise to acridan.[81]

Sodium amalgam,[82] sodium in ethanol,[83] 4-methylthiophenol,[84] and zinc and hydrochloric acid[83] have all been used to reduce acridine to acridan, but substantial amounts of 9,9-biacridan were also obtained. No dimeric products were observed in the reduction of acridine using lithium in ammonia buffered with ammonium acetate,[85] nor when aqueous ethanolic sodium tetrathionate[86] or lithium aluminium hydride[87] were used. The latter technique gave a yield of 70%.

The concomitant production of 9,9-biacridans is not a problem when a 9-substituent is present, probably as a result of the added steric hindrance. 9-Methyl-[88] and 9-phenyl-acridine[83] were both quantitatively reduced to the acridans with sodium amalgam in boiling ethanol.

9-Phenylacridine was also quantitatively reduced with zinc and hydrochloric acid,[83] and sodium tetrathionate has also been used. [75] The reductive methylation of acridine to 9,10-dimethylacridan was brought about in unstated yield by treatment with lithium and methyl iodide in liquid ammonia-tetrahydrofuran.[85] The nmr spectrum of the product was reported and is mentioned in Chapter XII (p. 704).

(2) *Reduction of 9-Acridanones*

Sodium and amyl alcohol have given good results in the reduction of 10-phenyl-9-acridanone[30] and the four monomethoxy-10-methyl-9-acridanones[27] to the respective acridans. Sodium in ethanol[32] has also been used for this type of reduction. These reductions were all continued until the fluorescence of the solutions disappeared. No biacridan derivatives appear to be formed in these alkaline conditions. ^{14}C-Acridan has been prepared by the sodium and amyl alcohol method.[89]

The reduction of 10-methyl-9-acridanone by zinc and hydrochloric acid gave a mixture of acridinium salts, 10-methylacridan, and biacridine derivatives, the amount of acridan increasing with the reaction time.[90a, b] The zinc and hydrochloric acid reduction of 2,7-diamino-10-methyl-9-acridanone is stated to be much preferable to the sodium amalgam reduction when the acridan is desired.[20]

(3) *Reduction of Acridinium Salts*

The reduction of acridinium salts to acridans is the reverse of the normal preparative procedure for the salts and has not been widely used. All but one of the monochloro-9,10-diphenylacridinium acetates have been reduced to the acridans by zinc and acetic acid.[34] The 3-chloro compound was dechlorinated and gave 9,10-diphenylacridan; but the desired 3-chloro-9,10-diphenylacridan was obtained in excellent yield by reduction of the acridinium formate with sodium formate and formic acid.[34] The same technique has been used for the reduction of 3,9,10-triphenylacridinium formate.[91] The reduction of 3,6-diamino-10-methylacridinium chloride by zinc and hydrochloric acid has been superficially examined.[51]

(4) *Reduction of Acridines with Metals and Organometallic Compounds*

Acridine,[92] 9-phenyl-,[93] and 9-methyl-acridine[94] reacted with sodium under ether to give deep red-violet mixtures containing the disodium derivatives. Treatment with ethanol gave the acridans, and the derivatives of acridine and 9-phenylacridine with carbon dioxide gave the acridan-9-carboxylic acids.

Considerable amounts of 9,9-biacridan were formed by dimerization in the reactions involving acridine and in the reduction of other acridines with a potassium-sodium alloy.[95] A 9-substituent seems to prevent this, probably for steric reasons.

Acridine reacted with phenyl[78] or benzyl[96] magnesium bromides to give addition compounds that were hydrolyzed with water to the corresponding 9-substituted acridans. Better results were obtained, however, with phenyl lithium,[78] and butyl lithium.[97] 4-Dimethylaminophenyl lithium[98] and diphenylmethyl sodium[96] behaved similarly. The initial products of the lithium alkyl reactions were colorless.[78, 99] 9-Methylacridine in n-butyl ether adds methyl magnesium iodide to give, after hydrolysis, 9,9-dimethylacridan,[100] but gives no acridan with phenyl lithium.[97] This failure could be a result of the formation of the N-metallic salt of the tautomeric 9-methyleneacridan (23), like those thought to be formed from 9-methyl- and 9-benzyl-acridines and sodium and potassium amides.[101]

23

A related migration of a 10-benzyl group was observed during the treatment of 10-benzyl-9-methylacridan with butyl lithium (see Section 2.B).[102]

Good yields of 9-alkylacridans (60–89%) have been obtained by the

Grignard reagent reduction of acridines in ether at 100° in a sealed tube,[103] and 9-allylacridan has been obtained in 81% yield from acridine and allyl magnesium bromide.[104] Dimethylaminopropyl magnesium chloride in tetrahydrofuran reacted with acridine[105] to give the 9-dimethylaminopropyl acridan (24). Compound 25, which has sedative properties, has been prepared by the reduction of 3-chloroacridine with the Grignard reagent derived from 2-(2-chloroethyl)-N-methylpiperidine.[106]

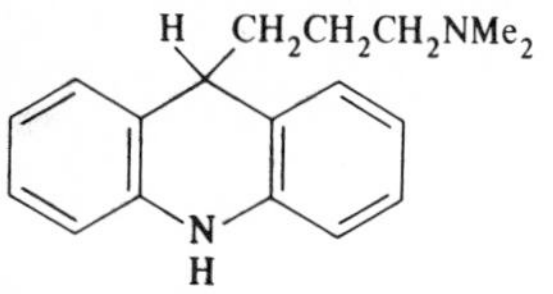

24

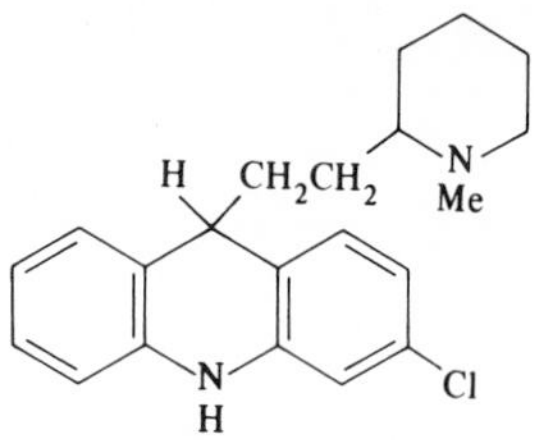

25

9-Chloroacridine reacted with 2 moles of dimethylaminopropyl magnesium chloride to give the 9,9-disubstituted acridan.[107]

9-Acridanylphosphines, such as 26, have been prepared by refluxing an ether solution of acridine and the lithium phosphine.[108] Compound 26 was obtained in 31% yield.

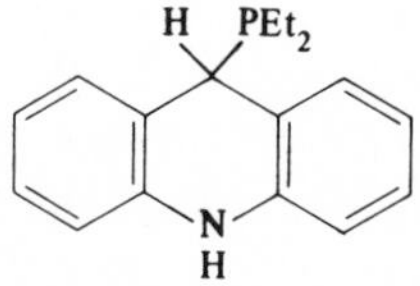

26

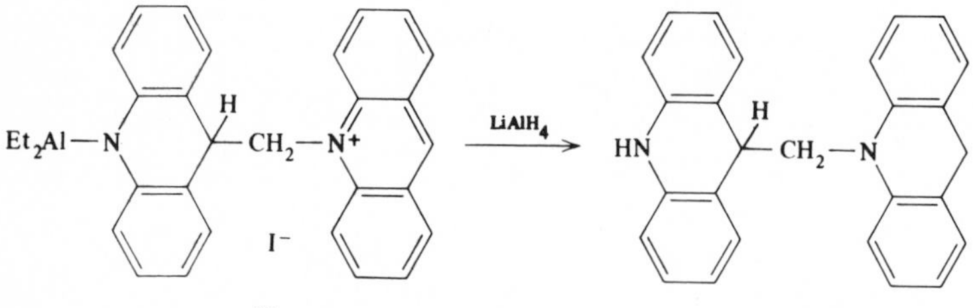

27

28

The organometallic reagent Et_2AlCH_2I reacted with acridine, forming the salt 27, which was reduced with lithium aluminium hydride[109] to 28.

(5) *Reductions of 9-Acridanones with Grignard Reagents*

Acridans such as **29**, possessing central nervous system activity, have been prepared by the reduction of 10-alkyl-9-acridanones with 1-dimethyl-aminopropyl magnesium chloride in tetrahydrofuran.[110] Sometimes the initial product was the 9-hydroxyacridan, which could be dehydrated to the 9-alkylideneacridan.[111] Further reduction of **29** with lithium aluminium hydride gave **30**, which shows the same type of activity. Other compounds derived from 9-alkylideneacridans, such as **31**, are coronary vasodilators.[112]

Similar reduction of 10-alkyl-9-acridanones with Grignard reagents derived from 4-chloro-*N*-alkylpiperidines gave 9-piperidylidene derivatives (**32**). On further reduction these derivatives gave the 9-piperidylacridans (**33**), which have tranquillizing properties.[113]

(6) Reductions of Acridinium Salts with Grignard Reagents

Acridinium salts, particularly those unsubstituted at position 9, have been converted to acridans by reduction with Grignard reagents. 10-Benzyl-acridinium bromide and methyl magnesium iodide[102] gave 10-benzyl-9-methylacridan (**34**); 3,7-diamino-10-methylacridinium chloride reacted with benzyl magnesium bromide to give the corresponding 9-benzylacridan.[114]

34

The reaction of 10-methylacridinium iodide with Grignard reagents, originally thought to give 10-methylacridans,[3] was reinvestigated; it was found that the product in the case of methyl magnesium iodide was a mixture of acridine and 9,10-dimethylacridan, together with a considerable amount of an insoluble material.[100] This may be 10,10'-dimethyl-9,9'-biacridan, since heat decomposed it to 10-methylacridan and acridine. 9,10-Dimethylacridinium iodide and methyl magnesium iodide gave 9,10-dimethylacridan with some 9,9,10-trimethylacridan.[100]

9,10-Diphenylacridinium chloride reacted with diphenyl magnesium to give 51–58% of 3,9,10-triphenylacridan, with 9,10-diphenylacridan as a minor product.[91] This ring substitution is an example of the reactivity of the 3-position to cationoid reagents. With phenyl magnesium bromide, however, the chloride gave 10% of 9,10-diphenylacridyl peroxide, but most of the starting material was recovered.

(7) Polarographic Reduction of Acridines

Polarographic reductions of acridine[115] and its derivatives [116–120] showed that two distinct stages are involved. The first step in the reduction of acridine in ethanolic solution is probably the reversible addition of a proton and an electron to give the semiquinone radical **35**, which is subsequently irreversibly reduced to the acridan.

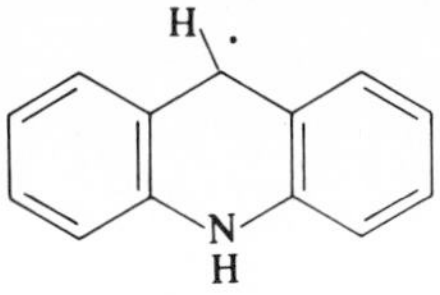

35

The theoretical aspects of the reduction have been studied.[121] The semiquinone **35** was comparatively stable and showed little tendency to disproportionate to acridine and acridan. A phase diagram of the acridine-acridan system gave no indication of the formation of a molecular compound,[119] although some interaction clearly took place because of the strong yellow color of the melt. The semiquinone can be little stabilized by resonance. In the presence of hydrochloric acid, however, a green-black addition compound was formed,[122] but it was largely dissociated into its components in solution.[123] Morgan's base (see also Chapter VII, p. 554), prepared by the polarographic reduction of dibenz[*aj*]acridine or by mixing equivalent quantities of the benzacridine and its 7,14-dihydro derivative,[120] is a similar semiquinone to **35**, but it is more stable, since the unpaired electron can be distributed over a much larger aromatic system. The uv absorption spectrum of Morgan's base was the sum of those of its constituents; ebullioscopic measurements showed that the radicals, or ion radicals, were not appreciably associated in solution. The semiquinone obtained in the polarographic reduction of 9-phenylacridine was bright red.[120] Polarographic reduction has been used to convert 9-(2-iodophenyl)acridine to the acridan or to 9-phenylacridan. The reduction potentials differ by only 0.3 V, and this shows the degree of control that is possible if the correct potential is used.[120]

(8) *Cyclization of Diphenylamine and Diphenylmethane Derivatives*

9,9-Dimethylacridan[124] has been prepared in 44–60% yield by heating acetone, hydrochloric acid, and diphenylamine at 260°. It is likely that the tertiary carbinol **36** is first formed and subsequently cyclized under the influence of the acid.[125] Acridine and 9-methylacridine occur as by-products.

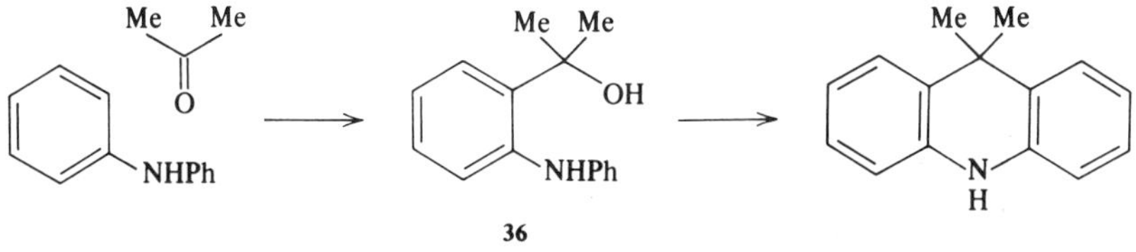

36

Many derivatives of 9,9-dimethylacridan have been prepared by this type of reaction, and a high-yield synthesis of compounds similar to dimethacrine, **38** (see also p. 459) involves the cyclization[126] with a catalytic amount of sulfuric acid of N-substituted o-aminoarylstyrenes (**37**).

The metabolism of dimethacrine has been investigated, and one of the metabolites (**39**) was synthesized from the appropriately substituted diphenylamine.[127]

9,9-Diphenylacridan has been made by a similar route for methyl diphenylamine-2-carboxylate and phenyl magnesium bromide. The carbinol was best cyclized with hot acetic acid containing a few drops of concentrated sulfuric acid,[128] although ethanolic hydrochloric acid was also effective.[129] The reaction is a general one and has been applied in other instances.[130] A large number of aminoacridans have been prepared by this method from aminodiphenylamine-2-carboxylic esters with phenyl, ethyl, and methyl magnesium halides.[128, 131, 132] Although methyl 4-aminodiphenylamine-2-carboxylate reacted normally,[133] the 5-amino compound failed to give the acridan,[133] as did methyl 4-nitrodiphenylamine-2-carboxylate.[133]

Several nitro-9,9-diphenylacridans have been obtained by heating the potassium salts of the products from 2-aminotriphenylmethanes with 2,4,6-trinitrochlorobenzenes or 2,6-dinitrochlorobenzene, in quinoline.[134] Attempts to make 2,4-dinitroacridans and 2,4-dinitro-9-phenylacridans from picryl chloride and 2-aminotoluene or 2-aminodiphenylmethane, and 2-nitro-9,9-diphenylacridan from 2,4-dinitrochlorobenzene and 2-aminotriphenylmethane, failed, [134] as did the attempted cyclization of N-(2,4-dinitro-1-naphthyl)-2-aminotriphenylmethane.[135]

9,10-Diaryl-9-hydroxyacridans have been obtained in yields exceeding 50% by the reaction of triphenylamine with aromatic carboxylic acids in the presence of polyphosphoric acid. The ketone **40** is an intermediate; *p*-acylated triphenylamines are also produced.[136]

(9) *Photoreduction of Acridines*

9-Hydroxymethylacridans have been obtained both by the photoreduction of acridine or 9-methylacridine in methanol[137] and by the reduction of acridine-9-carboxylic acid[138] or its ethyl ester[139] with lithium aluminum hydride. When the photoreduction was carried out in cyclohexane or dioxan, 70–90% yields of **41** or **42** were obtained.[140]

Photoreduction of acridines with aliphatic carboxylic acids gave up to 68% of the 9-alkylacridan with about 10% of biacridan by-product.[141] 4-(9-Acridanyl)valeric acid cyclized to give 10% of the spiro compound **43**.

Acridine 10-oxide added ethanol under the influence of uv light [142] to give a 40–45% yield of 9-ethoxy-10-hydroxyacridan (**44**).

43   **44**

The nmr spectrum (CDCl$_3$) showed the expected signals, including a low-field singlet at τ 3.95, due to the methine proton.

(10) *Other Reductions of Acridines Leading to Acridans*

Graebe found that acridine with sulfurous acid gave two products, one colorless and one deep red.[143] Further investigation showed that acridine hydrochloride and sodium sulfite gave the colorless sodium acridan-9-sulfonate (45), decomposed by alkalis to acridine.[144] Its structure was demonstrated by conversion to 9-cyanoacridan (46) with potassium cyanide in boiling alcohol, followed by aerial oxidation to 9-cyanoacridine. The oxidation of 45 by air gave Graebe's deep red compound, considered to be acridinium acridan-9-sulfonate, as it could be synthesized from acridine hydrochloride and sodium acridan-9-sulfonate. The deep red color suggests that it may have a semiquinone type of structure. Ethyl phenyl sulfone was added to acridine[145] to give a 54% yield of a compound thought to be 47.

45   **46**   **47**

Hydrogen cyanide was added to acridine in ethanol, giving 46, which was oxidized by air to 9-cyanoacridine.[144a,b] In a similar way, diketene[146] gave 9-acridanyl acetone (48); propionaldehyde[147] in 10 days gave 2-(9-acridanyl)-propionaldehyde (49); and malonic acid[148] in pyridine yielded 9-acridanyl-acetic acid (50).

Chloranil was used to oxidize 48[146] and 49[148] to acridine and 9-acetylacridine, respectively.

Acridine failed to react with maleic anhydride[149, 150] but combined with dimethylketene to give an unidentified product[150] now considered to be the

48   49   50

2:1 adduct **51**.[151] 9-Methylacridine reacted with dimethylketene to give the 9-methyl analogue of **51** and the bridged compound **52**, which was reduced to **53**.[152]

51   52   53

Acridine reacts at the activated methylene groups of nitromethane, malondinitrile, acetylacetone, deoxybenzoin, ethyl acetoacetate, and even diethyl malonate at room temperature or on a steam bath to give the corresponding acridans (e.g., **53a**), which were oxidized by lead tetraacetate to the corresponding acridines.[345]

53a

The 10-methyl derivative of **53a** has been obtained from 10-methylacridinium iodide, potassium hydroxide, and nitromethane,[59] and lead tetraacetate oxidation gave 10-methyl-9-nitromethyleneacridan.[345]

Acridine was reported to react with dimethyl acetylenedicarboxylate in methanol to give **54**, a structure that is inconsistent with the data in the paper.[153] Another investigation has shown that the product of the reaction was largely **55** (or the conjugate acridinium methoxide), along with some uni-

solated *cis* isomers.[154] Recrystallization of **55** from ethanol replaced one
methoxyl group with an ethoxyl group, and both alkoxy compounds gave the
same orange 9-acridanone (**56**) on gentle oxidation.

Compound **55** had a very similar uv spectrum to that of the corresponding
acridinium hydrochloride, prepared by the addition of dilute hydrochloric
acid, and had a typical acridan spectrum in the presence of excess alkali. All
these reactions are characteristic of 10-methylacridinium salts unsubstituted
at position 9.

The benzoxazine **57** rearranged on heating to the acridan **58**, which was
independently synthesized from 9-phenylacridan and diphenylacetaldehyde.[155]

9-Chloromethyl-10-phenylacridan was obtained in poor yield by a benzyne-
induced chloromethylation of acridine with methylene chloride.[156]

B. General Properties and Reactions of Acridans

Acridan or 9,10-dihydroacridine is a colorless compound, mp 172°, and was at one time referred to as carbazine (9,9-diphenylacridan was analogously known as 9,9-diphenylcarbazine). It occurs in traces in coal tar.[157a-c] Acridan is a very weak base (pK_a = −0.93)[158]; in this fact and in many of its properties, it shows a resemblance to diphenylamine (pK_a = +0.86).[158] The uv spectrum [119] and refractivity[159] are very similar to those of diphenylamine and indicate that the two benzene rings are electronically isolated. Acridan is insoluble in water and only slightly soluble in dilute acids, but dissolves in ether, benzene (the heat of solution has been measured),[160] and it can be crystallized from ethanol. Many 9- and 10-substituted acridans exhibit pharmacological activities of various types; an account of a clinical trial of an "antipsychotic" acridan has been published.[161]

Acridan has been partially dehydrogenated[143] by heating at 300° and is hardly affected by hydrogen and 30% palladium-charcoal.[162] Oxidation to acridine is brought about by many reagents; it has been observed that aerial oxidation of acridan and 6-chloro-2-methoxyacridan in aqueous-ethanol proceeded rapidly in the presence of acid and was much slower in the presence of alkali.[163] The addition of acid caused the solution to fluoresce, and an acridinium ion absorption band appeared at 355 nm. It seems that the oxidation occurs via an acridinium cation. The method has been exploited by using 78% sulfuric acid, in which acridan is soluble.[164] An efficient conversion is brought about by dilute chromic acid[165] or by potassium ferricyanide in alkali (> 80% yield).[103] Boiling nitric acid oxidizes 10-substituted acridans to the corresponding acridinium salts, and more powerful oxidizing agents give the 9-acridanones.[40, 166]

The rates of oxidation of 10-methylacridan by 2,6-dichlorophenolindophenol at various pH's have been examined; it appeared probable from the results that oxidation took place by hydrogen atom transfer, rather than by electron transfer.[167] The oxidation of ethyl 2-(2,7,10-trimethyl-9-acridanyl)-acetoacetate (59) with lead tetraacetate gave 60, or 2,7,10-trimethylacridinium acetate, depending on whether the solvent was benzene or acetic acid.[168]

59   60

Acylation of acridan with acetyl chloride gave 10-acetylacridan,[169] which was also obtained by the reductive acetylation of acridine using acetic anhydride in the presence of zinc dust.[119] A Friedel-Crafts acetylation of this compound with bromoacetyl bromide gave 3-bromoacetyl-10-acetylacridan.[170] Numerous 10-acetylacridans have been prepared and shown to possess pharmacological activity. Compounds such as 10-acetyl-9-dimethylaminopropylacridan (61), prepared from the acridan with acetic anhydride, show antidepressant activity,[105] and the amide 62 has been claimed as an antiviral and antitumor agent.[171, 172]

Acridan with propyl magnesium iodide gave 10-acridanyl magnesium iodide,[160] and with ethyl lithium gave ethane and the 10-lithio derivative, which itself adds methyl iodide to give 10-methylacridan.[78] Similarly, 9,9-dimethylacridan gave 9,9,10-trimethylacridan, when treated successively with sodamide and methyl iodide.[100] Lithium amide has been used in the alkylation of 9-methylacridan.[173] 9,10-Dimethylacridan (the nmr spectrum of which was reported; see also Chapter XII, p. 704) on successive treatment with butyl lithium and N,N-dimethylcarbamoyl chloride gave an 80% yield of the amide 63, which was converted to the starting material with lithium aluminium hydride.[102]

However, an interesting rearrangement occurred when 10-benzyl-9-methylacridan was treated with butyl lithium and 9-benzyl-9-methylacridan 64 was obtained.[102]

The product is considered to arise from displacement of the 10-benzyl group, which subsequently attacks position 9. The intermediate step is analogous to the attack of methyl magnesium iodide at the 9-position of 9-methylacridine.[100] 10-Methylacridans have been demethylated with hydrobromic acid.[174]

The reduction of acridan with phosphorus and hydriodic acid afforded *trans*-octahydroacridine,[82] and catalytic reduction with Raney nickel gave octa- and dodeca-hydroacridines.[76] The deuterium-hydrogen exchange of 10-methylacridan with deuterium oxide in the presence of acids and bases has been investigated.[175] Acridan nitroxides have been prepared and examined by esr spectroscopy.[176]

The production of acridan from aniline hydrochloride and compounds containing methylenedioxy groups was used as a test for these groups,[177] but subsequently many substances containing primary alcoholic groups were also found to give the reaction.[178]

9,9-Dimethylacridan, mp 125–126° or 92–93° (metastable form), resembles acridan in most of its properties.[100] It was precipitated unchanged from its solution in concentrated hydrochloric acid by water and was stable to boiling aqueous potassium permanganate,[100] but it was oxidized to 9-methylacridine on heating with concentrated hydrochloric acid or with 10-acridanyl sodium.[100] It formed addition compounds with acridine and 9-methylacridine,[100] although acridan itself forms no complex with acridine, but does with the hydrochloride. 9,9-Dimethylacridan finds use in conjunction with diphenylamine as a rubber antioxidant.[179] Both 9,9-dimethylacridan and its 10-isopropyl derivative are components of tobacco smoke.[180]

9,9-Dimethylacridan derivatives are of great pharmacological interest, and a number of different types of activity have been found. 9,9-Dimethyl-10-dimethylaminopropylacridan or "dimethacrine" (65) is a rapid-acting antidepressant in man. It was found that the drug (administered as the hydrogen tartrate) was very rapidly transported to the brain of rats; this may explain its fast action in man.[181, 182]

Dimethacrine and analogous compounds have been prepared from the acridan with an alkyl halide in the presence of sodamide,[183–185] or by treatment of the acridan with phosgene and then sodamide and the appropriate alcohol to give the ester 66, which was then decarboxylated.[183, 186]

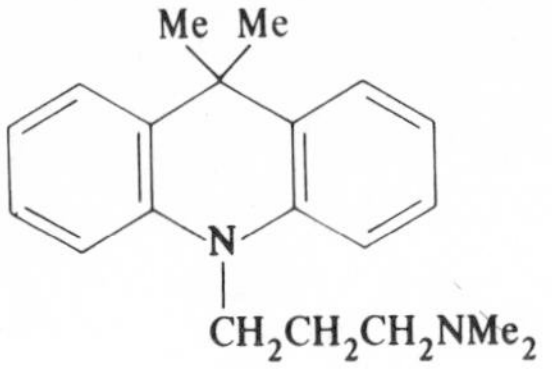

Me Me

CH₂CH₂CH₂NMe₂

65

Me Me

COOCH₂CH₂CH₂NMe₂

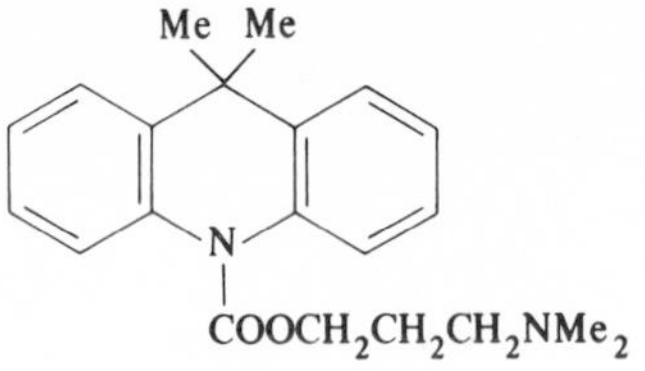

66

Lower alkyl analogues of **65** prepared similarly[187] showed anti-reserpine and thymoleptic activity.[188, 189] An alternative synthesis is the cyanoethylation of the acridan, followed by reduction to the amine with lithium aluminum hydride.[186] Esters of 9,9-dialkylacridan-10-propionic acid have been reported as showing spasmolytic activity.[190] A high-yield synthesis of dimethacrine that avoids the necessity of dealing with the unsubstituted acridan is the cyclization of the appropriately substituted *o*-aminoarylphenylcarbinol with 80% sulfuric acid.[126] The metabolism of dimethacrine has been studied (see p. 452).[127]

9,9-Diphenylacridan forms a 10-acetyl derivative[129] that on nitration furnished a series of nitro compounds containing 2-, 3-, 4-, and 6-nitro groups.[134] The positions of the nitro groups were determined by unambiguous syntheses and in other ways.[131, 133, 191] 9,9-Diphenyl-10-methylacridan behaved similarly on nitration. All the 9,9-diphenyl-2-nitroacridans gave deep red solutions in alcoholic alkali, presumably as a result of the acidic amino hydrogen atom and the formation of ions such as **67**. The presence of a 10-methyl group prevents this; such compounds are colorless. The uv spectra of these compounds have been recorded.[192]

All the 9,9-diphenylnitroacridans and their 10-methyl derivatives have been reduced to the corresponding amines; some of the nitroamines have been obtained directly by the Grignard synthesis. 4-Amino-2,7-dinitro-9,9-diphenylacridan diazotized normally[193] and gave the cyclic product **68**.

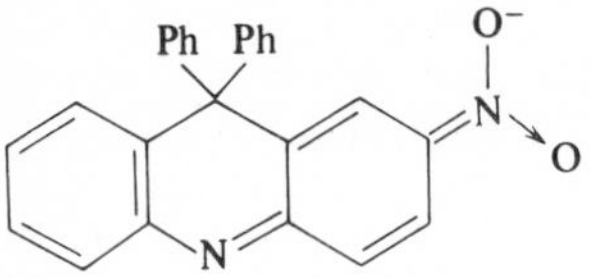

Ph Ph

67

Ph Ph

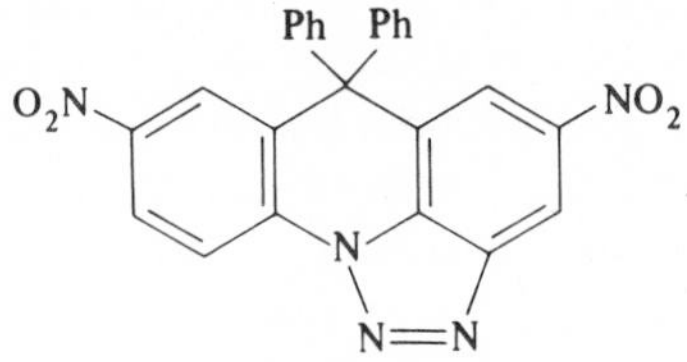

68

10-Acyl-9,9-diarylacridans have been reported as biologically active.[194]

The oxidation of 2-amino-9,9-disubstituted acridans by air or ferric chloride proceeded easily and gave rise to 9,9-disubstituted carbazims (69) (see Ref. 195 and preceding papers). During the oxidations the solutions frequently darkened, perhaps as a result of the intermediacy of semiquinone radicals. The products vary in color from yellow to violet and are occasionally fluorescent in solution, but they are insufficiently stable for use as dyes since they are easily hydrolyzed to carbazons (70).

This hydrolysis is generally carried out with sodium carbonate.[196a, b] Neither carbazims nor carbazons (better described as 2,9-dihydro-2-iminoacridine and 2,9-dihydro-2-acridinones) have been investigated further in recent years.

Acridans provide a convenient route into the 5H-dibenz[bf]azepine series of pharmacologically interesting compounds, since phosphorus pentoxide causes 9-hydroxymethylacridans to undergo Wagner-Meerwein rearrangement to this ring system.[50, 138, 197] In the rearrangement[50] of 71, the main product 72 was accompanied by an unrearranged compound 73, formed presumably via 9-methyleneacridan by isomerization.

It is interesting that in the case of the as-octahydroacridine analogue, 74, the tetrahydroacridine, 75, was the exclusive product.[198] 9,9-Dichloro-10-methylacridan has been converted with liquid ammonia to the imide 76 in 90% yield. [199]

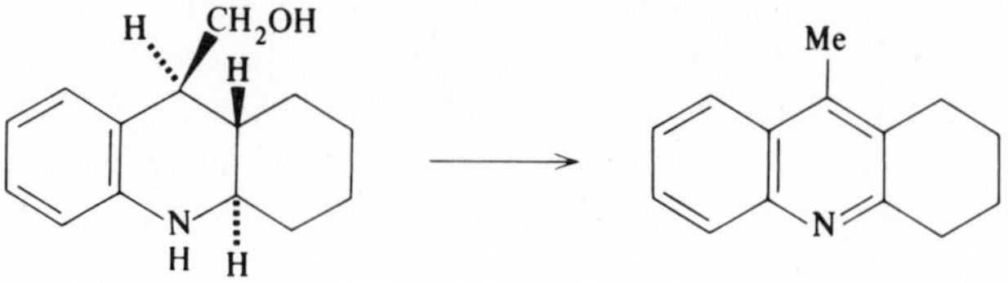

74   75

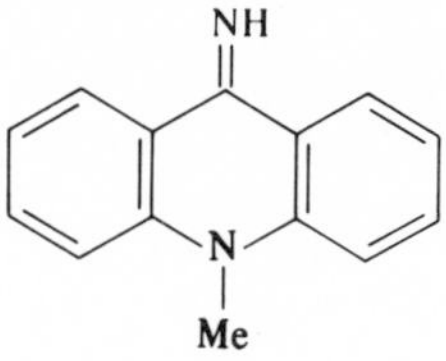

76

The nmr spectrum of the product was reported. An alternative preparation is given in Section 1.B.

3. 1,2,3,4-Tetrahydroacridines

A. Methods of Preparation

1,2,3,4-Tetrahydroacridines are almost invariably prepared by the condensation of suitably substituted aromatic and alicyclic six-membered ring compounds, using either acidic or basic conditions. The preparation of 9-halo- and 9-amino-1,2,3,4-tetrahydroacridines is discussed in Section 3.B and that of 1,2,3,4-tetrahydro-9-acridanone in Section 4.A.

(1) *Base-catalyzed Preparations of 1,2,3,4-Tetrahydroacridines*

The original preparation[200] of 1,2,3,4-tetrahydroacridine from cyclohexanone and 2-aminobenzaldehyde has been applied in other instances,[201, 202] e.g., 4-piperidinomethyl-1,2,3,4-tetrahydroacridine (**77**) has been prepared from 2-piperidinomethylcyclohexanone and 2-aminobenzaldehyde.[202]

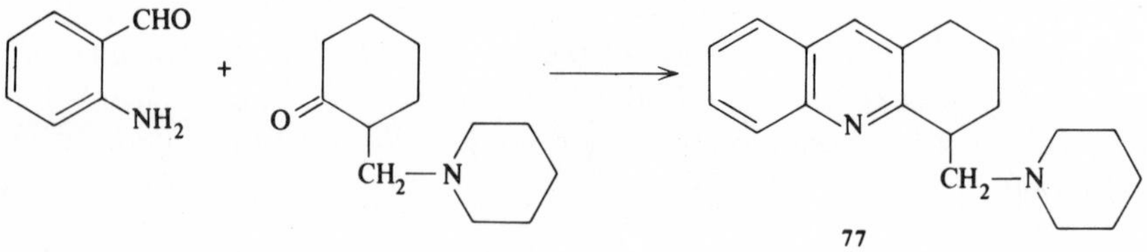

77

This technique, however, has been used much less frequently than the base-catalyzed Pfitzinger reaction, which is generally more satisfactory.[203] The preparation of 1,2,3,4-tetrahydroacridine (**79**) by the latter procedure involves two steps: the reaction of isatin and cyclohexanone (which proceeds via the salt **78**) and subsequent decarboxylation of the 1,2,3,4-tetrahydroacridine-9-carboxylic acid.[198, 204]

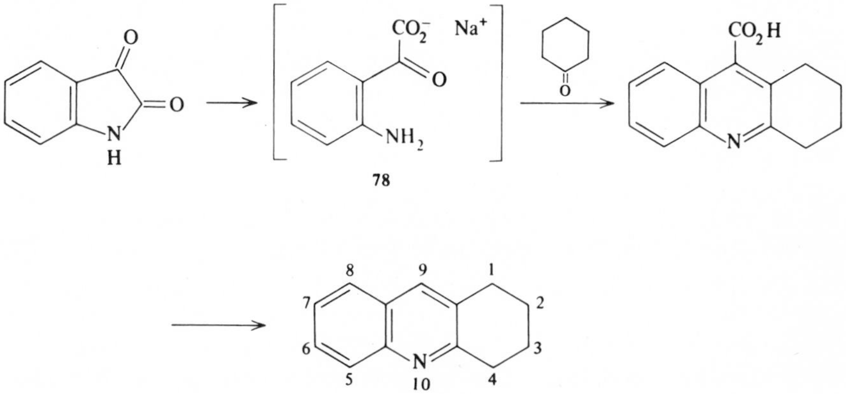

78

79

The reaction provides a general route to 1,2,3,4-tetrahydroacridines[204, 205] and condensed derivatives[206a–c] unsubstituted at position 9, since the carboxylic acid group is quantitatively removed by heating the acid above its melting point.

A number of alkyl-substituted 1,2,3,4-tetrahydroacridine-1,9-, -2,9-, and -3,9-dicarboxylic acids (e.g., **80** and **81**) have been prepared from the appropriate cyclohexanonecarboxylic acids and isatin in 10% potassium hydroxide.[207]

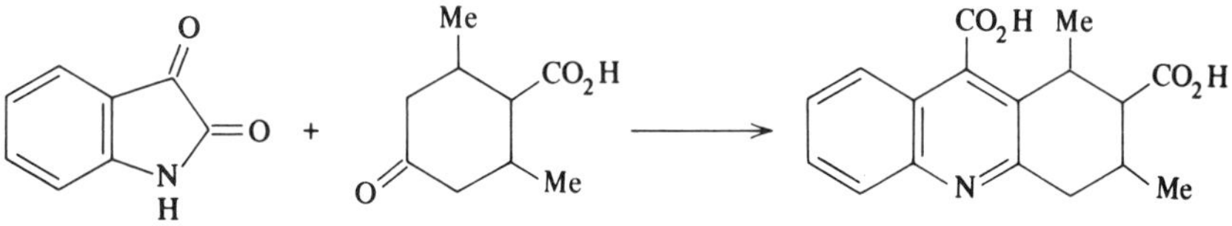

80

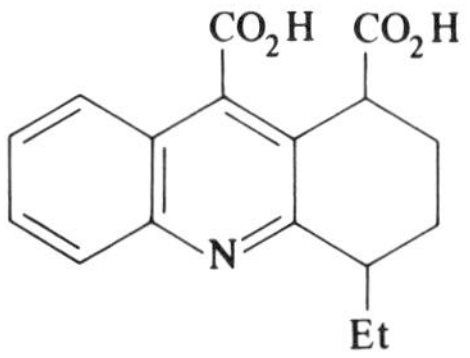

81

Unfortunately, the Pfitzinger reaction appears to be very sensitive to steric factors[208] and a number of surprisingly simple alkanones, such as pulegone,[209] isopulegone,[205] dihydropulegone,[205, 209] tetrahydrocarvone,[205] camphor,[209] norcamphor,[209] and menthone,[209] do not form acridines with isatin despite the normal reactions of their —COCH$_2$— groups with benzaldehyde and ketonic reagents. Tetrahydrocarvone (2-methyl-5-isopropylcyclohexanone) stands in strong contrast to 2-methylcyclohexanone, which reacts in good yield, and this suggests that the presence of a substituent destined to appear *peri* to the 9-carboxylic acid group prevents reaction. α- And β-naphthisatin[210a, b] and α-acenaphthisatin[205] did not react with cyclohexanone, although the 1-substituted products **80** and **81** were obtained in yields[207] exceeding 50%. Another case in which a steric effect seems to be operative is the preparation of 3- (not 1-) methyl-1,2,3,4-tetrahydroacridine-9-carboxylic acid from 3-methylcyclohexanone.[201, 204] (Decarboxylation of the product gave a tetrahydroacridine, mp 73°, and dehydrogenation gave a methylacridine, mp 125–126°, which agrees with the mp of authentic 3-methylacridine prepared at a later date.)[22] However, the reaction of 3-methylcyclohexanone with 2-aminobenzaldehyde[201] gave exclusively a tetrahydroacridine of mp 70–71°, i.e., 3-methyl-1,2,3,4-tetrahydroacridine; yet in this reaction there can be very little steric hindrance to the formation of the 1-isomer. This suggests that the 6-position of 3-methylcyclohexanone is inherently more reactive than position 2. The optical enantiomers of 3-methyl-1,2,3,4-tetrahydroacridine have been prepared from the (+)- and (−)-3-methylcyclohexanones.[204]

The Pfitzinger reaction has been used to prepare 1,2,3,4-tetrahydroacridines, substituted in the benzenoid ring with halogen atoms, as potential carcinogens.[204, 208, 211] 2-Cyclohexylidenecyclohexanone and 2-cyclohexenyl-

82   83

84

cyclohexanone gave the expected products **82** and **83** in about 30% yield; these were decarboxylated by heating on a Wood's metal bath.[212]

In a variation of the 2-aminobenzaldehyde synthesis, the amine **84** was condensed with cyclohexanone to give 6,7-dimethoxy-1,2,3,4-tetrahydroacridine.[213]

(2) *Acid-catalyzed Preparations of 1,2,3,4-Tetrahydroacridines*

The acid-catalyzed Friedlander synthesis provides an alternative route to 1,2,3,4-tetrahydroacridines. Borsche, Tiedtke, and Schmidt[200] prepared 3,9-dimethyl-1,2,3,4-tetrahydroacridine by condensation of 2-aminoacetophenone and 3-methylcyclohexanone to give an anil that cyclized with sulfuric acid. The product was identified by dehydrogenation to the acridine. More recently, 9-phenyl-1,2,3,4-tetrahydroacridine has been prepared in 83% yield from 2-aminobenzophenone and cyclohexanone, using concentrated sulfuric acid in acetic acid,[214] and the dihydrobenzacridine (**85**) has been obtained from β-tetralone and *o*-aminoacetophenone hydrochloride.[215]

Petrow prepared 1,2,3,4-tetrahydroacridine by cyclization of the anil **86** (formed in situ from formylcyclohexanone and the amine) using aniline hydrochloride and zinc chloride in boiling ethanol.[216]

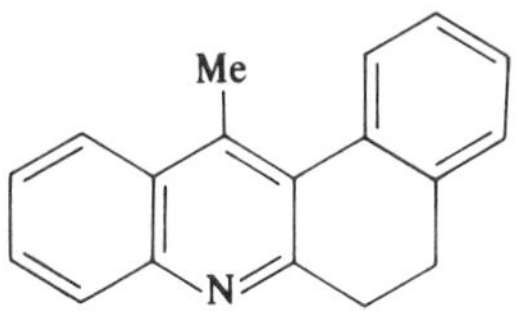

85

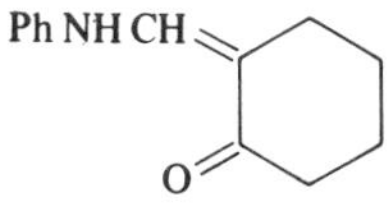

86

Yields of about 65% were obtained in a series of analogous preparations involving 22 different amines and three formylcyclohexanones; the following mechanism was proposed:

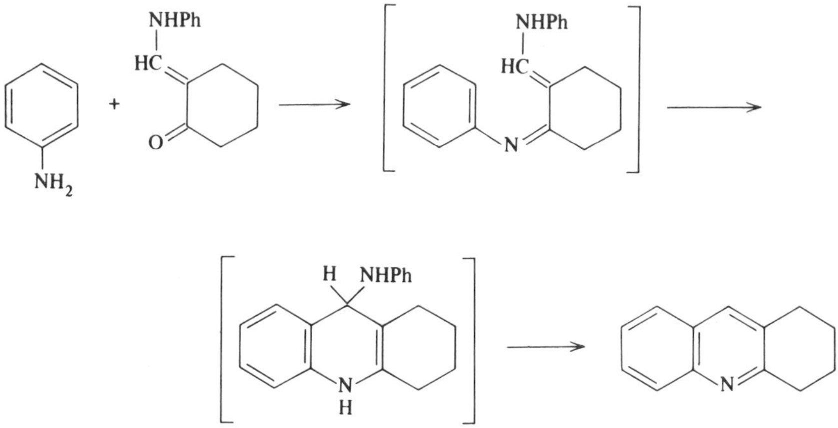

The reaction of *m*-toluidine with formylcyclohexanone gave only one of the two possible isomers, **87** and not **88**, and the structure was proved to be 3-methyl-1,2,3,4-tetrahydroacridine by dehydrogenation to 3-methylacridine.

The 6- or 8-acetyl-1,2,3,4-tetrahydroacridine referred to in the same paper[216] has since been reduced and dehydrogenated to 1-ethylacridine,[217] establishing that the precursor was the 8-isomer.

A simple one-step preparation of 1,2,3,4-tetrahydroacridine in 54% yield is the reaction of the β-chlorovinyl aldehyde (**89**) with aniline in acetic acid under reflux.[218] The aldehyde **89** was obtained by formylation of cyclohexanone with *N,N*-dimethylformamide and phosphoryl chloride.

cis-2-Hydroxy- or 2-*p*-tosyloxy-methylenecyclohexanones react with arylamines to give 2-arylaminomethylenecyclohexanones (**90**), which have been cyclized to 1,2,3,4-tetrahydroacridines[219] with the following reagents in ethanol: arylamine hydrochloride and zinc chloride, arylamine and lactic acid, lactic acid, and zinc chloride.

Reaction A	90	$\xrightarrow[\text{ZnCl}_2, \text{EtOH}]{}$	**91**	60%
Reaction B	90	$\xrightarrow[\text{ZnCl}_2, \text{EtOH}]{}$	**92** + **91**	60%
Reaction C	90	$\xrightarrow[\text{lactic acid, EtOH}]{}$	**91**	95%
Reaction D	90	$\xrightarrow[\text{lactic acid, EtOH}]{}$	**91** + **92**	95%
Reaction E	90	$\xrightarrow[\text{lactic acid, EtOH}]{}$	**91**	95%
Reaction F	90	$\xrightarrow[\text{ZnCl}_2, \text{EtOH}]{}$	**91**	20%

Reaction A was considered to proceed by attack of the second arylamine on the carbonyl group of **90** to give the intermediate **93**, which on cyclization and loss of arylamine gives the tetrahydroacridine **91**.

A similar cyclization occurs in the conversion of the enamine **94** to the 9-amino-10-methyl-1,2,3,4-tetrahydroacridinium tosylate.[220] (See Section 3.B.) If the two arylamines involved were different (Reaction B), aryl exchange occurred and both the possible products **91** and **92** were obtained. The mechanism by which the original aryl moiety is retained appeared to involve the reversible formation of **95** from which either aryl group could be lost to give a 2-arylaminomethylenecyclohexanone capable of cyclization in the manner shown for **93**.

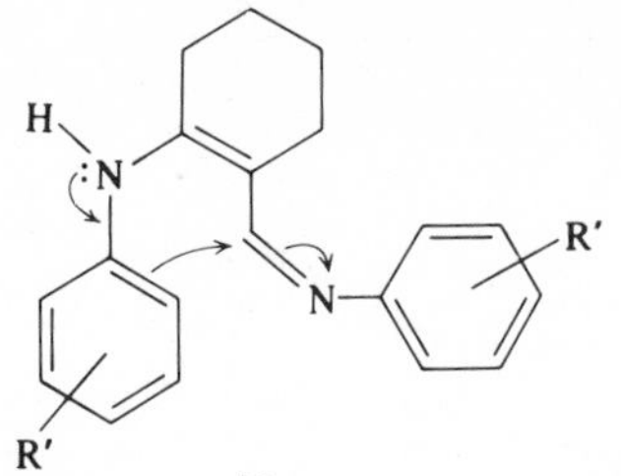

93

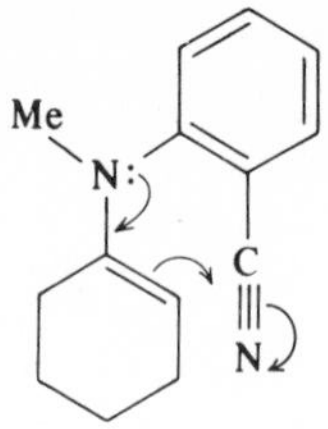

94

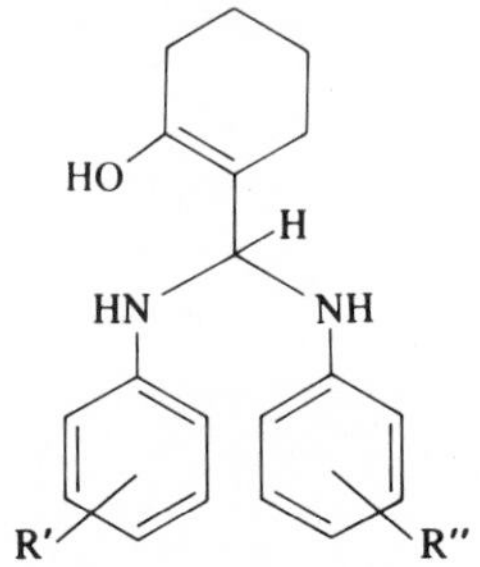

95

Under mild conditions the "exchanged" 2-arylaminomethylenecyclohexanone could in fact be isolated.

The best yields were obtained using lactic acid (Reactions C and D) and its particular efficacy was considered due to the anchimeric assistance that can be provided by the hydroxyl group in the formation of the proposed intermediate **96**. The intermediate **96** was prepared independently and gave a 97% yield of **97** on reaction with *m*-anisidine.

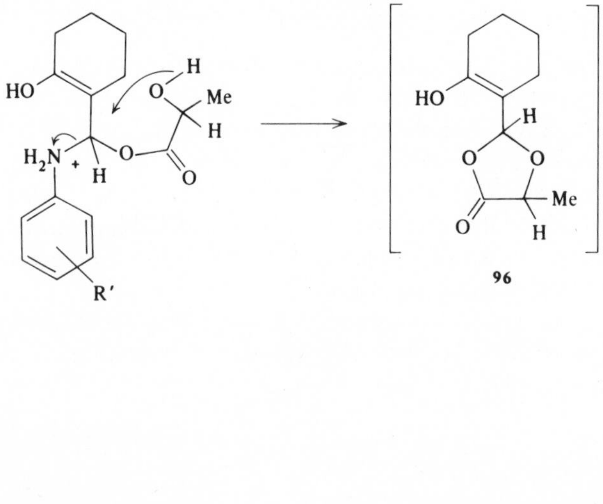

96

In the cyclization of **90** with lactic acid alone (Reaction E), the aryl moiety is eliminated during formation of **96**, which it then attacks to give the tetrahydroacridine. Cyclodehydration of **90** also occurred when zinc chloride and ethanol were used (Reaction F), giving a low yield of the tetrahydroacridine, perhaps via an azetidine intermediate.

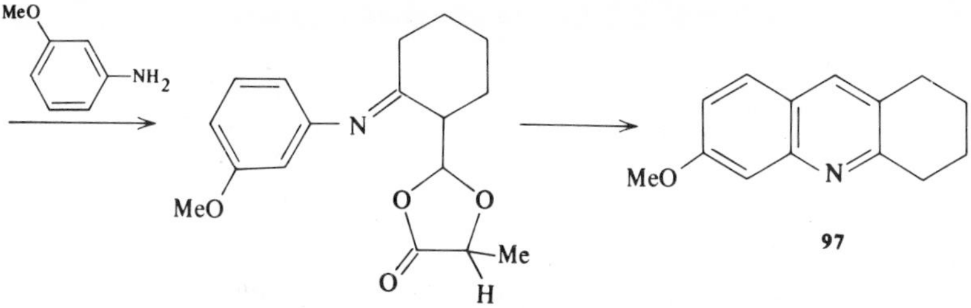

A different mechanism was proposed earlier for a case similar to Reaction A above.[221] The formation of 8,9,10,11-tetrahydrobenz[c]acridine (**99**) from the arylaminomethylenecyclohexanone **98** was explained by the following rearrangement:

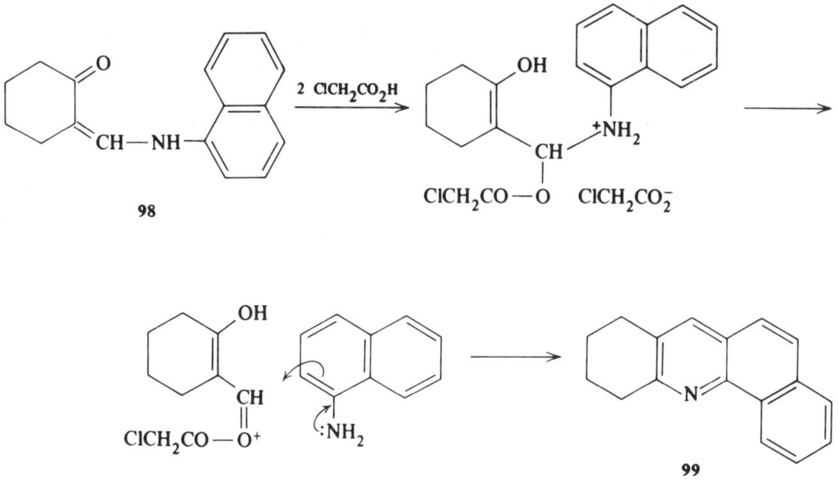

However, no products of *para* substitution by the amine have been isolated,[219, 221] even though the occurrence of aryl exchange would, in the case of this mechanism, imply complete separation of the amine and carbonium ion. It has also been pointed out that the nitrogen atom of **98** would be only weakly basic, since it is a part of a vinylogous amide system.[219]

9-Methyl-1,2,3,4-tetrahydroacridine was obtained when 9-hydroxymethyl-1,2,3,4,4a,9,9a,10-*trans*-octahydroacridine was treated with phosphorus pentoxide.[198]

B. General Properties and Reactions of 1,2,3,4-Tetrahydroacridines

1,2,3,4-Tetrahydroacridine closely resembles the 2,3-dialkylquinolines in smell and physical and chemical properties.[222] It forms a yellow picrate (mp 222°, dec)[216, 223] and a methiodide.[203]

Nitration of 1,2,3,4-tetrahydroacridine gave a mixture of the 5- and 7-nitro derivatives[203, 204, 216]; the former has been oxidized with permaleic acid to the 10-oxide.[224] Sulfonation at 100° gave a monosulfonic acid. At 130–140° the chief product was an isomeric monosulfonic acid accompanied by a small proportion of the disulfonic acid,[204] but the positions of the substituents have not been determined. Bromine in the cold with 1,2,3,4-tetrahydroacridine gave what appeared to be a perbromide ($C_{13}H_{14}Br_3N$), as treatment with ammonia regenerated the base.[204] At higher temperatures, the reaction gave an intractable mixture,[204] and bromo-1,2,3,4-tetrahydroacridines are best prepared by the general methods with the bromine atom incorporated at the start.[204] The vapor of the 6-bromo compound attacks the mucous membrane, and the solid is a powerful skin irritant.

1,2,3,4-Tetrahydroacridine shows the same type of reactivity in its 4-methylene group as the 4- and 5-methylene groups of *sym*-octahydroacridine and the methyl group of 2-methylquinoline. It reacted with benzaldehyde and *p*-nitro- and *p*-dimethylamino-benzaldehyde in the presence of zinc chloride to give 4-benzylidene derivatives.[204, 216] These compounds were not quaternized by methyl sulfate in boiling nitrobenzene, but 10-methyl-1,2,3,4-tetrahydroacridinium iodide reacted with *p*-dimethylaminobenzaldehyde in acetic anhydride in 6 min to give the bright red iodide (**100**). Ethanol and piperidine were ineffective solvents. Boiling the reaction mixture for 90 min caused decomposition[225] to **101**.

10-Methyl-1,2,3,4-tetrahydroacridinium chloride reacted with salicylaldehyde[226] in the presence of acetic acid to give the 4-salicylidene derivative **102**. The treatment of **102** with aqueous ammonia gave a purple precipitate that slowly became colorless; this change was rapid in the presence of organic matter. The product[226] is probably best represented as **103**.

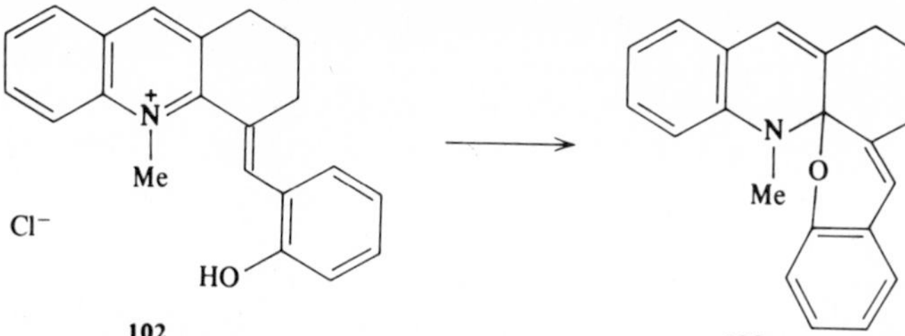

102   103

At least one cyanine dyestuff derived from position 4 of 1,2,3,4-tetrahydro-acridine is known.[227]

1,2,3,4-Tetrahydroacridine reacted with ethyl oxalate in the presence of potassium[228] to give the ester 104. Alkaline hydrolysis gave the acid and heating the oxime of the acid with acetic anhydride in pyridine gave the deep blue 105, the color of which resembles that of pyrrolo[1,2-a]quinoline-1,2-dione.[229]

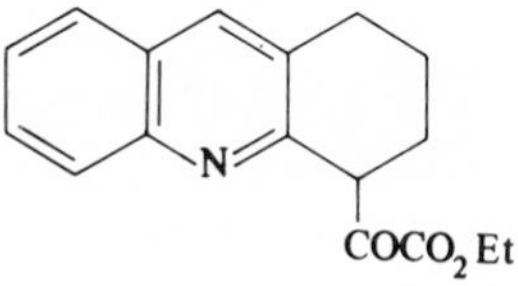

104 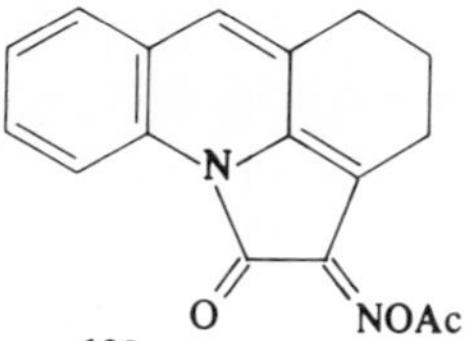   105

Vacuum distillation of 105 gave 4-cyano-1,2,3,4-tetrahydroacridine, which was hydrolyzed to the ester with ethanolic hydrogen chloride. 1,2,3,4-Tetrahydroacridine was obtained on alkaline hydrolysis of this ester, or the cyano compound, with the loss of carbon dioxide. 4-Methoxyphenyldiazonium

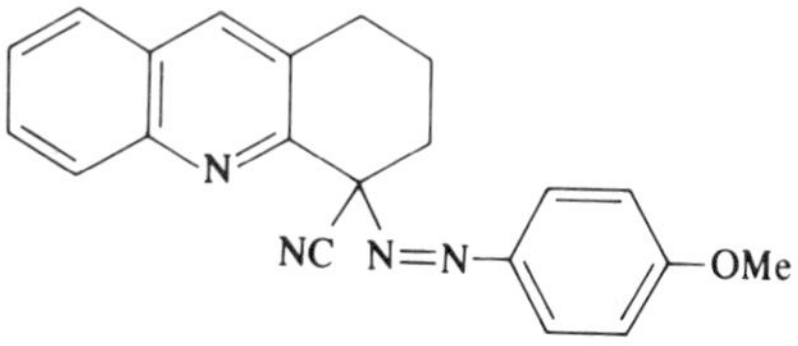

106 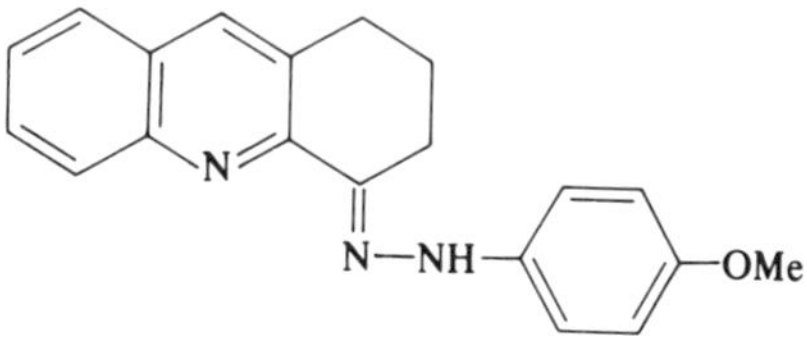   107

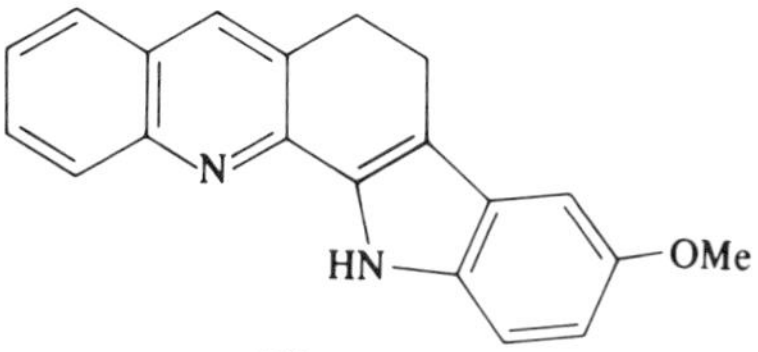

108

chloride reacted with 4-cyano-1,2,3,4-tetrahydroacridine to give **106**, and with the glyoxalate **104** in pyridine to form the phenylhydrazone **107**, which cyclized to the indole **108** with alcoholic hydrogen chloride.[228]

4-Dialkylaminoalkyl-1,2,3,4-tetrahydroacridines, such as **109**, have been synthesized from 1,2,3,4-tetrahydroacridine with 3-dimethylaminopropyl chloride in the presence of sodamide in toluene.[230] 4-Diethylaminomethyl-1,2,3,4-tetrahydroacridine(**110**) was prepared from 1,2,3,4-tetrahydroacridine, diethylamine hydrochloride, and formalin[230] by a Mannich reaction; both **109** and **110** showed weak antidepressant activity.

Oxidation of 1,2,3,4-tetrahydroacridine and the 9-methoxycarbonyl ester with peracetic acid gave the respective 10-oxides.[231] Acetic anhydride converted 1,2,3,4-tetrahydroacridine 10-oxide to the 4-acetoxy derivative **111**, which could be hydrolyzed to 1,2,3,4-tetrahydroacridin-4-ol with hydrochloric acid. Dehydration of the latter leads to the 1,2-dihydroacridine (see Section 8).

9-Chloro-1,2,3,4-tetrahydroacridine 10-oxide (**112**) has been obtained by direct oxidation of the chloroacridine with permaleic acid[232] and also by nitration of 1,2,3,4-tetrahydroacridine 10-oxide and subsequent treatment with concentrated hydrochloric acid.[232]

The photolysis of 9-methyl-, 9-phenyl-, and 9-methoxycarbonyl-1,2,3,4-tetra-
hydroacridine 10-oxides has been reported[233] and the main products were
benz[d]-1,3-oxazepines (113).

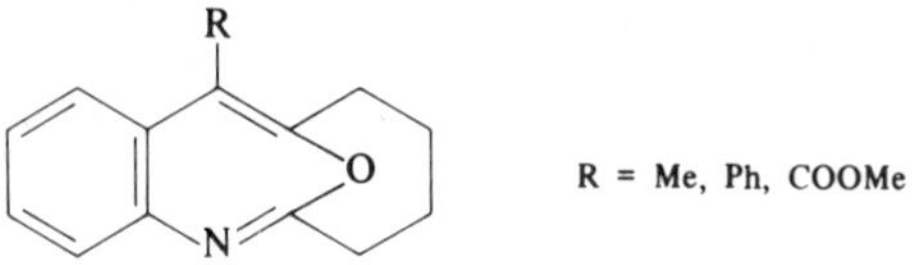

R = Me, Ph, COOMe

113

Attempted oxidation of 1,2,3,4-tetrahydroacridines with selenium dioxide
gave very low yields of the tetrahydro-4-acridinones isolated as their 2,4-
dinitrophenylhydrazones.[234] Dehydrogenation was the main reaction. 9-
Methyl-1,2,3,4-tetrahydro-4-acridinone has been obtained by reaction of
cyclohexane-1,2-dione with o-aminoacetophenone hydrochloride.[235] The 1,3-
dione gave rise to the expected mixture of isomers **114** and **115**.

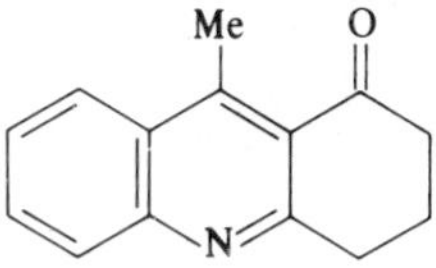

114 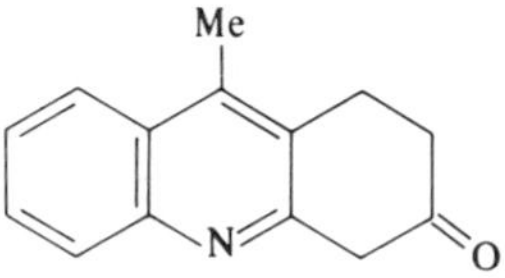   **115**

3,3-Dimethyl-1,2,3,4-tetrahydro-1-acridinone (**116**) has been prepared from
2-aminobenzaldehyde and 5,5-dimethylresorcinol in the presence of boiling
alcoholic caustic potash,[236, 237] and from 5,5-dimethyldihydroresorcinol and
the p-toluidine anil of the aldehyde[238] with piperidine at 100°. This keto-
acridine formed a semicarbazone,[238] a 2,4-dinitrophenylhydrazone,[238] a
picrate[236, 238] and a methiodide.[236] Reaction with phenylhydrazine is stated

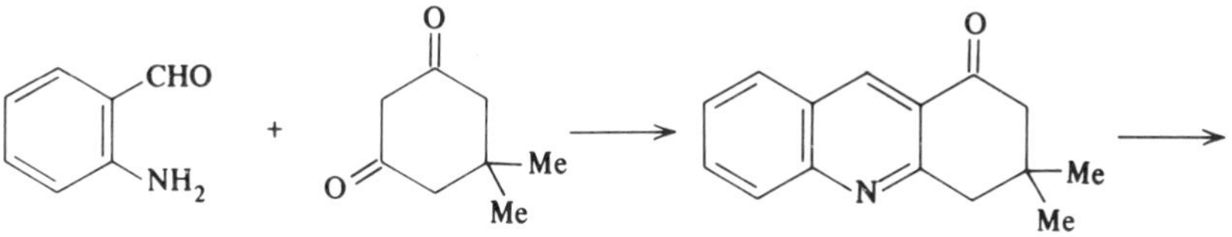

116

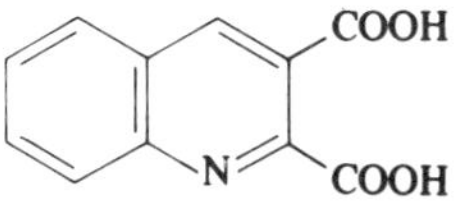

117

to give the phenylhydrazones of dimedone and 2-aminobenzaldehyde.[236] The oxidation of 116 with acid or alkaline permanganate gave acridinic acid (117).

2-Acetamidobenzaldehyde and 5,5-dimethyldihydroresorcinol in alcoholic potash also gave 116, but in absolute alcohol 118 was formed.[236] Compound 118 lost water at 105–110° with the formation of 119 (colorless), which cyclized to 116 with aqueous alcoholic potash. Absolute alcoholic potash in the last reaction gave the isomer 120; on standing or on treatment with acids, it also gave 116.

1,2,3,4-Tetrahydroacridine-9-carboxylic acid, colorless plates mp 284–286° (dec) from ethanol or acetic acid, like acridine-9-carboxylic acid, very easily loses the carboxylic acid group and proved difficult to esterify.[204] The ethyl and methyl esters were prepared from the alkyl halides and the silver salt of the acid.

The catalytic reduction of 1,2,3,4-tetrahydroacridines can give perhydroacridines and is a major approach to the *asym*-octahydroacridines. These reductions are considered in the appropriate preparative sections. A tetrahydroacridine, alleged to be 121, has been prepared by reducing the condensation product of 2-dimethylaminomethylcyclohexanone and 2-aminobenzaldehyde with tin and hydrochloric acid; on zinc dust distillation it yielded 4-methylacridine.[239]

9-Chloro-1,2,3,4-tetrahydroacridine has been prepared from 1,2,3,4-tetrahydro-9-acridanone (see Section 4.A) by boiling with phosphorus oxychloride,[240] addition of the pentachloride being unnecessary.[241] This procedure is a general one and has been used in the preparation of three alkyl-substituted 9-chloro-1,2,3,4-tetrahydroacridine-2- or -3-carboxylic acids[207] from the appropriate tetrahydro-9-acridanones. 7,9-Dichloro-1,2,3,4-tetrahydroacridine has been prepared from 7-chloro-1,2,3,4-tetrahydro-9-acridanone and phosphorus oxychloride.[242] 4,9-Dichloro-1,2,3,4-tetrahydroacridine (122) has been prepared in 85% yield in one step by refluxing anthranilic acid, 2-chlorocyclohexanone, and phosphorus oxychloride.[243] 9-Chloro-1,2,3,4-tetrahydroacridines containing alkyl, chloro or methoxyl groups in the carbocyclic aromatic ring have been prepared from the 9-acridanones,[242] but 9-chloro-5-nitro-1,2,3,4-tetrahydroacridine was best prepared by nitration of the 9-chloro-1,2,3,4-tetrahydroacridine.[243]

The chlorine atom of 9-chloro-1,2,3,4-tetrahydroacridines is rather unreactive, compared with that of 9-chloroacridine; there is a closer similarity to 4-chloroquinoline in this respect. The chlorine atom remained unaffected when 9-chloro-7-nitro-1,2,3,4-tetrahydroacridine was reduced with iron and ethanolic hydrochloric acid to the amine,[244] and treatment of 4,9-dichloro-1,2,3,4-tetrahydroacridine (122) with 4-diethylamino-1-methylbutylamine gave the 1,2-dihydroacridine 123 with the 9-chloro group intact.[243]

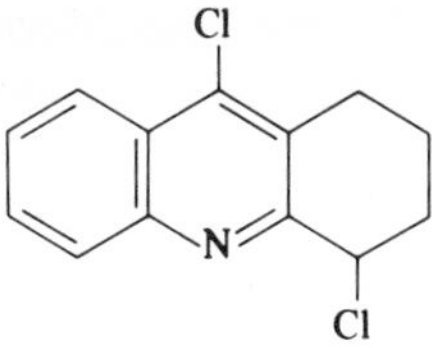

122 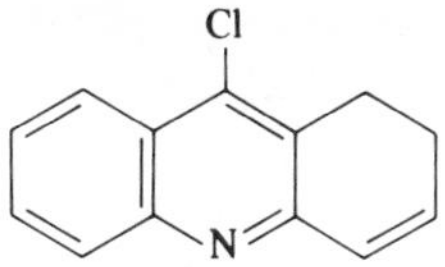   123

The resistance of the 9-chloro group toward hydrolysis is probably associated with the greater loss of aromaticity that occurs in the event of nucleophilic attack at position 9 (124), as compared with 9-chloroacridine (125).

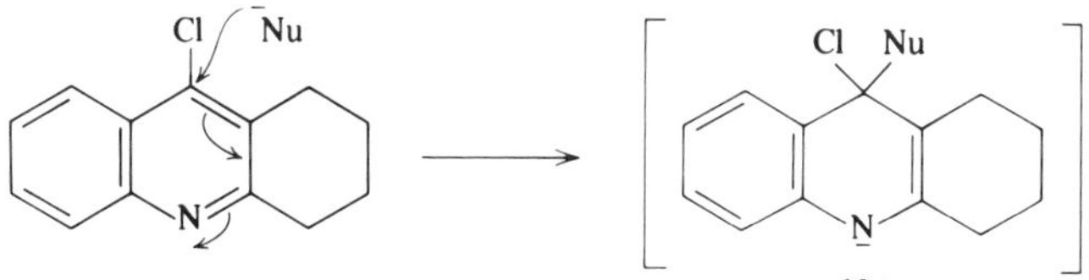

124

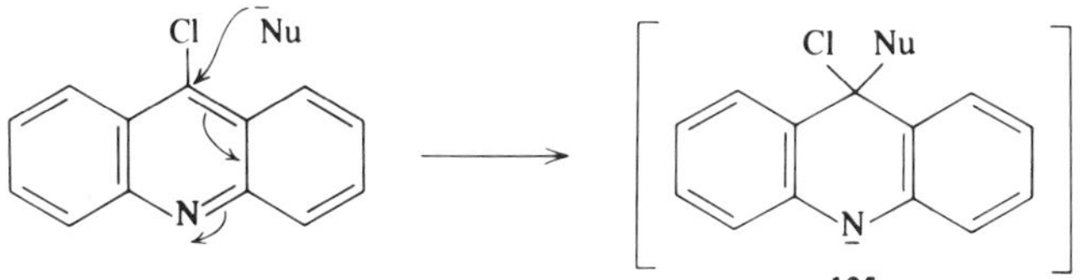

125

Nu = nucleophile

In certain circumstances, however, the 9-chloro group can be selectively displaced. Treatment of 7,9-dichloro-1,2,3,4-tetrahydroacridine with *p*-cresol and a stream of anhydrous ammonia left the 7-chlorine atom unaffected. Organic phase extraction of the product gave **126**; extraction with acid, followed by basification, gave **127**.

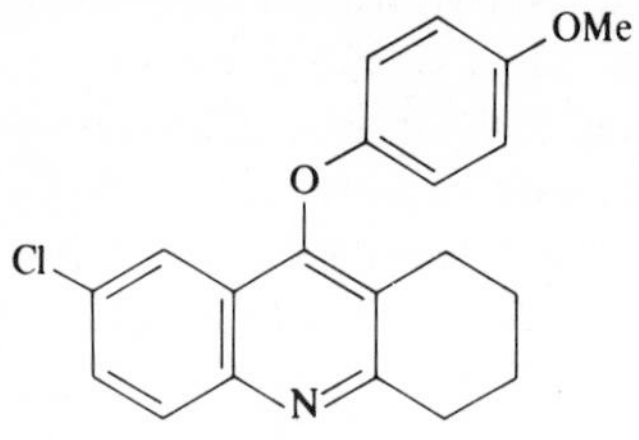

126

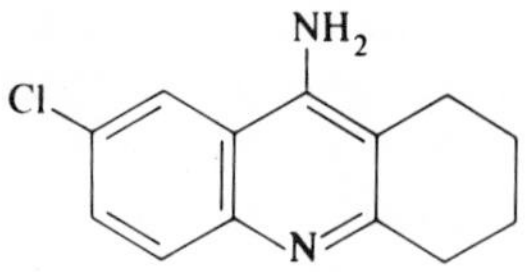

127

9-Bromo-1,2,3,4-tetrahydroacridine has been obtained from the 9-acridanone and phosphorus tribromide.[240]

9-Amino-1,2,3,4-tetrahydroacridine ("tacrine"), colorless crystals mp 178–180° from benzene, is a comparatively strong base and forms mono- and diacetyl derivatives.[244] It is a compound of considerable pharmacological interest that has been prepared in a variety of ways. The chlorine atom of 9-chloro-1,2,3,4-tetrahydroacridine has been replaced by an amino group, using alcoholic ammonia and a copper salt catalyst.[245] Heating 2-amino-benzonitrile with cyclohexanone in the presence of 1 mole of zinc chloride gave a 96% yield of 1:1 complex of zinc chloride and tacrine that was decomposed to the acridine with alkali.[334] The difficulty of obtaining 2-amino-benzonitrile led to a variation on this technique, in which 2-aminobenzamide (anthranilamide), cyclohexanone, and a catalytic amount of zinc chloride were heated to form the anil **128**, which was cyclized at 220–250° with zinc

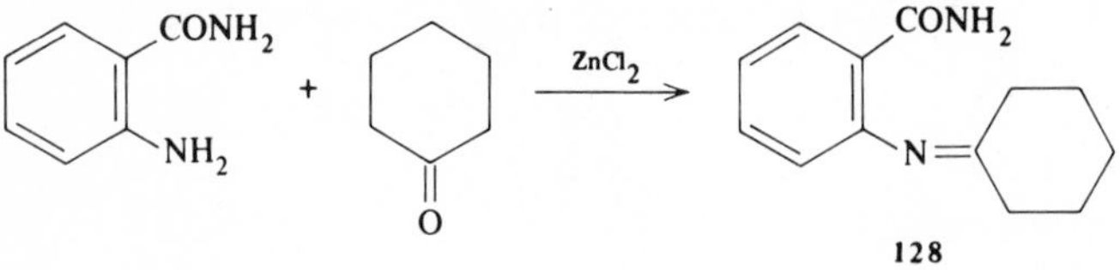

128

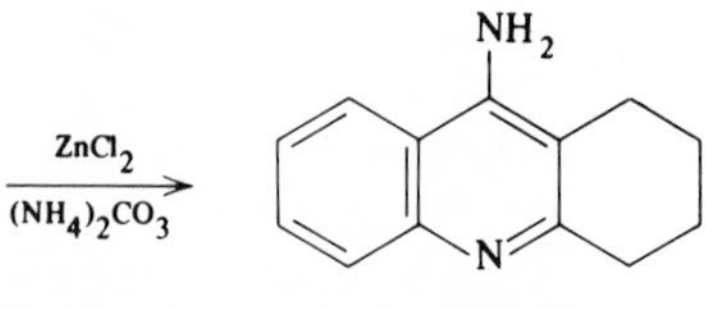

129

chloride in ammonium carbonate[334] to give a 55% yield of the tetrahydro-acridine **129**.

Tacrine finds use as an analeptic; it inhibits acetylcholinesterase,[247] and has also been shown to possess 10% of the anticholinesterase activity of neo-stigmine.[336] Its effect on neuromuscular transmission has been investigated.[337] Tacrine is a morphine antagonist,[338] and mixtures of morphine and tacrine have been claimed as nonhabit-forming analgesics.[251] The quaternary com-pounds obtained by the reaction of tacrine with alkyl halides or p-toluene-sulfonates have been claimed to show bacteriostatic and fungistatic activity.[339]

9-Amino-10-methyl-1,2,3,4,-tetrahydroacridinium p-toluenesulfonate (**130**) has been obtained in high yield by the cyclization of 2-methylaminobenzo-nitrile and cyclohexanone with p-toluenesulfonic acid.[220] The reaction pro-ceeds via the enamine; subsequent hydrolysis of the product[220] gives 10-methyl-1,2,3,4-tetrahydro-9-acridanone (**131**).

9-Arylamino-1,2,3,4-tetrahydroacridines have been prepared[340] by the intramolecular condensation of the arylamides of cyclohexylideneanthranilic acids (**132**). These azomethines were obtained in about 85% yield by heating the arylamides of anthranilic acid with cyclohexanone. Heating **132** in benzene with phosphoryl chloride gave the compounds **134** in yields up to 95%. The cyclization was considered to proceed via the imidochloride **133**.

Alternative procedures by which 9-phenylamino-1,2,3,4-tetrahydroacridine (**135**) may be prepared have been reported: heating N,N'-diphenylthiourea with cyclohexanone[341] and heating pimelic dianilide with phosphorus penta-chloride.[245]

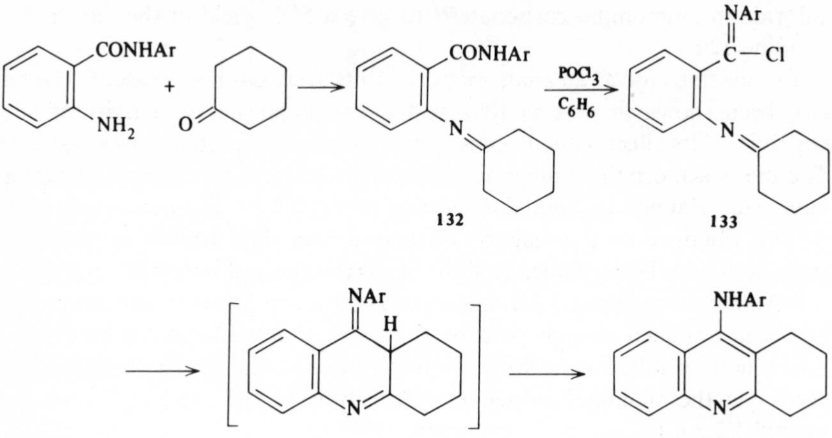

132　　　　　　　133

134

Compound **135** is hydrolyzed to the 9-acridanone and aniline with boiling concentrated hydrochloric acid.[245]

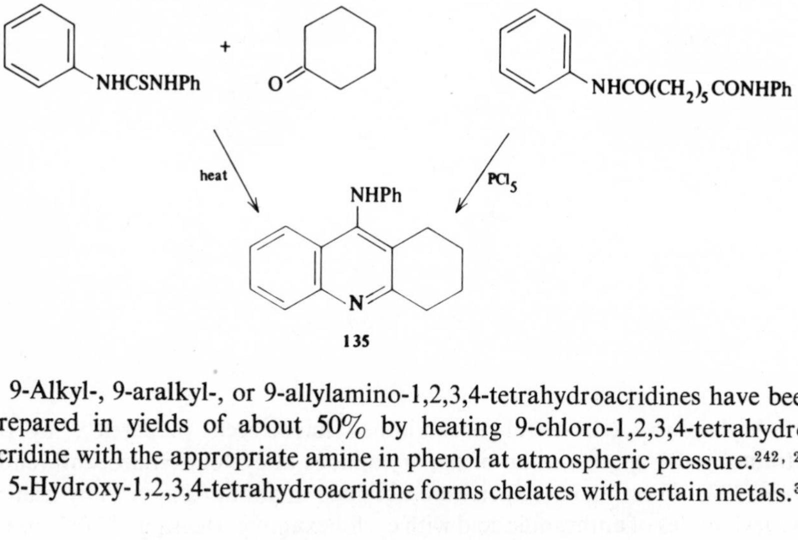

135

9-Alkyl-, 9-aralkyl-, or 9-allylamino-1,2,3,4-tetrahydroacridines have been prepared in yields of about 50% by heating 9-chloro-1,2,3,4-tetrahydro-acridine with the appropriate amine in phenol at atmospheric pressure.[242, 246] 5-Hydroxy-1,2,3,4-tetrahydroacridine forms chelates with certain metals.[335]

4. 1,2,3,4-Tetrahydro-9-acridanones

A. Methods of Preparation

The original preparation of 1,2,3,4-tetrahydro-9-acridanone (**137**) is a general one [248] and has frequently been employed for the preparation of the parent compound,[248, 249] and its alkyl-substituted analogues[250, 251] and alkyl-substituted 1-, 2-, or 3-carboxylic acid derivatives.[207] The method consists of the condensation of anthranilic acid with either a cyclohexanone, an alkyl-cyclohexanone, or an alkylcyclohexanonecarboxylic acid. The reaction[248] may be conducted in oil at 220° or without any solvent[207] for 1 hr at 125°, followed by a further 2-hr heating at 220°, or in diphenyl ether[249] for 1 hr at 120–130°. A somewhat better yield has been claimed when a water collector is used.[240] The reaction proceeds via the anil **136**.

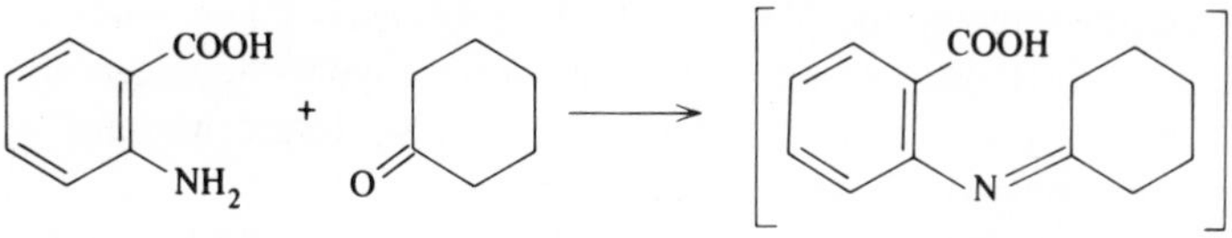

136

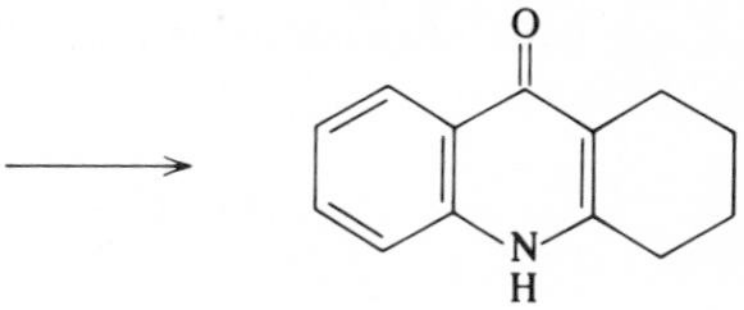

137

Considerable decarboxylation occurred with certain acids, e.g., with 4-methoxyanthranilic acid[252] and 5-nitroanthranilic acid.[203]

Heating *N*-methylanthranilic acid with cyclohexanone similarly gave 10-methyl-1,2,3,4-tetrahydro-9-acridanone,[250] although no anil can be formed as an intermediate. In this and similar cases, the enamine may be the inter-mediate. A series of 10-alkyl-1,2,3,4-tetrahydro-9-acridanones has been pre-pared with substituents ranging from ethyl to heptyl.[253, 254]

In some instances, more than one product might be expected: anthranilic acid with 3-methylcyclohexanone, for example, could give either **138** or **139**.

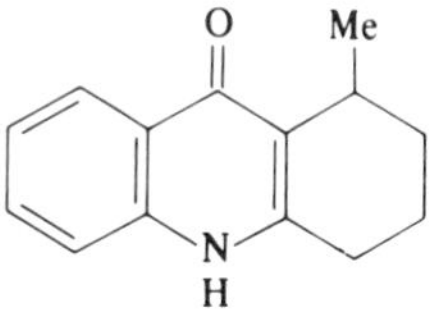

138

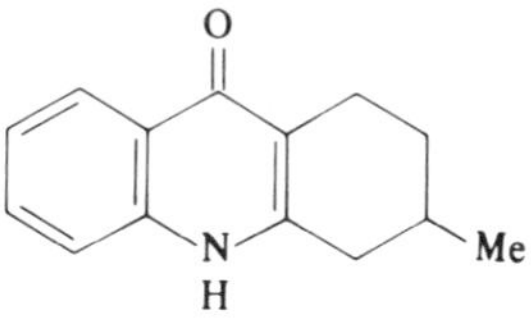

139

In fact, **139** appeared to be the exclusive product: dehydrogenation of the product gave 3-methyl-9-acridanone, which was identical with a sample prepared from 4-methylanthranilic acid and cyclohexanone, followed by dehydrogenation.[250]

An alternative procedure is the reaction of ethyl cyclohexanone-2-carbox-ylate with an aromatic amine at room temperature in the presence of a trace of acid, to give an anil that may be cyclized to the tetrahydro-9-acridanone by heating at 250°. It should be noted that simple heating of the reagents without prior formation of the anil gives the corresponding amides, which are cyclized with sulfuric acid to tetrahydrophenanthridones.[255, 256] The anils have been cyclized by heating alone to 255°, but this gave poor yields.[257, 258] Dropping the crude anils into liquid paraffin at 250–280° gave much improved results,[259] but the best technique is to use boiling diphenyl ether, since no charring of the product occurs and many 1,2,3,4-tetrahydro-9-acridanones have been obtained by this latter procedure.[252] The optimum amount of solvent depends on the individual cyclization.[260]

Cyclization of the anil derived from *m*-anisidine and ethyl cyclohexanone-2-carboxylate gave the two possible products, **140** and **141** in 2:1 ratio.[252] In the case of *m*-chloroaniline, the two isomers were obtained in equal amounts.[261]

A synthesis of 10-methyl-1,2,3,4-tetrahydro-9-acridanones (**144**) under much milder conditions than the above-mentioned techniques involves the reaction of the aminonitrile **142** with cyclohexanone in the presence of toluenesulfonic acid in refluxing toluene.[220] The intermediate **143** was isolated in high yield and converted to the product with alkali in 96% yield.

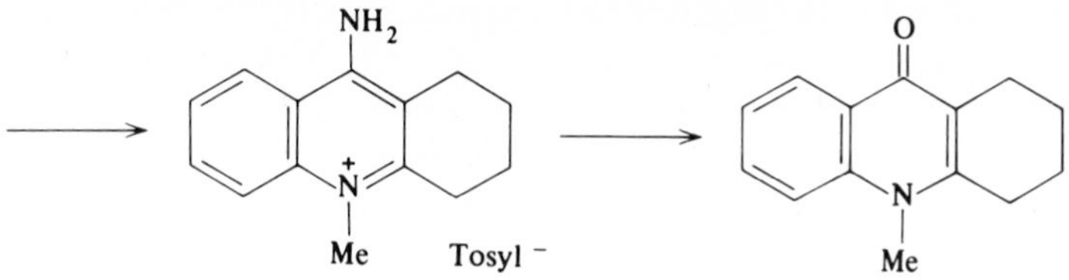

143   144

This synthesis appears to be a potentially versatile one.

B. General Properties and Reactions of 1,2,3,4-Tetrahydro-9-acridanones

1,2,3,4-Tetrahydro-9-acridanone is a colorless solid, mp 358°, very sparingly soluble in most organic solvents but which crystallizes from ethanol in needles.[248] It is much more basic than 9-acridanone, being soluble in dilute acids. This may be a reflection of a greater tendency of the tetrahydro-9-acridanone to form the hydroxyacridine tautomer (145).

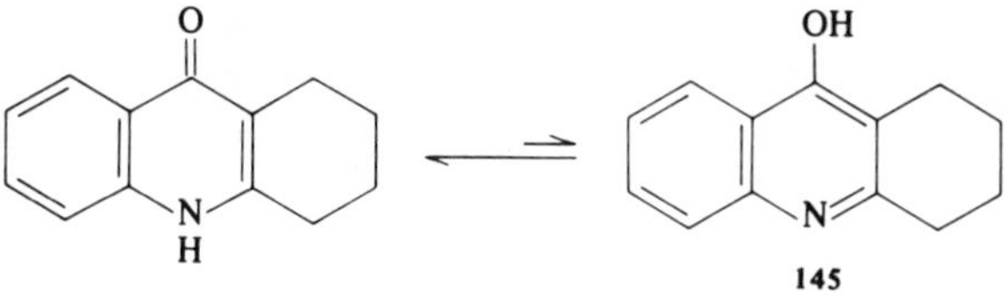

145

The hydrochloride is hydrolyzed by water, but the picrate can be crystallized from acetic acid. 1,2,3,4-Tetrahydro-9-acridanone has been dehydrogenated[248] to 9-acridanone by dry air at 280°, or better, in the presence of copper[250] at 360°. A number of other tetrahydroacridanones have been satisfactorily dehydrogenated by this procedure.[262] The distillation of tetrahydro-9-acridanone over zinc gave acridine.[248] In general, solutions of 1,2,3,4-tetrahydro-9-acridanones do not fluoresce in daylight but do so when exposed to uv light.

The reduction of 1,2,3,4-tetrahydro-9-acridanones with a variety of reagents gives cis and/or trans-1,2,3,4,4a,9,9a,10-octahydroacridines, 1,2,3,4,-5,6,7,8-octahydroacridines, or 1,2,3,4,4a,9a-hexahydro-9-acridanones. These are discussed in Sections 6.A, 5.A, and 8, respectively.

The reaction of 1,2,3,4-tetrahydro-9-acridanones with phosphoryl chloride, a widely used technique for the preparation of 9-chloro-1,2,3,4-tetrahydroacridines, is discussed in Section 3.B.

1,2,3,4-Tetrahydro-9-acridanone with bromine in acetic acid gave, according to the conditions, the 7-bromo or 5,7-dibromo derivative.[263] These compounds were insoluble in dilute aqueous acids and were independently synthesized from the appropriate anthranilic acids. The bromination of 5,7-dimethyl-1,2,3,4-tetrahydro-9-acridanone gave a mono-substitution product, but the bromine appeared to be aliphatic in character, since the compound reacted with pyridine.[263] Also, 5-, 6-, 7-, and 8-nitro-1,2,3,4-tetrahydro-9-acridanones have been prepared from nitroanthranilic acids, and the nitration of 1,2,3,4-tetrahydro-9-acridanone was shown to give a mixture of the 5- and 7-nitro derivatives.[203] All these nitro compounds have been reduced to the corresponding amines that acetylate and diazotize normally.[203]

5. 1,2,3,4,5,6,7,8-Octahydroacridines

A. Methods of preparation

Sym-Octahydroacridines have been prepared most frequently by the cyclization of 2,2′-methylenedicyclohexanone (146) with amino compounds. A convenient preparation of the 1,5-diketone 146 is the β-acylethylation of cyclohexanone with the Mannich base 2-dimethylaminocyclohexanone.[265] The yield is 63%. A lower yield is obtained by the condensation of formaldehyde and cyclohexanone.[267] An efficient cyclization of 146 to *sym*-octahydroacridine (147) is brought about by hydroxylamine hydrochloride in ethanol in 95% yield.[265]

146   147

This reaction does not lead to the *asym*-octahydroacridine, as reported elsewhere.[266] A similar reaction,[267,268] using formamide instead of hydroxylamine hydrochloride, is less satisfactory, since α- and β-perhydroacridine are also formed simultaneously. The perhydroacridines may be removed by vacuum distillation[268] or by treatment of the reaction mixture with sodium nitrite and acetic acid, which causes precipitation of the 10-nitrosoperhydroacridines and leaves *sym*-octahydroacridine acetate in solution.[267] The cyclization of 146 to 147 has also been achieved with anhydrous ammonia in toluene at 150° under pressure in 54% yield.[269]

Sym-octahydroacridine and its 9-phenyl derivative have been prepared by pyrolysis of the appropriate adducts (148), obtained by the reaction of ethanolamine with 146 or 2,2′-phenylmethylenedicyclohexanone.[270]

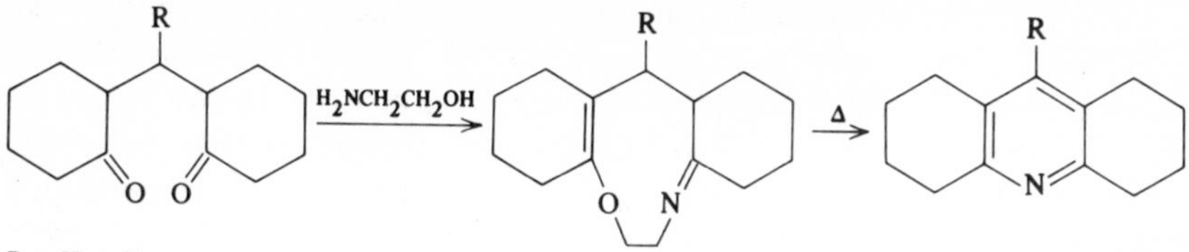

R = H or Ph

148

Sym-octahydroacridinium chlorides, substituted at position 10 with a variety of aryl, aralkyl, and chloroalkyl groups, have been prepared in yields of 80–85% by cyclization of the 1,5-diketone **146** with the appropriate primary amine in benzene-carbon tetrachloride.[271] The preparation of the 1-naphthyl derivative **149** by this technique suggests that the reaction is not sensitive to steric effects.

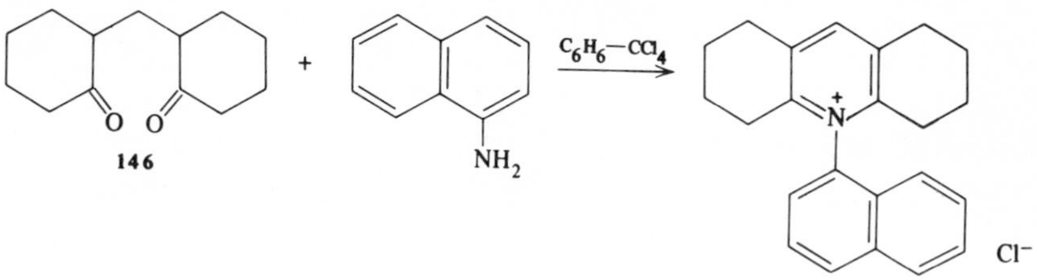

146   **NH₂**   **149**

The 4,5-disubstituted compound **151** was prepared by the aminoalkylation of the diketone **146** with paraformaldehyde and dimethylamine hydrochloride,[272] followed by cyclization of **150** with hydroxylamine hydrochloride.

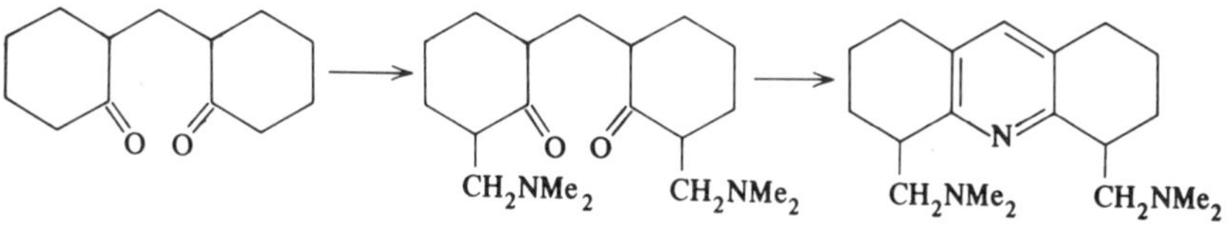

146   **150**   **151**

A related reaction of 2,2′-methylenedicyclohexanone with aniline in 95% acetic acid gave rise to 10-phenyl-*sym*-octahydroacridinium perchlorate after removal of 10-phenyl-$\Delta^{4a,10}$-dodecahydroacridine and treatment with sodium perchlorate.[273] *Sym*-octahydroxanthylium perchlorate was prepared by the condensation of oxymethylenecyclohexanone and cyclohexanone in about 55% yield; this has been converted to *sym*-octahydroacridinium perchlorate[274–276] with anhydrous ammonia, or to 10-phenyl-*sym*-octahydroacridinium perchlorate[273] with aniline, in unstated yields.

Hydrogenation and dehydrogenation of other acridines or polyhydro-acridines do not constitute good preparative routes to *sym*-octahydroacridines. Acridine has been hydrogenated over Raney nickel at 100° to give a 16% yield of *sym*-octahydroacridine.[76] *Trans-asym*-octahydroacridine has been iso-merized at 200° over palladium charcoal to give the symmetrical compound exclusively, but at higher temperatures tetrahydroacridine was also formed.[162] The pyrolysis of 4a,10a-dicyanoperhydroacridine (see Section 7.A) with calcium oxide-calcium carbonate gave *sym*-octahydroacridine in unstated yield.[277, 278]

Fused ring analogues of *sym*-octahydroacridine have been obtained by the pyrolysis of *N,N,N*-trimethylhydrazonium tetrafluoroborate derivatives of fused ring cyclohexanones.[279] 4-Methyl-α-tetralone, for example, gave **152**, which was converted to **153** on pyrolysis.

152   **153**

B. General Properties and Reactions of *sym*-Octahydroacridines

1,2,3,4,5,6,7,8-Octahydroacridine (*sym*-octahydroacridine) colorless nee-dles, mp 69–71° from petroleum ether,[265, 267, 268, 278] forms a picrate mp 198–200°, which was recrystallized from ethanol.[265, 268, 278]

The 4- and 5-methylene groups of *sym*-octahydroacridine show the same type of reactivity as the 4-methylene group of 1,2,3,4-tetrahydroacridine and the methyl group of 2-methylpyridine. The treatment of *sym*-octahydroacri-dine (**154**) with acetic anhydride and benzaldehyde under reflux[280] gave the dibenzylidene derivative **155** in 6 hr.

154   **155**

A number of other hydroxy-, acetoxy-, mono-, and dinitro-benzylidene derivatives were prepared in the same way and in high yield in sterically favored cases. The analogous 9-phenyl-*sym*-octahydroacridine derivatives were prepared from the 9-phenyloctahydroacridine.[280] Other 4-substituted derivatives have been prepared by the reaction of *sym*-octahydroacridine with phenyl lithium and subsequent treatment with either benzaldehyde, formaldehyde, cyclohexanone, or benzyl chloride to give 4-phenylhydroxymethyl, 4-hydroxymethyl-, 4-(1-hydroxycyclohexyl)-, or 4-benzyl-*sym*-octahydroacridines.[281] Dehydration of the 4-hydroxymethyl compound **156** gave a 45% yield of the exocyclic olefin **157**, which was reduced to 4-methyl-*sym*-octahydroacridine.[282]

The oxidation of *sym*-octahydroacridine with peracetic acid gave the 10-oxide (**158**) in 76.5% yield,[283] and this on nitration afforded 9-nitro-*sym*-octahydroacridine 10-oxide (**159**). The reduction of **159** with sodium tetrathionate gave 9-amino-*sym*-octahydroacridine (**160**) in high yield[284]; this gave octahydro-9-acridanone (**161**) on diazotization.

10-Alkyl-substituted octahydroacridanones have been prepared by the catalytic hydrogenation of 1,2,3,4-tetrahydro-9-acridanones with Adam's catalyst. These compounds find use as sun-screening materials.[253]

The addition of **159** to acetyl chloride gave a 68% yield[284] of 9-chloro-*sym*-octahydroacridine 10-oxide (**162**). If the order of addition is reversed, **162** is accompanied by 36.5% of the 4-acetoxyl derivative (**163**).

Both the 9-nitro and 9-chloro compounds **159** and **162** were deoxygenated at position 10 normally by phosphorus trichloride.[284]

The treatment of *sym*-octahydroacridine 10-oxide with acetic anhydride gave a 90% yield of the 4-acetoxyl derivative (**164**), which on hydrolysis[285, 286] afforded the alcohol **165**. Reoxidation[285] gave *sym*-octahydroacridin-4-ol 10-oxide (**166**).

These last two reactions have been performed in one step by hydrolysis of the 4-acetoxyl derivative **164** with peracetic acid.[287] Both **165**[285, 286] and **166**[287] have been dehydrated with polyphosphoric acid, giving 1,2,3,4,7,8-hexahydroacridine and its 10-oxide, respectively (see Section 8). Acetylation of **166** gave 4,5-diacetoxy-*sym*-octahydroacridine,[287] which on hydrolysis and dehydration gave *sym*-tetrahydroacridine (see Section 8). The intermediate *sym*-

octahydroacridin-4,5-diol (167) gave the corresponding dichloro compound with thionyl chloride and was converted to the keto-alcohol 168 in 58% yield by activated manganese dioxide.[287]

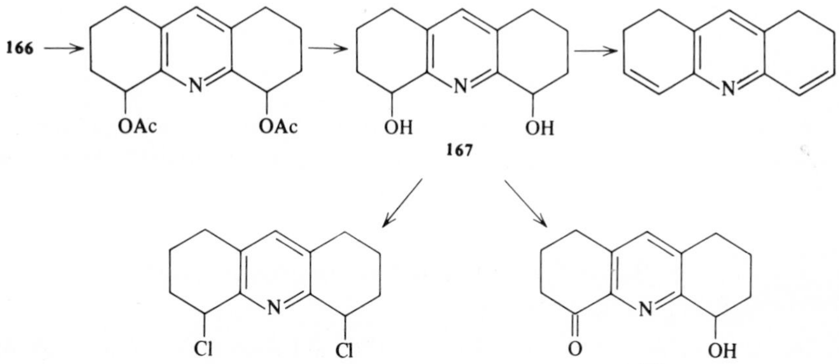

166 →

167

168

Similar oxidation of 165 gave the 4-oxo derivative; thionyl chloride gave 4-chloro-*sym*-octahydroacridine.[285]

Similar reactions[288] have been observed to occur with 9-chloro-*sym*-octa-hydroacridine oxide (162). The treatment of 162 with acetic anhydride gave the 4-acetoxyl derivative 169, which was converted to the alcohol 170 on hydrolysis. Thionyl chloride converted this to 4,9-dichloro-*sym*-octahydro-acridine (171), yielding 9-chloro-1,2,3,4,7,8-hexahydroacridine with poly-phosphoric acid. 9-Chloro-*sym*-octahydroacridine reacted quantitatively with ethanolamine,[288] giving 172.

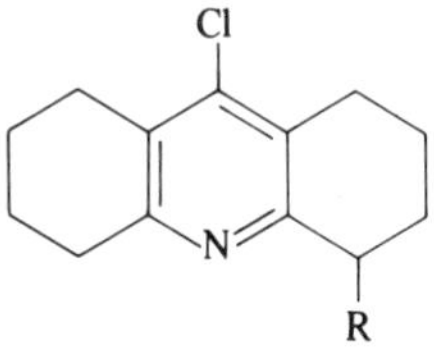

169, R = OAc
170, R = OH
171, R = Cl

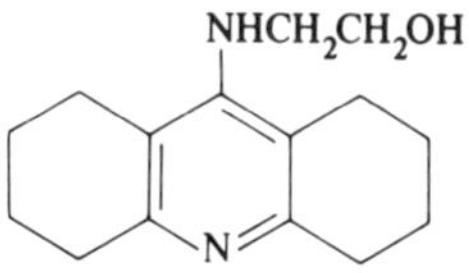

172

Simultaneous treatment of *sym*-octahydroacridine 10-oxide with *p*-nitro-benzaldehyde and acetic anhydride gave a 50% yield of 173, which was hydrolyzed subsequently to the alcohol; the latter gave a hexahydroacridine on dehydration (see Section 8).[289]

173

Catalytic dehydrogenation of *sym*-octahydroacridine with palladium on charcoal at 200-250° gave a mixture of acridine and acridan.[162] Attempted reduction with platinum oxide in acetic acid was unsuccessful.[290]

6. 1,2,3,4,4a,9,9a,10-Octahydroacridines

Asym-octahydroacridine exists in *cis* (**174**) and *trans* (**175**) forms, both of which have been prepared and characterized.

cis m.p. 72°   trans m.p. 82°

174   **175**

A. Methods of Preparation

Both *cis* and *trans* octahydroacridines are obtained by the reduction of 1,2,3,4-tetrahydroacridine with tin and hydrochloric acid[203, 291]; an improved separation procedure has given a yield of 33% of the *cis* isomer from the crude product.[292] The yield of the *trans* isomer is lower, and this is better prepared by the sodium amalgam reduction of 1,2,3,4-tetrahydro-9-acridanone in ethanol, which gives a 24% yield.[203, 292] *Trans*-octahydroacridine has been prepared in much higher yield (51.6%) by the reduction of 1,2,3,4-tetrahydroacridine with sodium formate in 95% formic acid.[293] The reaction required 20–22 hr at 170–175°, and the *trans*-isomer was extracted with ether from an aqueous solution of the crude product adjusted to pH 9–9.5. Further treatment gave 19% of the *cis* isomer. Other techniques that give the *trans* isomer are the reduction of acridine with red phosphorus and iodine[143] and the reduction of 9-phenylamino-1,2,3,4-tetrahydroacridine with sodium in ethanol.[245]

Cis-octahydroacridine has been prepared by catalytic hydrogenation of 1,2,3,4-tetrahydroacridine with platinium oxide in glacial acetic acid,[290] but the *cis* isomer was accompanied by smaller amounts of *trans*- and *sym*-octahydroacridine. The *cis* isomer appears to be the exclusive product when palladium/platinum/charcoal in methanol is used[231]; the same reduction of 1,2,3,4-tetrahydroacridin-4-ol gave an octahydroacridin-4-ol. Hydrogenation of acridine at 100° in dioxan in the presence of Raney nickel gave a mixture of reduced acridines containing 38% of *asym*-octahydroacridine, consisting mostly of the *trans*-isomer.[76] Copper chromite as the catalyst gave a mixture of *asym*-octahydroacridines in 70% yield.[76]

The reduction of 1,2,3,4-tetrahydroacridine-9-carboxylic acid with sodium amalgam gave two octahydroacridine-9-carboxylic acids, mp 180 and 221°, which were decarboxylated to give the *trans* and *cis asym*-octahydroacridines, respectively.[203]

B. Stereochemistry, General Properties, and Reactions of 1,2,3,4,4a,9,9a,10-Octahydroacridines

An assignment of stereochemical structure to the two isomers of *asym*-octahydroacridine, based on the observation that the melting points of one isomer and its derivatives were for the most part higher than those of the other isomer and its derivatives,[294] has been confirmed by nmr spectroscopy. [198, 295] The conclusion that the higher melting isomer has the *trans* configuration is consistent with the behavior of the methohydroxides towards Hofmann elimination.[292, 296] The original assignments are now considered to be incorrect.[297]

The nmr evidence is based on the magnitude of the coupling constants between the protons at positions 9 and 9a, and 4 and 4a. In the spectrum of the isomer, mp 72°,[198, 295] the two values of $J_{9,9a}$ were found to be 4.3 and 2.2 Hz; these values were consistent with dihedral angles of 40–50° and 50–60°, respectively, suggesting conformation **176** for the *cis* isomer. This conclusion was confirmed by the narrow band-width of the 4a-proton signal (9 Hz), which cannot include a *trans*-diaxial coupling. In the case of the rigid *trans* isomer (**177**), the 4a proton must be in a *trans*-diaxial relationship with one of the 4-protons, as well as the 9a-proton.

176   177

The 60 MHz spectrum[295] of the isomer mp 82° was rather intractable, but at 100 MHz the 4a-proton signal[198] was identified as a broadened multiplet with a band-width of 22 Hz. This was consistent with the two *trans*-diaxial couplings (to the proton at position 9a and to one of the protons at position 4), which are required by the *trans* structure **177**. In addition, one of the 9-proton signals showed a *trans*-diaxial coupling af 12 Hz.

Similar studies were made of the 10-methyl and 10-acetyl derivatives of the *cis* isomer, and the spectral parameters suggested that in these cases the alternative conformation **178** was involved.[295]

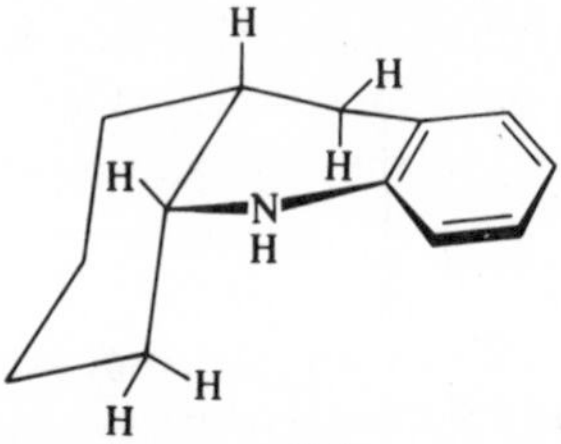

178

The 60 MHz spectrum of the benzoyl derivative of the *trans* isomer[295] showed a broadened triplet ($J = 9$ Hz) for the 4a-proton that is consistent with the 100 MHz spectrum of the unsubstituted *trans* isomer.[198]

The 10,10-dimethyl-*cis*- and -*trans*-octahydroacridinium hydroxides have been prepared and decomposed[292, 296] under Hofmann degradation conditions at 100°. In both cases, simple extrusion of methanol occurred with the regeneration of the base; this implies that in neither case were any of the hydrogen atoms β to nitrogen *trans*-coplanar with the nitrogen atom. This is inevitably so for the rigid *trans* methohydroxide, but it implies that the *cis* methohydroxide is in conformation **179**. This was shown to be the case by nmr spectroscopy.[292, 295]

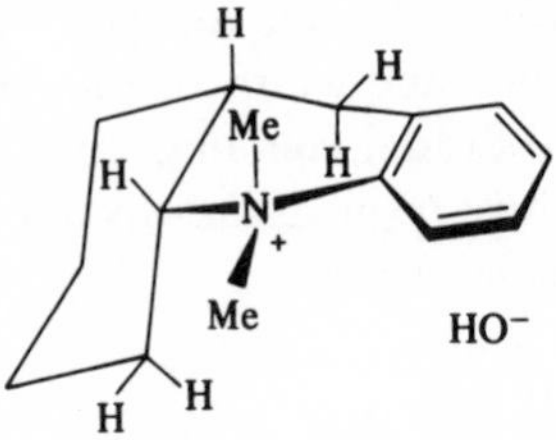

179

The nmr spectrum of the methiodide showed a 4a-proton signal possessing one *trans*-diaxial coupling, whereas a conformation like **176** would give rise to no such coupling.[295] Presumably, the methohydroxide is in the same conformation (**179**).

Both *cis* and *trans* octahydroacridine have been resolved into their enantiomeric forms with camphorsulfonic acid.[201] Racemization of the *cis* enantiomorphs occurred slowly at the melting point and was complete on distillation.

9-Hydroxymethyl-*as*-octahydroacridine (182) has been prepared[198] by the reduction of the ester 180 with anhydrous formic acid[293] to 181, which was further reduced with lithium aluminium hydride.

The structure of 182 was found by nmr spectroscopy[198] to be that of the *trans*-octahydroacridine with the 9- and 9a-protons in a *trans* relationship (183).

183

The nmr spectrum showed a broad multiplet for the 4a-proton with a band width of 27 Hz, which implies the presence of two *trans* couplings. The *trans* ring junction was also demonstrated by the broadened quartet structure ($J = 12$ Hz) of the 9a-proton signal, which results from three *trans* couplings and one (weak) axial-equatorial coupling.

The oxymethylene protons were magnetically nonequivalent, probably because of the asymmetry of the molecule. It seems likely that the acid 181 would have had the same configuration and reduction of the tetrahydroacridine with anhydrous formic acid led preferentially to the *trans*-octahydroacridine. The reduction of the unsubstituted tetrahydroacridine by a similar technique (see Section 6.A) also favored the *trans* over the *cis* octahydroacridine.[293]

A 9-methoxycarbonyl-*asym*-octahydroacridine[231] has been obtained by the reduction of 180 with hydrogen over platinum/palladium/charcoal in methanol, but the stereochemistry was not specified. However, the application of the same technique to tetrahydroacridine gave *cis*-octahydroacridine (see Section 6.A).

The treatment of **183** with phosphorus pentoxide gave 9-methyl-1,2,3,4-tetrahydroacridine (nmr spectrum recorded), presumably by the following sequence.[198]

The corresponding dehydration in the acridan series occurs with rearrangement leading to dibenzo[*bf*]azepines (see Section 2.B).

N-Dialkylaminoalkyl derivatives of *cis* and *trans* octahydroacridine have been prepared by reaction of the acridine with sodamide and the appropriate alkyl halide in toluene.[293] The products, administered as their maleates, were found to possess antidepressant activity of the same kind as imipramine but were inferior to imipramine. No significant difference in activity[298] was discerned between the *cis* and *trans* forms of **184**.

Asym-octahydroacridines may be converted to their *N*-acyl derivatives with acyl halides; also certain amino derivatives of the amides (e.g., **185**) are claimed to be long-acting analgesics, as well as having sedative, cough-suppressant, and hypotensive properties.[299]

10-Carbamoyl and 10-phenylcarbamoyl derivatives of *trans*-octahydroacridine have been prepared using sodium cyanate or phenylisocyanate in ice-cold acetic acid and were found to be anticonvulsants.[230] The 10-isonicotinoyl derivative was obtained from the acridine and the acid chloride, but no pharmacological activity was reported.[230]

7. Perhydroacridines

Tetradecahydroacridine or perhydroacridine (186) is, in theory, capable of existing in five geometric configurations.

186

Three of these, α-, β-, and γ-perhydroacridines, have been prepared, characterized, and assigned structures; the 10-methyl derivative of a fourth, δ-perhydroacridine, has been obtained but not further investigated.[300]

A. Methods of Preparation

Perhydroacridines are prepared either by the reduction of acridine or partially reduced acridines, or, directly or indirectly by the cyclization of 2,2'-methylenedicyclohexanone. A summary of the reactions leading to perhydroacridines is given in Table I.

The most convenient preparation of α-perhydroacridine is the catalytic reduction of acridine,[76, 290] using Raney nickel at 240°. At lower temperatures, dodecahydroacridine was obtained.[76] Alternatively, 11,14-dicyanoperhydroacridine, the preparation of which is described below, affords a rather lower yield of α-perhydroacridine on reduction with a sodium/potassium alloy in butanol.[278]

The only preparative procedures reported to give β-perhydroacridine alone are the catalytic reduction of 4a,10a-dicyanoperhydroacridine, using platinum oxide in dioxan-acetic acid,[301] and a similar reduction of 4a-cyano-Δ[10]-dodecahydroacridine.[301] All the other perhydroacridine syntheses gave mixtures of isomers, sometimes with partially reduced acridines.

β-Perhydroacridine has been prepared by the reductive cyclization of 2,2'-methylenedicyclohexanone with formamide.[267] The optimum yield of the β-isomer was 43%, but this was accompanied by 36% of the α-isomer and some sym-octahydroacridine. The perhydroacridines were separated from the sym-octahydroacridine, either by vacuum distillation[268] or by formation of their 10-nitroso derivatives with nitrous acid in acetic acid.[267] The nitroso derivatives separated as a yellow oil, leaving sym-octahydroacridine acetate in solution. The ir spectra of the derivatives of both pure perhydroacridine

TABLE I Syntheses of Perhydroactidines

Starting material	Reactions and conditions	Products	mp	Yield (%)	Ref.
Acridine	Raney nickel/hydrogen, 240°	α-Perhydroacridine	90.5-91.5°	81	76
Acridine, tetrahydroacridine, cis or trans-octahydro-acridine	Raney nickel/hydrogen 2.2-4.5 hr, 240-270°, 75-100 atm	α-Perhydroacridine	—	70-80	290
4a,10a-Dicyanoperhydro-acridine	2% Potassium in sodium alloy/butanol, 150°	α-Perhydroacridine	88°	31.5	278
2,2'-Methylenedicyclo-hexanone	Formamide/formic acid, 165°	α-Perhydroacridine	92-92.5°	36	267
		β-Perhydroacridine	49.5-50°	43	268
		sym-Octahydroacri-dine	71°	21	
trans-Octahydroacridine	Platinum oxide/acetic acid/hydrogen, normal tem-perature and pressure	α-Perhydroacridine	90.5-91.5°	60	290
		β-Perhydroacridine	48-51°	15	
$\Delta^{4a,10}$- Dodecahydro-acridine	Platinum oxide/dioxan/hydrogen, normal temperature and pressure	α-Perhydroacridine	89-90°	30.5	278
		β-Perhydroacridine	50-51°	28	
4a,10a-Dicyanoperhydro-acridine	Platinum oxide/acetic acid/dioxan/hydrogen, room tem-perature and pressure	β-Perhydroacridine	—	42.5	301
4a-Cyano-$\Delta^{5a,10}$- dodeca-hydroacridine	Platinum oxide/acetic acid/hydrogen, 3 hr at room temperature	β-Perhydroacridine	50-51°	49	301

494

Table I. (Continued)

Starting material	Reaction and conditions	Products	mp	Yield (%)	Ref.
cis-Octahydroacridine	Platinum oxide/acetic acid/hydrogen, room temperature and pressure	β-Perhydroacridine	48-51°	66	290
		γ-Perhydroacridine	72.5-74.5°	9	
		Dodecahydroacridine	61-63°	20	
sym-Octahydroacridine	Sodium/ethanol/nickel formate, room temperature	Unidentified perhydroacridines	87-88.5°	—	268
Acridine	Ruthenium oxide/hydrogen, 180°, 110 atm	Unidentified perhydroacridines	—	60	77

isomers and seven intermediate mixtures of these derivatives have been published[267] and it is possible to assess the isomeric composition of a mixture of 10-nitroso derivatives of α- and β-perhydroacridine by comparison of the ir spectrum with these published spectra.

It has been reported that the reduction of *sym*-octahydroacridine with sodium in ethanol in the presence of a nickel formate catalyst gives perhydroacridines identical with those obtained by vacuum distillation of the reaction product from the formamide reduction.[268] This suggests that direct reduction, by this technique, of the crude, undistilled formamide reduction product would give a mixture consisting exclusively of α- and β-perhydroacridine. These two isomers have been separated by treatment with benzoyl chloride in the presence of aqueous potassium carbonate.[267, 290] A precipitate of α-perhydroacridine hydrochloride was formed, and the β-isomer remained in solution as the benzoyl derivative, from which it was obtained by hydrolysis.

A detailed study of this cyclo-amination of 2,2'-methylenedicyclohexanone and its isomer tricyclohexanolone has been made. The yields obtained under various conditions have been tabulated.[267]

γ-Perhydroacridine has been obtained as a by-product (9%) in the hydrogenation of *cis*-octahydroacridine over platinum oxide.[290] The major product was the β-isomer.

Unspecified perhydroacridines were obtained when acridine was hydrogenated over ruthenium oxide.[77]

Substituted perhydroacridines have been prepared by cyclo-aminocyanation[273, 277, 278, 302] and cyclo-hydrazinocyanation[303] of 2,2'-methylenedicyclohexanone; this reaction was the starting point in the synthesis of various octa-, deca-, dodeca-, and per-hydroacridines. 4a,10a-Dicyanoperhydroacridine (187) was obtained in 76% yield from the 1,5-diketone, potassium cyanide and ammonium chloride.[278] A similar reaction employing primary amines, potassium cyanide, and 60–80% acetic acid gave yields in excess of 70% of the corresponding 10-substituted compounds (188).[302] The compounds 188 were converted to the 10-substituted perhydroacridines 189 with 85% formic acid, and the perhydroacridinylacetic acid was decarboxylated at 200° in glycerol to the 10-methylperhydroacridine.[302]

187

188   R = Me, CH$_2$CH$_2$OH,
Ph, CH$_2$COOH,
CH$_2$COOEt
    189

10-Methylperhydroacridines have also been prepared by direct methylation of the different stereoisomers using formalin and formic acid,[300, 304] and by catalytic hydrogenation of 10-methyl-*cis*- and -*trans*-octahydroacridines.[300] The reaction of 2,2′-methylenedicyclohexanone with hydrazine, potassium cyanide, and acetic acid[303] gave a 90% yield of 10-amino-4a,10a-dicyanoperhydroacridine (**190**). This reaction parallels the formation of a 1-amino-2,6-dicyanopiperidine from 2,6-heptanedione.[305]

4a,10a-Dicyanoperhydroacridine was obtained by treatment of **190** with nitrous acid, and the 10-formyl derivative (**191**) could be prepared either by reaction of **190** with formic acid, or directly by cyclization of the 1,5-diketone with potassium cyanide and formhydrazide in dilute acetic acid.

190   **191**

B. Stereochemistry

The three common stereoisomers of perhydroacridine have been assigned the following stereochemical structures:[300]

α, *trans, syn, trans*   β, *trans, anti, cis*

192   **193**

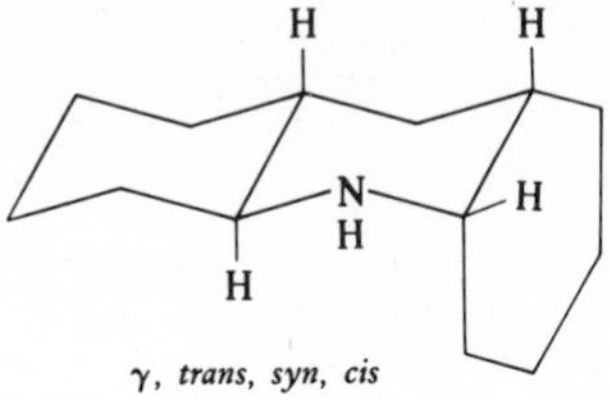

γ, *trans, syn, cis*

194

These assignments supersede earlier ones[306] and are based on the following physical and chemical evidence.

(1) *Kinetic Evidence*

The rates of quaternization of the 10-methyl derivatives of α-, β-, and γ-perhydroacridine with methyl iodide have been determined, along with the rates for a number of structurally analogous compounds.[300] It was found that the reaction rate was related to the basicity by a similar proportionality constant in all cases, except that of 10-methyl-β-perhydroacridine, which showed a much lower reaction rate for its basicity than the others. This markedly lower nucleophilicity was attributed to the steric interaction that would occur in the transition state between an axial β-methylene group and one of the 10-methyl groups. (It is assumed that the transition state resembles a quaternary ammonium cation.)

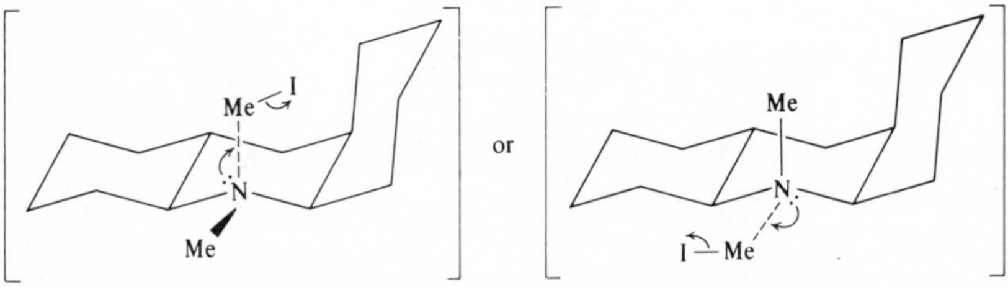

None of the other compounds used in the study possessed axial β-methylene groups, so that this result suggests a structure like **193** for the β-isomer. The presence of two α-methylene groups in *N*-methyl α- and γ-perhydroacridine resulted in a marginally lower rate of reaction for these isomers, compared with such compounds as 1-methyl-*trans*-perhydroquinoline and 5-methyl-*trans*-octahydrophenanthridine (see below), which have only one α-methylene group.

(2) *Bohlmann Bands*

The ir spectra of compounds containing a nitrogen lone electron pair with an anticoplanar α C—H bond show absorptions at 2820–2700 cm^{-1}, known as Bohlmann bands[307, 308] which are considered due to delocalization of charge from the lone pair to the C—H bond. The intensity of these bands is roughly proportional to the number of α C—H bonds anticoplanar to the lone pair.[308] The studies made in the perhydroacridine series were primarily intended to establish the conformational preference of the nitrogen lone pair in piperidine derivatives,[309] but the results for the N-methyl derivatives can be given a stereochemical interpretation if the assumption that the lone pair will adopt an axial conformation in these cases is accepted.[310, 311a, b] The Bohlmann band intensities for 10-methyl-α-, -β-, and -γ-perhydroacridines[309] were 10.0, 10.7, and 4.7 (l/cm^2. mole) $\times$ 10^{-3}. These figures reflect the fact that the N-methyl derivatives of **192, 193,** and **194** have, respectively, two, two, and one C—H bonds oriented anticoplanar to the axial nitrogen lone pair.

(3) *Hydrogen Bonding*

It has been established that there is an approximately linear relationship between the length of a hydrogen bond and the shift of the ir frequency of the associated hydroxyl group from that of the unassociated hydroxyl group frequency.[312–314a, b] In an ir study of the bonding of n-butanol to 10-methylperhydroacridines,[300] it was shown that the β-isomer gave rise to a longer hydrogen bond than the α- or γ-isomers. If it is accepted that the lone pair will adopt an axial conformation, then this result can be interpreted as a consequence of the steric interaction[300] of the n-butyl group with the axial β-methylene group found in structure **193**. Corresponding experiments with the unmethylated α- and β-isomers[310] showed that the hydrogen-bond lengths were approximately the same in these cases, which was taken to mean that inversion of the nitrogen atom in the β-isomer occurred, enabling an equatorial bond to form, of about the same length as the axial bond of the α-isomer.

(4) *Hofmann Degradation*

Pyrolysis of the 10,10-dimethyl-α- and -γ-perhydroacridinium hydroxides caused reversion to the original 10-methylperhydroacridine,[300] but the β-analogue gave 3-(2-dimethylaminocyclohexylmethyl)cyclohexene. This indicates that the β-isomer possesses a β-hydrogen atom *trans* coplanar to the nitrogen atom.[315] This is provided for in structure **193** but is absent in structures **192** and **194**. Simple extrusion of methanol occurred on heating the 10,10-dimethyl hydroxide derivatives of these compounds.

(5) *NMR Spectroscopic Evidence*

The nmr spectra of α- and β-perhydroacridine and some of their derivatives,[342] and of γ-perhydroacridine and some of its derivatives,[343] have been recorded; the chemical shifts of the 4a- and 10a-protons and the multiplicities of the signals support the assignments **192–194**

The spectrum of α-perhydroacridine[342] contained a two-proton multiplet at τ 7.95 (carbon tetrachloride as solvent) with a half-height band width of about 20 Hz, which was ascribed to the axial 4a- and 10a-protons (of **192**). The large band width is consistent with the two *trans*-diaxial couplings involved, and the chemical shift was in accord with the corresponding protons of other alkylpiperidines.[344]

The spectrum of β-perhydroacridine[342] included 2 one-proton signals at τ 7.15 and τ 7.95. The lower-field signal was a narrow multiplet free of *trans*-diaxial couplings and was ascribed to the 4a-proton of **193**, which is equatorial with respect to the cyclohexane ring. This assignment was confirmed by the strong upfield shift that occurred on *N*-methylation of the perhydroacridine. A similar observation has been made in the case of the 2- and 6-axial protons of certain piperidines.[344] The τ 7.95 signal was considered due to the axial 4a(10a)-proton.

γ-Perhydroacridine[343] in carbon tetrachloride gave an nmr spectrum containing a double triplet centered on τ 7.17, assigned to the 4a-proton of **194** (equatorial with respect to the piperidine ring) and a broader multiplet at τ 7.85, considered due to the 10a-proton. The major coupling of the double triplet was 11 Hz, which is consistent with the single *trans*-diaxial coupling expected for such a proton.

C. General Properties and Reactions of Perhydroacridines

Unsubstituted perhydroacridines are unreactive, apart from the normal reactions of secondary amines such as quaternization.[300] By contrast, the reactions of 4a,10a-dicyanoperhydroacridine are numerous and lead to various other hydroacridines such as 4a-cyano-Δ^{10}-dodecahydroacridine and tetracyclic derivatives of perhydroacridine.[278, 301]

Hydrolysis of the dicyano compound **195** with 60% sulfuric acid at 120° afforded the lactone of 4a-hydroxyperhydroacridine-10a-carboxylic acid (**196**), presumably by elimination, hydrolysis, and cyclization.[278, 301] The lactone could be titrated with ethanolic caustic soda, reflecting the ease of hydrolysis to the free acid. At 180° the lactone lost carbon dioxide to give $\Delta^{4a,10}$-dodecahydroacridine (**197**), which easily autoxidized in the manner of $\Delta^{1,8a}$-octahydroquinolines.[316]

Concentrated sulfuric acid converted **195** to the iminoimide of 4a,10a-perhydroacridine dicarboxylic acid (**198**). This was easily hydrolyzed to the imide **199**, which was itself very resistant to hydrolysis.[278, 301] However, long treatment with 70% sulfuric acid gave the lactone **196**. The reduction of **199**

with lithium aluminium hydride gave 4a,10a-dimethyleneiminoperhydro-
acridine (200), but with sodium in butanol the lactam of 4a-aminomethylene-
perhydroacridine-10a-carboxylic acid (201) was obtained. Treatment of the
dicyano compound 195 with 4% aqueous-ethanolic caustic soda gave the
lactam 202. These facile cyclization reactions were taken to demonstrate
the cis-diaxial relationship of the two cyano groups, and the reduction of 195
with sodium in butanol to α-perhydroacridine suggested the configuration[278]
shown below.

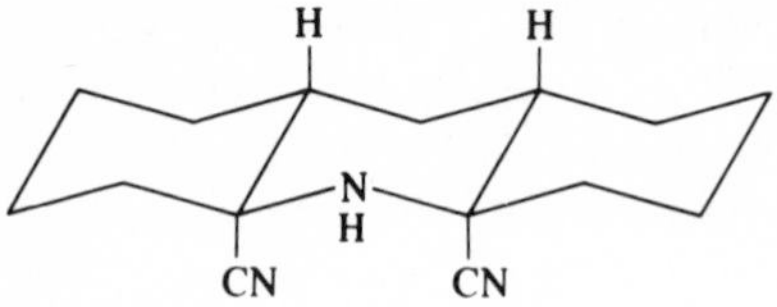

195 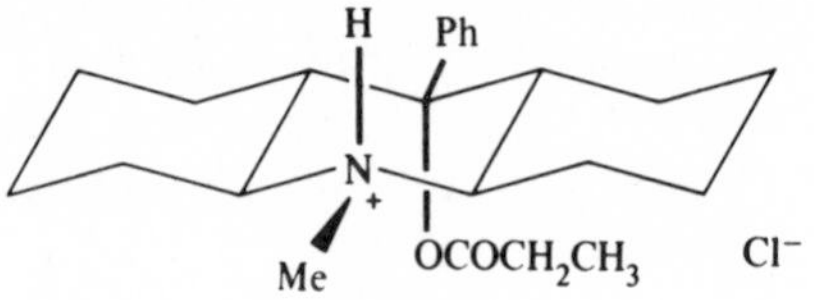   203

   In an investigation of the nature of the analgesic receptor site the compound 203 was prepared as a rigid analogue of prodine[304] but proved inactive
as an analgesic. The precursor alcohol was prepared by phenyl lithium reduction of α-perhydroacridanone. The structure of the latter was confirmed by
the presence of Bohlmann bands in the ir spectrum.
   A polymer containing perhydroacridine units has been prepared and had
insecticidal properties.[317]

8. Less Common Reduced Acridines

A. Dihydroacridines

   1,2-Dihydroacridine (204), an isomer of acridan, has been prepared by the
dehydration of 1,2,3,4-tetrahydroacridin-4-ol with polyphosphoric acid. The
alcohol was obtained by oxidation of the lithium salt of tetrahydroacridine
with gaseous oxygen.[318]

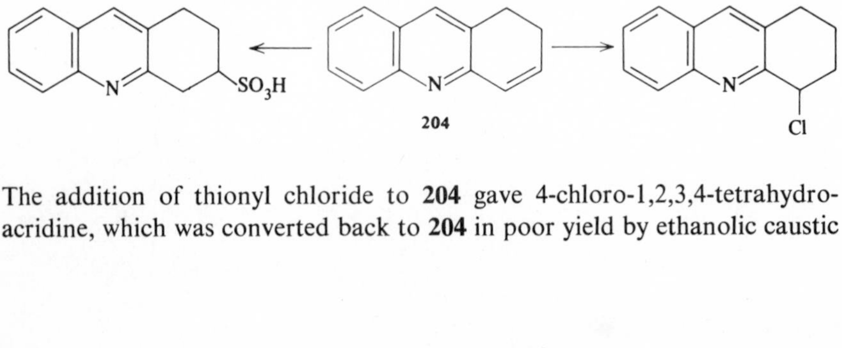

204

The addition of thionyl chloride to 204 gave 4-chloro-1,2,3,4-tetrahydro-
acridine, which was converted back to 204 in poor yield by ethanolic caustic

potash. The treatment of **204** in aqueous ethanol with sulfur trioxide gave the 3-sulfonic acid. [318]

A general single-stage synthesis of 1,2-dihydroacridines has been described that involves the condensation of 2-aminoacetophenone hydrochloride with aliphatic ketones.[235] A number of heterocyclic derivatives, such as **205** and **206**, have been prepared in this way.

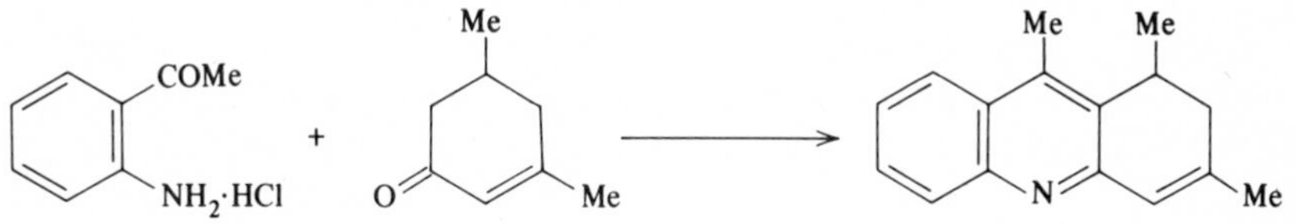

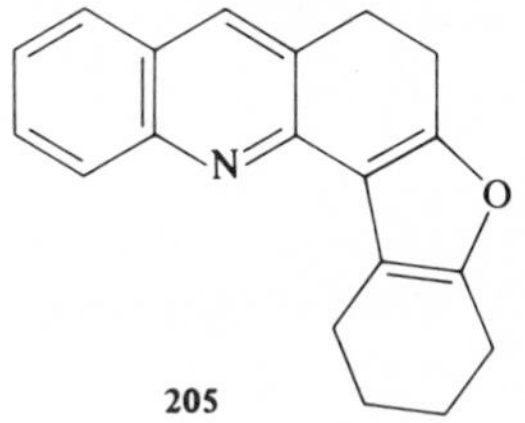

205 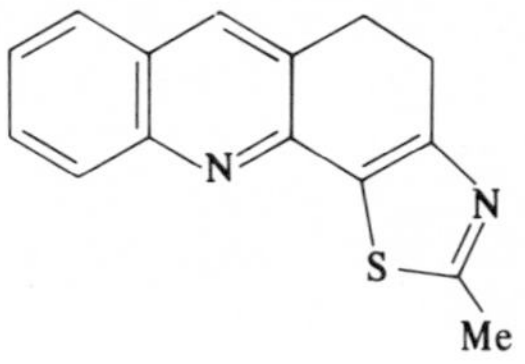   206

4,9-Dichloro-1,2,3,4-tetrahydroacridine, when heated under reflux with *N*-methylaniline and hydrated sodium carbonate, gave 1,2-dihydro-9-acridanone (**207**). Heating the dichlorotetrahydroacridine with anhydrous potassium carbonate and 4-amino-1-diethylaminopentane[243] gave 1,2-dihydro-9-chloroacridine (**208**).

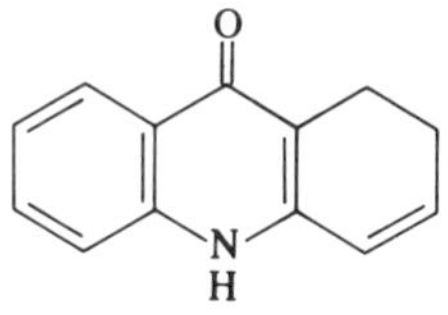

207 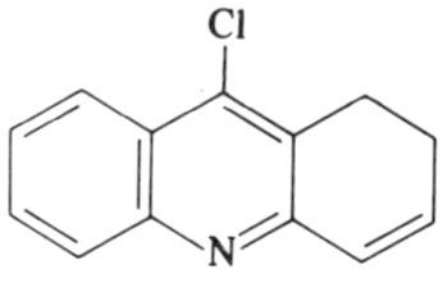   208

B. Sym-Tetrahydroacridines

1,2,7,8-Tetrahydroacridine (**210**), a cyclic analogue of 2,6-divinylpyridine, has been prepared in 67% yield by dehydration of the glycol **209** at 180° with polyphosphoric acid.[319, 320]

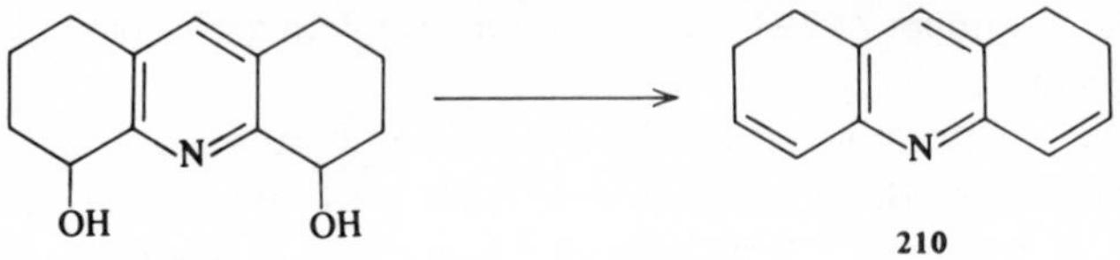

209 → 210

Reduction with platinum in acetic acid gave *sym*-octahydroacridine.[320] Acetoacetic ester in the presence of sodium[320] was added at position 3.

Another *sym*-tetrahydroacridine, 1,4,5,8-tetrahydroacridine (211) was obtained by the reduction of acridine with a large excess of lithium in ethanol[85]; the structure of the product was confirmed by its nmr, ir, and uv spectra, as well as by its reduction to *sym*-octahydroacridine with tris(triphenylphosphine)rhodium. Because anthracene gave 1,4,5,8,9,10-hexahydroanthracene under these conditions,[321] it seemed surprising that the pyridine ring should remain unreduced, and it was found that acridan also gave rise to 211. The elimination of metal hydride[85] appears to account for the aromatization of the dihydropyridine ring (cf. 212).

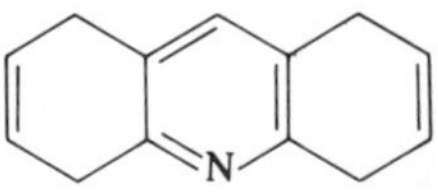

211

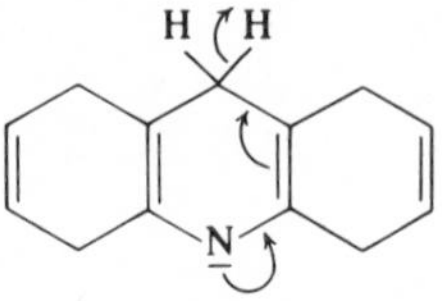

212

C. Hexahydroacridines

1,2,3,4,7,8-Hexahydroacridine (213) has been obtained by dehydration of 1,2,3,4,5,6,7,8-octahydroacridin-4-ol with polyphosphoric acid[285] and by reduction of the hexahydroacridine 10-oxide (214) with iron and acetic acid.[320] The *N*-oxide was prepared by dehydration of *sym*-octahydroacridin-4-ol 10-oxide.[320] The addition of hydrogen cyanide to 213 gave 3-cyano-*sym*-octahydroacridine.[285, 286] The chlorination of 9-chloro-*sym*-octahydroacridin-4-ol with thionyl chloride gave the 4,9-dichloro derivative, which was converted to 9-chloro-1,2,3,4,7,8-hexahydroacridine (215) on treatment with polyphosphoric acid.[322]

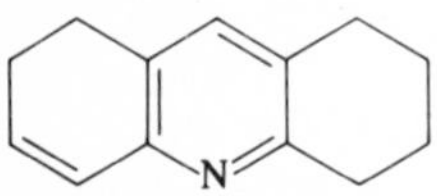

213

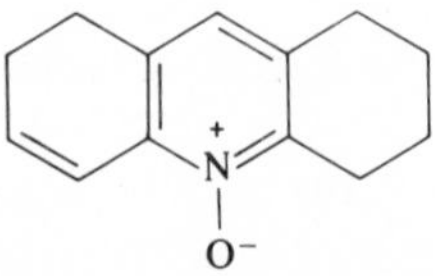

214

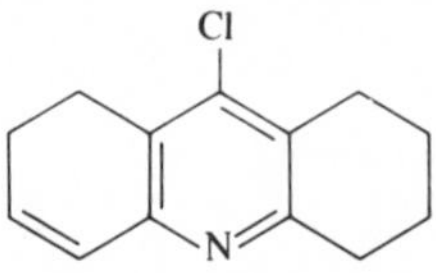

215

The addition of hydrogen cyanide gave 9-chloro-3-cyano-*sym*-octahydro-acridine.[322]

A number of 1,2,3,4,4a,9a-hexahydro-9-acridanones have been prepared by the lithium aluminium hydride reduction of 1,2,3,4-tetrahydro-9-acridanones[254] and are used as sun-screening materials. Some of the compounds (such as **216**) possess hypotensive activity.

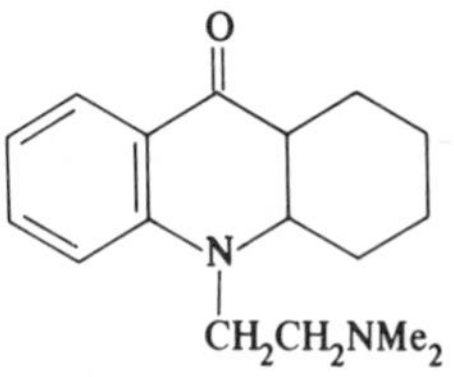

216

D. Decahydroacridines

A general method for the preparation of 10-substituted 1,2,3,4,5,6,7,8,9,10-decahydroacridines has been described[323] that involves the heating of a mixture of 2,2'-methylenedicyclohexanone and the appropriate primary amine, either at atmospheric pressure or in a bomb at 140–160°, with or without an inert solvent. The 10-phenyl derivative **217** was obtained quantitatively by this technique and has also been prepared by the dry distillation of 10-phenyl-4a,10a-dicyanoperhydroacridine.[302]

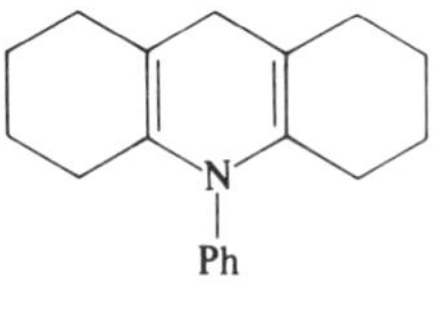

217

The reaction was reversed by hydrogen cyanide.[302] These compounds were unstable and absorbed oxygen from the air; they are reported to show activity as fungicides and herbicides.[323]

1,2,3,4,5,6,7,8,8a,10a-Decahydro-9-acridanone (**218**) has been prepared in 46% yield by the condensation of *cis*-hexahydroanthranilic acid with cyclohexanone and its nmr spectrum was recorded.[304] Apparently the same compound was obtained from the *trans* acid.[203] Hydrogenation of **218** with 5% rhodium on aluminium gave 9-hydroxyperhydroacridine,[304] which was methylated and oxidized to **219**.

218   219

The configuration of **219** was inferred from the presence of Bohlmann bands in the ir spectrum (see Section 7.B) and the lack of low-field equatorial methine proton resonances in the nmr spectrum.[304]

The 10-methyl derivative of **218** has been prepared by the reaction of cyclo-hexanone with 2-cyano-*N*-methylcyclohexylamine in the presence of *p*-toluenesulfonic acid[220] to give an 80% yield of the salt **220**, which on alkaline hydrolysis afforded the decahydro-9-acridanone **221** as a 77:23 mixture of the *trans* and *cis* isomers.

220   221

A large number of decahydro-1,8-acridindiones have been prepared by the condensation of aromatic aldehydes with cyclohexane-1,3-dione in the presence of ammonium acetate.[324] The use of aromatic amines,[325] instead of ammonium acetate, gave 9,10-diaryldecahydro-1,8-acridindiones, and other analogues with a variety of substituents have been synthesized.[326-328] These compounds may be oxidized to the octahydro analogues with chromium trioxide[325] or nitrous acid[264]; the reverse reduction is accomplished by hydrogenation over platinum.[283]

The compound **222** has been prepared from methylene-bis-dihydro-resorcinol and is stated to react with alcoholic ammonia[264] at 100° to give **223**.

223   222

The oxidation of **223** with nitrous acid gave the *sym*-octahydro-1,8-acridin-dione.[264] Zinc dust distillation of **223** gave acridine.[329] 3-, 9-, and 10-substituted analogues of **223** have also been prepared.[237, 329]

The ir spectra of a range of these decahydro-1,8-dioxo compounds have been recorded and compared in the 6 μ region with the spectra of other *trans*-rigid aminovinyl carbonyl compounds, but no assignments were made.[330]

E. Dodecahydroacridines

The dodecahydroacridine **224** has been isolated in 22% yield from the product of hydrogenation of acridine or acridan over Raney nickel[76] at 100°. The 10-phenyl derivative **225** resulted from the cyclization of 2,2'-methylenedicyclohexanone with aniline in 98% acetic acid.[273]

The structure was confirmed by hydrogenation with platinum oxide to the known 10-phenylperhydroacridine. *cis*-Octahydroacridine on hydrogenation over platinum oxide in glacial acetic acid gave a 20% yield of dodecahydroacridine.[290]

4a-Cyano-Δ^{10}-dodecahydroacridine (**227**) was proposed as an intermediate[278] in the acid hydrolysis of 4a,10a-dicyanoperhydroacridine to the lactone **228** and was subsequently prepared by treatment of **226** with concentrated acid and ether.[301]

The ir spectrum was recorded and, on the basis of a comparison with the spectrum of 1,4-dihydropyridine, it was concluded that some of the tautomeric enamine **229** was present.[301] The compound was easily autoxidized to **230**, like 3,4,5,6-tetrahydropyridines.[316, 331]

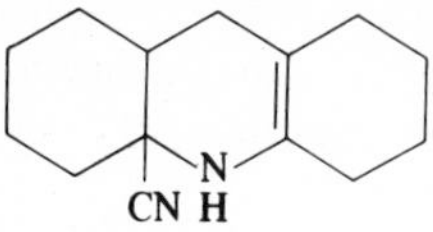

229

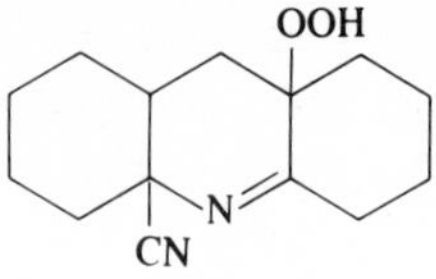

230

The unsubstituted $\Delta^{4a,10}$-dodecahydroacridine **231** was obtained[278, 332] in 69% yield by heating the lactone **228** to 180°. Apparently, the ring opens and decarboxylation occurs. The ir spectrum showed no NH absorption but did show a C=N band at 1670 cm^{-1}. Hydrogenation of **231** led to α- and β-perhydroacridines, and autoxidation occurred very easily, as was the case with $\Delta^{1,8a}$-octahydroquinoline.[316] The oxidation of **231** in heptane with atmospheric oxygen gave the unstable crystalline hydroperoxide **232**, which was reduced to the 9a-hydroxy-$\Delta^{4a,10}$-dodecahydroacridine (**233**) with sodium bisulfite.[332, 333]

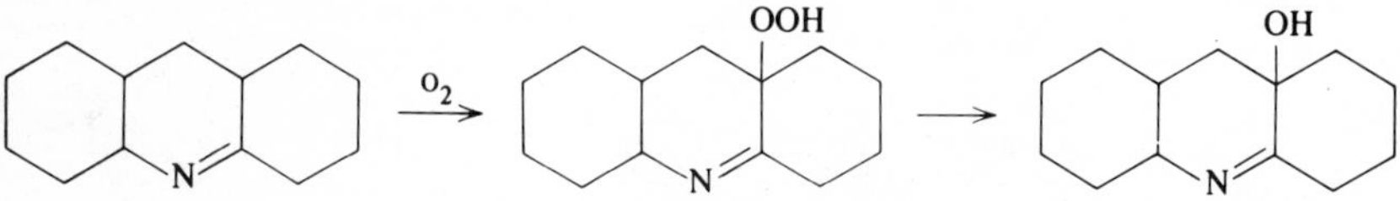

231

232

233

The addition of a little hydrochloric acid to a water-dioxan solution of **232** gave[332] the macrocyclic keto-lactam 5-azabicyclo[0.4.8]tetradecan-6,11-dione **234**.

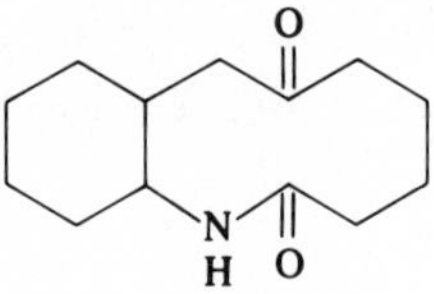

234

References

1. H. C. Longuet-Higgins and C. A. Coulson, *Trans. Faraday Soc.*, **43**, 87 (1947).
2. A. Albert and B. Ritchie, *J. Chem. Soc.*, 458 (1943).
3. M. Freund and G. Bode, *Chem. Ber.*, **42**, 1746 (1909).
4. F. Kroehnke and H. L. Honig, *Chem. Ber.*, **90**, 2215, 2226 (1957).

5. H. Decker and T. Hock, *Chem. Ber.*, **37**, 1564 (1904).
6. A. Schmid and H. Decker, *Chem. Ber.*, **39**, 933 (1906).
7. R. M. Acheson and C. W. C. Harvey, in press.
8. A. Kaufmann and A. Albertini, *Chem. Ber.*, **42**, 1999 (1909).
9. E. Grandmougin and K. Smirous, *Chem. Ber.*, **46**, 3425 (1913).
10. (a) F. Ullmann and P. Wenner, *Justus Liebigs Ann. Chem.*, **327**, 120 (1903); (b) F. Ullmann and P. Wenner, *Chem.-Ztg.*, *Chem. App.*, **44**, 329 (1913).
11. A. E. Dunston and T. P. Hilditch, *J. Chem. Soc.*, 1659 (1907).
12. C. H. Browning, J. B. Cohen, R. Gaunt, and R. Gulbransen, *Proc. Roy. Soc.*, *Ser. B.*, **93**, 329 (1922).
13. L. Benda, *Chem. Ber.*, **45**, 1787 (1912).
14. F. Ullmann and A. Maric, *Chem. Ber.*, **34**, 4307 (1901).
15. F. Ullmann and E. Naef, *Chem. Ber.*, **33**, 2470 (1900).
16. A. Weizmann, *J. Amer. Chem. Soc.*, **69**, 1224 (1947).
17. A. Albert and B. Ritchie, *J. Soc. Chem. Ind.*, **60**, 120 (1941).
18. A. Albert and W. H. Linnell, *J. Chem. Soc.*, 22 (1938).
19. J. Biehringer, *J. Prakt. Chem.*, **54**, 217 (1896).
20. K. Lehmstedt and H. Hundertmark, *Chem. Ber.*, **64**, 2386 (1931).
21. P. Kraenzlein, *Chem. Ber.*, **70**, 1785 (1937).
22. R. A. Reed, *J. Chem. Soc.*, 679 (1944).
23. B. L. Van Duuren, B. M. Goldschmidt, and H. H. Seltzman, *J. Chem. Soc.*, *B*, 814 (1967).
24. B. Loew and K. M. Snader, *J. Org. Chem.* **30**, 1914 (1965).
25. (a) F. Ullmann, *Chem. Ber.*, **36**, 1017 (1903); (b) H. Decker and G. Dunant, *Chem. Ber.*, **39**, 2720 (1906).
26. (a) W. H. Perkin and G. R. Clemo, British Patent 214,756; *Chem. Abstr.*, **18**, 2715 (1924); (b) S. M. Sherlin, G. I. Braz, A. I. Jakubovich, E. I. Worobjowa, and A. P. Ssergejef, *Justus Liebigs Ann. Chem.*, **516**, 218 (1935); (c) E. R. Blout and R. S. Corley, *J. Amer. Chem. Soc.*, **69**, 763 (1947).
27. S. Nitzsche, *Chem. Ber.*, **76**, 1187 (1943).
28. H. Buenzly and H. Decker, *Chem. Ber.*, **37**, 575 (1904).
29. H. Decker and R. Pschorr, *Chem. Ber.*, **37**, 3396 (1904).
30. F. Ullmann and R. Maag, *Chem. Ber.*, **40**, 2515 (1907).
31. M. Gomberg and L. H. Cone, *Justus Liebigs Ann. Chem.*, **370**, 203 (1909).
32. O. Eisleb, *Chem. Ber.*, **74**, 1433 (1941).
33. K. Lehmstedt and F. Dostal, *Chem. Ber.*, **72**, 1071 (1939).
34. M. Gomberg and D. L. Tabern, *J. Amer. Chem. Soc.*, **48**, 1345 (1926).
35. (a) O. Fischer and K. Demeler, *Chem. Ber.*, **32**, 1307 (1899); (b) K. Gleu and A. Schubert, *Chem. Ber.*, **73**, 757 (1940).
36. N. S. Drozdov and V. A. Sklyarov, *Trudy Kafedry Biokhim. Moskov. Zootekh. Inst. Konevodstva*, **33**, 1944 (1945); *Chem. Abstr.*, **41**, 764 (1947).
37. K. Gleu, S. Nitzsche, and A. Schubert, *Chem. Ber.*, **72**, 1093 (1939).
38. N. S. Drozdov, *Zh. Obshch. Khim.*, **9**, 1373, 1456 (1939); *Chem. Abstr.*, **34**, 1667, 2847 (1940).
39. M. Bockmuehl and L. Stein, German Patent 644,076; *Chem. Abstr.*, **31**, 5382 (1937).
40. A. Kaufmann and A. Albertini, *Chem. Ber.*, **42**, 1999 (1909).
41. A. Hantzsch and M. Kalb, *Chem. Ber.*, **32**, 3109 (1899).
42. J. G. Aston, *J. Amer. Chem. Soc.*, **53**, 1448 (1931).
43. J. J. Dobbie and C. K. Tinkler, *J. Chem. Soc.*, 269 (1905).
44. H. Decker, *J. Prakt. Chem.*, **45**, 161 (1892).

45. P. Ramart-Lucas, *C. R., Acad. Sci., Paris, Ser.* **211,** 436 (1940).
46. C. K. Tinkler, *J. Chem. Soc.,* 856 (1906).
47. J. G. Aston and C. W. Montgomery, *J. Amer. Chem. Soc.,* **53,** 4298 (1931).
48. A. Albert and R. Goldacre, *J. Chem. Soc.,* 454 (1943).
49. R. Goldacre and J. N. Phillips, *J. Chem. Soc.,* 1724 (1949).
50. E. D. Bergmann, M. Rabinovitz, and A. Bromberg, *Tetrahedron,* **24,** 1289 (1968).
51. P. Ehrlich and L. Benda, *Chem. Ber.,* **46,** 1931 (1913).
52. H. Decker and P. Becker, *Chem. Ber.,* **46,** 969 (1913).
53. (a) J. Gadamer, *Arch. Pharm.,* **243,** 12 (1905); (b) J. Gadamer, *J. Prakt. Chem.,* **84,** 817 (1911).
54. O. N. Chupakhin, V. A. Trofimof, and Z. V. Pushkareva, *Khim. Geterotsikl. Soedin.,* **5,** 954 (1969); *Chem. Abstr.,* **72,** 111'270u (1970).
55. A. Hantzsch and A. Horn, *Chem. Ber.,* **35,** 877 (1902).
56. A. Hantzsch, *Chem. Ber.,* **42,** 68 (1909).
57. J. F. Brunet, C. F. Hauser, and K. V. Nahabedian, *Proc. Chem. Soc.,* 305 (1961).
58. O. Dimroth and R. Criegee, *Chem. Ber.,* **90,** 2207 (1957).
59. N. J. Leonard, G. W. Leubner, and E. H. Burk, *J. Org. Chem.,* **15,** 979 (1950).
60. R. W. Hoffmann and J. Schneider, *Chem. Ber.,* **99,** 1899 (1966).
61. H. Decker and T. Hock, *Chem. Ber.,* **37,** 1002 (1904).
62. H. Decker and C. Schenk, *Chem. Ber.,* **39,** 748 (1906).
63. C. Schenk, *Chem. Ber.,* **39,** 2424 (1906).
64. R. Wizinger and H. Wenning, *Helv. Chim. Acta,* **23,** 247 (1940).
65. E. Bergmann and H. Corte, *Chem. Ber.,* **66,** 39 (1933).
66. J. H. Wilkinson and I. L. Finar, *J. Chem. Soc.,* 115 (1946).
67. O. Fischer and K. Demeler, *Chem. Ber.,* **32,** 1307 (1899).
68. J. J. Fox and J. T. Hewitt, *J. Chem. Soc.,* 529 (1904).
69. J. T. Hewitt and J. J. Fox, *J. Chem. Soc.,* 1058 (1905).
70. A. E. Dunstan and J. T. Hewitt, *J. Chem. Soc.,* 482, 1472 (1906).
71. S. Nitzsche, *Chem. Ber.,* **77,** 337 (1944).
72. F. Kehrmann and A. Stepanoff, *Chem. Ber.,* **41,** 4133 (1908).
73. C. Baezner and A. Gardiol, *Chem. Ber.,* **39,** 2623 (1906).
74. H. Baumann, German Patent 1,232,295; *Chem. Abstr.,* **66,** 76933q (1957).
75. R. M. Acheson, T. G. Hoult, and K. A. Barnard, *J. Chem. Soc.,* 4142 (1954).
76. H. Adkins and H. L. Coonradt, *J. Org. Chem.,* **63,** 1563 (1941).
77. A. A. Ponomarev, A. S. Chegolya, and V. N. Dyukareva, *Khim. Geterotsikl. Soedin., Akad. Nauk. Latv. SSR,* **00,** 239 (1966).
78. E. Bergmann, O. Blum-Bergmann, and A. von Christiani, *Justus Liebigs Ann. Chem.,* **483,** 80 (1930).
79. R. R. Burtner and J. W. Cusic, *J. Amer. Chem. Soc.,* **65,** 1582 (1943).
80. A. L. Tarnoky, *Biochem. J.,* **46,** 297 (1950).
81. K. Ito, *Yakugaku Zasshi,* **86,** 1166 (1966); *Chem. Abstr.,* **66,** 75899w (1967).
82. C. Graebe and H. Caro, *Justus Liebigs Ann. Chem.,* **158,** 265 (1871).
83. A. Bernthsen and F. Bender, *Chem. Ber.,* **16,** 1971 (1883).
84. J. L. Towle, *Iowa State J. Sci.,* **26,** 308 (1952).
85. A. J. Birch and H. H. Mantsch, *Aust. J. Chem.,* **22,** 1103 (1969).
86. R. Scholl and W. Neuberger, *Monatsh. Chem.,* **39,** 238 (1918).
87. F. Bohlmann, *Chem. Ber.,* **85,** 390 (1952).
88. S. G. Sastry, *J. Chem. Soc.,* 270 (1916).
89. A. P. Wolf and R. C. Anderson, *J. Amer. Chem. Soc.,* **77,** 1608 (1955).
90. (a) H. Becker and G. Dunant, *Chem. Ber.,* **39,** 2720 (1906); **42,** 1176 (1909); (b) K. Gleu and S. Nitzsche, *J. Prakt. Chem.,* **153,** 233 (1939).

91. G. M. Kosolapoff and C. S. Schoepfle, *J. Amer. Chem. Soc.* **76**, 1276 (1954).
92. E. Bergmann and O. Blum-Bergmann, *Chem. Ber.*, **63**, 757 (1930).
93. W. Schlenk and E. Bergmann, *Justus Liebigs Ann. Chem.*, **463**, 300 (1928).
94. O. Blum, *Chem. Ber.*, **62**, 881 (1929).
95. K. Lehmstedt, W. Bruns, and H. Klee, *Chem. Ber.*, **69**, 2399 (1936).
96. E. Bergmann and W. Rosenthal, *J. Prakt. Chem.*, **35**, 267 (1932).
97. K. Zeigler and H. Zieser, *Justus Liebigs Ann. Chem.*, **485**, 174 (1931).
98. H. Gilman and D. A. Shirley, *J. Amer. Chem.*, *Soc.* **72**, 2181 (1950).
99. A. Senier, P. C. Austin, and R. Clarke, *J. Chem. Soc.*, 1469 (1905).
100. W. L. Semon and D. Craig, *J. Amer. Chem. Soc.*, **58**, 1278 (1936).
101. F. W. Bergstrom and W. C. Fernelius, *Chem. Rev.*, **12**, 163 (1933).
102. G. A. Digenis, *J. Pharm. Sci.*, **58**, 335 (1969).
103. E. Hayashi, S. Ohsumi, and T. Maeda, *Yakugaku Zasshi*, **79**, 967 (1959); *Chem. Abstr.*, **53**, 21947 (1959).
104. H. Gilman, J. Eisch and T. Soddy, *J. Amer. Chem. Soc.*, **79**, 1245 (1957).
105. Smith, Kline, and French Laboratories, French Patent 1,530,413; *Chem. Abstr.*, **72**, 3397g (1970).
106. C. L. Zirkle, U.S. Patent 3,449,334; *Chem. Abstr.*, **71**, 49800r (1969).
107. A. Marxer, *Helv. Chim. Acta*, **49**, 572 (1966).
108. K. Issleib and L. Bruesehaber, *Z. Naturforsch.*, **206**, 181 (1965).
109. H. Hoberg, *Justus Liebigs Ann. Chem.*, **709**, 123 (1967).
110. Smith, Kline, and French Laboratories, French Patent M3550; *Chem. Abstr.*, **64**, 8157c (1966).
111. C. L. Zirkle, U.S. Patent 3,131,190.
112. N. V. Koninklijke Pharmaceutische Fabrieken voorheen Brocades-Stheeman and Pharmacia, Belgian Patent 638,839; *Chem. Abstr.*, **62**, 13131b (1965).
113. C. Kaiser and C. L. Zirkle, U.S. Patent 3,391,143; *Chem. Abstr.*, **69**, 96501n (1968).
114. P. Karrer, *Helv. Chim. Acta*, **6**, 402 (1923).
115. R. C. Kaye and H. I. Stonehill, *J. Chem. Soc.*, 27 (1951).
116. R. C. Kaye and H. I. Stonehill, *J. Chem. Soc.*, 2368 (1951); A. Blazek, R. Kalvoda, and J. Zyka, *Casopis Ceskeho Lekarniktva*, **63**, 138 (1950); *Chem. Abstr.*, **46**, 7707 (1952).
117. D. L. Hammick and S. F. Mason, *J. Chem. Soc.*, 345 (1950).
118. B. Breyer, G. S. Buchanan, and H. Duewell, *J. Chem. Soc.*, 360 (1944).
119. E. R. Blout and R. S. Corley, *J. Amer. Chem. Soc.*, **69**, 763 (1947).
120. J. J. Lingane, C. G. Swain, and M. Fields, *J. Amer. Chem. Soc.*, **65**, 1348 (1943).
121. N. S. Hush, *J. Chem. Phys.*, **20**, 1660 (1952).
122. K. Lehmstedt and H. Hundertmark, *Chem. Ber.*, **62**, 414 (1929).
123. P. Ramart-Lucas, M. Grumez, and M. Martynoff, *Bull. Soc. Chim. Fr.*, **8**, 228 (1941).
124. D. Craig, *J. Amer. Chem. Soc.*, **60**, 1458 (1938).
125. D. P. Craig, *Chem. Rev.*, **38**, 494 (1946).
126. I. Molnar, *Swiss Patent* 450,415; *Chem. Abstr.*, **69**, 96500m (1968).
127. I. Molnar and T. Wagner-Jauregg, *Helv. Chim. Acta*, **52**, 401 (1969).
128. H. Goldstein and G. Huser, *Helv. Chim. Acta*, **27**, 616 (1944).
129. A. Baeyer and V. Villiger, *Chem. Ber.*, **37**, 3192 (1904).
130. M. E. Konshin and P. A. Petyunin, *Khim. Geterotsikl. Soedin.*, **310** (1967).
131. H. Goldstein and J. Vaymatchar, *Helv. Chim. Acta*, **11**, 245 (1928).
132. H. Goldstein and W. Kopp, *Helv. Chim. Acta*, **11**, 478 (1928).
133. H. Goldstein and W. Rodel, *Helv. Chim. Acta*, **9**, 772 (1926).
134. F. Kehrmann, H. Ramm, and C. Schmajewski, *Helv. Chim. Acta*, **4**, 538 (1921).
135. F. Kehrmann and F. Brunner, *Helv. Chim. Acta*, **9**, 221 (1926).
136. B. Staskun, *J. Org. Chem.*, **33**, 3031 (1968).

137. F. Hoffmann-La Roche and Co., A-G, French Patent 1,374,544; *Chem. Abstr.*, **62,** 4040h (1965).

138. E. D. Bergmann and M. Rabinovitz, *J. Org. Chem.*, **25,** 827 (1960).

139. A. Campbell and E. N. Morgan, *J. Chem. Soc.*, 1711 (1958).

140. V. Zanker and E. Erhardt, *J. Thies. Ind. Chim. Belges,* **32,** 24 (1967); *Chem. Abstr.,* **70,** 87538y (1969).

141. R. Noyari, M. Kato, M. Kawanisi, and H. Nazaki, *Tetrahedron,* **25,** 1125 (1969).

142. H. Mantsch and V. Zanker, *Tetrahedron Lett.*, 4211 (1966).

143. C. Graebe, *Chem. Ber.*, **16,** 2828 (1883).

144. (a) K. Lehmstedt and F. Dostal, *Chem. Ber.*, **72,** 804 (1939); (b) N. S. Drozdov and O. M. Cherntsov, *Zh. Obshch. Khim.*, **21,** 1918 (1951).

145. H. Nozaki, Y. Yamamoto, and T. Nisimura, *Tetrahedron Lett.*, 4625 (1968).

146. T. Kato and Y. Yamamoto, *Chem. Pharm. Bull.* (Tokyo), **15,** 1426 (1967).

147. E. Hayashi and T. Nakura, *Yakugaku Zasshi*, **87,** 745 (1967); *Chem. Abstr.* **68,** 21815 p (1968).

148. C. Van der Steldt, P. S. Hofmann, and W. T. Nauta, *Arzneim. Forsch.*, **15,** 1081 (1965).

149. E. deB. Barnett, N. F. Goodway, A. G. Higgins, and C. A. Lawrence, *J. Chem. Soc.*, 1224 (1934).

150. H. Staudinger, H. N. Klever and P. Kober, *Justus Liebigs, Ann. Chem.*, **374,** 8 (1910).

151. S. A. Proctor and G. A. Taylor, *J. Chem. Soc.*, 5877 (1965).

152. S. A. Proctor and G. A. Taylor, *Chem. Commun.*, 569 (1965).

153. O. Diels and W. E. Thiele, *Justus Liebigs, Ann. Chem.*, **543,** 79 (1940).

154. R. M. Acheson and M. L. Burstall, *J. Chem. Soc.*, 3240 (1954).

155. F. Eiden and H. Wiedermann, *Tetrahedron Lett.*, 1111 (1970).

156. B. H. Klandermann, *Tetrahedron Lett.*, 6141 (1966).

157. (a) H. Decker and G. Dunant, *Chem. Ber.*, **42,** 1178 (1909); (b) G. Kraemer and A. Spilker, *Chem. Ber.*, **29,** 561 (1896); (c) O. Kruber and A. Räilhel, *Chem. Ber.*, **85,** 327 (1952).

158. R. Reynaud and R. Rumpf, *Bull. Soc. Chim. Fr.*, 1805 (1963).

159. K. von Auwers and R. Kraul, *Chem. Ber.*, **58,** 543 (1925).

160. V. V. Chelintsev and B. V. Tronov, *J. Russ. Phys. and Chem. Soc.*, **46,** 1886 (1914); *Chem. Zent.*, **III,** 540 (1915)

161. J. L. Claghorn, J. C. Schoolar, and J. Kinross-Wright, *Psychosomat, Med.*, **8,** 212 (1967).

162. T. Masamune and G. Homma, *J. Fac. Sci., Hokkaido Univ., Ser. III,* **5,** 64 (1957); *Chem. Abstr.*, **52,** 14581 (1958).

163. A. L. Tarnoky, *Biochem. J.*, **46,** 297 (1950).

164. J. Boes, *Apoth. Zeitg.*, **30,** 406 (1915); *Chem. Zent.*, **II,** 711 (1915).

165. A. Albert and J. B. Willis, *J. Soc. Chem. Ind.*, **65,** 26 (1946).

166. I. G. Farbenindustrie, French Patent 773,918; *Chem. Abstr.*, **29,** 2177 (1935).

167. S. J. Leach, J. H. Baxendale, and M. G. Evans, *Aust. J. Chem.*, **6,** 409 (1953).

168. R. Criegee, Ph.D. Dissertation, University of Würzburg, Germany, 1925; in *Methods of Preparative Organic Chemistry*, Interscience, New York-London, 1940, p. 4.

169. L. J. Sargent and L. F. Small, *J. Org. Chem.*, **13,** 447 (1948).

170. L. J. Sargent, *J. Org. Chem.*, **22,** 1494 (1957).

171. A. Meisels and A. Stomi, Swiss Patent 460,009; *Chem. Abstr.*, **70,** 5786c (1969).

172. A. Meisels and A. Stomi, Swiss Patent 460,010; *Chem. Abstr.*, **70,** 47320q (1969).

173. J. Mills, U.S. Patent 2,586,370; *Chem. Abstr.*, **46,** 8153 (1952).

174. F. Kehrmann and J. Tschui, *Helv. Chim. Acta*, **8,** 24, 27 (1925).

175. W. G. Brown and N. J. Letang, *J. Amer. Chem. Soc.*, **63,** 358 (1941).

176. L. A. Kalashnikova, M. B. Neiman, E. G. Rozantsev, and L. A. Skripko, *Zh. Org. Khim.*, **2**, 1529 (1966).
177. K. Freudenberg, F. Klinck, E. Flickinger, and A. Sobek, *Chem. Ber.*, **72**, 217 (1939).
178. K. Freudenberg and E. Plankenhorn, *Chem. Ber.*, **80**, 149 (1947).
179. N. A. Shepard, *Ind. Eng. Chem.*, **28**, 281 (1936).
180. R. L. Miller, W. J. Chamberlain, and R. L. Stedman, *Tobacco Sci.*, **13**, 21 (1969); *Chem. Abstr.*, **70**, 94068y (1969).
181. F. Schatz, U. Jahn, R. W. Adrian, and I. Molnar, *Arzneim.-Forsch.*, **18**, 862 (1968); *Chem. Abstr.*, **69**, 58148v (1968).
182. U. Jahn and G. Haeusler, *Wien. Klin. Wochschr.*, **78**, 21 (1966); *Chem. Abstr.*, **64**, 11737h (1966).
183. I. Molnar and T. Wagner-Jauregg, *Helv. Chim. Acta*, **48**, 1782 (1965); *Chem. Abstr.*, **64**, 3476e (1966).
184. M. Haering, I. Molnar, and T. Wagner-Jauregg, U.S. Patent 3,284,454; *Chem. Abstr.*, **66**, 28684n (1967).
185. J. R. Geigy A.-G., Netherlands Patent Applied 6,603,826; *Chem. Abstr.*, **66**, 55412d (1967).
186. H. Kataoka, French Patent 1,438,357; *Chem. Abstr.*, **66**, 10856j (1967).
187. I. Molnar and T. Wagner-Jauregg, Swiss Patent 455,772; *Chem. Abstr.*, **69**, 86836m (1968).
188. I. Molnar and T. Wagner-Jauregg, Swiss Patent 451,144; *Chem. Abstr.*, **69**, 96502p (1968).
189. M. Haering, I. Molnar, and T. Wagner-Jauregg, U.S. Patent 3,431,342; *Chem. Abstr.*, **71**, 49799x (1969).
190. I. Molnar, Swiss Patent 467,263; *Chem. Abstr.*, **71**, 38827d (1969).
191. H. Goldstein and M. Piolino, *Helv. Chim. Acta*, **10**, 334 (1927).
192. F. Kehrmann and H. Goldstein, *Helv. Chim. Acta*, **4**, 26 (1921).
193. F. Kehrmann and M. Rohr, *Helv. Chim. Acta*, **10**, 596 (1927).
194. M. E. Konshin and P. A. Petyunin, *Biol. Aktiv. Soedin., Akad. Nauk SSSR*, 155 (1965); *Chem. Abstr.*, **63**, 16348g (1965).
195. H. Goldstein and W. Kopp, *Helv. Chim. Acta*, **11**, 486 (1928).
196. (a) F. Kehrmann and J. Tschui, *Helv. Chim. Acta*, **8**, 23 (1925); (b) H. Goldstein and W. Kopp, *Helv. Chim. Acta*, **11**, 478 (1928).
197. P. N. Craig, B. M. Lester, A. J. Saggiomo, C. Kaiser, and C. L. Zirkle, *J. Org. Chem.*, **26**, 135 (1961).
198. V. G. Ermolaeva, N. P. Kostyuchenko, V. G. Yashunskii, and Y. N. Sheinker, *Khim-Farm. Zh.*, **3**, 19 (1969).
199. A. Schoenberg and K. Junghaus, *Chem. Ber.*, **99**, 1015 (1966).
200. W. Borsche, H. Tiedtke, and R. Schmidt, *Justus Liebigs Ann. Chem.*, **377**, 78 (1910).
201. W. H. Perkin and W. G. Sedgwick, *J. Chem. Soc.*, 438 (1926).
202. W. O. Kermack and W. Muir, *J. Chem. Soc.*, 3089 (1931).
203. W. H. Perkin and W. G. Sedgwick, *J. Chem. Soc.*, 2437 (1924).
204. W. Borsche and W. Rottsieper, *Justus Liebigs Ann. Chem.*, **377**, 101 (1910).
205. N. P. Buu-Hoi, *J. Chem. Soc.*, 795 (1946).
206. (a) J. von Braun and P. Wolff, *Chem. Ber.*, **55**, 3675 (1922); (b) N. P. Buu-Hoi, N. Hoan, N.-H. Khoi, and N.-D. Xuong, *J. Org. Chem.*, **15**, 511 (1950); (c) N. P. Buu-Hoi, N. Hoan and N.-D. Xuong, *J. Chem. Soc.*, 279 (1952).
207. M. Artico and V. Nacci, *Ann. Chim.*, **55**, 1085 (1965).
208. N. P. Buu-Hoi, R. Royer, N.-D. Xuong, and P. Jacquinon, *J. Org. Chem.*, **18**, 1209 (1953).

209. W. Borsche, *Justus Liebigs Ann. Chem.*, **377**, 70 (1910).
210. (a) W. Borsche and M. Wagner-Roemmich, *Justus Liebigs Ann. Chem.*, **544**, 272 (1940); (b) E. A. Robinson and M. T. Bogert, *J. Org. Chem.*, **1**, 65 (1936).
211. V. Q. Yen, N. P. Buu-Hoi, and N.-D. Xuong, *J. Org. Chem.*, **23**, 1858 (1958).
212. F. L. Al-Tai, G. Y. Sarkis, and J. M. Al-Janabi, *Bull. Coll. Sci.*, **9**, 55 (1966); *Chem. Abstr.*, **68**, 39436A (1968).
213. W. Borsche and J. Barthenheimer, *Justus Liebigs Ann. Chem.*, **548**, 50 (1941).
214. E. A. Fehnel, *J. Org. Chem.*, **31**, 2899 (1966).
215. M. Scholz, H. Limmer, and G. Kempter, *Z. Chem.*, **5**, 154 (1965); *Chem. Abstr.* **64**, 4905b (1966).
216. V. A. Petrow *J. Chem. Soc.*, 693 (1942).
217. L. J. Sargent and L. F. Small, *J. Org. Chem.*, **19**, 1400 (1954).
218. J. M. F. Gagan and D. Lloyd, *Chem. Commun.*, 1043 (1967).
219. B. D. Tilak, H. Berde, V. N. Gogte, and T. Ravindranathan, *Indian J. Chem.*, **8**, 1 (1970).
220. S. Singh and A. I. Meyers, *J. Heterocyclic Chem.*, **5**, 737 (1968).
221. G. E. Hall and J. Walker, *J. Chem. Soc.*, 2237 (1968).
222. A. Albert and R. Goldacre, *J. Chem. Soc.*, 706 (1946).
223. G. E. Calf and E. Ritchie, *Proc. Roy. Soc. New South Wales*, **83**, 117 (1949).
224. E. Hayashi and R. Goto, *Yakugaku Zasshi*, **85**, 645 (1965); *Chem. Abstr.*, **63**, 9913a (1965).
225. V. Petrow, *J. Chem. Soc.*, 18 (1945).
226. C. F. Koelsch, *J. Org. Chem.*, **16**, 1362 (1951).
227. L. G. S. Brooker, F. L. White, R. H. Sprague, R. H. Dent, and G. van Zandt, *Chem. Rev.*, **41**, 325 (1947).
228. W. Borsche and R. Manteuffel, *Justus Liebigs Ann. Chem.*, **534**, 56 (1938).
229. W. Borsche and R. Manteuffel, *Justus Liebigs Ann. Chem.*, **526**, 22 (1936).
230. V. G. Ermolaeva and V. G. Yashunskii, *Khim-Farm. Zh.*, **3**, 13 (1969). *Chem. Abstr.* **72**, 78832d (1970).
231. E. Hayashi and T. Nagao, *Yakugaku Zasshi*, **84**, 198 (1964); *Chem. Abstr.* **61**, 3071 (1964).
232. T. Okano and T. Takahashi, *Yakugaku Zasshi*, **89**, 1305 (1969); *Chem. Abstr.*, **72**, 12534b (1970).
233. C. Kaneko, I. Yokoe, S. Yamada, and M. Ishikawa, *Chem. Pharm. Bull.*, **17**, 1290 (1969); *Chem. Abstr.*, **71**, 81119m (1969).
234. W. Borsche and H. Hartmann, *Chem. Ber.*, **73**, 839 (1940).
235. G. Kempter and G. Moebius, *J. Prakt. Chem.*, **34**, 298 (1966); *Chem. Abstr.*, **66**, 46359q (1967).
236. B. H. Iyer and G. C, Chakravarti, *J. Indian Inst. Sci. A*, **14**, 157 (1932).
237. A. Sonn and H. Schreiber, *J. Prakt. Chem.*, **155**, 65 (1940).
238. W. Borsche, M. Wagner-Roemmich, and J. Bartenheimer, *Justus Liebigs Ann. Chem.*, **550**, 160 (1942).
239. C. Mannich and B. Reichert, *Arch. Pharm.*, **271**, 116 (1938).
240. L. J. Sargent and L. F. Small, *J. Org. Chem.*, **11**, 259 (1946).
241. U. Basu and S. J. Das-Gupta, *J. Indian Chem. Soc.*, **14**, 468 (1937).
242. M. V. Sigal, B. J. Brent, and P. Marchini, U.S. Patent 3,232,945; *Chem. Abstr.*, **64**, 14174h (1966).
243. W. H. Linnell and L. K. Sharp, *J. Pharm. Pharmacol.*, **2**, 145 (1950).
244. V. Petrow, *J. Chem. Soc.*, 634 (1947).
245. J. von Braun, A. Heymons, and G. Manz, *Chem. Ber.*, **64**, 227 (1931).

246. W. P. Brian and B. L. Souther, *J. Med. Chem.*, **8**, 143 (1965).
247. A. K. S. Ho and S. E. Freeman, *Nature*, **205**, 1118 (1965).
248. H. Tiedke, *Chem. Ber.*, **42**, 621 (1909).
249. R. W. Hinde and W. E. Matthews, Australian Patent 266,105; *Chem. Abstr.*, **68**, 105024t (1968).
250. R. A. Reed, *J. Chem. Soc.*, 425 (1944).
251. A. W. Woods Proprietary, Ltd., British Patent 974,022; *Chem. Abstr.*, **62**, 3891d (1965).
252. J. M. L. Stephen, I. M. Tonkin, and J. Walker, *J. Chem. Soc.*, 1034 (1947).
253. J. A. Meschino, U.S. Patent 3,244,720; *Chem. Abstr.*, **64**, 17558f (1966).
254. J. A. Meschino, U.S. Patent 3,278,539; *Chem. Abstr.*, **65**, 18567d (1966).
255. H. K. Sen and U. Basu, *J. Indian Chem. Soc.*, **6**. 309 (1929).
256. B. K. Blount, W. H. Perkin and S. G. P. Plant, *J. Chem. Soc.*, 1977 (1929).
257. W. Bukhsh and R. D. Desai, *Proc. Indian Acad. Sci.*, **10A**, 262 (1939).
258. H. K. Sen and U. Basu, *J. Indian Chem. Soc.*, **7**, 435 (1930).
259. G. K. Hughes and F. Lions, *Proc. Roy. Soc. N. S. Wales*, **71**, 458 (1937).
260. C. C. Price and R. M. Roberts, *J. Amer. Chem. Soc.*, **68**, 1204 (1946).
261. L. J. Sargent and L. F. Small, *J. Org. Chem.*, **12**, 571 (1947).
262. R. A. Reed, *J. Chem. Soc.*, 186 (1945).
263. H. P. W. Huggill and S. G. P. Plant, *J. Chem. Soc.*, 784 (1939).
264. D. Vorländer and F. Kalkow, *Justus Liebigs Ann. Chem.*, **309**, 356 (1899).
265. N. S. Gill, K. B. James, F. Lions, and K. T. Potts, *J. Org. Chem.*, **5**, 4923 (1952).
266. A. Albert, *The Acridines*, 2nd ed., Arnold, London, 1966, p. 366.
267. N. Barbulescu and F. Potmischil, *Rev. Roum. Chim.*, **14**, 1427 (1969).
268. M. N. Tilichenko and V. I. Vysotskii, *Dokl. Akad. Nauk SSSR*, **119**, 1162 (1958).
269. L. R. Freimiller and J. W. Nemec, U.S. Patent 3,325,498; *Chem. Abstr.*, **67**, 73533u (1967).
270. N. Barbulescu and L. Ivan, *An. Univ. Bucuresti Ser. Stiint. Natur. Chim.* **15**, 47 (1966); *Chem. Abstr.*, **70**, 87130j (1969).
271. V. A. Kaminskii, A. N. Saverchenko, and M. N. Tilichenko, *Zh. Org. Khim.*, **6**, 404 (1970).
272. M. N. Tilichenko and G. V. Pavel, *Zh. Org. Khim.*, **1**, 1992 (1965).
273. V. A. Kaminskii and M. N. Tilichenko, *Zh. Org. Khim.*, **5**, 186 (1969).
274. G. N. Dorofeenko, G. V. Lazurev'skii, and G. I. Zhungietu, *Dokl. Akad. Nauk SSSR*, **161**, 355 (1965).
275. G. N. Dorofeenko and G. I. Zhungietu, *Zh. Obshch. Khim.*, **35**, 589 (1965).
276. A. T. Balabon and N. S. Barbulescu, *Rev. Roum. Chim.*, **11**, 109 (1966).
277. V. A. Kaminskii and M. N. Tilichenko, *Zh. Org. Khim.*, **1**, 612 (1965).
278. V. A. Kaminskii and M. N. Tilichenko, *Khim. Geterotsikl. Soedin.*, **4**, 708 (1967).
279. G. R. Newkome and D. L. Fishel, *J. Heterocyclic Chem.*, **4**, 427 (1967).
280. V. I. Vysotskii and M. N. Tilichenko, *Khim. Geterotsikl. Soedin.*, **6**, 1080 (1968).
281. M. N. Tilichenko, V. A. Stonik, and V. I. Vysotskii, *Khim. Geterotsikl. Soedin.*, **3**, 570 (1968).
282. V. A. Stonik, V. I. Vysotskii, and M. N. Tilichenko, *Khim. Geterotsikl. Soedin.*, **4**, 763 (1968).
283. E. I. Stankevich and G. Vanags, *Khim. Geterotsikl. Soedin., Akad. Nauk Latv. SSR*, **2**, 305 (1965); *Chem. Abstr.*, **63**, 6974d (1965).
284. G. A. Klimov and M. N. Tilichenko, *Zh. Org. Khim.*, **2**, 1507 (1966).
285. G. A. Klimov and M. N. Tilichenko, *Khim. Geterotsikl. Soedin.*, **1**, 306 (1967).

286. E. I. Stankevich and G. Vanags, *Khim. Geterotsikl. Soedin.*, *Akad. Nauk Latv. SSR*, **2**, 305 (1965); *Chem. Abstr.* **63**, 6974c (1965).

287. G. A. Klimov, M. N. Tilichenko, and E. S. Karaulov, *Khim. Geterotsikl. Soedin.*, **2**, 297 (1969).

288. G. A. Klimov and M. N. Tilichenko, *Khim. Geterotsikl. Soedin.*, **3**, 572 (1969).

289. G. A. Klimov and M. N. Tilichenko, *Khim. Geterotsikl. Soedin.*, **1**, 175 (1969).

290. T. Masamune and S. Wakamatsu, *J. Fac. Sci., Hokkaido Univ., Ser. III*, **5**, 47 (1957); *Chem. Abstr.*, **52**, 11850 (1958).

291. W. G. Perkin and S. G. P. Plant, *J. Chem. Soc.*, 2583 (1928).

292. D. A. Archer, H. Booth, P. C. Crisp, and J. Parrick, *J. Chem. Soc.*, 330 (1963).

293. V. G. Ermolaeva, V. G. Yashunskii, A. I. Polezhaeva, and M. D. Mashkovskii, *Khim.-Farm. Zh.*, **2**, 20 (1968).

294. M. M. Cartwright and S. G. P. Plant, *J. Chem. Soc.*, 1898 (1931).

295. H. Booth, *Tetrahedron*, **19**, 91 (1963).

296. T. Masamune, M. Takasugi, H. Suginome, and M. Yokoyama, *J. Org. Chem.*, **29**, 681 (1964).

297. W. H. Perkin and W. G. Sedgwick, *J. Chem. Soc.*, 441 (1926).

298. M. D. Mashkovskii, A. I. Polezhaeva, V. G. Ermolaeva, and V. G. Yashunskii, *Farmakol. Toksikol.*, **31**, 427 (1968).

299. K. Landgraf and E. Seeger, U.S. Patent 3,282,943; *Chem. Abstr.*, **66**, 28681j (1967).

300. T. Masamune, M. Ohno, K. Takemura, and S. Ohuchi, *Bull. Chem. Soc. Japan*, **41**, 2458 (1968).

301. V. A. Kaminskii and M. N. Tilichenko, *Khim. Geterotsikl. Soedin.*, **2**, 298 (1968).

302. V. A. Kaminskii, L. N. Donchak, and M. N. Tilichenko, *Khim. Geterolskil. Soedin.*, **6**, 1134 (1969).

303. V. A. Kaminskii, V. K. Gamov, and M. N. Tilichenko, *Khim. Geterotsikl. Soedin.*, **1**, 181 (1969).

304. E. E. Smissman and M. Steinman, *J. Med. Chem.*, **10**, 1054 (1967).

305. C G. Overberger and B. S. Marks, *J. Amer. Chem. Soc.*, **77**, 4097 (1955).

306. T. Masamune and S. Wakamatsu, *J. Chem. Soc. Jap.*, **77**, 1145 (1956).

307. F. Bohlmann, *Chem. Ber.*, **91**, 2157 (1958).

308. M. Wiewiorowski, O. E. Edwards, and M. D. Bratek-Wiewiorowska, *Can. J. Chem.*, **45**, 1447 (1967).

309. T. Masamune and M. Takasugi, *Chem. Commun.*, **625** (1967).

310. T. Masamune, *Chem. Commun.*, **244** (1968).

311. (a) N. L. Allinger, J. G. D. Carpenter, and F. M. Karkowski, *J. Amer. Chem. Soc.*, **87**, 1232 (1965); (b) J. B. Lambert, K. G. Keske, R. E. Carhart and A. P. Jovanovich, *J. Amer. Chem. Soc.*, **89**, 3761 (1967).

312. L. P. Kuhn, *J. Amer. Chem. Soc.*, **80**, 1318 (1958).

313. K. Nakamoto, M. Margoshes, and R. E. Rundle, *J. Amer. Chem. Soc.*, **77**, 6480 (1955).

314. (a) R. M. Badger, *J. Chem. Phys.*, **8**, 288 (1940); (b) R. C. Lord and R. E. Merrified, *J. Chem. Phys.*, **21**, 166 (1953).

315. A. C. Cope, *Org. Reactions*, **11**, 317 (1960).

316. L. Cohen and B. Witkop, *J. Amer. Chem. Soc.*, **77**, 6595 (1955).

317. M. N. Tilichenko, R. G. Soboleva, T. M. Domanyuk, and B. K. Gavrilova, *Soobshch. Dal'nevost. Filiala, Sibirsk. Otd., Akad. Nauk SSSR*, 113 (1963); *Chem. Abstr.*, **62**, 9719g (1965).

318. V. A. Stonik, G. A. Klimov, V. I. Vysotskii, and M. N. Tilichenko, *Khim. Geterotsikl. Soedin.*, **5**, 953 (1969).

319. G. A. Klimov and M. N. Tilichenko, *Zh. Org. Khim.*, **2**, 1526 (1966).

320. G. A. Klimov, M. N. Tilichenko, and E. S. Karaulov, *Khim. Geterotsikl. Soedin.*, **2**, 297 (1969).
321. A. J. Birch, P. Hextall, and S. Sternhell, *Aust. J. Chem.*, **7**, 256 (1954).
322. G. A. Klimov and M. N. Tilichenko, *Khim. Geterotsikl. Soedin.*, **3**, 572 (1969).
323. L. R. Freimiller and J. W. Nemec, U.S. Patent 3,326,917; *Chem. Abstr.*, **68**, 49469c (1968).
324. H. Antaki, *J. Chem. Soc.*, 4877 (1963).
325. H. Antaki, *J. Chem. Soc.*, 2263 (1965).
326. J. Lielbriedis and E. Gudriniece, *Latv. P.S.R. Zinat. Akad. Vestis. Kim. Ser.*, **3**, 318 (1967); *Chem. Abstr.*, **68**, 12821k (1968).
327. W. Kirkor and S. Rychter, *Lodz. Towarz. Nauk Wydz. III Acta Chim.*, **11**, 91 (1966); *Chem. Abstr.*, **66**, 65379n (1967).
328. D. Sveics, E. Stankevics, and O. Neilands, *Latv. P.S.R. Zinat. Akad. Vestis Kim. Ser.*, **2**, 213 (1968); *Chem. Abstr.*, **69**, 51964z (1968).
329. D. Vorländer and O. Strauss, *Justus Liebigs Ann. Chem.*, **309**, 375 (1899).
330. E. I. Stankevich and G. Vanags, *Zh. Org. Khim.*, **1**, 809 (1965).
331. R. F. Parcell and F. R. Hauck, *J. Amer. Chem. Soc.*, **85**, 346 (1963).
332. V. A. Kaminskii, V. I. Vysotskii and M. N. Tilichenko, *Khim. Geterotsikl. Soedin.*, **2**, 373 (1969).
333. V. I. Vysotskii and M. N. Tilichenko, *Khim. Geterotsikl. Soedin.*, **4**, 751 (1969).
334. J. A. Moore and L. D. Kornreich, *Tetrahedron Lett.*, **1277** (1963).
335. H. Irving, E. J. Butler, and M. F. Ring, *J. Chem. Soc.*, 1498 (1949).
336. M. J. Bleiberg, *Life Sci.* (Oxford), **4**, 449 (1965).
337. J. H. Kario, W. L. Nastuk and R. L. Katz, *Brit. J. Anaesth.*, **38**, 762 (1966); *Chem. Abstr.*, **66**, 45397p (1967).
338. H. W. Woods, Australian Patent 263,669; *Chem. Abstr.*, **68**, 16137q (1968).
339. C. H. Boehringer Sohn, Netherlands Patent 6,611,621; *Chem. Abstr.*, **68**, 49470w (1968).
340. P. A. Petyunin, M. E. Konshin, and Yu. V. Kozhevnikov, *Zh. Vses. Khim. Obshchest.*, **12**, 238 (1967); *Chem. Abstr.*, **67**, 54025p (1967).
341. K. Dziewonskii and J. Schoen, *Bull. Intern. Acad. Polonaise, Classe Sci. Math. Nat.*, A, 448, 1934; *Chem. Abstr.*, **29**, 2958 (1935).
342. N. Barbulescu and F. Potmischil, *Tetrahedron Lett.*, 2309 (1969).
343. N. Barbulescu and F. Potmischil, *Tetrahedron Lett.*, 5275 (1969).
344. H. Booth and J. H. Little, *Tetrahedron*, **23**, 291 (1967).
345. F. Kröhnke and H. K. Honig, *Justus Liebigs Ann. Chem.*, **624**, 97 (1959).

CHAPTER VI

Biacridines*

FRANK McCAPRA

*The Chemical Laboratory, University of Sussex,
Brighton, England*

*The term "biacridine" includes all compounds with two directly linked, acridine ring systems. At the time of publication of the first edition of "The Acridines" in this series, only 9,9', 10,10', and 9,10' linkages were known. The situation is as yet unchanged.

1. 9,9'-Biacridines

A. 9,9'-Biacridine

These compounds, directly analogous to bipyridyls and related dimeric heterocycles, are named in *Chemical Abstracts* as 9,9'-biacridines. The dearth of other positional isomers is no doubt because of the unique reactivity of the position *para* to the nitrogen atom, the methods of synthesis often being reminiscent of those in the pyridine and quinoline series. 9,9'-Biacridine (1) is easily prepared in high yield by adding ethereal 9-chloroacridine to phenyl magnesium bromide, biphenyl being a by-product.[1] Earlier workers used 9-chloroacridine dichlorophosphate,[2] and reported that the use of ethyl magnesium bromide gave a mixture of 9,9'-biacridine and $\Delta^{9,9'}$-biacridan,

519

which were not easily separated. Other syntheses include treatment of 9-chloroacridine with Raney nickel in boiling methanol (66–75% yield),[3] and Raney copper and zinc dust,[4] both of which gave reduced yields. Metal-catalyzed reductions of 9-chloroacridine of various sorts have, in fact, been successfully used, e.g., hydrogen and palladium on barium sulfate at 130° in xylene,[5] copper (92% yield),[6] zinc and concentrated hydrochloric acid (96% yield).[7] The reduction of 9-acridanthione by copper is less effective (60% yield).[8]

9,9'-Biacridine is formed when 10,10'-dimethyl-9,9'-biacridinium iodide or bromide is heated, the corresponding methyl halide being evolved,[9, 10] and was first prepared in this way.[9]

A biacridine derivative has also been isolated as a metabolite of acridine 10-oxide in rabbits, the injection of this compound causing the excretion of 9,9'-biacridine 10,10'-dioxide as a glucuronide.[10]

9,9'-Biacridine is a pale yellow crystalline solid, sparingly soluble in most solvents, giving nonfluorescent solutions. Purification is achieved by Soxhlet extraction with pyridine or toluene and crystallization from these solvents. It crystallizes as small needles, with a high melting point[2, 5, 11] given as 400, 383, and 382°. A yellow hydrochloride and picrate may be formed.[8] Alkylation proceeds in stages, as might be expected, with methyl sulfate in boiling toluene methylating one nucleus,[11] and boiling methyl sulfate alkylating both nuclei.[4, 11] Precipitation of the dibromide or diiodide (a red, very insoluble salt) from aqueous solution by addition of the potassium halide[4, 6,11] is a convenient means of purifying the diquaternary methosulfates. The ir spectra of 1,1'-, 2,2'-, 3,3'-, and 4,4'-dihydroxy-9,9'-biacridines have been recorded,[12, 13] together with those of the corresponding methyl ethers,[12] and the varying tendencies toward inter- and intra-molecular hydrogen bonding have been noted. It is interesting that the uv spectra of highly concentrated acridine solutions begin to resemble that of 9,9'-biacridine, with the appearance of a new band[14] at 390 nm. This is in accord with the absence of effective conjugation between the two nuclei (discussed later).

1

2

3

B. 9,9'-Biacridans

Other names for $\Delta^{9,9'}$-biacridan (2) are 9,9'-biacriden, 9,9'-diacriden, and 9,9'-biacridylidene. The compound has been obtained by the reduction of 9,9'-biacridine with zinc and hydrochloric acid, followed by Soxhlet extraction of the insoluble residue with benzene.[11] It separated from the benzene solution, which had a yellow-green fluorescence, as yellow leaflets, mp 392°. It can also be prepared by heating 9-(9-acridanyl)-10-methylacridinium bromide to 350°, eliminating methyl bromide.

$\Delta^{9,9'}$-Biacridan formed a yellow hydrochloride on heating with hydrochloric acid, although there was no reaction in the cold. Since biacridans are insufficiently basic to form hydrochlorides, it was suggested that the analogue of 3 had been formed.[11] The exact nature of the isolated compound is unconfirmed.

10-Methyl-$\Delta^{9,9'}$-biacridan, or 9-(10-methyl-9-acridanyl)acridine (3) was obtained from the reduction of 10-methyl-9-(9-acridinyl)acridinium bromide with zinc in benzene.[11] Its solutions are fluorescent and it forms a picrate easily. A red compound, though to be a free radical, was formed by allowing a solution in benzene to stand in air.[11]

10,10'-Dimethylbiacridylidene (4) was first obtained in poor yield, together with 10-methylacridan and 10-methylacridinium acetate, by the reduction of 10-methyl-9-acridanone with zinc in acetic acid.[9] The reduction appears to take place through the intermediate pinacol 5, which has been isolated from the reaction.[6] 3,6-Diamino-10-methyl-9-acridanone, when reduced with zinc and hydrochloric acid, gives only the 3,3',6,6'-tetraamino-9,9'-biacridinium chloride. The dication 6 is apparently too stable to allow either the isolation of a pinacol or further reduction.

4   5   6

9-Phenylimino-10-substituted acridans are also reduced to $\Delta^{9,9'}$-biacridans. The reduction of 10,10'-dialkylacridinium salts to the corresponding biacridylidene can also be achieved by alkaline sodium sulfite, by shaking an aqueous solution with zinc and benzene,[11] or by zinc in methanolic hydrochloric acid.[15]

A useful route to 10,10'-disubstituted $\Delta^{9,9'}$-biacridans is to convert the appropriate 9-acridanone to the 9-chloro-10-substituted acridinium dichlorophosphate, using boiling phosphorous oxychloride. Reduction by zinc and acetone[16] or by phenyl magnesium bromide then affords the $\Delta^{9,9'}$-biacridan in good yield.

The 10,10'-dialkyl-$\Delta^{9,9'}$-biacridans are sparingly soluble, yellow crystalline compounds giving highly fluorescent solutions. (See Chapter IX, p. 615, concerning chemiluminescence for further details of this and related properties.) 10,10'-Dimethyl-$\Delta^{9,9'}$-biacridan (4) is best recrystallized from pyridine, and has mp 350°; with hydrogen chloride in benzene it gives 10-methyl-9-(10-methyl-9-acridanyl)acridinium chloride.

The biacridylidenes belong to a general class of highly substituted ethylenes, many of which are thermo- and photo-chromic. However, the biacridylidenes are not particularly noteworthy in these respects, although thermochromism is observed when 10,10'-dimethyl-$\Delta^{9,9'}$-biacridan is adsorbed on magnesium oxide.[17] Although the stereochemistry of biacridylidenes has not been examined in detail, by comparison with related substituted ethylenes (e.g., bianthronylidene) for which X-ray data exist,[18a, b] some information is obtainable. It would seem that the double bond remains planar (length 1.31 Å) and the benzene rings twist out of plane by about 40°, while themselves remaining planar. The whole structure is still centrosymmetric, and the bonds from the central double bond are single in character.

C. 10,10'-Dialkyl-9,9'-biacridinium salts

A large number of 10,10'-disubstituted-$\Delta^{9,9'}$-biacridans and 10,10'-disubstituted-9,9'-biacridans have been oxidized to the corresponding biacridinium nitrates by boiling with dilute nitric acid, and this is probably the most common route. (For recent examples, see Refs. 19a and b.) The biacridinium iodides are also obtained directly from the 10-alkyl-9-acridanones by boiling with magnesium and magnesium iodide in anisole.[20] As might be expected, these biacridinium salts usually give the corresponding carbinol bases (pseudobases) with alkali.[6] However, 3,3'-dimethoxy-10,10'-dimethyl-9,9'-biacridinium dinitrate (7) gives 8 with aqueous sodium hydroxide, presumably by the addition of the hydroxyl ion at the 3- and 3'-positions, with the elimination of methanol.[16]

The reaction can be reversed by treatment[21] with methyl sulfate at 100°. The uv spectrum of **8** is very little different from that of 10-methyl-3,9-dihydroacridine-3-one, except for a slight uniform shift of all maxima to longer wavelengths. This suggests that there is very little conjugation between the two ring systems. It is likely that all 9,9'-biacridines and biacridinium salts are not planar and that as in other very hindered systems related to the biphenyls, there is considerable twisting around the connecting single bond.

In boiling phosphorous oxychloride, **8** gave 3,3'-dichloro-10,10'-dimethyl-9,9'-biacridinium dichlorophosphate (**9**), convertible into 3,3'-dichloro-10,10'-dimethyl-$\Delta^{9,9'}$-biacridan with zinc in acetone. The identity of the latter compound was confirmed by its synthesis from 3,9-dichloro-10-methylacridinium dichlorophosphate using zinc in acetone.[21] As expected, the demethylation of the 1,1'-, 2,2'-, and 4,4'-dimethoxy-10,10'-dimethylbiacridinium dinitrates is achieved not by alkali but by the more usual ether cleavage reagent, hydrobromic acid. 10,10'-Dimethyl-9,9'-biacridinium dinitrate is commonly known as "Lucigenin" because of the bright chemiluminescence that occurs on treatment with alkaline hydrogen peroxide. This is probably the most noteworthy chemical reaction of the biacridinium salts and is considered in detail in Chapter IX (Section 2).

D. 9,9'-Biacridans

These compounds, similar to diphenylamine in chemical behavior, are white nonbasic solids. 9,9'-Biacridan (**10**) decomposes at its melting point (265°) into acridine and acridan, and can be obtained by treating the sodium adduct of acridine with ethanol[23, 24] or by direct reduction of 9,9'-biacridine with sodium.[24] The dianion formed by the dimerization of the sodium adduct of acridine can be trapped either by carbon dioxide[25] or sulfur dioxide to form, respectively, *N,N'*-tetrahydrobiacridyldicarboxylic acid and the related disulfamic acid.[26]

9,9'-Biacridan formed a dibenzoyl derivative, mp 305°, decomposed by boiling 30% potassium hydroxide to acridine (presumably by disproportionation and oxidation) and potassium benzoate.[23] This biacridan can also be prepared by sodium amalgam reduction of 9-cyanoacridan and has a melting point of 249° under specified conditions, although varying with the rate of heating.[8]

An isomeric 9,9'-biacridan was obtained on reducing 9,9'-biacridine with zinc in acetic acid[27] and by boiling 9-cyanoacridan with ethanol, cyanogen being evolved.[28] It was nonbasic and decomposed to acridine and acridan, even in the presence of a trace of alkali, which hindered the decomposition of the higher melting isomeride.[8]

Both 9,9'-biacridans were converted[8] by sulfuric acid at 100° to the same glycol (11), which was easily dehydrated to the oxide (12). This oxide cyclized to the 9,9'-biacridan of mp 214°.

A 1:1 mixture of the isomeric biacridans separated from benzonitrile in colorless crystals melting at 215°, whereas a 2 (mp 214°):1 mixture had a melting point of 195°. The 1:1 mixture and the low- and high-melting isomer had similar X-ray diffraction patterns, although there were differences. It is not clear what these relatively small differences denote; but if simple polymorphism is to be excluded, then the most reasonable explanation of the isomerism is in the *cis* or *trans* arrangement of the hydrogen atoms in **10**.

Other syntheses of 9,9'-biacridan involve the reduction of acridine by Raney nickel[29] or *p*-thiocresol[30] in hot xylene by analogy with bipyridyl synthesis (yield 15%); the treatment[31] of acridine with sodium carbonate in ethylene glycol at 150° (17% yield); and reduction with zinc in acetic anhydride.[32] In the last case, the compound had a melting point of 247–248° and it leaves the status of the isomers discussed above in a rather doubtful state, since the isomer with a melting point of 214° was not found. Fair yields of biacridans were found in the reaction of o-fluorobromobenzene with acridine in the presence of magnesium using tetrahydrofuran as the solvent.[33] Benzyne is the likely intermediate, and the isolation of reduced biacridines suggests a radical chain process involving the solvent.

The photolysis of acridine usually results in 9,9′-biacridan formation, with other products derived from the solvent. Early work[34a, b] has been re-examined and shown to require hydrogen-donating solvents for reaction.[35a, b, 36] Acridan is also formed,[37a, b] together with products of radical combination with the solvent (e.g., **13**).

The irradiation of acridine in ethanol at 253 and 366 nm, probably producing an n,π^* excited state,[38] leads to hydrogen abstraction from the solvent to give **13**. At concentrations of acridine greater than $10^{-3}M$, dimerization to 9,9'-biacridan takes place.[38, 39] In air-saturated *iso*-propanol, the quantum yield[40] for biacridan formation is 0.032. 10-Alkylacridinium salts also undergo photoreduction to give 10,10'-dialkyl-9,9'-biacridans.[37, 39, 41] With 10-ethyl-acridinium salts, a photostationary state is achieved at 40% conversion to the 9,9'-biacridan.[42]

2. 9,10'-Biacridines

No recent preparations of this type of compound have been reported, and only three 9,10'-biacridines are known. 10-(9-Acridyl)-9-acridanone (**15**) was synthesized[43] in unspecified yield by heating 9-acridanone and 9-chloro-acridine to 300° and in 33% yield by treating *N*-(2-phenylaminobenzoyl)-diphenylamine-2-carboxylic acid (**14**) with concentrated sulfuric acid at 100° for 90 min.

3-Chloro-10-(3-chloro-9-acridinyl)-9-acridanone was similarly prepared by both methods.[43] The unsubstituted compound (**15**) was insoluble in hydrochloric acid and alcoholic sodium hydroxide and had a melting point of 383–384° after crystallization from pyridine. 6-Chloro-10-(6-chloro-2-methoxyacridinyl)-2-methoxy-9-acridanone, mp 350–352°, is an impurity in commercial 6,9-dichloro-2-methoxyacridine and was prepared by fusing the latter compound with the corresponding 9-acridanone.[44]

3. 10,10'-Biacridines

The only compound of this class known, 10,10'-bi-9-acridanonyl (**16**), was obtained from 9-acridanone by chromic acid oxidation and was earlier isolated from the oxidation products of acridine.

16

It has also been obtained in small yield from similar oxidations of 10-methyl-9-acridanone,[45] most of which was unattacked, and of 4-methylacridine when the methyl group was eliminated.[46]

10,10'-Bi-9-acridanonyl, also called 10,10'-diacridonyl, is a yellow solid, mp 251°, easily soluble in chloroform, less so in acetic acid and sparingly soluble in ethanol. It is reported as being nonfluorescent in solution, which is surprising, since *N*-substituted 9-acridanones are generally highly fluorescent. This is an interesting observation if the structure is correct. It is reduced to acridan by sodium amalgam and gives acridine on distillation from zinc dust. Its molecular weight (determined in phenol) is in agreement with the structure, but it has not been investigated further and derivatives are as yet unknown.

References

1. R. M. Acheson and M. L. Burstall private communication.
2. K. Gleu and A. Schubert, *Chem. Ber.*, **73**, 805 (1940).
3. A. M. Grigorovski, *J. Gen. Chem.* (*USSR*) **17**, 1124 (1947). A. M. Grigorovski and V. S. Fedorov, *J. Gen. Chem.* (*USSR*) **21**, 529 (1948); *Chem. Abstr.*, **43**, 646 (1949).
4. A. M. Grigorovski and A. A. Simeonov, *Zh. Obshch. Khim.*, **21**, 589 (1951); *Chem. Abstr.*, **46**, 114 (1952).
5. A. Ya. Yakubovich and M. Nevyadomskii, *J. Gen. Chem.* (*USSR*), **18**, 887 (1948).
6. K. Lehmstedt and H. Hundertmark, *Chem. Ber.* **62**, 1065 (1929).
7. R. Royer, *J. Chem. Soc.*, 1663 (1949).

8. K. Lehmstedt and H. Hundertmark, *Chem. Ber.*, **63**, 1229 (1930).
9. H. Decker and G. Dunant, *Chem. Ber.*, **42**, 1176 (1909).
10. R. Ito, H. Otaka, A. Sasaki, and Y. Hashimoto, *Nippon Univ. J. Med.*, **3**, 93 (1961).
11. H. Decker and P. Petsch, *J. Prakt. Chem.*, **143**, 211 (1935).
12. A. I. Gurevich, *Opt. Spektrosk.*, **12**, 42 (1962); *Chem. Abstr.*, **57**, 293 (1962).
13. A. I. Gurevich and Yu. N. Sheinker, *Khim. Nauk. Prom.*, **3**, 129, (1958); *Chem. Abstr.*, **52**, 11566 (1958).
14. L. V. Levshin, *Zh. Eksp. Teor. Fiz.*, **28**, 213 (1955); *J. Exp. Theoret. Phys.* **1**, 244 (1955).
15. R. M. Acheson and M. L. Burstall, *J. Chem. Soc.*, 1954, 3240.
16. K. Gleu and S. Nitzsche, *J. Prakt. Chem.*, **153**, 233 (1939).
17. O. Kortium, W. Theilacker, and G. Schreyer, *Z. Phys. Chem.* (Frankfurt am Main), **11**, 182 (1957).
18. E. Harnik, F. H. Herbstein, G. M. J. Schmidt, and F. L. Hirshfeld, *J. Chem. Soc.*, 3288 (1954); (b) E. Harnik and G. M. J. Schmidt *J. Chem. Soc.*, 3295 (1954).
19. (a) A. Braun, *Rocz. Chem.*, **36**, 151 (1961). (b) A. Chrzasziezewska and A. Braun, *Lodz Towarz. Nauk. Wydzial III, Acta Chim.* **9**, 189 (1964); *Chem. Abstr.*, **62**, 9103 (1965).
20. K. Gleu and R. Schaarshmidt, *Chem. Ber.*, **73**, 909 (1940).
21. S. Nitzsche, *Chem. Ber.*, **77**, 337 (1944).
22. K. Gleu and P. Petsch, *Angew. Chem.*, **48**, 57 (1935).
23. W. Schlenk and E. Bergmann, *Justus Liebigs Ann. Chem.* **463**, 300 (1928).
24. E. Bergmann and O. Blum-Bergmann, *Chem. Ber.*, **63**, 757 (1930).
25. W. E. Kramer, U. S. Patent 3,147,262; *Chem. Abstr.*, **61**, 13291 (1965).
26. W. E. Kramer, U.S. Patent 3,118,891; *Chem. Abstr.*, **60**, 9255 (1964).
27. K. Lehmstedt and H. Hundertmark, *Chem. Ber.*, **62**, 1742 (1929).
28. K. Lehmstedt and E. Wirth, *Chem. Ber.*, **61**, 2044 (1928).
29. G. M. Badger and W. H. F. Sasse, *J. Chem. Soc.*, 616 (1956).
30. H. Gilman, J. L. Towle, and R. K. Ingham, *J. Amer. Chem. Soc.*, **76**, 2920 (1954).
31. A. Albert and G. Catterall, *J. Chem. Soc.*, 4657 (1965).
32. I. W. Elliott and R. B. McGriff, *J. Org. Chem.*, **22**, 514 (1957).
33. G. Wittig and K. Niethammer, *Chem. Ber.*, **93**, 944 (1960).
34. (a) W. R. Orndoff and F. K. Cameron, *Amer. Chem. J.*, **17**, 658 (1895); (b) C. Dufraisse and J. Houpillart, *Bull. Soc. Chim. Fr.*, **5**, 626 (1938).
35. (a) A. Kellmann, *J. Chim. Phys.*, **54**, 468 (1957); **56**, 574 (1959); **57**, 1 (1960); (b) *Bull Soc. Chim. Belges*, **71**, 811 (1962).
36. V. Zanker and P. Schmid, *Z. Phys. Chem.* (Frankfurt am Main), **17**, 11 (1958).
37. (a) M. Giurgea, G. Mihai, V. Topa, and M. Musa, *J. Chem. Phys.* **61**, 619 (1964); (b) H. Goeth, P. Ceruthi, and H. Schmid, *Helv. Chim. Acta*, **48**, 1395 (1965).
38. A. Kellmann, *J. Chim. Phys.*, **63**, 936 (1966).
39. F. Mader and V. Zanker, *Chem. Ber.*, **97**, 2418 (1964).
40. S. Nüzuma and M. Koizumi, *Bull. Chem. Soc. Jap.*, **36**, 1629 (1963).
41. V. Zanker, E. Erhardt, and J. Thies, *Ind. Chem. Belge*, **32**, 24 (1967); *Chem. Abstr.*, **70**, 87538 (1969).
42. V. Zanker, E. Erhardt, F. Mader, and J. Thies, *Z. Naturforsch, B*, **21**, 102 (1966).
43. A. M. Grigorovski, *J. Gen. Chem. (USSR)*, **19**, 1744 (1949).
44. A. M. Girgorovski and N. S. Milovanov, *J. Appl. Chem., (USSR)* **23**, 197 (1950); *Chem. Abstr.*, **45**, 1596 (1951).
45. C. Graebe and K. Lagodzinski, *Justus Liebigs Ann. Chem.*, **276**, 35 (1893).
46. C. Graebe and H. Locher, *Justus Liebigs Ann. Chem.*, **279**, 275 (1894).

CHAPTER VII

Benzacridines and Condensed Acridines

D. A. ROBINSON

Molecular Pharmacology Unit,
Medical Research Council,
Cambridge, England

It is essential at the outset to state quite clearly that this chapter is not a complete record of all the ring systems contained by its title. This would be an unenviable task for even an encyclopedic work and is neither feasible nor desirable for the purposes of this book. The Ring Index (2nd ed.) and its supplements[1a, b] (I, II, and III, the last covering the literature to the end of 1963) list over 170 systems whose correct generic name contains the suffix "acridine," while 46 more systems that can be envisaged structurally as being benzacridines or condensed acridines are officially named with titles in which the word "acridine" is not present. Only those nuclei whose chemistry has been extensively studied or which otherwise show particularly interesting properties will be discussed here. The names and numberings laid down by *Chemical Abstracts* (also used by the Ring Index) are employed throughout. Some of the difficulties described by the author of the first edition[2] with regard to nomenclature regrettably are still present when the literature of the years 1953–1970 is searched. The *Chemical Abstracts* system is still not used by all authors, so that special care must be taken if a literature search is to be effective.

1. Benzacridines and Condensed Acridines with Four Rings

Compounds of this type, containing more than one heterocyclic atom, are dealt with elsewhere in this series.[3] Benz[a]acridine (1), benz[b]acridine (2), benz[c]acridine (3), and a great many of their derivatives have been made by standard methods that require no further consideration here, since they have been discussed in some detail with reference to the simple acridines in Chapter I (p. 13–57).

The most convenient methods for preparing the three parent benzacridines in optimum yields and key references are presented in Table I.

A. Syntheses of Benz-[a]-, -[b]- and -[c]-acridines

A great number of substituted benz[a]- and benz[c]-acridines, including fluorobenzacridines,[11] have been made by the Bernthsen reaction (Chapter I, Section 2.B) and by its now more often used modification in which the acid is largely replaced by its anhydride. (See Refs. 12a, b; 13a, b. There are a total of 16 papers in J. Chem. Soc.; see especially Refs. 14–18.) By the use of the appropriate acids (or anhydrides), ethyl, n-propyl, n-pentyl, phenyl, benzyl, and cyclohexyl, substituents[5, 14, 15a, b] have been incorporated at position 12 in benz[a]acridine and/or position 7 in benz[c]acridine (the so-called *meso* positions), while attempts to place *t*-butyl (or other tertiary alkyl)[16] or perfluoroalkyl[17] groups at these positions has failed. The Bernthsen reaction of trialkylacetic acids with N-arylnaphthylamines leads to benzacridines having no *meso*-substituent, suggesting a greater lability for tertiary alkyl groups at this position, since their location elsewhere in the rings has been achieved.[18]

TABLE I

Compound	Appearance, solvent, and mp ($^{\circ}$C)	Preparation	Ref.
Benz[a]acridine 1	Colorless needles ex. aq ethanol, 131°	2-Chlorobenzoic acid is condensed with 2-naphthylamine, the product cyclized to 12-chlorobenz[a]acridine with POCl$_3$, hydrogenated to 7,12-dihydrobenz[a]-acridine (60-90% yield this step) and this last oxidized with chromic acid (quantitative yield).	4, 5
Benz[b]acridine 2	Orange crystals ex. ethanol, 223°	Benz[b]acridan-12-one, obtained by condensation of aniline with 2-hydroxynaphthalene-3-carboxylic acid, followed by cyclization with POCl$_3$, on reduction to the acridan (Na/Hg), followed by oxidation with FeCl$_3$, gave benz[b]acridine (70% yield).	6
	225-226°	Benz[f]isatin is condensed with cyclohexanone in the presence of alkali to give 1,2,3,4-tetrahydrobenz[b]acridine-12-carboxylic acid (70% yield), which with PbO at ca.300° gave benz[b]acridine (27%).	7, 8

(Table Continued)

531

Table I. (Continued)

Compound	Appearance, solvent, and mp (°C)	Preparation	Ref.
Benz[c]acridine 3	Brilliant yellow needles, ex. benzene-petroleum, ether, 108°	Heating 2-tolyl-1-naphthylamine with lead oxide gave benz[c]-acridine (41%) isolated as its picrate.	9
		2-Chlorobenzoic acid is condensed with 1-naphthylamine, the product cyclized and the resulting acridanone reduced by distillation over zinc.	10
	107-108°	7-Chlorobenz[c]acridine hydrogenated (Raney Ni) to the acridan, which gave benz[c]acridine (75%) on chromic acid oxidation.	5

The intermediacy of aminoketones in the modified Bernthsen reaction has been supported by the isolation and subsequent facile cyclodehydration of compounds such as **4**, which yields 12-methylbenz[a]acridine on heating in an acetic/sulfuric acid medium.[19] Parallel treatment of **5** for only 5 min gives benz[a]acridine (**1**), which lends support to the postulate that intermediate aldehydes (e.g., **5**) may be formed in the Ullmann-Fetvadjian reaction (Chapter I, p. 41–48).[19]

Many substituted benzacridines (and condensed acridines) have been synthesized by this reaction procedure from the appropriate phenols and aromatic amines, the universal use of paraformaldehyde in the procedures giving rise to products unsubstituted at their *meso* positions[20, 21] (i.e., *para* to nitrogen in the heterocyclic ring).

The somewhat unusual synthesis of benz[c]acridine[22] from 1-tetralone and diphenylthiourea first described in 1934 has been revived recently; its utility has been extended to include the synthesis of benz[a]acridine, dibenz[ah]-, and dibenz[aj]-acridines by variation of the aryl substituent of the urea and the use of other cyclic ketones.[23, 24] Thus benz[a]acridine has been synthesized by two such routes: 1,3-di(2-naphthyl)thiourea heated with cyclohexanone gave (6), which when autoclaved with ethanolic potash yielded 1,2,3,4-tetrahydrobenz[a]acridan-12-one (7) [or its hydroxyacridine tautomer; cf. (9)] from which benz[a]acridine (1) could be obtained by pyrolysis with zinc dust under carbon dioxide; a parallel set of reactions[23, 24] starting from 2-tetralone and diphenylthiourea proceeds through intermediates 8 and 9 to the same product (1).

Pfitzinger-type condensations giving initially reduced benzacridines, bearing a carboxyl group in the *meso* position, have been used to prepare a variety of substituted benz[*a*]-, benz[*b*]-, and benz[*c*]-acridines by the reaction of suitably substituted isatins or benzisatins with substituted tetralones, *cis*-2-decalone or cyclohexanone, followed by decarboxylation and dehydrogenation of the initial products.[8, 25–35] For instance, in the presence of strong alkali, benz[*f*]isatin and cyclohexanone[8] gave 10, isatin and *cis*-2-decalone[31] gave both 11 and 12, the latter product interestingly arising from cyclization across the 2,3-positions of the decalone. In the 1-tetralone series, 5,8-dimethyl-1-tetralone (13) failed to give a benz[*c*]acridine intermediate with 4,7-, 5,7-, 6,7-, and 5,6-dimethylisatin,[32] presumably because of steric restrictions caused by the 8-methyl group of the tetralone.

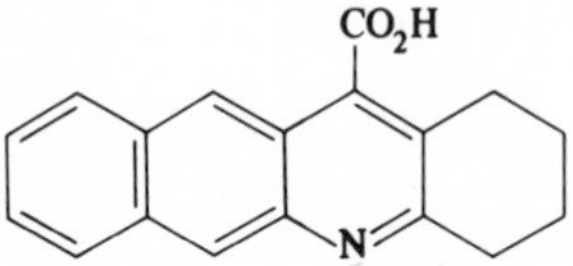

10

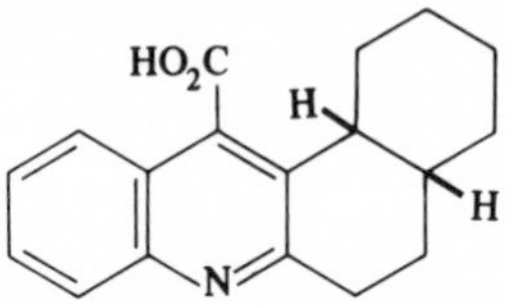

11

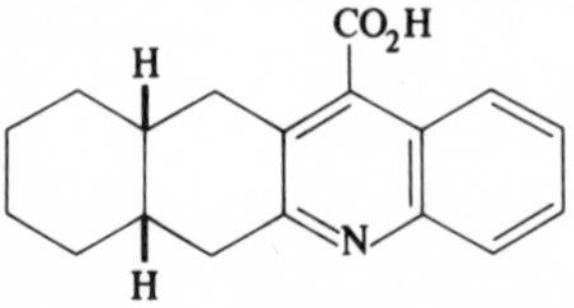

12

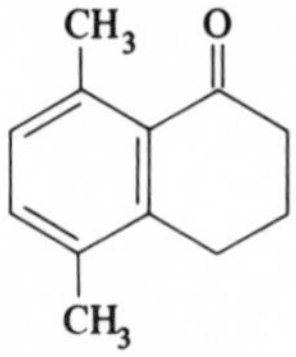

13

A closely related reaction is the cyclization of *N*-arylbenz[*e*]isatins (14) to substituted benz[*a*]acridine-12-carboxylic acids[36a, b]; the parent benz[*a*]-acridine-12-carboxylic acid is formed from naphtho[2,1-*b*]furan-1,2-dione (15), and aniline in boiling acetic acid.[33] Substituted *N*-(*o*-acetylphenyl)-1-naphthylamines have been cyclized under acid conditions to substituted 7-methylbenz[*c*]acridines in a reaction reminiscent of the Bernthsen procedure.[37] The thermal condensation of anthranilic acid with various tetralones leads to 5,6-dihydrobenz[*a*]-[38] and benz[*c*]-acridanones[39] in good yields. In an improved version of the Friedländer reaction, heating *o*-aminoacetophenone hydrochloride with various tetralones afforded dihydrobenzacridines, bearing a *meso* methyl group[40, 41]; on steric grounds the benz[*b*]acridine derivative 16 is preferred to the benz[*a*]acridine isomer, which would be the expected product from 2-tetralone in this reaction if the usual pathway of cyclization across the 1,2-positions were followed.[41]

14   15   16

The cyclization of compound **17** by means of aniline and anhydrous zinc chloride to the benz[c]acridine derivative **18**, with retention of the methyl substituent,[42] repudiated the earlier mechanism[43] that postulated the formation of the 1-anil, followed by cyclization with the elimination of aniline from the 2-methylene group in the formation of 5,6-dihydrobenz[c]acridine from the anilinomethylene analogue of compound **17**.

17   18

19

Further light on this reaction was provided by a study[44] of the cyclization of isomeric naphthylamine derivatives (e.g., **19**) in the presence of various acids or amine hydrochlorides, when a Hofmann-Martius type mechanism (see reaction sequence below) explained most of the results in this[44] and the earlier studies.[42, 43]

The condensation of *o*-nitrobenzaldehyde under acid conditions with *gem*-dimethyltetralones leads to intermediates **20** and **21** ,which after reduction by iron in acetic acid are cyclized to the dihydrobenzacridines, **22**[45] and **23**,[46] respectively.

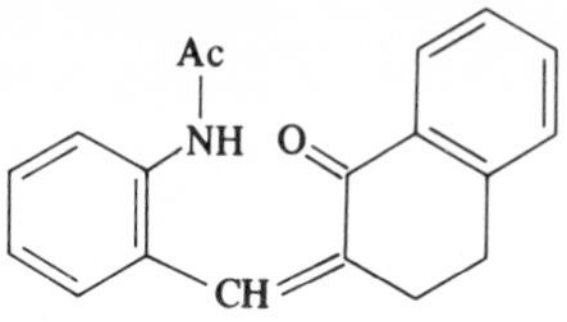

24

The reaction has been extended to include additional substituents in both the o-nitrobenzaldehyde and the tetralone moieties,[47a, b] while in a related modified reaction,[48] o-acetamidobenzaldehyde was condensed with 1-tetralone under mild alkaline conditions, yielding 24, which was cyclized to 5,6-dihydro-benz[c]acridine by heating with a sodium sulphate/potassium bisulfate mixture. The condensation of o-nitrobenzaldehyde and naphthalene by means of concentrated sulfuric acid yielded only intractable tars, but in the presence of 85% polyphosphoric acid benz[c]acridan-7-one (25) was obtained.[49] The N-(1-naphthyl)benzimidate (26) undergoes extrusion of methyl benzoate on heating to yield 11-methylbenz[c]acridan-7-one[50] by rearrangement and cyclization (cf. Chapter II, p. 171); the N-(2-naphthyl) analogue of 26 similarly yields 8-methylbenz[a]acridan-12-one.

In all of the synthetic routes to benzacridines considered thus far, the new ring formed in the process of synthesis has been the one containing the nitrogen atom. In concluding this section on the preparation of the benz-acridines, brief mention is made of two syntheses from quinoline derivatives in which the new ring produced by the reaction procedure is carbocyclic. In an Elbs-type reaction, heating 3-o-toluoylquinoline to 380–400° gave a 10% yield of benz[b]acridan, a free radical mechanism being suggested,[51] while the 3-phenylquinoline derivatives 27 and 28 have been cyclized by means of polyphosphoric acid to give, respectively, benz[a]acridine-5-carboxylic acid and 5-phenylbenz[a]acridine.[52]

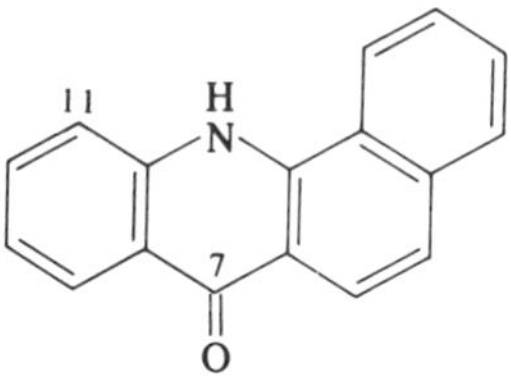

25

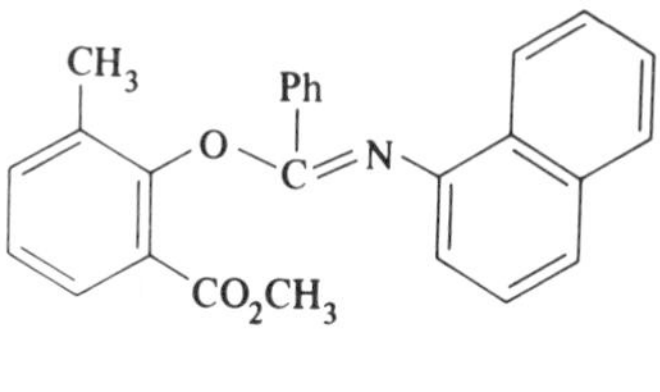

26

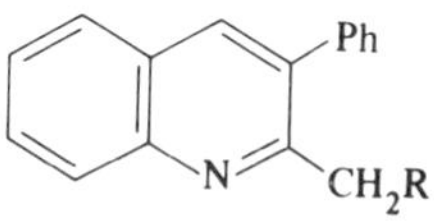

27   R = CO·CO₂C₂H₅

27   R = CO·CO$_2$C$_2$H$_5$

28   R = CH$_2$Ph

Mention has been made earlier (by way of Table I; syntheses described by Refs. 4, 5, and 6) of the synthesis of benz[a]- and benz[c]-acridines, bearing a chloro substituent at the *meso* position, by the cyclization of N-(1- or 2-naphthyl)anthranilic acids using phosphorus oxychloride. This route, originated by Bachmann and Picha,[53] has had more recent usage for the synthesis of a variety of substituted 12-chlorobenz[a]acridines and 7-chlorobenz[c]acridines, since these compounds are the precursors of benzacridines bearing complex alkylamino substituents at the *meso* position; a large number of these last-named benzacridines have been prepared and tested for amebicidal and antimalarial properties (Chapter XVIII, p. 829).

B. Properties and Reactions of Benz-[a]-, -[b]-, and -[c]-acridines

The simple benzacridines are very similar to acridine in their general properties and give strongly fluorescent solutions. The much deeper color of benz[b]acridine than benz[a]acridine or benz[c]acridine corresponds to the colors of the analogous hydrocarbons and indicates that the linear compounds have greater reactivity at the *meso* positions than their isomers. Solutions of the salts of benz[b]acridine are violet in color. The photochemistry of benz[b]-acridine under a variety of conditions is now partly understood. In carbon disulfide or benzene solution sunlight causes the formation of an unstable photooxide **29** with a *meso*-peroxide bridge; this compound cannot be isolated (though the analogous product from dibenz[bh]acridine has been characterized), the only material obtained from the irradiation being the 6,11-dione (quinone) in 34% yield.[7, 8]

These postulates received the support of a second group of investigators,[59] who extended the study of the photochemistry to a wider range of solvents, and whose confirmation of structure **29** and that of the quinone were based on spectroscopic evidence. In photooxidation benz[b]acridine therefore resembles anthracene and differs from pyridine in its behavior. Three days irradiation of an ethereal solution of benz[b]acridine by sunlight under anaerobic conditions gives a photodimer (mp 369–370°) of unspecified structure in 75% yield, reverting to the monomer on heating.[8] The yield of dimer is around 40% if air is not excluded, presumably the formation of the photooxide being a competitive process under these condtions. In both methanol and carbon tetrachloride solutions, the three simple benzacridines resemble acridine in their photochemical behavior, although the quantum yield determinations (at 366 nm) show a weakening in the acridine character, particularly for benz[b]- and benz[c]-acridine.[60] In the former solvent, hydrogen-abstraction by the electronically excited benzacridine is the first step in the photoreaction, the nature of the final product being unknown, although on spectroscopic evidence for the case of benz[b]acridine, benz[b]acridan or a photodimer are postulated.

Molecular orbital calculations support this mechanism and predict the ionic pathway,[60] followed by the reaction in carbon tetrachloride solution, in which solvent all three benzacridines give major products exemplified by **30**, the quantum yields for the disappearance of the precursors (0.2–0.8) being one to two orders of magnitude greater than their corresponding values in methanol solution. Irradiation under Pyrex of 12-methylbenz[a]-acridine 7-oxide gives a single product, the [10]annulene **31**, in 80% yield, whose structure has been elucidated by nmr techniques[61] and comparison with the behavior of acridine 10-oxide.[62]

29

30

31

The nmr spectra of benz[a]acridine and benz[c]acridine have been partially interpreted (see also Chapter XII, Section 6), and comparisons with the carbocyclic isosteres and solvent effects on the chemical shifts have been discussed.[63, 64] Noteworthy is the fact that the proton on C-1 in benz[c]-acridine, which is angularly opposed to the nitrogen atom is found at very low field (τ 0.5 in CDCl$_3$) being more deshielded (by 0.55 ppm) than the corresponding proton in benz[a]anthracene. The mass spectra of 14 alkyl-substituted benz[a]- and benz[c]-acridines have been subjected to a detailed analysis (see Chapter XIII, p. 714),[65] the main features that emerge being (1) the presence of strong peaks corresponding to several stages of dehydrogenation; (2) very little fragmentation of the polycyclic frame with, in ethyl-benzacridines, a noteworthy splitting off of methyl groups; and (3) the importance of doubly charged ions, some of which contribute more to the ionic current than do the corresponding singly charged ones, fragmentation with the loss of two electrons being energetically favored.

The separation of a very large number of substituted and parent benz- and dibenz-acridines, related aza heterocycles, and polycyclic aromatic compounds by means of thin-layer- and paper-chromatography and electrophoresis has been the subject of exhaustive studies,[66–71] stimulated by investigations in the fields of coal-tar products, air pollution, and carcinogens. The uv absorption data given in one of these reports[66] supplements earlier recordings of such data for benzacridines.[72a–d] When slow neutrons collide with benz[a]acridine, 7-^{14}C-benz[a]anthracene is produced by a [14]N(n,p)[14]C nuclear reaction, together with randomly labeled ^{14}C-benz[a]acridine produced by the recoil effect of the ^{14}C atom.[73a–e]

The cationoid substitution of these three benzacridines remains little investigated; a chlorine atom can be introduced at position 6 of both 3,12-dichlorobenz[b]acridine and 2,12-dichlorobenz[b]acridine by phosphorus pentachloride.[74] The nitro group in 9-nitrobenz[a]acridan-12-one behaves as in nitrobenzene, undergoing reduction to the amino derivative, diazotization, and thereafter the Sandmeyer reaction to give the 9-cyano derivative.[75] Chlorine atoms and amino groups present *para* to the heterocyclic nitrogen are readily hydrolyzed, as in the case of the simple acridines, giving the corresponding benzacridanones. In the reduced benzacridine series, aryl-amino groups at this position can be displaced hydrolytically under autoclave conditions, yielding di- and tetra-hydrobenzacridanones (see reaction sequence, p. 533).[23, 24] The *meso* chlorine atoms in benzacridines are also easily replaced by heating with amines in phenol; a number of quite complex aminobenzacridines have been prepared in this way as potential chemo-therapeutic agents.[6, 53, 55, 56, 58, 76–80]

The syntheses of 7-(substituted-amino)benz[c]acridines bearing sulfur- and nitrogen-mustard functions appended to the amino group have been under-taken, and the products possess antitumor activity.[81] When this substitution reaction is carried out in phenol solution, the *meso*-phenoxybenzacridines are the first-formed intermediates. The substitution of 7-chlorobenz[c]-acridine under basic conditions by various thiols gave the 7-alkyl(aryl)thio derivatives. The 7-substituents, like the phenoxy group at a *meso* position, were readily displaced by amines in anhydrous pyridine.[82a, b] The placement of the dicyanomethyl group at position 12 in benz[a]acridine results from treating the 12-chloro derivative with malonodinitrile and sodium butoxide in refluxing xylene solution[83]; the product was a useful intermediate for placing other functional groups at position 12. A photodimer of 12-aminobenz[b]acri-dine has been reported[6]; the isomeric 12-aminobenz[a]acridine and 7-amino-benz[c]acridine were little affected by light.

The relative reactivity of the methyl group at the *meso* positions in acridine and various benz- and dibenz-acridines has been estimated in terms of the extent of their (condensation) reaction with *m*-nitrobenzaldehyde[84]; 9-methyl-acridine is the most reactive member of the series, 7-methylbenz[c]acridine being slightly more reactive than 12-methylbenz[a]acridine, while the dibenzacridine derivatives reacted to an extent of 11% or less, occupying the bottom of the activity scale. These methyl groups in benz[a]acridine and benz[c]-acridine are relatively highly active, nevertheless, and condensation and derived products have been characterized from their reaction with aromatic aldehydes, nitroso compounds, and diazonium salts.[85-87] Similar condensation reactions have been studied using 5,6-dihydro-7,12-dimethylbenz[c]-acridinium methosulfate as the source of the 7-methyl group.[88] 12-Methyl-benz[a]acridine and aqueous formaldehyde heated in ethanolic solution yield the 12-β-hydroxyethyl derivative if hydrochloric acid is the catalyst,[89] while triethylamine hydrochloride (or secondary alicyclic amine salts)[89] in this capacity led directly to the 12-vinylbenz[a]acridine[90]; 7-methylbenz[c]-acridine undergoes the Mannich condensation in very high yields[89] in contrast with 12-methylbenz[a]acridine, which is inert, the distinction here presumably arising from steric hindrance at position 12 in the benz[a]acridines.

Selenium dioxide oxidation proves to be specific for the conversion of a methyl group at the *meso* position into a formyl group in excellent yield[33,91,92]; for instance, 7-formyl-9,10-dimethylbenz[c]acridine is the product from the 7,9,10-trimethyl precursor.[33] The carbonyl functions in benz[a]acridine-12-carboxaldehyde and benz[c]acridine-7-carboxaldehyde undergo the usual condensation reactions with hydrazines, aromatic amines, and active methyl groups.[91,93] Benz[a]acridine-12-carboxaldehyde failed to react with the Wittig reagents $ArCH=PPh_3$, but gave the expected products 32 on treatment with the phosphonium salts[94] $ArCH_2^+PPh_3$ X^- and lithium ethoxide; in the inverse reaction sequence, 7-chloromethylbenz[c]acridine gave a phosphonium salt with triphenylphosphine, which with benzaldehyde in the presence of lithium ethoxide yielded mainly the expected *trans* isomer[95] of compound 33.

32   33

The reaction of free benzyl radicals with benz[a]acridine and benz[c]-acridine has been studied[5]; both reactants yielded considerably more dibenzyl than was obtained with acridine, and thus must be less reactive as radical acceptors than acridine, a result in accord with the predictions from calculations of bond localization energies at the *meso* carbon centers in the isoconjugate hydrocarbons.[96] The initial attack of the benzyl radical is at the *meso* positions, and steric factors affect any subsequent reaction; thus benz[*a*]acridine gives a very low yield (7.5%) of 7,12-dibenzylbenz[*a*]acridan, the first step here being subject to steric inhibition, the second addition process being relatively facile. Also, benz[*c*]acridine gives a good yield (64%) of the monosubstituted product, 7-benzylbenz[*c*]acridine, and only 1% of a dibenzylbenz[*c*]acridine as would be expected; the difficulty of reaction of benz[*a*]acridine is further emphasized by a higher yield of dibenzyl and a 70% recovery of unchanged material. Benz[*b*]acridine resembles anthracene in forming Diels-Alder type adducts bridged across the 6,11-positions with maleic anhydride and dimethyl azodicarboxylate, but with dimethyl acetylenedicarboxylate, a Michael-type addition occurs[97] to give the equilibrium pair **34** (cf Chapter V, p. 455).

34

The reactions of 5,6-dihydro-5,5-dimethylbenz[*c*]acridine and 6,11-dihydro-6,6-dimethylbenz[*b*]acridine (whose syntheses were described on p. 536) have been studied in depth.[39,45–47] The former compound brominates at position 6, using *N*-bromosuccinimide with a trace of organic peroxide as the reagent.[45] The bromine atom at position 6 can be replaced by various nucleophiles, but by far the most interesting reaction of 6-bromo-5,6-dihydro-5,5-dimethylbenz[*c*]acridine (**35**) occurs on heating it briefly under nitrogen at 160–170°, when an α-elimination of hydrogen bromide occurs, accompanied by a Wagner-Meerwein rearrangement[45] to give 5,6-dimethylbenz[*c*]-acridine (**36**).

35   **36**

This reaction can also be achieved at a slightly higher temperature (to achieve concurrent decarboxylation in situ), when the precursor bears in addition a 7-carboxylic acid group.[47a, b] A variety of other reaction conditions have caused the rearrangement to occur for the 6-bromo-5,5-dimethyl precursor, a related 6-hydroxy derivative, and interestingly, for the case of 5,6-dihydro-5,5-tetramethylenebenz[c]acridine, which was brominated and rearranged in situ to give the predicted 1,2,3,4-tetrahydrodibenz[ac]acridine in 76% yield.[47a, b] 6,11-Dihydro-6,6-dimethylbenz[b]acridine under similar conditions with N-bromosuccinimide gives the 11-bromo derivative, which is likewise susceptible to nucleophilic attack at the 11-position.[46]

The formation of acridine-acridan complexes, which are very much more stable than in the case of the simple acridines, has been taken as an example to indicate the more pronounced reactivity at the *meso* positions in benzacridines, as compared with acridine itself. The reduction of benz[a]acridine with lithium aluminium hydride gave a quantitive yield of an orange complex.[4] A similar complex, along with benz[a]acridan, which was separated chromatographically, was also formed in the hydrogenation of 12-chlorobenz[a]acridine over Raney nickel.[4] Treatment of the orange complex (mp 140°) with concentrated hydrochloric acid gave the pale yellow benz[a]acridan, mp 158°, and a solution of benz[a]acridine hydrochloride. The orange complex was formed on mixing equimolar proportions of benz[a]acridine and benz[a]acridan, but uv absorption data indicated that it was largely dissociated in solution.[4] Its structure is similar to that of Morgan's base. The interpretation of the significance of these results requires some care in view of the work of Waters's group on the radical substitution of benzacridines.[5] Thus while theory[96] and experiment[5] support the proposition that the overall energetics for radical substitution at the *meso* position are less favorable for benzacridines as compared with acridine, these results for the formation of the benz[a]acridan–benz[a]acridine complex and the related Morgan's base in the dibenz[aj]acridine series relate only to the overall energetics for the one-electron transfer in these two systems to give radical anion–radical cation pairs. These energetics, in contrast to the above, are least favorable for acridine itself.

The oxidation of 5,6-dihydrobenz[c]acridine with chromic acid[25] gave the *ortho* quinone **37**; this product has since been obtained by the same oxidation procedure from benz[c]acridine, benz[a]acridine yielding no dione when treated similarly.[98] The oxidation of 7,9-dimethylbenz[c]acridine with osmium tetroxide gave 5,6-dihydro-5,6-dihydroxy-7,9-dimethylbenz[c]acridine.[99] A number of benzacridanones have been synthesized by the methods described in Chapter III, and they have already been described in this section as intermediates in a number of synthetic pathways to the parent benzacridines. Benz[b]acridan is oxidized by potassium dichromate in glacial acetic acid[51] to benz[b]acridan-12-one; chromic acid oxidation of benz[b]acridine gave the *para* quinone **38** in 48% yield.[8]

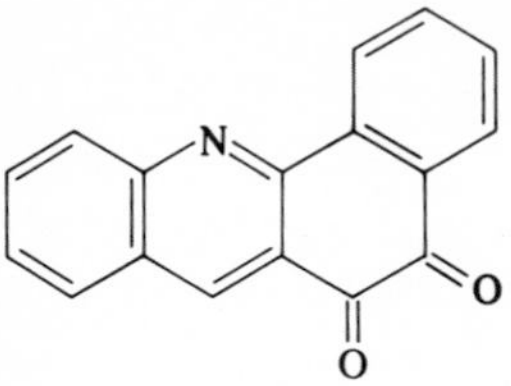

37

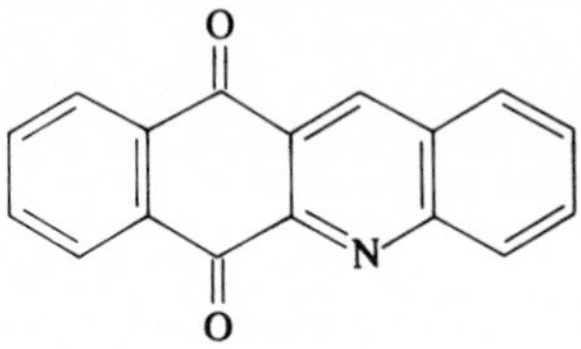

38

Lithiation of N-phenyl-β-naphthylamine, followed by treatment with solid carbon dioxide and cyclization, is stated to yield benz[b]acridan-12-one.[100] The chromic acid oxidation of 12-methyl- and 12-ethyl-benz[a]acridine gave[11] benz[a]acridan-12-one, converted to benz[a]acridine by distillation with zinc dust in a typical procedure. The benzacridanones have been little investigated but appear to be very similar to 9-acridanone itself.

The conversion of the carbonyl group to a *meso* chlorine substituent by treatment with phosphorus pentachloride or phosphoryl chloride is a widely used synthetic procedure. Sodium in amyl alcohol reduction of 9-methyl-benz[c]acridan-7-one gave 9-methylbenz[c]acridan.[50] 6,11-Dihydro-6,6-di-methyl-11-oxo-benz[b]acridine (**39**) gave the 11-oxime with hydroxylamine hydrochloride and the expected 11-carbinol **40** with methyl magnesium iodide[46]; however, the Grignard reagent from bromobenzene reacted to give the 12-phenylacridan derivative **41** by conjugate addition[101] (cf. 3-benzoylquinoline yields 3-benzoyl-1,4-dihydro-4-phenyl-quinoline in a parallel reaction.[102]

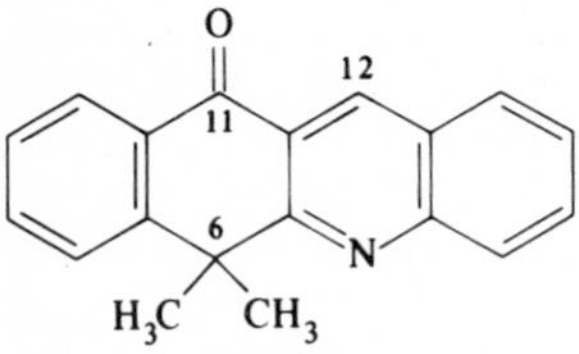

39

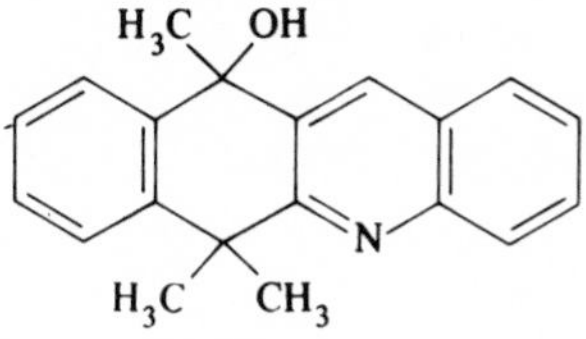

40

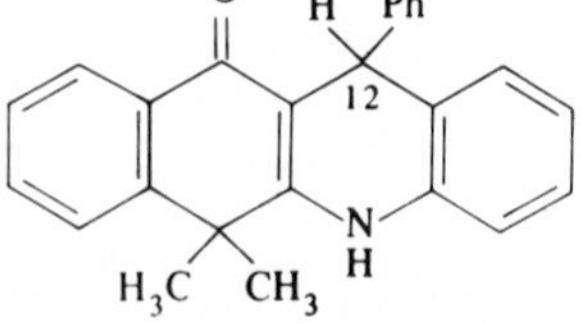

41

C. Other Benzacridines

7H-Benz[kl]acridine (**44**), originally called *ms*-benzacridan, was first obtained[103] by heating the triazine **43** in naphthalene. A more recent variant[104] on this reaction consists in heating the naphthalene **42** with *N*-nitrosodiphenylamine in naphthalene solution until nitrogen evolution ceases; the benzacridan **44**, mp 125°, is obtained in 69% yield. Heating the 2'-nitro derivative of the triazine **43** with triethyl phosphite gave the 8-nitro derivative of compound **44** in only 2% yield,[105] while the 2',4'-dinitro analogue of precursor **42** gave[104] a 93% yield of 8,10-dinitro-7H-benz[kl]acridine on treatment with *N*-nitrosodiphenylamine in acetic acid at 70° for 1 hr.

The first derivative of pyrido[3,2,1-*de*]acridine, which was briefly described in the literature, is the ketone **45**, synthesized by reductive cyclization, as indicated in the reaction sequence.[106]

Ziegler's group have discovered a novel access to this ring system: acridan and malonic acid heated with phosphorus oxychloride in naphthalene gave

1H,7H-pyrido[3,2,1-de]acridine-1,3(2H)-dione (**46**; R = R′ = H) in good yield.[107] Heating acridan with the 2,4-dichlorophenyl diester of benzylmalonic acid to 150° for 1 hr gave **46** (R = CH$_2$Ph; R′ = H) in good yield.[107] Compounds **46** (R′ = H) are in tautomeric equilibrium with their enol forms (**47**); this synthetic route has been generalized for various substituted malonic acids.[108]

46   **47**

Reaction of the sodium salt of the enol **47** with alkyl halides gave the 3-ether derivatives,[109] while various chlorinating reagent systems and the parent compound **46** (R = R′ = H) gave the 2,2-dichloro derivative **46** (R = R′ = Cl) in almost quantitive yield, which on reduction with zinc dust in an organic medium gave the 2-chloro compound **47** (R = Cl).[110] This last derivative was rapidly nitrated in an acetic acid-concentrated nitric acid mixture[111] at 50° to give the 2-chloro-2-nitro compound **46** (R = Cl; R′ = NO$_2$). The benzyl derivative **46** (R = CH$_2$Ph, R′ = H) gave the monobromo compound **46** (R = CH$_2$Ph; R′ = Br), while bromine in aqueous dioxan under a nitrogen atmosphere converted the parent compound **46** (R = R′ = H) into an unstable 2,2-dibromo derivative **46** (R = R′ = Br) in 76% yield.[112]

Acronycine, an acridanone alkaloid (see Chapter IV, p. 412) with broad spectrum antitumor activity against experimental neoplasms, is a member of the pyrano[2,3-c]acridine system, having the structure **48**; it has recently been synthesized by three interrelated routes.[113] Earlier workers had at first been unable to distinguish between structure **48** and the corresponding linear structure for acronycine,[114a–d] although degradation studies favored the former.[115a, b] The synthetic work of Beck's group,[113] an X-ray crystallographic study[116] of a bromo derivative of dihydroacronycine, and an elucidation by nmr spectroscopy[117] of the structure of the Diels-Alder adduct **49** of acronycine and dimethyl acetylenedicarboxylate established structure **48** as the correct one. Very recently Ritchie's group has provided a further synthetic pathway to acronycine,[118] utilizing the Claisen rearrangement of the 9-acridanone **50**.

Their observations of nuclear Overhauser effects in acronycine (**48**), noracronycine, and the adduct **49** established the angular structure unequivocally.

2. Dibenzacridines and Condensed Acridines with Five Rings

It remains true that comparatively little work has been done on the dibenzacridines, although a number of substituted derivatives have been made by general methods detailed in Chapter I and in Section A of this chapter, and tested as carcinogens; a number are quite active in this respect (see Chapter XVII). The known unsubstituted dibenzacridines and their methods of preparation are listed in Table II or discussed individually below. They have been little investigated from a chemical standpoint, and several of the possible dibenzacridines are as yet unknown.

The procedures [66, 67, 69, 70] for the thin layer and paper chromatographic and electrophoretic separation of the benzacridines (see p. 540) included data for a number of substituted dibenz[ah]-, dibenz[ch]-, and dibenz[aj]-acridines. The uv spectra of five dibenzacridines, namely, dibenz-[ac]-, -[ah]-, -[ai]-, -[aj]-, and -[ch]-acridines have been measured at −183° for both the free bases and their cations,[72a−c] and the results analyzed in terms of theoretical predictions, photochemical stabilities, and comparison with the carbocyclic isosteres. In an earlier paper, Zanker predicted some of the spectral properties of the as yet unknown dibenz[bi]acridine (**56**), a molecule expected to have low stability because of a small energy separation between the first excited singlet and triplet states, with the latter state being accessible by thermal pathways.[72a−c]

TABLE II. Dibenzacridines

Compound	Structure	Appearance and mp (°C)	Preparation	Ref.
Dibenz[ac]acridine	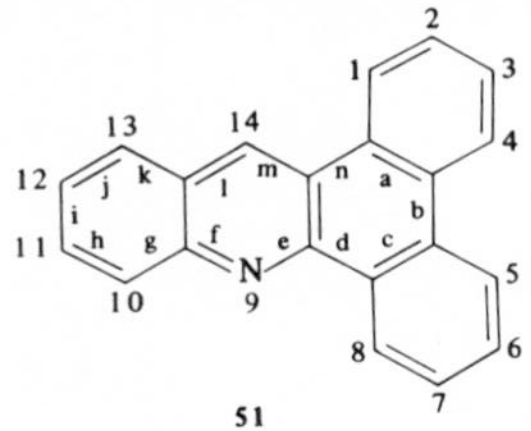 51	Yellow needles ex. toluene, 204°	Phenanthraquinone, 2-nitrobenzyl chloride, and stannous chloride are heated in concd. aq HC1.	119
		204-205°	Thermal rearrangement of 6-bromo-5,6-dihydro-5,5-tetramethylenebenz-[c]acridine, followed by Pd-C dehydrogenation.	47
Dibenz[ah]acridine	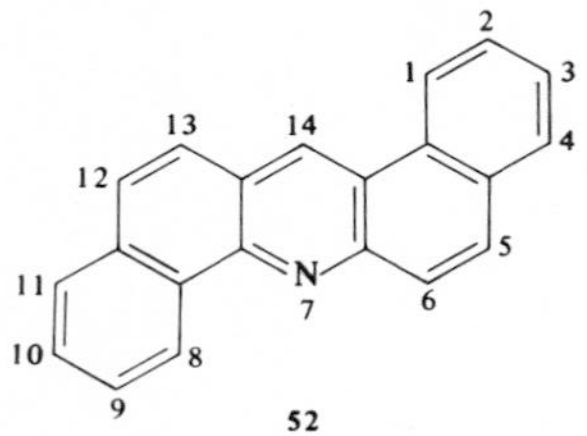 52	Yellow crystals ex. toluene, 228°	1-Naphthylamine, 2-naphthol, and paraformaldehyde are heated.	20, 127
		228°	1- and 2-naphthylamines, and dichloromethane are heated in a sealed tube at 240°.	121
		228°	1,3-di(2-naphthyl)thiourea and 1-tetralone are heated. Hydrolysis, zinc reduction of the acridanone and oxidative aromatization follow.	23

548

Table II. (Continued)

Compound	Structure	Appearance and mp (°C)	Preparation	Ref.
Dibenz[*ai*]acridine	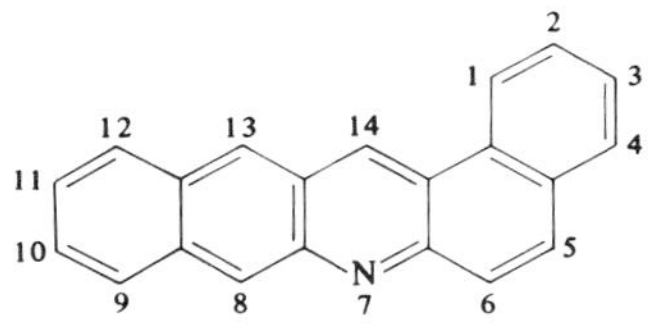 **53**	Yellow needles ex. pyridine, 205.5-206°	Dibenz[*ai*]acridan-14-one, obtained from 2-naphthylamine and 2-hydroxynaphthalene-3-carboxylic acid, is distilled with zinc.	120, 121
			Dibenz[*ai*]acridan-14-one, also from Chapman rearrangement of 3-methoxycarbonyl-2-naphthyl *N*-β-naphthyl benzimidate.	50
		207-208°	5,6-Dihydrobenz[*ai*]acridine-14-carboxylic acid is heated with PbO or in vacuo to c.250°	7,8
Dibenz[*ai*]acridine	 **54**	Straw yellow needles ex. acetone alcohol, 216° Very pale yellow leaflets ex. benzene or toluene, 220-221° (after chromatography).	2-Naphthylamine and formaldehyde are heated or 2-naphthylamine, 2-naphthol, and formaldehyde are heated optimally at 250-260°.	122, 127, 128
				20, 123

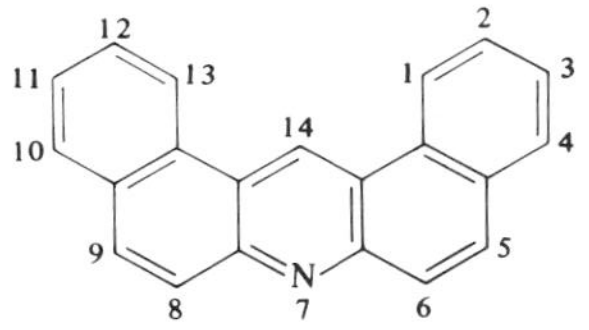

(Table Continued)

549

Table II. *(Continued)*

Compound	Structure	Appearance and mp (°C)	Preparation	Ref.
Dibenz[*aj*]acridine		215.5°	2-Naphthylamine, diiodomethane and potassium carbonate are heated to 150-160°.	121
		217°	Isolated from benzene insoluble/picrate, from 2-naphthol and formamide heated 20 hr at ca. 200°	124
		216-217°	1,3-Di(2-naphthyl)-thiourea and 2-tetralone heated. Hydrolysis, zinc reduction of the acridanone and dehydrogenation follow.	24
		216°	Di-2-naphthylamine, *N*-methylformanilide, and phosphoryl chloride are heated at 100°. The crude aldehyde is cyclized by acid at 100°.	19

Table II. (Continued)

Compound	Structure	Appearance and mp (°C)	Preparation	Ref.
Dibenz[*bh*]acridine	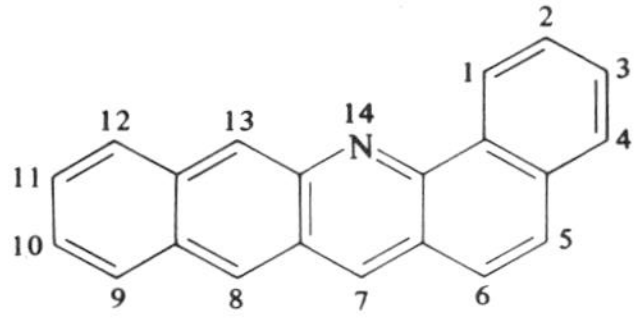 **55**	Yellow needles ex. benzene, 194-195°	Decarboxylation and dehydrogenation of 5,6-dihydrodibenz[*bh*]-acridine-7-carboxylic acid with PbO at 300° or in vacuo at 290°	7
Dibenz[*bi*]acridine	**56**		For derivatives, see refs. For predicted properties, see ref.	125, 126 72
Dibenz[*ch*]acridine	**57**	Pale cream needles from acetone or benzene, 189° Pale white solid, 188°	1-Naphthylamine and dichloromethane are heated in a sealed tube at 220-230° or 1-naphthylamine, diiodomethane, and potassium carbonate are heated at 150-160°. Pyrolysis of the *N,N,N*-trimethylhydrazonium fluoroborate of 1-tetralone, followed by dehydrogenation (Pd-C) at 250° in vacuo.	121, 127 129

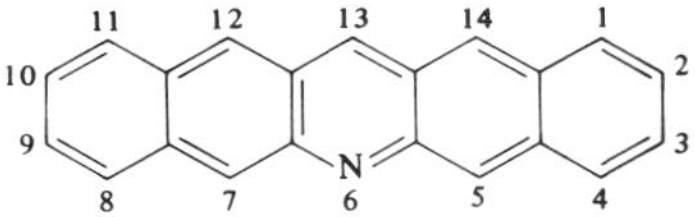

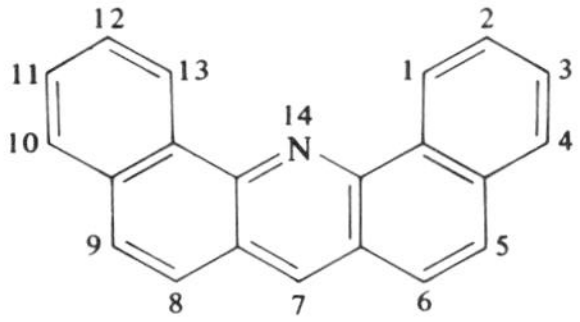

(Table Continued)

Table II. (Continued)

Compound	Structure	Appearance and mp (°C)	Preparation	Ref.
Naphth[2,3-c]acridine	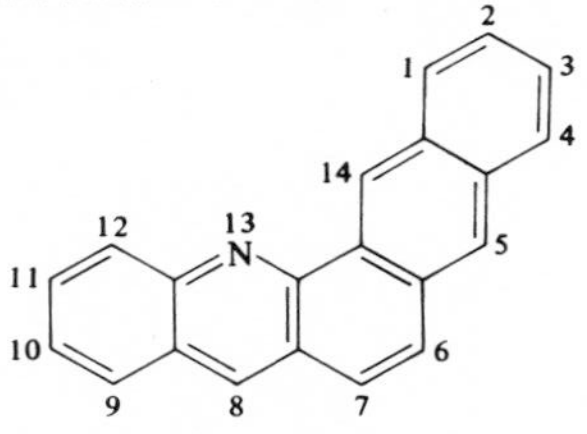 **58**	Yellow leaflets from benzene or ethanol 223-224°	Naphth[2,3-c]acridan-5,8,14-trione (available readily from suitable anthraquinone precursor), heated at 500-750° with zinc dust under hydrogen (1 atm)	130
		223-224°	Acridine with o-xylylene generated in situ gives six products, a major component (7% yield) being naphth[2,3-c]acridine.	130
Naphth [2,3-a]acridine	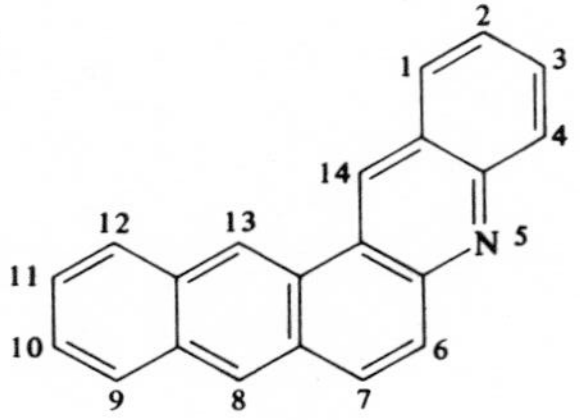 **59**	—	For derivatives see refs.	131, 132

552

Table II. (Continued)

Compound	Structure	Appearance and mp (°C)	Preparation	Ref.
Naphth[2,1-c]acridine	**60**	Colorless needles ex. ethanol, 202°	Decarboxylation and dehydrogenation of the product from isatin and 4-oxo-1,2,3,4-tetrahydrophenanthrene.	25, 133
7H-Naphth[1,8-bc]-acridine	**61**	Colorless needles ex. ethanol, 137°	Decarboxylation and dehydrogenation of the product from isatin and 2,3-dihydro-1H-phenalen-1-one.	133
Naphth[1,2-c]-acridine	**62**	Yellow crystals ex. ethanol, 184°	Decarboxylation and dehydrogenation of the product from isatin and 1-oxo-1,2,3,4-tetrahydrophenanthrene.	134
Naphth[2,3-b]-acridine	**63**	—	For derivatives, see refs.	132, 135, 136

Both dibenz[*ai*]acridine and dibenz[*bh*]acridine form photooxides on exposure to sunlight in carbon disulfide or benzene solution.[8] These initial products are relatively unstable, although the photooxide from dibenz[*bh*]-acridine has been isolated as very pale yellow needles, mp 195–200°, and is thought to possess an *endo* peroxide bridge across the 8,13-positions. This product is converted to the 8,13-dione when heated, or diluted in solution; 8,13-dihydrodibenz[*ai*]acridine-8,13-dione was the only product isolated from the photooxidation of the former acridine. Both acridines slowly form high-melting photodimers on irradiation in ethereal solution, the yields being greater when air is excluded.[8] The molecular parameters of dibenz[*ch*]-acridine (**57**) have been determined by a least squares refinement of the complete two-dimensional X-ray data, and the bond lengths and charge densities in this molecule have been calculated by a molecular orbital theory treatment.[137]

The nmr spectra of dibenz[*ah*]acridine (**52**), dibenz[*aj*]acridine (**54**), and dibenz[*ch*]acridine (**57**), have been investigated.[138, 139] For dibenz[*aj*]acridine the spin tickling technique has made possible a complete analysis of the nmr spectrum, and the actual and computed spectra have been matched.[138] Nuclear magnetic double resonance techniques have given unequivocal evidence of inter-ring proton spin-spin coupling in all three dibenzacridines[139]; for instance, in dibenz[*aj*]acridine (**54**), nonzero values have been found for $J_{1,5}$, $J_{1,14}$, and $J_{6,14}$. Certain protons in these systems show remarkably low field chemical shifts[139]; thus the H-14 resonance in dibenz[*aj*]-acridine occurs at τ 0.11 (CDCl$_3$), while the H-1 (and H-13) resonances in dibenz[*ch*]acridine are found at τ 0.27 (CDCl$_3$); similarly, the signal at τ 0.10 in the nmr spectrum of naphth [2,3-*c*]acridine (**58**) has been assigned[130] to H-14.

Three routes to the dibenz[*aj*]acridine system omitted from Table II merit brief mention at this point. Trimethyl-2-naphthylammonium iodide, when refluxed in a mixture of glacial acetic acid and anhydrous sodium acetate with 2,4,6-triamino-5-nitrosopyrimidine, this last nitroso compound behaving as an oxidizing agent in the system, gave the methiodide of dibenz[*aj*]acridine; other oxidizing agents (e.g., chromic acid) failed to achieve this synthesis.[140] Catalytic conversion with ammonia in benzene over alumina at 470° gave 14-methyldibenz[*aj*]acridan from the oxygen isostere 14-methyl-14*H*-dibenzo[*aj*]xanthen.[141] Mention has been made earlier (see p. 530) that trialkylacetic acids in the Bernthsen synthesis of benzacridines failed to give *meso*-tertiary-alkyl-substituted products; thus pivalic acid and di-2-naphthylamine in the Bernthsen procedure gave unsubstituted dibenz[*aj*]acridine.[16]

Morgan's base has been referred to earlier (see p. 543). Early attempts to synthesize dibenz[*aj*]acridine from 2-naphthylamine and either formaldehyde or diiodomethane (see Table II) gave, among other compounds, an orange complex, known as Morgan's base, whose structure was uncertain.[121, 122a, b] Later work showed that the orange compound could be obtained in good yield from 2-naphthylamine, 2-naphthol, and formaldehyde in boiling toluene or xylene, or by the zinc and hydrochloric acid reduction of dibenz[aj]acridine.[123] Dibenz[aj]acridan, mp 153°, was also formed in both reactions, but the latter reduction gave a second dihydrobenz[aj]acridine of mp 183°. The former dihydro compound was the acridan, being nonbasic and with a uv absorption spectrum unlike that of dibenz[aj]acridine; the latter dihydro derivative may be 5,6-dihydrodibenz[aj]acridine, since it was basic, had a uv absorption spectrum very similar to that of dibenz[aj]acridine, and was oxidized to this compound by potassium permanganate.

The admixture of equimolar quantities of dibenz[aj]acridan and dibenz[aj]-acridine gave Morgan's base, and a phase diagram investigation confirmed the 1:1 ratio of the components. Various oxidizing agents or repeated recrystallization from acetone or acetic acid converted Morgan's base, mp 248–249°, to dibenz[aj]acridine of mp 220–221°, while reduction gave color-less dibenz[aj]acridan. The green salts that it gives with acids may be com-pared with the green acridine–acridan–hydrochloric acid complex (Chapter V, p. 451). Both ebullioscopic and spectral measurements showed that Morgan's base was completely dissociated in solution. Polarographic reduc-tion of dibenz[aj]acridine occurred by two successive one-electron transfers. The orange Morgan's base has therefore been represented by two ion radi-cals,[123] structures **64** and **65** portraying one of the possible canonical forms contributing to the overall resonance structure in each case, or as the radical form **66**.

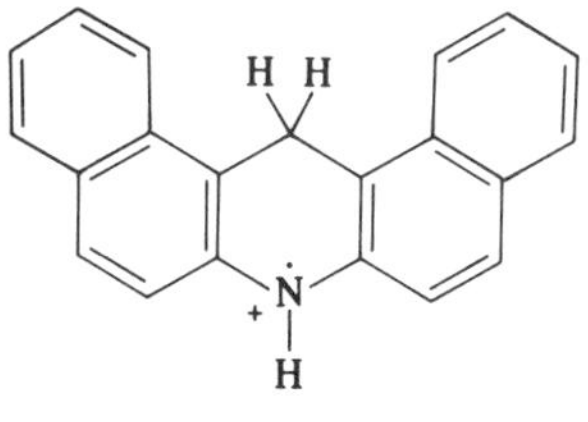

64

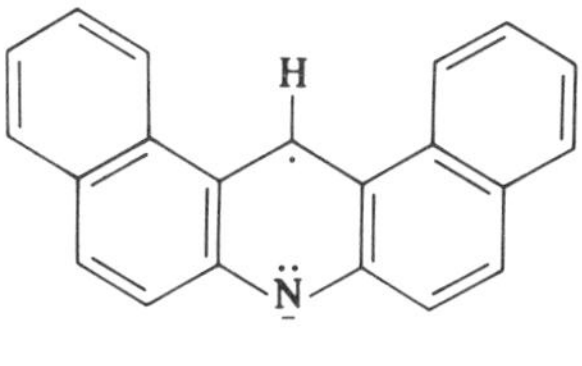

65

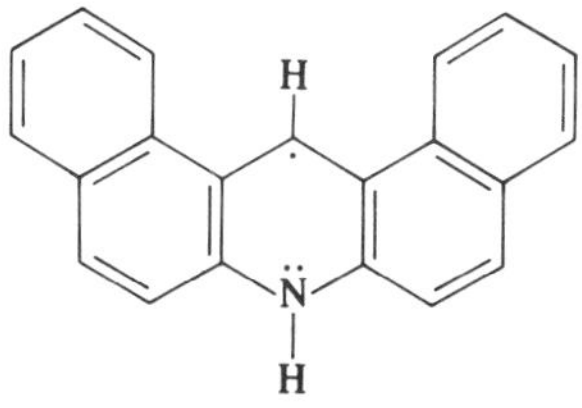

66

Structures derived in an equivalent manner were subsequently proposed for the anion radical-cation radical pair that comprise the similar orange benz[a]-acridine–benz[a]acridan complex.[4] The complex can be pictured as being built up of successive layers of planar molecules having positive or negative charges, the layers being held together by electrostatic attraction.

Mention has been made of the fact that the *meso*-methyl groups in dibenzacridines are the least reactive (toward *m*-nitrobenzaldehyde) of their kind among acridine, benzacridines, and dibenzacridines.[84] Both dibenz[ai]-acridine (**53**) and dibenz[bh]acridine (**55**) are oxidized in high yield (76 and 90%, respectively) to their respective 8,13-diones by chromium trioxide, both compounds existing in two crystalline forms of differing melting points.[8] Dibenz[ch]acridan-7-one, obtained from the Chapman rearrangement of the benzimidate **67**, forms a 7-chloro derivative with phosphorus oxychloride which proved to be unusually slow to aminate.[50]

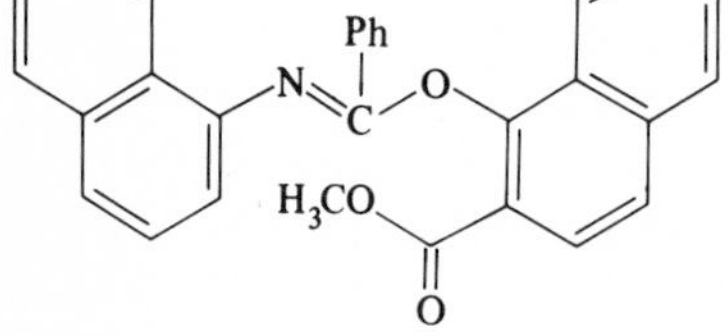

67 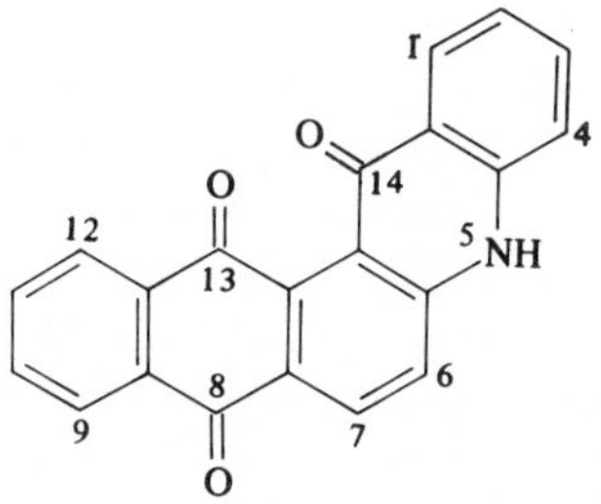   68

The most frequently encountered representatives of the naphth[2,3-a]-acridines, naphth[2,3-b]acridines, and naphth[2,3-c]acridines, are the triones **68, 69,** and **70,** often referred to as phthaloylacridones.

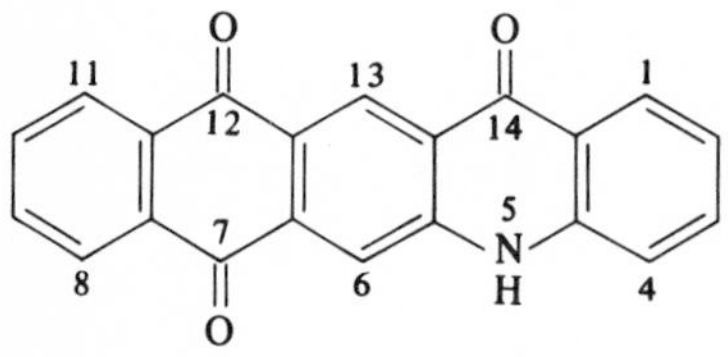

69 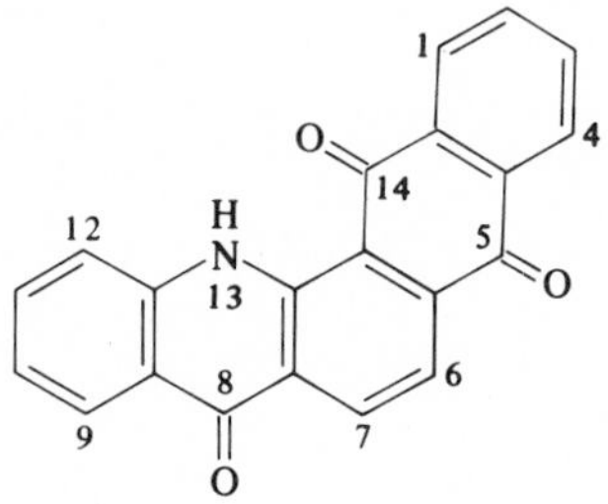   70

Of these three, 5,14-dihydronaphth[2,3-c]acridan-5,8,14-trione (**70**) (and its derivatives) has been the most investigated. On the evidence of their uv spectra, lower solubility (in chlorobenzene), and higher melting points, the triones **68** and **69** form intermolecular hydrogen bonds, while the more soluble lower melting compound, **70,** forms intramolecular hydrogen bonds. *N*-Methylation of the acridanones **68** and **69** prevents the hydrogen bonding which results in increased solubility and lower melting points; the effects of N-alkylation and ring substitution on the uv spectral properties, solubilities, and melting points[132] have been analyzed for the isomer **70**.

Syntheses of the trione **70** and its derivatives rely on two main routes: (1) nucleophilic substitution by the amino group of an anthranilic acid at position 1 of an anthraquinone derivative bearing a labile chloro or nitro substituent at position 1, followed by cyclization, e.g., by heating with sulfuric acid; or (2) a parallel sequence of reactions involving the amino group of a suitable aromatic amine and an anthraquinone-2-carboxylic acid derivative bearing a labile 1-substituent.[132, 142a, b, 143] Nitration of the trione **70** by fuming nitric acid/concentrated sulfuric acid[144] gave in excellent yield a 1:3 mixture of, respectively, the 6,10,12-trinitro derivative and a tetranitro derivative, in which the location of the fourth nitro group is unspecified among positions 1–4. Chloromethylation of precursor **70** can be controlled to give either the 12-chloromethyl or the 10,12-bis(chloromethyl) derivatives.[145] A halogen (chlorine or bromine) at position 7 in the trione **70** or its derivatives can be replaced by an amino or an alkoxyl function; the yields in this nucleophilic displacement are high when the reaction is carried out in refluxing nitrobenzene. Halogens at positions 6, 9, 10, and 12 are inert under these conditions.[143, 146, 147] A sulfonic acid moiety at the 7-position has been replaced by a nitrile group by treatment with an aqueous potassium cyanide/potassium carbonate mixture at 120° in a pressure vessel.[148]

8H-Dibenz[c,mn]acridin-8-one (**73**) has been obtained by three different routes; in the earliest of these,[149] the amide **71** cyclized as shown to the phenanthridine **72**, which underwent cyclic dehydration to the required product **73**.

In another synthesis, 1,4-naphthoquinone and 1-o-nitrophenylbuta-1,3-diene undergo a Diels-Alder condensation to give a mixture of the tetrahydroanthraquinone 74 and a related dihydroanthraquinone, produced by dehydrogenation.[150] The latter was the only product isolated when the reaction was carried out in benzene at 100°; however, at 150° compound, 75 resulted. Boiling crude 74 with methanolic potash caused an intramolecular redox reaction leading to 8H-dibenz[c,mn]acridin-8-one (73). The phenanthraquinone derivative 75 was unaffected by this treatment, but with alkaline sodium dithionite gave 73, presumably via reduction to the amino compound and cyclic dehydration. The condensation of o-aminothiophenol with 1-chlorophenanthraquinone gave the thiazepine derivative 76, from which sulfur could be extruded by a variety of reagents (e.g., copper-bronze or Raney nickel) to give 8H-dibenz[c,mn]acridin-8-one (73) in 70–80% yield.[151]

74   75   76

The acridinone 73 obtained as yellow needles either from xylene,[149] mp 221–223°, or from ethyl acetate,[150] mp 218°, has a uv absorption spectrum very similar to that of its carbocyclic analogue and did not form a picrate or a 2,4-dinitrophenylhydrazone under normal conditions.[150]

Reduction of the acridinone 73 with alkaline sodium dithionite solution, followed by neutralization with acetic acid, precipitated black-violet needles, mp 197°, which were readily oxidized to the original compound. This product, although stable in the solid state and slightly soluble in aqueous sodium hydroxide, gave a violet, readily oxidized solution in toluene. Boiling with acetic anhydride or treatment with benzoyl chloride in pyridine gave a mixture of the original acridinone (73) and a desoxy-compound, 8H-dibenz[c,mn]-acridine. The latter was also prepared from 73 by reduction with stannous chloride in acetic/hydrochloric acid but could not be obtained pure because of its aerial oxidation; when freshly prepared it was completely soluble in 10% aqueous hydrochloric acid. The structure of the black-violet solid remains a matter of conjecture; it may be related to Morgan's base and may have a radical formulation.

9*H*-Naphth[3,2,1-*kl*]acridin-9-one (**78**), also known variously as ceramidine, ceramidone, and ceramidonine, has been prepared by the cyclization of 1-anilinoanthraquinone (**77**), using 70% sulfuric acid,[152, 153] and from 9-(2′-carboxyphenyl)acridine (**79**) in a similar manner.[154]

A number of derivatives has been prepared,[153, 155-158] for the most part by the anthraquinone route, and 1,2,3,4-tetrahydro-9*H*-naphth[3,2,1-*kl*]acridin-9-one was obtained from 1-aminoanthraquinone and cyclohexanone in the presence of alkali.[159] Powdered zinc and concentrated hydrochloric acid, acting on 1,4-bis(*p*-tolylamino)anthraquinone, gave as one product the 2-methyl-8-*p*-tolylamino derivative[160] of **78**. The reduction of this last product with tin and acetic/hydrochloric acid[157] gave the acridan derivative **80**, demonstrating also the facility with which hydrolytic cleavage of the 8-substituent occurs; the reduced form **80** is oxidized to the 2-methyl-8-hydroxy derivative of **78** by heating in nitrobenzene. An 8-amino function in derivatives of **78** has also been converted to a hydroxyl group by prolonged heating in 70% sulfuric acid; on the other hand, the 8-amino-9*H*-naphth[3,2,1-*kl*]-acridin-9-one behaves as a typical aromatic amine on diazotization.[158] Conversely, heating the 8-hydroxy derivatives under a variety of conditions with ammonia or aromatic amines gave the 8-(substituted)amino compounds.[153, 158]

Derivatives of **78**, variously substituted at positions 2 and/or 7, on heating at 120–150° with aromatic amines, gave mainly the 6-arylamino product with lesser quantities of the 8-isomer. The more basic cyclohexylamine and methylamine preferentially entered position 8 under milder conditions (20–60°); cupric acetate (for the arylamines) or pyridine (for the alkylamines) was present in trace amounts.[161] Direct amination of 7-chloro-2-methyl-9*H*-naphth-[3,2,1-*kl*]acridin-9-one with 25% aqueous ammonia, finally heating the system to 170° in the presence of cupric acetate, gave 29% of the 6-amino derivative, while a mixture of sodamide and liquid ammonia in dry dimethylaniline at 60° gave 15% of the 8-amino isomer.[162]

Indolo[3,2,1-*de*]acridan-8-one (**81**) has been synthesized from 9-phenylcarbazole by successive treatment with butyl lithium, carbon dioxide, water, phosphorus pentachloride, and stannic chloride,[163] and from carbazole by reaction with 2-iodobenzoic acid, followed by cyclization.[164] After crystallization from xylene, it had a melting point of 180–181°, and was stated to give

80   **81**

an oxime, mp 175–176°, with hydroxylamine.[163] On the assumption that this last product had been correctly formulated, this reaction provided the first recorded case of an acridanone behaving as a ketone with such a reagent. (For a more recent instance of such a reaction,[46] see Section 1.B concerning the 11-oxobenz[b]acridine derivative **39**.)

In the last decade, there has been an enormous growth, primarily in the patent literature, in the number of papers on the subject of linear quinacridones, the essential motivation for this expansive research being their importance to the dyestuffs chemist. Their detailed chemistry will consequently be dealt with in Chapter VIII; therefore, only the fundamentals of their synthesis and reactions are presented here. Quinacridones are the subject of a recent review article.[165] By far the most frequently used synthetic route to derivatives of quin[2,3-b]acridine-7(12H),14(5H)-dione (**83**) or lin-quinacridone, involves the cyclization of 2,5-diarylaminoterephthalic acid derivatives, **82**.

82   **83**

In an early report,[166] precursor **82**, heated with boric acid to 320°, gave lin-quinacridone (**83**) in 87% yield, as a violet-red powder carbonizing above 400°. A great variety of cyclizing agents has been discovered (e.g., aluminium chloride,[167] polyphosphoric acid,[168] sulfuric acid,[169] sulfonic acids,[170] acid chlorides in nitrobenzene,[171] phosphorus (oxy)halides,[172] phthalic anhydride and Lewis acids,[173] benzotrichloride,[174] etc.) and a range of substituents has been located in the terminal benzene rings of **83** by selection of the appropriate arylamine components of precursors **82**. The complementary cyclization of precursors of type **84** to derivatives of **83** has been effected with boiling phosphorus oxychloride,[175] or with acid chlorides or anhydrides in refluxing nitrobenzene.[176] Many 6,13-dihydro-*lin*-quinacridones (derivatives of **86**) have been synthesized directly, in a similar manner, [177a, b] from the appropriate 2,5-diarylamino-3,6-dihydroterephthalic acid diester precursors (**85**).

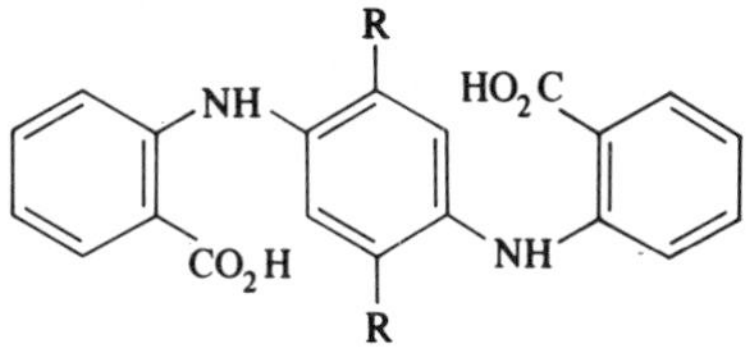

84

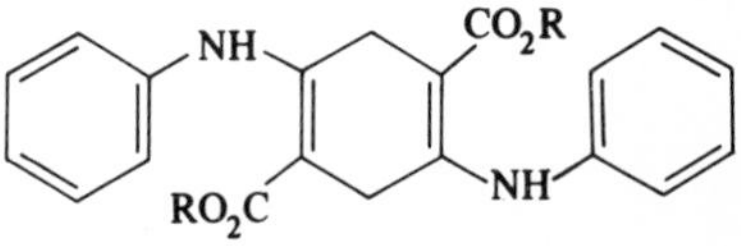

85

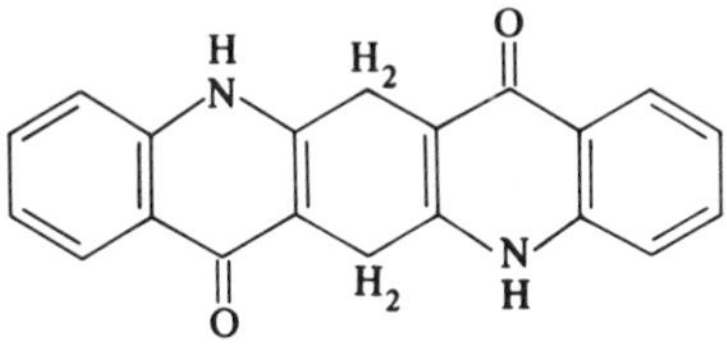

86

Both routes[178a, b, 179a, b] have been employed for the direct synthesis of quinacridonequinones (derivatives of **89**), using precursors derived either from **87** or from **88**.

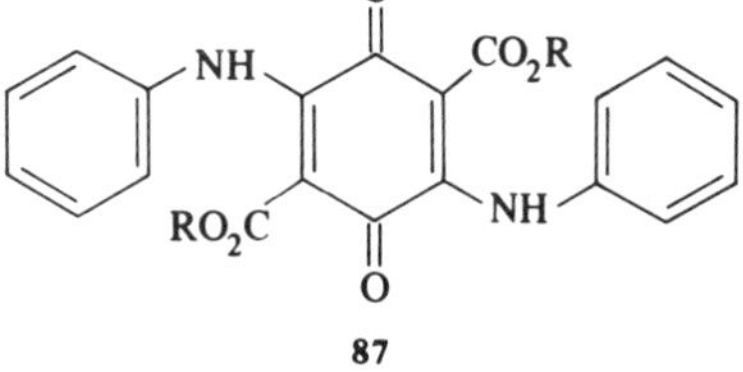

87

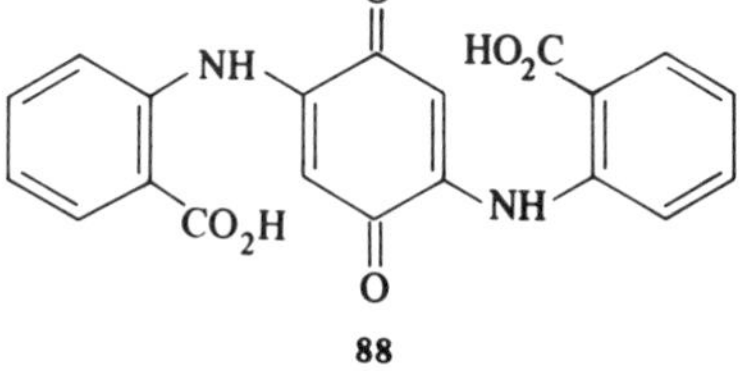

88

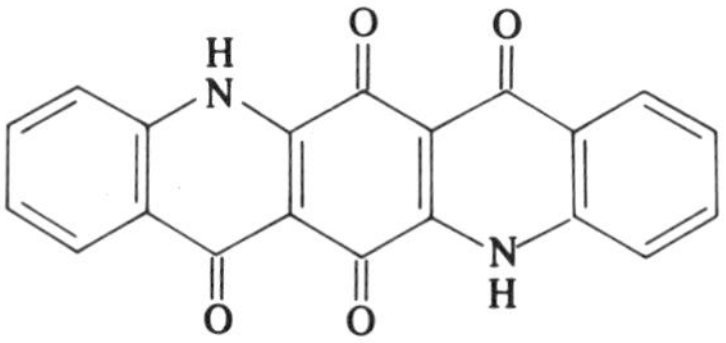

89

Again a range of cyclization agents[178a, b, 179a, b] has been devised for the synthesis of derivatives of **86** and **89**. Condensation[180] by means of sulfuric acid of cyclohexanone and 2,5-diaminoterephthalic acid gave **90**, while cyclohexanone and 2,5-dibenzoyl-*p*-phenylenediamine in a reaction catalyzed by aniline hydrochloride[181] yielded **91**. Then the *p*-benzoquinone derivative **92** was cyclized by polyphosphoric acid,[182] and the product reduced in situ by zinc dust to give *lin*-quinacridone (**83**).

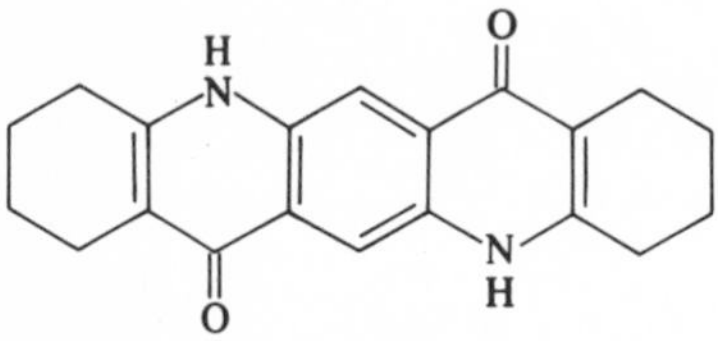

90

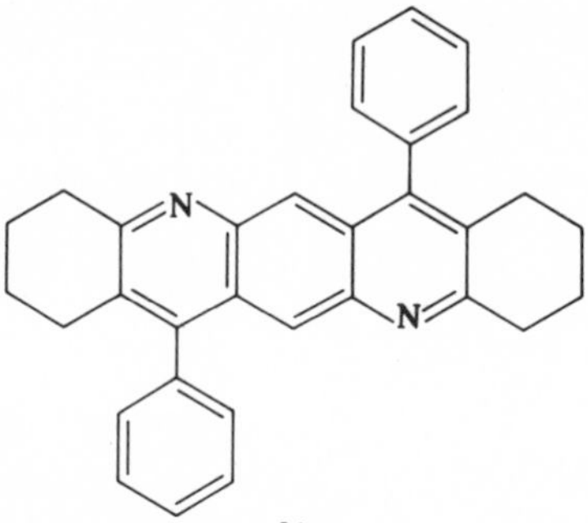

91

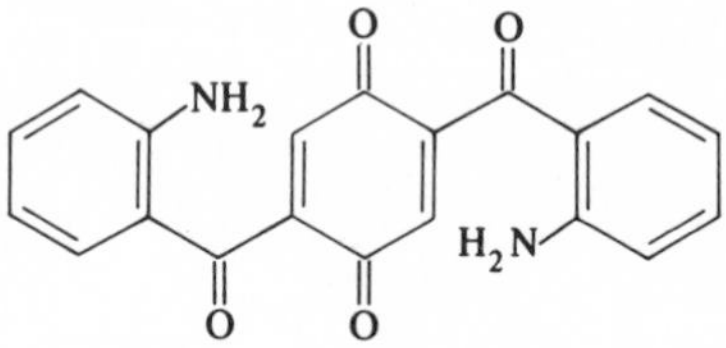

92

The study of the reactions of the quin[2,3-*b*]acridine system is mainly confined to the oxidation-reduction interchange of derivatives of the four systems, the dione **83**, its 6,13-dihydro reduction product **86**, the 6,13-dihydroxy derivative **93**, and the quinone **89**. For example, the quinone **89** is reduced to the leuco 6,13-dihydroxy compound **93** by tin and 6*N* hydrochloride acid,[178a,b,183] to the dione **83** by zinc, aluminium chloride, and urea at 80°,[184] and to the 6,13-dihydro product **86** by tin and polyphosphoric acid at 130°.[185–187] This last reaction system[186, 187] also reduces the dione **83** to compound **86**. The 6,13-dihydro compound **86** can be oxidized to the dione **83** by, e.g., sodium *m*-nitrosulfonate in aqueous alcoholic caustic soda solution[179a, b, 185, 187–189] or by chloranil.[177a, b, 189] Inorganic oxidants,[190] such as chromium trioxide in aqueous acetic acid at 100°, give the quinone **89**. Phosphorus pentachloride converts derivatives of the quinol **93** to the 6,13-dichloro compounds in good yield.[191a, b] Quinacridone **83** and the quinone **89** have been chlorinated and brominated to varying extents at unspecified locations with reagents such as sulfur dichloride-aluminium chloride,[192] thionyl chloride-aluminium chloride-bromine,[192] and chlorine-titanic chloride.[193] The 6,13-dichloro derivative of **83** can be dechlorinated, either by zinc dust and acetic acid or by tin and potassium hydroxide in dimethylformamide.[194a, b]

Whereas the cyclization[176] of **84** (R = Me) gave mainly the *lin*-quinacridone derivative **83** (6,13-Me₂), the unsubstituted precursor **84** (R = H) gave the isomeric quin[3,2-*a*]acridine-13(8*H*),14(5*H*)-dione (**94**) by the alternative mode of cyclization.[195a, b] The minor product from **84** (R = Me) was the 7-methyl derivative of **94**, cyclization having occurred with the extrusion of a methyl group.[176] Precursors of type **84** (R = H), bearing a variety of substituents in the two anthranilic acid rings, have been cyclized to derivatives[196, 197] of system **94**. Reduction of the dione **94** by sodium and alcohol to the acridan, followed by nitric acid oxidation, gave the parent aromatic compound quin[3,2-*a*]acridine or 5,8-diazapentaphene (**95**), whose uv absorption spectrum was very similar to that of pentaphene but not to that of pentacene.[195a, b]

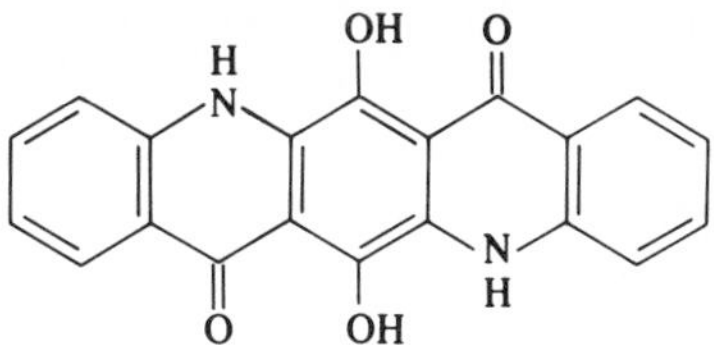

93

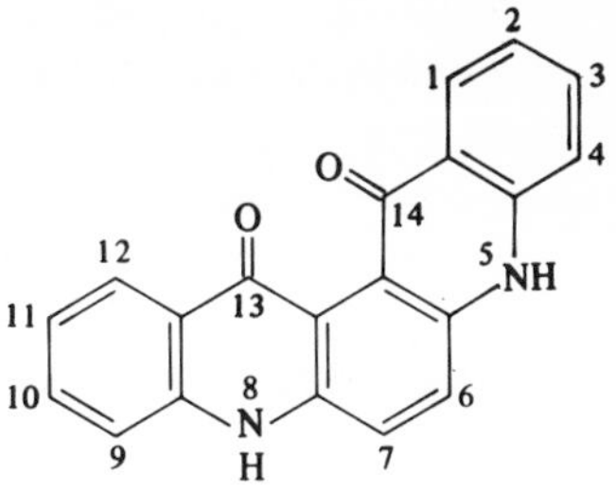

94

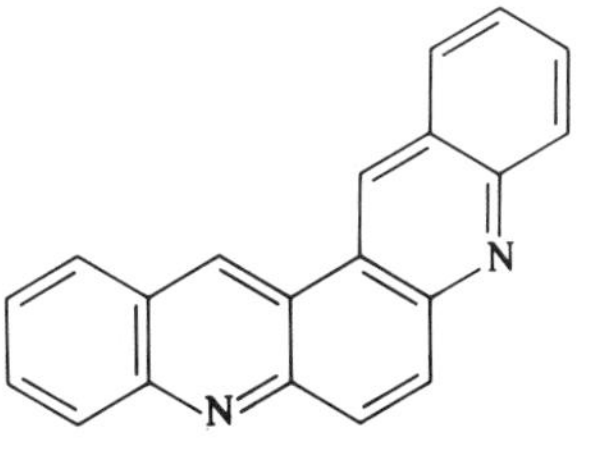

95

A number of *iso*phthalic acids of the general formula **96** (Ar, Ar' = aryl groups) has been cyclized under a variety of acidic conditions to give derivatives[198] of quin[3,2-*b*]acridine-12(7*H*),14(5*H*)-dione (**97**). The condensation of 4,6-diamino*iso*phthalic acid with cyclohexanone[180] in the presence of concentrated sulfuric acid gave the reduced derivative **98**.

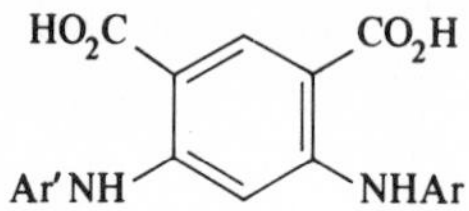

96

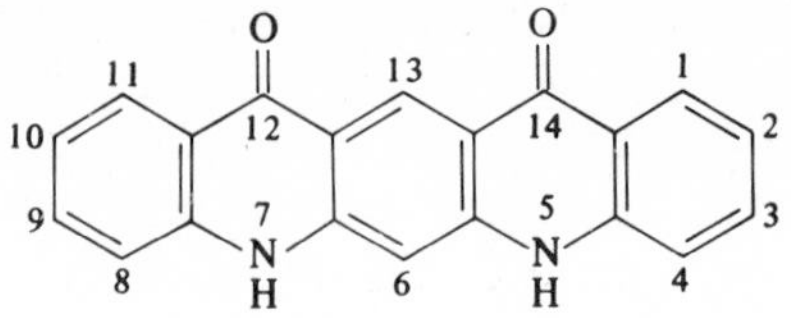

97

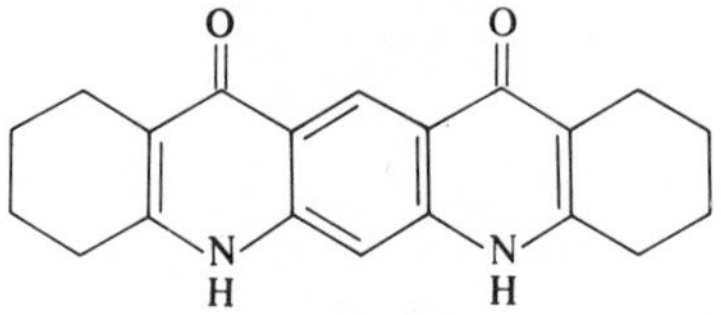

98

Phloroglucinol (cyclohexan-1,3,5-trione) reacted with two molecules of *o*-aminobenzaldehyde[199] to give 7-hydroxyquin[2,3-*a*]acridine (**99**; R = OH), oxidized to the *ortho* quinone by chromic acid. This quinone gave a phenazine with *o*-phenylenediamine. In a similar way, phloroglucinol and anthranilic acid at 150° are reported to give a 90% yield of 7-hydroxyquin[2,3-*a*]-acridine-8(13*H*),14(5*H*)-dione,[200] which on distillation with zinc gave the parent quin[2,3-*a*]acridine (**99**; R = H), which crystallized as colorless plates, mp 221°, with a blue fluorescence in solution.

8-Methyl-5*H*-quin[2,3,4-*kl*]acridine (**100**) was prepared by cyclizing 1-anilino-4-methyl-9-acridanone with phosphorus oxychloride[201]; it is a deep blue compound melting above 355°, giving blue solutions in aqueous hydrochloric acid. 2,2'-Diaminobenzophenone and dimedone[202] cyclize in acetic acid containing a trace of concentrated hydrochloric acid to give a related compound, 7,8-dihydro-7,7-dimethyl-6*H*-quin[2,3,4-*kl*]acridine (**101**), described as pale yellow prisms, mp 170–171°.

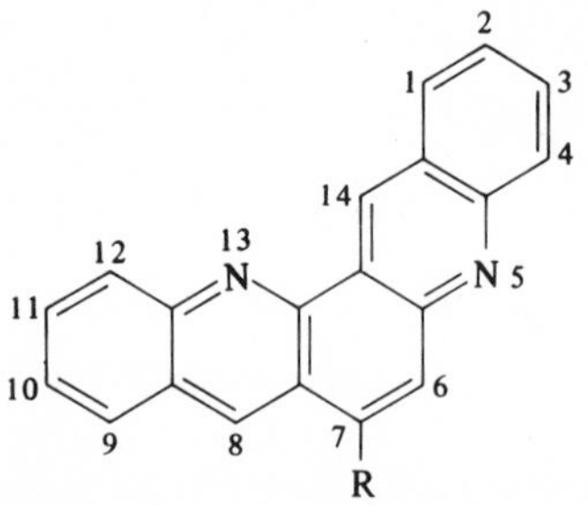

99

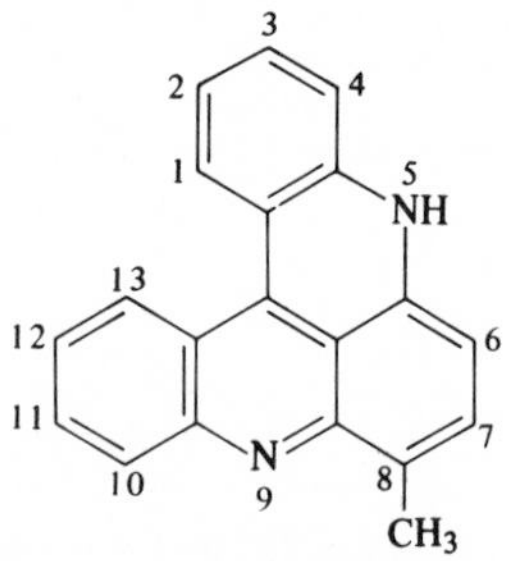

100

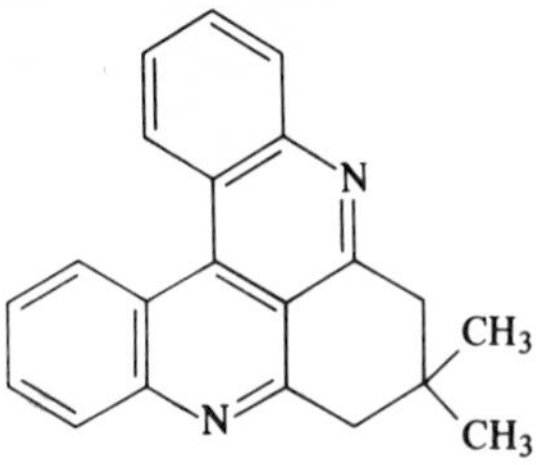

101

3,6-Diaminoacridine treated with potassium thiocyanate in acetic acid, followed by bromine in that acid, gave the acridine **102**, which with 20% hydrochloric acid,[203] followed by basification with ammonia, gave 2,10-diamino-*bis*-thiazolo[4,5-*c*: 5′,4′-*h*]acridine (**103**, R = NH$_2$), mp > 350°. The acridine **102**, treated with sodium sulfide in aqueous ethanol, gave the salt **104**; heating this with 99% formic acid gave the parent compound **103** (R = H), mp 340–341°, while **104** and acetic anhydride or benzoyl chloride gave, respectively, **103** (R = Me) and **103** (R = Ph).

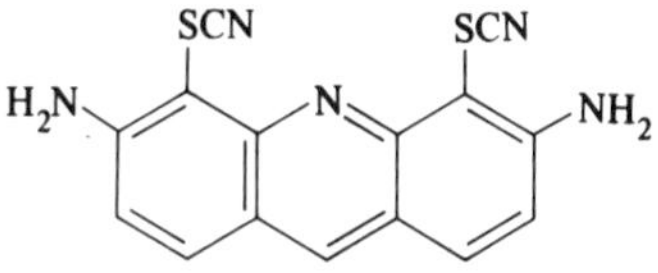

102

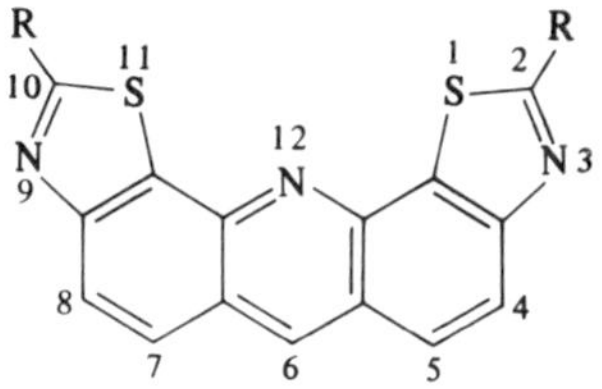

103

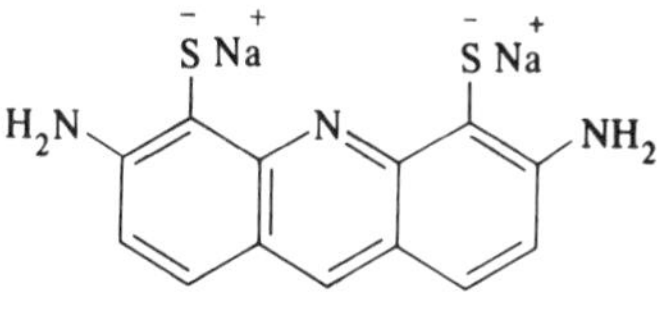

104

Derivatives of four isomeric systems, benzo[*a*]thiazolo[4,5-*j*]-, benzo[*a*]-thiazolo[5,4-*j*]-, benzo[*h*]thiazolo[5,4-*a*]-, and benzo[*h*]thiazolo[4,5-*a*]-acridine have been prepared from the appropriate aminobenzothiazoles, isomeric naphthols, and paraformaldehyde, and tested for their carcinogenicity in mice.[204]

3. Condensed Acridines with Six or More Rings

6-Methyl-8H,12H-phenaleno[2,1,9,8-$klmn$]acridine-8,12-dione (**106**) has been prepared from 1-chloroanthraquinone and 3,5-dimethylaniline.[156] An Ullmann condensation, followed by cyclization of the product with sulfuric acid, gave 1,3-dimethyl-9H-naphth[3,2,1-kl]acridin-9-one (**105**), which with silver oxide at 180° gave **106** in quantitative yield.

Paraformaldehyde, 1,2,3,4-tetrahydro-9-phenanthrol (**107**), and 1-naphthyl-amine, or 2-naphthylamine, or 9-aminophenanthrene, followed by dehydro-genation of the initial products with 5% palladium charcoal, gave, respec-tively, tribenz[ach]acridine (mp 216°), tribenz[acj]acridine (mp 253°), and tetrabenz[$achj$]acridine.[205]

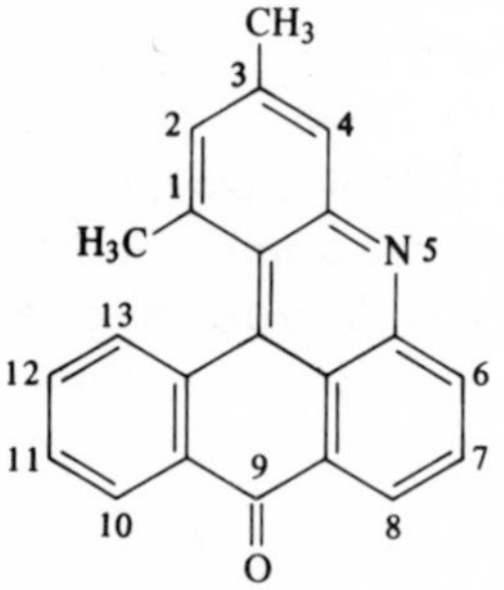

105 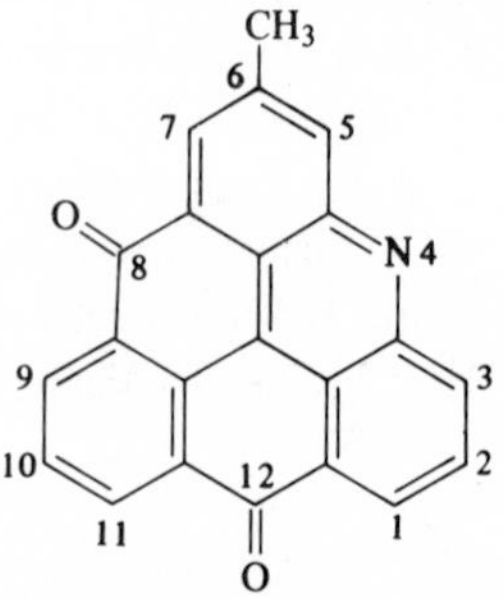   106 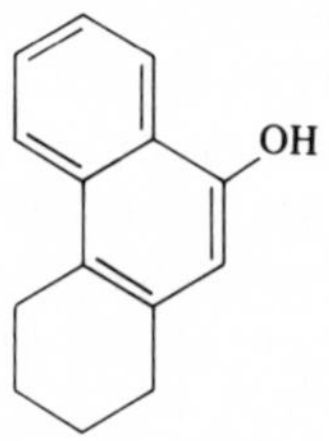   107

In the complementary syntheses, paraformaldehyde, and 9-aminophen-anthrene with 1-naphthol and 2-naphthol, gave, respectively, as minor products only,[206] the expected tribenz[ach]acridine and tribenz[acj]acridine; in both instances the major product was tetrabenz[$achj$]acridine, mp 463°. The Ullmann-Fetvadjian synthesis with 2-aminophenanthrene and 1- and 2-naphthol proceeds exclusively with cyclization to position 1 of the phen-anthrene, the products being, respectively,[206] benzo[h]naphth[2,1-a]acridine (**108**), mp 309–310°, and its isomer **109**, mp 233°.

3-Aminofluoranthene, 2-naphthol, and paraformaldehyde yielded benzo-[a]fluoren[1,9-hi]acridine (**110**), mp 282°; the reaction fails with 1-naphthol, probably, it was suggested, for steric reasons.[207]

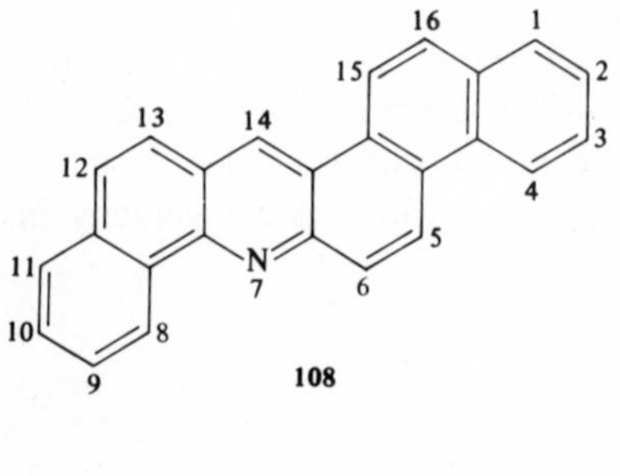

108 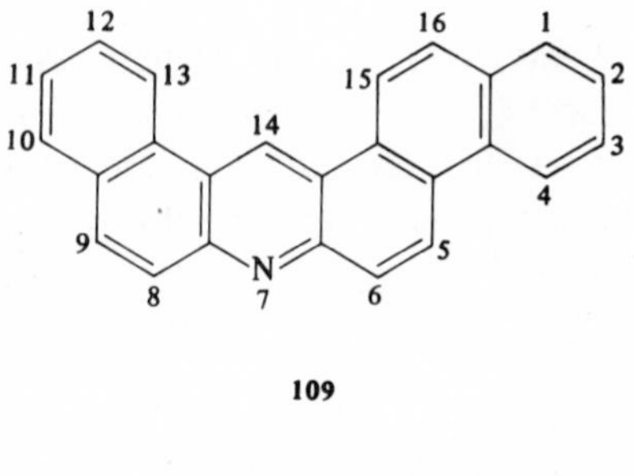   109

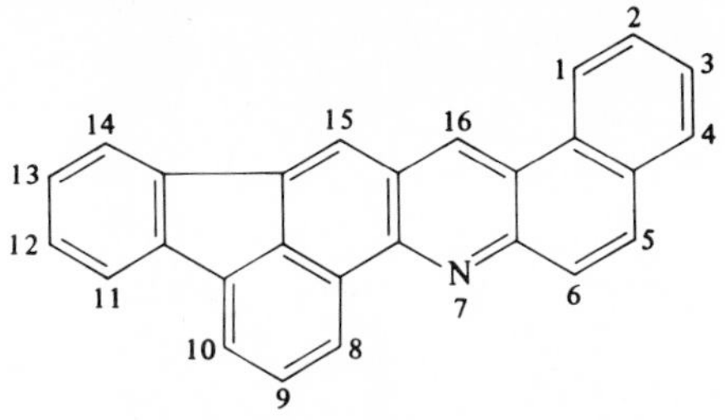

110

1-Aminoanthracene in this reaction gave benzo[c]naphth[2,3-h]acridine (**111**), with 1-naphthol, and the benzo[a]-isomer with 2-naphthol; 2-aminoanthracene in parallel reactions cyclizes only to position 1, giving benzo[h]naphth-[2,3-a]acridine (**112**) and the isomer **113** with 1- and 2-naphthol, respectively.[208]

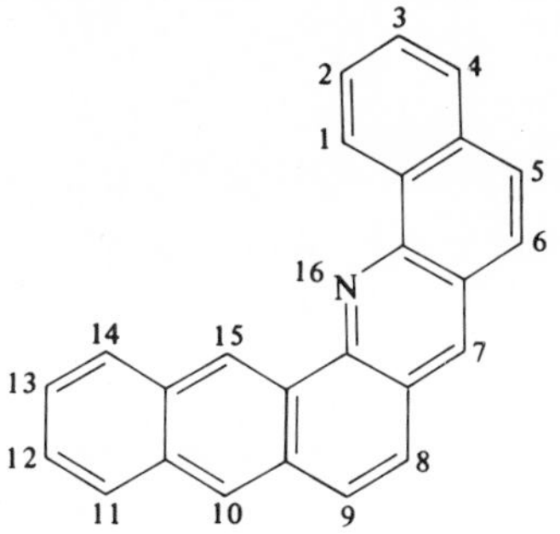

111

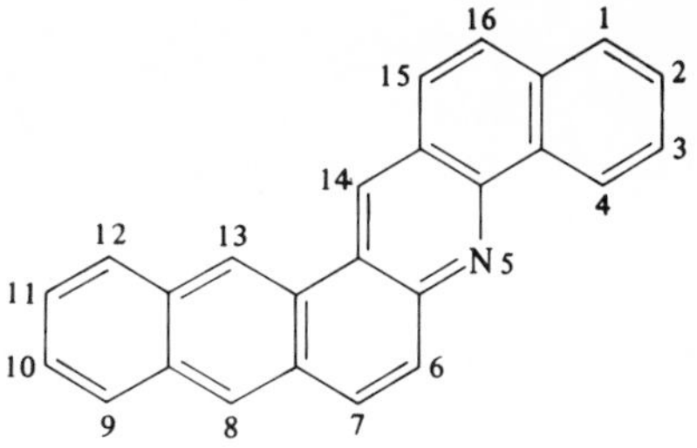

112

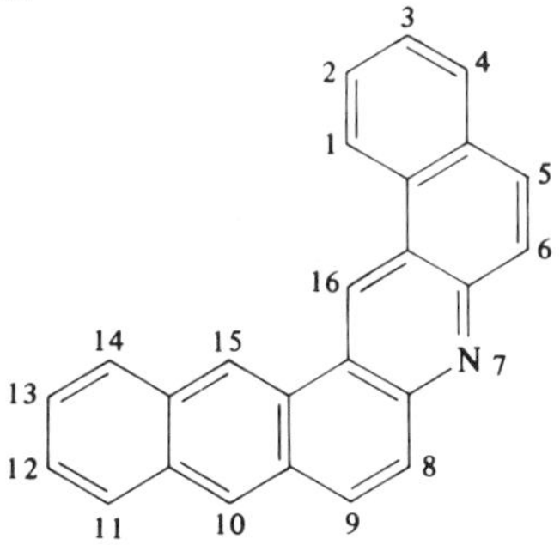

113

This agrees with the earlier results of Ullmann et al.,[209] who found that 2-aminoanthracene, 2-anthrol, and formaldehyde gave only the bis-angular product **114**. In contrast, 2-aminofluorene cyclizes to position 3 in its reaction with 1-naphthol and paraformaldehyde, yielding the linear benz[h]indeno-

[3,2-*b*]acridine (**115**); with 2-naphthol,[210] cyclization occurs in both modes (to positions 1 and 3), giving the isomers **116** (80%) and **117** (20%). The struc-

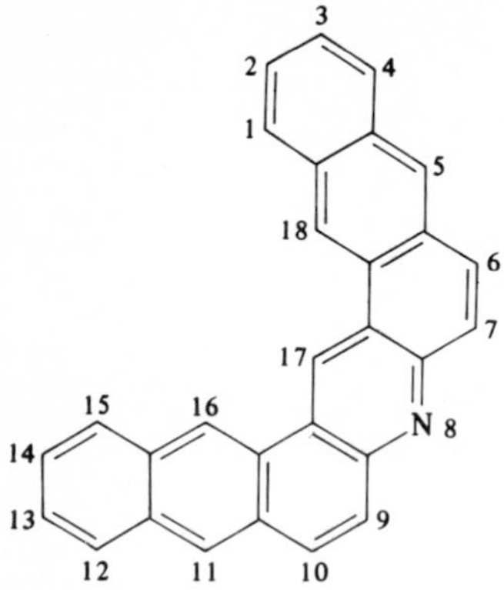

114

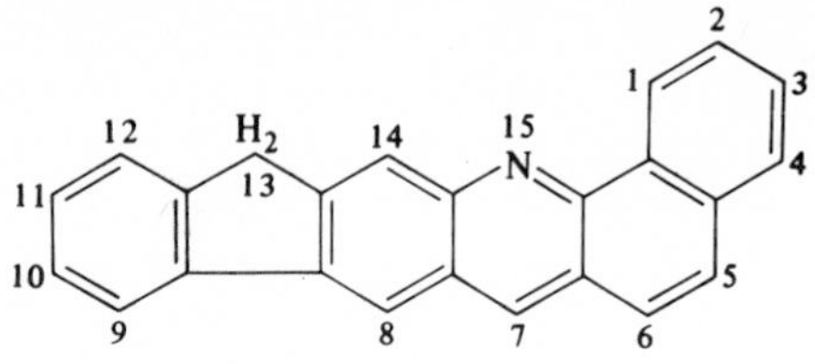

115

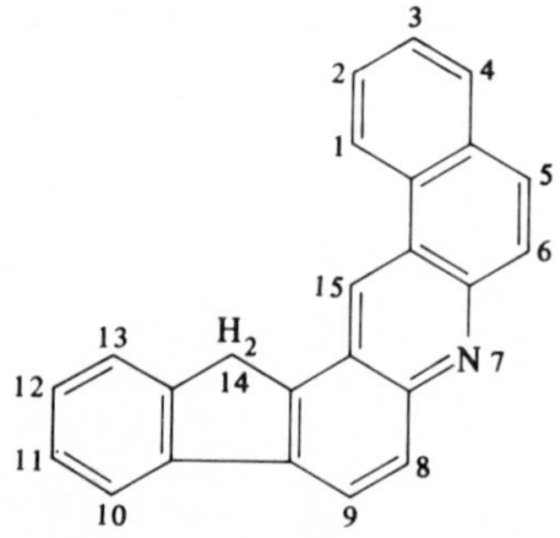

116

tures were elucidated by nmr spectroscopy by comparison with the spectrum of the acridine **118** (obtained from 4-aminofluorene, 1-naphthol, and para-formaldehyde),[206] whose structure was unequivocal.

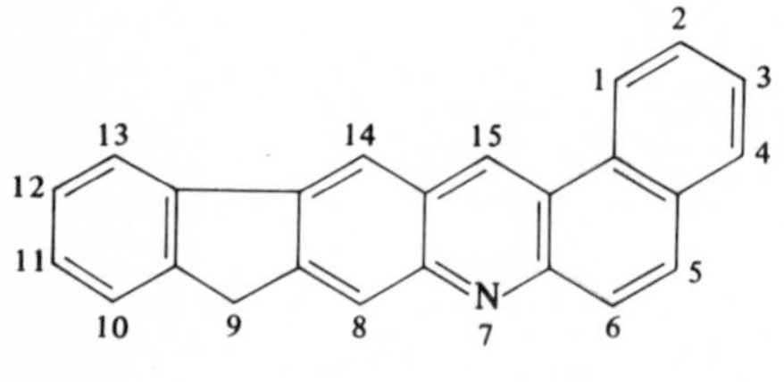

117

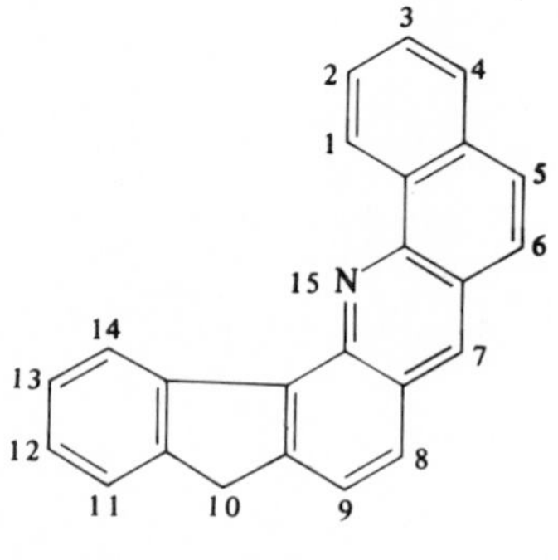

118

Thus 2-aminofluorene, like 3-aminophenanthrene,[211] disobeys Marckwald's rule[212] when participating in the Ullmann-Fetvadjian reaction, the suggestion being that obedience to the rule would result in sterically overcrowded molecules.[210, 211]

1-Aminoanthraquinone and an equimolar amount of 1-aminoanthra-
quinone-2-sulfonic acid, condensed in the presence of aluminium chloride in
anhydrous pyridine,[213a, b] gave the complex acridine **119**.

Oxidation of 1,5-dianilinonaphthalene, in the presence of aluminium
chloride,[214] gave acridino[2,1,9-*mna*]acridine (**120**), chrome yellow needles,
mp 362°; this acridine could be sulfonated (by chlorosulfonic acid) but was
inert to bromine or nitric acid.

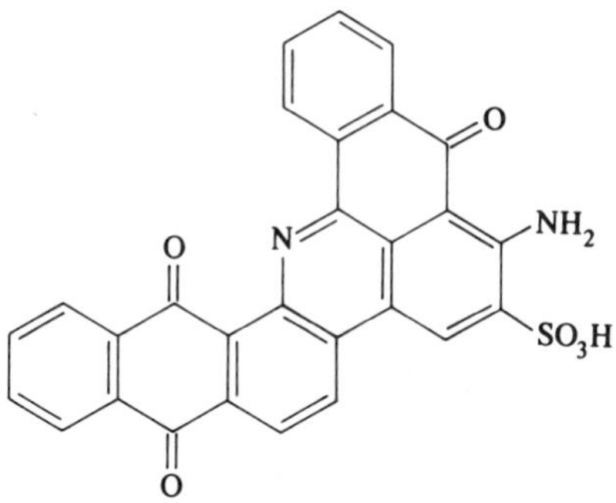

119

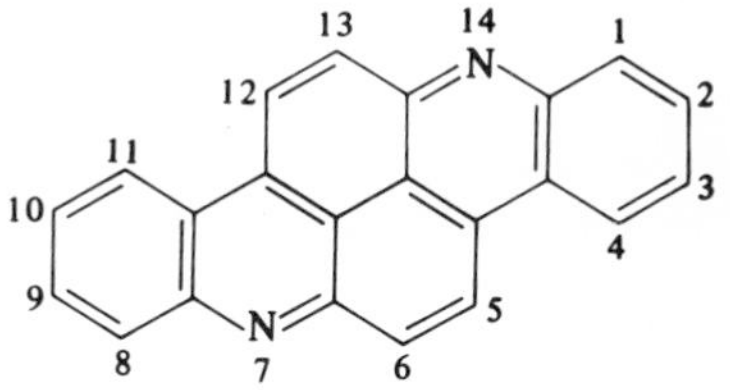

120

Acridino[2,1,9,8-*klmna*]acridine (**122**), mp 312°, has been obtained by the
oxidation of 8-acetamido-2-naphthol[215] to **121**, followed by hydrolysis and
cyclization with sulfuric acid at 130°. Dibromo and mononitro derivatives
have been obtained by direct substitution, and an anilino derivative by boiling with aniline in the presence of air. The structure of **122** has been confirmed by its conversion to *peri*-xanthenoxanthene (**123**) by water, hydrogen,
and alumina at 300°; attempts to reverse this reaction failed.

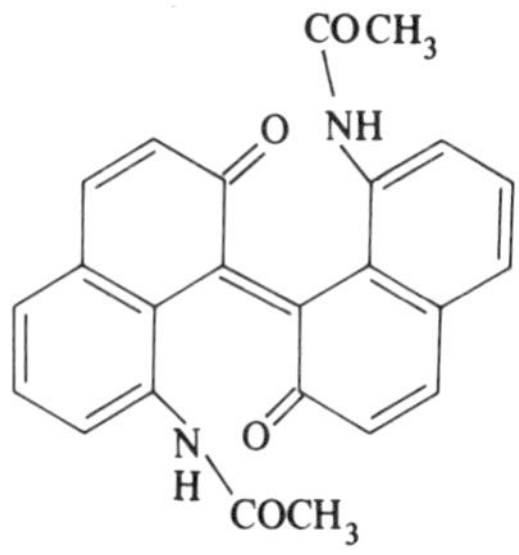

121

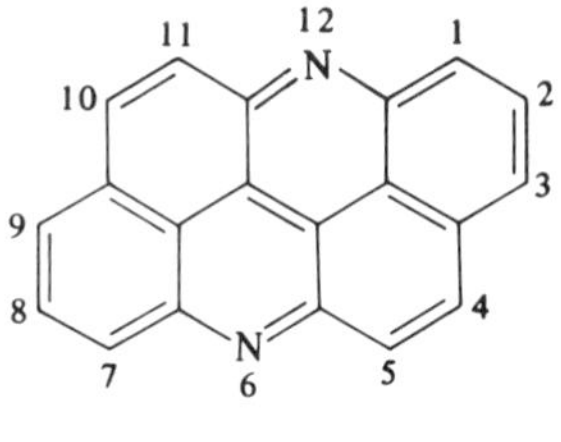

122

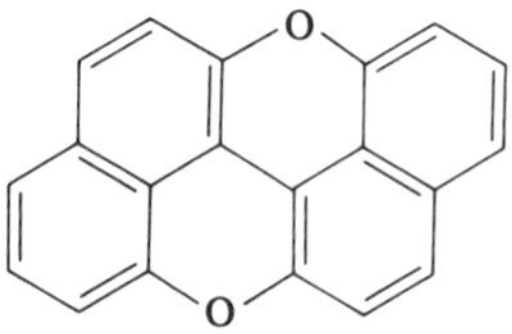

123

1,5-Di(*o*-nitrophenyl)anthraquinone,[150] on reduction with sodium dithionite in boiling aqueous ethanol cyclized to 8,16-diazadibenzo[*bk*]perylene (124), a pale yellow solid almost insoluble in the usual solvents. 1,5-Dichloroanthraquinone and 1-chloro-4-nitroanthraquinone with *o*-aminothiophenol gave, respectively, compounds 125 and 126, from which sulfur was extruded under a variety of reaction conditions (cf. p. 558) to give, respectively, the perylene 124 and dibenzo[*ci*]naphtho[1,2,3,4-*lmn*][2,9]-phenanthroline (127), previously named 11,16-diazatribenzo[*aei*]pyrene.[151]

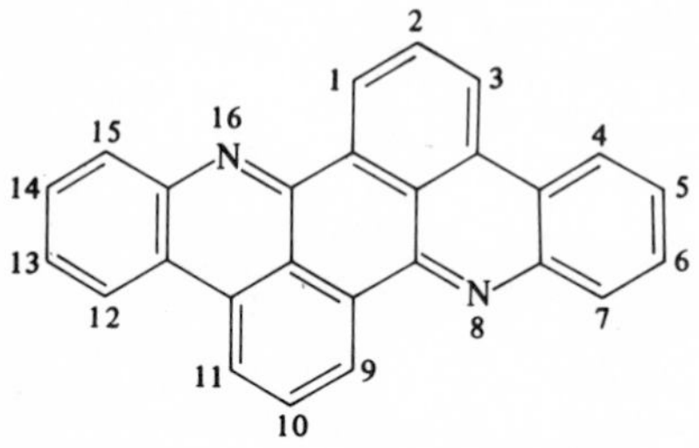

124

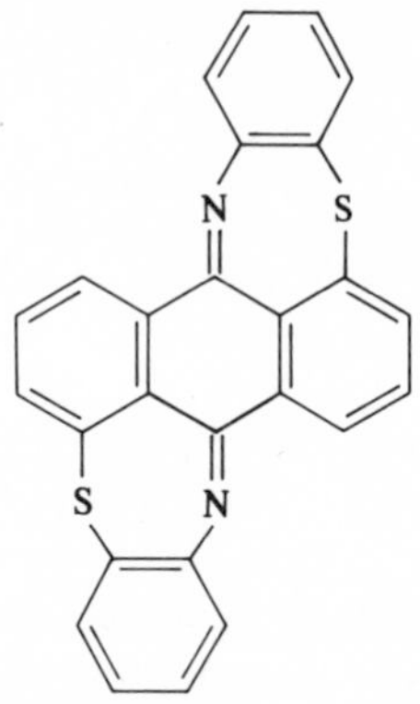

125

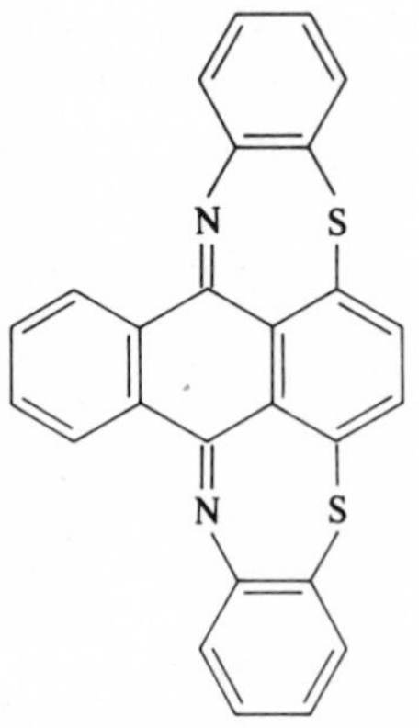

126

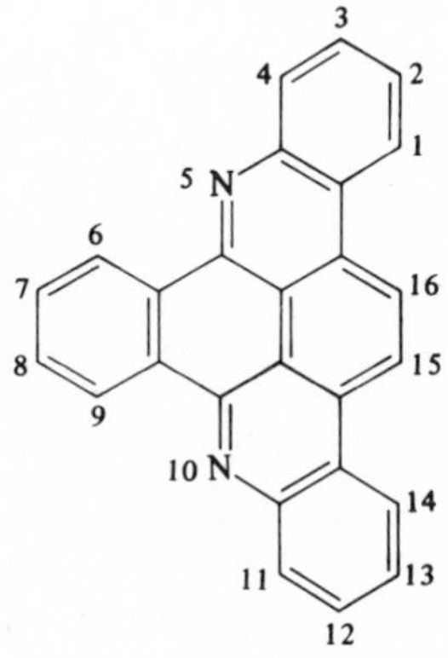

127

Diquin[2,3-*a*:2′,3′-*c*]acridine (128) has been obtained[199] by condensing phloroglucinol with excess 2-aminobenzaldehyde at 120–150°. It is extremely stable, crystallizes from nitrobenzene as yellow-brown needles, mp 403°, and is effectively insoluble in all solvents, except acids. Although inert to alkaline permanganate or chromic-acetic acid, it was oxidized to pyrido[2,3-a]quin-[2,3-c]acridine-7,8-dicarboxylic acid (129) by a mixture of sodium dichromate and nitric acid.[216]

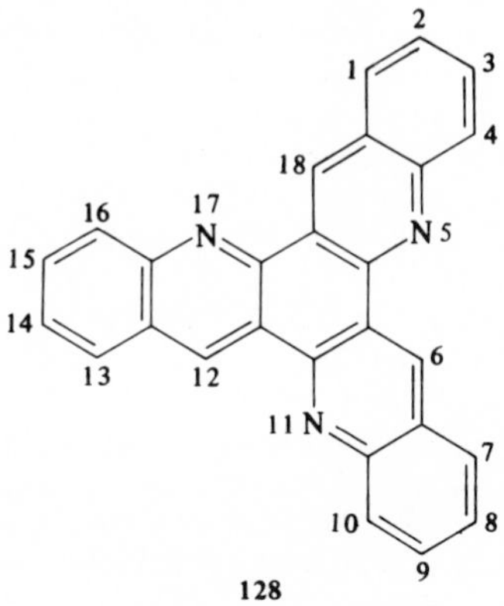

128

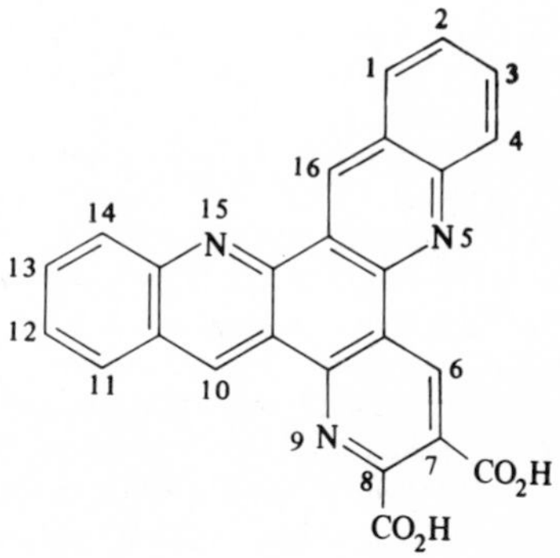

129

This acid formed an anhydride, which lost one carboxyl group (probably at position 8) on distillation and a second on subsequent distillation with lime.

A number of related complex acridines have been synthesized by the Ull-mann-Fetvadjian route from formaldehyde, the isomeric naphthols, and the amines p-phenylenediamine, and 1,4- and 1,5-diaminonaphthalene; certain ambiguities remain with the structures of some of the products.[217]

References

1. (a) A. M. Patterson, L. T. Capell, and D. F. Walker, *The Ring Index*, 2nd ed., American Chemical Society, 1960. (b) L. T. Capell, and D. F. Walker, Supplement I, (1957-1959), American Chemical Society, 1963; Supplement II (1960-1961), published 1964; Supplement III, (1962-1963), published 1965.
2. R. M. Acheson, *Acridines*, Interscience, New York-London, 1956, p. 307.
3. C. F. H. Allen, *Six-Membered Heterocyclic Nitrogen Compounds with Four Condensed Rings*, Interscience, New York-London, 1951.
4. G. M. Badger, J. H. Seidler, and B. Thomson, *J. Chem. Soc.*, 3207 (1951).
5. W. A. Waters and D. H. Watson, *J. Chem. Soc.*, 2082 (1959).
6. A. Albert, D. J. Brown, and H. Duewell, *J. Chem. Soc.*, 1284 (1948).
7. A. Etienne and A. Staehelin, *C. R. Acad. Sci., Paris*, 234, 1433 (1952).
8. A. Etienne and A. Staehelin, *Bull. Soc. Chim. Fr.*, 748 (1954).
9. F. Ullmann and A. La Torre, *Chem. Ber.*, 37, 2922 (1904).
10. F. Ullmann, *Justus Liebigs Ann. Chem.*, 355, 312 (1907).
11. I. Ya. Postovskii and B. N. Lundin, *J. Gen. Chem. USSR*, 10, 71 (1940); *Chem. Abstr.*, 34, 4738[8] (1940).

12. (a) N. P. Buu-Hoï and J. Lecocq, *C. R. Acad. Sci., Paris*, **218**, 792 (1944); (b) N. P. Buu-Hoï et al., *J. Chem., Soc.*, 670 (1949); *J. Chem. Soc., C*, 662 (1967).
13. (a) E. D. Bergmann, J. Blum, S. Butanaro, and A. Heller, *Tetrahedron Lett.*, 15, (1959); (b) E. D. Bergmann and J. Blum, *J. Org. Chem.*, **27**, 527 (1962).
14. N. P. Buu-Hoï, R. Royer, M. Hubert-Habart, and P. Mabille, *J. Chem. Soc.*, 3584 (1953).
15. (a) N. P. Buu-Hoï, L. C. Binh, T. B. Loc, N. D. Xuong, and P. Jacquignon, *J. Chem. Soc.*, 3126 (1957); (b) M. Marty, N. P. Buu-Hoï, and P. Jacquignon, *J. Chem. Soc.*, 384 (1961).
16. N. P. Buu-Hoï, P. Jacquignon, M. Dufour, and M. Mangane, *J. Chem. Soc., C*, 1792 (1966).
17. T. Thu-Cuc, N. P. Buu-Hoï, and N. D. Xuong, *J. Chem. Soc., C*, 87 (1966).
18. N. P. Buu-Hoï, *J. Chem. Soc.*, 1146 (1950).
19. N. P. Buu-Hoï, R. Royer, and M. Hubert-Habart, *J. Chem. Soc.*, 1082 (1955).
20. F. Ullmann and A. Fetvadjian, *Chem. Ber.*, **36**, 1027 (1903).
21. N. P. Buu-Hoï, M. Mangane, and P. Jacquignon, *J. Chem. Soc., C*, 662 (1967).
22. K. Dziewonski and J. Schoen, *Bull. Int. Acad. Polonaise, Classe Sci. Math. Nat.*, 448 (1934A); *Chem. Abstr.*, **29**, 2958[9] (1935).
23. J. Schoen and K. Bogdanowicz, *Rocz. Chem.* **34**, 1339 (1960); **36**, 1493 (1962).
24. J. Schoen and W. Laskowska, *Rocz. Chem.*, **39**, 1633 (1965); **40**, 1315 (1966).
25. J. von Braun and P. Wolff, *Chem. Ber.*, **55**, 3675 (1922).
26. N. P. Buu-Hoï and P. Cagniant, *Bull. Soc. Chim. Fr.*, **11**, 343 (1944).
27. N. P. Buu-Hoï, *J. Chem. Soc.*, 792 (1946).
28. N. P. Buu-Hoï, P. Cagniant, and C. Mentzer, *Bull. Soc. Chim. Fr.*, **11**, 127 (1944).
29. J. von Braun, et al., *Justus Liebigs Ann. Chem.*, **451**, 1 (1926).
30. J. von Braun and A. Stuckenschmidt, *Chem. Ber.*, **56**, 1724 (1923).
31. N. P. Buu-Hoï, P. Jacquignon, and D. Lavit, *J. Chem. Soc.*, 2593 (1956).
32. N. P. Buu-Hoï, G. Saint-Ruf, P. Jacquignon, and M. Marty, *J. Chem. Soc.*, 2274 (1963).
33. N. P. Buu-Hoï, M. Dufour, and P. Jacquignon, *J. Chem. Soc.*, 5622 (1964).
34. M. Sy and G. A. Thiault, *Bull. Soc. Chim. Fr.*, 1308 (1965).
35. F. A. Al-Tai, A. M. El-Abbady, and A. S. Al-Tai, *J. Chem. UAR*, **10**, 339 (1967).
36. (a) A. Martinet, *Ann. Sci. Univ. Besançon, Chim.*, **2**, No. 3, 27 (1957); *Chem. Abstr.*, **53**, 5268c (1959); (b) A. Martinet, *C. R. Acad. Sci., Paris*, **243**, 278 (1956).
37. M. J. Sacha and S. R. Patel, *J. Indian Chem. Soc.*, **34**, 821 (1957).
38. L. N. Lavrischeva, G. A. Fedorova, and V. N. Belov, *Zh. Obshch. Khim.*, **33**, 3961 (1963); *Chem. Abstr.*, **60**, 9244g (1964).
39. N. H. Cromwell and L. A. Nielsen, *J. Heterocycl. Chem.*, **6**, 361 (1969).
40. G. Kempter, P. Andratschke, D. Heilmann, H. Krausmann, and M. Mietasch, *Z. Chem.*, **3**, 305 (1963).
41. G. Kempter, P. Andratschke, D. Heilmann, H. Krausmann, and M. Mietasch, *Chem. Ber.*, **97**, 16 (1964).
42. F. Boyer and J. Décombe, *Bull. Soc. Chim. Fr.*, 2373 (1967).
43. F. Boyer and J. Décombe, *C. R. Acad. Sci., Paris*, **255**, 1945 (1962).
44. G. E. Hall and J. Walker, *J. Chem. Soc., C*, 2237 (1968).
45. V. L. Bell and N. H. Cromwell, *J. Org. Chem.*, **23**, 789 (1958).
46. N. H. Cromwell, and J. C. David, *J. Amer. Chem. Soc.*, **82**, 1138 (1960).
47. (a) N. H. Cromwell and V. L. Bell, *J. Org. Chem.*, **24**, 1077 (1959); (b) J. L. Adelfang and N. H. Cromwell, *J. Org. Chem.*, **26**, 2368 (1961).
48. M. Gindy and I. M. Dwidar, *Egypt. J. Chem.*, **2**, 119 (1959).

49. I. Tănăsescu, M. Ionescu, I. Goia, and H. Mantsch, *Bull. Soc. Chim. Fr.*, 698 (1960).
50. J. Cymerman-Craig and J. W. Loder, *J. Chem. Soc.*, 4309 (1955).
51. G. M. Badger and R. Pettit, *J. Chem. Soc.*, 2774 (1953).
52. C. R. Hauser and J. G. Murray, *J. Amer. Chem. Soc.*, **77**, 3858 (1955).
53. G. B. Bachmann and G. M. Picha, *J. Amer. Chem. Soc.*, **68**, 1599 (1946).
54. D. P. Spalding, E. C. Chapin, and H. S. Mosher, *J. Org. Chem.*, **19**, 357 (1954).
55. E. F. Elslager, F. W. Short, and M-J. Sullivan, U.S. Patent 2,773,064; *Chem. Abstr.*, **51**, 6707i (1957).
56. E. F. Elslager, A. M. Moore, F. W. Short, M-J. Sullivan, and F. H. Tendick, *J. Amer. Chem. Soc.*, **79**, 4699 (1957).
57. K. G. Yekundi and S. R. Patel, *J. Indian Chem. Soc.*, **35**, 285 (1958).
58. E. N. Morgan and D. J. Tivey, British Patent 844,818; *Chem. Abstr.*, **55**, 12431e (1961).
59. V. Zanker and F. Mader, *Chem. Ber.*, **93**, 850 (1960).
60. A. Kellmann, *J. Chim. Phys.*, **63**, 949 (1966).
61. C. Kaneko, S. Yamada, and M. Ishikawa, *Tetrahedron Lett.*, 2329 (1970).
62. C. Kaneko, S. Yamada, and M. Ishikawa, *Chem. Pharm. Bull.* (Tokyo), **17**, 1294 (1969).
63. E. vander Donckt, R. H. Martin, and F. Geerts-Evrard, *Tetrahedron*, **20**, 1495 (1964).
64. R. H. Martin, N. Defay, F. Geerts-Evrard, and D. Bogaert-Verhoogen, *Tetrahedron*, Suppl. 8, Pt. I, 181 (1966).
65. N. P. Buu-Hoï, C. Orley, M. Mangane, and P. Jacquignon, *J. Heterocycl. Chem.*, **2**, 236 (1965).
66. E. Sawicki, T. W. Stanley, J. D. Pfaff, and W. C. Elbert, *Anal. Chim. Acta*, **31**, 359 (1964).
67. E. Sawicki, T. W. Stanley, and W. C. Elbert, *Occupational Health Rev.*, **16**, 8 (1964); *Chem. Abstr.*, **62**, 5108d (1965).
68. A. M. Luly and K. Sakodynsky, *J. Chromatogr.*, **19**, 624 (1965).
69. E. Sawicki, M. Guyer, and C. R. Engel, *J. Chromatogr.*, **30**, 522 (1967).
70. C. R. Engel and E. Sawicki, *J. Chromatogr.*, **31**, 109 (1967).
71. M. Lederer and G. Roch, *J. Chromatogr.*, **31**, 618 (1967).
72. (a) A. Pacault, *Bull. Soc. Chim. Fr.*, 1270 (1950). (b) A. Cheutin, et al., *C. R. Acad. Sci.*, *Paris*, **241**, 52 (1955). (c) V. Zanker and W. Schmid, *Chem. Ber.*, **90**, 2253 (1957); (d) V. Zanker and P. Schmid, *Chem. Ber.*, **92**, 615 (1959).
73. (a) R. Muxart, *C. R. Acad. Sci.*, *Paris*, **242**, 2457 (1956); (b) R. Muxart and G. Pinte, *Bull. Soc. Chim. Fr.*, 1675 (1956); (c) R. Muxart, *Bull. Soc. Chim. Fr.*, 1857 (1956).
74. B. Cairns and W. O. Kermack, *J. Chem. Soc.*, 1322 (1950).
75. B. S. Joshi, N. Parkash, and K. Venkataraman, *J. Sci. Ind. Res.*, **14B**, 325 (1955); *Chem. Abstr.*, **50**, 11341i (1956).
76. (a) B. Stefanska and A. Ledóchowski, *Rocz. Chem.*, **42**, 1535 (1968). (b) D. Shortridge, R. Turner and H. N. Green, *Brit. J. Cancer*, **23**, 825 (1969); *Chem. Abstr.*, **72**, 100464j (1970).
77. G. B. Bachmann and F. M. Cowen, *J. Org. Chem.*, **13**, 89 (1948).
78. J. Dobson, W. C. Hutchison, and W. O. Kermack, *J. Chem. Soc.*, 123 (1948).
79. E. F. Elslager, F. H. Tendick, L. M. Werbel, and D. F. Worth, *J. Med. Chem.*, **12**, 970 (1969).
80. (a) A. K. Chatterjee, *J. Org. Chem.*, **24**, 856 (1959). (b) *Science and Culture (Calcutta)*, **24**, 90 (1958); *Chem. Abstr.*, **54**, 9928a (1960). (c) *J. Org. Chem.*, **24**, 2067 (1959). (d) *J. Indian Chem. Soc.*, **38**, 333 (1961); **39**, 565 (1962).

81. R. M. Peck, A. P. O'Connell, and H. J. Creech, *J. Med. Chem.*, **9**, 217 (1966).
82. (a) A. M. Moore, E. F. Elslager, and F.W. Short, U.S. Patent 2,915,523; *Chem.Abstr.*, **54**, 5707b (1960). (b) U.S. Patent 2,981,731; *Chem. Abstr.*, **55**, 25994e (1961).
83. A. Campbell and E. N. Morgan, *J. Chem. Soc.*, 1711 (1958).
84. B. M. Mikhailov and G. S. Ter-Sarkisyan, *Izv. Akad. Nauk SSSR, Otdel. Khim. Nauk*, 846 (1954); *Chem. Abstr.*, **49**, 13994i (1955); **54**, 5653d (1960).
85. (a) A. E. Porai-Koshits and G. S. Ter-Sarkisyan, *Izv. Akad. Nauk SSSR, Otdel. Khim. Nauk*, 601, 771 (1951); *Chem. Abstr.*, **46**, 8116dg (1952); (b) B. M. Mikhailov and G. S. Ter-Sarkisyan, *Izv. Akad. Nauk SSSR, Otdel. Khim. Nauk*, 656 (1954); *Chem. Abstr.*, **49**, 10953c (1955).
86. O. Tsuge, M. Nishinohara, and M. Tashiro, *Bull. Chem. Soc. Jap.*, **36**, 1477 (1963).
87. N. P. Buu-Hoï, J. P. Hoeffinger, and P. Jacquignon, *J. Chem. Soc.*, 5383 (1963).
88. G. Kempter, H. Dost, and W. Schmidt, *Chem. Ber.*, **98**, 945 (1965).
89. B. S. Tanaseichuk and I. Ya. Postovskii, *Khim. Geterotsikl. Soedin., Akad. Nauk Latv. SSR*, 390 (1965); *Chem. Abstr.*, **63**, 14810b (1965).
90 B. S. Tanaseichuk and I. Ya. Postovskii, *Metody Poluch. Khim., Reaktivov Prep.*, No. 14, 28 (1966); *Chem. Abstr.*, **67**, 64227v (1967).
91. B. S. Tanaseichuk and I. Ya. Postovskii, *Zh. Org. Khim.*, **1**, 1279 (1965); *Chem. Abstr.*, **63**, 13207c (1965).
92. B. S. Tanaseichuk and I. Ya. Postovskii, *Metody Poluch. Khim., Reaktivov Prep.*, No. 17, 158 (1967); *Chem. Abstr.*, **71**, 30342w (1969).
93. B. S. Tanaseichuk, I. Ya. Postovskii, and L. F. Lipatova, *Zh. Org. Khim.*, **2**, 293 (1966); *Chem. Abstr.*, **65**, 2218d (1966).
94. O. Tsuge, T. Tomita, and A Torii, *Nippon Kagaku Zasshi*, **89**, 1104 (1968); *Chem. Abstr.*, **70**, 96595s (1969).
95. I. Ya. Postovskii and B. S. Tanaseichuk, *Zh. Org. Khim.*, **1**, 1276 (1965); *Chem. Abstr.*, **63**, 13173f (1965).
96. M. J. S. Dewar, *J. Amer. Chem. Soc.*, **74**, 3357 (1952).
97. R. M. Acheson and C. W. Jefford, *J. Chem. Soc.*, 2676 (1956).
98. K. Tada, R. Takitani, and S. Iwasaki, *Kyoritsu Yakka Daigaku Kenkyu Nempo*, **5**, 16 (1961); *Chem. Abstr.*, **55**, 15488c (1961).
99. G. M. Badger, *J. Chem. Soc.*, 1809 (1950).
100. N. S. Narasimhan and A. C. Ranade, *Indian J. Chem.*, **7**, 538 (1969); *Chem. Abstr.*, **71**, 49743z (1969).
101. N. H. Cromwell and J. C. David, *J. Amer. Chem. Soc.*, **82**, 2046 (1960).
102. R. C. Fuson and J. J. Miller, *J. Amer. Chem. Soc.*, **79**, 3478 (1957).
103. H. Waldmann and S. Back, *Justus Liebigs Ann. Chem.*, **545**, 52 (1940).
104. H. Sieper, *Chem. Ber.*, **100**, 1646 (1967).
105. H. Sieper and P. Tavs, *Justus Liebigs Ann. Chem.*, **704**, 161 (1967).
106. F. Lions, *J. Proc. Roy. Soc. New South Wales*, **71**, 192 (1938).
107. E. Ziegler, H. Junek, E. Nölken, K. Gelfert, and R. Salvador, *Monatsh. Chem.*, **92**, 814 (1961).
108. E. Ziegler and F. Litvan, U.S. Patent 3,052,678; *Chem. Abstr.*, **58**, 3437d (1963).
109. E. Ziegler, U. Rossmann, F. Litvan, and H. Meier, *Monatsh. Chem.*, **93**, 26 (1962).
110. E. Ziegler and T. Kappe, *Monatsh. Chem.*, **94**, 447 (1963).
111. T. Kappe and E. Ziegler, *Monatsh. Chem.*, **95**, 415 (1964).
112. E. Ziegler, R. Salvador, and T. Kappe, *Monatsh. Chem.*, **94**, 941 (1963).
113. J. R. Beck et al., *J. Amer. Chem. Soc.*, **90**, 4706 (1968).
114. (a) F. N. Lahey and W. C. Thomas, *Aust. J. Sci. Res. (A)*, **2**, 423 (1949); (b) R. D. Brown, L. J. Drummond, F. N. Lahey, and W. C. Thomas, *Aust. J. Sci. Res. (A)*, **2**,

622 (1949); (c) L. J. Drummond and F. N. Lahey, *Aust. J. Sci. Res.* (*A*), **2**, 630 (1949); (d) R. D. Brown and F. N. Lahey, *Aust. J. Sci. Res.* (*A*), **3**, 593 (1950).

115. (a) P. L. Macdonald and A. V. Robertson, *Aust. J. Chem.*, **19**, 275 (1966); (*b*) T. R. Govindachari, B. R. Pai, and P. S. Subramaniam, *Tetrahedron*, **22**, 3245 (1966).

116. J. Z. Gougoutas and B. A. Kaski, private communication to J. R. Beck.

117. J. A. Diment, E. Ritchie, and W. C. Taylor, *Aust. J. Chem.*, **22**, 1721 (1969).

118. J. Hlubucek, E. Ritchie, and W. C. Taylor, *Aust. J. Chem.*, **23**, 1881 (1970).

119. P. C. Austin, *J. Chem. Soc.*, **93**, 1760 (1908).

120. E. Strohbach, *Chem. Ber.*, **34**, 4146 (1901).

121. (a) A. Senier and W. Goodwin, *J. Chem. Soc.*, **81**, 280 (1902); (b) A. Senier and P. C. Austin, *J. Chem. Soc.*, **89**, 1387 (1906).

122. G. T. Morgan, *J. Chem. Soc.*, **73**, 536 (1898).

123. E. R. Blout and R. S. Corley, *J. Amer. Chem. Soc.*, **69**, 763 (1947).

124. N. Saito, C. Tanaka, and M. Okubo, *J. Pharm. Soc. Jap.*, **76**, 359 (1956).

125. J. Pajak, *Rocz. Chem.*, **12**, 507 (1932).

126. M. A. Mikhaleva and V. P. Mamaev, *Izv. Sib. Otdel. Akad. Nauk SSSR, Ser. Khim. Nauk*, 106 (1969); *Chem. Abstr.* **71**, 91140a (1969).

127. W. O. Kermack, R. H. Slater, and W. T. Spragg, *Proc. Roy. Soc. Edinburgh*, **50**, 243 (1930).

128. R. Möhlau and O. Haase, *Chem. Ber.*, **35**, 4164 (1902).

129. G. R. Newkome and D. L. Fishel, *J. Heterocycl. Chem.*, **4**, 427 (1967).

130. K. Sisido, K. Tani, and H. Nozaki, *Tetrahedron*, **19**, 1323 (1963).

131. F. Ullmann, *Chem. Ber.*, **43**, 536 (1910).

132. W. Bradley and H. Kaiwar, *J. Chem. Soc.*, 2859 (1960).

133. N. P. Buu-Hoï and P. Cagniant, *C. R. Acad. Sci.*, *Paris*, **216**, 447 (1943).

134. N. P. Buu-Hoï and P. Cagniant, *C. R. Acad. Sci.*, *Paris*, **215**, 144 (1942).

135. F. Ullmann and I. C. Dasgupta, *Chem. Ber.*, **47**, 553 (1914).

136. G. Wittig, H. Härle, E. Knauss, and K. Niethammer, *Chem. Ber.*, **93**, 951 (1960).

137. R. Mason, *Proc. Roy. Soc., Ser. A*, **258**, 302 (1960).

138. B. Clin and B. Lemanceau, *C. R. Acad. Sci., Paris, Ser. C*, **270**, 598 (1970).

139. B. Clin and B. Lemanceau, *C. R. Acad. Sci., Paris, Ser. D*, **271**, 788 (1970).

140. D. G. I. Felton and G. M. Timmis, *J. Chem. Soc.*, 2881 (1954).

141. V. L. Vaiser, V. D. Ryabov, and A. K. Ostroumova, *Dokl. Akad. Nauk SSSR*, **125**, 799 (1959); *Chem. Abstr.*, **53**, 20045e (1959).

142. (a) J. Haase, Czechoslovakian Patent 91429; *Chem. Abstr.*, **55**, 6502h (1961); (b) J. Haase and A. Boehmova, Czechoslovakian Patent 97989; *Chem. Abstr.*, **58**, 6809f (1963).

143. D. C. Eaton and F. Irving, British Patent 944,513; *Chem. Abstr.*, **60**, 10846a (1964).

144. Y. Hosada and O. Teramachi, Japanese Patent 25,847; *Chem. Abstr.*, **60**, 14646d (1964).

145. S. Nakazawa, *Yuki Gosei Kagaku Kyokai Shi*, **20**, 661 (1962); *Chem Abstr.*, **58**, 13913c (1963).

146. A. Schuhmacher and A. Ehrhardt, Belgian Patent 629,607; *Chem. Abstr.*, **60**, 14485e (1964).

147. French Patent 1,420,726; *Chem. Abstr.*, **65**, 13669b (1966).

148. W. Zerweck and E. Heinrich, German Patent, 1,228,355; *Chem. Abstr.*, **66**, 30026t (1967).

149. C. F. Koelsch, *J. Amer. Chem. Soc.*, **58**, 1325 (1936).

150. E. A. Braude and J. S. Fawcett, *J. Chem. Soc.*, 3117 (1951).

151. R. H. B. Galt, J. D. Loudon, and A. D. B. Sloan, *J. Chem. Soc.*, 1588 (1958).

152. German Patent 120,193.
153. N. I. Grineva, V. V. Puchkova, and V. N. Ufimtsev, *Khim. Geterotsikl. Soedin.*, 744 (1966); *Chem. Abstr.*, **66,** 76899h (1967).
154. H. Decker and C. Schenck, *Justus Liebigs Ann. Chem.*, **348,** 242 (1906).
155. A. H. Cook and W. Waddington, *J. Chem. Soc.*, 402 (1945).
156. R. Weiss and W. Knapp, *Sitzungsber. Akad. Wiss. Wien*, IIb, **135,** 459 (1926).
157. N. I. Grineva, V. V. Puchkova, and V. N. Ufimtsev, *Zh. Obshch. Khim.*, **33,** 597 (1963); *Chem. Abstr.*, **59,** 572e (1963).
158. E. P. Fokin and R. P. Shishkina, *Zh. Obshch. Khim.*, **33,** 3674 (1963); *Chem. Abstr.*, **60,** 7997b (1964).
159. C. Weinand, German Patent 566,473; *Chem. Abstr.*, **27,** 1198 (1933).
160. Z. I. Krutikova, *Zh. Org. Khim.*, **4,** 1673 (1968); *Chem. Abstr.*, **70,** 3813y (1969).
161. R. P. Shishkina and E. P. Fokin, *Khim. Geterotsikl. Soedin., Akad. Nauk Latv. SSR*, 420 (1966); *Chem. Abstr.*, **65,** 13780e (1966).
162. E. P. Fokin, R. P. Shishkina, and I. V. Fomicheva, *Khim. Geterotsikl. Soedin., Akad. Nauk Latv. SSR*, 467 (1966); *Chem. Abstr.*, **65,** 15320g (1966).
163. H. Gilman, C. G. Stuckwisch, and A. R. Kendall, *J. Amer. Chem. Soc.*, **63,** 1758 (1941).
164. A. Eckert, F. Seidel, and G. Endler, *J. Prakt. Chem.*, **104,** 85 (1922).
165. S. S. Labana and L. L. Labana, *Chem. Rev.*, **67,** 1 (1967).
166. H. Liebermann, et al., *Justus Liebigs Ann. Chem.*, **518,** 245 (1935).
167. A. Schuhmacher and A. Ehrhardt, German Patent 1,112,517; *Chem. Abstr.*, **56,** 4769c (1962).
168. A. Caliezi, German Patent 1,112,597; *Chem. Abstr.*, **56,** 5940a (1962).
169. W. Deutschel and K. Schrempp, German Patent 1,151,081; *Chem. Abstr.*, **60,** 9396c (1964).
170. British Patent 896,803; *Chem. Abstr.*, **59,** 10277h (1963).
171. German Patent 1,136,438; *Chem. Abstr.*, **59,** 632f (1963).
172. Belgian Patent 632,114; *Chem. Abstr.*, **61,** 1987f (1964).
173. K. Hashizume, A. Kashiwaoka, and T. Ishii, Japanese Patent 5414; *Chem. Abstr.*, **67,** 44848v (1967).
174. A. Schuhmacher and A. Ehrhardt, German Patent, 1,190,125; *Chem. Abstr.*, **63,** 712c (1965).
175. W. Lesnianski and K. Dziewonski, *Przem. Chem.*, **13,** 401 (1929); *Chem. Abstr.*, **24,** 120 (1930).
176. H. Vollmann, W. Burmeleit, and W. Hohmann, German Patent 1,136,040; *Chem. Abstr.*, **58,** 9263c (1963).
177. (a) W. S. Struve, U.S. Patent 2,821,529; *Chem. Abstr.*, **52,** 10215d (1958); (b) N. S. Corby, E. D. Harvey, and D. G. Wilkinson, British Patent 894,610; *Chem. Abstr.*, **57,** 15290f (1962).
178. (a) R. M. Acheson and B. F. Sanson, *J. Chem. Soc.*, 4440 (1955); (b) Y. Nagai, H. Nishi, N. Goto, and H. Hasegawa, *Kogyo Kagaku Zasshi*, **67,** 2099 (1964); *Chem. Abstr.*, **62,** 14644d (1965).
179. (a) E. Anton, German Patent 1,140,300; *Chem. Abstr.*, **59,** 791f (1963); French Patent, 1,365,773; *Chem. Abstr.*, **63,** 7148f (1965). (b) W. Braun and R. Mecke, *Chem. Ber.*, **99,** 1991 (1966).
180. A. L. Nelson, U.S. Patent 3,133,071; *Chem. Abstr.*, **61,** 7018g (1964).
181. D. A. Kinsley and S. G. P. Plant, *J. Chem. Soc.*, 1 (1958).
182. Belgian Patent 627,375; *Chem. Abstr.*, **60,** 12150a (1964).

183. Y. Nagai, H. Nishi, and H. Hasegawa, *Kogyo Kagaku Zasshi*, **68**, 1910 (1965); *Chem. Abstr.*, **64**, 19582b (1966).
184. W. Braun, W. Ruppel, and R. Mecke, German Patent 1,178,159; *Chem. Abstr.*, **62**, 1775b (1965).
185. H. Nishi, Y. Nagai, and H. Hasegawa, *Kogyo Kagaku Zasshi*, **68**, 1717 (1965); *Chem. Abstr.*, **64**, 9697g (1966).
186. Y. Nagai, H. Nishi, and H. Hasegawa, *Kogyo Kagaku Zasshi*, **69**, 669 (1966); *Chem. Abstr.*, **66**, 85708x (1967).
187. Y. Nagai, H. Nishi, and K. Morikubo, *Kogyo Kagaku Zasshi*, **70**, 2199 (1967); *Chem. Abstr.*, **69**, 11400f (1968).
188. W. S. Struve, U.S. Patent 2,821,530; *Chem. Abstr.*, **52**, 10216a (1958).
189. British Patent 909,602; *Chem. Abstr.*, **58**, 6832e (1963).
190. E. E. Jaffe, U.S. Patent 3,251,845; *Chem. Abstr.* **65**, 9069d (1966).
191. (a) Y. Nagai, Japanese Patent 9,274; *Chem. Abstr.*, **61**, 16208a (1964); (b) Y. Nagai, H. Nishi, and N. Gotoh, *Kogyo Kagaku Zasshi*, **71**, 386 (1968); *Chem. Abstr.*, **69**, 52918z (1968).
192. R. L. Sweet, U.S. Patent 3,272,822; *Chem. Abstr.*, **66**, 11865z (1967).
193. French Patent 1,448,922; *Chem. Abstr.*, **67**, 22892c (1967).
194. (a) Y. Nagai, Japanese Patent 29,449; *Chem. Abstr.*, **62**, 11817f (1965); (b) Y. Nagai, H. Nishi, S. Nagai, and K. Morikubo, *Kogyo Kagaku Zasshi*, **71**, 717 (1968); *Chem. Abstr.*, **69**, 106516n (1968).
195. (a) F. Ullmann and R. Maag, *Chem. Ber.*, **40**, 2515 (1907); (b) G. M. Badger and and R. Pettit, *J. Chem. Soc.*, 1874 (1952).
196. J. H. Cooper, U.S. Patent 3,107,248; *Chem. Abstr.*, **60**, 2931d (1964).
197. W. Rauner and F. Wolf, *Z. Chem.*, **8**, 304 (1968).
198. H. Bohler and F. Kehrer, U.S. Patent 3,124,581; *Chem. Abstr.*, **61**, 13462f (1964).
199. S. von Niementowski, *Chem. Ber.*, **39**, 385 (1906).
200. S. von Niementowski, *Chem. Ber.*, **29**, 76 (1896).
201. R. Weiss and J. L. Katz, *Sitzungsber. Akad. Wiss. Wien*, IIb, **137**, 701 (1928).
202. M. W. Partridge and H. J. Vipond, *J. Chem. Soc.*, 632 (1962).
203. A. Fravolini, G. Grandolini, and A. Martani, *Ann. Chim.*, **58**, 533 (1968).
204. N. P. Buu-Hoï, A. Martani, A. Ricci, M. Dufour, P. Jacquignon, and G. Saint-Ruf, *J. Chem. Soc.*, C, 1790 (1966).
205. N. P. Buu-Hoï, J. C. Perche, and G. Saint-Ruf, *Bull. Soc. Chim. Fr.*, 627 (1968).
206. D. C. Thang, E. K. Weisburger, P. Mabille, and N. P. Buu-Hoï, *J. Chem. Soc.*, C, 665 (1967).
207. N. P. Buu-Hoï, P. Mabille, and J. Brasch, *J. Chem. Soc.*, 3920 (1964).
208. N. P. Buu-Hoï, M. Dufour, and P. Jacquignon, *J. Chem. Soc.*, C, 1337 (1969).
209. F. Ullmann and D. Urményi, *Chem. Ber.*, **45**, 2259 (1912).
210. N. P. Buu-Hoï, M. Mangane, and P. Jacquignon, *J. Heterocycl. Chem.*, **7**, 155 (1970).
211. N. P. Buu-Hoï, D. C. Thang, P. Jacquignon, and P. Mabille, *J. Chem. Soc.*, C, 467 (1969).
212. W. Marckwald, *Justus Liebigs Ann. Chem.*, **274**, 331 (1893).
213. (a) French Patent 1,509,260; *Chem. Abstr.*, **70**, 115027b (1969); (b) A. K. Wick, *Helv. Chim. Acta*, **49**, 1748 (1966).
214. G. R. Clemo and E. C. Dawson, *J. Chem. Soc.*, 1114 (1939).
215. A. Rieche, W. Rudolph, and R. Seifert, *Chem. Ber.*, **73**, 343 (1940).
216. L. T. Bratz and S. von Niementowski, *Chem. Ber.*, **51**, 366 (1918).
217. N. P. Buu-Hoï and P. Cagniant, *Bull. Soc. Chim. Fr.*, **11**, 406 (1944).

CHAPTER VIII

Acridine Dyes

B. D. TILAK and N. R. AYYANGAR

National Chemical Laboratory,
Poona, India

A number of synthetic dyes contain the acridine (1) or "acridone" (2) ring systems. The name "acridone" has been changed to the more systematic name "9-acridanone" by *Chemical Abstracts* and the new nomenclature has been used here for the simple acridones. However, for the more complicated systems the existing generally accepted nomenclature is employed.

The chemistry of the acridine and 9-acridanone dyes has been discussed in books on dyes and/or heterocyclic compounds.[1-8] In this chapter, an attempt is made to describe the important dyes containing the acridine and "acridone" ring systems; emphasis will be laid on more recent literature. Although

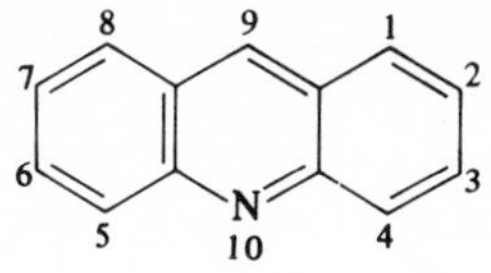

Acridine Dyes

1

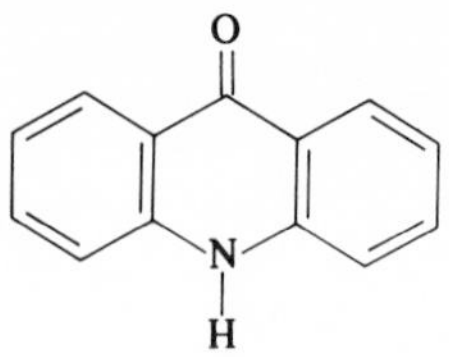

2

acridine (1) and 9-acridanone (2) are practically colorless or pale yellow, the synthetic dyes containing these ring systems range in shades from yellow to green and grey. These dyes may be classified on the basis of their application into groups such as basic, disperse, vat dyes, and pigments. Presently, these dyes are described under four major groups: (1) acridine dyes, (2) simple acridone dyes, (3) complex acridone dyes, (4) quinacridones, which are considered in this order in this Chapter.

Acridine dyes include well-known basic dyes, which contain amino or alkylamino substituents in *meta* (3 or 3,6) positions to the nitrogen atom of the acridine nucleus, in comparison with oxazine and thiazine dyes, which contain amino substituents in the *para* position with respect to the heterocyclic nitrogen atom. The simple 9-acridanone dyes are of interest as disperse dyes for synthetic fibers. The 9-acridanone ring system also forms part of quinonoid polycyclic compounds. These complex acridone dyes are generally used as vat dyes for cotton. Some of them can also be used as disperse dyes for hydrophobic fibers, such as cellulose acetate and polyester fibers. Although quinacridones have been known for quite some time, their commercial use as high-grade pigments is comparatively recent. They were first marketed by E. I. du Pont de Nemours and Company in 1958.

Some of the naturally occurring coloring matters such as acronycine (3), a yellow alkaloid of Australian *Rutaceae* (see Chapter IV), are derivatives of acridine[9]; but they do not appear among the natural dyes listed in the *Colour Index* (C.I.).

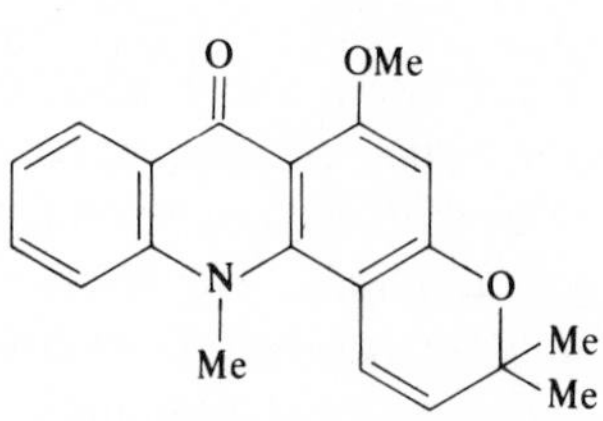

3

1. Acridine Dyes

The water-soluble basic acridine dyes form one of the oldest group of synthetic dyes; they were discovered in the late nineteenth or early twentieth century. Chrysaniline is the oldest acridine dye, reported as far back as 1862. The simple acridine dyes usually yield yellow, orange, and orange-brown shades. The *Colour Index* (1956) lists 17 acridine dyes, many of which are obsolete now. No recent addition to this class of dyes seems to have been made.[10] These dyes now have only a limited use for the dyeing of silk, bast fibers, and tannin-mordanted cotton. The United States sales of acridine dyes in 1962 were only 29,000 lb, valued at $80,000, out of the total dye sales of 178 million lb, valued at $227.2 million. Among the nontextile uses of these dyes may be mentioned the coloration of leather, paper, and lacquer; they are also used in spirit inks. These dyes exhibit strong blue-violet to orange fluorescence in dilute solutions, and some of them are used in biological staining techniques. Acridine derivatives are usually noncarcinogenic; in fact, some of them are anticarcinogenic.[4] Although the use of metal-free acridine orange NO for the coloration of cheese has been reported,[11] acridine dyes are not included in the usual range of permissible dyes for coloring foods and drugs.

Like other basic dyes, acridine dyes are also fugitive to light. Their light fastness is seldom 2 or more on a scale of 1 to 8 (8 denotes the maximum light fastness). The light fastness of acridine dyes such as acridine orange NO or rhoduline orange NO (basic orange 14, C.I. 46005)* (4) is comparatively inferior to that of the related pyran, thiopyran, thiazine, and oxazine (5) dyes.[12] The oxazine (5) has much better light fastness when dyed on acetate rayon.

4

5

The above acridines are usually prepared by the condensation of appropriate diamines and aldehydes. Proflavine or 3,6-diaminoacridine sulfate is the simplest acridine dye. It is prepared from *m*-phenylenediamine[13a, b] (see Chapter I, Section C). Proflavine is better known for its antiseptic properties (see Chapter XVI, p. 791). Methylation of proflavine with dimethyl sulfate or

Colour Index, generic name (Part I) and number (Part II).

methyl *p*-toluenesulfonate gives the well-known antiseptic agent, acriflavine (C.I. 46000) (see also Chapter XVI, Section 2.C).

Proflavine   Acriflavine

The above methylating reagents introduce the methyl group on the ring-nitrogen forming acridinium compounds. Under these conditions, the primary amino groups are not methylated. However, if the alkylation is carried out with an alcohol and mineral acid or alkyl halides, the free amino groups get alkylated.

Acridine orange NO (**4**) is prepared from 4,4′-dimethylaminodiphenyl-methane.[14] The latter on nitration and reduction with iron and sulfuric acid gives **6**. The aqueous acid solution of **6** when heated at 140° for 6 hr (3.5-atm pressure) gives the leuco base **7** which on oxidation yields **4**.

6   7

Methylation of **4** with dimethyl sulfate gives the methosulfate **8**. The 2,7-dibromo derivative of **4**, obtained by bromination in sulfuric acid or nitrobenzene, is the obsolete dye, acridine scarlet J (C.I. 46015). It dyes tannin-mordanted cotton a yellowish red. Auracine G (basic yellow 6, C.I. 46030) (**9a**) is made by reacting formaldehyde with 2,4-diaminotoluene. The resulting acridan (**10**) is oxidized and the formate prepared by treating the base (140 parts) with 85% formic acid (21 parts).[11]

Acridine yellow G (C.I. 46025), which was once used in colored discharge prints, was the corresponding chloride (**9b**). Methylation of **9b** with dimethyl sulfate or methyl *p*-toluenesulfonate gives diamond phosphine GG (basic orange 4, C.I. 46035). There are several similar alkylated dyes (basic orange 5-11). Baking the acridan (**10**) with sodium polysulfide at 280° for 24 hr gives the sulfur dye, immedial brown FR extra concentrated (sulfur brown 20,

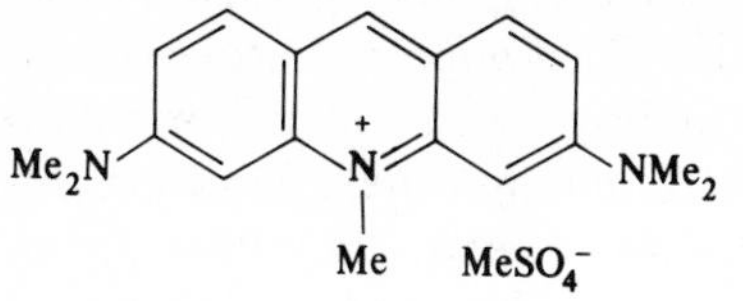

8   MeSO$_4^-$

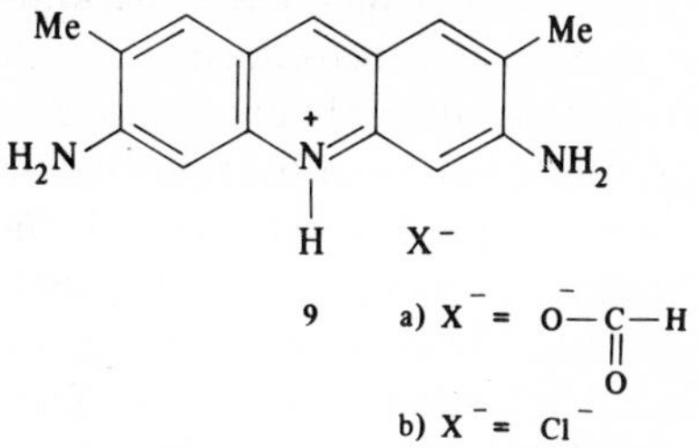

9   a) X$^-$ = O—C—H
             ‖
             O
     b) X$^-$ = Cl$^-$

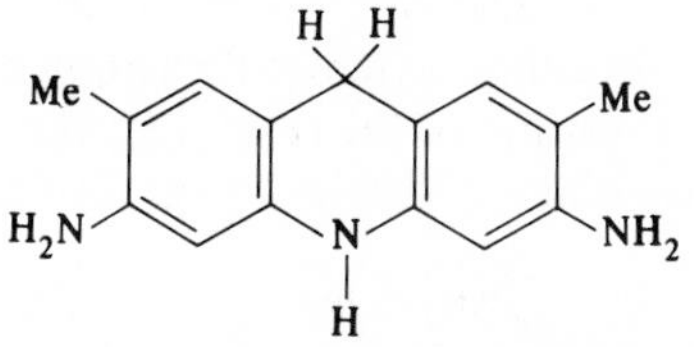

10

C.I. 53680). This dye gives brown shades on cotton with good wash and light fastness. The condensation of 2,4-diaminotoluene with acetaldehyde,[11] instead of formaldehyde, gives euchrysin GGNX (basic yellow 9, C.I. 46040). This 9-methyl analogue of **9b** gives a yellow shade with greenish fluorescence on tannin-antimony mordanted silk.

The unsymmetrically substituted coriphosphine BG (basic yellow 7, C.I. 46020) (**11**), which is made by the action of formaldehyde on a mixture of 2,4-diaminotoluene and *N,N*-dimethyl-*m*-phenylenediamine, followed by oxidation, has found use in India for the preparation of artificial gold thread ("Jari").[2]

A few derivatives of 9-phenylacridines have also been used as dyes. The nitric acid salt (**12**) of chrysaniline or phosphine (basic orange 15, C.I. 46045), formed as a by-product in the synthesis of magenta (C.I. 42510), is readily isolated since it is sparingly soluble. Its 2-methyl analogue is phosphine E.

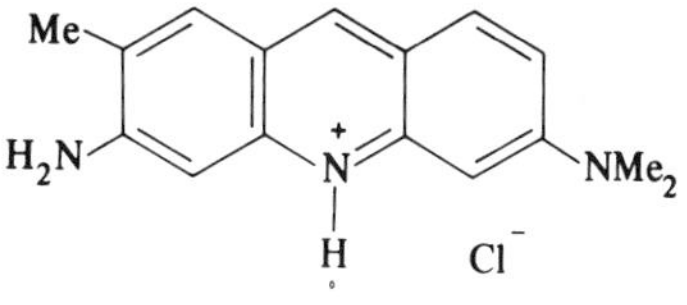

11

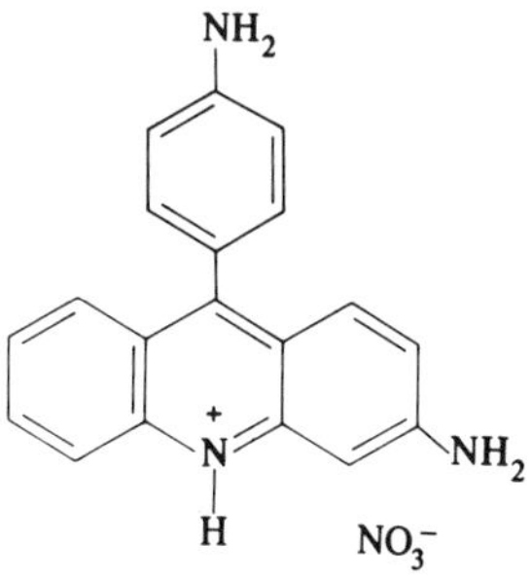

12

The 9-phenyl analogue of acridine orange NO (C.I. 46055) (4) is obtained by condensing benzaldehyde with *N*,*N*-dimethyl-*m*-phenylenediamine in ethanol containing hydrochloric acid, followed by cyclization and oxidation. The condensation of benzaldehyde with 2,4-diaminotoluene, cyclization, and oxidation gave benzoflavine (C.I. 46065) (13), which is used in leather dyeing. The zinc chloride double salt (basic orange 18, 46070) of 14, prepared from 4-amino-2-dimethylaminotoluene and benzaldehyde gives a reddish orange shade on cotton and leather. The interaction of *m*-phenylenediamine, its hydrochloride, and 4,4′-bis-dimethylaminobenzophenone (Michler's ketone) at 200° for 5 hr, and heating of the resulting product at 100° with 30% hydrochloric acid, gives rheonine A (basic orange 23, C.I. 46075) (15), which

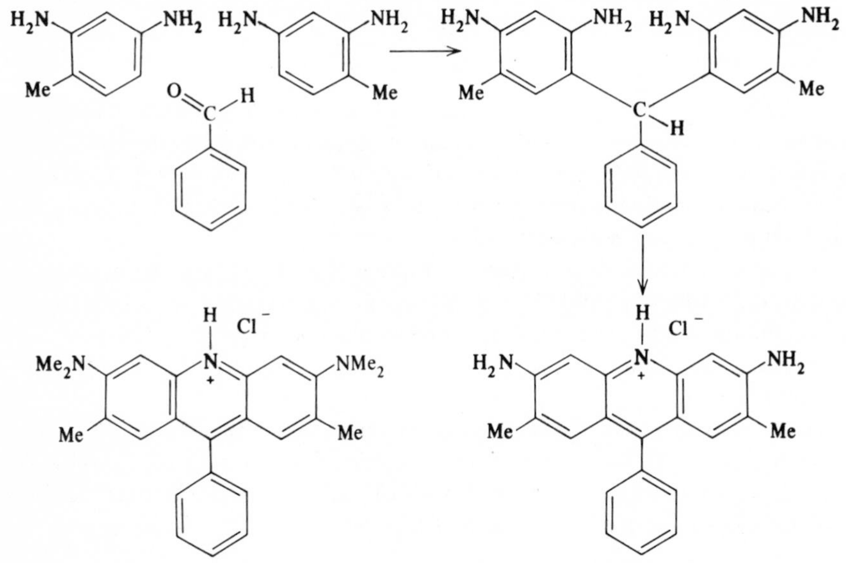

14

13

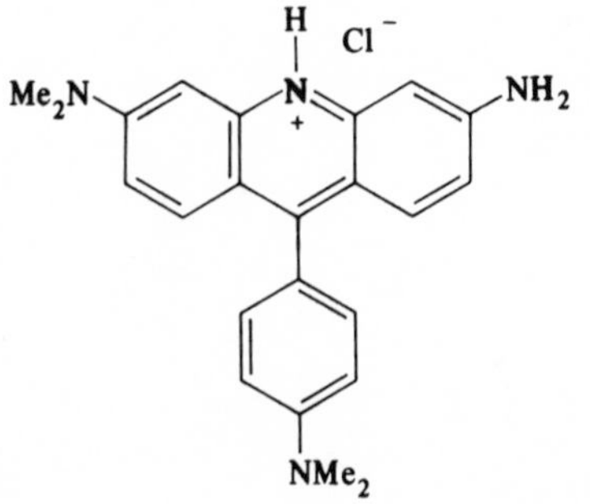

15

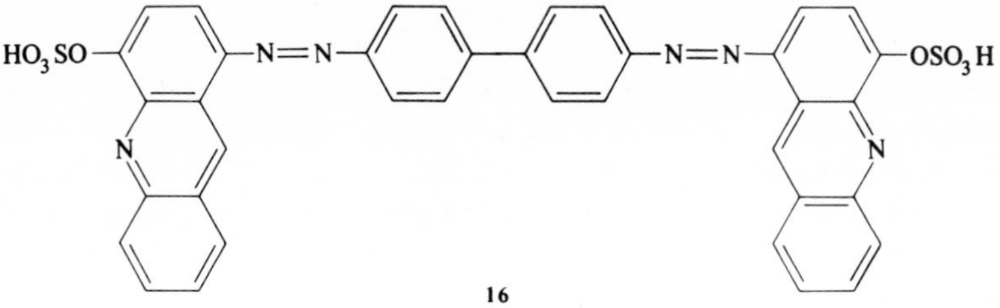
HO3SO
N=N
OSO3H
N
N

16

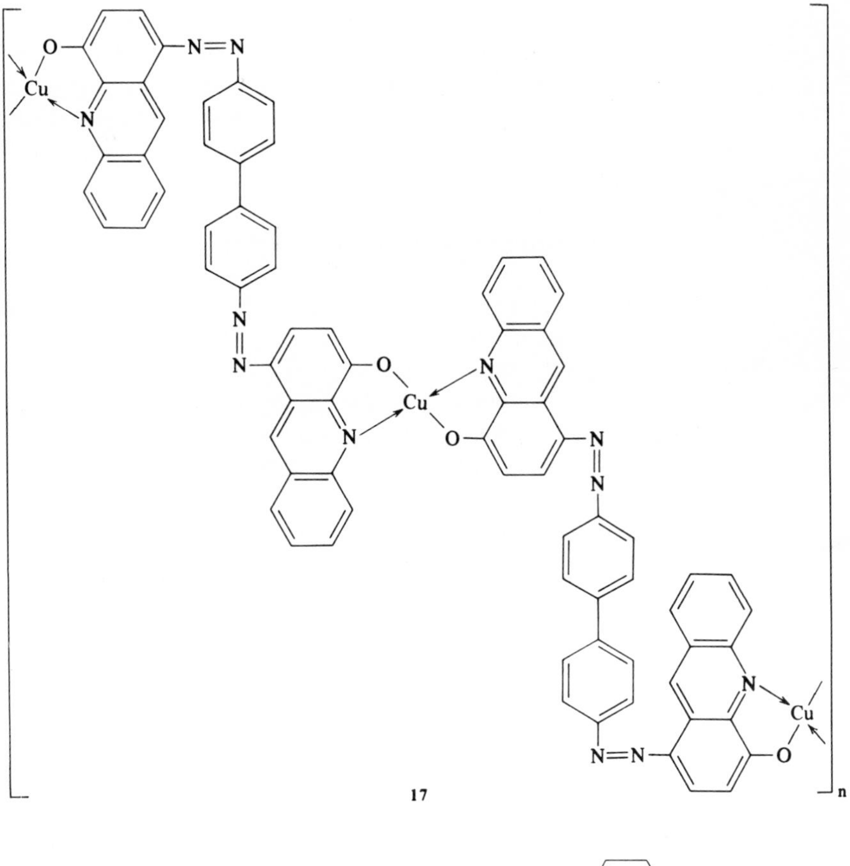
O
Cu
N
N=N
N
N
O
Cu
N
O
N
N
N
N
Cu
O
N=N
n

17

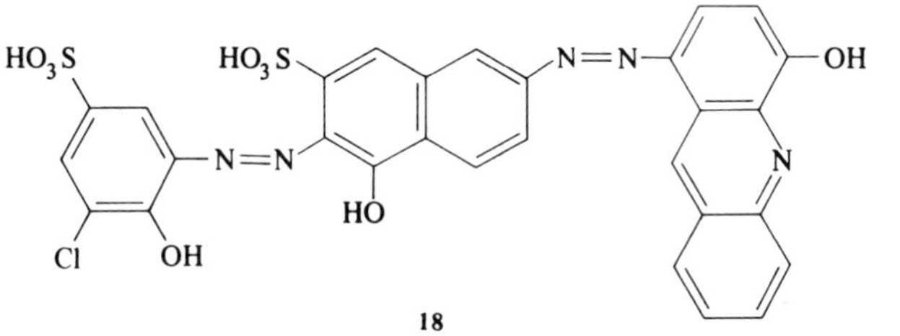
HO3S
HO3S
N=N
OH
N=N
N
HO
Cl
OH

18

is used for dyeing cotton, silk, wool, and leather.[15] A potassium titanoxalate is usually used as a mordant for the application of acridine dyes to leather.[1]

The use of 4-hydroxyacridine and 8-hydroxyquinoline as coupling components in the preparation of bisazo dyestuffs has been patented. These dyes give bright dyeings on cotton, and the fastness properties can be considerably improved by an after-coppering treatment. The water-soluble dyes containing sulfate ester groups (e.g., 16) form polymeric-insoluble copper complexes (e.g., 17) and thus improve the fastness properties.[16] A bisazo dye 18 is obtained by coupling 6-amino-1-naphthol-3-sulfonic acid (J-acid) under alkaline conditions with diazotized 2-hydroxy-3-chloroaniline-5-sulfonic acid, followed by diazotization of the amino group of the J-acid moiety and coupling with 4-hydroxyacridine. When dyed on cotton, and after being treated with a copper salt, (18) gives light and wash-fast blue shades.[17]

2. Simple 9-Acridanone Dyes

Although several nitro-9-acridanone derivatives have been patented as disperse dyes, none seems to have been marketed commercially. However, celanthrene fast yellow GL (disperse yellow 2), (du Pont) is probably 4-nitro-9-acridanone (19),[1, 18] although its structure is not disclosed in the Colour Index and the subsequent supplements. 4-Nitro-9-acridanone (19), prepared by the cyclization of 2'-nitrodiphenylamine-2-carboxylic acid, gives bright yellow shades of good light and wash fastness on cellulose acetate and polyester fibers. Compound 19 and its derivatives are reported to be useful for the dyeing of polypropylene yellow shades which are fast to light.[19]

The condensation of 1-chloro-4-nitro-9-acridanone with 4-chlorothiophenol in the presence of potassium carbonate and dimethyl sulfoxide yields 20, which dyes a greenish yellow shade on polyester fibers.[20] The corresponding 7-amino-9-acridanone (21) gives a brown-red shade. The 4-nitro-9-acridanone-2-sulfonamide (22), prepared by the condensation of 4-chloro-3-nitrobenzenesulfonamide with anthranilic acid and the cyclization of the condensation product, gives a fast greenish yellow shade on cellulose acetate.[21] Chlorosulfonation of 4-nitro-9-acridanone and interaction of the resulting sulfonyl chloride with phenol, in the presence of a base, gives 23, which dyes polyester fibers a fast yellow shade.[22]

19

20

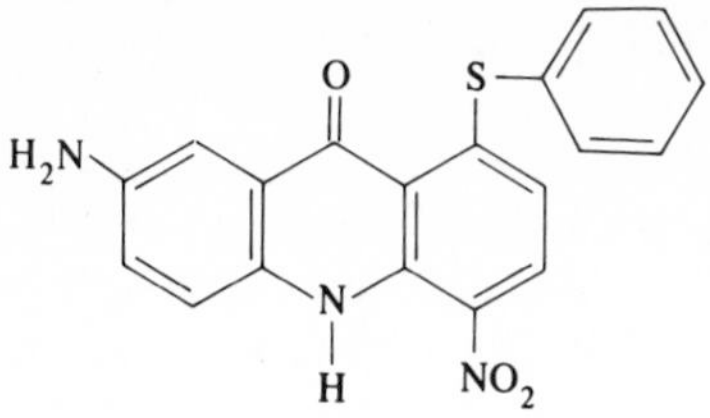

21

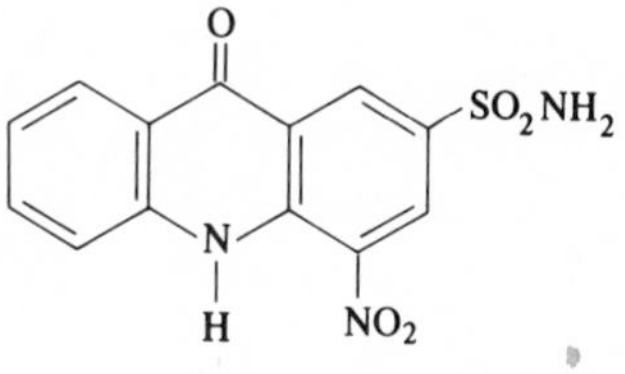

22

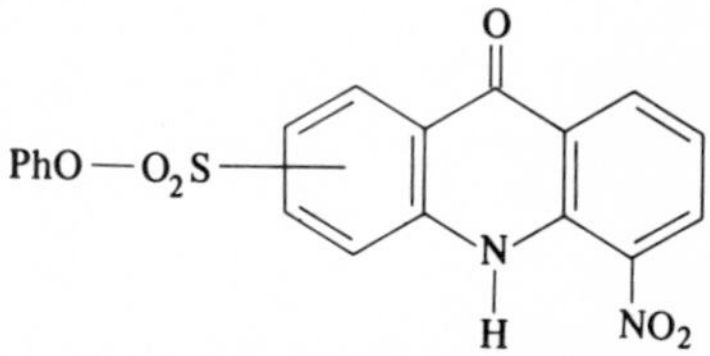

23

Attempts have also been made to prepare yellow to orange pigments from 9-acridanone derivatives. Thus the azo compound **24**, obtained by coupling diazotized 2-chloro-4-nitroaniline with 2-(acetoacetylamino)-9-acridanone, when boiled with methanol and dimethylformamide, gives a stable golden orange pigment.[23] The 2-hydroxy-3-naphthoic acid arylide **25** exemplifies an-

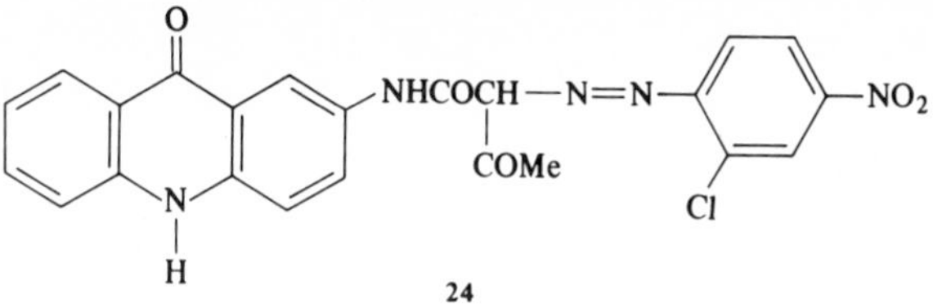

24

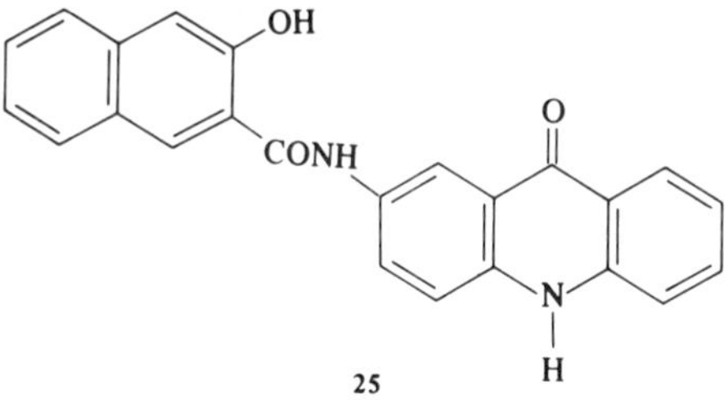

25

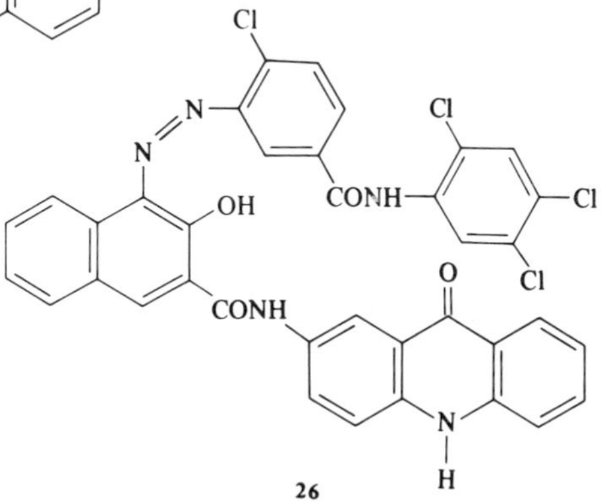

26

other 9-acridanone-containing coupling component. The azo pigment **26** colors polyvinyl chloride and lacquers brown.[24]

The 2-amino-1,4-dimethoxy-9-acridanone (**27**) has been suggested as a diazo component for deep shades.[25] Thus cotton fabrics padded with an alkaline solution of 2-hydroxy-3-naphthoic anilide (naphthol AS) (**28**), and then treated with diazotized **27**, are dyed fast blue shades.

Compound **27** is prepared by the condensation of 2,5-dimethoxy-4-amino-acetanilide with *o*-bromobenzoic acid, followed by cyclization and hydrolysis of the acetamido group.

3. Complex 9-Acridanone Dyes

The dyes of this group are complex polycyclic compounds containing quinonoid groups, in addition to the acridone ring system. They are mostly derivatives of 3,4-phthaloylacridones and benzanthrone. The importance of these dyes is indicated from their large patent coverage and available production statistics. Structures of 16 phthaloylacridones and 14 benzanthrone derivatives (three of which are solubilized vat dyes) are reported in the *Colour Index*. A noteworthy feature of these dyes is their wide color range. Thus they include orange, red, violet, blue, green, and gray dyes.[1-4]

Complex acridone dyes are generally used as vat dyes; the leuco sulfuric esters of some of the benzanthrone derivatives have been used as solubilized vat dyes. The alkali-dithionite vat solutions of these dyes are generally wine red to violet, and they dye cotton shades of excellent light, wash, and bleach fastness. A recent trend appears to be the use of phthaloylacridones as disperse dyes for polyester fibers.

A. Phthaloylacridones

Although three isomeric phthaloylacridones are possible, only the derivatives of 3,4-phthaloylacridone (**29**) are valuable as dyes. The naphth[2,3-*c*]-acridan-5,8,14-trione (**29**) is generally prepared by cyclization of either **30a** or **30b**.

29

30 a) X_1 = COOH; X_2 = H
    b) X_1 = H; X_2 = COOH

The cyclization of **30a** and **30b** may be effected by interaction with several reagents, such as concentrated sulfuric acid,[26a, b] chlorosulfonic acid,[27] phosphorus pentachloride or thionyl chloride,[28] and acetyl or benzoyl chloride.[29] The acid chlorides of **30a** and **30b** can be cyclized by the action of aluminium chloride.[30] The alkyl esters of **30a** and **30b** have also been cyclized by treating with zinc and ammonia or alkaline dithionite.[31] Compound **30a** is prepared by the Ullman condensation of 1-chloro- or 1-nitro-anthraquinone-2-carboxylic acid with an aromatic amine or by the condensation of 1-amino-anthraquinone-2-carboxylic acid with bromobenzene. Compound **30b** is synthesized by the condensation of 1-chloroanthraquinone and anthranilic acid or 1-aminoanthraquinone and o-chlorobenzoic acid.[26, 32, 33] Several other methods for the preparation of **29** have also been reported.[2, 3]

The red-violet-colored acridone **29** is not a commercial dye, but many of its derivatives are valuable dyes. The hydrogen atom attached to nitrogen in **29** is conveniently located to form an intramolecular hydrogen bond with the carbonyl oxygen (14-position). This property may be partly responsible for the dyeing and fastness properties of these dyes. Thus the *N*-methyl derivative of (**29**) is relatively less substantive to cotton and it also exhibits a hypso-chromic shift.[34] The phthaloylacridones with amino and substituted amino groups at 6-position are blue; otherwise, they are red. Phthaloylacridone dyes, derived from the parent structure **29**, may be divided broadly into the following three groups: (1) 3,4-phthaloylacridones carrying substituents/fused rings at 9–12 positions, (2) 6-amino derivatives of 3,4-phthaloyl-acridones, and (3) carbazole phthaloylacridone dyes.

(1) 3,4-Phthaloylacridones Carrying Substituents/Fused Rings at 9-12 Positions

Indanthrene violet RRK (C.I. 67800) and indanthrene red RK new (C.I. 67805), which are now obsolete, were 11-chloro- and 10,12-dichloro deriva-tives of (**29**). Indanthrene red violet RRK (vat violet 14, C.I. 67895) is essentially 6,10,12-trichlorophthaloylacridone,[35, 36] prepared either by the chlorination of (29) or the uncyclized 30b. The 6,9,10,11,12-pentachloro derivative[33] of 29 is marketed as indanthrene brilliant pink BBL (vat red 39, C.I. 67900). Indanthrene pink B (C.I. 67905) is the 6,7,9,10,11,12-hexachloro derivative. The labile chlorine at 7-position probably gets removed during vat dyeing.[33]

Indanthrene brilliant pink BL (vat red 38, C.I. 67810) (31) is prepared by the condensation of 3,5-dichloro-2-phenoxyaniline with methyl ester of 1-chloroanthraquinone-2-carboxylic acid, followed by hydrolysis of the ester and cyclization of the resulting acid with benzoyl chloride in o-dichlorobenzene.[37] Indanthrene Red RK (vat red 35, C.I. 68000) (32) is prepared from 1-nitroanthraquinone-2-carboxylic acid and β-naphthylamine with arsenic trichloride in o-dichlorobenzene.[38]

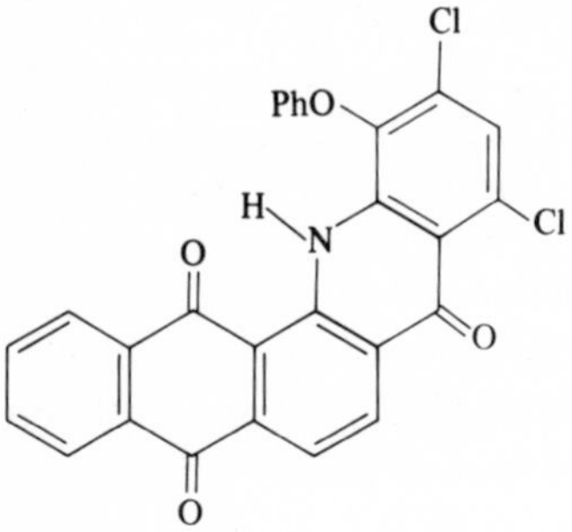

31

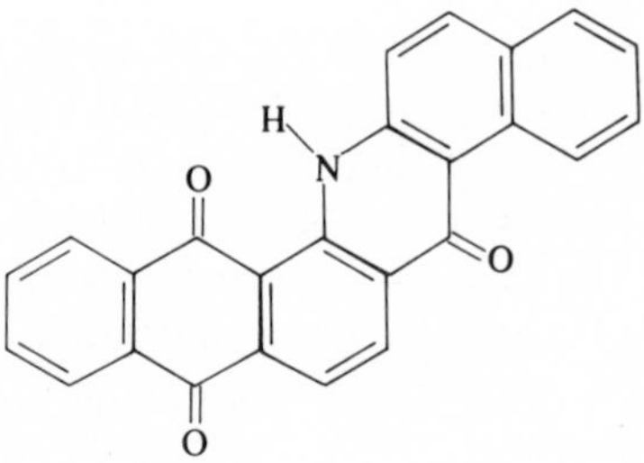

32

The carcinogenic β-naphthylamine can be replaced by 2-naphthylamine-1-sulfonic acid (Tobias acid).[39] The phthaloylacridone 33a gives wine red shades on cotton.[40] Vat dyes such as 33b, containing a sulfonic acid group, are reported to possess improved penetration and level dyeing properties.[41]

Indanthrene orange RR (vat orange 13, C.I. 67820) (34) is prepared by the condensation of 1-nitroanthraquinone-2-carboxylic acid with p-amino-benzoic acid, cyclization, convertion to acid chloride and final condensation with 1-amino-5-benzamidoanthraquinone.[33, 42]

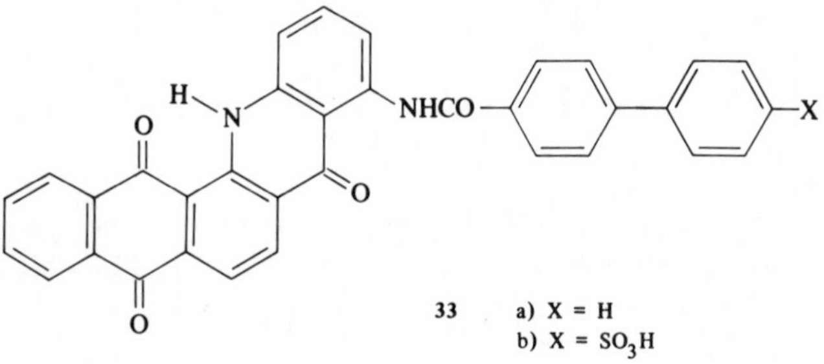

33   a) X = H
          b) X = SO₃H

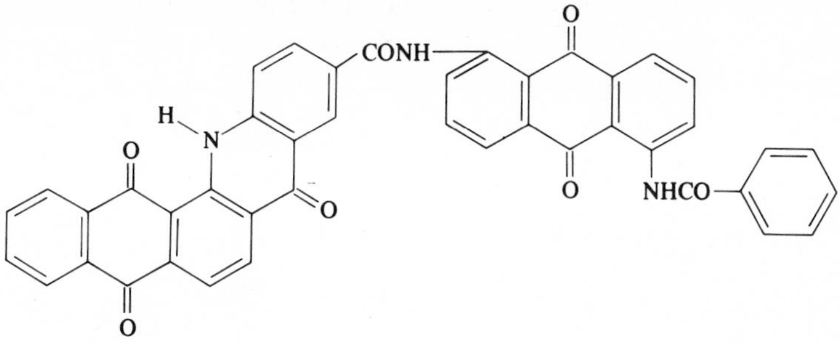

34

The acridone derivative, obtained starting from 1-chloroanthraquinone-2-carboxylic acid and 2-aminoanthracene, is red brown.[43] By using more complex arylamines such as 3-aminopyrene, dyes that absorb at longer wavelengths are obtained. Thus the dye 35, derived from 3-aminopyrene, is green.[44] The hypochlorite oxidation of 10-aminophthaloylacridone gives 35a, which dyes cotton a fast brown shade.[45]

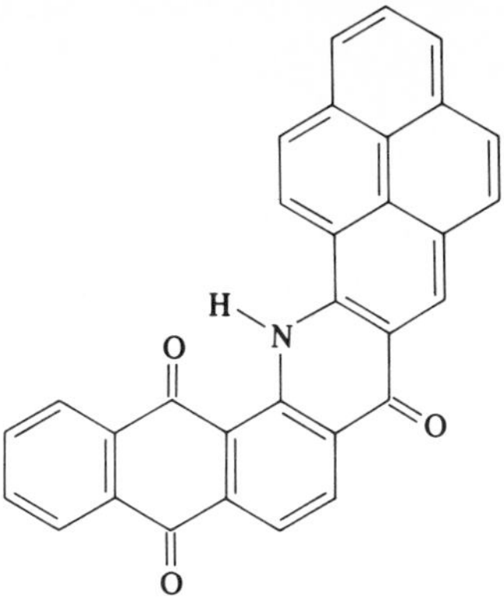

35

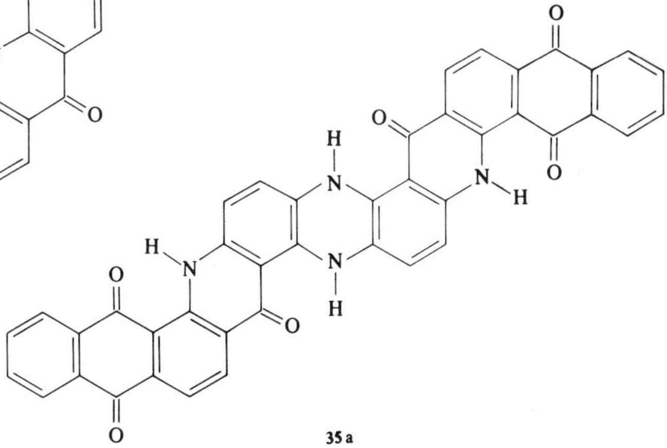

35a

Indanthrene red brown R (vat brown 31, C.I. 70695) is prepared by the interaction of 2 moles of 1-nitroanthraquinone-2-carboxylic acid with 1 mole of benzidine, followed by cyclization of the condensation product. Decarboxylation of one of the carboxyl groups which is still present, by heating with alkali under pressure, yields indanthrene red brown R[46] which is a mixture of the acridonylceramidone **36b** (70%), **36a** (23%), and the diceramidonyl **36c** (7%).

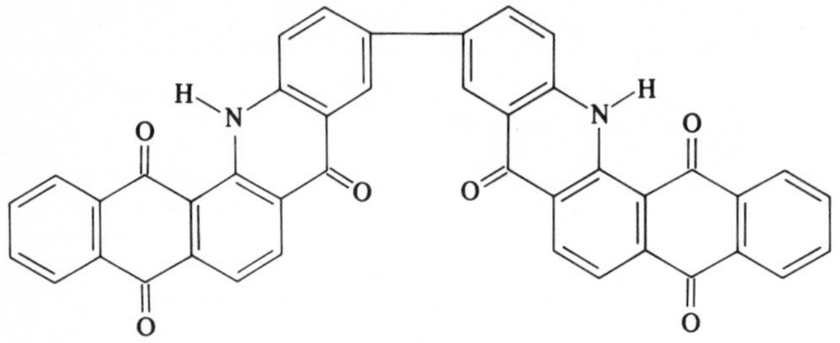

36a

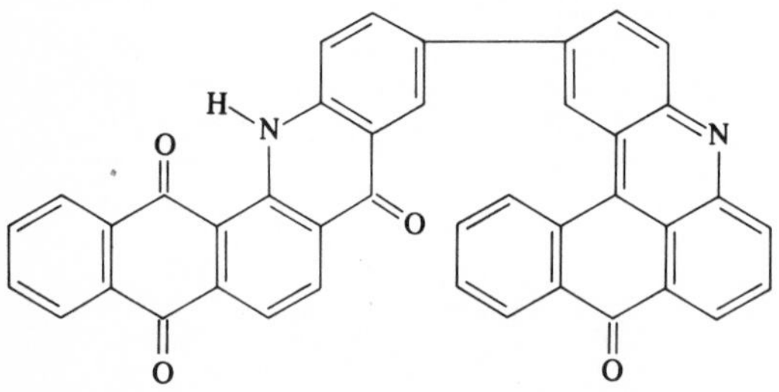

36b

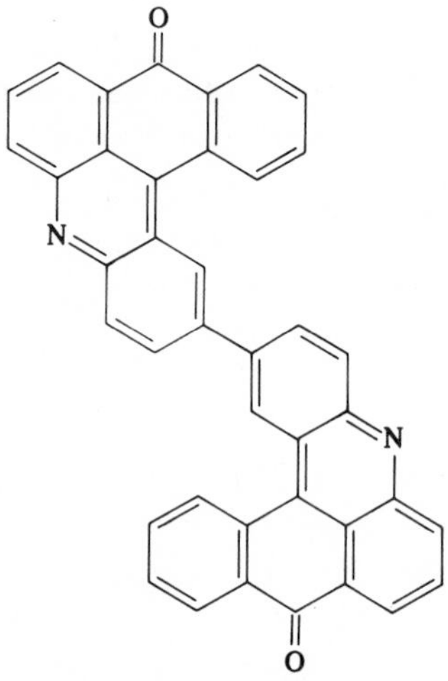

36c

The condensation of 1-chloroanthraquinone-2-carboxylic acid with 2-aminoanthraquinone and cyclization[47] gives indanthrene orange F3R (vat orange 16, C.I. 69540) (**37**). The bisacridone, indanthrene violet FFBN (vat violet 13, C.I. 68700) (**38**), which dyes bluish violet shades, is prepared by the condensation of 1,5-dichloroanthraquinone and potassium anthranilate, followed by cyclization.[11, 48]

Phthaloylacridones are usually stable to nitric acid and other oxidizing agents. However, they break down when fused with caustic potash. Thus the parent acridone (**29**) on alkali fusion yields benzoic acid and 9-acridanone-3-carboxylic acid, whereas **38** gives only 9-acridanone-3-carboxylic acid.[49]

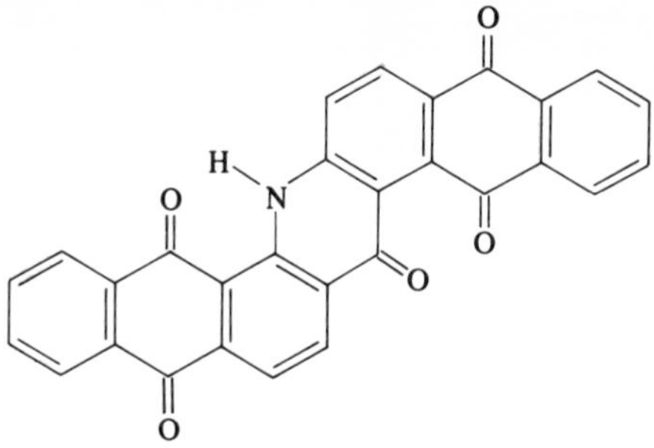

37

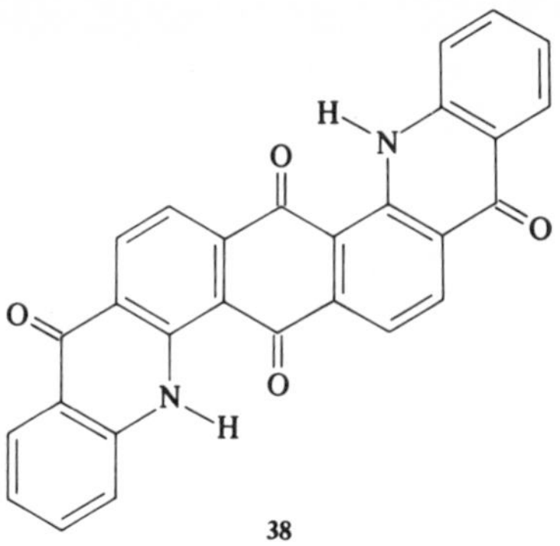

38

(2) 6-Amino Derivatives of 3,4-Phthaloylacridones

Derivatives of (29) having amino substituents in the 6-position are blue. Thus indanthrene turquoise blue GK (vat blue 32, C.I. 67910) and indanthrene turquoise blue 3GK (vat blue 33, C.I. 67915) are 39 and 40, respectively.[50]

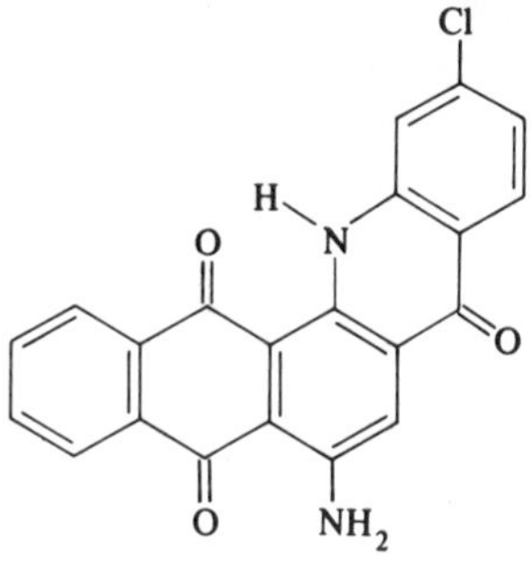

39

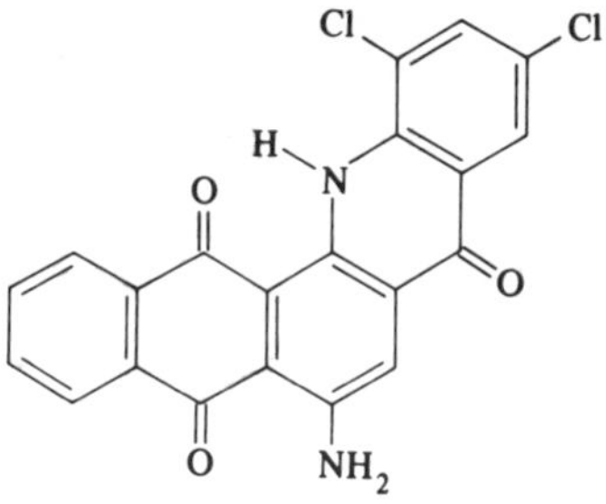

40

They are prepared by the condensation of 1-amino-4-bromoanthraquinone-2-sulfonic acid (bromamine acid) with 3-chloro- and 2,4-dichloro-aniline, respectively, followed by cyclization with simultaneous elimination of the sulfonic group.[51] The nitration and reduction of (29) gives a mixture of tri- and tetra-amino derivatives, which dyes cotton in a bluish green shade.[52] The 11-trifluoromethyl-6-benzamido derivative of (29) is indanthrene printing blue HFG (vat blue 21, C.I. 67920). It is used for printing direct and resist styles on cotton.[11, 53] Apart from their use as vat dyes for cotton, dyes such as 39 and their acylamino derivatives can also be used as disperse dyes for turquoise blue prints and dyeings of excellent fastness on polyester fibers.[54]

Indanthrene blue CLN (vat blue 39) (41) is prepared by acylation of 6-aminophthaloylacridone with p-chlorobenzoyl chloride[3] or by starting from 1-(p-chlorobenzoylamido)-4-chloroanthraquinone and anthranilic acid.[55]

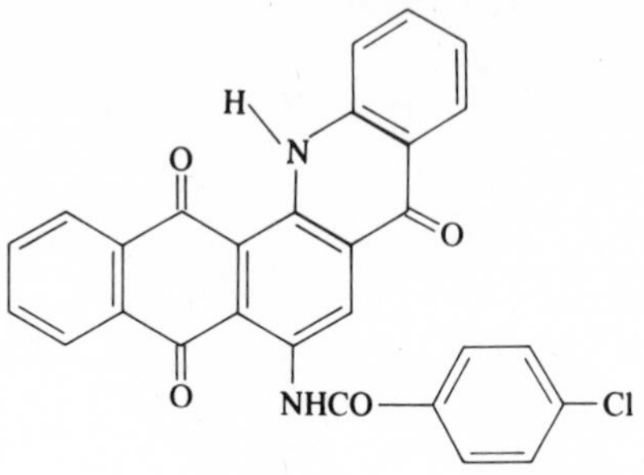

41

    The dyes **42** and **43** are other examples of disperse dyes that give brilliant bluish green shades on polyester fibers. The dye **42** is prepared by the inter-action of 1-amino-4-bromoanthraquinone with acrylamide, hydrolysis of the amide group, condensation with anthranilic acid, followed by cyclization and esterification.[56] The treatment of appropriate acridone with ethylene glycol monoethylether, dimethylformamide, *m*-nitrobenzenesulfonic acid, and caustic soda[57] is reported to give **43**. However, the structure of this com-pound requires confirmation. The bluish green disperse dye **44** is prepared either by the condensation of 4-bromo-1-(ω-methoxypropylamino)anthra-quinone and anthranilic acid or by the interaction of ω-methoxypropylamine with 6-aminophthaloylacridone in the presence of boiling aqueous sodium dithionite and oxidation of the leuco derivative.[58]

    The condensation of thiosalicylic acid with 6-bromophthaloylacridone, in the presence of potassium carbonate and amyl alcohol, gives **45**, which dyes polyester fibers a bright blue shade by mass coloration.[59]

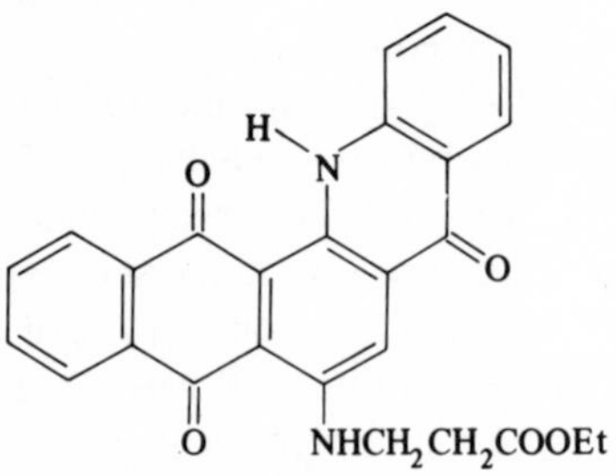

42

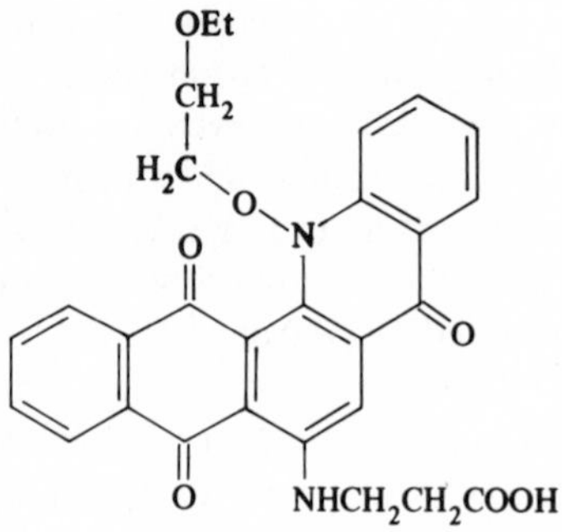

43

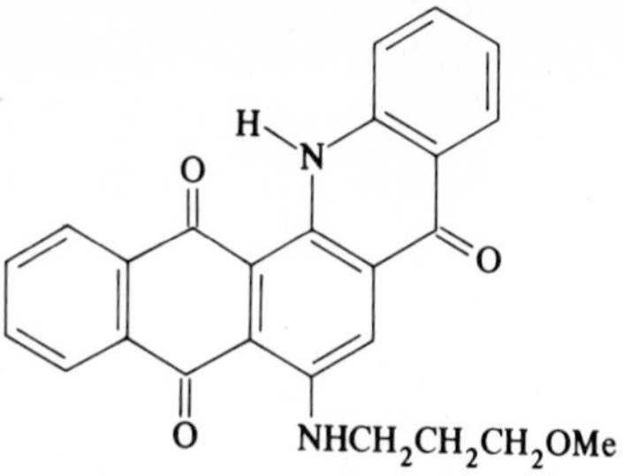

44

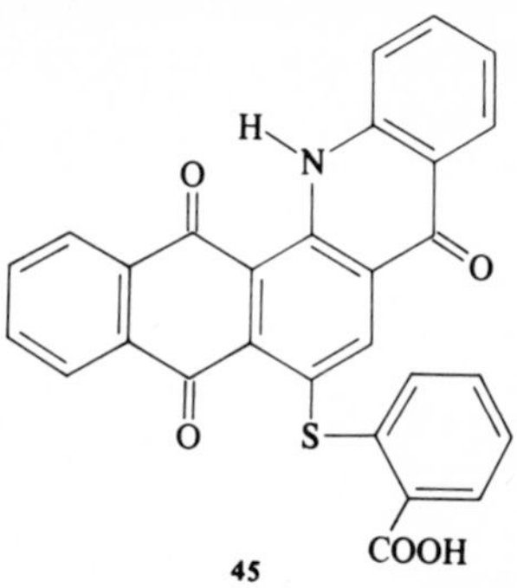

45

The phthaloylacridone **46a**, prepared by the condensation of 1-amino-2,4-dibromoanthraquinone with anthranilic acid followed by cyclization with phosphoric acid, reacts with potassium or copper cyanide to give the 7-nitrile **46b**. The latter on hydrolysis[60] yields the amide **46c**.

The interaction of **46a** with sodium phenoxide[61] gives **46d**. These disperse dyes, **46b, c,** and **d,** give bright bluish green shades on polyester fiber.

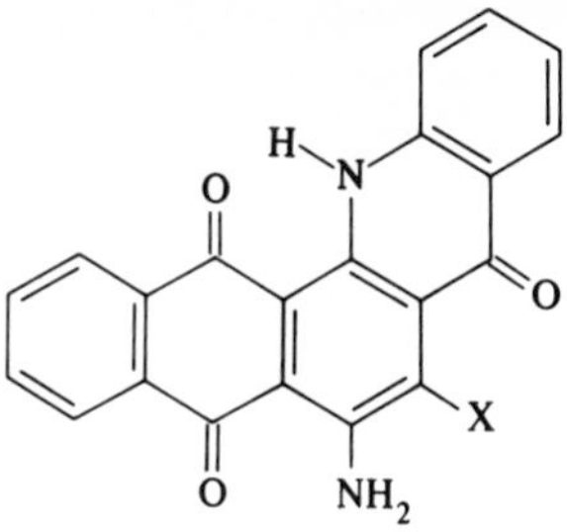

46 a) X = Br
b) X = CN
c) X = CONH₂
d) X = OPh

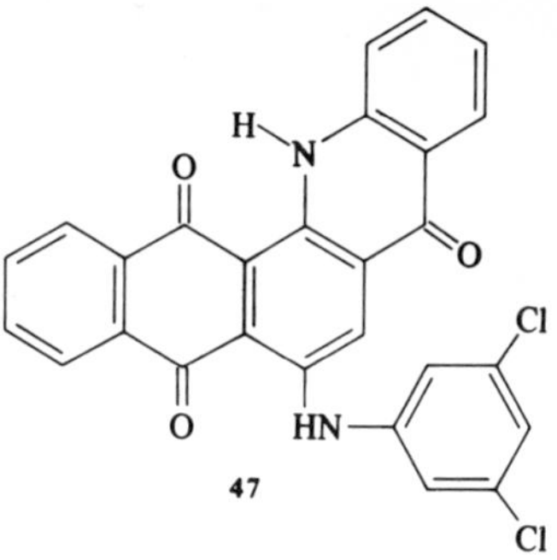

47

By introducing appropriate arylamino substituents at the 6-position of phthaloylacridone (**29**), it is possible to produce useful green vat dyes. Thus the condensation of 1,3,5-trichlorobenzene with 6-aminophthaloylacridone, in the presence of potassium carbonate and copper chloride,[62] yielded a brilliant green dye (**47**). The yellowish green vat dye, indanthrene green 4G (vat green 12, C.I. 70700) (**48**), has good fastness properties.[63] Similar dyes, in which sulfonic acid groups are introduced for better penetration and level dyeing, have been reported in recent patents.[64] A green dye similar to **48**, containing a biphenyl substituent in place of phenyl, has also been reported.[65]

The condensation of 6-aminophthaloylacridone with 1-(*m*-bromobenzamido)-anthraquinone gives **49**, which dyes cotton yellowish green.[66]

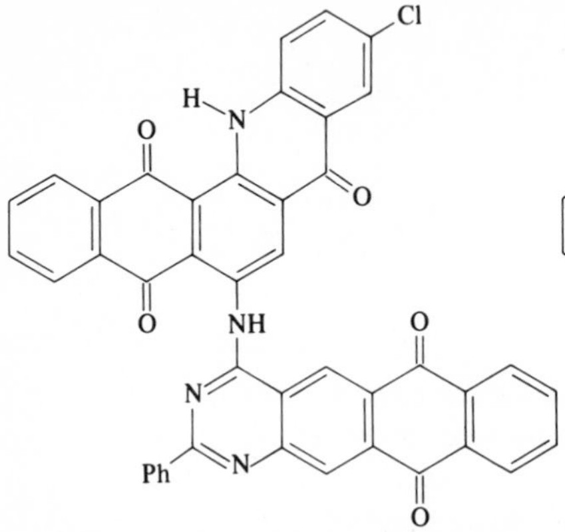

48

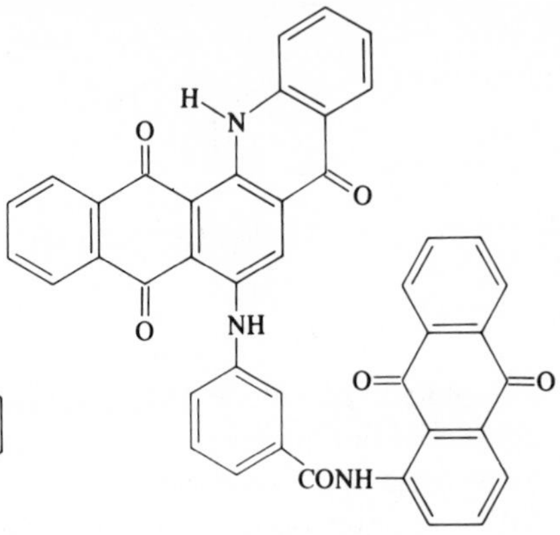

49

The green dye **50** is prepared by the condensation of terephthaloyl chloride with one molecule each of 6-aminophthaloylacridone and 1,2-diamino-anthraquinone.[67] The condensation of 6-aminophthaloylacridone with

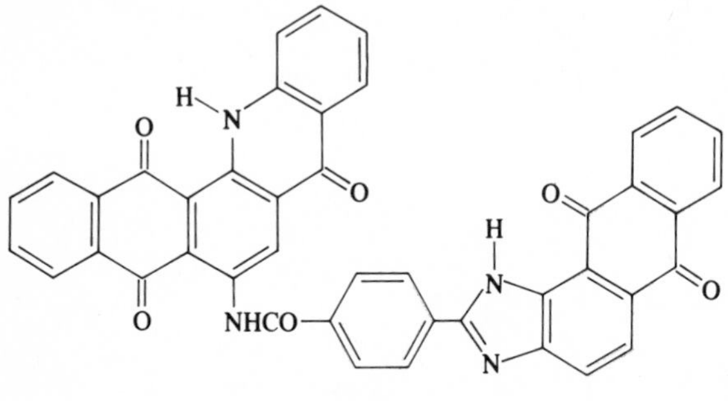

50

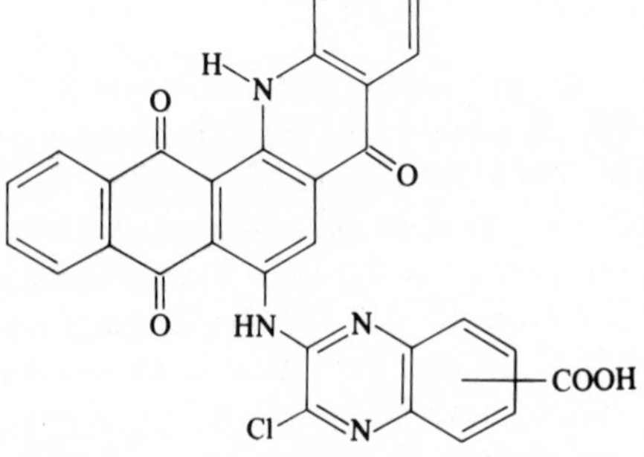

51

2,3-dichloroquinoxaline-6-carboxylic acid gives the dye **51**, which dyes cotton an olive green shade from a wine red vat.[68]

The condensation of 2 moles of 10-chloro-6-aminophthaloylacridone with 1 mole of a monosubstituted cyanuric chloride gives (**52**), which dyes blue shades on cotton that are fast to bleaching and soda boil.[69, 70]

52

(3) *Carbazole Phthaloylacridone Dyes*

Carbazole phthaloylacridone derivatives dye fast brown shades on cotton. Indanthrene khaki GR (vat brown 16, C.I. 70910) (**53**) is a mixed carbazole acridone, prepared by the condensation of 6,7,9,12-tetrachlorophthaloyl-acridone with two molecules of α-aminoanthraquinone, followed by cycliza-tion with aluminium chloride.[71] It gives brownish olive shade of good, all-round fastness. Indanthrene brown 3GT (vat brown 26, C.I. 70510) (**54**) is another carbazole-acridone dye prepared by the interaction of 1-amino-5-benzamidoanthraquinone with 6,10,12-tri-chlorophthaloylacridone, followed by carbazolization.[72]

Indanthrene brown LG (vat brown 46, C.I. 70905) (**55**) is obtained when 2 moles of 1-amino-5-benzamidoanthraquinone are condensed with 1 mole of 6,10,12-trichlorophthaloylacridone and the product carbazolized.[33] The carbazole-acridone **56**, prepared by the condensation of 2 moles of 6-benz-amido-10-bromophthaloylacridone and 1 mole of 1,5-diaminoanthraquinone and carbazolization is a good brown vat dye.[73]

Several similar complex carbazole-acridone brown and gray vat dyes have been reported in recent literature.[74]

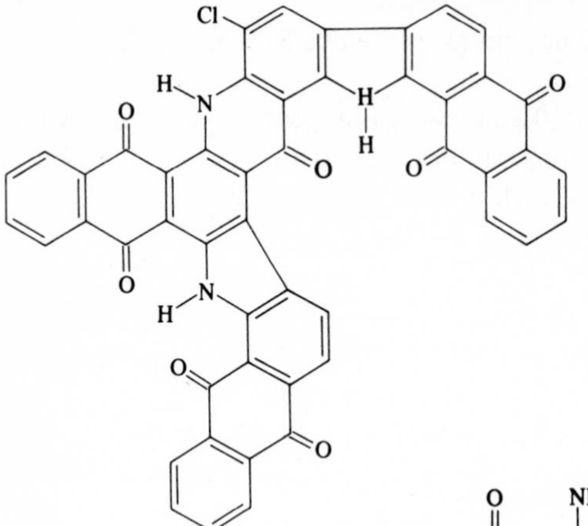

53

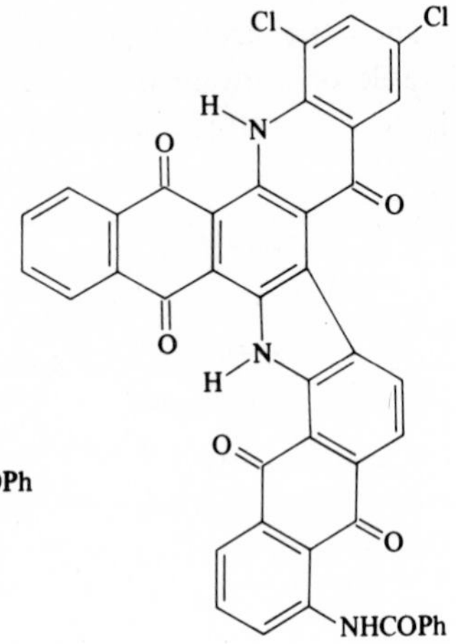

54

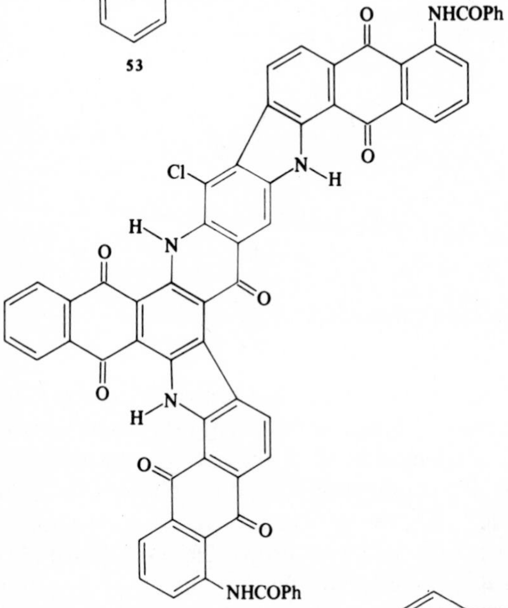

55

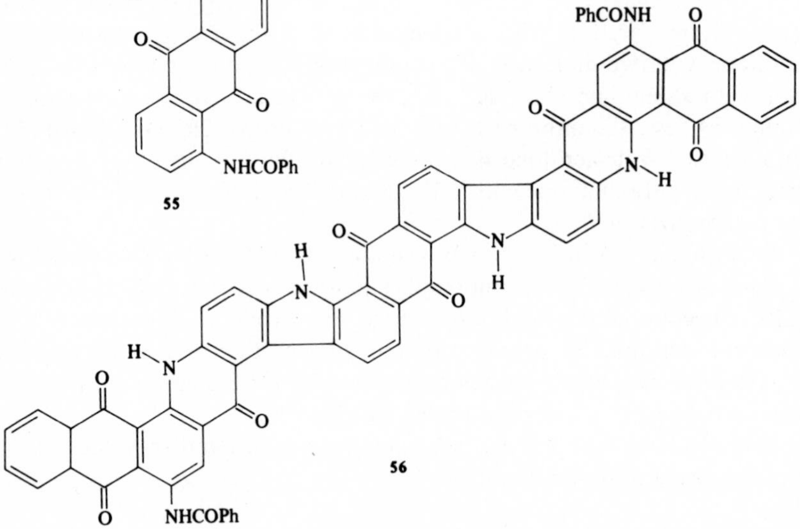

PhCONH

56

598

B. Benzanthrone Acridone Dyes

Vat dyes containing benzanthrone and acridone moieties dye shades of excellent fastness (light fastness around 8). Because of their outstanding fastness, they are used for the dyeing of cotton curtains, furnishings, and shirting. The controlled bromination of benzanthrone gives high yields of either 3-bromobenzanthrone or 3,9-dibromobenzanthrone.[33] Both are valuable intermediates in the preparation of benzanthrone-acridone vat dyes.

Indanthrene olive green B (vat green 3, C.I. 69500) (57) is prepared by the condensation of 3-bromobenzanthrone with α-aminoanthraquinone in the presence of sodium carbonate and copper oxide in boiling naphthalene. The resulting 3-(α-anthraquinonylamino)-benzanthrone on cyclization by heating with potassium hydroxide in isobutanol[33, 75] gives 57. Following similar reaction conditions and using 2 moles of α-aminoanthraquinone per mole of 3,9-dibromobenzanthrone, indanthrene olive T (vat black 25, C.I. 69525) (58) is obtained.[11, 33, 76]

The carbazolization of 58 by heating at 100° in a melt containing urea and aluminium chloride results in an olive green vat dye.[77] Both 57 and 58, which give fast olive green and brownish gray shades, are widely used.

An interesting method of preparing **57** and **58** consists in the cyclization of the sulfonated intermediates **59a** and **59b** by treatment (vatting) with alkaline dithionite[78] at 100°.

59 a; X = H
    b; X = α-anthraquinonylamino

When **59a** and **59b** are used for vat dyeing, the dyes **57** and **58** are formed on the fiber by desulfonation and cyclization. The dyeings have the same shades (green and gray) and are equally fast. Indanthrene olive green GG (vat green 5, C.I. 69520) (**60**) is prepared by the interaction of 3-bromobenzanthrone with 1,5-diaminoanthraquinone, followed by cyclization, benzoylation, and chlorination.[33]

60

Derivatives of **57** having benzamido or α-anthraquinonylamino groups in 6-position also give greenish olive shades on cotton.[79] Of special interest are the derivatives of **57**, containing substituted benzamido group in the 4'-position.[80] They are olive green dyes that show low ir reflectance properties and

give fast dyeings on cotton. These dyes have been recommended for camou-
flaging military uniforms and equipments, since conventional ir detectors are
incapable of detecting dyes of ir reflectance below 25%. An example of such a
dye is **61**, which is prepared by the interaction of *m*-dimethylsulfonamido-
benzoyl chloride with 4'-amino derivative of **57**. The condensation of 1-amino-
4-chloroanthraquinone with thiosalicylic acid, followed by cyclizing to a
thioxanthone and then further condensation with 3-bromobenzanthrone
and cyclization, yields a brown dye (**62**) that is similar to **61** in low ir reflectance.[81]

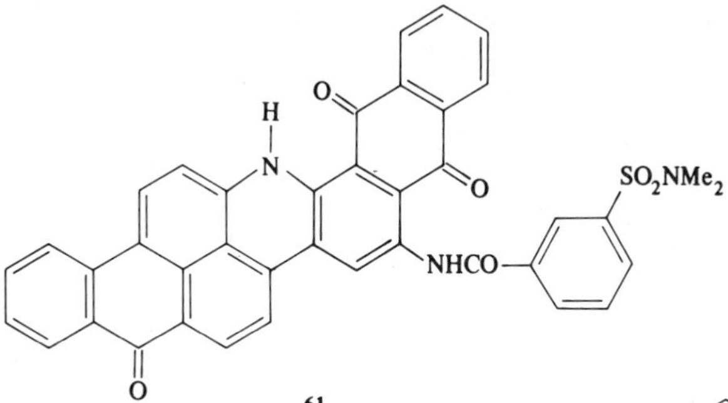

61

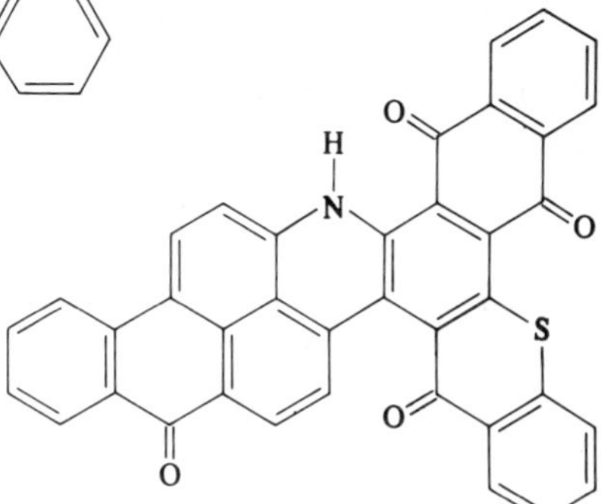

62

If the acylation of 4'-amino derivative of **57** is carried out with 1-amino-
anthraquinone-2-carboxyl chloride, the resultant dye gives a maroon shade
on cotton that shows good fastness, weatherability, and printability.[82]

The condensation of 4-bromobenzanthrone and α-aminoanthraquinone in
the presence of caustic potash and dimethyl sulfoxide, followed by cyclization, yields **63**, an isomer of **57**.

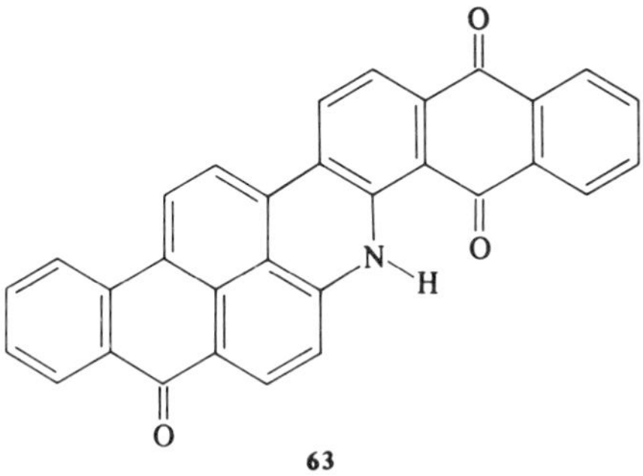

63

Such dyes prepared from 1-amino-4-benzamidoanthraquinone and 1-amino-5-benzamidoanthraquinone gave olive dyeings that are fast to light and alkali boil.[83] Their chloro derivatives give olive green shades on cotton.[84]

C. Miscellaneous Complex Acridone Dyes

An example of such a dye is the fast yellowish green dye **64**, which is obtained by interaction of 4,5-dichloronaphthalene-1,8-dicarboxylic acid, o-phenylenediamine, and 1-aminoanthraquinone.[85] Other examples are the anthrapyridone pigments **65** and **66**, which dye polypropylene reddish purple shades by melt spinning.[86] The condensation of anthranilic acid with 6-bromo-N-methylanthrapyridone and 8-chloro-N-methylanthrapyridone and cyclization of the condensation products yields, respectively, **65** and **66**. The anthrapyridone derivative **67** has also been mentioned as a violet dye.[87]

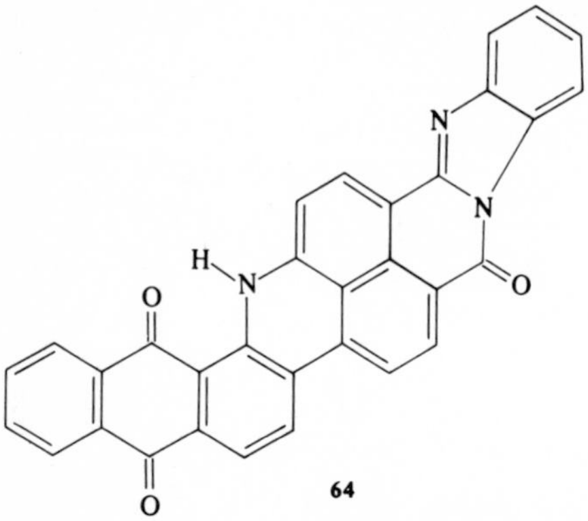

64

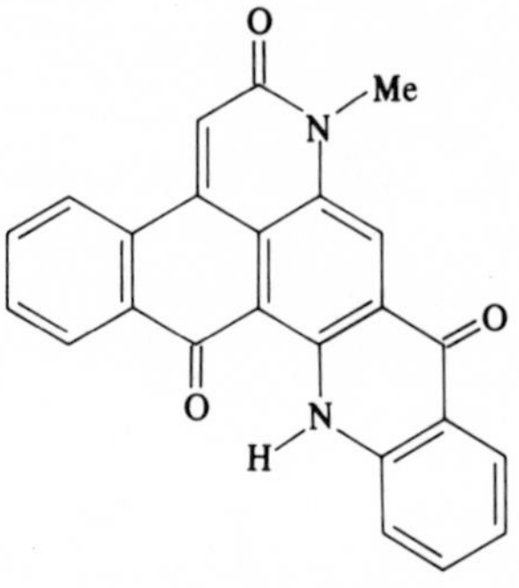

65

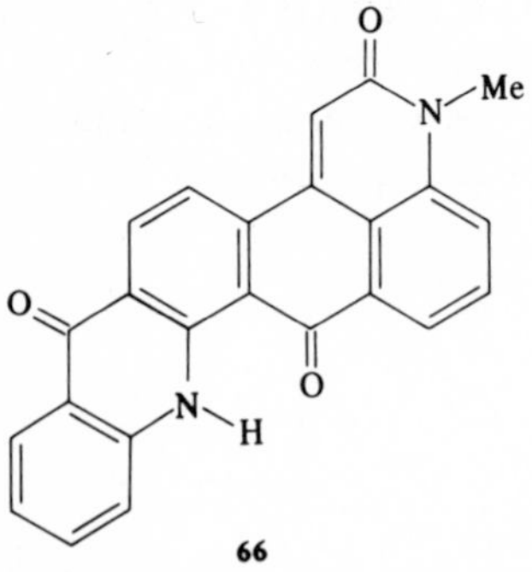

66

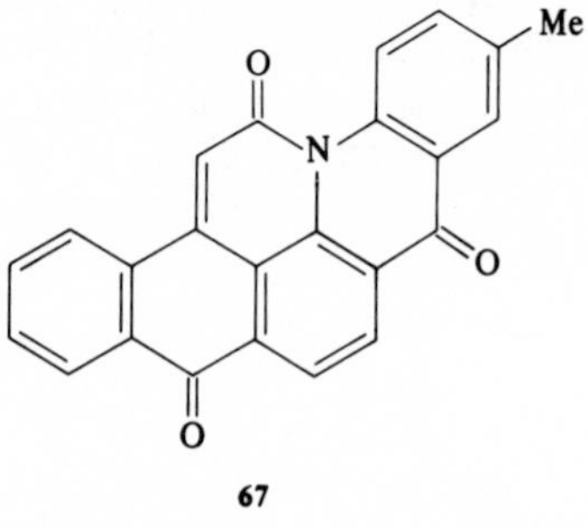

67

Pyrazolanthrone derivatives, indanthrene navy blue R (vat blue 25, C.I. 70500) **(68)** and indanthrene gray M, MG (vat black 8, C.I. 71000) **(69)** are widely used fast vat dyes. The former (68) is prepared by the condensation of 3-bromobenzanthrone with anthrapyrazole, followed by cyclization by treatment with alcoholic caustic potash.[88] The latter (69) is prepared similarly, starting from 3,9-dibromobenzanthrone, anthrapyrazole, and α-aminoanthraquinone.[11, 33] The well-known phototropic vat dye, flavanthrene (vat yellow 1, C.I. 70600), (70) may also be looked upon as a complex acridone dye.

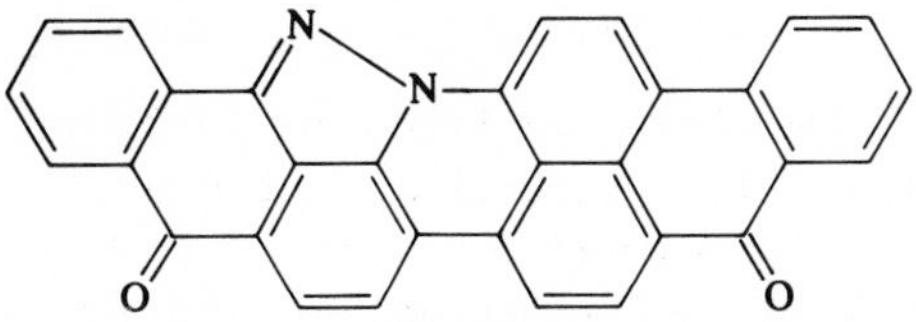

68

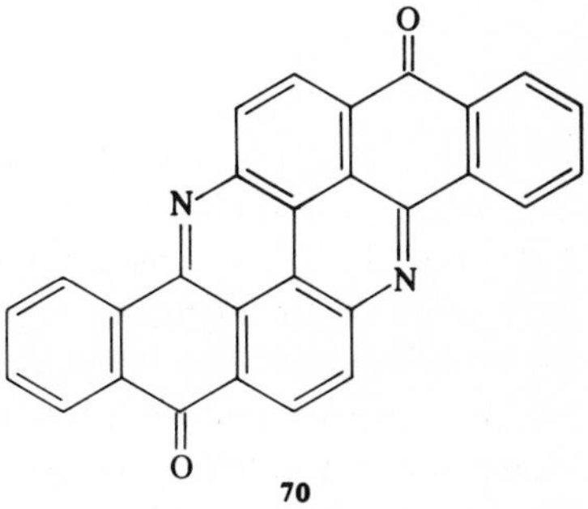

70

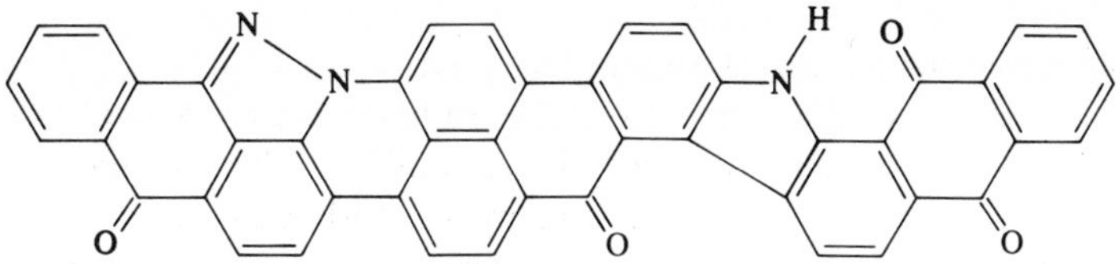

69

It is prepared by treating 2-aminoanthraquinone with antimony pentachloride in nitrobenzene or by simultaneous dehydration and cyclization of 2,2'-diphthalimido-1,1'-bianthraquinonyl under alkaline conditions.[11, 33]

4. Quinacridones

Diketoquinolinoacridines are designated by the trivial name "quinacridones." They can be linear *trans* (71) and *cis* (72) or angular (73) and (74).

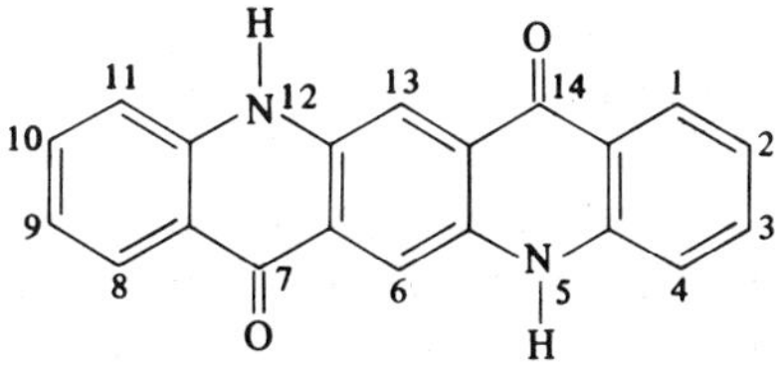

71

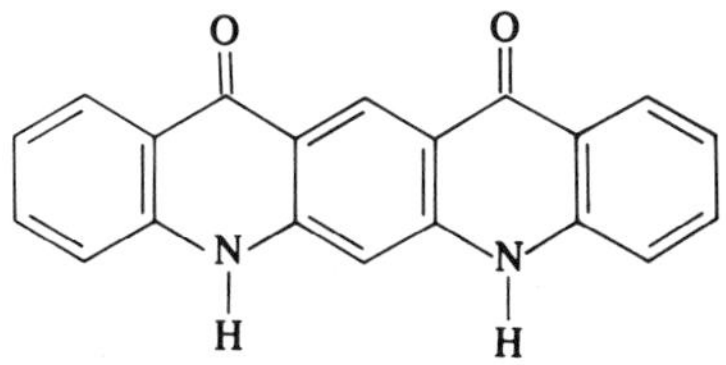

72

73   74

All four compounds and their derivatives have been known for a long time. The linear *trans*-quinacridone (**71**) was first prepared by Leibermann[89] in 1935, but the other three compounds were known even earlier.[90-92] It was only when their pigmentary properties were recognized in 1955 by the du Pont chemists that they attracted wide attention. The compound **71**, which is usually referred to as quinacridone, and its derivatives are red or violet and are valuable as high-quality pigments. The other three isomers **72**, **73**, and **74** are yellow and are comparatively much less valuable as pigments.

The linear *trans*-quinacridone **71** and its derivatives are insoluble in organic solvents and dispersion media. They are stable to heat up to 165° and have good light fastness. These properties, which are similar to those of phthalo-cyanines, make them extremely valuable as pigments. Although they have low molecular weight (312), the intermolecular hydrogen bonding between the imino and carbonyl groups of quinacridone molecules in crystal lattice is responsible for their high stability and insolubility.[93a, b] The *N*-methylated derivatives of **71** are soluble in alcohol.[94] This compound and the unsubstituted **71** have similar spectral characteristics, thus indicating the ketonic structure for the quinacridone.[93a, b]

The quinacridones show polymorphism. The bluish red α form of linear quinacridone lacks stability in solvents. The commercial quinacridone pigments are in the stable, reddish violet β form and the red γ form. The β form was first marketed in 1958 as cinquasia violet R or monastral violet R (pigment violet 19, C.I. 46500) (du Pont). Cinquasia red B (bluer) and Red Y (yellower) have the γ form. The transparent red B and opaque red Y differ only in particle size,[93a, b] the red B having a particle size of 1 μ or bigger and the red Y having a particle size of 1 μ or smaller. Resination of the pigment with rosin salts increases the transparency.[95] The brand names of this pigment by some of the other firms are quindo violet RV-6902 (Allied Chemical Corporation), paliogen red BB (BASF), helio fast red E5B (FBY), hostaperm red E3B (Hoechst), and quinacridone violet. Other commercial derivatives of **71** include the 2,9-dimethyl derivative, quindo magenta RV-6803 (Allied

Chemical Corporation) or hostaperm pink E (Hoechst) (pigment red 122), indazin red 2B (Cassella) (pigment violet 30) and sandorin brilliant red 5BL (Sandoz) (pigment red 192). These pigments can be used in alkyl resin enamels, vinyl and acrylic lacquers, printing inks, vinyl and plastic products, and textile printing.

A. Linear trans-Quinacridone

The linear *trans*-quinacridone is prepared by the dehydrogenation of the dihydro (75) or the octahydro (76) derivatives, or the cyclization of 2,5-diarylaminoterephthalic acid or ester (77).

The preparation of 75 and its dehydrogenation were covered by du Pont's original patents[95] in 1955. The condensation of two molecules of diethyl succinate in sodium ethoxide gave diethyl succinoylsuccinate or 2,5-bis-ethoxycarbonylcyclohexane-1,4-dione (78). The condensation of 78 with aniline in the presence of aniline hydrochloride in Dowtherm A yields 79, which on cyclization by heating at 250° under nitrogen atmosphere gives 75. Oxidation of the dihydro derivative 75 to quinacridone 71 can be effected by heating with sodium *m*-nitrobenzenesulfonate in alcoholic solvents,[95] or by heating its benzene solution at 200° in an autoclave.[96] The enolic form (78b) can be condensed with aniline and cyclized to 75 in one step by heating it with aniline in polyphosphoric acid.[97] Direct conversion of 79 to 71 can be achieved by one-step cyclization and oxidation, using polyphosphoric acid and chloranil.[98] Some of the other methods of oxidation of 75 to 71 include air oxidation in the presence of tetramethylene sulfone as the solvent and

2-chloroanthraquinone as the catalyst.[99] Alternatively **79** may be first oxidized with chloranil and then cyclized.[100]

Condensation of 2,5-diaminoterephthalic acid with cyclohexanone in the presence of sulfuric acid gives **76**, which can be dehydrogenated under pyrolytic conditions.[101]

The reaction of excess aniline with 2,5-dichloroterephthalic acid or its ester[102] gives **77**. The cyclization of compound **77** to quinacridone **71** can be effected either by heating in sodium chloride-aluminium chloride melt[103] or by heating with aluminium chloride in trichlorobenzene.[104] A good method for cyclization[105] appears to be the heating of **77** (R = H) with polyphosphoric acid at 140–170°. The resulting quinacridone is generally in the pigmentary form. The cyclization, if effected by heating with concentrated sulfuric acid, yields a partially sulfonated product. Subsequent desulfonation can be effected by heating the cyclization product with dilute sulfuric acid at higher temperatures under pressure.[106] Some of the other cyclizing agents that have been suggested are boric acid in the presence of phthalic anhydride as the solvent or fused zinc chloride in o-dichlorobenzene,[107] benzoyl chloride,[108] and phosphorus oxychloride in nitrobenzene.[109] The condensation of two molecules of anthranilic acid with p-dibromobenzene and cyclization[93] yields the angular compound **73**. However, 6,13-dimethyl derivative of linear trans-quinacridone can be obtained from anthranilic acid and 2,5-dibromo-p-xylene.[110]

B. Quinacridonequinone

The quinacridone quinone **80** is the oldest-known linear quinacridone derivative. This brownish yellow pigment has high thermal stability, but its fastness to light is comparative less. The light fastness, however, can be improved by incorporation of other suitable chemicals.[93] Derivatives of **80** proved to be valueless as vat dyes. The quinone **80** was prepared in 1918 by condensing anthranilic acid with benzoquinone and cyclization.[91] In this oxidative condensation, part of the benzoquinone is reduced to hydroquinone. Other reagents, such as sodium chlorate and vanadium pentoxide, can be used in place of an excess of benzoquinone.[111] The condensation product (**81**) can be cyclized by treatment with polyphosphoric acid,[112] sulfuric acid,[113] or thionyl chloride in nitrobenzene.[114] Vat dyes similar to **81** such as helindon yellow CG (vat yellow 5, C.I. 56005), have been prepared starting from *p*-chloroaniline and benzoquinone.[11] A more attractive method, which can be used for the preparation of **80** (and substituted derivatives) is the cyclization of **82**. The reaction of aniline with 2,5-dichloro-3,6-*bis*-ethoxycarbonyl-1,4-benzoquinone[115] or 2,5-dihydroxyterephthalic acid in the presence of an oxidizing agent (sodium chlorate and ammonium vanadate)[116] gives **82b**. The chlorination of **78b**, followed by condensation with aniline,[117] also gives **82b**. The cyclization of **82a** or **82b** can be effected by heating with sulfuric acid. Thermal cyclization of **82b** by heating in Dowtherm A at 230–270° is also known. If the heating is done at a lower temperature (below 200°), formation of the dioxazine derivative **83** is facilitated.

To avoid this, the benzoquinone derivative **82** can be reduced with sodium dithionite to the corresponding hydroquinone, which can then be cyclized by heating with chloronaphthalene[118] at 250°. The 6,13-dihydroxyquinacridone (leuco derivative), thus obtained, can then be oxidized with sodium *m*-nitrobenzenesulfonate. By this process, the quinacridonequinone **80** is exclusively formed.

Quinacridone (**71**) and its dihydro derivative (**75**), on oxidation with chromic acid,[119] give a good yield of quinacridonequinone (**80**). Conversely, it is also possible to reduce the latter to quinacridone (**71**) by heating it with zinc dust in a fused mixture of aluminium chloride, sodium chloride, and potassium chloride.[120] The reduction of **80** with copper powder in sulfuric acid gives the violet-red-colored 6-hydroxy derivative and the red-violet 6,13-dihydroxyquinacridone.[121] The quinonoid carbonyl and the >NH groups of **80** are well situated to form metal chelates; such chelates with copper, zinc, and nickel have been reported.[122]

C. Other Quinacridone Derivatives

Several quinacridone derivatives have been prepared, varying in shade from yellowish red to violet. Quindo magenta RV-6803 (pigment red 122) is probably 2,9-dimethylquinacridone, obtained by using *p*-toluidine in place of aniline in the preparation of **71**.[93, 123-125] The chloro derivatives of quinacridone are orange and red colored. The 2,9-dichloroquinacridone has also been found to exist in α, β, and γ forms.[126] The 4,11-dichloroquinacridone is a reddish orange pigment.[127] The red 6,13-dichloroquinacridone is prepared by the action of phosphorus pentachloride[128] on the 6,13-dihydroxy derivative of **71**. Another route to **71** is from 2,5-dianilino-3,6-dichloroterephthalic acid.[129] Yellowish red octa- and decachloro-quinacridones have been prepared by the direct chlorination of quinacridone with sulfur monochloride in the presence of anhydrous aluminium chloride.[130] The 3,10-trifluoromethyl derivative of **71**, obtained by using *m*-trifluoromethylaniline, is purple.[131] The 2,9-difluoroquinacridone is red-violet.[132] The stable β and γ forms of 2,9-dimethoxyquinacridone are reddish violet.[133] Several other similarly substituted quinacridones have also been reported. The aluminium salt of quinacridone-2,9-disulfonic acid, prepared by using *p*-aminobenzenesulfonamide in place of aniline for condensation with **78**, can be mixed with the β form of **71** (15:85) to give a pigment that has superior tinctorial and rheological properties.[134] The derivative of **71** having phenylazo substituents at the 2,9-positions is a red pigment,[135] whereas that with phenylthio substituents is violet.[136]

D. Pigmentary Forms

As mentioned earlier, quinacridones show polymorphism. The crude quinacridones, obtained by the different processes described above, cannot be directly used as pigments. They usually exist in the solvent unstable α form. Special pigmentary conditioning methods have to be used to convert the crude pigment into the stable β and γ forms, which are valuable as pigments. The bluish red α form of **71** is usually unaffected if the product is subjected to ball milling with sodium chloride or dissolved in concentrated sulfuric acid at low temperature and reprecipitated.[95] However, if the α form of **71** is ball-milled in the presence of traces of inert solvents such as xylene and o-dichlorobenzene, it is converted into the β form. The use of a small quantity of dimethylformamide during ball milling converts the pigment into the γ form.[95] The quinacridones obtained by cyclization of the relevant intermediate with polyphosphoric acid generally have the desired pigmentary properties.[105] If crude quinacridone is dissolved in polyphosphoric acid and is then reprecipitated at 45° by pouring the polyphosphoric acid solution *rapidly* into ethanol, it is converted into the β form; *slow* addition gives the γ form.[137] This process also gives the quinacridones in smaller particle size. The γ form can also be obtained by heating the crude quinacridone in dimethylformamide,[95] quinoline,[138] ethanol in an autoclave,[139] or 2-pyrrolidone.[140] If quinacridone (**71**) is treated with N-methylpyrrolidone at 200°[141] or dimethylsulfoxide at 150°,[142] a yellower and brighter modification of the γ form is obtained. Sublimation of **71** at 405–425° under high vacuum has been reported to give the red δ form, a fourth variety.[143] The *cis*-quinacridone (**72**) also exists in α, β, and γ forms, and the latter two can be used as yellow pigments.[144]

E. Solid Solutions

Quinacridones have another interesting property. A mixture of two different kinds of quinacridones, when refluxed in dimethylformamide or ball-milled with salt and then treated with cold dimethylformamide, form solid solutions.[145] These solid solutions have their own characteristic X-ray diffraction patterns; their shades are different from those of the individual quinacridone components. The solid solutions are usually brighter and possess higher light fastness. The solid solution between **71** and its 2,9-dichloro derivative is scarlet, whereas those of **71** and the quinone **80** are maroon. It is also possible to prepare solid solutions between two different forms of the same quinacridone derivative.[146]

References

1. R. M. Acheson, *Acridines, The Chemistry of Heterocyclic Compounds*, Vol. 9, Interscience, New York, 1956.
2. K. Venkataraman, *Chemistry of Synthetic Dyes*, Vol. 2, Academic Press, New York, 1952.

3. H. A. Lubs, *The Chemistry of Synthetic Dyes and Pigments*, Reinhold, New York, 1955.
4. A. Albert, *The Acridines*, Arnold, London, 1966.
5. F. M. Rowe, *The Development of the Chemistry of Commercial Synthetic Dyes* (1956-1938), Institute of Chemistry, London, 1938.
6. H. E. Fierz-David, *Kuenstliche Organische Farbstoffe*, Ergangsband, Springer, Berlin, 1935.
7. J. T. Hewitt, *Dyestuffs Derived from Pyridine, Quinoline, Acridine and Xanthene*, Longmans, Green, London, 1922.
8. J. F. Thorpe and R. P. Linstead, *The Synthetic Dyestuffs*, Griffin, London, 1933.
9. P. L. Macdonald and A. V. Robertson, *Aust. J. Chem.*, **19**, 275 (1966).
10. N. R. Ayyangar and B. D. Tilak in *Chemistry of Synthetic Dyes*, Vol. 4, K. Venkataraman, ed., Academic Press, New York, 1971.
11. FIAT, 1313 II.
12. J. Wegmann, *Melliand Textilber.*, **39**, 408 (1958).
13. (a) A. Albert, *J. Chem. Soc.*, 121,484 (1941); (b) Poulene Freres, British Patent 137,214.
14. BIOS, 959.
15. BASF, U.S. Patent 546,177.
16. I.C.I., U.S. Patent 2,794,797; British Patent 785,038.
17. I.C.I., British Patent 760,595.
18. E. I. du Pont de Nemours and Co., British Patent 432,360; U.S. Patent 2,005,303; French Patent 765,001.
19. Inter Chemical Corporation, U.S.Patent 3,188,164.
20. J. R. Geigy A. G., British Patent 1,127,721; French Patent 1,509,386.
21. CIBA, German Patent 1,816,990.
22. Nippon Kayaku Co., Japanese Patent 14,706 (1969).
23. F. Hoechst, British Patent 1,130,283.
24. CIBA, German Patent 1,816,990.
25. American Cyanamid Co., U.S. Patent 2,725,375.
26. (a) F. Ullmann and P. Ochsner, *Justus Liebigs Ann. Chem.*, **381**, 1 (1911); (b) F. Ullmann, German Patent 221 853; U.S. Patent 961 047; BASF, German Patent 237,546.
27. MLB, German Patent 243,586; I. G. Farbenindustrie, German Patent 531,013.
28. F. Ullmann and M. Sone, *Justus Liebigs Ann. Chem.*, **380**, 336 (1911); BASF, German Patent 237,236-7; BASF, US Patent 1,011,068.
29. BASF, German Patent 248,170; General Aniline Works, U.S. Patent 2,097,112; E. I. du Pont de Nemours and Co., U.S. Patent 2,100,532.
30. F. Ullmann, *Chem. Ber.*, **43**, 536 (1910); German Patent 237,236; F. Ullmann and I. C. Dasgupta, *Chem. Ber.*, **47**, 553 (1914).
31. BASF, German Patent 246,966 (1912); Ullmann and Dootson, *Chem. Ber.*, **51**, 9 (1918).
32. BASF, German Patent 229,394; 272,297; I. G. Farbenindustrie, German Patent 534,934.
33. BIOS, 987.
34. W. Bradley and H. Kaiwar, *J. Chem. Soc.*, 2859 (1960).
35. FIAT, 1313, III.
36. BASF, German Patent 221,853; 237,236.
37. BIOS, 1088.
38. BASF, German Patent 237,236-7; 242,063; 248,170.

39. I.C.I. British Patent 691,118; U.S. Patent 2,689,247.
40. CIBA, British Patent 961,690.
41. CIBA, German Patent 1,185,747 (1965).
42. I. G. Farbenindustrie, German Patent 534,934.
43. General Aniline Works, U.S. Patent 1,785,801.
44. W. Kern, U.S. Patent 2,189,503; T. Holbro, *J. Appl. Chem.*, **3,** 1 (1953).
45. J. Haase and J. Filipi, Czechoslovakian Patent 98,220.
46. BIOS, 1493; BIOS, 987; BASF, German Patent 237,236.
47. BASF, German Patent 237,236-7.
48. BIOS, 1493; E. I. du Pont de Nemours and Co., U.S. Patent 2,726,242.
49. B. S. Joshi, N. Parkash, and K. Venkataraman, *J. Sci. Ind. Res.*, (India) **14B,** 325 (1955).
50. B. S. Joshi, B. D. Tilak, and K. Venkataraman, *Proc. Indian Acad. Sci.*, **32,** 201 (1950).
51. BASF, German Patent 287,614; I. G. Farbenindustrie, German Patent 531,013.
52. Mitsui Chemical Industry, Japanese Patent 25,847 (1963).
53. BIOS, 1493; E. I. du Pont de Nemours and Co., U.S. Patent 2,061,186.
54. Cassella, German Patent 1,135,415; 1,171,101; British Patent 919,270; F. Bayer, British Patent 775,802.
55. Ya. E. Berezin and V. Vanifat'ev, Russian Patent 183,857.
56. Sandoz, French Patent 1,493,903 (1967).
57. Sandoz, Belgian Patent 658,475 (1965).
58. I.C.I., British Patent 944,513.
59. I.C.I., British Patent 1,127,704 (1968).
60. Toms River Corp., U.S. Patent 3,299,071; British Patent 1,027,576; Cassella, German Patent 1 228 355.
61. I.C.I., British Patent 944,722.
62. BASF, German Patent 1,112,800; British Patent 892,402.
63. BIOS, 1493; BIOS, Misc. 55.
64. CIBA, British Patent 1,027,567; Belgian Patent 618,825.
65. CIBA, Belgian Patent 635,078 (1964).
66. Cassella, British Patent 767,004; German Patent 955,173; German Patent 1,081,584; F. Hoechst, German Patent 955,084; F. Bayer, German Patent 1,056,306; I.C.I. British Patent 737,586.
67. CIBA, Belgian Patent 668,789.
68. CIBA, Swiss Patent 419,399 (1967).
69. BASF, British Patent 829,699; F. Bayer, British Patent 993,257; CIBA, British Patent 915,266; F. Hoechst, British Patent 837,298.
70. CIBA, Belgian Patent 632,847.
71. BIOS 1088; BIOS, Misc. 20.
72. BIOS 1493; Berliner, U.S. Patent 1,994,033 (1935).
73. F. Bayer, U.S. Patent 3,167,557 (1965).
74. I.C.I. U.S. Patent 2,414,155; CIBA, Swiss Patent 298,380-8; BASF French Patent 1,371,077; Cassella, German Patent 1,084,404.
75. BASF, German Patent 212,471; U.S. Patent 995,936. I. G. Farbenindustrie, German Patent 499,352; 504,016; Hironaka and Tono, Japanese Patent 172,744; CIBA, British Patent 613,836; General Aniline, U.S. Patent 2,392,794; American Cyanamid Co., U.S. Patent 2,483,238; T. Maki and T. Mine, *J. Soc. Chem. Ind. Jap.*, **51,** 13 (1948); T. Maki and A. Kikuti, *J. Soc. Chem. Ind. Jap.*, **42,** 316B (1939).
76. I. G. Farbenindustrie, British Patent 337,741; U.S. Patent 184,569; German Patent 517,442.

77. BASF, British Patent 1,016,665 (1965).
78. E. I. Dupont de Nemours and Co., U.S. Patent 3,254,935 (1966).
79. F. Bayer, U.S. Patent 3,134,781.
80. American Cyanamid Co., U.S. Patent 2,993,901; 3,004,029; 3,027,369; 3,027,373-4; 3,027,376-7; 3,030,369.
81. American Cyanamid Co., U.S. Patent 3,027,375 (1962).
82. Cassella, German Patent 1,066,304.
83. Cassella, German Patent 1,171,441.
84. Cassella, German Patent 1,197,182.
85. F. Hoechst, British Patent 1,036,273; 1,079,109.
86. I.C.I., U.S. Patent 3,324,132 (1967).
87. A. M. Lukin and P. M. Aronowich, *J. Gen. Chem. (USSR)*, **19**, 358 (1949).
88. BIOS 1493; I. G. Farbenindustrie, German Patent 468,986.
89. H. Liebermann, *Justus Liebigs Ann. Chem.*, **518**, 245 (1935).
90. A. Eckert and F. Seidel, *J. Prakt. Chem.*, **102**, (2), 354 (1921).
91. W. S. Lesnianski, *Chem. Ber.*, **51**, 701 (1918); *Rocz. Chem.*, **6**, 881 (1926).
92. S. Neimentowski, *Chem. Ber.*, **29**, 76 (1896).
93. (a) J. Lenoir, in *Chemistry of Synthetic Dyes*, Vol. V. K. Venkataraman. ed.. Academic Press, New York, 1971; (b) *Peintures, Pigments, Vernis*, **39**, 545 (1963).
94. Y. Nagai, H. Nishi, and K. Morikubo, *Kogyo Kagaku Zasshi*, **70**, 2199 (1967).
95. E. I. du Pont de Nemours and Co., U.S. Patent 2,821,529-30; 2,844,484-5.
96. Tekkosha Co., British Patent 1,106,397.
97. E. I. du Pont de Nemours and Co., French Patent 1,264,480.
98. I.C.I., British Patent 894,610.
99. E. I. du Pont de Nemours and Co., U.S. Patent 3,475,436; British Patent 1,091,027; French Patent 1,496,960; Cassella, French Patent 1,328,160.
100. I.C.I. British Patent 868,361.
101. E. I. du Pont de Nemours and Co., U.S. Patent 3,133,071.
102. E. I. du Pont de Nemours and Co., French Patent 1,264,157; Sandoz, French Patent 1,233,785.
103. CIBA, French Patent 1,277,183.
104. Allied Chemical Corp., U.S. Patent 3,020,279; F. Hoechst, German Patent 1,127,019.
105. Allied Chemical Corp., French Patent 1,255,770; U.S. Patent 3,257,405; 3.342.823; CIBA, French Patent 1,226,825; Cassella, French Patent 1,258,551; I.C.I. British Patent 868,360; F. Hoechst, German Patent 1,184,881; German Patent 1,199,906; Sandoz, Swiss Patent 413,181; Swiss Patent 430,915; 431,767.
106. E. I. du Pont de Nemours and Co., U.S. Patent 3,261,836; CIBA, British Patent 896,803.
107. Toyo Ink Manufacturing Co., Japanese Patent 5414 (1967).
108. CIBA, Swiss Patent 392,737; BASF, French Patent 1,295,839.
109. Sandoz, Swiss Patent 404,034; 419,393; 419,396.
110. F. Bayer, German Patent 1,136,040.
111. BASF, German Patent 1,254,269.
112. BASF, Belgian Patent 627,375.
113. E. I. du Pont de Nemours and Co., U.S. Patent 3,185,694.
114. BASF, German Patent 1,254,269.
115. BASF, German Patent 1,140,300.
116. BASF, German Patent 1,195,425.
117. E. I. du Pont de Nemours and Co., U.S. Patent 3,124,582 (1964).

118. J. R. Geigy A.G., Swiss Patent 419,395; A. Pugin and J. Von der Crone, *Chimia*, **19,** 242 (1965).
119. E. I. du Pont de Nemours and Co., U.S. Patent 3,251,845.
120. Y. Nagai, Japanese Patent 21,007 (1967); BASF, German Patent 1,178,159; 1,184,440; 1,187,755.
121. BASF, British Patent 1,051,450; Belgian Patent 609,423; French Patent 1,411,752.
122. E. I. du Pont de Nemours and Co., U.S. Patent 3,121,718.
123. Dainippon Ink and Chemical Industry, Japanese Patent 22,419 (1969).
124. Sandoz, Swiss Patent 413,181.
125. American Cyanamid Co., British Patent 1,030,216.
126. BASF, British Patent 923,069; Sandoz; Swiss Patent 405,560.
127. E. I. du Pont de Nemours and Co., U.S. Patent 3,301,856.
128. Y. Nagai, Japanese Patent 9274 (1964); Y. Nagai et al., *Kogyo Kagaku Zasshi*, **71,** 386 (1968).
129. Sandoz, French Patent 1,233,785.
130. E. I. du Pont de Nemours and Co., U.S. Patent 3,272,822.
131. Sandoz, Swiss Patent 408,944; Dainichiseika Color Chemicals Manufacturing, Japanese Patent 22,417 (1969).
132. N. Ishikawa and T. Tanable, *Kogyo Kagaku Zasshi*, **72,** 1146 (1969).
133. E. I. du Pont de Nemours and Co., U.S. Patent 3,317,539.
134. E. I. du Pont de Nemours and Co., U.S. Patent 3,386,843.
135. BASF, German Patent 1,236,696.
136. ACNA, Italian Patent 703,319.
137. E. I. du Pont de Nemours and Co., U.S. Patent 3,265,699.
138. F. Hoechst, German Patent 1,268,586.
139. ACNA, French Patent 1,379,970.
140. Tokkosha, French Patent 1,489,908; 1,480,493; British Patent 1,110,997.
141. BASF, German Patent 1,183,884.
142. ACNA, French Patent 1,397,723.
143. Eastman Kodak, U.S. Patent 3,272,821; French Patent 1,352,663.
144. BASF, Belgian Patent 625,666.
145. E. I. du Pont de Nemours and Co., U.S. Patent 3,160,510; 3,298,847; I.C.I., British Patent 896,916.
146. CIBA, French Patent 1,395,204.

CHAPTER IX

Chemiluminescent Reactions of Acridines

FRANK McCAPRA

The Chemical Laboratory, University of Sussex,
Brighton, England

1. Introduction

Although bioluminescence[1] represents the oldest-known class of chemiluminescence in organic compounds, the realization of the source of the light is of very recent origin. The earliest example of luminescence from a solution of a particular, known organic substance is that of lophine,[2] described in 1877. Luminol,[3] discovered in 1928 and lucigenin (**1**), whose luminescence was first observed in 1935, can be considered the other "classical" examples of organic chemiluminescence. Although the principles of chemiluminescent mechanism are described in detail elsewhere,[4, 5] a brief introduction may be helpful.

Chemiluminescence occurs when the product of an exothermic reaction is formed in an electronically excited state. Direct light emission from this product is observed if it is fluorescent; if other fluorescent products or impurities are present, then energy transfer to these, with subsequent light emission from the acceptor molecules, may occur. Such energy transfer can take place by any of the processes known from photochemical studies.[6]

The first requirement for the assignment of mechanism is that the excited product should be identified. If this is the primary product of the reaction, then exact correspondence of the chemiluminescence emission spectrum and the fluorescence spectrum of the pure product must be obtained. It is a fair assumption that in an oxygenated solution at room temperature, the only state to radiate in either case will be a short-lived singlet in thermal equilibrium with its surroundings. In such circumstances phosphorescence can be excluded and chemiluminescence considered as a chemically generated fluorescence. In addition, the reaction suspected of yielding the excitation must produce in the excitation step energy at least as great as that of the spectroscopic energy of the radiating molecule, as determined by the lowest observed wavelength of emission. (This O-O level is not easily identified in the broad band spectra typical of most organic molecules.)

Early workers usually considered that a radical species must be generated as the emitter, probably owing to too close an identification of excited states with molecules of a radical nature. Some modern investigations have shown that radical reactions, as exemplified by electron transfer sequences, do indeed lead to excited states.[7–9] If we consider that acridines are readily reduced (to the radical anion), then this is in principle a likely mechanism. In outline, a reduced polynuclear aromatic molecule ($Ar^{\cdot-}$) is oxidized by the removal of a bonding electron to form the first excited singlet state that radiates in the usual way.

$$Ar + e \rightarrow Ar^{\cdot-}$$
$$Ar^{\cdot-} + ox \rightarrow Ar^* + h\nu + red$$
$$Ar^* \rightarrow Ar + h\nu$$

where ox and red are the oxidized and reduced forms of a suitable oxidant. Until recently, chemical reactions of a more usual kind, i.e., those involving bond making and breaking in a conventional transition state, were not thought to be a major pathway to excitation. However, there is no doubt that the decomposition of peroxides is a powerful and important chemiluminescence mechanism.[4, 10, 11] Although other routes are possible (particularly that involving singlet oxygen formation), these two provide a suitable basis for the discussion of acridine chemiluminescence. The efficiency of a chemiluminescent process is measured by the quantum yield: a yield of 1.0 (or 100%) occurring when every molecule of a substrate reacting produces one photon. Yields of 1% are normally considered high, although still higher yields (20–30%) are possible.

2. Biacridinium Salts—Lucigenin

Although it is a difficult compound to investigate, lucigenin (**1**) is the archetypal chemiluminescent acridine derivative. Standing in relative isolation (with lophine and luminol) for so long, it attracted a fair amount of experimental effort but with many contradictory results. Nevertheless, it is proper that we should examine it first, with a reminder that more recent work on related, but simpler, compounds will assist considerably in understanding the basic mechanism. These compounds are considered later in this chapter.

The reaction is very sensitive to medium effects but in essence requires a base and hydrogen peroxide.[12] It proceeds best in hydroxylic solvents and is rather sensitive to the structure of the alcohol used[13, 14] (see the discussion of its use as an analytical reagent).[13, 14]

The nature of the base is also relevant, some amines being considered more effective.[13,15] However, it is probable that these effects are the result of changes in the medium and rate caused by the added compounds (e.g., a change to general base catalysis; the quantum yield is probably not affected.[13]) In fact, the influence of a variety of additives[13, 16a, b, 17] has been noted but seldom investigated in a way that would lead to any conclusion about the involvement of the additive in the mechanism. The reaction is complex, and the major product, 10-methylacridanone, is not formed quantitatively. There is some indication that the initial intensity of the reaction has a linear relationship to the hydrogen peroxide concentration, but that the order in base is not simple.[15, 17] In spite of the fairly large amount of work on the compound, there is no thorough and convincing examination of the mechanism of the reaction in its simplest form. However, the growth in understanding of organic chemiluminescence and recent work allow a plausible mechanism to be written.

A final difficulty that has caused confusion lies in the identification of the primary excited product. The reaction, as usually performed with alkaline hydrogen peroxide in aqueous alcohol, displays a green emission. Since lucigenin itself is green fluorescent (λ_{max}, ca. 500 nm) it was suggested[18, 19] that lucigenin is re-formed (e.g., as a biradical) with subsequent emission. A related view has been expressed recently.[20] However, there is now conclusive evidence[21—23] that the primary emitting molecule is 10-methylacridanone (λ_{max}, 442 nm in ethanol); since this is the most energetic emission observed, then energy transfer to lucigenin or a degradation product is occurring. Such transfer to added fluorescent molecules is easily demonstrated.[19, 24]

It is then necessary to provide a mechanism that produces 10-methylacridanone directly in an excited state. The details of this process are not entirely clear, but it must occur by a route similar to the scheme on page 618. Although a radical mechanism (involving a radical derived from lucigenin (1) or its pseudobase 2 is possible, recent extensive investigations[23,25a, b] have shown that such intermediates do not lead to efficient chemiluminescence. For example, Janzen and co-workers[25a, b] have identified DBA$^{\cdot+}$ 3 and 4 by esr, formed by the addition of hydroxyl ion to lucigenin. However, the subsequent reaction with oxygen produced a quantum yield 10^{-4} times that of the

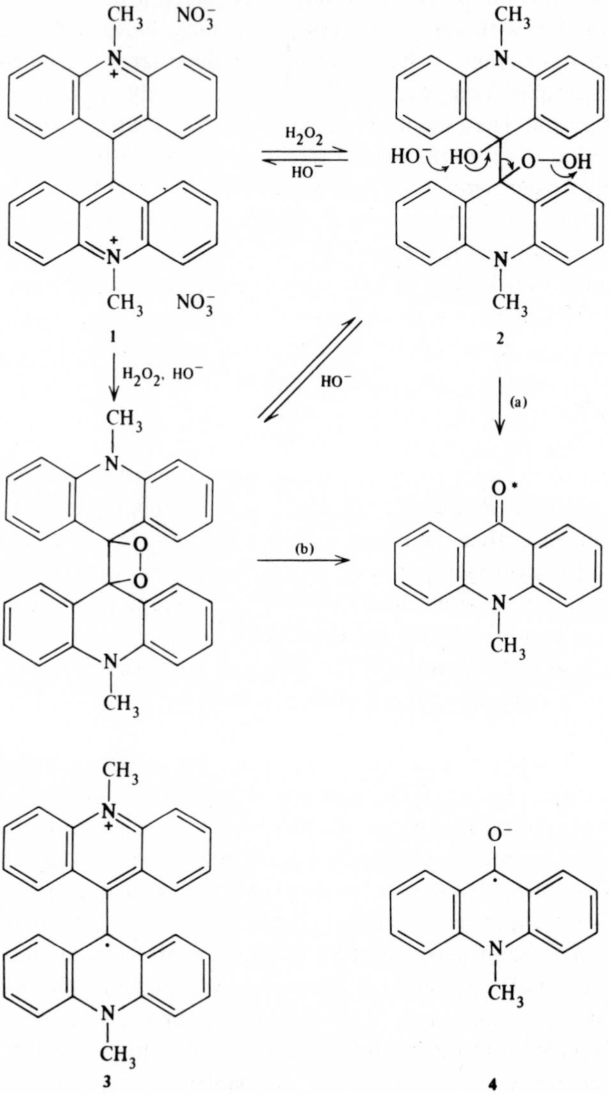

CH_3 NO_3^-

CH_3

$\dfrac{H_2O_2}{HO^-}$

CH_3 NO_3^-

1

HO^- HO $O{-}OH$

CH_3

2

$\Big\downarrow H_2O_2, HO^-$

HO^-

(a)

CH_3

O

O

CH_3

(b) →

O•

CH_3

CH_3

CH_3

3

O^-

CH_3

4

standard reaction using hydrogen peroxide. Moreover, no radicals were detectable during the latter reaction. Weak chemiluminescence, also requiring oxygen, was also observed by adding cyanide ion to lucigenin. Hercules and co-workers repeated earlier work which seemed to indicate that electrogenerated chemiluminescence was possible from lucigenin.[16] However, it is apparent[23] that direct reduction in the absence of oxygen, at a platinum electrode, merely reduces lucigenin to 10,10′-dimethyl-$\Delta^{9,9'}$-biacridan (6) and that a mercury electrode, at potentials more negative than -0.15 V, reduces oxygen, which then reacts with lucigenin, giving light. At pH 7 in water, in fact, the reduction of oxygen gives both the reagents of the classical chemiluninescence.

$$O_2 + 2H_2O + 2e \rightarrow H_2O_2 + 2HO^-$$

Using nonaqueous solvents, the superoxide ion O_2^- is produced from oxygen, and lucigenin is also reduced to 10,10′-dimethyl-$\Delta^{9,9'}$-biacridan. With the solvents ethanol, dimethylformamide, dimethylsulfoxide, and acetonitrile, it is possible to have varying solubilities and fluorescence efficiencies for the two principal energy acceptors. These are lucigenin itself (λ_{max}, 500 nm) and 10,10′-dimethyl-$\Delta^{9,9'}$-biacridan (λ_{max}, 510 nm). Because of this flexibility, it can be shown that 10-methylacridanone is the primary emitter, and at concentrations of acceptor greater than $10^{-4}M$, singlet to singlet energy transfer of the Förster type [6, 26] occurs. Thus the arguments in favor of nucleophilic attack of hydrogen peroxide or its conjugate base as a principal step in all cases of lucigenin luminescence are reinforced.

Various other mechanisms[18–20, 27a–d] have been advanced, some accompanied by considerable experimental work. However, almost all can be excluded on the grounds that the excitation step either does not form 10-methylacridanone, or more important, does not release the large amount of energy required to populate the observed excited state.

To date, it has not been possible to distinguish between route (a) or (b); but insofar as alkyl peroxides are considerably less efficient (if effective at all), then route (b) is implied. It is a relatively efficient compound; a quantum yield of 1.6% obtained [21, 22] on the basis of the yield of 10-methylacridanone is very variable and the quantum yield based on lucigenin consumed may be only 0.08%.

A variety of lucigenin analogues have also been prepared.[28–31] By alkylating the potassium salt of 9-acridanone and treating the 10-alkylacridanone with zinc in methanolic hydrochloric acid, the 10,10′-dialkyl-$\Delta^{9,9'}$-biacridans are formed.[30] Oxidation with nitric acid gives the dialkyl, dipropyl, and dibutyl analogues of lucigenin. Some N-aryl derivatives have also been prepared via the 10-arylacridanone, made by the condensation of an N-arylanthranilic acid with cyclohexanone. They are all chemiluminescent, and it is not expected that important differences in behavior will be observed. However, the electron-withdrawing effect of the aryl group may have some influence on the efficiency of the light reaction.[31] Substituents on the nucleus have a more pronounced effect,[29a–e] since the reactions may now take a different course. Changes in color of the emission are also observed, but no detailed work has been done.

3. Other Acridines

10,10'-Dimethyl-9,9'-biacridan-9,9'-oxide (**5**) and 10,10'-dimethyl-$\Delta^{9,9'}$-biacridan (**6**) are said to be spontaneously chemiluminescent on being dissolved in pyridine or methanol.[32] However, hydrogen peroxide is required for both reactions.[32, 33] The luminescence is weak, and the former compound requires radical initiation to produce a visible intensity. A direct approach to the problem, with implications for lucigenin chemiluminescence, was made by oxidizing the biacridan **6** by singlet oxygen. The reagent can be the triphenylphosphite-ozone complex, bromine and alkaline hydrogen peroxide, or singlet oxygen gas generated by a radio-frequency discharge. It is an efficient reaction, and it can be shown that water is not involved. The emitter is undoubtedly 10-methylacridanone; since an increase in its concentration does not increase the light yield, it is probably not excited by energy transfer. Thus we can assume that the emitter is excited during its formation and singlet oxygen may be expected to be added, as shown in the following scheme.

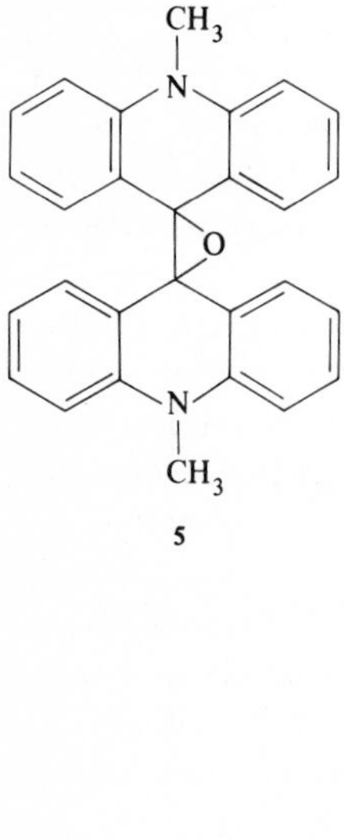

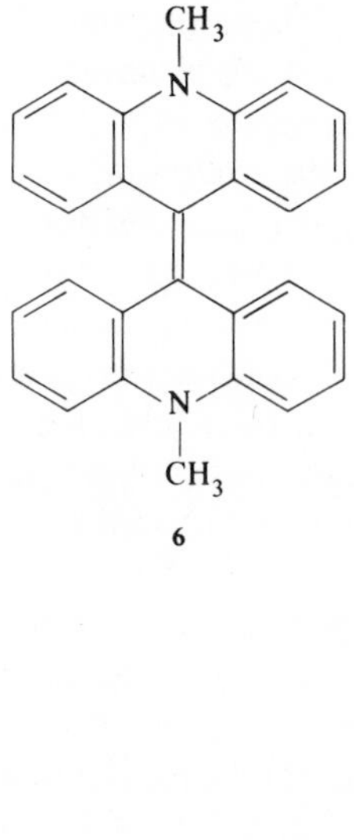

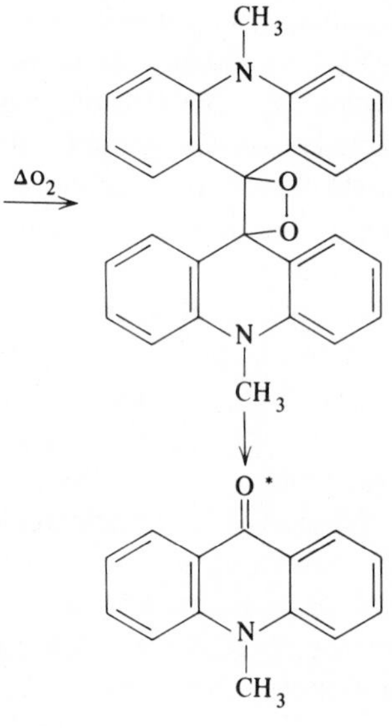

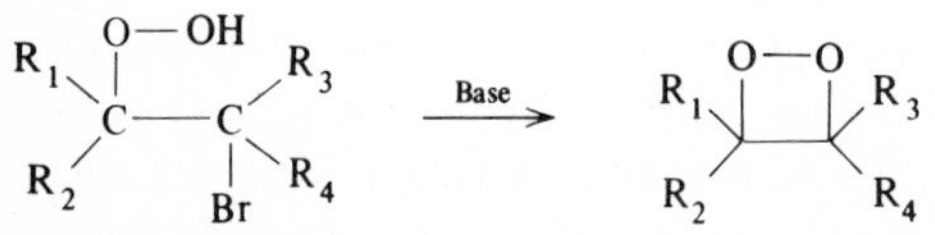

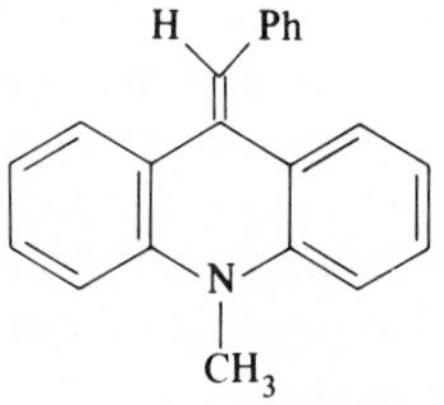

HO—O CHBrPh

Ph O
H   O

O *

HO⁻ →

CH₃

CH₃

CH₃

9

+ PhCHO

Recently,[34a–c] the addition of singlet oxygen to simple olefins has resulted in the isolation of less complex dioxetanes (alkyl-substituted dioxetanes can be made by treating a bromoperoxide such as 7 with a base[35]). The yields are very high and the dioxetanes surprisingly stable. In addition, almost quantitative population of the carbonyl product excited state occurs on decomposition.[34a–c] The state involved is not surprisingly the lowest triplet, which does not radiate efficiently in solution. The more complex (unisolated) dioxetanes, made by either of the methods above (see **6-9**), do chemiluminesce strongly since one product, 10-methylacridanone has a lowest radiative singlet state. Preliminary studies[36] suggest that the dioxetane derived from 6 has a half-life of 5 sec at $-70°C$, while for that derived from the benzylidene acridan **8**, the value is about 90 min at the same temperature. The treatment of the acridan **9** with a base also leads to very rapid efficient light emission. Sterically accelerated decomposition and an effect related to the catalysis involved in the chemiluminescent reaction of active oxalates may be involved.[10]

There is some dispute[37] as to whether ozonolysis of the $\Delta^{9,9'}$-biacridan **6** is particularly chemiluminescent. If the reaction is carried out quickly (by adding a saturated solution of ozone to a solution of **6**, then the quantum yield is about 5×10^{-7}. This is around 10^{-5} less than for the singlet oxygen reaction and may fairly be considered a negligible value. The yield of 10-methylacridanone is about the same in each case. If ozone is slowly passed into the solution of **6**, as a gas, then the quantum yield increases. A reasonable interpretation is that solvent ozone reactions occur (the reaction is markedly solvent-dependent[37]) involving radical species (not rare in ozonolysis), and, where unreacted **6** is still present, the already noted radical initiated chemiluminescence ensues.

4. Acridinium Salts

This series,[38] which resembles lucigenin in outline but which is free from most of the difficulties was first discovered in 1964. The nitrile **10** is the first example of a chemiluminescent compound made with the expectation of strong luminescence, and not found by accident.

The reaction is formulated as involving an intermediate four-numbered peroxide ring because the sequence of events includes the initial fast addition of peroxide to the acridinium nucleus. The (slow) addition of hydrogen peroxide to the nitrile group is unlikely, since the rate of reaction is not particularly sensitive to peroxide concentration. Moreover, the attack of hydroxide on the nitrile, besides being notoriously slow, is not a likely chemiluminescent route, since this must form the amide as an intermediate. The amide (**11**; R = CONH$_2$) is not chemiluminescent.[39] There is no question that 10-methylacridanone is the excited product, since correspondence of the chemiluminescence and fluorescence spectra is exact and it is virtually the only product. It is not difficult to construct other systems on this basis, with the added advantage that the dark route does not occur.

The acid chloride **12** was simultaneously investigated by two groups,[40, 41] and the Cyanamid workers favored a mechanism involving the peroxide **13**. The main observations in support of this are that the dilution of the initial solution of **12** in 90% hydrogen peroxide with water markedly increased the intensity of emission. Other peroxides, such as *t*-butyl peroxide and perlauric acid, also produced light but not as efficiently as hydrogen peroxide itself. Both groups commented on the failure of the peroxyacid **13** to give light.

The reaction is not quite as straightforward as this, however. The increase of emission with the addition of water is not simply a result of the addition of water to the acridinium nucleus, since the intensity increases proportionately with a volume increase from 0.01 to 3 l. Obviously, changes in pH are occurring, an interpretation supported by the rate increase caused by base. Furthermore, the *t*-butylperoxide reaction does not emit light on simple dilution with water.

A related, more stable series,[40, 42] with greater opportunity for experimental investigation, is provided by the phenyl esters **14**.

Again, the major product (90% yield) is 10-methylacridanone, the light emission from this series occurring with a quantum yield of about 2%. The most significant observations[42, 43] are:

1. The reaction is first order in base (over a limited pH range), first order in acridinium salt, and zero order in hydrogen peroxide (at pH values above 8).

2. Hydrogen peroxide is essential, *t*-butyl and methyl hydroperoxide being considerably less effective (the small amount of light emitted almost certainly being caused by traces of hydrogen peroxide).

3. The quantum yield is not significantly affected by substitution on the phenyl group, providing that the conjugate acid of the leaving group is stronger than hydrogen peroxide. The reaction constant (ρ) in aqueous ethanol is $+4.6$.

4. The order of addition of the reagents, base and hydrogen peroxide, is important. In either case the acridinium absorption disappears virtually instantaneously, with the appearance of an absorption characteristic of an acridan. If hydrogen peroxide is added after the base, then the fast bright reaction is replaced by a slow dim emission.

The peroxide **15** can, in fact, be isolated. The addition of a base then gives bright chemiluminescence. Alkyl esters (**11**; R = CO$_2$R) are considerably less efficient and can only just be considered chemiluminescent. A reasonable explanation is that the alkoxy group (p$K_a \sim 16$) is not easily expelled by the peroxide (p$K_a \sim 12$). Three routes to the products may be considered.

Route (a) certainly does not lead to light emission, since the peroxy acid can be made and shown to decompose quantitatively to 10-methylacridanone in a fast, dark, reaction. The tetrahedral intermediate in route (b) might be expected to decompose by the expulsion of phenoxide ion, thus merging with route (a). If this route does give rise to emission, then we might expect the quantum yield to fall with the increasing acidity of the leaving (phenoxide) group. This is not observed. In addition, the inefficient reaction with alkyl peroxides is hard to explain [there should be no difference in the final quantum yield between the compounds **15** (R = H) and **15** (R = alkyl)]. Finally, the value of ρ (+4.6) is much higher than that for the hydrolysis of phenyl esters (+2.1), a result probably owing to the need to shift the equilibrium in route (c) to the right, offsetting the developing strain in the four-membered ring.

It is not surprising that route (c) is most likely, given the high reactivity of peroxide anions toward carbonyl groups[44a, b] and the intramolecular nature of the reaction. The extra strain energy (20–30 kcal mole^{-1}) of the dioxetane may, in fact, be required to populate the excited state (about 70 kcal mole^{-1}). If this is so, then intermediate dioxetanes are implicated in all chemilumines- cent acridine reactions that produce 10-methylacridanone. It would be of interest to add substituents to the nucleus in order to lower the energy of the first excited singlet of the 9-acridanone, perhaps allowing open chain routes, such as (a) or (b), to populate the excited state efficiently. Route (b), in particular, expelling the resonance-stabilized carbonate grouping, is about 15 kcal more exothermic than route (a).

5. Acridan-9-carboxylic Esters

Reduction of the acridinium esters in acetic acid with zinc gives the corre- sponding acridan. These react in a strong base in polar aprotic solvents to give bright chemiluminescence.[36]

The quantum yield is higher, perhaps because of direct formation of the peroxide, excluding competition with other nucleophiles. These are among the brightest of all chemiluminescent organic compounds, ($\phi = 10\%$) and serve as useful models for bioluminescent processes. After the autoxidation, the sequence is identical to that of the acridinium salts.

Acridan itself, on oxidation with benzoyl peroxide, is weakly chemi-luminescent.[45] The quantum yield is very low (3.0×10^{-7}), and there must be doubts about the identification of the reaction responsible. However, the emission seen corresponds to the fluorescence of the acridinium cation, whose lowest excited singlet state is of significantly lower energy (58.6 kcal mole^{-1}) than that of acridine itself (67.9 kcal mole^{-1}). For this reason, the reaction is formulated as a two-electron oxidation as in the following scheme:

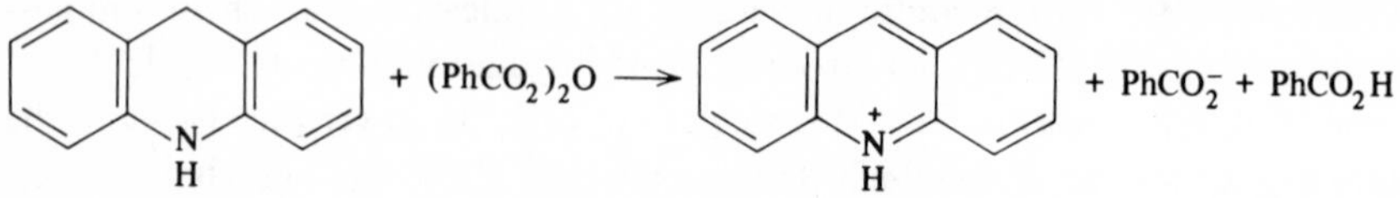

6. Acridines

Acridine-9-carboxylic acid derivatives[41] are also chemiluminescent, but in contrast to the behavior of acridinium salts, would not be expected to add either water or hydrogen peroxide readily at C_9 of the nucleus. Accordingly, the luminescence is weaker. No investigation of the mechanism has been made, but it should be related to that of the acridinium salts. The oxidation of 9-methylacridine (16) is chemiluminescent and may also be considered to react in a similar fashion.

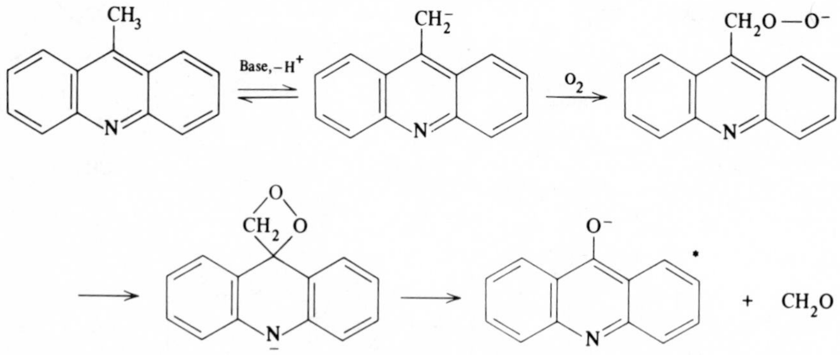

The acridine-9-carboxhydrazide (11; R = CONHNH$_2$) is chemiluminescent,[46] but it is not yet clear whether the reaction is to be related to hydrazide chemiluminescence (which is poorly understood) or whether it falls within the class of acridine luminescence already considered. It is probable that both are operating.[47]

7. Reaction of Lucigenin with Enzymes

The luminescence of lucigenin induced by xanthine oxidase has been investigated as a model for the bioluminescent luciferin-luciferase reaction.[21, 48-50] In the presence of the natural substrate hypoxanthine as reductant, lucigenin and oxygen are both reduced; but it is not clear whether the ensuing reactions, which lead to light emission, are related to the more usual oxidation by alkaline hydrogen peroxide. The bacterium *Serratia marcescens*[51] also induces light emission, but the interpretation is even less clear.

8. The Use of Lucigenin in Analysis

Since the light emission from a chemiluminescent reaction is easily measured by modern devices, it is not surprising that lucigenin has been used in a variety of analytical techniques. These uses are virtually confined to Eastern European countries, where most work has been done in recent times. Three main uses for lucigenin have been investigated: (1) in titration as an indicator, (2) in transition metal analysis, and (3) in semiquantitative analysis of the lower alcohols.

Erdey and co-workers[52] have examined lucigenin as a general-purpose chemiluminescent indicator, both alone[53] and with fluorescent materials as energy acceptors.[54] Other workers have also suggested their value in acid-base titrations in opaque solutions of natural and industrial materials.[55, 56a-c]

Lucigenin chemiluminescence is very sensitive to medium effects, as mentioned previously; the addition of alcohols shifts the pH of maximum emission to lower values.[57a-c] All the alcohols up to and including the butanols can be distinguished by their effect on the rate of reaction.

It is claimed[58a, b] that the error for detection of methanol in ethanol (equimolar solution) is less than 1%. It would seem that the effect of the alcohol is to increase the activity of the base, the higher alcohols being most effective. The nerve gas tabun also induces chemiluminescence (probably acting as a base). There is no obvious reason why this reagent should act in a specific manner, and it is therefore a little odd that oximes (known *in vivo* antagonists) should reverse the effect.[59]

Transition metals catalyze lucigenin chemiluminescence to some extent, although the reasons for this have never been sought. Several workers have devised techniques for the analysis of a variety of cations,[60] but there is no clear indication of the relative worth of the methods. Among the metals analyzed are cobalt,[61] osmium[62] (as the tetroxide),[12, 16] silver,[63] and palladium.[64] If we consider the extraordinary sensitivity of modern light detection devices, it would seem that further exploration of this technique of analysis would be useful.

References

1. (a) E. N. Harvey, *Bioluminescence*, Academic Press, New York, 1952; (b) J. W. Hastings, *Ann. Rev. Biochem.*, **37**, 597 (1968).
2. B. Radziszewski, *Chem. Ber.*, **10**, 70, 321 (1877).
3. H. O. Albrecht, *Z. Phys. Chem.*, **136**, 321 (1928).
4. F. McCapra, *Quart. Rev.*, *(London)*, **20**, 485 (1966).
5. K. D. Gunderman, Chemilumineszenz Organische Verbindungen, Springer-Verlag, Berlin, 1968.
6. F. Wilkinson, *Advan. Photochem.*, **3**, 241 (1964).
7. (a) E. A. Chandross and F. I. Sonntag, *J. Amer. Chem. Soc.*, **86**, 5350 (1964); (b) D. M. Hercules, *Science*, **145**, 808 (1964); (c) M. M. Rauhut, D. L. Maricle, G. W. Kennedy, and J. P. Mohns, American Cyanamid Co., "Chemiluminescent Materials," Tech. Rep. No. 5 (1964).
8. A. Zweig, *Advan. Photochem.*, **6**, 425 (1968).
9. D. M. Hercules, *Accounts Chem. Res.*, **2** 301 (1969).
10. M. M. Rauhut, *Accounts Chem. Res.*, **2**, 80 (1968).
11. F. McCapra, *Chem. Commun.*, 155 (1968).
12. K. Gleu and P. Petsch, *Angew Chem.*, **48**, 57 (1935).
13. K. Weber, *Z. Phys. Chem.*, **b50**, 100 (1941).
14. L. Erdey, J. Tackacs, and I. Burzas, *Acad. Sci. Acta Chim. Hung.*, **39**, 295 (1963).
15. O. Shales, *Chem. Ber.*, **72**, 1155 (1939).
16. (a) B. Tamamushi and H. Akiyama, *Trans. Faraday Soc.*, **35**, 491 (1939). (b) J.Matkovic and K. Weber, *Arch. Toxikol.*, **21**, 355 (1966); *Chem. Abstr.*, **65**, 10851 (1966).
17. K. Weber and W. Ochsenfeld, *Z. Phys. Chem.*, **51**, 63 (1942).
18. A. V. Kariakin, *Opt. Spectrosc.*, **7**, 75 (1959).
19. B. D. Ryzhikov, *Bull. Acad. Sci. URSS, Phys. Ser.*, **20**, 487 (1956).
20. K. Maeda and T. Hayashi, *Bull. Chem. Soc. Jap.*, **40**, 169 (1967).
21. J. R. Totter, *Photochem. Photobiol.*, **3**, 231 (1964).
22. (a) C. J. Spruit and A. Spruit van der Burg, in *The Luminescence of Biological Systems*, F. H. Johnson, ed., Washington, D.C., 1955; (b) A. Spruit van der Burg, *C. R. Acad. Sci.*, *Paris*, **69**, 1525 (1950); (c) B. Ya. Sveshnikov, *Izv. Akad. Nauk USSR, Ser. Fiz.* **9**, 341(1945).
23. K. D. Legg and D. M. Hercules, *J. Amer. Chem. Soc.*, **91**, 1902 (1969).
24. L. Erdey, *Acta Chim. Acta. Sci. Hung.*, **3**, 81 (1953).
25. (a) E. G. Janzen, J. B. Pickett, J. W. Happ, and W. DeAngelis, *J. Org. Chem.*, **35**, 88 (1970); (b) J. W. Happ and E. G. Janzen, *J. Org. Chem.*, **35**, 96 (1970).
26. T. Förster, *Z. Electrochem.*, **53**, 93 (1949).
27. (a) H. Kautsky and H. Kaiser, *Naturwissenschaften*, **31**, 505 (1943); (b) L. Erdey, *Acta Chim. Acad. Sci. Hung.*, **7**, 95 (1953); (c) J. R. Totter and G. Philbrook, *Photochem.*

Photobiol., **5,** 177 (1966); (d) J. R. Totter, in *Bioluminescence in Progress* p. 25; F. H. Johnson and Y. Haneda, eds., Princeton University Press, Princeton, N.J., 1965.

28. K. Gleu and A. Schubert, Chem. Ber., **73B,** 805 (1940).
29. (a) K. Gleu and R. Schaarschmidt, *Chem. Ber.*, **73B** 909 (1940); (b) K. Gleu and S. Nitzsche, *J. Prakt. Chem.*, **153,** 233 (1939); (c) Crow and Price, *Aust. J. Sci. Res.*, **A2,** 282 (1949).
30. K. Kormendy, *Acta Chim. Acad. Sci. Hung.*, **21,** 83 (1959); *Chem. Abstr.*, **54,** 18524 (1960).
31. A. Braun, A. Dorabialska, and W. Reimschuessel, *Rocz. Chem.*, **40,** 247 (1966); *Chem. Abstr.*, **65,** 1620 (1966). A. Chrzaszczewska, A. Braun, and M. Nowaczyk, *Soc. Sci. Lodziensis Acta Chim.*, **3,** 93 (1958); *Chem. Abstr.*, **53,** 13148 (1959).
32. A. M. Grigorovskii and A. A. Simeonov, *Zh. Obshch. Khim.*, **21,** 589 (1951).
33. F. McCapra and R. A. Hann, unpublished work.
34. (a) P. D. Bartlett and A. P. Schaap, *J. Amer. Chem. Soc.*, **92,** 3223 (1970); (b) S. Mazur and C. S. Foote, *J. Amer. Chem. Soc.*, **92,** 3225 (1970); (c) T. Wilson and A. P. Schaap, *J. Amer. Chem. Soc.*, **93,** 4126 (1971).
35. K. R. Kopecky and C. Mumford, *Can. J. Chem.*, **47,** 709 (1969).
36. F. McCapra and R. A. Hann, unpublished work.
37. E. G. Janzen, I. G. Lopp, and J. W. Happ, *Chem. Commun.*, 1140 (1970).
38. F. McCapra and D. G. Richardson, *Tetrahedron Lett.*, 3167 (1964).
39. F. McCapra and D. G. Richardson, unpublished work.
40. F. McCapra, D. G. Richardson, and Y. C. Chang, *Photochem. Photobiol.*, **4,** 1111 (1965).
41. M. M. Rauhut, D. Sheehan, R. A. Clarke, B. G. Roberts, and A. M. Semsel, *J. Org. Chem.*, **30,** 3587 (1965).
42. F. McCapra, *Pure and Appl. Chem.*, **24,** 611 (1970).
43. F. McCapra, D. G. Richardson and R. A. Hann, unpublished work.
41. (a) W. P. Jencks, *Catalysis in Chemistry and Enzymology*, McGraw-Hill, New York, 1968); (b) C. A. Bunton in *Peroxide Reaction Mechanisms*, J. D. Edwards, ed., Interscience, New York, 1962.
45. S. Steenken, *Photochem. Photobiol.*, **11,** 279 (1970).
46. E. H. White and D. F. Roswell, *Accounts Chem. Res.*, **3,** 54 (1970).
47. E. H. White and D. F. Roswell, Joint Conference of Chemical Institute of Canada and American Chemical Society, Toronto, Canada, May 25-29, 1970.
48. (a) J. R. Totter, V. J. Medina, and J. L. Scoseria, *J. Biol. Chem.*, **235,** 238 (1960); (b) J. R. Totter, J. L. Scoseria, and V. J. Medina, *Anal. Fac. Med. Montevideo*, **44,** 1467 (1959); *Chem. Abstr.*, **55,** 23609 (1961).
49. L. Greenlee, G. Fridovich, and P. Handler, *Biochemistry*, **1,** 779 (1962).
50. W. J. DeAngelis and J. R. Totter, *J. Biol. Chem.*, **239,** 1012 (1964).
51. W. S. Oleniacz, M. A. Pisano, and R. V. Isnolera, *Photochem. Photobiol.*, **6,** 613 (1967).
52. L. Erdey and I. Buzas, *Anal. Chem. Acta* **22,** 524 (1960).
53. L. Erdey and I. Buzas, *Magyar Tudomanyos Akad. Kem. Tudomanyok Osztalyanak Kozlemenyli*, **5,** 299 (1954); *Chem. Abstr.*, **49,** 9431 (1955).
54. L. Erdey, J. Takacs, and I. Buzas, *Acta. Chim. Acad. Sci. Hung.*, **39,** 295 (1964); *Chem. Abstr.*, **60,** 9878 (1964).
55. G. N. Kosheleva and E. Ya. Yarovenko, *Sb. Statei Uses. Nauchn. Issled. Inst. Khim. Reacktivov Osoba Chistykh Khim. Veshchestv*, **4,** 91 (1961); *Chem. Abstr.*, **57,** 4041 (1962).
56. (a) R. Parizek and V. Mouchka, *Chem. Listy*, **48,** 626 (1954). (b) E. Michalski and I. Adolf. *Lodz. Towarz. Nauk Wydzial III*, **2,** 33 (1956); *Chem. Abstr.*, **52,** 7930 (1958).

(c) F. B. Martinez, A. Bardinas, and A. P. Benza, *Inform. Quim. Anal.* (Madrid), **14**, 151 (1960); *Chem. Abstr.*, **55**, 15225 (1961).

57. (a) E. Michalski and M. Turowska, *Rocz. Chem.*, **30**, 985 (1956); *Chem. Abstr.*, **51**, 14468 (1957). (b) *Chem. Anal.* (*Warsaw*), **3**, 599 (1958); *Chem. Abstr.*, **54**, 3047 (1960). (c) *Chem. Anal.* (*Warsaw*), **4**, 651 (1959), *Chem. Abstr.* **54**, 5345 (1960).

58. (a) M. Turowska, *Chem. Anal.* (Warsaw), **6**, 1051 (1961); *Chem. Abstr.*, **57**, 6615 (1962); (*b*) E. Michalski and M. Turowska, *Chem. Anal.* (Warsaw), **5**, 625 (1960), *Chem. Abstr.*, **55**, 4254 (1961).

59. J. Matkovic and K. Weber, *Arch. Toxikol*, **21**, 355 (1966); *Chem. Abstr.*, **65**, 10851 (1966).

60. A. K. Babko, and L. I. Dubovenko, and A. V. Terletskaya, *Ukr. Khim. Zh.*, **32**, 1326 (1966); *Chem. Abstr.*, **66**, 80637 (1967).

61. J. Bognar and L. Sipos, *Mikrochim. Ichnoanal. Acta*, 442 (1963); *Chem. Abstr.*, **59**, 8119 (1963).

62. J. Bognar and L. Sipos, *Mikrochim. Ichnoanal. Acta*, 1066 (1963); *Chem. Abstr.*, **60**, 6209 (1964).

63. A. K. Babko, A. V. Terletskaya, and L. I. Duboveno, *Zh. Anal. Khim.*, **23**, 932 (1968); *Chem. Abstr.*, **69**, 64389 (1968).

64. A. K. Babko and A. V. Terletskaya, *Otkrytiay, Izobret., Prom. Obratsy, Tovarnyl Znaki*, **47**, 128 (1970); *Chem. Abstr.*, **73**, 52089 (1970).

CHAPTER X

Ultraviolet and Visible Absorption Spectra

MARGARET L. BAILEY

Chemistry Department, Victoria University of Wellington,
Wellington, New Zealand

1. Introduction

The study of the ultraviolet and visible absorption spectra of molecules has led to the relation of spectra to structure. Since certain functional groups lead to characteristic bands in a spectrum, spectra can be a useful aid in determining the composition of unknown molecules. For example, alternative structures can be distinguished, enabling the study of keto-enol and other forms of tautomerism, and of the type of species present in solution.

An absorption spectrum is obtained when a molecule absorbs radiation of the required energy to be promoted from its ground state to an excited state. The wavelength of the absorbed light is determined by the energy difference between the excited and ground states, and the extinction coefficient is related to the probability of the transition. Ultraviolet and visible light results in transitions between electronic energy levels; however, the electronic excitation is accompanied by vibrational excitations that cause the broad absorption bands seen in the spectra of molecules in solution. For aromatic molecules such as the acridines, these spectra arise from transitions between π-electronic energy levels. The dependence of the electronic energy levels on the geometry of the molecule accounts for the usefulness of absorption spectra in determining the structure of molecules.

In this chapter, the relation between absorption spectra and structure will be discussed for several classes of acridines. After a brief section on acridine itself, the spectra of various amino derivatives will be examined. The next section will cover the spectra of hydroxyacridines, acridanones, and acridine 10-oxides, including the effects of solvents on keto-enol tautomerism. Various other derivatives of acridine, such as the acridans and 9-substituted acridines, will be discussed next. The final section will consider the calculation of spectra and other properties of the acridines using molecular-orbital (MO) theory, and in particular π-electron MO theory.

2. Acridine

Acridine (1) is a heteroaromatic molecule, differing from its parent hydrocarbon, the linear three-ringed anthracene (2), in the replacement of a C-H group in the central ring by a nitrogen atom. It is known theoretically[1] that a nitrogen atom introduced into an alternant hydrocarbon produces rather small changes in the π-electronic energy levels, resulting in similar spectra for the heteroaromatic molecule and its parent hydrocarbon. This is illustrated by the spectra of acridine and anthracene[2] (Fig. 1). The strong band at 250 nm is almost identical in the two spectra, and the weaker bands that lie in the range of 300 to 400 nm differ only in their fine structure.

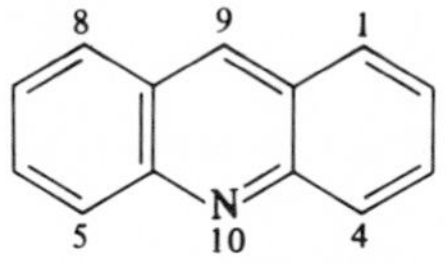

1

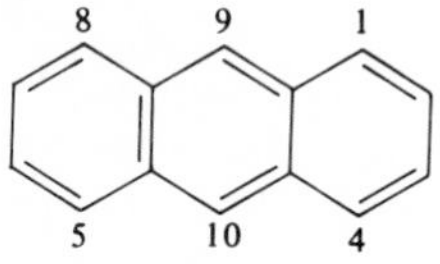

2

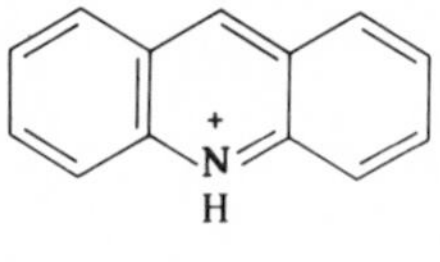

3

In contrast, the formation of acridinium cation (3) by the protonation of the ring nitrogen leads to a much more pronounced shift in the spectrum than that caused by the replacement of a C-H group by a nitrogen atom because

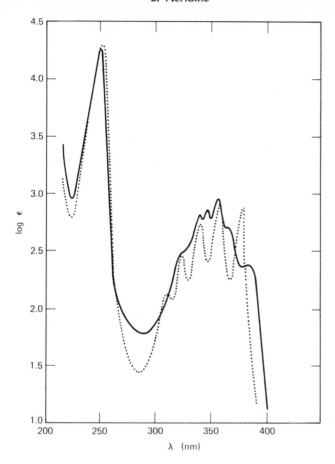

Fig. 1. (———) Acridine; (. . .) anthracene.[2]

the positively charged N-H group produces a much greater change in the π-electronic energy levels than does the neutral nitrogen atom. The main difference between acridine and its cation, as shown in Figure 2, is the extension to 440 nm of the weak long-wavelength band[3]; the other bands shift only slightly. This long-wavelength band is characteristic of acridine derivatives in which the ring nitrogen is protonated. Therefore, if more than one ionic species is possible in solution, the appearance of this band is indicative of the presence of this type of cation. It should be noted that the extinction coefficients for acridine in Figures 1 and 2 differ by one unit. Those for Figure 1, which is taken from the original paper,[2] are probably one unit too low, for when this correction is made the curves are in accord.

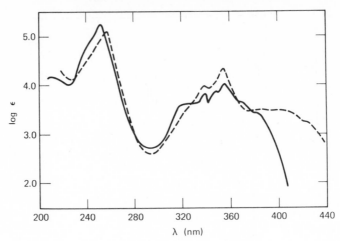

Fig. 2. Acridine.[3] (——) Neutral species in water at pH 8; (– – –) cation in water at pH 3.

3. Aminoacridines

The aminoacridines are perhaps the most important derivatives of acridine and have been studied extensively.[3–6] Substitution of one amino group produces five different monoaminoacridines; many diaminoacridines can be formed, but the two symmetric derivatives, 3,6- and 4,5-diaminoacridine, have been of most interest. Since the amino group contributes two π electrons to the conjugated system, the spectra of aminoacridines are expected to be different from that of acridine.

The spectra of 1-, 2-, 3-, and 4-aminoacridine (those formed by substituting an amino group in the outer ring) are shown in Figures 3 to 6. Each spectrum shows two strong bands at about 240 and 260 nm, a series of weaker bands with maxima in the range of 320 to 360 nm, and a shallow broad band at about 410 nm. A comparison of each spectrum with that of acridine shows that the spectra of the aminoacridines have their bands at longer wavelengths, the shift being most pronounced for the long-wavelength band. This displacement is the result of the perturbation of the π-electronic energy levels caused by the addition of a substituent, such as the amino or hydroxy group, which contributes a lone-pair of electrons to the π-electron system. Another feature is that the spectra of 2- and 3-aminoacridine (Fig. 7) are almost identical with one another[3] and with the spectrum of 2-aminoanthracene[7]; the spectra of 1- and 4-aminoacridine (Fig. 8) are also similar. This shows that the close resemblance between the spectra of acridine and anthracene is also

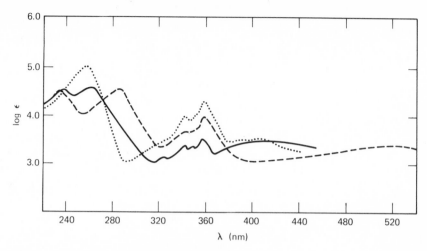

Fig. 3. 1-Aminoacridine.[3] (——) Neutral species in water at pH 11; (– – –) monocation in water at pH 2.5; (. . .) dication in 5N-HCl.

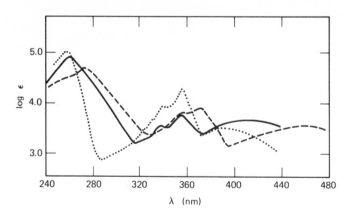

Fig. 4. 2-Aminoacridine.[3] (——) Neutral species in water at pH 11; (– – –) monocation in water at pH 2.5; (. . .) dication in 5N-HCl.

seen in the spectra of their derivatives; in other words, the 9- and 10-positions in acridine are almost equivalent.

In acid solution, the aminoacridines are converted into the monocations. As there are two sites available for protonation, the ring-nitrogen and the amino-nitrogen atoms, a careful study is required to elucidate which nitrogen atom is protonated first. First, it has been shown that although primary

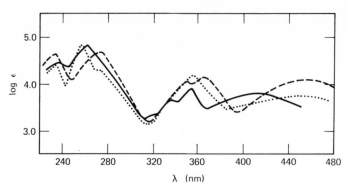

Fig. 5. 3-Aminoacridine.[3] (——) Neutral species in water at pH 11; (– – –) monocation in water at pH 2.5; (. . .) mixture of mono- and dication in 5N-HCl.

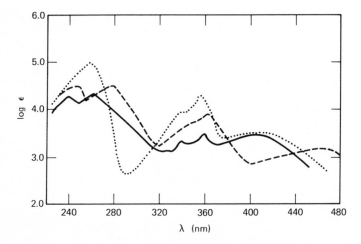

Fig. 6. 4-Aminoacridine.[3] (——) Neutral species in water at pH 11; (– – –) monocation in water at pH 2.5; (. . .) dication in 5N-HCl.

aromatic amines absorb at longer wavelength than the hydrocarbons from which they are derived, protonated amines have spectra very similar to those of the parent hydrocarbon.[8a, b] The quaternary amino-nitrogen atom no longer supplies electrons to the π-electron system, which reverts to that of the parent-conjugated molecule. Second, it has been shown that the long-wavelength band in acridine[3] shifts to longer wavelength on protonation of the ring-nitrogen atom (cf.3). Therefore, if the amino-nitrogen were protonated

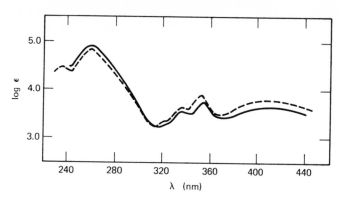

Fig. 7. (———) 2-Aminoacridine; (– – –) 3-aminoacridine neutral species. From Figs. 4 and 5.

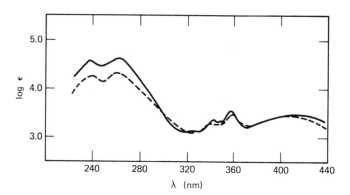

Fig. 8. (———) 1-Aminoacridine; (– – –) 4-aminoacridine neutral species. From Figs. 3 and 6.

to form the monocation, the spectra of all four aminoacridinium cations would resemble the spectrum of acridine; whereas, if the ring nitrogen were protonated, the spectra would be displaced to longer wavelengths.

The spectra of the four aminoacridinium cations are shown in Figures 3 to 6, together with those of the neutral molecules. The most noticeable change is the shift to longer wavelength of the shallow broad band that appears at 410 nm in the neutral molecules, this band appearing at about 460 nm for 2-, 3-, and 4-aminoacridinium cations, and at 520 nm for 1-aminoacridinium cation. Hence it is deduced that the ring nitrogen is protonated to form the monocation. Protonation causes changes in the positions and intensities of the other bands in the spectra, leading to differences among all four molecules. This again demonstrates the stronger influence that the protonated nitrogen, as opposed to the ordinary nitrogen, has upon the spectrum.

On the basis of the above discussion, the spectra of the dications, formed in more highly acidic solution, would be expected to resemble the spectrum of acridinium cation because the second proton must be added to the amino-nitrogen atom. The spectra of 1-, 2-, and 4-aminoacridinium dications (Figs. 3, 4, and 6) do revert to the spectrum of acridinium cation in $5N$ hydrochloric acid. However, 3-aminoacridine is only about half converted to the dication in $5N$ hydrochloric acid, its spectrum (Fig. 5) being a mixture of the mono-cation and dication. In concentrated sulfuric acid, the dication is formed completely, the spectrum becoming that of acridinium cation.[5] The stability of the monocation has been attributed to its high degree of resonance[9]; this theory will be discussed later in this section, together with other possible explanations.

The spectra of 9-aminoacridine (4) and its cation (Fig. 9) are somewhat different from those of the other monoaminoacridines.[3] In the neutral mole-cule, two strong bands occur at 218 and 260 nm, and weaker bands occur in the range of 320 to 440 nm under a broad envelope; the shallow broad band at 410 nm in the other aminoacridines is not seen. The spectrum of the mono-cation differs very little from that of the neutral molecule, two weak bands appearing at 311 and 326 nm, and the structure of the weaker bands being more pronounced. This deviation from the general pattern is believed to occur

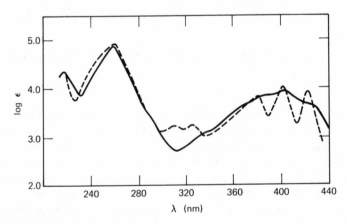

Fig. 9. 9-Aminoacridine.[3] (——) Neutral species in water at pH 12; (– – –) cation in water at pH 6.

because the position of substitution is at the carbon atom opposite the ring-nitrogen atom. Since this atom is the most π-electron deficient,[10] the lone-pair electrons from the amino-nitrogen atom are strongly delocalized. The 9-amino group is also situated along the short axis of the molecule and cannot add to conjugation along the long axis, unlike amino groups in all other positions. This is believed to account for the small differences between the spectra of the neutral molecule and its cation, since 4-aminopyridine (5) and 4-aminoquinoline (6) both having an amino group opposite a ring-nitrogen atom, also have similar spectra for neutral and cationic species.

because the position of substitution is at the carbon atom opposite the ring-nitrogen atom. Since this atom is the most π-electron deficient,[10] the lone-pair

The difference between the spectrum of 9-aminoacridine and the spectra of the other aminoacridines has led to some controversy about the structure of 9-aminoacridine in solution. A comparison of the spectra[11] of 9-dimethyl-aminoacridine and 9-imino-10-methylacridan with 9-aminoacridine (Fig. 10), together with the resemblance of the spectrum of 9-acridanone (Fig. 22) to that of 9-aminoacridine, led to support for the "imino-acridan" structure.[12] (See Chapter II, p. 121.) Chemical evidence from ionization constants in aqueous solution supported the amino form,[13] and this was confirmed by ir spectra in nonaqueous media.[14] More recently, the ir spectra of 9-amino- and 9-dideuterioamino-acridine have provided further support for the amino form.[15] (See Chapter XI, p. 666; Chapter XII, p. 696.) The basis for the difference of the spectrum of 9-aminoacridine has already been discussed, this explaining the similarity in the spectra of 9-acridanone (7) and 9-aminoacridine. The differences between the spectra of 9-aminoacridine (4) and 9-dimethylaminoacridine (8) have been attributed to the loss of conjugation in the latter because of steric interference between the methyl groups and the hydrogen atoms in the 1- and 8-positions.[16]

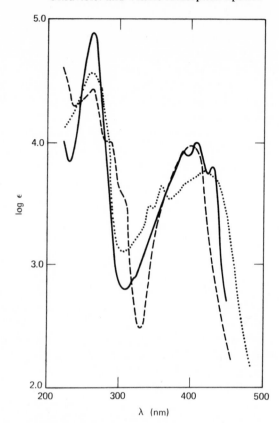

Fig. 10. (——) 9-Aminoacridine.[11] (— — —) 9-imino-10-methylacridan; (. . .) 9-dimethyl-aminoacridine. All in methanol.

The dication is not formed from 9-aminoacridine in $5N$ hydrochloric acid but requires concentrated sulfuric acid; its spectrum resembles that of acridinium cation.[17] This stability of the monocation, like that of 3-aminoacridinium cation, has been explained by the use of resonance structures (see Chapter II, p. 110).[9] An alternative explanation is offered by MO theory. The 9-position in acridinium cation is highly π-electron-deficient, the 3 position moderately so.[10] A substituent such as the amino group is strongly delocalized in these positions, increasing the stability of the molecule. Thus the cations of 3- and 9-aminoacridine are more stable than the cations of the other aminoacridines and require severe conditions to be converted into the dications.

N-Alkylation of the ring nitrogen produces much the same effect on the spectra of aminoacridines as protonation. The spectrum of 3-amino-10-

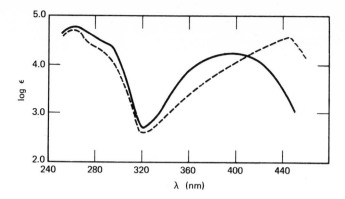

Fig. 11. 3,6-Diaminoacridine.[3] (——) Neutral species in water at pH 12; (– – –) cation in water at pH 7.

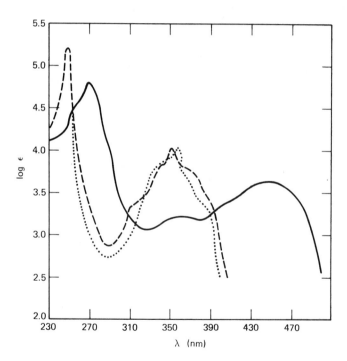

Fig. 12. 4,5-Diaminoacridine.[19] (——) Neutral species in absolute alcohol; (– – –) cation in 5N-HCl; (. . .) acridine, neutral species at pH 11.

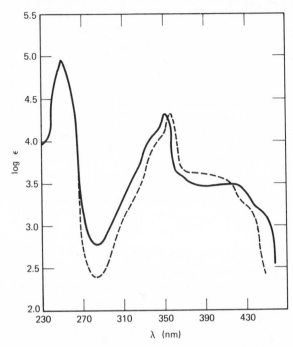

Fig. 13.   4,5-Diaminoacridine.[19] (——) Cation in 18N-H_2SO_4; (– – –) acridine cation in 5N-HCl.

methylacridinium chloride[4] is practically identical to that of 3-aminoacridinium hydrochloride.[5] Studies on other aza-aromatics, namely, pyridine, quinoline, and isoquinoline, have shown that these also have similar spectra for hydrochlorides and methochlorides.[18]

The spectra of several of the diaminoacridines have been studied.[3, 4, 6, 19] The general effect on the spectrum of adding an additional amino group is to intensify the long-wavelength band, and sometimes to shift this band.[3] This can be seen in the spectrum of 3,6-diaminoacridine, proflavine (Fig. 11), in which the long-wavelength band is quite strong (log ϵ 4.25), compared with that in 3-aminoacridine (log ϵ 3.79).[3] The effect of protonation is to shift this band to longer wavelengths, as would be expected for the protonation of the ring-nitrogen atom. Conversion to the monocation produces this bathochromic shift for most diaminoacridines.[3]

The spectra of 4,5-diaminoacridine in various media, however, do not follow the usual pattern.[19] In 5N hydrochloric acid, the spectrum reverts almost exactly to that of acridine itself (Fig. 12), not to that of the acridinium

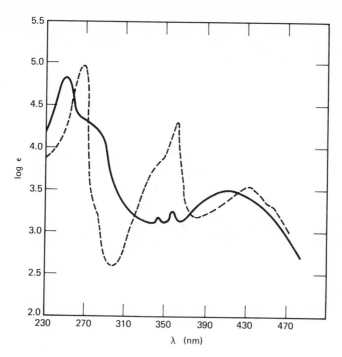

Fig. 14. 4-Amino-5-methylacridine.[19] (——) Neutral species in absolute alcohol; (– – –) dication in 5N-HCl.

cation. From the principles discussed above, it is deduced that the two amino groups are protonated first and the ring-nitrogen atom is not protonated. The spectrum of 4,5-diaminoacridine in $18N$ sulfuric acid confirms this (Fig. 13), as it is almost identical with that of acridinium cation. This anomalous behavior is attributed to steric hindrance, the two amino groups preventing the approach of the protonating hydrogen to the ring nitrogen. The ionization of the amino groups further hinders the approach to the ring nitrogen, the latter being protonated only under the severe conditions of $18N$ sulfuric acid.

The spectra of 4-amino-5-methylacridine in various media also show some interesting features.[19] Although the spectra of the neutral molecule in alcohol and the dication in $5N$ hydrochloric acid (Fig. 14) show no anomalies, two monocations are seen to exist in suitable media. In alcoholic $0.07N$ hydrochloric acid, the spectrum reverts to that of acridine (Fig. 15), indicating that the amino group has been protonated, and not the ring nitrogen. On the other hand, in aqueous $0.25N$ hydrochloric acid, the spectrum resembles that

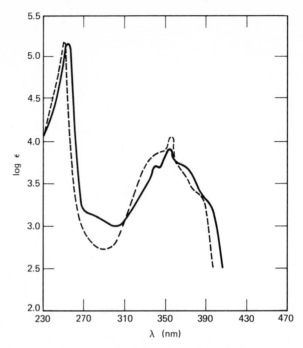

Fig. 15. (——) 4-Amino-5-methylacridine in alcoholic 0.07N-HCl; (– – –) acridine in water at pH 11.[19]

of the monocation of 4-aminoacridine (Fig. 16), in which the ring nitrogen is protonated, and not the amino group. In this case, a change in solvent has caused a change in the relative basic strengths of two possible ionizing processes, a most unusual feature.[3]

Another spectrum which deviates from the expected pattern because of steric hindrance[11] is that of 9-dimethylaminoacridine, **8** (Fig. 10). This spectrum has already been mentioned in the discussion of the structure of 9-aminoacridine. The usual effect of extranuclear *N*-alkylation is to displace the long-wavelength band in the spectra of both the cation and its neutral species to longer wavelengths.[3] This can be seen by comparing the spectra of 3,6-diaminoacridine (proflavine) and 3,6-bisdimethylaminoacridine (**9**) (acridine orange) cations.[20]

Me₂N⟨...⟩N⟨...⟩NMe₂

9

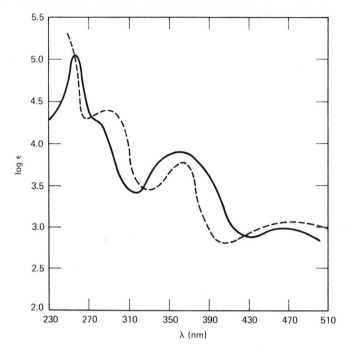

Fig. 16. (———) 4-Amino-5-methylacridine in 0.25N-HCl; (- - -) 4-aminoacridinium cation at pH 2.5.[11]

Acridine orange has been studied extensively[21-23] because of the tendency of its cation to form association dimers. As the concentration of the solution is raised, a new peak, due to the dimer absorption, appears at 451 nm (Fig. 17). 3,6-Diaminoacridine (proflavine) does not show the same tendency to aggregate. The absorption spectra of both of these molecules[24a, b] have also been studied in connection with their binding to nucleic acids (Chapter XIV, p. 770).

4. Hydroxyacridines, 9-Acridanones, and Acridine 10-Oxides

The other major derivatives of acridine, the hydroxyacridines, have also been studied in detail.[25-27] The hydroxy group, like the amino group, contributes a lone-pair of electrons to the π-system, and is therefore expected to cause a displacement of the spectrum to longer wavelength. However, the displacement caused by the hydroxy group is known to be less than that

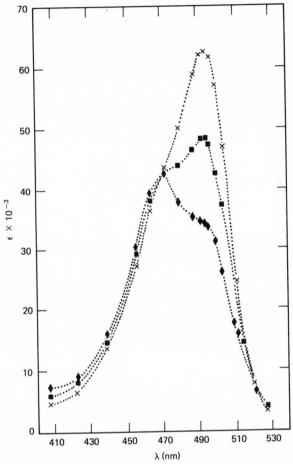

Fig. 17. Bis-3,6-dimethylaminoacridine (acridine orange) hydrochloride.[22] (♦ ♦) 1.80 x $10^{-4}M$; (| |) 3.90 x $10^{-5}M$; (xx) 2.50 x $10^{-6}M$. In water.

caused by the amino group. Hence the spectrum of a hydroxyacridine should be intermediate between that of acridine and the corresponding amino-acridine.

The spectra of 1-, 2-, 3-, and 4-hydroxyacridine have been measured in dioxan[26] and in alcohol and alcohol/water mixtures,[25] the latter spectra being given in Figures 18 through 21. The 2- and 4-hydroxyacridines give a strong band at about 255 nm and a series of weaker bands with maxima at about 340, 360, and 390 nm, these spectra being similar in all the media used.

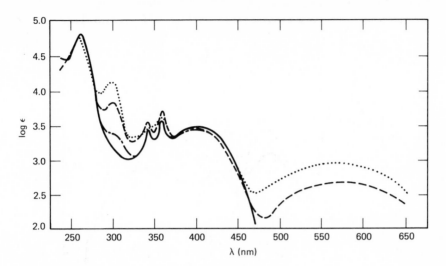

Fig. 18. 1-Hydroxyacridine.[25] (——) In absolute alcohol; (– · – ·) 67% aqueous alcohol; (– – –) 33% aqueous alcohol; (. . .) 20% aqueous alcohol.

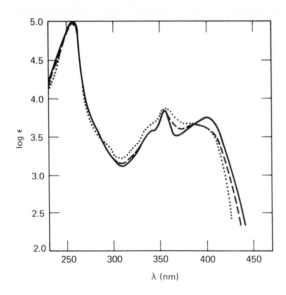

Fig. 19. 2-Hydroxyacridine.[25] (——) In absolute alcohol; (– – –) 67% aqueous alcohol; (. . .) 20% aqueous alcohol.

647

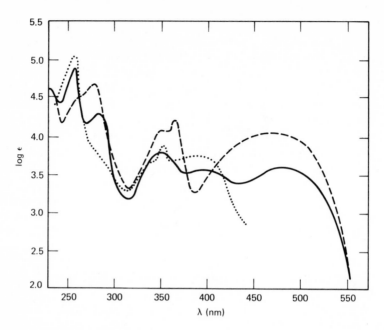

Fig. 20. 3-Hydroxyacridine.[25] (——) In absolute alcohol; (– – –) 20% aqueous alcohol; (. . .) 90% ether/10% alcohol.

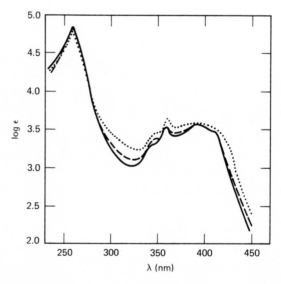

Fig. 21. 4-Hydroxyacridine.[25] (——) In absolute alcohol; (– – –) 67% aqueous alcohol; (. . .) 33% aqueous alcohol.

In absolute alcohol and dioxan, 1-hydroxyacridine has a spectrum similar to that of 4-hydroxyacridine, but in aqueous alcohol two new bands appear at 300 and 570 nm. 3-Hydroxyacridine also shows this complicated behavior; a spectrum similar to that of 2-hydroxyacridine is obtained in 9:1 (v/v) ether/alcohol mixture and in dioxan, but in aqueous, and even absolute, alcohol two more bands appear at 280 and 465 nm. The changes in the spectra in different media are a result of the formation of new species that need to be identified.

The spectra of all four hydroxyacridines in dioxan are virtually identical with those of the corresponding methoxyacridines in alcohol.[26] As the spectra of hydroxy and methoxy derivatives are similar, this confirms that the species present in dioxan is the hydroxy form. This is further shown by the position of the long-wavelength band, which is at 390 nm, between the positions for acridine (380 nm) and the aminoacridines (410 nm), as expected for a hydroxy derivative. It is also known from dissociation constants that neither the anionic nor the cationic species will be present in the solvents used.[9] Comparison of the spectrum of 1-hydroxyacridine (10) in aqueous alcohol with that of 1,10-dihydro-10-methyl-1-acridinone[28] (11) reveals that the new bands result from the presence of the acridinone form.

10

11

Therefore, the changes that occur in the spectra of 1- and 3-hydroxyacridines are caused by the formation of the tautomeric keto forms. An increase in dielectric constant, such as occurs in going from absolute to aqueous alcohol, shifts the position of the tautomeric equilibrium toward the ketonic form because it is more polar than the enolic form. The ketonic form of 3-hydroxyacridine (12) is so stable that it exists even in absolute alcohol, and it is probably the only species present in 20% aqueous alcohol.[25] The stability of the ketonic form can be explained either by resonance with its zwitterion (13),[27] or because of the π-electron deficiency of the 1- and 3-positions leading to strong delocalization.

12

13

Neither 2- nor 4-hydroxyacridine can form such tautomers, so that their spectra are independent of alcohol concentration.

The four hydroxyacridines also form cations and anions in acid and alkali, respectively. Both ionic species give spectra that are displaced to longer wavelength in comparison with the spectra of the neutral species.[26]

The fifth member of the hydroxy derivatives, 9-acridanone (14), which is tautomeric with 9-hydroxyacridine (16), again shows different behavior from the other four derivatives. A comparison[11] of the spectra of 9-acridanone (14), 10-methyl-9-acridanone (15), and 9-methoxyacridine (17) (Fig. 22) shows that the hydroxy form (16) is present to a very small extent if at all.

14; R = H
15; R = Me

16; R = H
17; R = Me

This has been confirmed by a study of the basic ionization constants of 9-methoxyacridine and 9-acridanone,[29] from which it was concluded that only one part in 10 million of 9-hydroxyacridine was in equilibrium with 9-acridanone in aqueous solution at 20°. Indeed, the hydroxy form has not even been detected in nonpolar solvents.[26] Its spectrum is independent of alcohol concentration and is different from the spectra of the other hydroxyacridines,[25] with major bands at 251, 295, 381, and 400 nm. A similarity between the spectrum of 9-acridanone and that of 9-aminoacridinium cation may be interpreted in terms of a contribution of the zwitterion of the former to the resonance hybrid.[3] The band at about 300 nm in 9-acridanone, which is not seen in the spectrum of acridine, also appears in the spectra of 1- and 3-hydroxyacridine under conditions in which the ketonic form is present.

The absorption spectrum of the 9-acridanone cation[27, 30] has two new bands at 345 and 416 nm. The other bands do not shift significantly from the positions for 9-acridanone, although their intensities change somewhat. The appearance of the two new bands is a result of the formation of the 9-hydroxy-acridinium cation, with the characteristic shift to longer wavelength of the band at 400 nm.

The sulfur analogue of 9-acridanone, 9-acridanthione, has also been studied to determine its tautomeric form. The spectra of 10-methyl-9-acridan-thione, 9-methylthioacridine, and 9-acridanthione (Fig. 23) have been interpreted as showing that the last-mentioned compound[11] is present in the thione form (cf. 14) Another study in heptane, ethanol, dioxan, and water-dioxan

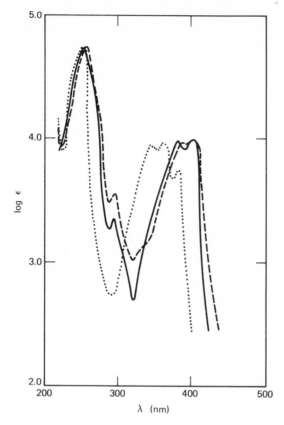

Fig. 22. (———) 9-Acridanone in methanol, (– – –) 10-methyl-9-acridanone in methanol; (. . .) 9-methoxyacridine in cyclohexane.[11]

solvents has led to the conclusion that 9-acridanthione exists in both the thiol and thione forms in neutral media.[31] The presence of the thiol form (cf. **16**) may account for the greater difference between the spectra of 9-acridanthione and its 10-methyl derivative than that between 9-acridanone and its 10-methyl derivative. The spectra of 9-acridanone and 9-acridanthione show similarities, the main difference being the displacement to longer wavelengths of the bands in the spectrum of 9-acridanthione; the two long-wavelength bands occur at 448 and 477 nm in 9-acridanthione, compared with 381 and 400 nm in 9-acridanone. In acidic media, the mercaptoacridinium cation is formed.[31]

Various derivatives of 9-acridanone have also been studied. A methoxy or hydroxy substituent in one of the benzene rings increases the intensity of the characteristic 9-acridanone band at 300 nm.[32] Indeed, this band in 1,3-dihydroxy-10-methyl-9-acridanone is more intense than the long-wavelength

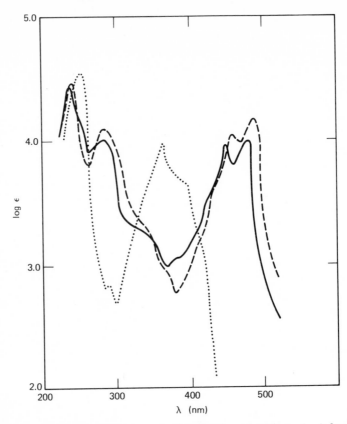

Fig. 23. (——) 9-Acridanthione; (– – –) 10-methyl-9-acridanthione; (. . .) 9-methylthio-acridine. In methanol.[11]

band. This has been attributed to the perturbation caused by a nonsymmetrical substitution making a symmetry-forbidden transition an allowed transition.[32] The long-wavelength band shifts to a longer wavelength for the 2-methoxy derivative, and to a shorter wavelength for the 1- and 3-methoxy derivatives. The methyl group exerts a similar but weaker influence than that of the methoxy group.[33]

Alkyl-substituted 9-acridanones have recently been studied.[30] Differences in the spectra in the region of 250 to 260 nm divide the compounds into two classes: (1) 2-alkyl-substituted 9-acridanones and 9-acridanone itself, and (2) 1- and 4-alkyl-substituted 9-acridanones. The wavelengths and extinction coefficients of some of these molecules in both dry methanol and acidified methanol are given in Table I.

TABLE I.   Spectra of Alkyl-Substituted 9-Acridanones

Compound														
9-Acridanone	M	211* (4.04)	215 (4.10)	250* (4.40)	255 (4.41)	260* (4.24)	269* (3.88)	295* (2.75)				365* (3.18)	382 (3.52)	400 (4.54)	
	MA	211* (3.86)	215 (3.94)	252* (4.45)	256 (4.49)		269* (3.75)	295* (2.15)			345 (2.92)	365* (3.18)	382 (3.44)	400 (3.51)	416* (2.75)
2-Et	M	211* (4.30)	215 (4.36)	252.5 (4.74)	257.5 (4.75)	265* (4.70)		298 (3.27)	310 (3.03)			367* (3.65)	385 (3.90)	405 (3.93)	
	MA	211* (4.18)	215 (4.24)	255* (4.82)	261 (4.90)			298 (3.27)	310 (3.03)		340 (3.50)		385 (3.82)	402 (3.85)	423* (3.30)
2-Bu^n	M	211* (4.32)	215 (4.39)	253 (4.75)	258 (4.76)	263 (4.71)	271* (4.39)	298 (3.30)	310* (3.00)			370* (3.66)	386 (3.86)	405 (3.86)	
	MA	211* (4.18)	215 (4.23)	255* (4.83)	262 (4.94)			298 (3.30)	310* (3.00)		340 (3.60)	370* (3.60)	386 (3.77)	405 (3.77)	425* (3.00)
2-(1-Me-Bu^n)	M	211* (4.31)	215 (4.35)	253 (4.74)	259 (4.75)	263 (4.71)	271* (4.38)	285* (3.30)	298 (3.29)	309* (3.03)		370* (3.66)	387 (3.83)	404 (3.85)	
	MA	211* (4.19)	215 (4.24)	256* (4.81)	262 (4.88)			285* (3.30)	298 (3.29)	309* (3.03)	340 (3.47)	370* (3.61)	387 (3.73)	404 (3.79)	423* (3.13)
1-Pr^i-4-Me	M	209* (4.02)	214* (4.10)	219 (4.19)	258 (4.68)		285* (3.37)	298 (3.32)	310 (3.21)				385 (3.84)	403 (3.83)	
	MA	207* (4.02)	214* (4.08)	219 (4.14)	258 (4.67)	266* (4.48)	285* (3.37)	298 (3.32)	310 (3.21)		340 (2.97)		385 (3.80)	403 (3.80)	425* (2.84)
4-Bu^{sec}	M	213* (4.19)	217.5 (4.29)		255 (4.77)			295 (3.27)	305* (3.00)			367* (3.70)	383 (3.89)	400 (3.90)	
	MA	213* (4.15)	217.5 (4.24)		256 (4.77)			295 (3.27)	305* (3.00)		338 (3.24)	367* (3.65)	383 (3.84)	400 (3.84)	425* (2.87)
4-Pr^i	M	214* (4.28)	217.5 (4.33)		255 (4.81)			295 (3.19)	305* (2.89)			367* (3.67)	380 (3.91)	400 (3.94)	

(Table Continued)

653

Table I. (Continued)

4-Pr^i (Continued)	MA	214* (4.20)	217.5 (4.28)	256 (4.83)	295 (3.19)	305 (2.89)	337 (3.29)	367* (3.59)	380 (3.80)	400 (3.85)	425* (3.20)

* = Inflection.

Note: λ_{max} and log ϵ (in brackets) of 9-acridanones in dry methanol solution at concentrations in the range of 0.003-0.007 mg ml^{-1} (M) and with the addition of one drop of 70% HClO$_4$ to both cells (MA).

Class 1 compounds show three maxima at about 255 nm; on the addition of perchloric acid, these almost double in intensity to give one band with a small shoulder. Class 2 compounds show one band near 255 nm, which hardly changes on the addition of perchloric acid. Additional evidence from solid ir spectra shows that hydrogen bonding is disfavored in the 1- and 4-substituted 9-acridanones (Chapter XI, p. 670). The bands in the regions of 210 to 220, 270 to 310, and 365 to 430 nm are not changed significantly, either by the substitution of an alkyl group or by a change from a neutral to an acidic medium. One possible explanation of the difference between the two classes of 9-acridanones is that substituents in the 1- and 4-positions may cause steric interference with the N—H and C—O groups, whereas substituents in the 2- and 3-positions are less likely to cause interference. Hydrogen bonding may also differentiate between these two classes of 9-acridanones.

Another class of compounds similar to the acridanones are the acridine 10-oxides. The spectra of acridine 10-oxide, its cation and its anion, are given in Figure 24. The spectrum of the neutral molecule[34] is similar to the spectra of 9-acridanone (Fig. 22) and 9-aminoacridine (Fig. 9); for all these molecules, the conjugated system is enhanced by a lone-pair of electrons along the short axis of the molecule. The spectrum of the cation resembles that of acridine.[34]

Two interesting derivatives of acridine 10-oxide are the 9-hydroxyacridine and 9-acridinethiol 10-oxides. Studies of the tautomeric structure of both these

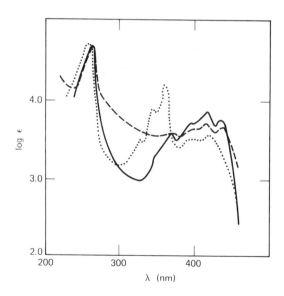

Fig. 24.   Acridine 10-oxide.[34] (——) Neutral species; (— — —) basic solution; (. . .) acidic solution. In methanol.

molecules have been made using spectra and basicity measurements.[35] In aqueous solution, the 9-hydroxyacridine 10-oxide **(18)** coexists with a comparable amount of the 10-hydroxy-9-acridanone **(19)**.

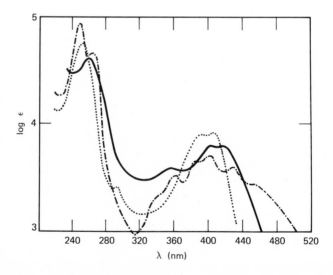

This can be seen from the spectra of these molecules, together with the spectrum of 9-methoxyacridine 10-oxide, all in neutral solution (Fig. 25). The changes that occur in these spectra in acidic medium are shown in Figure 26. 9-Acridinethiol 10-oxide also exists in solution in equilibrium with a comparable amount of 10-hydroxy-9-acridanthione. This pattern differs from that of the "non-10-oxides," in which the oxo or thione forms are more stable than the hydroxy of thiol forms; an explanation of this has been given in terms of resonance structure.[35]

Fig. 25. (———) 9-Hydroxyacridine 10-oxide; (–·–·) 9-methoxyacridine 10-oxide; (. . .) 10-methoxy-9-acridanone. In 50% EtOH-AcONa aq-HCl buffer of pH 4.8.[35]

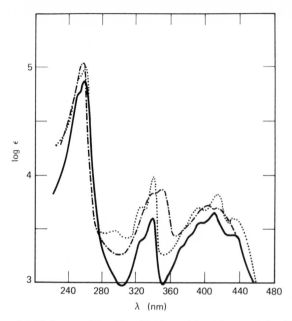

Fig. 26.   (———) 9-Hydroxyacridine 10-oxide; (– · – ·) 9-methoxyacridine 10-oxide; (. . .) 10-methoxy-9-acridanone. In 15N-H$_2$SO$_4$.[35]

5. Miscellaneous Acridines

Acridan (9,10-dihydroacridine) **(20)** has a much simpler spectrum than that of acridine,[36] having a band at 290 nm with two inflections at 315 and 335 nm (see Figure 27).

This spectrum is similar to that of diphenylamine (Fig. 27), which differs from acridan only in a methylene group. The substitution of a methyl group in a 9-position causes small shifts and an enhancement of the transition of weak intensity in acridan polarized perpendicularly to the long molecular axis.[37] For phenyl substitution, the changes in the spectrum[37] can be explained in terms of the independent absorption by the benzene ring (Fig. 27). The spectra

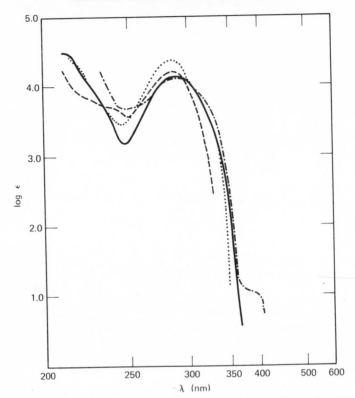

Fig. 27. (——) Acridan; (– – –) diphenylamine; (. . .) 9-methylacridan; (– · – ·) 9-phenyl-acridan. In ethanol.[36]

for 10-substituted acridans (Fig. 28) show much the same pattern.[36] Whereas the methyl and ethyl groups have only a small effect on the spectrum of acridan, the phenyl group leads to the appearance of a new band in the region of 340 to 370 nm.

The absorption spectra of 10,10′-dialkyl- (or diaryl-) -9,9′-biacridans have also been studied.[36] In Figure 29, the spectra of the methyl, n-butyl, and phenyl derivatives are given. Beside the band at 290 nm and the inflections between 310 and 335 nm found in the monoacridans, two new bands appear in the region of 340 to 360 nm, and 240 to 250 nm. These two new bands show more clearly at lower temperatures.

9-Imino-10-methylacridan has already been mentioned because of its use in determining the form of the amino group in 9-aminoacridine (Fig. 10). The spectrum of this molecule differs from that of 9-aminoacridine[11] in having the characteristic acridan band at 290 nm. However, it is necessary to use

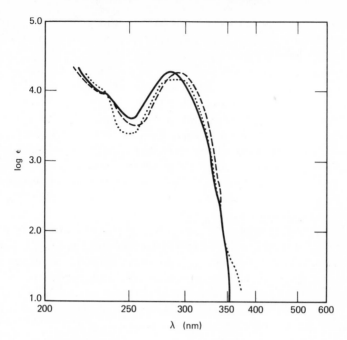

Fig. 28. (——) 10-Methylacridan; (– – –) 10-ethylacridan; (. . .) 10-phenylacridan. In ethanol.[36]

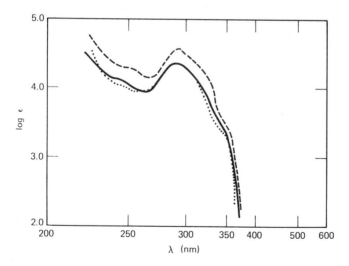

Fig. 29. (——) 10,10′-dimethyl-9,9′-biacridan; (– – –) 10,10′-di-*n*-butyl-9,9′-biacridan; (. . .) 10,10′-diphenyl-9,9′-biacridan. In ethanol.[36]

anhydrous solvents in order to see this band, as the presence of water causes the cation to form; this has a spectrum similar to that of 9-aminoacridine and almost identical with the cation of the latter.[17]

Saturated aliphatic groups substituted in the 9-position cause no change in the spectrum of acridine.[38] Even unsaturated aliphatic or aromatic substituents do not greatly influence the spectrum of acridine.[39] The results of dipole moment studies[40] are consistent with the idea that the benzene and acridine rings make an angle of at least 40° to one another in 9-phenylacridine, because of steric interference between the hydrogens in the 1- and 8-positions and the phenyl group. As the phenyl group makes little contribution to the conjugated system, the spectrum of 9-phenylacridine would be expected to be similar to that of acridine. However, if an unsaturated aliphatic or aromatic substituent is further substituted with an electron-donor group such as NH_2, NMe_2, OH, or OMe in the ω-, o- or p-positions relative to the point of attachment of acridine, a new long-wavelength band appears.[39] In aromatic substituents, a further new band appears in the middle ultraviolet, usually in the position where acridine has a minimum. No such bands occur when the electron-donating group is in the m-position.

In glacial acetic acid or acidified ethanol, 9-p-aminoarylacridines give highly colored solutions of the monocations (Fig. 30).[41] When strongly acidified, the solutions revert to the spectrum of acridinium cation, unless the basicity of the amino group is low, showing that the ring nitrogen is the first nitrogen protonated. If the basicity of the acridine nucleus is lowered by several bromine substituents, the colors are not observed (Fig. 30), since the amino group is then relatively more basic than the ring-nitrogen.[41] Two explanations have been proposed for the appearance of the long-wavelength band in these monocations (see also Chapter II, p. 129). In one, the color is considered to arise from an intramolecular charge-transfer effect[38, 41, 42]; an electron initially largely localized on the donor aminoaryl group transfers to the positively charged acridine ring during light absorption. The other explanation considers that the color is caused by the coplanar quinonoid resonance structure.[3] The coplanar state is believed to be reached only after absorption in the long-wavelength, charge-transfer band.[43]

The spectra of angular and linear benzacridines and dibenzacridines have been studied at low temperatures.[44] Linear benzacridine (benz[b]acridine) has its spectrum at longer wavelength than that of acridine. This is an example of the well-known phenomenon, in which the addition of a ring to a linear hydrocarbon causes its spectrum to be displaced to longer wavelength. The angular benzacridines are not displaced to longer wavelength.[44]

The spectra of 1-, 2-, 3-, and 4-bromoacridine have been studied in ethanol and in 1N sulfuric acid.[45] A bromo substituent lowers the basic strength of acridine, as has already been seen for 9-p-dimethylaminophenylacridine.[41]

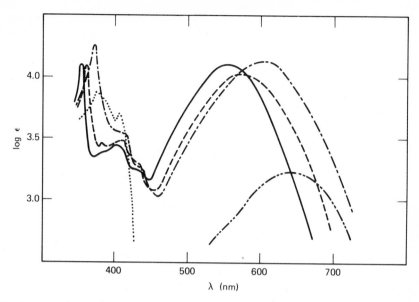

Fig. 30. 9-*p*-dimethylaminophenylacridinium monocations.[41] (——) Parent molecule; (– – –) 2-bromo derivative; (– · – ·) 2,5-dibromo derivative; (– · · – · ·) 2,4,7-tribromo deriv-ative; (. . .) 2,4,5,7-tetrabromo derivative. In methanol.

This has been accounted for by the inductive effect[45]; however, in the 1- and 3-positions, this is opposed by the mesomeric effect. Bromo substitution causes small shifts in the positions of the bands, compared with the spectrum of acridine; again, the spectra of the 1- and 4- derivatives are similar, as are the 2- and 3- derivatives. Protonation causes bathochromic shifts of the bands; in particular, the long-wavelength band appears in the region of 400 to 430 nm, as is expected for the protonation of the ring nitrogen. 9-Bromo-acridine, together with the 9-chloro- and 9-iodo- derivatives, has been studied in ethanol.[46]

6. Theoretical Calculations

In the past few years, many of the spectra of the acridines have been inter-preted by theoretical calculations, giving an understanding of the relation of structure to spectra. A brief mention will therefore be made of the type of calculations that have been made, although details of the methods used will not be given. An introduction to the theory of spectra may be found in the book by Murrell.[1] All the calculations discussed here have used some variant of semiempirical MO theory; most calculations of spectra use Pariser-Parr-Pople (PPP) self-consistent-field, π-electron MO theory, although a few calculations have been performed using the much simpler, and less accurate, Hückel π-electron MO theory.

Acridine has been studied by most variants of these two methods, mainly because it is an aromatic molecule with only one heteroatom, and as such is useful in testing new variants. Recent PPP calculations of its spectrum have included those by the variable-beta method[47] with limited configuration interaction, and by the Linderberg-beta method with full configuration interaction.[10] The spectrum of acridinium cation has also been studied,[10, 48] its spectrum being more difficult to calculate because of the need to obtain parameters for the positively charged nitrogen atom. In order to gain an understanding of the effect of substitution, a study of acridine and its cation has been performed using the CNDO all-valence-electron, self-consistent-field MO theory to obtain the σ- and π-electron densities.[10] These have proved useful in explaining some of the features of the derivatives of acridine.

Most of the spectra of acridine derivatives have been calculated, mainly by Zanker and his co-workers. Using the variable-beta method,[47] they have calculated the transition energies, oscillator strengths, polarization directions, and other properties of the aminoacridines,[49] the hydroxyacridines,[50] the benzacridines,[51] 9-substituted acridines,[52] acridine 10-oxide,[53] acridan[54] and the monohydroxyacridinium cations.[55] Hückel calculations for the cations of acridine, 3-aminoacridine and 3,6-diaminoacridine (proflavine) have been performed[56]; the cation of 3,6-diaminoacridine has also been calculated by the PPP method.[10, 20, 57] The Linderberg-beta method has also been used to calculate the spectra of all the monoaminoacridines and proflavine.[58]

It is likely that further MO calculations will be performed in order to interpret the spectra of the acridines. These calculations will employ more sophisticated π-electron theory or even all-valence-electron theory, so that nonplanar acridines may be studied.

Acknowledgments

The author gratefully acknowledges the granting of permission to use the following: Figures 2, 3, 4, 5, 6, 7, 8, 9, and 11, Edward Arnold (Publishers) London; Figures 10, 22, 23, 24, and 30, the Chemical Society and Dr. R. M. Acheson; Figures 12, 13, 14, 15, and 16, the Chemical Society and Professor D. P. Craig; Figure 17, the American Chemical Society (copyright 1965); Figures 18, 19, 20, and 21, the Chemical Society and Professor A. Albert; Figures 25 and 26, Pergamon Press and Professor A. R. Katritzky; and Figures 27, 28, and 29, Akademische Verlagsgesellschaft.

References

1. J. N. Murrell, The Theory of the Electronic Spectra of Organic Molecules, Methuen, London, 1963, p. 197.
2. D. Radulescu and G. Ostrogovich, *Chem. Ber.*, **64**, 2233 (1931).
3. A. Albert, The Acridines, 2nd ed., Arnold, London, 1966, Chap. 11.
4. D. P. Craig and L. N. Short, *J. Chem. Soc.*, 419 (1945).
5. N. H. Turnbull, *J. Chem. Soc.*, 441 (1945).
6. A. Wittwer and V. Zanker, *Z. Phys. Chem. Neue Folge*, **22**, 417 (1959).
7. R. N. Jones, *Chem. Rev.*, **41**, 353 (1947).
8. (a) C. L. Harberts, P. M. Heertjes, L. J. N. Van der Hulst, and H. I. Waterman, *Bull. Soc. Chim. Fr.*, **3**, 643 (1936); (b) C. de Borst, P. M. Heertjes, and H. I. Waterman, *Bull. Soc. Chim. Fr.*, **5**, 888 (1938).
9. A. Albert and R. Goldacre, *J. Chem. Soc.*, 454 (1943).
10. M. L. Bailey and J. P. M. Bailey, *Theor. Chim. Acta*, **16**, 303 (1970).
11. R. M. Acheson, M. L. Burstall, C. W. Jefford, and B. F. Sansom, *J. Chem. Soc.*, 3742 (1954).
12. L. E. Orgel, in R. M. Acheson, *The Acridines*, John Wiley, New York, 1956, Chap. 7.
13. S. J. Angyal and C. L. Angyal, *J. Chem. Soc.*, 1461 (1952).
14. S. F. Mason, *J. Chem. Soc.*, 1281 (1959).
15. N. Bacon, A. J. Boulton, R. T. C. Brownlee, A. R. Katritzky, and R. D. Topsom, *J. Chem. Soc.*, 5230 (1965).
16. A. Albert and R. Goldacre, *J. Chem. Soc.*, 706 (1946).
17. A. R. Sukhomlinov, *Zh. Obshch. Khim.*, **28**, 1038 (1958); *Chem. Abstr.*, **52**, 16350g (1958).
18. E. Spinner, *Aust. J. Chem.*, **16**, 174 (1963).
19. D. P. Craig, *J. Chem. Soc.*, 534 (1946).
20. H. Ito and Y. J. l'Haya, *Int. J. Quantum Chem.*, **2**, 5 (1968).
21. V. Zanker, *Z. Phys. Chem.*, **199**, 225 (1952).
22. M. E. Lamm and D. M. Neville, Jr., *J. Phys. Chem.*, **69**, 3872 (1965).
23. R. E. Ballard and C. H. Park, *J. Chem. Soc.*, A, 1340 (1970).
24. (a) A. Blake and A. R. Peacocke, *Biopolymers*, **6**, 1225 (1968); (b) B. J. Gardner and S. F. Mason, *Biopolymers*, **5**, 79 (1967).
25. A. Albert and L. N. Short, J. Chem. Soc., 760 (1945).
26. A. I. Gurevich and Yu. N. Sheinker, *Zh. Fiz. Khim.*, **33**, 883 (1959); *Chem. Abstr.*, **54**, 8285b (1960).
27. V. Zanker and A. Wittwer, *Z. Phys. Chem. Neue Folge*. **24**, 183 (1960).
28. S. Nitzsche, *Angew. Chem.*, **52**, 517 (1939).
29. A. Albert and J. N. Phillips, *J. Chem. Soc.*, 1294 (1956).
30. R. Bolton, D. Phil. Thesis, Oxford University, Oxford, England, 1970.
31. V. P. Maksimets and O. N. Popilin, *Khim. Geterotsikl. Soedin.*, 191 (1970); *Chem. Abstr.*, **73**, 13681e (1970).
32. R. D. Brown and F. N. Lahey, *Aust. J. Sci. Res.*, A3, 593 (1950).
33. H. Brockmann and H. Muxfeldt, *Chem. Ber.*, **89**, 1379 (1956).
34. R. M. Acheson and B. Adcock, with G. M. Glover and L. E. Sutton, *J. Chem. Soc.*, 3367 (1960).
35. M. Ionescu, A. R. Katritzky, and B. Ternai, *Tetrahedron*, **22**, 3227 (1966).
36. V. Zanker, E. Erhardt, F. Mader, and J. Thies, *Z. Phys. Chem. Neue Folge*, **48**, 179 (1966).
37. V. Zanker and B. Schneider, *Z. Phys. Chem. Neue Folge*, **68**, 19 (1969).

38. V. Zanker and G. Schiefele, *Z. Elektrochem.*, **62,** 86 (1958); *Chem. Abstr.*, **52,** 9762b (1958).
39. V. Zanker and A. Reichel, *Z. Elektrochem.*, **63,** 1133 (1959); *Chem. Abstr.*, **54,** 5243b (1960).
40. C. W. N. Cumper, R. F. A. Ginman, and A. I. Vogel, *J. Chem. Soc.*, 4525 (1962).
41. R. M. Acheson and M. J. T. Robinson, *J. Chem. Soc.*, 484 (1956).
42. L. E. Orgel, *Quart. Rev.*, **8,** 422 (1954).
43. V. Zanker and A. Reichel, *Z. Elektrochem.*, **64,** 431 (1960); *Chem. Abstr.*, **54,** 13855i (1960).
44. V. Zanker and P. Schmid, *Chem. Ber.*, **92,** 615 (1959).
45. G. S. Chandler, R. A. Jones, and W. H. F. Sasse, *Aust. J. Chem.*, **18,** 108 (1965).
46. V. Zanker and W. Flügel, *Z. Naturforsch.*, **19b,** 376 (1964); *Chem. Abstr.*, **61,** 10559h (1964).
47. K. Nishimoto and L. S. Forster, *Theor. Chim. Acta*, **4,** 155 (1966).
48. K. Nishimoto, K. Nakatsukasa, and R. Fujishiro, *Theor. Chim. Acta*, **14,** 80 (1969).
49. W. Seiffert, V. Zanker, H. Mantsch, and B. Schneider, *Tetrahedron Lett.*, **54,** 5655 (1968).
50. W. Seiffert, V. Zanker and H. Mantsch, *Tetrahedron*, **25,** 1001 (1969).
51. W. Seiffert, V. Zanker, and H. Mantsch, *Tetrahedron Lett.*, **30,** 3437 (1968).
52. W. Seiffert, V. Zanker, and H. Mantsch, *Tetrahedron Lett.*, **40,** 4303 (1968).
53. (a) H. Mantsch, W. Seiffert, and V. Zanker, *Tetrahedron Lett.*, **27,** 3161 (1968); (b) *Rev. Roum. Chim.*, **14,** 125 (1969).
54. V. Zanker, B. Schneider and W. Seiffert, *Tetrahedron Lett.*, **19,** 1497 (1969).
55. W. Seiffert, H. H. Limbach, V. Zanker, and H. Mantsch, *Tetrahedron*, **26,** 2663 (1970).
56. H. Lang and G. Löber, *Tetrahedron Lett.*, **46,** 4043 (1969).
57. L. L. Ingraham and H. Johansen, *Arch. Biochem. Biophys.*, **132,** 205 (1969).
58. M. L. Bailey, D. Phil. Thesis, Oxford University, Oxford, England, 1969.

CHAPTER XI

The Infrared Spectra of Acridines

R. M. ACHESON

The Department of Biochemistry
and
The Queen's College,
University of Oxford, England

The ir spectra of a relatively small number of acridines are scattered through the literature, and most are of value only for comparison with the spectra of new samples of the same acridines as an aid to identification. Spectra of this type are not considered.

1. Acridine

The ir spectrum of acridine (**1**) (Fig. 2) has been compared with that of the much more symmetrical anthracene (**2**) in the 400–3000 cm^{-1} region; above 700 cm^{-1} the compounds have almost identical spectra arising from overtones, combination tones, and planar interactions.[1]

1   **2**

Below 700 cm^{-1}, acridine has many more peaks than anthracene, mainly due to nonplanar vibrations. There are skeletal vibration frequencies that appear to correspond to the C—C stretching of benzene and pyridine, and which can be related to the various spectral regions between 1625 cm^{-1} and about 1475 cm^{-1} discussed by Bellamy.[2] Other absorption frequencies corresponding to single C—H bonds (900–986 cm^{-1}), two adjacent C—H bonds (700–750, 800–860 cm^{-1}) three adjacent C—H bonds (750–810, 680–725 cm^{-1}), and four adjacent C—H bonds (735–770 cm^{-1}) have been allocated.[3] The suggestion[4] that the 810 cm^{-1} frequency is associated with the 9-hydrogen atom is difficult to sustain, as a band appears at this position in the spectra of several of the 9-substituted acridines which are reproduced. Three acridine N-oxides possess strong absorption in the 1320–1370 cm^{-1} region which has been assigned to the N–O vibration.[5] However, this is a poor criterion for the presence, or absence, of the N-oxide group in an unidentified acridine, since many acridines which are not N-oxides possess absorption maxima in this range.

2. Amino- and Hydroxy-acridines

The ir spectra of all five possible monoaminoacridines have been examined,[7] and all show quite similar stretching (3115–3450 cm^{-1}) and bending (1619–1661 cm^{-1}) frequencies normal for aromatic primary amino groups. Studies[8] on corresponding N-deuteriated acridines support these results; it can be concluded that the primary amino group is present as such, and not as the tautomeric imino form, in all these compounds. Partial deuteriation of the amino group of 9-aminoacridine has been achieved,[9] and the scissors vibration mode of the hydrogen and deuterium atoms attached to the nitrogen atom observed. The presence of much of the tautomeric form **4** in chloroform and the other solvents used is therefore excluded. However, in aqueous solvents, the pK_a of 9-aminoacridine (**3**) and 9-imino-10-methylacridan (**5**) can be considered to suggest that a 9:1 equilibrium mixture of 9-aminoacridine (**3**) and the imino tautomer (**4**) exists under these conditions.[7, 10]

3   4   5

The ir spectra of the mono- and diacetyl derivatives of 9-aminoacridine show that the acyl groups are substituents of the 9-amino group, and there is no maximum in the 1600–1650 cm^{-1} region expected[11] of a 9-acylimino-10-acylacridan because of the presence of the N=C bond. In contrast, both 9-trichloro- and 9-trifluoro-acetamidoacridine show maxima at 1619 cm^{-1} not shown by any of the other trichloroacetamidoacridines; this indicates[11] that they possess the iminoacridan structure (6).

12-Aminobenz[b]acridine in chloroform and carbon tetrachloride,[7] in contrast to 9-aminoacridine, exists apparently entirely as the imino tautomer (7). It shows the 10-hydrogen (N—H) frequency at 3448 cm^{-1}, close to that of diphenylamine (3428 cm^{-1}) and the 9-imino hydrogen atom at 3298 cm^{-1}.

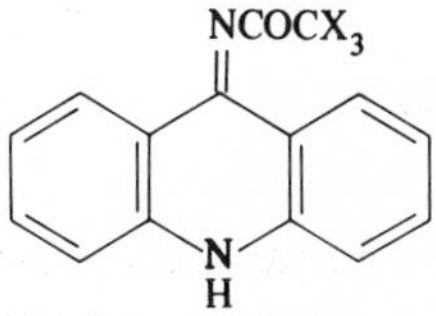

6,   X = Cl or F

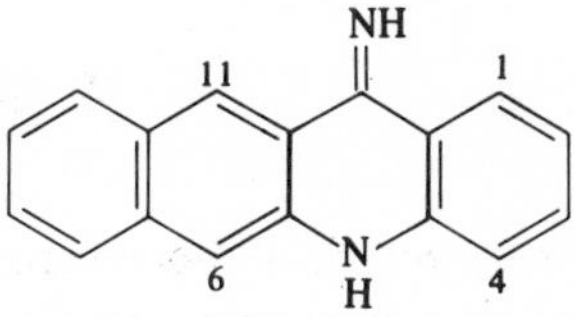

7

The N—H-bond-stretching frequencies for the monoaminoacridines in chloroform and carbon tetrachloride are similar, but a substantial lowering of the frequencies in dioxan and pyridine (ca. 150 cm^{-1}) was observed for all the acridines except the 4-amino derivative, where the lowering in pyridine was only 50 cm^{-1}. This is attributed to the presence of an intramolecular hydrogen bond (8). Similar hydrogen bonding is revealed by 4-hydroxyacridine (9) in carbon tetrachloride, as the hydroxyl group appears as a broad band at 3398 cm^{-1}.

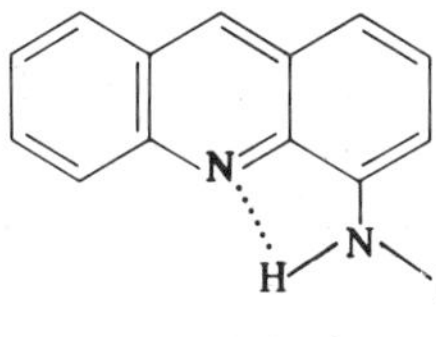

8

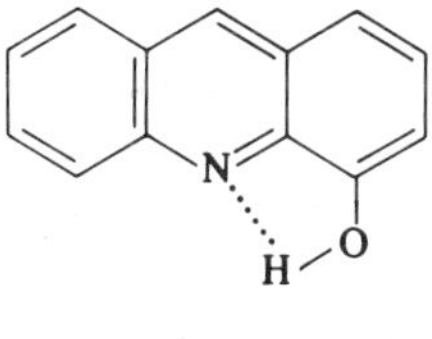

9

The hydroxyl groups of 1- and 2-hydroxyacridine in the same solvent appear as sharp peaks at 3607 and 3616 cm^{-1}, respectively.[12]

3. Acridans

Acridan (10), and 9-alkyl and 9-phenyl derivatives, in chloroform showed an N—H stretching frequency at 3424–3430 cm^{-1}, similar to that of diphenylamine (3428 cm^{-1}), which changes to 2548–2553 cm^{-1} on deuteriation.[13]

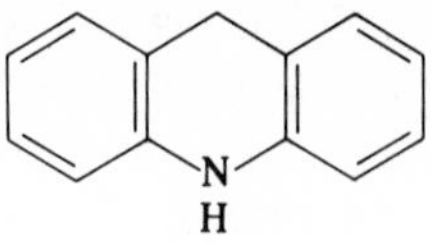

10

The N—H frequency for acridan, both in the solid phase (KBr) and in the tetrahydrofuran solution, shows a reduction of about 50 and 100 cm^{-1}, which is attributed to association involving the 10-hydrogen atom.[13] The ir spectra of some more highly reduced acridines are mentioned in Chapter V (p. 499).

4. 9-Acridanones

The ir spectrum of 9-acridanone (**11**) (Fig. 10) in the solid phase shows a carbonyl stretching frequency at 1645 cm^{-1} and a characteristic broad N—H stretching region indicating strong intermolecular hydrogen bonding (Chapter III, p. 200).

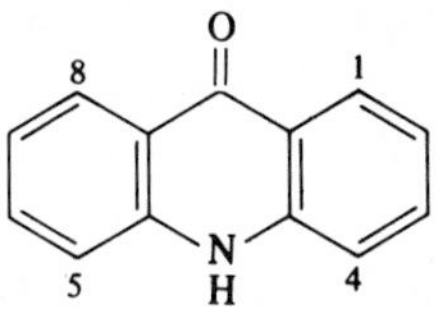

11

The carbonyl frequencies for 9-acridanone and a series of 10-methyl-9-acridanones have been reported at 1632 ± 6 cm^{-1}, showing that the carbonyl frequency is little affected by hydrogen bonding,[14] as is the case for anthraquinones.[15]

There is no indication of the presence of the possible 9-hydroxyacridine tautomer in 9-acridanone itself. Replacement of the N—H by N—[²H] gives the corresponding N—[²H] frequencies with a reduced band width, but the fine structure is retained, which suggests that the spectra have a common origin.[13] The complexity of the N—H absorption has been attributed to the superimposition of peaks, due to different molecular species present in the lattice.[13] Alkyl[16] and halogen[13] substituents at position 2 have almost no effect on the N—H region and the carbonyl frequency (1640–1645 cm^{-1}). 1- and 4-Mono-alkyl-9-acridanones have a different N—H region, indicating a new arrangement of the molecules in the crystals,[16] but the carbonyl frequency is

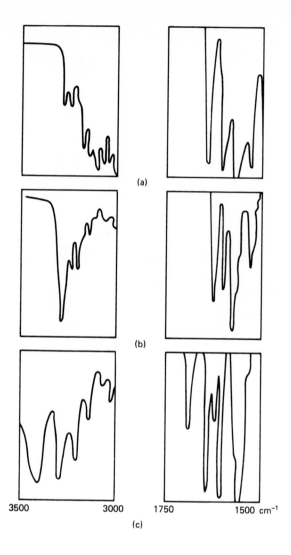

| 3500 | 3000 | 1750 | 1500 cm⁻¹ |

Fig. 1. 9-Acridanones in nujol[16]: (a) 2-ethyl-9-acridanone; (b) 4-isopropyl-9-acridanone; (c) 1-isopropyl-4-methyl-9-acridanone.

virtually unaltered (1630–1640 cm^{-1}). However, 1-isopropyl-4-methyl-9-acridanone[16] (Fig. 1) shows a very different N—H region and carbonyl peaks at both 1640 and 1690 cm^{-1}. It therefore appears that the steric effect of substituents at positions 1 and 4, particularly when both types are present, can reduce intermolecular hydrogen bonding in the lattice. This is consistent with the remarkably low mp (75°) for this compound, since it is unsubstituted at position 10 (Chapter III, p. 200). However, at 90–100°, it resolidifies; it remelts at 208–209°. The ir spectrum of the high melting form is identical with those of the 1- and 4-alkyl-9-acridanones in the N—H and carbonyl regions (Fig. 1), thereby indicating a change in crystal lattice. Cooling of the melt from 209° to room temperature again gave material of mp 75°. The maximum at 1690 cm^{-1} cannot be attributed to non-hydrogen-bonded carbonyl groups, as 10-methyl-9-acridanone shows its carbonyl absorption at 1640 cm^{-1}. The N—H region of the spectrum for 2,4,5,7-tetrachloro-9- acridanone, measured in a nujol mull,[13] is very much simpler than that of 9-acridanone itself and shows only two peaks at about 3355 and 3065 cm^{-1}. Octachloro-9-acridanone shows only one N—H frequency at about 3370 cm^{-1}. These results can be attributed to the steric effects of the peri substituents that reduce intermolecular interaction in the crystal lattice.[13]

The ir spectrum of 9-acridanthione (Fig. 12) has a similar N—H region to that of the 9-acridanone. In dilute chloroform solution, this region simplifies to give a single band at 3420 cm^{-1} that is similar to those of diphenylamine and acridan. Little association, therefore, seems to take place under these conditions.

Investigations of the structures of the 9-acridanone alkaloids (Chapter IV, p. 386) provided the first ir spectra of 9-acridanones and showed that hydroxyl groups present at the 2-, 3-, and 4-positions of the ring could be detected at 3250–3280 cm^{-1}, while those at position 1 (e.g., 12) showed no "free hydroxyl" absorption in this region.[17]

12

This indicates strong hydrogen bonding and agrees with Flett's observations that the hydroxyl absorption is also absent in anthraquinones possessing only hydroxyl groups peri to the carbonyl groups.[15]

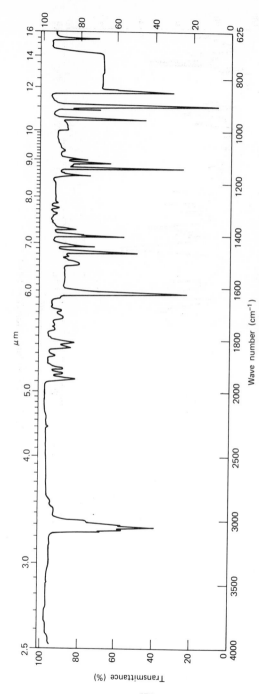

Fig. 2. Acridine in carbon tetrachloride.[18]

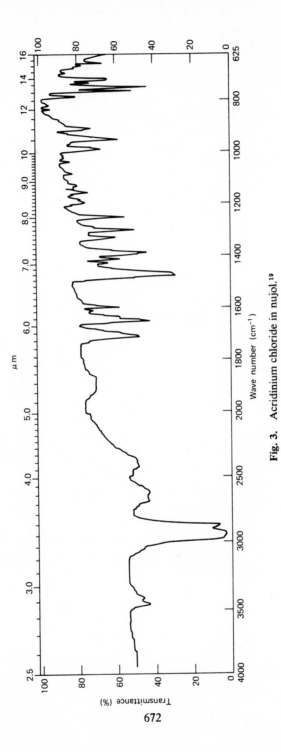

Fig. 3. Acridinium chloride in nujol.[19]

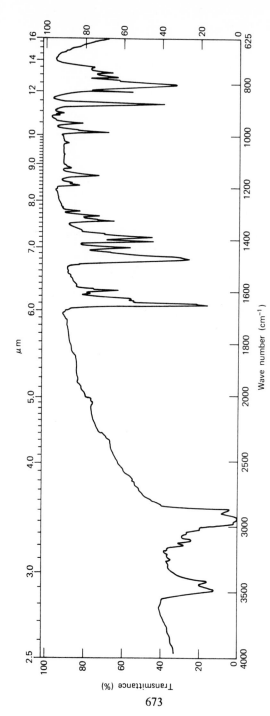

Fig. 4. Acridinium bromide in nujol.[19]

673

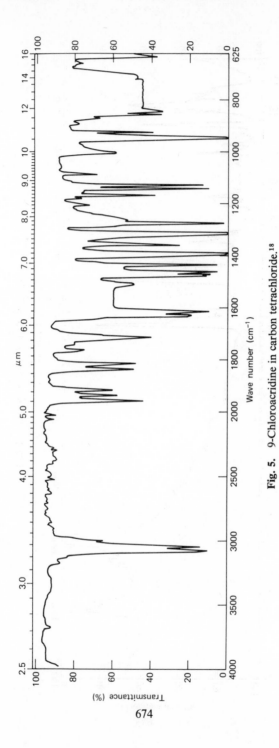

Fig. 5. 9-Chloroacridine in carbon tetrachloride.[18]

674

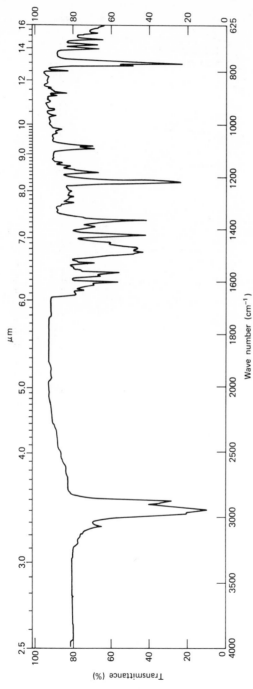

Fig. 6. 9-Phenoxyacridine in nujol.[19]

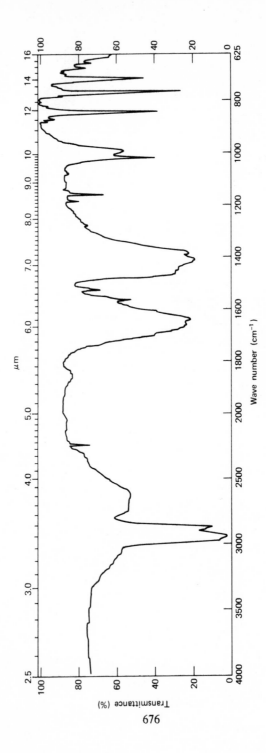

Fig. 7. 9-Cyanoacridine in nujol.[19]

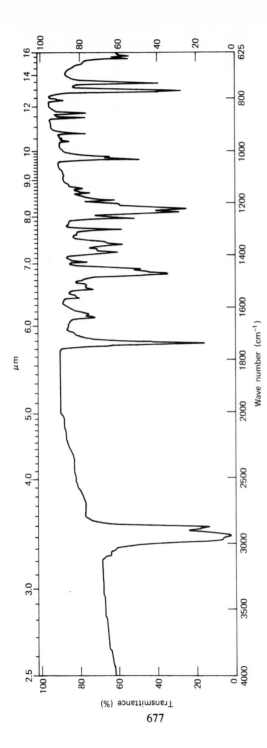

Fig. 8. Methyl acridine-9-carboxylate in nujol.[19]

677

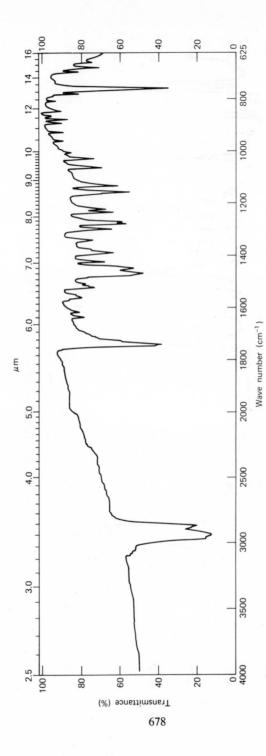

Fig. 9. Methyl 3-(9-acridinyl)propionate in nujol.[19]

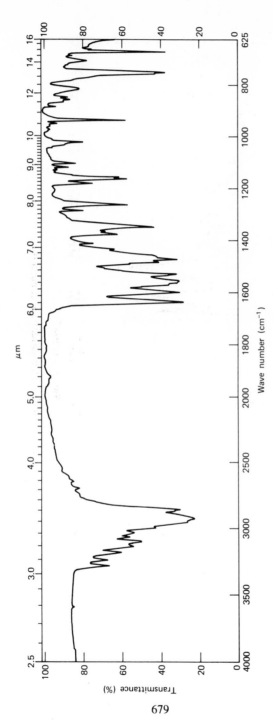

Fig. 10. 9-Acridanone in nujol.[19]

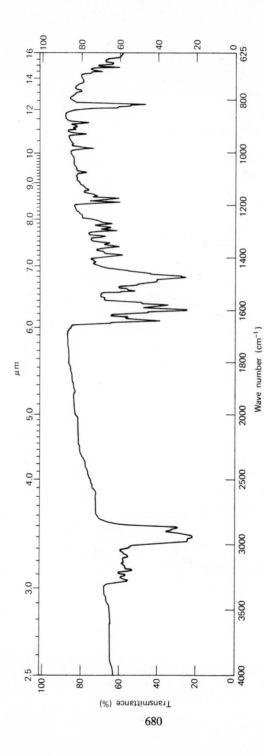

Fig. 11. 2,7-Dibromo-9-acridanone in nujol.[19]

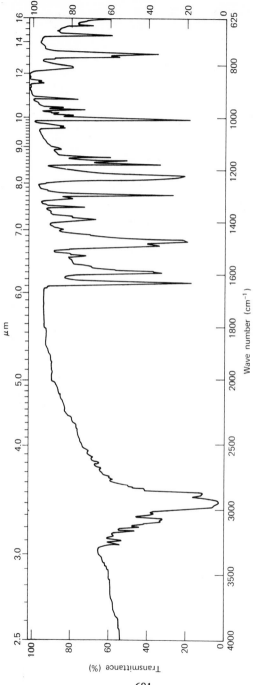

Fig. 12. 9-Acridanthione in nujol.[19]

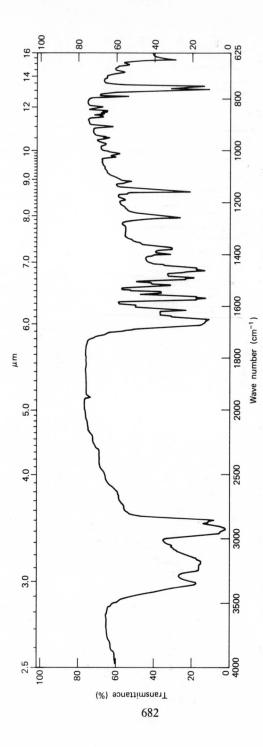

Fig. 13. 9-Aminoacridine in nujol.[18]

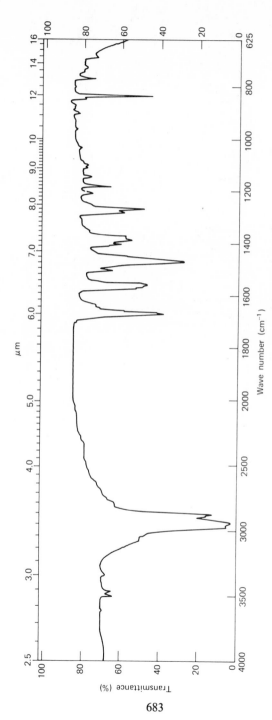

Fig. 14. 9-Amino-2,7-di-*tert*-butylacridine in nujol.[19]

683

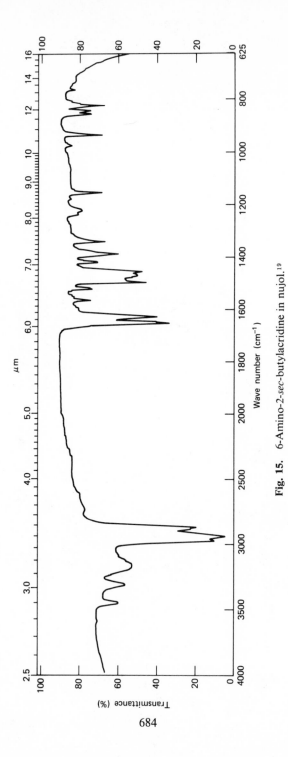

Fig. 15. 6-Amino-2-*sec*-butylacridine in nujol.[19]

References

1. O. U. Fialkovskaya and A. U. Nefedov, *Opt. Spektrosk.*, **25**, 766 (1968).
2. L. J. Bellamy, *Infra-red Spectra of Complex Molecules*, Methuen, London, 1954.
3. H. H. Perkamstus and E. Z. Baumgarten, *Z. Elecktrochem.*, **64**, 951 (1960).
4. A. Albert, *The Acridines*, 2nd ed., Arnold, London, 1966, p. 211.
5. I. Ionescu, H. Mantsch, and I. Goia, *Chem. Ber.*, **96**, 1726 (1963).
6. R. M. Acheson and B. Adcock, *J. Chem. Soc.*, C, 1045 (1968).
7. S. F. Mason, *J. Chem. Soc.*, 1281 (1959); 4584 (1952).
8. Y. N. Sheinker and E. M. Peresleni, *Dokl. Akad. Nauk, SSSR*, **131**, 1366 (1960); *Chem. Abstr.*, **54**, 21094 (1960).
9. N. Bacon, A. J. Boulton, R. T. C. Brownlee, A. R. Katritzky, and D. T. Topson, *J. Chem. Soc.*, 5230 (1965).
10. R. M. Acheson, M. L. Burstall, C. W. Jefford, and B. F. Sansom, *J. Chem. Soc.*, 3742 (1954).
11. A. I. Gurevich and Y. N. Sheinker, *Zh. Fiz. Khim*, **36**, 734 (1962); *Chem. Abstr.*, **57**, 7231a (1962).
12. S. F. Mason, *J. Chem. Soc.*, 4874 (1957).
13. V. Zanker, H. H. Mantsch, and E. Erhardt, *An. Quim.* (Spain), **64**, 659 (1968).
14. J. R. Price and J. B. Willis, *Aust. J. Chem.*, **12**, 589 (1959).
15. M. St. C. Flett, *J. Chem. Soc.*, 1441 (1948).
16. R. G. Bolton, D. Phil. Thesis, University of Oxford, England, 1970.
17. W. D. Crow and J. R. Price, *Aust. J. Sci. Res.*, **A2**, 282 (1949).
18. J. K. Stubbs, private communication.
19. C. W. Harvey, private communication.

CHAPTER XII

The Nuclear Magnetic Resonance Spectra of Acridines

R. M. ACHESON

The Department of Biochemistry
and
The Queen's College,
University of Oxford, England

1. Acridines

The nmr spectrum[1] of acridine (**1**) was first examined at 60 MHz in 1963 in both dimethyl sulfoxide and deuteriochloroform solutions (Table I). The observed spectrum was computer-simulated, using the assumption that there would be no coupling between hydrogen atoms at positions 1 to 4 with those at positions 5 to 8 and that a first-order interaction between the hydrogen atoms of the carbocyclic rings and the 9-hydrogen atom was possible.

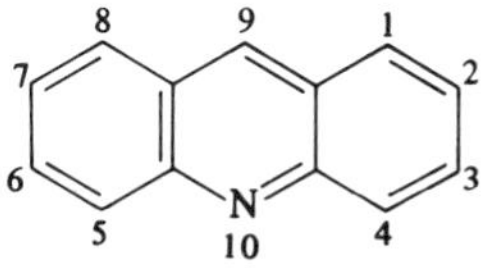

1

Very satisfactory agreement (± 0.1 Hz) was obtained between the observed and calculated spectra, and almost as good agreement has now been obtained using the same parameters for spectra run at 100 MHz in deuteriochloroform.[2] In spite of the fact that such good agreement has been obtained at two frequencies, it should be noted that alternative parameters, giving as good a fit with the experimental spectra, have not been sought and the accuracy of the parameters is only reasonably certain to a few tenths of a Hertz (cf. Ref. 3).

Although it was clearly established from these results that the carbocyclic ring protons could be placed in a definite order, ABCD, the assignment of protons A to position 1 and 8, or 4 and 5, of the rings could not be made with certainty. The assignment actually made, on the basis of the electron densities calculated for the various positions both by MO[4a−c] and valence bond[5] methods, have been established as correct by comparison with the unambiguous assignments that can be made for 2,7,9-[^{2}H]$_3$-acridine[6] and other substituted acridines (Table I). The simplification of the spectrum of acridine, caused by the replacement of the 9-hydrogen atom by deuterium, is exactly in accord with the expected elimination of the long range 1,9- and 4,9-couplings. The 4,9-coupling is remarkably strong. The other coupling constants are normal for aromatic-type protons, but it is interesting that $J_{1,2}(J_{7,8})$ is a little less than $J_{3,4}(J_{5,6})$ and is markedly greater than $J_{2,3}(J_{6,7})$. Although assignments of the protons for the 4-methyl- and 2-ethyl-acridines have been made for the 100 MHz spectra, superimposed resonance lines make the task difficult and a little uncertain. The spectra at 220 MHz,* however, are almost first order (e.g., Fig. 1) and have confirmed the analysis.

The effect of the concentration of acridine on the chemical shifts of the protons in dimethyl sulfoxide is negligible, but in deuteriochloroform there are interesting changes.[1] On dilution of 10% solutions, the resonance positions of all the protons, except those at positions 4 and 5, move upfield, the biggest shift being shown by the 9-proton.

The effect of the bromine atoms of 2,7-dibromoacridine (Table I) on the chemical shifts of the protons in the same ring is less than that observed for benzene rings,[7] and even a 9-ester group, probably because of its lack of planarity with the ring, exerts only a small deshielding effect on the 1 and 8 protons. In contrast, these protons show a marked downfield shift in 9-methylthioacridine, presumably caused by the magnetic effects of the sulfur atom.

The assignment of the protons in 9-substituted acridines, not possessing other substituents, is difficult but can be done assuming that the 2-protons will appear at the highest field of the aromatic group. For most acridines so far examined, it appears that $J_{3,4} > J_{1,2}$. This could be used as an assignment

*Recorded for us by M. S. Sunley.

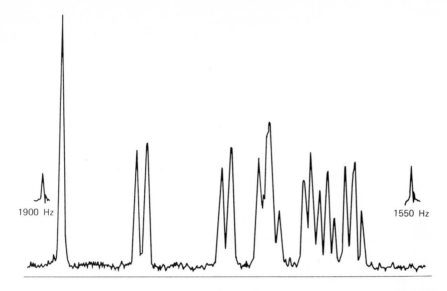

1900 Hz   1550 Hz

Fig. 1.   The nmr spectrum of acridine in carbon tetrachloride, measured at 220 MHz.

criterion, but as these coupling constants may differ only by a few tenths of a Hertz, and are consequently difficult to determine with precision, it would not appear to be a reliable method. For the 9-alkylacridines the spectral analogy with acridine itself is adequate, but for methyl acridine-9-carboxylate (2) and dimethyl acridine-9-phosphonate (Table I) the situation is less certain. However, if the ester group of 2 can be presumed not to move the 4- and 5-protons upfield, as compared with acridine, then all the protons of this compound can be assigned as shown in Table I and have similar shifts to those of acridine itself. As no marked deshielding of the 1- and 8-protons can occur, it is clear that the steric requirements of the ester group require it to be oriented at roughly 90° to the plane of the ring.

The nmr spectrum of acridine 10-oxide (3) has been measured (Table I) and computer-simulated, although the accuracy with which this has been done is not stated.[8]

CO$_2$Me

2

3

TABLE I. Nuclear Magnetic Resonance Spectra of Acridine and Simple Derivatives Given in τ, J in Hertz

Substituents	MHz	Solvent	Proton resonances										Ref.
			1	2	3	4	5	6	7	8	9	10	
The Bases													
None	60	DMSO[a]	1.82 $J_{1,2}$ 8.2	2.36 $J_{1,3}$ 1,4	2.11 $J_{1,4}$ 0.6	1.78 $J_{1,9}$ 0.4	$J_{2,3}$ 6.6	$J_{2,4}$ 1.2	$J_{3,4}$ 9.0	$J_{4,9}$ 0.9	0.90		1
	60	$CDCl_3$[b]	1.97	2.46	2.20	1.75					1.20		1
	60	$CDCl_3$[c]	2.19	2.61	2.33	1.79					1.47		1
2,7,9-[^{2}H]$_3$	100	$CDCl_3$	1.96		2.20	1.74	$J_{3,4}$ 9.0 ± 0.2						6
2-CHBrMe	100	$CDCl_3$	ca. 2.1	CHMe, 4.63; CHMe, 7.92; J 7.0	2.2	1.8	1.8	2.3	2.52	2.16	1.60		6
9-Bu[n]	100	$CDCl_3$	1.82	2.56	2.32	1.82	[MeC_3H_7, 9.02t J 7; $C_3H_7CH_2$, 6.48t, J 7.5; $Me(CH_2)_2CH$, 8.05-8.70;						6
9-CN	100	$CDCl_3$	1.6-1.8	2.0-2.4	2.0-2.4	1.6-1.8	$J_{1,3}$ 2.0						2
2,7-Br$_2$	100	$CDCl_3$	1.77		2.11	1.84	1.82	$J_{3,4}$ 9.3 ± 0.1			1.37		6
2-Et	100	CCl_4	2.41	$MeCH_2$, 8.65; $MeCH_2$, 7.15; J 8	2.46	1.90		2.35	2.59	2.14	1.50		6
2-Et	220	CCl_4	2.47 $J_{3,4}$ 8.5		2.49	1.93 $J_{5,6}$ 8.5	1.83 $J_{5,7}$ ca. 1.5	2.37 $J_{6,7}$ 7.0	2.66 $J_{6,8}$ ca. 1.0	2.25 $J_{3,8}$ 8.0	1.58		6
4-Et	100	$CDCl_3$	2.15	2.50	2.37	$MeCH_2$, 8.54; $MeCH_2$, 6.53; J 8	1.70	2.32	ca. 2.45	2.02	1.30		6
9-CO_2Me	100	$CDCl_3$	1.99	2.44	2.22	1.74	$J_{1,2}$ 8.9 1.69	$J_{1,3}$ 1.8 2.24	$J_{1,4}$ 0.7 2.52	$J_{2,3}$ 6.6 2.04	$J_{2,4}$ 1.4 1.32	$J_{3,4}$ 8.6	2
4-Me	100	$CDCl_3$	2.19	2.60	2.39	7.06[o]	$J_{5,6}$ 8.7 1.79[k]	2.35[k]					6
4-Me	220	CCl_4	2.31[k] $J_{1,2}$ 8.5	2.71	2.51[k] $J_{2,3}$ 6.5 2.33				2.59[k]	2.15[k]	1.45		6
9-Me	100	$CDCl_3$	1.93 $J_{1,2}$ 8.5	2.58	$J_{1,3}$ 6.3 2.33	1.79 $J_{2,4}$ 1.2	$J_{5,6}$ 8.5	$J_{5,7}$ ca. 1.5	$J_{6,7}$ 6.5	$J_{7,9}$ 9	7.05		2
9-SMe	100	DMSO	1.25 $J_{1,2}$ 1.4	1.95-2.35	1.95-2.35	1.75	$J_{3,4}$ 9 $J_{1,3}$ ≃ $J_{2,4}$ ≃ 1.8	$J_{1,3}$ ≃ $J_{2,4}$ ≃ 1.8	$J_{1,2}$ 8.4	$J_{3,4}$ 8	e		2
10-Oxide	60	$CDCl_3$	2.12 $J_{1,2}$ 8.2	2.50	2.28	1.16	$J_{2,3}$ 6.6	$J_{2,4}$ 1.0	$J_{3,4}$ 9.0	$J_{4,9}$ 1.0	1.88		8

Table I. (Continued)

Substituents	MHz	Solvent	Proton Resonances										Ref.
			1	2	3	4	5	6	7	8	9	10	
The Bases													
4-Ph	100	$CDCl_3$	1.95-2.34[f]		2.36-2.60[g]		1.81				1.21		6
4-Pr^i	100	$CDCl_3$	2.15	2.56	ca. 2.4	[h]	1.70	2.24	ca. 2.4	2.02	1.30		6
9-Pr^i	100	$CDCl_3$	1.84	2.59	2.34	1.84					[i]		6
9-PO(OEt)$_2$	60	$CDCl_3$	0.65m	2.26m	2.26m	1.70	1.70m	2.26m	2.26m	0.65m	[r]		10
The Salts													
None	60	CF_3CO_2H	1.4-1.74m	2.02[j]	1.4 ———— 1.75m		$J_{1,2}$ 8				0.26		*
None, HI	100, 100	DMSO[a]	1.42[k]	2.0[j]	1.61[l]	1.61[l]					-0.1		2
2,7-Br$_2$	and 60	CF_3CO_2H	1.33		1.55	1.73					1.38		*
2-Et	100	CF_3CO_2H	$J_{1,2}$ 1.7; 1.70[n]	$J_{3,4}$ 9.4; m	1.6 ————			—1.8m	1.98[j]	1.53[d]; $J_{1,8}$ 8.5	0.33		†
4-Me	100	CF_3CO_2H	1.4	2.2m	2.2m	6.96[o]	1.4			2.2m	0.21		†
10-Me, MeSO$_4$	60	DMSO[a]	1.1-1.8m	2.02t[k]	1.1-1.8m	1.1-1.8m					-0.11	5.20[o]	*
10-Pr^n, I	60	D_2O	1.7	1.91t[k,p]	1.48[k,p]	2.4m					0.77	5.12[o]	†
	100	DMSO[a]	1.30[d,k,p]; $J_{1,2}$ 8.5	$J_{1,3}$ ca. 6	$J_{3,4}$ 9.0	1.15[d,k,p]					-0.19	[q]	*

Multiplicities are not given where it follows from the coupling constants, which are reported whenever possible.

[a] Fully deuteriated. [b] 10% solution. [c] Calcd for infinite dilution. [d] CO_2Me, γ 5.82. [e] Obscured by solvent. [f] 6 protons. [g] 5 protons. [h] MeC_2H_4, 8.90, J 7; $MeCH_2CH_2$, 8.07m; $EtCH_2$, 6.60t, J 8. [i] MeC_2H_4, 8.93t, J 7.0; $MeCH_2CH_2$, 8.18m; [j] 6 lines. [k] Apparent doublet with little additional splitting. [l] Small further splitting. [m] $MeCH_2$, 6.88q, J 7.5. [n] Apparent singlet standing out from complex multiplet. [o] Me. [p] Assignments uncertain. [q] $Me(C_2H_4)$, 8.80t, J 7.3; $MeCH_2CH_2$, 7.9m; $EtCH_2$, 4.60t, J 7.7. [r] $MeCH_2$, 5.75m, J 7.
* Measured by P. J. Abbott; † measured by Mrs. E. E. Richards.

The large inter-ring couplings are similar to those for acridine itself, and the protons have been assigned on the assumption that the 10-oxide group, like a carbonyl group, will cause a marked downfield shift of the *peri* (4 and 5) protons. The resonance of the 9-proton appears at higher field than that of acridine itself and is consistent with the relatively high electron density at this position as shown by dipole moment studies and the chemistry of the compound (see Chapter I, p. 75).

Comparison of the nmr spectra of 2,7-dibromoacridine in deuteriochloroform and in trifluoroacetic acid, where the ring is protonated at position 10, shows unequivocally that the downfield shift of the 9-, 3,7-, 1,8-, and 4,5-hydrogen atoms are τ 0.99, 0.56, 0.44, and 0.11, respectively. It is impossible to say how much of this is a result of the change in the solvent. The large shift of the 9-hydrogen atom is similar to that observed for the 4-proton of 2-methylquinoline under the same conditions.[9] On the assumption that similar shifts occur, the spectra of acridinium iodide and trifluoroacetate can be assigned; this interpretation is consistent with that made for 2-ethylacridinium trifluoroacetate.

2. 9-Chloroacridines

The nmr spectra of eighteen 9-chloroacridines[6] have been examined, and first-order analyses of the spectra (in most cases), have enabled unambiguous proton assignments to be made (Table II). Only two cases, 9-chloroacridine (4) and its 4-*sec*-butyl derivative, have been examined sufficiently closely to enable accurate coupling constants and chemical shifts to be obtained.

4

The spectrum for the 4-*sec*-butyl compound was measured at both 60 and 100 MHz and measured to ± 0.2 Hz. Iteration assuming independent ABCD and ABC systems, and using LACON III, yielded parameters that gave calculated spectra agreeing with the observed spectra within the errors of experimental measurement (Fig. 2). The coupling constants for the corresponding positions of the two carbocyclic rings are not exactly equal; it is interesting that the values for $J_{5,6}$ and $J_{7,8}$ for both of these acridines are closer together than the values for acridine itself, although $J_{5,6}$ is the greater, as it is for most acridines so far examined carefully.

TABLE II. Nuclear Magnetic Resonance Spectra of 9-Chloroacridines Recorded in τ^*, Measured at 100 MHz in Deuteriochloroform

Substituent	Proton resonances								Ref.
	1	2	3	4	5	6	7	8	
None	1.76	2.54	2.32	1.88	$J_{2,4}$ 1.3	$J_{3,4}$ 8.9			2
	$J_{1,2}$ 8.6	$J_{1,3}$ 1.6	$J_{1,4}$ 0.7	$J_{2,3}$ 6.4					
2-CHBrMe	1.61	a	2.05	1.74	1.74	2.17	2.36	1.58	6
4-CHBrMe	1.60	ca. 2.4	1.88	b	1.72	2.18	ca. 2.4	1.60	6
4-CHBrEt	1.67	ca. 2.4	1.99	c	1.75	2.23	ca. 2.4	1.67	6
2-Bu^n	1.92	d	2.40	1.87	1.80	2.28	2.48	1.67	6
2-Bu^t	1.67	8.51^e	2.07	1.80	1.76	2.12	2.40	1.57	2
4-Bu^{sec}	$J_{1,3}$ ca. 2 1.68	$J_{3,4}$ 9 2.42	2.32	f	1.72	2.21	2.38	1.58	6
	$J_{1,2}$ 8.4; $J_{7,8}$ 9.0	$J_{1,3}$ 1.7	$J_{2,3}$ 7.3	$J_{5,6}$ 9.1	$J_{5,7}$ 1.2	$J_{5,8}$ 0.65	$J_{6,7}$ 6.7	$J_{6,8}$ 1.2	
2-CBr$_2$Me	1.48	6.80^g	1.67	1.67	1.73	2.13	2.34	1.59	6
2,7-diBu^t	1.67	8.51^e	2.08	1.80	$J_{1,3}$ ca. 2	$J_{3,4}$ ca. 8.5			2
1,3-(OMe)$_2$- 2-CO$_2$Me^m	5.99^g	5.97^g	5.95^g	2.73	1.5 ————			2.5m	11
2-Et	1.87	h	2.35	1.87	1.80	2.25	2.45	1.65	6
4-Et	ca. 1.77	2.53	ca. 2.41	i	ca. 1.77	2.27	ca. 2.45	1.65	6
1-Me	6.89^g	2.69	ca. 2.43	1.94	1.84	2.24	2.43	1.55	6
2-Me	1.83	7.42^g	2.37	1.87	1.75	2.22	2.41	1.58	6
3-Me	1.79	2.57	7.41^g	2.01	1.71	2.21	2.42	1.60	6

(Table Continued)

693

Table II. (Continued)

Substituent	Proton Resonances								Ref.
	1	2	3	4	5	6	7	8	
4-Me	1.74	2.54	ca. 2.4	7.10g	1.74	2.22	ca. 2.4	1.62	6
4-Me-1-Pr^i	j	ca. 2.50	ca. 2.50	7.18g	1.79	2.27	2.44	1.47	6
4-Pr^n	1.69	2.49	ca. 2.37	k	1.72	2.23	ca. 2.4	1.58	6
4-Pr^i	1.72	2.44	ca. 2.35	l	1.74	2.24	2.42	1.60	6

* $J_{1,2} \simeq J_{3,4} \simeq 9$, $J_{2,3} \simeq 7$, $J_{1,3} \simeq J_{2,4} \simeq 1.5$ Hz unless cited.

a Me, 7.82d, J 7.5; CH, 4.54q. b Me, 7.76d, J 7.1; CH, 2.98q. c Me, 8.86t, J 8.0; $MeCH_2$, 7.60 (six lines clear with further splitting); CH, 3.23t, J 8.5. d $Me(CH_2)_3$, 9.05t, J 7; $MeCH_2(CH_2)_2$, 8.60m; EtCH_2CH$_2$, 8.30m; PrCH_2, 7.20t, J 7.5. e Bu^t resonance. f $Me(C_3H_7)$, 9.00t, J 7; $MeCH_2(C_2H_4)$, 8.09m; EtCHMe, 8.51d, J 7.5. g Me resonance. h Me, 8.63t, J 8; CH_2, 7.13q. i Me, 8.58t, J 8; CH_2, 6.57q. j Me_2CH, 8.65d, J 7; Me_2CH, 5.28m. k Me, 8.95t, J ca. 8; $MeCH_2CH_2$, 8.10 (six lines); EtCH_2, 6.61t, J ca. 8. l Me_2CH, 8.58d, J 8; Me_2 CH, 5.50m. m Measured at 60 MHz.

694

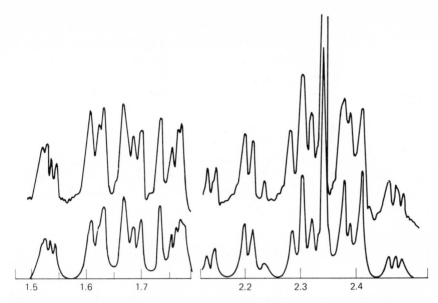

Fig. 2. Upper curve: nmr spectrum of 4-*sec*-butyl-9-chloroacridine in deuteriochloroform, measured at 100 MHz. Lower curve: spectrum calculated by LACON III at a theoretical line width of 0.75 Hz at half height.

TABLE III.   Upfield Proton Shifts (τ) Caused by Alkyl Substitution of 9-Chloroacridines[6]

Substituent	Proton shifts			
	1,2- or 3,4-	2,3-	1,3- or 2,4-	1,4-
2-Bu^n	0.25	0.12	0.07	—
2-Bu^t	0.17	0.01	0.18	—
4-Bu^{sec}	0.13	—	0.02	0.08
2-Et	0.22	0.10	0.08	—
1-Me	0.26	—	0.19	0.12
2-Me	0.25	0.15	0.12	—
3-Me	ca. 0.2	0.15	0.19	—
4-Me	0.30	—	0.12	0.10

695

The deshielding effect of the chlorine atom on the 1 and 8 (*peri*) protons is clearly shown, since they appear at a lower field than in acridine itself, and usually at a lower field than the 4,5-protons. 1-Alkyl groups are affected similarly. The shielding constants (Table III) were assessed[6] on the assumptions that the solvation or association which causes the concentration dependence of the chemical shift for acridine in deuteriochloroform[1] would affect both carbocyclic rings equally, and that the substituent would have no effect on the protons of the unsubstituted ring. A methyl group shields an *ortho* proton across the 1,2- or 3,4-bond by about τ 0.25, and across the 2,3-bond by only τ 0.15. This correlates with the lower proton coupling constant across the 2,3-bond than the 1,2- or 3,4-bonds.

3. Aminoacridines

The nmr spectrum of 9-aminoacridine (**5**) (Table IV) has been recorded at 60 MHz and computer simulated,[1] but the four-spin system was assigned the wrong way around.

5

This became clear when the spectrum of 9-amino-2,7-di-*tert*-butylacridine was examined,[2] for which unambiguous proton assignments can be made if it is assumed that $J_{3,4} > 1$–2 Hz. The low-field protons must then be at positions 1 and 8; in contrast to acridine, the highest-field proton is at position 3.

The lowest-field aromatic protons of 9-amino-2,7-di-*tert*-butylacridinium chloride are again at positions 1 and 8, and can be definitively assigned; consequently the highest field protons in the spectrum of 9-aminoacridinium chloride must be at positions 2 and 7. It is interesting that the downfield shifts of the ring hydrogen atoms on protonation of the molecule are 3 > 1 > 4 > 2, the protons at positions 2 and 7 hardly changing position. The 9-aminoacridinium cation can be written in the resonance forms **6** and **7**, accounting for its high basicity, where the positive charge is close to the 4- and 1-protons. The positive charge can also be placed at positions 1 and 3 (**8** and **9**), where

TABLE IV. Aminoacridines

			Proton resonances										
Substituent	MHz	Solvent	1	2	3	4	5	6	7	8	9	10	Ref.
The Bases													
9-NH$_2$[a]	60	DMSO[b]	1.56 $J_{1,2}$ 8.8 2.1-2.7m	2.35	2.66 $J_{1,4}$ 0.6 2.1-2.7m	2.13 $J_{2,3}$ 6.0 2.1-2.7m	$J_{2,4}$ 1.2 3.13 $J_{5,7}$ 2	$J_{3,4}$ 9.0			c		1
6-NH$_2$-2-Bu^{sec}	60	DMSO[b]		$J_{1,3}$ 1.7 d				4.10[e]	2.99 $J_{7,8}$ 8.6	2.1-2.7m	1.48		*
9-NH$_2$-2,7-Bu$_2$^{t}	60	DMSO[b]	1.83[f]	8.60[g]	2.38[f]	2.38[f]					g		2
3,6-(NH$_2$)$_2$	100	DMSO	2.37	3.12	4.19[e]	3.17	$J_{1,2}$ 8.3	$J_{2,4}$ 2			1.70		2
3,6-(NH$_2$)$_2$-2,7-Me$_2$[a]	60	DMSO[b]	2.51	7.40	4.39[e]	3.09	$J_{1,4}$ <0.2	$J_{4,9}$ 0.1-0.2	$J_{1,9}$ 0.8		1.81		1
3,6-(Me$_2$N)$_2$	60	DMSO[b]	2.12[i]	2.80	6.82	3.39					1.21		1
The Salts													
9-NH$_2$,HCl	60 and 100	DMSO[b]	1.14	2.4 m	1.9 m	1.9 m	$J_{1,2}$ 8.3				h		2, *
9-NH$_2$-2-Bu^{sec},HCl	100	DMSO[b]	1.30	8.55[g]	1.75				2.05 m	1.15	h		2
9-NH$_2$-2,7-Bu$_2$^{t},HCl	100	DMSO[b]	1.35	8.56[g]	1.85	2.03	$J_{1,3}$ 1.5	$J_{3,4}$ 9.1			-0.15[e]		2
3,6-(NH$_2$)$_2$,H$_2$SO$_4$	60	DMSO	2.22 $J_{1,2}$ 8.8	3.07 $J_{2,4}$ 1.7	5.8[e]	3.31					1.28[c]		*

[a] Assignments given in reference have been altered. [b] Deuteriated. [c] Typical broad NH$_2$ peak. [d] MeCH, 9.23t, J 7; MeCH$_2$, 8.40 apparent pentuplet; CH$_2$ CHMe, 7.34, apparent pentuplet: MeCH, 8.77d, J 6.8. [e] NH$_2$ resonance, broad. [f] Apparent singlet but showing signs of splitting. [g] Me$_3$C group. [h] Not observed, possibly exchanged out with the DMSO. [i] Too insoluble for J to be measured accurately.
* Measured by P. J. Abbott.

hydrogen atoms show the greatest downfield shift when the molecule is protonated.

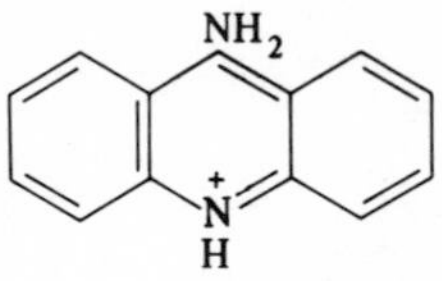

6

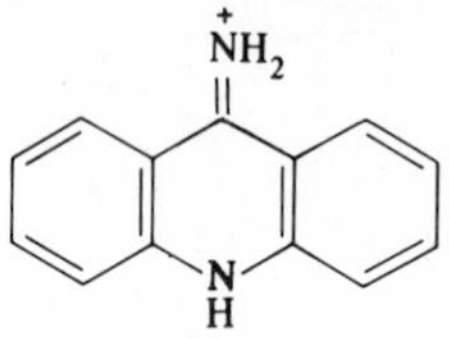

7

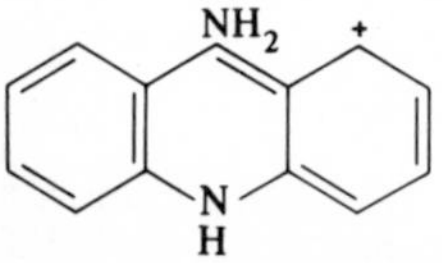

8

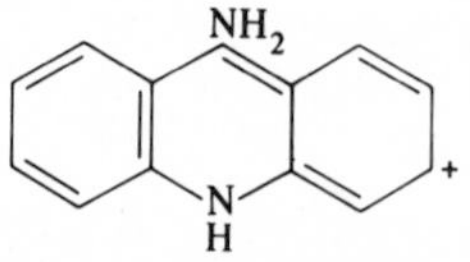

9

Formal charges cannot be placed on the 2 and 7 positions, which show a negligible shift.

The protons of 3,6-diaminoacridine and its sulfate are readily assigned, the largest downfield shift on protonation being shown by the 9-hydrogen atom. Assignments for acridine orange, 3,6-bisdimethylaminoacridine, are similar, while those given[1] for acridine yellow have been revised (Table IV), so that the shielding effects of the methyl groups enter the normal range.

The aggregation of acridine orange in aqueous solution has been studied in the 10^{-2} to $5 \times 10^{-4} M$ concentration range by measuring the changes in its nmr spectrum. It was found that the most favored configuration for the aggregates occurs when the flat molecules are packed in nearly parallel array.[12]

4. 9-Acridanones

The nmr spectrum of 9-acridanone (10) has been simulated and assigned on the assumption that the lowest field protons would be those *peri* to the carbonyl-type oxygen atom.[1]

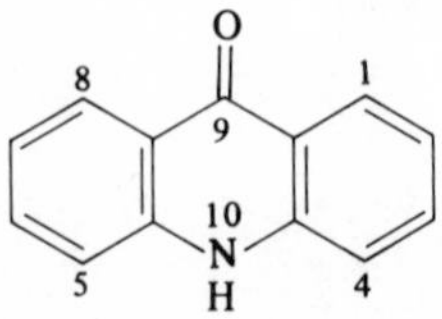

10

TABLE V. Nuclear Magnetic Resonance Spectra of 9-Acridanones

Substituents	MHz	Solvent	Proton resonances										Ref.
			1	2	3	4	5	6	7	8	9	10	
None	60	DMSO	1.73 $J_{1,2}$ 8.3	2.73 $J_{1,3}$ 1.4	2.26 $J_{1,4}$ 0.4	2.43 $J_{1,10}$ 0.4	$J_{2,3}$ 7.0	$J_{2,4}$ 1.0	$J_{3,4}$ 8.6			-1.7	1
Acronycine[a]	60	CDCl$_3$	6.23[b]	3.72			2.4———		3.1m	1.47		6.03[c]	11
3-PhCH$_2$O-1,2-(MeO)$_2$-10-Me	60	CDCl$_3$	5.95[b]	6.10[b]	4.8[d]	3.50	e	e	e	1.60[d]		6.45[c]	15
3-PhCH$_2$O-1-HO-2-MeO-10-Me	60	CDCl$_3$	4.8	6.10[b]	4.8[d]	3.80	e	e	e	1.70[d]		6.40[c]	15
1,3-(HO)$_2$-2-MeO-10-Me	60	DMSO[f]	5.1[e]	6.10[b]	broad[g]	3.30	e	e	e	1.65[d]		6.25[c]	15
1,3-(MeO)$_2$-2-CO$_2$Me-10-Me	60	CDCl$_3$	6.17[bh]	6.05[bh]	6.39[bh]	3.65	2.3———		3.1m	1.70		6.02[ch]	11
1,3-(MeO)$_2$-4-CO$_2$Me-10-Me	60	CDCl$_3$	6.06[b]	3.68	6.32[b]	5.97[bh]	2.2———		2.9m	1.62		6.06[ch]	11
1,2-(MeO)$_2$-3,4-OCH$_2$O-10-Me	60	CDCl$_3$	6.01[b]	6.01[b]	3.95[i]		2.6m	2.6m	2.6m	1.63[d]		6.08[c]	14
1,4-(MeO)$_2$-2,3-OCH$_2$O-10-Me	60	CDCl$_3$	5.93[b]	3.95[i]		6.13[b]	2.6m	2.6m	2.6m	1.67[d]		6.11[c]	14
1,5-(MeO)$_2$-2,3-OCH$_2$O-10-Me	60	CDCl$_3$	5.88[b]	4.10[i]		3.40	6.20[b]	2.81——— $J_{6,8}$ 3.3	2.94 $J_{7,8}$ 6.6	2.00		6.10[c]	14

(Table Continued)

Table V. (Continued)

Substituents	MHz	Solvent	1	2	3	4	5	6	7	8	9	10	Ref.
							Proton resonances						
1,5-(MeO)$_2$-2,3-OCH$_2$,O-10-Me	60	CF$_3$CO$_2$H	5.40^b				5.85^b					5.47^c	14
2,4-(MeO)$_2$-10-Me	60	CDCl$_3$	2.35	5.97^{bh}	3.10^g	6.03^{bh}	e	e	e	1.55^d		6.15^{ch}	15
1,3-(MeO)$_2$-10-Me-4-CO$_2$H	60	CF$_3$CO$_2$D	5.65^{bh}	2.87	5.58^{bh}	-1.80^j	1.0		2.5m			5.43^{ch}	11
1,3-(MeO)$_2$-2-O-pentenyl	60	CDCl$_3$	6.10^b	k	6.20^b	3.85	e	e	e	1.90^d		6.65^c	15
4-CHO-1,3-(MeO)$_2$-10-Me	60	CDCl$_3$	6.00^{bh}	3.64	6.37^{bh}	-.50	2.2		2.9m	1.60		5.92^{ch}	11
4-CHO-3-HO-1-MeO-10-Me	60	CF$_3$CO$_2$D	5.45^b	3.07	-1.80^{gi}	-0.40	0.9			2.3m		5.34^c	11
2-HO-1,3-(MeO)$_2$-10-Me	60	CDCl$_3$	6.20^b	3.85	6.10^b	3.85	e	e	e	1.55^d		6.55^c	15
1-HO-3-MeO-2-CO$_2$Me	60	DMSO	-5.9^{gi}	6.13^{bh}	6.05^{bh}	3.52	2.0		2.9m	1.80		-2.1	11
1-HO-2MeO-3,4-OCH$_2$,O-10-Me	60	CDCl$_3$	-5.05	5.97^b	4.00^i		e	e	e	1.65^d		6.00^c	15
1-HO-2MeO-10-Me-3-O-pentenyl	60	CDCl$_3$	4.65^g	6.15^b	m	3.95	e	e	e	1.80^d		6.45^c	15

700

Table V. *(Continued)*

Substituents	MHz	Solvent	Proton resonances										Ref.
			1	2	3	4	5	6	7	8	9	10	
1-HO-3-MeO-10-Me-2-O-pentenyl	60	CDCl$_3$	4.7^g	k	6.10^b	4.00	e	e	e	1.85^d		6.45^c	15
2-HO-1-MeO-10-Me-3-O-pentenyl	60	CDCl$_3$	5.90^b	6.3^g	n	3.45	e	e	e	1.50^d		6.30^c	15
1-HO-2,3-(MeO)$_2$-10-Me	60	CDCl$_3$	4.5^g	6.15^b	6.15^b	4.15	e	e	e	2.00^d		6.55^c	15
1-MeO-2,3-OCH$_2$O-10-Me	60	DMSO	6.03^b	3.85^b	3.95^i	3.65	2.52m	2.52m	2.52m	1.77^d		6.18^c	14-16
2-MeO-3,4-OCH$_2$O-10-Me	60	CDCl$_3$	2.35	6.00^b	3.95^i		e	e	e	1.75^d		6.10^c	15
1-MeO-10-Me	60	CDCl$_3$	5.95^b	2.25 ————————					3.50m	1.50^d		6.25^c	15
2-MeO-10-Me	60	CDCl$_3$	2.2-3.6m	6.20^b	2.2 ————————				3.6m	1.65^d		6.50^c	15
10-Me	60	CDCl$_3$	1.60^d	2.2 ————————					3.0m	1.60^d		6.30^c	15
1,2,3-(MeO)$_3$-10-Me	60	CDCl$_3$	5.95^b	6.15^{bh}	6.10^{bh}	3.60	e	e	e	1.60^d		6.50^c	15

701

^{a}See Chapter IV, p. 412, 11-H, 3.42d; 12-H, 4.44d; $J_{11,12}$ 9; Me$_2$, 8.50. ^{b}OMe. ^{c}N-Me. ^{d}CH$_2$ resonance, Ph 2.6.
^{e}Not given. ^{f}DMSO-d$_6$. ^{g}OH. ^{h}Assignments should possibly be interchanged. ^{i}O-CH$_2$-O. ^{j}Exchanged with D$_2$O.
^{k}CH$_2$, 5.65d; CH$_2$CH, 4.45t; CH$_2$CH = CMe$_2$, 8.30, 8.25. ^{l}NH, broad. ^{m}CH$_2$, 5.35d; CH$_2$-CH, 4.45t; CH$_2$CH = CMe$_2$,
8.20 broad. ^{n}As ^{m} but CH$_2$CH = CMe$_2$ 8.15.

This assignment is confirmed by the many spectra obtained for the 9-acridanone alkaloids (Chapter IV), which are included in Table V. Methoxyl groups at positions 1 and 8 are also significantly deshielded by the 9-carbonyl group and appear at τ 5.95–5.98. Hydroxyl groups at positions 1 and 8 are strongly hydrogen-bonded to the 9-carbonyl, and therefore appear at much lower field than when present at other positions.[13] N-Methyl resonances, τ 6.3–6.5, are usually broader than methoxyl resonances[13] and show a greater downfield shift when the solvent is changed from chloroform to trifluoroacetic acid.[14] The irradiation of both N- and O-methyl groups (nuclear Overhauser effect) has been used to place nearby protons by the sharpening of their resonance lines (see also Chapter IV, p. 418).

An interesting aspect of the spectrum of 9-acridanone is the appreciable coupling (0.4 Hz) reported for the 1 (and 8) protons with that at position 10 and attached to the nitrogen atom.[1] Long-range coupling through a nitrogen atom does not appear to have been noted before, and proof that the 10-hydrogen atom is involved, for instance, by observing the loss of the coupling on 10-deuteriation or by examination of the spectrum of 9-acridanone measured at a higher frequency, is needed.

5. Acridans and Reduced Acridines

The nmr spectrum of acridan (11) has been measured and simulated,[1] and the assignments made have been confirmed by an inspection of the spectrum of 4-phenylacridan (Table VI).

11

An interesting feature of the last spectrum is the upfield shift, in comparison with acridan, of the 5-proton, presumably a result of shielding by the 4-phenyl group.

Agreement between the observed and calculated spectra for acridan is not as good as in the case of acridine, and the assumption of coupling between the 1 (and 8) and 10-hydrogen atoms though the nitrogen atom to the extent of 0.8 Hz is needed for the best results. This coupling is greater than the corresponding coupling for 9-acridanone (0.4 Hz), and needs confirmation.

TABLE VI. Nuclear Magnetic Resonance Spectra of Other Acridines

Substituent	MHz	Solvent	Proton resonances										Ref.
			1	2	3	4	5	6	7	8	9	10	
9,10-H₂	60	DMSO^a	2.97 $J_{1,2}$ 7.6	3.26 $J_{1,3}$ 1.4	2.97 $J_{1,4}$ 0.6	3.22 $J_{1,9}$ O	$J_{2,3}$ 7.0	$J_{2,4}$ 1.4	$J_{3,4}$ 8.0	$J_{4,9}$ O	6.04	1.4	1,6
	100	CCl₄	2.4			3.7					6.01	ca. 4.1	17
9,10-H₂-4-Ph	100	CDCl₃	2.94	3.10	3.00	2.52^b	3.51	2.98	3.17	2.87	5.91	3.71^c	6
1,2,3,4-H₄-7-Et	100	CDCl₃	7.10^d	8.0	8.2m	6.90^e	2.08 $J_{5,6}$ 9	2.53 $J_{6,8}$ 2	f	2.58	2.31		6
9,10-H₂-10-Me-9-PO(OEt)₂^g	60	CDCl₃	2.6							3.2m	h	6.68	10
9,10-H₂-9-MeO-10-trans-1',2'-di-CO₂Me-vinyl-	60	CDCl₃	2.45q^i	2.6	3.1	3.25^i	$J_{1,2} \simeq J_{3,4} \simeq 8, J_{1,3} \simeq 8, J_{2,4} \simeq 2$				4.09 MeO 7.08	i	*
9,10-H₂-9-EtO-10-HO	?	CDCl₃	2.33							3.27	3.27 k	3.87	23
9,10-H₂-9-imino-10-Me	60	CDCl₃	1.75	2.57					2.57	1.75	?	?	24
1,4,5,8-H₄	?	CDCl₃	6.60^i	4.1	4.1	6.55^i	6.55^i	4.1	4.1	6.60^i	2.89		25

^a Deuteriated. ^b Ph apparent singlet. ^c Exchanges with D₂O. ^d Broadened triplet, J ca. 6. ^e Broadened triplet, J ca. 7.

f MeCH₂, 8.70; MeCH₂, 7.22; J 7.5. ^g The spectrum of the 10-H compound was similar, the 10-H appearing in the aromatic region.

^h MeCH₂, 0, 8.87t, J 7; MeCH₂O, 6.17m; CH, 5.5d, $J_{H,P}$ 26. ^i These assignments and coupling constants could be exchanged.

^j Vinyl-H, 2.42, ester-Me, 6.29, 5.46. ^k MeCH₂, 8.76t, MeCH₂, 6.00q.

*Measured by Mrs. E.E.Richards.

The nmr spectra of a series of 9-alkyl- and 9,9-dialkyl-acridans have been investigated for deuteriochloroform and carbon tetrachloride solutions.[17] The 10-H absorption was between τ 4.05 and 4.25, and the aromatic protons between τ 2.4 and 3.7. Where the 9,9-substituents are both methyl, ethyl, or isopropyl, only one type of alkyl group could be detected for each compound. The chemical shift of the methyl group of 9-methylacridan (τ 8.66) drops steadily downward to τ 8.25, as the 9-H is replaced successively by methyl, ethyl, *iso*-propyl, and *tert*-butyl. This is interpreted as indicating that the larger substituent adopts a pseudoaxial conformation from the central boat-shaped ring.

The aliphatic-type resonances in the nmr spectra of some complex acridans, obtained from 9-methyl-, 9-ethyl-, and 9-phenyl-acridine and dimethyl ketene, have been used to elucidate the structures of these compounds (e.g., 12 and 13).[18] The 9-H and 9-Me groups of 12 appear at τ 5.87 and 8.75, respectively, and show the expected multiplicities (*J* 7 Hz). The temperature dependence of the spectra of 14 (R = Me or Ph) is attributed to variations in the populations of the conformers, possibly because of rotation about the *N*-vinyl bond.

12   13   14

The nmr spectra of more highly reduced acridines are discussed in Chapter V (p. 489–491) in connection with their stereochemistry.

6. Benzacridines

The nmr spectra of benz[*b*]acridine* (15) and benz[*c*]acridine[19a, b] (16) have been recorded at 60 MHz.

*Measured by P. J. Abbott.

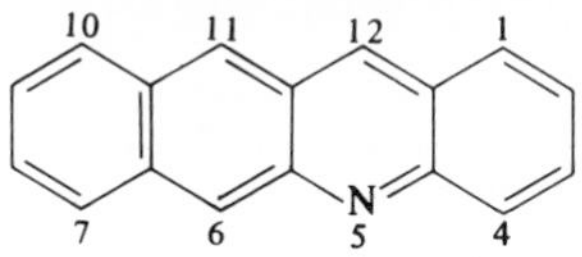

15

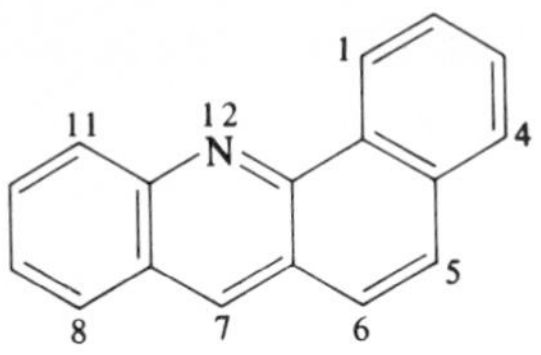

16

In hexadeuteriodimethylsulfoxide, **15** shows a one-proton singlet at τ 1.01, assigned to the 12-hydrogen atom, an apparent two-proton doublet at τ 1.68 (J ca. 8); the remaining protons appear as a complex multiplet, extending to about τ 3. Benz[c]acridine in carbon tetrachloride shows a single proton (position 7) at τ 1.66; a multiplet assigned to H-1 at τ 0.53; and a partially obscured doublet, due to H-11 centered on τ 1.72. The other protons are not resolved. The particularly lowfield resonance for H-1, in part because of the relative position of the nitrogen atom, is noteworthy.

The nmr spectrum of dibenz[aj]acridine (**17**) has been investigated at 100 MHz in deuteriochloroform, using double resonance and spin tickling techniques, and the observed spectrum has been matched (±0.1 Hz) by a computed spectrum calculated from the parameters deduced from the experiments.[20]

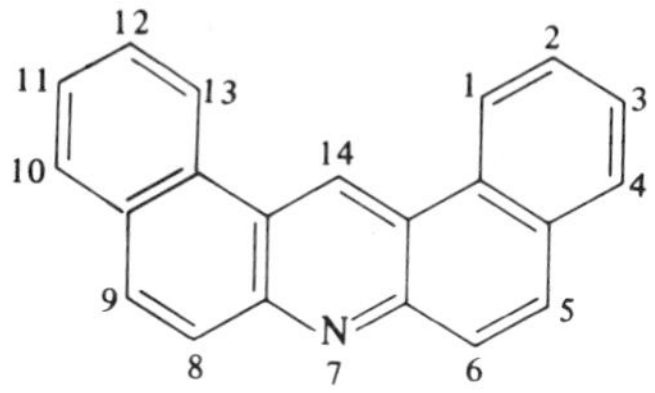

17

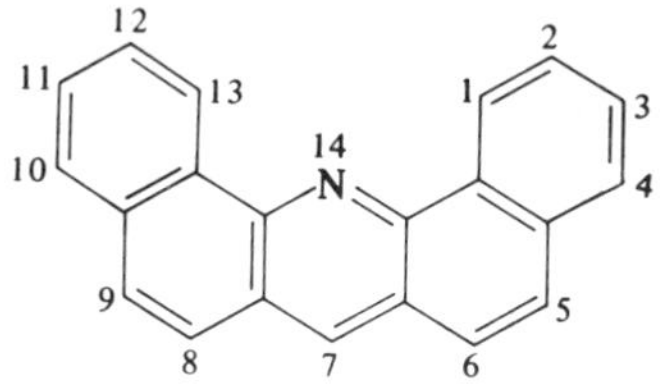

18

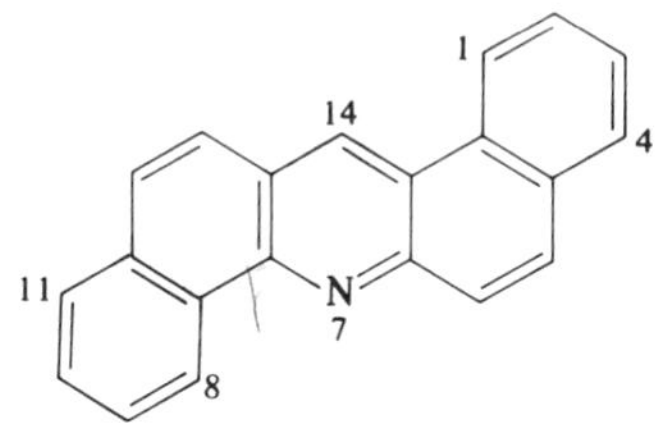

19

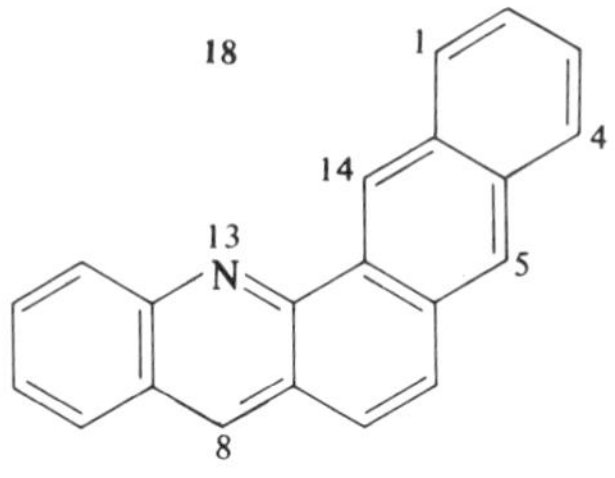

20

The complete second-order analysis for the spectra of **18** in both deuteriochloroform and hexadeuteriobenzene also appears to have been carried out, but only the chemical shift data given in Table VII are reported.[21] A partial analysis of the spectrum of dibenz[ah]acridine (19) has also been made.[21]

TABLE VII.   Nmr data for Benz- and Dibenz-acridines

	Proton chemical shifts in τ, J in Hz						
	Proton Position						
Compound	1(13)	2(12)	3(11)	4(10)	5(9)	6(8)	14(7)
17*	1.43	2.45	2.53	2.31	2.24	2.10	0.11
18	0.27	2.21	2.29	2.14	2.30	2.30	1.55
19	1.59	?	?	?	2.30	1.57	
19	(2.45)	(2.45)	?	?	?	(0.30)	(0.04)

* $J_{1,2}$ 8.51; $J_{1,3}$ 0.87; $J_{1,4}$ 0.76;
$J_{2,3}$ 7.28; $J_{2,4}$ 1.62; $J_{3,4}$ 8.17

The low-field singlet (τ 0.10) of naphth[2,3-c]acridine (20) has been assigned to the 14-H atom,[22] and its low-field position is doubtless a result of the proximity of the nitrogen atom (compare with the 1-H of 16). It is surprising that the 8-H proton is not assigned, but it perhaps contributes with the 5-proton to the two-proton multiplet recorded at τ 1.70.

Inter-ring coupling has been reported for several dibenzacridines (see Chapter VII, p. 554).

References

1. J. P. Kokko and J. H. Goldstein, *Spectrochim. Acta*, **19,** 1119 (1963).
2. C. W. C. Harvey, D. Phil. Thesis, University of Oxford, England, 1970.
3. R. J. Abraham and S. Castellano, *J. Chem. Soc.*, B, 49 (1970).
4. (a) D. A. Brown and M. J. S. Dewar, *J. Chem. Soc.*, 2406 (1953); (b) H. C. Longuet-Higgins and C. A. Coulson, *Trans. Faraday Soc.*, **43,** 87 (1947); (c) P. J. Black, R. D. Brown, and M. L. Hefferman, *Aust. J. Chem.*, **20,** 1305 (1967).
5. B. Pullman, *Bull. Soc. Chim. Fr.*, 533 (1948).
6. R. G. Bolton, D. Phil. Thesis, University of Oxford, England, 1970.
7. P. Diehl, *Helv. Chim. Acta*, **44,** 829 (1961).
8. H. H. Mantsch, W. Seiffert, and V. Zanker, *Rev. Roum. Chim.*, **12,** 1137 (1967).
9. R. M. Acheson, J. M. F. Gagan, and D. R. Harrison, *J. Chem. Soc.*, C, 362 (1968).
10. D. Redmore, *J. Org. Chem.*, **34,** 1420 (1969).

11. P. C. Macdonald and A. V. Robertson, *Aust. J. Chem.*, **19**, 275 (1966).
12. D. J. Blears and S. S. Danyluk, *J. Amer. Chem. Soc.*, **89**, 21 (1967).
13. J. A. Diment, E. Ritchie, and W. C. Taylor, *Aust. J. Chem.*, **20**, 1719 (1967).
14. K. H. Pegel and W. G. Wright, *J. Chem. Soc.*, *C*, 2327 (1969).
15. J. A. Diment, Ph.D. Thesis, University of Sydney, Australia, 1968.
16. R. H. Prager and H. M. Thredgold, *Aust. J. Chem.*, **22**, 1493 (1969).
17. G. A. Taylor and S. A. Procter, *Chem. Commun.*, 1379 (1969); *J. Chem. Soc.*, *C*, 2537 (1971).
18. S. A. Procter and G. A. Taylor, *J. Chem. Soc.*, *C*, 1937 (1967).
19. (a) E. V. Doncelet, R. H. Martin, and F. Geerts-Evrard, *Tetrahedron*, **20**, 1495 (1964); (b) R. H. Martin, N. Defay, F. Geerts-Evrard, and D. Bogaert-Verhoogen, *Tetrahedron*, *Supp.* **8**, Part 1, 181 (1967).
20. B. Clin and B. Lemanceau, *Acad. Sci.*, *Paris, Ser. C*, **270**, 598 (1970).
21. B. Clin and B. Lemanceau, *C. R. Acad. Sci.*, *Paris, Ser. D*, **271**, 788 (1970).
22. K. Sisido, K. Tani, and H. Nozaki, *Tetrahedron*, **19**, 1323 (1963).
23. H. Mantsch and V. Zanker, *Tetrahedron Lett.*, 4211 (1966).
24. A. Schönberg and K. Junghans, *Chem. Ber.*, **99**, 1015 (1966).
25. A. J. Birch and H. H. Mantsch, *Aust. J. Chem.*, **22**, 1103 (1969).

CHAPTER XIII

The Mass Spectra of Acridines

R. G. BOLTON

I. C. I. Pharmaceuticals Division Research Laboratories,
Cheshire, England

A so-called "mass spectrum" of an organic compound is more accurately an analysis of the products of electron-impact-induced degradation of the compound. Fragmentation is directed either by fission of the weakest bonds or by the stability of the products formed. The abundance of each ion, as in a chemical reaction, is a steady-state concentration determined by the relative rates of formation and decomposition of that species. Consequently, the mass spectra of aromatic compounds and, more particularly, those of polycyclic aromatic compounds such as acridine are characterized by molecular ions of high abundance and fragment ions of low abundance, due to the stability of the aromatic ring even under electron impact. This feature contrasts with the spectra of aliphatic compounds, in which small fragments are usually the more stable and therefore more abundant. Polycyclic aromatic systems are so suited to carrying positive charge that doubly charged ions are an important feature of their mass spectra.[1] This stability toward skeletal fragmentation is responsible for the third distinctive feature of the polycyclic aromatics, the process of dehydrogenation.[2] Fission of C—H bonds, normally an unfavored process, effects expulsion of one or more atoms of hydrogen from the molecular ion, resulting in the formation of a group of peaks for each subsequent skeletal fragment ion.

Where such a molecule carries substituents that fragment readily, abundant ions corresponding to this fragmentation may be observed. It is still a characteristic of the spectra of acridines, and polycyclic aromatic compounds in general, that once fragmentation of the substituent is completed, subsequent fragmentation gives rise to ions which are of low abundance by comparison. These general features are revealed in the mass spectra of all the acridines that will subsequently be discussed.

1. Acridine

The mass spectrum of acridine (1) reflects the great stability to fragmentation of both the singly and doubly charged molecular ions.[3]

1

All fragment ions, save that for dehydrogenation, are of less than 10% abundance relative to the molecular ion (100%). Predominant are groups of ions corresponding to successive losses of 26 and/or 27 u (atomic mass units) from the molecular ion and its dehydrogenated species. Doubly charged ions are absent at nominal electron energies below 30 eV; below 13.5 eV, dehydrogenation, the most facile fragmentation, is suppressed. Metastable peaks are extremely weak, and the only two recorded in the published spectrum are actually the triply charged molecular ion and $(M-1)^{3+}$.

Studies of the metastable transitions in the fragmentation of acridine in the first and second field-free regions of a double-focusing mass spectrometer have established the relationship between daughter ions, and those of their parents which decompose at the detectable rate.[4] Coupled with exact mass determinations, which supply the elemental compositions of fragment ions, these have enabled a scheme for the fragmentation of acridine under electron impact to be built up (see Scheme 1, page 711). The evidence of metastable transitions in the second field-free region is that the doubly charged ions fragment according to a similar scheme.

From a study of the metastable transitions of deuterated acridines, it would appear that an intermediate degree of scrambling of hydrogen atoms precedes the loss of HCN from the molecular ion at 70 eV. As a result of this scrambling, the mechanism of the expulsion of HCN is still not elucidated.

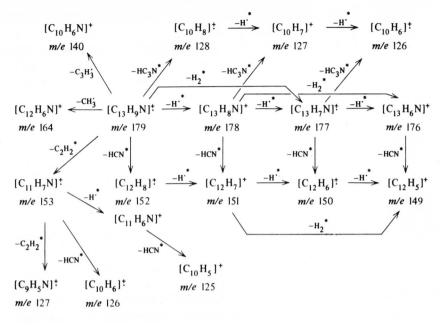

Acridine also displays marked differences in behavior from the fragmentation of four other $C_{13}H_9N$ isomers (phenanthridine, 2-cyanobiphenyl, 2-phenylethynylpyridine, and fluoren-9-imine), as reflected in the relative abundances of singly and doubly charged fragment ions at normal- and low-electron voltage and also the metastable transitions observed.

2. Substituted Acridines

Comparison of the spectra of a number of alkylacridines with those of the corresponding alkyl-9-chloroacridines indicates that in the latter the two substituents fragment independently and the loss of Cl· or HCl is simply superimposed on the alkyl fragmentation.[4] Possible exceptions are the 1-alkyl-9-chloroacridines that appear to display enhanced loss of the halogen atom. The fragmentation of the alkylacridines resembles that of analogous alkyl-quinolines.[5, 6]

Alkylacridines, except the 4-substituted compounds, fragment simply by β-cleavage, the ease of cleavage as observed in the methylacridines being 9 ≫ 1 > 2, 3, 4. The lability of the hydrogen atoms in 9-methylacridine, even in the absence of electron impact, is displayed by the exchange of deuterium label between the 9-methyl group and the residual water vapor in the mass spectrometer source. That the β-cleavage peak in the mass spectrum of 2-[α-d_2]ethyl-[9-d_1]acridine does not shift from (M-15)$^+$ is a demonstration that β-cleavage does not involve the α-CH$_2$ in any way.

Alkyl chains longer than C$_2$ undergo, in addition, McLafferty rearrangements to an ortho-position on the ring (e.g., Scheme 2 below); thus for 2-butyl-9-chloroacridine the ion (M-C$_3$H$_6$)$^+$ is approximately half as abundant as (M-C$_3$H$_7$)$^+$.

In addition to two successive β cleavages, 2-t-butyl-9-chloroacridine displays the fragmentation (M-Me$^{\cdot}$-C$_2$H$_4$), for which a "cationated cyclopropane" intermediate was postulated in the case of t-butyl benzene.[7]

The fragmentation of 4-alkylacridines is directed by the heteroatom in the peri position. Thus cyclization to the nitrogen [Scheme 3, (a) → (d)] leads to γ cleavage, and transfer of hydrogen to the heteroatom from the β or γ positions initiates McLafferty rearrangements [(a) → (b)] and loss of α-H. For example, in the spectrum of 4-ethylacridine, the base peak results from γ cleavage and the β-cleavage ion (M-Me$^{\cdot}$)$^+$ is of relatively low abundance. The alternative major fragmentation is a McLafferty rearrangement with expulsion of the side chain as ethylene. Similarly, the (M-Me$^{\cdot}$)$^+$ ion in the spectrum of 4-propylacridine is of much greater abundance than the (M-Et$^{\cdot}$)$^+$ ion resulting from β cleavage. Scheme 3, page 713, representing the fragmentation of 4-propylacridine, illustrates the variety of processes observed in the fragmentation of 4-alkylacridines.

Deuterium-labeling studies[4] indicate that the (M-H)$^+$ ion in 4-ethylacridine arises 20% from α-H loss, possibly initiated by β-H transfer, and 80% from γ cleavage through cyclization. The (M-Me$^{\cdot}$)$^+$ ion from 4-[α-d_1]propyl-acridine arises with no loss of label, but the process (M$^+$-C$_2$H$_4$) involves the α-CH$_2$ to 25%. It appears that while the majority of the (M-C$_2$H$_4$)$^+$ ion can arise through a 7-center McLafferty rearrangement[8] involving γ-H transfer with β cleavage [Scheme 3, (a) → (c)] a rearrangement with transfer of Me$^{\cdot}$ to the heteroatom [Scheme 3, (a) → (e)] must be postulated for the remaining 25%. The formation of the (M-C$_3$H$_6$)$^+$ ion from the same labeled compound involves no transfer of α-H to the nitrogen. Finally, it has been shown that the elimination of H$_2$ from the (M-1)$^+$ ion of 4-ethylacridine, envisaged as a cyclized species (2), is not totally random, but, allowing for a deuterium isotope effect, appears to involve specifically one α-H and one β-H.

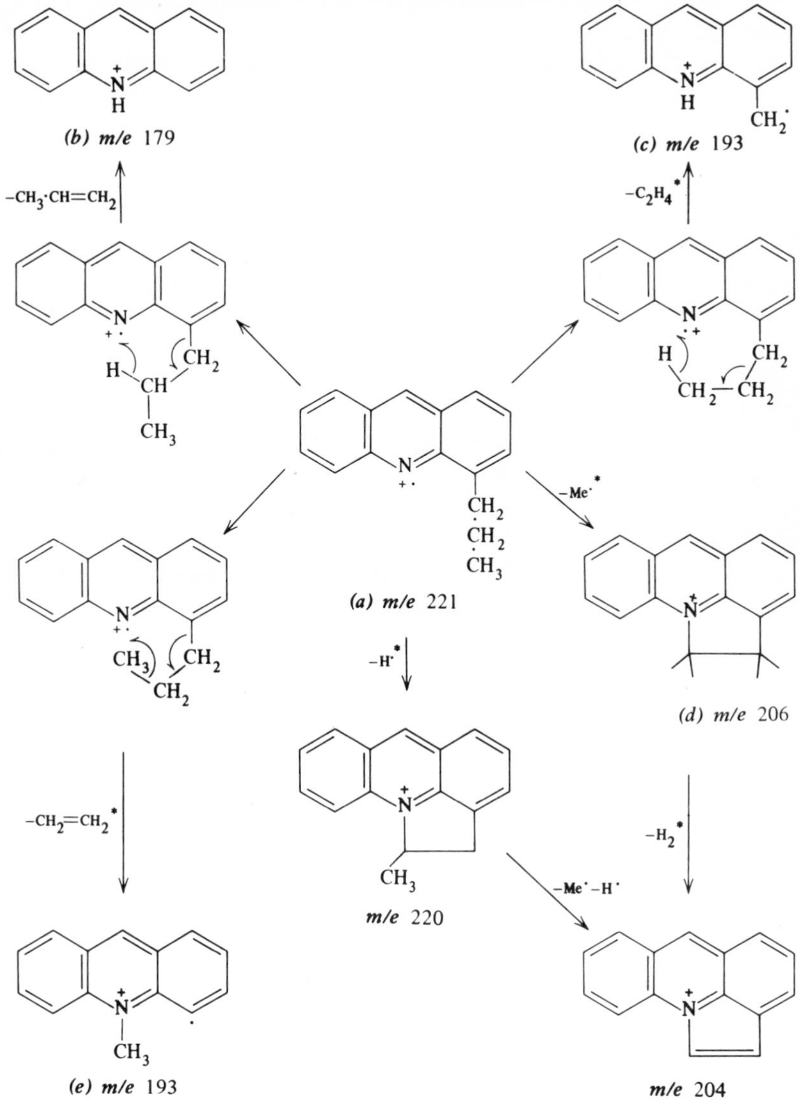

(b) m/e 179

$-CH_3 \cdot CH=CH_2$

(c) m/e 193

$-C_2H_4^{\bullet}$

(a) m/e 221

$-Me^{\bullet}$

(d) m/e 206

$-H^{\bullet}$

m/e 220

$-CH_2=CH_2^{\bullet}$

$-Me^{\bullet}-H^{\bullet}$

$-H_2^{\bullet}$

(e) m/e 193

m/e 204

Similar fragmentation routes with enhanced β-cleavage are displayed by α-branched 4-alkylacridines. In addition, the elimination of C_2H_4 from **3** (R = 4-Pr^i) and of C_3H_6 from **3** (R = 4-Bu^{sec}) suggests that β-hydrogen transfer to the heteroatom generates a species visualized as a "cationated cyclopropane,"[7] analogous to that formed by β cleavage of a tertiary alkyl substituent. The fact that (M-H)$^+$ is the base peak in the spectrum of 4-phenylacridine suggests that cyclization to the heteroatom occurs even for this compound.

By analogy with the peri effect postulated for 4-alkylquinoline by Djerassi[5] and MacLean,[6] and later retracted by Djerassi,[9] γ cleavage might be expected for those 9-alkylacridines that may cyclize to the 1- or 8-position to give an ion represented as **4**.

Such γ cleavage is observed, in fact, together with McLafferty rearrangement with hydrogen transfer possibly to a peri position, but is much less favored than β cleavage. This is the inverse of the tendency noted in 4-alkylacridines, where the heteroatom is in the peri position.

Save for the exception mentioned above, the halogen atom in substituted 9-chloroacridines fragments independently, being lost as Cl· or HCl. Precisely how this fragmentation is superimposed on the fragmentation of the other substituent depends on the relative rates of the two fragmentation processes. Thus the loss of Cl· precedes any β cleavage in all the 9-chloromethylacridines, occurs at approximately the same rate as β cleavage in 9-chloro-2-ethyl-acridine, and is always preceded by the γ cleavage and McLafferty rearrangement fragmentations of 9-chloro-4-ethyl- and -4-propyl-acridines. Poly-haloacridines fragment by successive losses of halogen, bromine in preference to chlorine, until only the skeleton remains, which fragments as a dehydrogenated acridine.

In accordance with the remarks made in the introduction, all ions below the molecular ion of abundance greater than about 10% in the spectra of substituted acridines so far examined result from fragmentation of the substituent, breakdown of the aromatic acridine nucleus being an unfavorable process. A possible exception is the doubly charged molecular ion of 4-phenyl-acridine, which has a relative abundance of about 20%. When no further facile fragmentation of the substituent is possible, then all the alkyl-9-chloro- and alkyl-acridines fragment further to a small extent after the fashion of acridine itself.

The main features[10] of the mass spectra of 14 benzo[a]- and benzo[c]-acridines have been described. As expected for the extended polycyclic aromatic heterocyclic system, ionization potentials were low and multiply charged ions a major feature of the spectra. Singly charged ions appeared at

8 eV and doubly charged ions at 22–23 eV, with some doubly charged ions of greater intensity than the corresponding singly charged ions at 70 eV. Extensive strong dehydrogenation peaks were observed but no noticeable fragmentation of the polycyclic skeleton. The possibility of α and β cleavage leading to the dehydrogenation peaks in the alkyl substituted compounds was discussed inconclusively, but examination of the tabulated spectra indicates that for most of the compounds dehydrogenation peaks are not substantially greater than in benzo[c]acridine itself (25%). Only in those ethylbenzacridines with the substituent peri to the nitrogen do the $(M-1)^+$ peaks greatly increase, as a result of the peri interaction. β Cleavage of ethyl substituents was noted, but several claims were made for α cleavage of ethyl and methyl substituents, particularly in the di- and trisubstituted compounds. In the dimethyl compounds, it would seem that as M-16, M-17, and M-18 are usually more intense than $(M-15)^+$, methyl expulsion is decreasingly likely to result from α cleavage. Expulsion of methyl from the nucleus of such a system is not without precedent, indeed loss of $CH_3^{\cdot}$ from stilbene has been studied in detail and the carbon atom expelled shown to be of random origin.[11] In the di- and triethyl compounds, all of which possess a peri-ethyl substituent, the $(M-28)^+$ ion is much more intense than $(M-29)^+$; again, expulsion of $Et^{\cdot}$ through simple α cleavage is less likely than ethylene loss from the peri substituent by β-H transfer and α cleavage. In each case, ethylene loss from a benzo[a]acridine was less favored than from a benzo[c]acridine. Unfortunately, no data on metastable transitions are available.

The aminoacridines display extensive dehydrogenation in the mass spectrometer.[4] The amino group is usually expelled as hydrogen cyanide but the single-step fragmentations $(M-15)^+$ and $(M-17)^+$ are also observed. For 3,6-diaminoacridine, the typical pathways $(M^+$-H-NH_3-HCN) and $(M^+$-HCN-H-HCN) are confirmed by the evidence of metastable transitions.

3. 9-Acridanones

The fragmentation of the 9-acridanone and 10-methyl-9-acridanone nuclei[12] was established as shown in Scheme 4, page 716. The simple monohydroxy-9-acridanones behave similarly, with an extra carbon monoxide loss occurring before expulsion of the hydrogen cyanide.[13]

The fragmentation of the alkoxy substituents of 9-acridanones related to the naturally occurring alkaloids (Chapter IV, p. 379) is directed by the presence of a 2- or 4-substituent, which fragments preferentially by alkyl-oxygen cleavage.[12] Thus 5 shows $(M-Et^{\cdot})^+$ as its base peak, with $(M-Me^{\cdot})^+$ only 15%. Deviations from this general pattern may occur for alkoxyl groups or where R is other than a simple alkyl group. Thus, when R = -$CH_2 \cdot CH$=

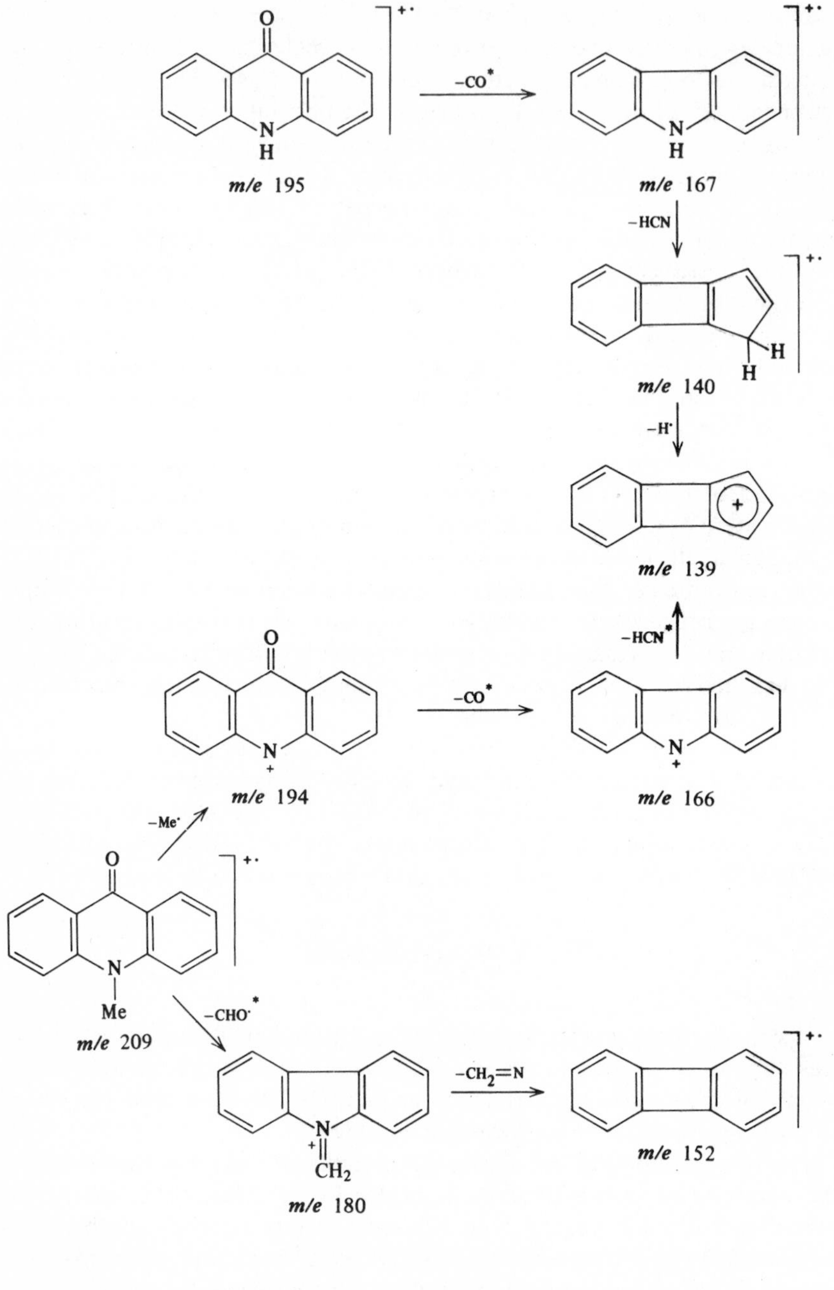

O
N
H
m/e 195
–CO*
N
H
m/e 167
–HCN
H
H
m/e 140
–H·
+
m/e 139
–HCN*
O
N
+
m/e 194
–CO*
N
+
m/e 166
–Me·
O
N
Me
m/e 209
–CHO·*
N
+
CH₂
m/e 180
–CH₂=N
m/e 152

$C(CH_3)_2$, the loss of this group, whether at the 2- or 3-position, is preferred[13]; the fragmentation of 3-OCH_2Ph groups is about as probable as that of 2- or 4-OCH_3 groups.

The study of **6** ($R_1 = CD_3$, $R_2 = CH_3$) and **6** ($R_1 = CH_3$, $R_2 = CD_3$) showed[12] that where competition can occur, 60% of the initial alkyl loss is from the 2-position and 40% from the 4-position.

Loss of carbon monoxide follows the alkyl-oxygen cleavage, but it has not yet been established which carbonyl grouping is lost, since attempts at exchanging the 9-carbonyl oxygen atom with oxygen-18 failed.

A remarkable process in the fragmentation of 1,2,3,4-tetra-substituted 9-acridanones, e.g., **7**, containing an -OH group actually or potentially (elision of C_2H_4 from an ethoxy substituent occurs readily) is the transfer of hydrogen from the hydroxyl to the central ring with associated cleavage to yield **8**, or **9** in the event of two hydroxyls being present. Labeling of the -OH has confirmed this process.

The general fragmentation pattern of the loss of R·, followed by carbon monoxide, is broken when there is a 1-ethoxy substituent. Even in the presence of 2- and 4-alkoxy groups, there is a prominent M-OH·-H_2 process represented by Scheme 5, page 718.

A similar interaction has been observed for 1-methoxy-10-methyl-9-acridanone,[14] where (M-17)$^+$ has been shown by peak matching and the presence of the appropriate metastable peak to be (M-OH·)$^+$ and is followed by loss of 29 u in the fragmentation scheme. It would seem that when there is no more facile cleavage possible, then this mode (not observed in the more complex 1-methoxy-9-acridanones) begins to operate.

High-resolution measurements were used to distinguish between the possible losses of CHO$\cdot$ or CO from the nucleus and $C_2H_5\cdot$ or C_2H_4 from the alkyl group of the alkyl-9-acridanones.[4] They indicated that only for the methyl-9-acridanones are the ions (M-CHO)$^+$ or (M-CO)$^+$ detectable. All other (M-28,29)$^+$ ions result from alkyl fragmentation, which is strikingly dominated by β cleavage for all substituents. Thus the McLafferty rearrangement observed in 2-butyl-9-chloroacridine is absent in the corresponding 9-acridanone, and the heteroatom-induced McLafferty rearrangements and γ cleavage of the 4-alkylacridines are not observed in the 4-alkyl acridanones. On the other hand, 1-isopropyl-4-methyl-9-acridanone displays losses of C_2H_4 and C_3H_6 that were absent in the 9-chloroacridines. These observations indicate strongly that the site of charge localization shifts in the 9-acridanones from nitrogen to the oxygen atom, which then initiates the H$\cdot$ transfers and the associated rearrangements when there is a 1-substituent larger than methyl.

An interaction related to those observed by Bowie[12] and Ritchie[13] and their co-workers for 1-alkoxy-9-acridanones was the expulsion of HO$\cdot$, observed for 1-isopropyl-4-methyl-9-acridanone. (M-17)$^+$ is the second most intense ion (80%) in this spectrum; high-resolution studies indicate that 90.5% of the ion has the composition $C_{17}H_{16}N$, (M-OH$\cdot$)$^+$, confirmed as a one-step loss by observation of the appropriate metastable ion.

As (M-H)$^+$ is much more intense for the methyl-9-acridanones than for the methylacridines, some N-H fission is implied. Enhanced hydrogen loss from 1-methyl-9-acridanone does point to β cleavage, however, assisted by interaction between the oxygen atom and the methyl substituent.

Following β cleavage, there are weaker ions in the spectra of all the 9-acridanones for loss of CHO˙ from the nucleus, followed by losses of 26 or 27 u, as typical of the acridine nucleus.

4. Reduced Acridines

Under electron impact, acridans expel one of the 9-substituents preferentially, to give rise to species analogous to protonated acridines.[4] Thus, for acridan (10), the base peak in the spectrum is $(M-1)^+$, further fragmentation resembling that of acridine; for 4-ethylacridan, the loss of a 9-H precedes any fragmentation of the alkyl group.

10

The loss of an alkyl group, rather than hydrogen, from the 9-position is favored; M^+, $(M-H)^+$, and $(M-Pr˙)^+$ for 9-propylacridan having relative abundances of, respectively, 7, 6, and 100%. An ion of low abundance, corresponding to β cleavage from the $(M-H)^+$ ion only, confirmed by an appropriate metastable ion, indicates the probable resemblance of $(M-H)^+$ to the corresponding acridine molecular ion. The $(M-Pr˙)^+$ ion for 9-propylacridan shows identical fragmentation to the $(M-1)^+$ ion from acridan, giving rise to the same series of metastable transitions, among which m/e 180 giving m/e 152 is prominent. Subsequent to β cleavage of the substituent, 7-ethyl-1,2,3,4-tetrahydroacridine loses C_2H_4 by a retro Diels-Alder mechanism, as observed for the 5,6,7,8-tetrahydroquinolines.[15]

5. Miscellaneous Acridine Derivatives

Acridine 10-oxide (11) displays two sequences of fragmentation under electron impact.[16] In addition to the expected route for N-oxides, the loss of oxygen to give the acridine molecular ion that fragments as such, the primary loss of carbon monoxide is observed. Although rearrangement via an intermediate oxaziridine might be expected for quinoline and isoquinoline N-oxides,[17] it is not feasible for 11. Supporting a postulated rearrangement is the observed photochemical rearrangement of 11 to cyclohept[b]indole-10-[5H]-

one (12),[18] and the near identity of the mass spectrum of **11**, neglecting the $(M-O^{\cdot})^{+}$ route of fragmentation, with those of 9-acridanone and **12**. Both of these fragment by carbon monoxide loss to give what might reasonably be expected to be the carbazole molecular ion.

11   12

The fragmentation of the pharmaceutically important acridine derivative, atebrin (mepacrine or quinacrine, **13**), is dominated by the rupture of the dialkylaminoalkylamino side chain.[4]

13

The base peak in the spectrum is m/e 86, corresponding to the ion $[CH_2:NEt_2]^{+}$, and is accompanied by abundant related fragments at m/e 73, 99, 112, and 126. Fragments in which the acridine ring is retained are, with the exception of the molecular ion, usually less abundant than the alkylamine fragments. Elimination of the elements of diethylamine is observed as a metastable transition, and the resulting species appears to fragment according

m/e 277 *(100%)*   *m/e* 199 *(8%)* $\xrightarrow{-Cl^{\cdot}}$ *m/e* 164 *(22%)*

$-Me^{\cdot}\downarrow$ $-Cl^{\cdot}\uparrow$

m/e 262 *(8%)* $\xrightarrow{-CO}$ *m/e* 234 *(57%)* $\xrightarrow{-HCl}$ *m/e* 198 *(7%)*

to either of the schemes $M^+ - HNEt_2 - \cdot CH_2CH{=}CH_2 - C_2H_2 - CH_3$) or
$M^+ - HNEt_2 - (Me - \cdot CH_2CH{=}CH_2 - H\cdot$. The fragmentation of the
acridine nucleus of atebrin is illustrated by that of its synthetic precursor
6,9-dichloro-2-methoxyacridine, shown in Scheme 6, page 720.

References

1. M. E. Wacks and V. H. Dibeler, *J. Chem. Phys.*, **31**, 1553 (1950).
2. E. J. Gallegos, *J. Phys. Chem.*, **72**, 3452 (1968).
3. "Catalogue of Mass Spectral Data," American Petroleum Institute Research Project No. 44, Carnegie Institute of Technology, Pittsburgh, Pa., Serial No. 639.
4. R. G. Bolton, D. Phil. Thesis, University of Oxford, England, (1970).
5. S. D. Sample, D. A. Lightner, O. Buchardt, and C. Djerassi, *J. Org. Chem.*, **32**, 997 (1967).
6. P. M. Draper and D. B. MacLean, *Can. J. Chem.*, **46**, 1487 (1968).
7. P. N. Rylander and S. Meyerson, *J. Amer. Chem. Soc.*, **78**, 5799 (1956).
8. N. C. Rol, *Rec. Trav. Chim. Pays-Bas*, **84**, 413 (1965).
9. T. S. Muraski and C. Djerassi, *J. Org. Chem.*, **33**, 2962 (1968).
10. N. P. Buu-Hoi, C. Orley, M. Mangane, and P. Jacquignon, *J. Heterocycl. Chem.*, **2**, 236 (1965).
11. P. F. Donaghue, P. Y. White, J. H. Bowie, B. D. Roney, and H. J. Rodda, *Org. Mass Spectrometry*, **2**, 1061 (1969).
12. J. H. Bowie, R. G. Cooks, R. H. Prager, and H. M. Thredgold, *Aust. J. Chem.*, **20**, 1179 (1967).
13. J. A. Diment, E. Ritchie, and W. C. Taylor, *Aust. J. Chem.*, **20**, 1719 (1967).
14. E. Ritchie, private communication.
15. P. M. Draper and D. B. MacLean, *Can. J. Chem.*, **46**, 1499 (1968).
16. H. H. Mantsch, *Rev. Roum. Chim.*, **14**, 549 (1969).
17. O. Buchardt, A. M. Duffield, and R. H. Shapiro, *Tetrahedron*, **24**, 3139 (1968).
18. M. Ishikawa, C. Kaneko, and S. Yamada, *Tetrahedron Lett.*, 4519 (1968).

CHAPTER XIV

The Interaction of Acridines with Nucleic Acids

A. R. PEACOCKE

St. Peter's College,
Oxford, England

The nature of the interaction of acridine derivatives, especially amino-acridines, with nucleic acids has attracted increasing attention since their earliest use as cellular stains. The widespread biological effects of acridine derivatives gradually came to be connected principally with their ability to interact with nucleic acids, but interest in this interaction was heightened by the unravelling of the main features of the binding processes involved[1] and the subsequent proposal of a detailed structure for the complex.[2] This proposal attempted to relate the structure at the molecular level to the known mutagenic effects of the acridine derivatives in question, aminoacridines, such as 2–9.

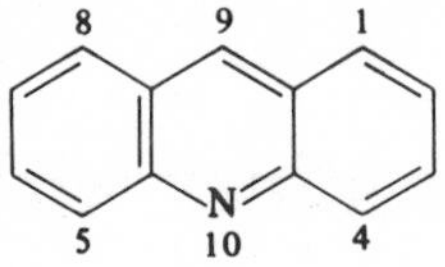

1, Acridine

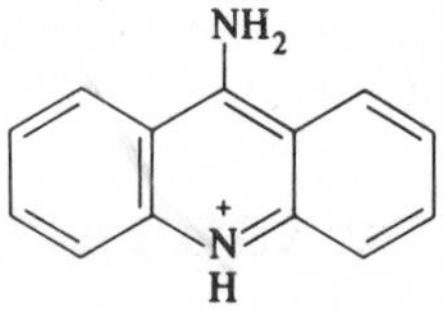

2, 9-Aminoacridinium cation

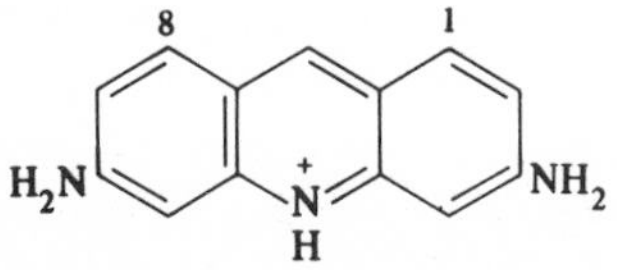

3, Proflavine, or 3,6-diaminoacridinium cation

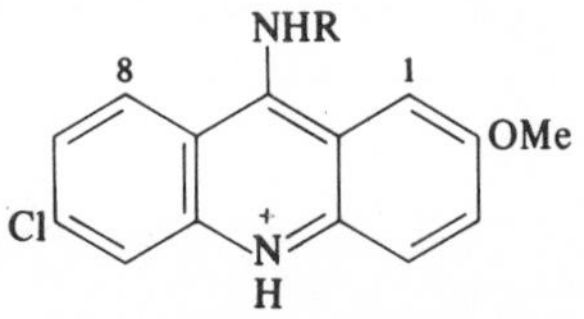

4, Atebrin (mepacrin or quinacrine) cation, R = CHMe(CH$_2$)$_3$NEt$_2$

5, Acranil cation, R = CH$_2$CH(OH)CH$_2$NEt$_2$

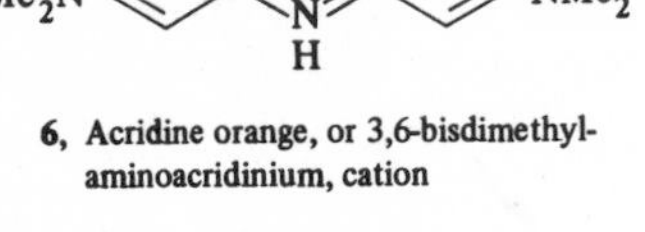

6, Acridine orange, or 3,6-bisdimethyl-aminoacridinium, cation

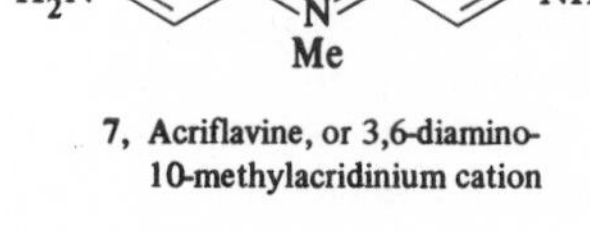

7, Acriflavine, or 3,6-diamino-10-methylacridinium cation

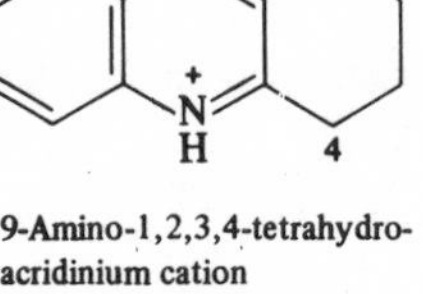

8. 9-Amino-1,2,3,4-tetrahydro-acridinium cation

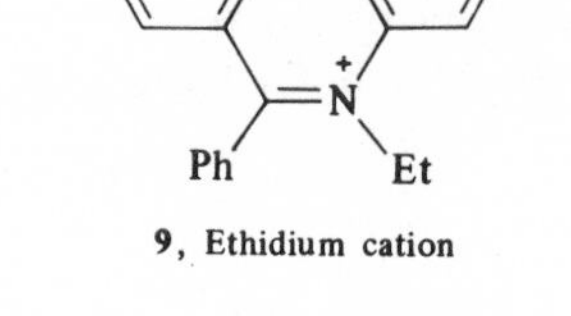

9, Ethidium cation

The intermittent trickle of publications in the 1950s on this interaction has grown into a broad river now, in the early 1970s, repeatedly bifurcating in many new directions, not least that leading to the study of the related antibiotic/nucleic acid interactions.

Concomitantly there has been a growing awareness in biochemistry, genetics, physiology and pharmacology of the significance of the interaction of small ligand molecules with biological macromolecules, e.g., the interaction of substrates and inhibitors with enzymes, carcinogens and antibiotics with nucleic acids, histological stains with macromolecules in the cell nucleus and cytoplasm and drugs with the components of cell membranes. The modes of investigation, the conceptual interpretative framework and the derived molecular structural situations often common to these different contexts and the much investigated acridine/nucleic acid system affords interesting examples and applications of more widespread features of small molecule/ macromolecule interactions. One of the motives in the study of such inter-actions has been to use the small molecule as a "probe" of subtle differences and changes in the structure of the macromolecules to which it is bound and the study of.the interaction of acridines with nucleic acids is no exception.

The number of publications is of such a magnitude that, in the account which follows, only the principal features and conclusions from these investi-gations can be outlined and only some typical examples of studies can be men-tioned specifically. The evidence for believing that much of the biological action of acridines is principally related to their ability to interact with nucleic acids will first be summarized (Section 1) and related to other inter-actions of nucleic acids. The quantitative study of the extent of the interaction, which yields "binding curves," will then be described (Section 2) since this is fundamental to all other structural studies. The effects on the interaction of structural modifications in the nucleic acid and of variations in the acridine structure are described in Section 3 and their effects on each other in Section 4 since, fortunately, the latter can be distinguished experimentally. A brief account (Section 5) is then given of studies on the kinetics of interaction. Certain diffraction and optical techniques are particularly relevant to the structural and geometrical relationships of acridine bound to nucleic acids and the results of the application of these methods are outlined in Section 6. The broad sweep of the evidence on the structure of the complexes is then assessed (Section 7) and, in conclusion, an attempt is made to indicate its relevance to the biological activity of the acridines (Section 8).

1. The Biological Role of the Interaction

Acridine derivatives have a wide variety of biological effects; not all of these can be attributed to the specific interaction that occurs between the acridines and the nucleic acids. Although this interaction is likely to be occurring when acridines act as antimalarial substances or carcinogens, the biological response is too complex and involves the disturbance of too many interlocking processes for this one type of interaction to be determinative, not even, it appears, when the inhibition is that of a DNA-primed RNA polymerase. Since these problems are discussed elsewhere in the appropriate chapters, the following section briefly summarizes some of the biological effects of acridines that seem to be closely linked with their ability to form complexes with nucleic acids.

The *staining properties* of diaminoacridines, such as proflavine (3,6-diaminoacridine, **3**) and acriflavine (3,6-diamino-10-methylacridinium chloride, **7**) and of acridine orange (bis-3,6-dimethylaminoacridine, **6**) have been much utilized because of two unusual features: these acridine derivatives stain the nuclei, and not the extranuclear nucleoproteins, of living cells without killing them,[3-6] and their complexes with DNA (deoxyribonucleic acid) and RNA (ribonucleic acid) in both fixed and living cells differ with respect to the color of their fluorescence (green and red, respectively).[7-10] In many of these investigations, a judicious use of the nuclease enzymes, combined with microscope photography, has unequivocally demonstrated that nucleic acids constitute the staining sites for acridines. As a consequence of these properties, the diaminoacridines already mentioned, as well as others, have been extensively used as vital stains, and acridine orange is also much employed for locating nucleic acids in nonliving fixed cells. More recently, the human Y chromosome has been shown to display an exceptional fluorescence when chromosomes are stained with atebrin (quinacrine) (**4**), and visualization and differentiation of heterochromatic chromosomes of a number of organisms have become possible.[11a, b, 12]

The action of acridines on *viruses* and *bacteriophage* includes, among other effects, the induction of mutations; since the nucleic acid (mainly DNA in the instances studied) is the unequivocal carrier of the genetic information, this biological effect must be a consequence of the specific nature of the interaction of the acridines with the nucleic acid moiety of the virus or phage. The mutagenic effects on viruses, which are described more fully in Chapter XV (p. 781), were first reported[13] in 1953 for the system proflavine-T2 bacteriophage (*Esterichia coli*) and have been observed in other viral systems, e.g., proflavine-polio virus,[14] and proflavine and other aminoacridines acting on the r_{II} region of the T4 bacteriophage.[15-17]

The last mentioned has been intensively studied, and it has been shown that when proflavine is mutagenic, it is not incorporated into the polynucleotide chains,[15, 18] as is 5-bromouracil, another mutagen toward the r_{II} region of T4 bacteriophage. The genetic evidence indicates rather that, when they are mutagenic, aminoacridines act by causing a deletion or insertion of a single nucleotide in DNA.[19] This hypothesis has been confirmed by a comparison of the amino acid sequences in lysozyme made under the control of the DNA of T4 bacteriophage which has, or has not, been subjected to treatment with proflavine.[20] Five particular amino acids in sequence in the lysozyme were changed as a result of the proflavine-induced mutation in the DNA of the T4-phage. These changes were of such a kind that on the basis of the known genetic code relating DNA and protein sequences, it was clear that at one extremity of the changed part of the DNA a nucleotide had been added and, at the other extremity, one had been removed. It was, of course, the ability of proflavine to induce mutants of this kind that made possible, by genetic experiments, the original characterization of the genetic code as triplet.[21] Aminoacridines inhibit the maturation and multiplication of phage progeny particles in *E. coli*[22-27] and their relative ability to do so follows the same order as their effectiveness as mutagens.[17]

Proflavine, acridine orange, and acriflavine accelerate the inactivation (i.e., loss of infectivity) of viruses which is caused by light and oxygen. The mechanism of this "photodynamic action" is still obscure.[28] However, with proflavine-sensitized polio virus, the action spectrum of the light inactivation is shifted in the direction of the changes in visible absorption that occur when proflavine binds to nucleic acid,[29] so that this latter interaction again appears to be a prerequisite of the biological effect.

The action of acridines on *bacteria* is described in Chapter XVI. In the present context, it must be asked if the nature of this action implicates nucleic acids in the bacteria as the principal biologically effective site of interaction. Briefly, it can be said that there is good circumstantial evidence that this is the case, but that specific evidence concerning the actual point at which the acridines dislocate the DNA/RNA/translation/replication sequence in bacteria has not yet been obtained. Albert, Rubbo, Goldacre, and their colleagues[30a-c, 31] convincingly demonstrated that (1) it was the cationic, not the anionic or zwitterionic, forms of acridines that were bacteriostatic (a cationic ionization of at least 50% was necessary, and preferrably 75% at pH 7.3 and 20°C); (2) the chemical nature of substituents usually had an influence on the acridine bacteriostatic action only through their effect on the cationic ionization, except where some specific chelating or dimensional effects (steric hindrance) were involved; (3) hydrogen ions competed very effectively with the acridinium cations for (presumably) anionic binding sites in the bacteria; and (4) the acridine should possess a minimum flat area of about 38 sq Å to be bacteriostatic, which implies that the binding site in the bacterium is also flat, so that attractive interactions are maximized. These factors strongly suggest nucleic acids as the principal binding site for bacteriostatic action by the acridines, since the nucleic acids are anionic, possess flat purine and pyrimidine rings as possible binding sites, and are, of course, deeply implicated in the control of protein synthesis, and so the growth rate, in bacteria. However, most of the proflavine absorbed by *Aerobacter aerogenes* has been shown to be absorbed at anionic sites which are not involved in bacteriostasis, as judged by the lack of influence of pH and of the acquisition of resistance on the net amount of proflavine bound to the whole bacteria.[32] Nevertheless, there is a considerable decrease (nearly 50%) in the amount of proflavine absorbed by whole *Clostridium welchii* cells when the nucleic acids are extracted from the cells,[33] so that nucleic acids form a substantial proportion of the binding sites in this Gram-positive organism. The importance of such interactions in bacteriostasis by proflavine, acriflavine, and ethacridine (or rivanol, 6,9-diamino-2-ethoxyacridine) had earlier been inferred from the McIlwain's discovery that the addition of RNA or DNA antagonized their bacteriostatic action.[34] This effect was always explicable in terms of a simple chemical removal of the acridine by insoluble complex formation, but our present knowledge of such complexes shows that, under the conditions of these experiments, only reversible complex formation would have occurred in the solution, so that an effect on the bacterial DNA/RNA metabolism would now seem to be implied by these observations.

Hinshelwood and his colleagues[35a, b] interpreted the kinetics of the inhibition of *A. aerogenes* by aminoacridines in terms of a dislocation of the delicately balanced kinetic organization of the cell, in particular of the rate of formation of an "autosynthetic enzyme," which we would undoubtedly now regard as the DNA/RNA system. This kinetic interpretation (which is more fully discussed in Chapter XVI, p. 804) therefore implicates nucleic acids as the site of acridine action on bacteria as much as does alternative interpretations in terms of a mutagenic action, again presumably on DNA. Acridines do induce mutations in bacteria,[13, 36] but whether or not the acquisition by bacteria of resistance to aminoacridines is to be attributed to the natural selection of preexisting resistant mutants is a subject of some debate (Chapter XV, p. 775).

The actions of acridines on *yeast* has revealed both an inhibitory effect on growth, which appears to be the result of interaction with the nucleoproteins of the yeast,[37] and, at much higher acridine concentration, an inhibition of respiration,[38] a purely enzymic inhibition phenomenon (Chapter XV, p. 760). The most-investigated mutational effects of acridines on yeast have been those that result in the formation of "petit" mutants,[39, 40] strains that produce small colonies and are deficient in certain enzymes.

Parallel with this action on yeast, aminoacridine derivatives interfere with the extranuclear DNA, *episomes* and *kinetoplasts*, which appear, respectively, in certain enteric bacteria and in trypanosomes. Thus acridine orange inhibits the multiplication of episomes in those particular strains of *E. coli* that are capable of transmitting this agent to other *E. coli* strains.[41a, b] and both acridine orange and acriflavine react with episomes so as to abolish their ability to transmit resistance to sulfonamides and antibiotics.[42] Acriflavine[43,44] induces certain trypanosomes to lose their kinetoplast, a specialized portion of the mitochondrion containing a large amount of DNA, which is self-replicating during interphase, distinct from the nuclear DNA of the same species and which appears to be involved in the coding of mitochondrial proteins. This loss is the result of a selective localization of the acriflavine in the DNA of the kinetoplast[44] and has been attributed to either the direct inhibition of the DNA replication process by the acridine and/or its causing incorrect replication of the DNA.[43] In any case, the biological effect of the acridine is clearly an effect of its interaction with the extranuclear DNA.

The action of acridines on *higher organisms* is often complex and it is not usually possible unequivocally to associate these effects with an interaction with nucleic acids, though this is clearly implied when, as is frequently the case, mutagenesis, inhibition or distortion of nuclear, or whole cell, division occurs.[45–47]

The foregoing accounts makes no pretense at completeness, but at least it is clear that the ability of acridines to interact strongly with nucleic acids is deeply implicated in some, though not all, of their major biological actions. This biological significance, especially the well-analyzed mutational effects on bacteriophage, have catalyzed a wide range of investigations into the molecular nature and basis of this interaction. The results of such studies must now be surveyed.

2. Binding Sites

Studies of the binding curves of acridines on nucleic acids have played a prominent part in developing ideas about the nature of the binding sites on the nucleic acids. The binding curve is the relation between the amount (r) of acridine bound per unit of nucleic acid (nucleotide phosphorus) and the concentration (c) of free, unbound acridine in the bulk solution and in equilibrium with the bound acridine. Such curves are also, or rather should be, an essential concomitant for the interpretation of other physicochemical studies of the complexes in solution.

The main *purpose* of determining such curves is to derive the number (n_J) of each of the P classes of binding sites (of intrinsic binding constant k_J) for the aminoacridine on the nucleic acid. In general,

$$r = \sum_{J=1}^{J=P} \frac{n_J k_J c}{1 + k_J c} \tag{1}$$

which simplifies to

$$r = \frac{n_I k_I c}{1 + k_I c} + \frac{n_{II} k_{II} c}{1 + k_{II} c} \tag{2}$$

for two classes (I, II) of binding sites; and to

$$r = \frac{nkc}{1 + kc} \tag{3}$$

or

$$\frac{r}{c} = kn - kr \tag{3a}$$

for one class of noninteracting binding sites.

In this last instance, a plot of r/c against r will be linear with an intercept at $r = n$ on the r axis and with a slope of $-k$ if this quantity, which is the intrinsic association constant of the group and is equal to [occupied sites]/ [unoccupied sites]·[free aminoacridine], is constant. Curvature of this plot can result from variation in one of the following factors: (1) an electrostatic free energy ($\Delta G_r°$), dependent on r so that $k = k' \exp(\Delta G_r°/RT)$; (2) the overlapping of the binding of more than one type of binding site, as expressed in Eq. (1); and (3) mutual, direct or indirect, interaction between bound molecules (see Refs. 48a–c and 49 for a fuller discussion). The electrostatic effects (1) can be suppressed by the addition of a neutral salt at sufficiently high concentration.[1]

When only two or three types of binding site are present, it is sometimes possible to discern distinct linear portions in the plot of r/c vs. r, from which the individual n_J and k_J can be determined. Even when there is a continuous overlap of a series of binding sites, the slope $d(r/c)/dr$ of this plot at a particular r is still an indication of the weighted, average association constant prevailing at that r value. The limiting value, as $r \to 0$, of the slope of the Scatchard plot [Eq. (3a): r/c vs. r] is the negative of the binding constant of the strongest binding sites, and the value of r as $r/c \to 0$ is

$$\sum_{J=1}^{J=P} n_J$$

Thus, by inspection of the binding curves (r vs. c) and analysis of the Scatchard plots, we hope to obtain information concerning the number and nature of the binding sites. However, quantitative analysis of the binding curves of acridines to nucleic acids is rarely straightforward. Even when electrostatic interaction effects have been suppressed by the addition of sufficient salt,[1] the Scatchard plots for these systems usually exhibit marked curvature, convex to the r axis. The existence of multiple binding sites and/or of mutual interaction between bound acridine molecules has therefore had to be postulated (see below). The treatment of these binding curves was at first relatively unsophisticated, for they were regarded simply as the sum of a few curves (usually two), each corresponding to a different type of binding.[1, 49] However, this approach, though giving broad useful interpretations, must soon give way to a more rigorous analysis of binding curves based on proper statistical mechanical treatments of multiple binding processes. There has been, over the years, a considerable literature in this field and the recent work of Schwarz and his colleagues[50a–c] seems to provide a suitable theoretical basis now for the analysis of empirical binding curves.

The *methods* of determining the binding curves of acridines on nucleic acids are not specific to these systems and have been critically surveyed and compared elsewhere.[49] They include spectrophotometric analysis,[1, 58, 59, 62, 63, 64] equilibrium dialysis,[1, 52, 58, 60, 61, 64, 75] sedimentation dialysis,[51] partition analysis[52] and fluorescence quenching.[52–57] The spectrophotometric method has been much employed but is restricted[49] to systems and conditions when, e.g., a good isosbestic point is observed in the family of spectra of the free and bound acridine molecules. Fortunately, this condition is fulfilled for many of the more interesting aminoacridines bound to double-helical DNA and facilitates the determination of binding curves. Equilibrium dialysis, although more tedious, has the virtue of being applicable, in principle, to all acridine/ nucleic acid systems, but it cannot reach to the very low r values available to fluorescence studies.

The binding curves of acridines, mainly aminoacridine derivatives, which have been obtained by these methods, exhibit, in summary, the following features.[51–64]

A. Heterogeneity of binding sites

Figure 1 shows the earliest determined,[1] but still typical, binding curve of proflavine on DNA. This form has frequently been observed and the implication of heterogeneity of binding sites is confirmed by the curvature of the corresponding Scatchard plot, even for very high ionic strength solvents in which electrostatic effects have been suppressed.[1] There is general agreement that the binding process can be divided into a process (I) by which up to about 0.2 molecules of acridine per nucleotide are bound strongly, with a ΔG of -6 to -9 kcal mole^{-1} aminoacridine and a weak process (II) by which further acridines are bound up to the electroneutrality limit of $r = 1.0$. This conclusion is based on the observation that the r/c vs. r plots fall into two distinct portions: one of high slope corresponding to process I up to $r \simeq 0.2$ and the other of low slope at r from 0.2 to 1.0. The inflected form of the binding curve suggests (e.g., Fig. 2, Ref. 49) that the process (II) of weaker binding of proflavine is cooperative, with a sigmoidal binding curve, so that the binding of one proflavine cation (3) facilitates the binding of the next,

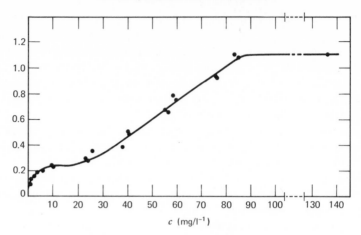

Fig. 1. The binding curve of proflavine on herring sperm DNA.[1]

especially as process II appears at high values of c at which self-aggregation occurs in proflavine solutions alone. Clearly, process II must also be occurring to some extent at the lower concentrations at which process I is dominant (see Section 5), so it is not surprising that the Scatchard plots, even in the range of r dominated by process I, are not linear; and moreover, the maximum number of aminoacridine molecules bound in process I (approximately corresponding to the value of r where the inflexion begins) varies with ionic strength.

These observations indicate that a number of binding constants are needed to characterize process I and/or that a continuous change occurs in binding constant(s) with increasing r consequent upon the continuous modification of the nucleic acid structure which occurs in binding (see Section 4). The heterogeneity of binding sites for proflavine and acridine orange on DNA has been analyzed by Armstrong, Kurucsev, and Strauss[64] by means of a combination of spectrophotometry and equilibrium dialysis. At an ionic strength of 0.2, the binding curves could be described in terms of a single, spectroscopically unique, bound ligand species (undoubtedly, monomer molecules), but at an ionic strength of 0.002 the spectra of the bound acridines denoted a stronger mode of binding (with a constant of 30×10^5 1 mole^{-1} for both proflavine and acridine orange) and a weaker mode (with constants of 7.3×10^5 and 1.5×10^5 1 mole^{-1}, respectively, for proflavine and acridine orange), both occurring within the c range, where the binding curve is concave to the c axis, i.e., the process I range. Because, *inter alia*, of the similarities and differences between the binding constants for these two acridines, the second, weaker mode was attributed to the binding of the acridine to the

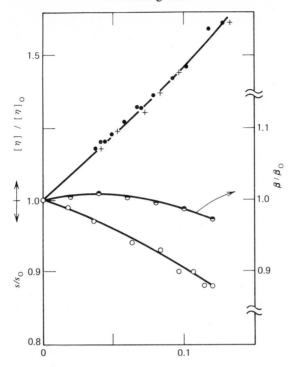

Fig. 2. Relative change of the intrinsic viscosity, sedimentation coefficient, and parameter β of rodlike DNA on binding proflavine.[89] ● Viscosity measurements, DNA mol. wt. = 395,000; + viscosity measurements, DNA mol. wt. = 492,000; ○ sedimentation measurements, DNA mol. wt. = 492,000; ◐ parameter β.[90] Subscript zero refers to DNA without proflavine.

exposed part of an acridine molecule already *partially* intercalated (see discussion in Section 7).

Some investigations, based on fluorescence quenching of the acridine/ nucleic acid interaction, have led to the conclusion, that at very low r values (<0.04), there is heterogeneity in the binding sites, adenine-thymine base pairs binding preferentially.[54, 55] However, it has also been maintained that because the quantum yield of bound acridines varies with r, conclusions from fluorescence measurements concerning heterogeneity of binding sites, and so of base-pair preferences, are invalid.[56, 57] However, this criticism of the earlier work[65] may itself be questioned on the ground that, at $r < 0.02$, any bound acridines are too distant for preferential energy transfer from higher-quantum-yield sites (A-T) to lower-quantum-yield sites (G-C) to be an explanation of the variable quantum yield at these low r values of the fluorescence of the bound acridine (acriflavine, in Refs. 54 and 57). Hence nonquenching sites have to be assigned to sequences of A-T, to which are attributed a greater association constant than G-C sequences, binding to which entirely quenches fluorescence.[65] Heterogeneity (two processes, A and B) within process I has also been inferred by Löber,[87b] both from Klotz plots[48b] deduced from studies of the fluorescence quenching of proflavine and trypaflavine (acriflavine or 3,6-diamino-10-methylacridinium, 7) by DNA; and from the observation that the spectra of bound proflavine, trypaflavine, 3,6-diamino-10-ethyl-acridinium chloride, and 3,6-diamino-10-amylacridinium chloride at very high ratios of DNA to acridine (at the extreme, $r \sim 0.14$) no longer passed through the same isosbestic point as at higher r within the process I range (private communication, 1970, and ref. 84b).

B. Effects of ionic strength (μ)

Increase in ionic strength diminishes the extent of binding, but binding by process II is decreased more by a given change in ionic strength than is binding by process I.[1, 58, 59] Hence it is concluded that electrostatic forms contribute greatly to both binding processes but are relatively more important in process II than in I. Gilbert and Claverie have made detailed calculations of the electrostatic free energy of the aminoacridine/DNA complexes in media of various ionic strengths.[66] They have shown that the strong binding curves, with its variation of association constant (the slope of r/c vs. r) and the variation of the inflexion point in the r vs. c binding curves, which represents approximately the upper limit of acridine bound by process I, could both be accounted for satisfactorily. They reached this conclusion on the basis of an intercalated model for the complex, in terms of the electrostatic attractions between the acridinium cation and DNA phosphates, the repulsions between bound acridinium cations, and the interaction of the ions in the bulk solutions (as represented by their ionic strength) with both the phosphate and acridinium ions. They stress the dominant contribution of these electrostatic interactions to the total binding energy and calculate that the mutual repulsion between the bound acridinium cations is the principal factor in limiting process I binding, independently of the relatively small differences in the stacking energies of acridines on A-T compared with G-C base pairs. Clearly, this dominance of the electrostatic interactions must not be forgotten when more detailed structural effects are under discussion.

C. Effect of temperature

The effect of temperature is complex. The binding of proflavine on DNA decreases with a rise in temperature,[58, 67, 68] and this can be explained partly by a decrease in the number (n) of binding sites. However, the ΔH of binding is not constant and also decreases with temperature.[58] Therefore, the effects are complex, even when electrostatic effects are suppressed at $\mu = 1.0$ and involve not only the different temperature coefficients of binding by processes I and II, but also the effects on binding (both n and k) of some change of state in the DNA, which occurs at a much lower temperature than its usual melting point, as well as the different binding abilities of native and heat-denatured DNA.[58] As a consequence of this last difference, a cooperative decrease[58] in binding occurs at a temperature that is higher than the actual melting temperature of DNA when binding proflavine at $\mu = 0.01$ and 0.1.

3. Structural Factors

A. Effect of Nucleic Acid Structure

The effect of the structure of a nucleic acid on its ability to bind acridines has been studied in connection with the problem of whether or not the intact double helix is necessary for strong binding. For example, it has long been known that changes in the secondary structure of DNA can have considerable influence on the binding process, as revealed by the observed metachromasy, the displacement of the absorption spectrum of the bound acridine.[69, 70] In particular, the effect on its ability to bind aminoacridine of heating solutions of DNA and rapidly cooling to 0°C has been especially examined because this procedure disorganizes the double-helical structure. Elucidation of this matter has been rendered difficult by its entanglement with the effects of temperature, as mentioned above, in the various binding processes on to any one form of DNA; in particular, the enhancement of proflavine binding by II relative to I on denaturation of DNA.[59, 63] However, there now seems to be general agreement that heat-denatured shock-cooled DNA binds aminoacridines, at moderate ionic strengths ($\sim$0.1) and temperatures of 20 to 25°C, by process I along almost the same binding curve as native double-helical DNA [see the survey in Ref. 49, pp. 1239–1240 and Refs. 62, 71 (Fig. 6), and 84a]. Yet, Chambron, Daune, and Sadron report a marked reduction in binding of proflavine at a critical temperature which is coincident, at $\mu = 1.0$, with the breakdown of the double-helical DNA structure.[58] This implies that denatured DNA does not bind proflavine at these high temperatures ($>$90°C). The whole matter has been much clarified by the work of Ichimura et al. who have, by extensive equilibrium dialysis studies, determined the operative thermodynamic factors in the binding of acridine orange (6) to both native and heat-denatured DNA.[60] Over the range $r = 0.04$ to 0.13, only one binding constant (somewhat surprisingly) was operative for both native and denatured DNA, so that a ΔG of binding ($= -RT$ in k) could be calculated, and hence the corresponding ΔH and ΔS from the temperature coefficients of k (Table I).

TABLE I. Thermodynamic Parameters of the Binding of Acridine Orange (bis-3,6-dimethylaminoacridine, 6) to Native and Heat-Denatured DNA ($\mu = 0.11$)[a]

	Temperature (°C)	Native DNA ($n = 0.2$)[b]	Denatured DNA ($n = 0.5$)
10^{-5}, k	25	3.1	2.1
	30	2.2	1.4
	40	1.5	0.6
	50	0.9	0.3
$-\Delta G$ (kcal mole^{-1})	25	7.5	7.3
	30	7.4	7.1
	40	7.4	6.9
	50	7.3	6.6
$-\Delta H$ (k cal mole^{-1})		9.5	14.5
ΔS (cal mole^{-1} deg^{-1})	25	-6.8	-24.3
	30	-7.3	-24.4
	40	-6.8	-24.4
	50	-6.8	-24.4

[a] From Ref. 60.

[b] n is the maximum value of r (binding sites per DNA phosphate) for strong binding (process I).

It is instructive to set out the thermodynameter parameters for acridine orange binding at 25°C ($\mu = 0.1$), in the following way:

	cal mole^{-1} ΔG		cal mole^{-1} ΔH		deg. e.u. mole^{-1} $T \cdot \Delta S$
Native DNA	$-7{,}500$	$=$	$-9{,}500$	$-$	$298(-6.8)$
Heat-denatured DNA	$-7{,}300$	$=$	$-14{,}500$	$-$	$298(-24.3)$

It is clear from the results of Ichimura et al. that a similar value of ΔG, and hence of effective k, at 25°C and $\mu = 0.11$ can result from the combination of very different ΔH and ΔS values. Acridine orange is, nevertheless, released much more steeply from denatured DNA than from native DNA as the temperature rises (more negative ΔH of binding) and this would explain how the binding curves of the two forms of DNA toward acridines can be very alike at 20–25°C but different at higher temperatures. [It must be noted that in the case of acridine orange, the n values for strong binding on the two forms of DNA are not the same (Table I) at any temperature, so that, unlike the situation with proflavine, the same binding curves are not obtained at 25°, even though the k's are so close. Because of the tendency of acridine orange to self-aggregate, it displays many other differences from the binding of the simpler aminoacridines.]

The effects on the interaction caused by denaturation with agents other than heat has not been much studied. At pH 2.8, DNA still binds proflavine, as has been demonstrated by observing, by means of absorption optics at $\lambda = 454$ nm (the usual isosbestic point), the sedimentation boundary at pH 2.8 of pro-flavine bound to DNA.[72] The series of spectra of proflavine, obtained when increasing amounts of DNA are added at pH 2.8, are like those of heat-denatured DNA at low ionic strength.[63] The limiting spectrum, at high concentrations of DNA and very low r values, is identical with the usual spectrum of strongly bound (process I) monomeric proflavine (isosbestic point at $\lambda = 454$ nm). As r increases, the spectrum includes an increasing contribution from aggregates, and this is also indicated by a rough r/c vs. r plot.[63] Thus acid-denatured DNA at acid pH (2.8) tends to bind proflavine more by process II than by I. This observation has been related to the loss on lowering the pH of the optical activity that is induced in proflavine when it is bound to DNA.[63] By means of fluorescence quenching studies, Löber[87a] has concluded from the change of the inferred binding curves with pH that the two distinct processes (A and B) which he distinguishes within process I (see above and Ref. 84b) are dependent in different ways on DNA structure, the stronger (A) requiring double helices but not the weaker (B).

Binding by the weak process II may be identified with binding as aggregates, to judge from the spectral shifts observed, especially when the acridine (e.g., acridine orange) self-aggregates with a distinctive change in spectrum.[73, 74] By such means, Bradley and Wolf studied the aggregation of acridine orange on nucleic acids and other polycleotrolytes, and introduced the concept of "stacking," which may be described as the binding of ligand molecules to ligand molecules already bound on an external surface of the macromolecule (by electrostatic forces, principally).[74] The ligands interact mutually in a direction perpendicular to their aromatic planes, so that they pile up on each other like a stack of coins. Heat-denaturation and shock-cooling of DNA caused an increase in the proportion of acridine orange bound in the stacked form, as indicated by the need to increase the ratio of DNA to acridine to cause complete unstacking,[74] in agreement with the relative enhancement of process II binding of proflavine relative to process I when DNA is denatured by heat or acid, as already described (see also the preliminary account of the effects on binding of DNA denaturation by pH changes and by organic sol-vents in Ref. 87a). However, such denatured DNA still contains entangled interacting chains, so these observations still leave as an open question the exact nature of the structural requirements in nucleic acids for the binding of acridines by process I.

Ribosomal RNA exhibits the same type of binding curve as does DNA, although it binds less proflavine and the inflection point, at $\mu = 0.1$, is at about $r = 0.09$ instead of at $r \simeq 0.2$, as with DNA.[1, 63] This suggests that the maximum binding of proflavine by process I is, for RNA, about half of its value with DNA. Since such RNA contains both single-stranded and helical regions, through the chain doubling back on itself in regions where the base sequences are complementary, it would be tempting to postulate that the acridine is binding only to the double-helical regions. However, polyadenylic acid at pH 6.2, in a form that is nonhelical though containing stacked bases, binds proflavine strongly by a markedly cooperative process, rather better than does its helical form at pH 5.0, so the preceding conclusion cannot be legitimately drawn.[61, 75]

There are marked similarities in the binding curves and spectral displacements of the acridine for DNA, double-helical poly- (A,U) and triple-helical poly- (A,2U)* interacting with proflavine,[63, 76] and for these same polynucleotides, and also poly(I,C), interacting with acriflavine, although in the latter instances acriflavine fluorescence is quenched by DNA but enhanced by the other polynucleotides.[57] On the other hand, the addition of poly-U to solutions of proflavine had only very slight depressive effects on the spectra, which suggested that little or no interaction was occurring.[63]

The various evidence is thus still somewhat confusing. It seems possible to rationalize it by asserting that, in order for aminoacridines to be bound strongly by process I to a system of polynucleotide chains, the conditions must be such that there is definite interaction, by either hydrogen bonding or by stacking forces, between the bases, which may be on the same or on different polynucleotide chains. These conditions are fulfilled not only by intact double-helical DNA but also by other, although not all, polynucleotide systems. Whether or not this generalization can be upheld will depend on further investigations, but it appears to represent the situation better than the apparently simpler assertion that the existence of strong binding is dependent on long runs of an intact double-helical structure. This generalization also seems to be consistent with the observations on the interaction of nucleic acids and of polynucleotide homopolymers and their mutual complexes with ethidium bromide, which behaves like proflavine in so many respects.[77a, b]

There is some evidence that the base composition of nucleic acids is also important to the interaction. The binding of acridine orange to polynucleotides, and probably RNA fractions, is dependent on the base composition[78]

*A = adenine, U = uracil, C = cytosine, I = inosine.

and fluorescence quenching studies suggest that binding is greatest to those sites with least quenching efficiency, which appear to be the adenine-thymine base pairs (see Section 2).[52–55, 73] This agrees with the observation that the T_m of complexes of DNA with acridine orange increases with the adenine-thymine content of the DNA,[79] with the observed ability of adenine and thymine mono- and polynucleotides to suppress the fluorescence of proflavine solutions[57, 80, 81] and with theoretical calculations.[67]

B. Effect of Acridine Structure

The effect of structure on the interaction of acridines with nucleic acids is clearly of great interest in relation to the various biological activities of different acridines.

The greater tendency of aminoacridine derivatives, such as acridine orange, which self-aggregate readily, to bind as aggregates on nucleic acids (stacking or process II binding) has already been mentioned. As with their ability to act as bacteriostatic agents,[30, 31] only those aminoacridines interact strongly which are in the fully cationic form at the pH (normally 6–8) of the binding experiments, which usually means that an amino group which may be substituted is present at the 3-, 6-, or 9-positions of the acridine ring.[82] It was also shown that a minimum planar area of 38 Å^2 (corresponding to the three rings of acridine) was required in the acridine ring for bacteriostatic activity.[30, 31] Hydrogenation of one ring of 9-aminoacridine (2) to form the quinoline derivative, 9-amino-1,2,3,4-tetrahydroacridine (THA) (8) practically eliminated its bacteriostatic activity[30] and reduced its maximum capacity to bind to DNA by process I by more than a half,[59] no doubt on account of its possession of a bulky, buckled ring in place of one of the aromatic rings of 9-aminoacridine. However, THA could still bind to DNA, presumably through possession of its two planar rings[83] and in this respect it is analogous to the quinoline derivative chloroquine (10) which probably (like the acridines) intercalates, at least partially, into DNA (see Section 7).

$$NHCHMe(CH_2)_3NEt_2$$

10, Chloroquine

When long and bulky side chains are attached to the 9-amino group of 9-aminoacridine (as in acranil 5 and atebrin) the binding is not reduced.[59] The structure of these complexes must be such that the 9-position is free to attach long side chains without detriment to the interaction. The dependence of ability to bind to nucleic acids on the basicity of the acridine has been confirmed by Löber,[84a–e] who also showed that alkylation of the ring nitrogen (e.g., 10-methylation of proflavine to acriflavine) enhances the binding ability of an aminoacridine to an extent not very dependent on the length of the alkyl chain.

A series of 9-aminoacridine derivatives (below), in which bulky alkyl groups were substituted at various positions in the framework [9-amino-4-ethyl-acridinium chloride (11); (+) and (−)-9-sec-butylaminoacridinium chloride (15); 9-amino-2-*tert*-butylacridinium chloride (13); 9-amino-2-*sec*-butyl-acridinium chloride (12); 9-amino-2,7-di-*tert*-butylacridinium chloride (14)] all showed evidence of strong interaction with DNA.[85, 86] Thus spectral shifts and, in the first three instances, extrinsic Cotton effects (see Section 4) have been observed, as well as that increase in DNA viscosity which is attributed to intercalation (see Section 7). However, although the 2,7-di-*tert*-butyl derivative of proflavine (16) binds readily to DNA, it does not cause the changes in viscosity and sedimentation velocity of the DNA which proflavine does and it is concluded that it does not intercalate.[87c]

11, 9-Amino-4-ethylacridinium
cation

12, 9-Amino-2-*sec*-butylacridinium
cation

13, 9-Amino-2-*tert*-butylacridinium
cation

14, 9-Amino-2,7-di-*tert*-butylacridinium
cation

15, (±)-9-*sec*-butylaminoacridinium
cation

16, 3,6-Diamino-2,7-di-*tert*-butylacridinium
cation

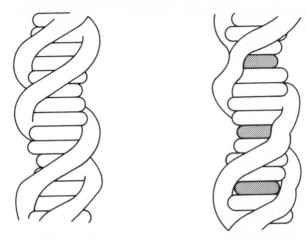

Fig. 3. The double-helical structure of DNA (left) and DNA-containing intercalated aminoacridine molecules (right) as proposed by Lerman.[91] The base pairs (open discs) and acridine molecules (black discs) are viewed edgewise and the sugar-phosphate backbone is depicted as a smooth tube.

4. The Effects of the Interaction

A. Effects on the nucleic acids

The combination of proflavine and acridine orange with DNA causes a marked increase in intrinsic viscosity and a decrease in sedimentation rate.[2, 51, 88] The variation of these properties with r has been determined[51, 88] and the best available hydrodynamic model for DNA of molecular weights higher than a million (the wormlike chain) was applied to interpret the results in terms of the change in the configuration of the DNA.[88] The change in viscosity corresponded, up to r values of about 0.2, to an increase in contour length for each bound aminoacridine (proflavine or 9-aminoacridine) of the order of the normal spacing (3.4 Å) between DNA bases.[88] This process of extension reached a limit when r attained a value corresponding to the completion of binding by process I, after which no further increase in viscosity occurred with increase in r (Fig. 2). The decrease in sedimentation coefficient with r also indicated an increase in contour length of the DNA double helix.[2, 51] Quantitative interpretation of these changes was somewhat indirect and obscured by the complexities of the wormlike chain mode. These difficulties were circumvented by Cohen and Eisenberg, who, by using sonically degraded rodlike DNA of molecular weight $4-5 \times 10^5$, were able to demonstrate unambiguously that the extension in the contour (i.e., rod) length of such DNA for each molecule of proflavine bound was equal to 84% of the characteristic 3.4 Å interbase spacing along the direction of the axis of the DNA.[89] The Scheraga-Mandelkern coefficient,[90]

$$\beta(= Ns°_{20,w}\{[\eta_0]^{1/3}/M^{2/3}(1 - \bar{\nu}\rho_0)\}$$

(where the subscripts zero refer to the solvent, N = Avogadro's number, and the other symbols refer to DNA and its solutions), depends only slightly on r for proflavine bound to the smaller rodlike DNA[89] but decreases with r with the larger, more coillike molecules,[51] which suggested that some increase in flexibility of the DNA accompanies the undoubted increase in contour length. This is now supported by more detailed viscosity measurements with acridine orange and the nonrodlike DNA.[64] A wider range of basic dyes with three or four fused rings, various antibiotics, benzacridines and a triphenylmethane dye have also been shown to enhance the viscosity of DNA.[91, 92a, b]

It is interesting to note that the viscosity of heat-denatured DNA,[88] which is, of course, much less that of native DNA, either does not change at all on complexing with proflavine and 9-aminoacridine at μ = 0.1, or is decreased along a flat curve at μ = 0.001. The last effect is indistinguishable from the effect of adding extra salt to a solution of μ = 0.001 and is an example of ion shielding on the viscosity of a polyelectrolyte. The absence of a specific increase in the viscosity of denatured DNA on interacting with amino-acridines suggests that the increase in contour length on strong binding of aminoacridines by process I is a specific feature of the interaction with the intact double-helical structure. Conversely, it is possible for such strong interaction by process I to occur, without there being an increase in the contour length, in polynucleotide structures other than the intact double helix.

The increase in contour length has been confirmed by autoradiography,[93] and light scattering[94] reveals a corresponding increase in the radius of gyration. This expansion is accompanied by a decrease in the mass per unit length, along the helical direction, according to low-angle X-ray measurements[95] and light scattering measurements,[94] and by a loss of the hypo-chromicity of the DNA at 260 mμ.[68, 96] X-Ray diffraction patterns[2] obtained from fibers of the DNA-proflavine complex were qualitatively consistent with a model in which the two DNA strands both untwisted and extended on interacting with DNA and similar conclusions have been drawn from such studies on the complexes of ethidium bromide and DNA.[97] A more detailed study has, in general, confirmed this view,[98] although the phenomena are much more complex than first appeared. For example, the X-ray evidence is consistent with an unwinding of between 12 and 45° from the original relative orientation of $+36°$ between successive base pairs, so that the final relative orientation may be between $+24$ and $-9°$.

The interaction of aminoacridines with native DNA produces not only the configurational changes mentioned above, but it also stabilizes the macromolecule toward thermal denaturation,[2, 67, 99, 100] mainly through its electrostatic effect.[58]

B. Effects on the acridines

The most characteristic effect of interaction with nucleic acids on acridine derivatives is the displacement of the electronic absorption spectrum to longer wavelengths, when bound as single cations (α spectrum), and the reversal of this trend when the acridine is bound as dimers and aggregates (β and γ spectrum).[73] The quenching of the fluorescence of the acridine on binding is also a notable change. Comparison of the spectra obtained in the case of acridine/acridine interaction with a related study of the change in the fluorescence spectra of aminoacridines bound to nucleic acids,[96] and the observation that the oscillator strength of proflavine bound to DNA is constant up to the point at which process II supersedes I,[101] lead to the conclusion that in strong binding by process I the absorption characteristics of the bound acridine are altered, at least partly, as a result of coupling with the nucleotide bases. In weak binding by II or stacking, however, the alterations are the result of coupling between bound acridines, as in concentrated acridine solutions.[102a, b] This interpretation is relevant to the problem of the structure of complexes I and II. A full theoretical treatment of the electronic and vibrational energy levels is only now developing (see Chapter X, p. 661), but this has not precluded the practical use of the changes in absorption spectra and in fluorescence in the determination of binding curves. It should be noted that the spectral displacements are not confined to the visible regions but also occur in the uv absorption bands.

When aminoacridines are bound to nucleic acids, they become optically active in the region of the (displaced) absorption bands of the bound molecules, i.e., they exhibit an extrinsic or induced Cotton effect in their optical rotation,[103–105] superimposed on the normal, wavelength dispersion curve, and in their circular dichroism in the same absorption bands.[71, 86, 105–109] These effects contain a contribution from the acridine bound at an isolated site (i.e., at low r values when bound molecules are far apart) and sometimes a more variable contribution which depends on the proximity of the bound acridines and which changes with r. The first effect must arise from the asymmetry of the acridine/nucleic acid interaction site but, even so, it is surprisingly sensitive in magnitude to external conditions, such as changes in salt concentration.[108] The other effect can be quite dramatic in its cooperativity, i.e., its steep increase with r, and appeared to result, statistically speaking, from the mutual effects on each other of acridines bound adjacent to each other by process I.[71, 104] This mutual effect was at first attributed to the effects of direct coupling between the bound acridine molecules[104] over distances of 6–7 Å, but subsequent more extended studies have shown that only acridines containing a 3-amino group (and ethidium bromide) display a steeply cooperative dependence on r of their induced circular dichroism on nucleic acids.[70, 86] Other aminoacridines showed only a very flat dependence on r of their induced circular dichroism per bound molecule, which tended to a finite value at zero r.

These and other observations and especially the congruence in the respect between the 3-aminoacridines and ethidium bromide (whose position of binding can be more closely located, because of its exacting structure[110]) have pointed to an interpretation of the cooperativity, in which one bound acridine affects the next through the changes it induces in the DNA structure that lies between them.[71] This conclusion has further implications for the structure of the complexes, since it suggests that the 3-amino group has a location and role in them which is distinct from that of amino groups at other positions on the acridine rings.[75, 86] When dimers of acridine orange (not itself optically active) are bound to nucleic acids, optical activity can also apparently be induced by coupling between the acridines in adjacently bound dimers, and the resulting optical rotation can show complex variations with wavelength and with the ratio of dye to nucleic acid.[107] These phenomena have been studied in some detail, and the possibility of optical activity being induced in the bound acridine has been found to depend in subtle ways on the structure, rigidity, and helicity of the polynucleotide chains.[104]

The chemical reactivity of the amino groups of the aminoacridines is reduced when they are bound to DNA: thus there is a 50-fold reduction in the rate of diazotization of proflavine and other aminoacridines on binding to DNA.[111] This reduction is much greater than when they interact with synthetic polymers, and it must be concluded that in the complex with DNA the amino groups are much less accessible to these nitrosating reagents.

5. Kinetics of the Interaction

The rate of processes involved in the interaction of some aminoacridines with DNA and polynucleotide have been studied by the temperature jump technique, coupled with the appropriate spectroscopic measurements in the absorption band of the bound acridine.[112, 113] The binding of proflavine on polyadenylic acid at pH 7.5 (single-chain-stacked, but not the double-helical form) involved a single relaxation process independent of the concentrations of the reactants and so corresponded to an intramolecular process subsequent to the formation of the proflavine-poly-A complex and probably resulting from the stacking of proflavine molecules along the poly-A chain.[112]

Within the process I region, proflavine has been shown[113] by the temperature jump method to bind to DNA in two kinetically distinguishable stages: (1) a rapid bimolecular binding, of low activation energy (4 kcal/mole^{-1} of proflavine) and with a rate constant (k_{12}, below) only one or two orders of magnitude smaller than that of an entirely diffusion-controlled process; and (2) a slower process of half-life of the order of milliseconds, i.e., its rate constant k_{23} is approximately 10^3 sec^{-1} and of a much larger activation energy (16 kcal mole^{-1} of proflavine), consistent with the occurrence of an intramolecular structural change in the DNA. The temperature jump technique also allowed the determination of the rate constants for the reverse processes (k_{21} and k_{32}, respectively) and the corresponding equilibrium constants ($K_{12} = k_{12}/k_{21}$, $K_{23} = k_{23}/k_{32}$). The processes have the relation:

$$\text{Pf} + \text{DNA} \underset{k_{21}}{\overset{k_{12}}{\rightleftharpoons}} \text{(bound Pf)}' \underset{k_{32}}{\overset{k_{23}}{\rightleftharpoons}} \text{(bound Pf)}'' \qquad (4)$$

$$\text{1st stage} \qquad\qquad \text{2nd stage}$$

The equilibrium constant for binding in the first stage (K_{12}) increased with pH, and so with increasing net negative charge; and increased even more steeply with a decrease in μ, which thereby implicates electrostatic interactions as principally involved, especially in view of the low activation energies corresponding to k_{12} and also to k_{21}. However, a decrease in μ lowers K_{21} for the second stage, while still increasing the actual rate constants, k_{23} and k_{32}— yet changes in pH had little effect on these parameters of the second stage. These observations led Li and Crothers[113] to argue that the second stage involves an intramolecular change in the DNA, in which base pairs are not opened (i.e., the hydrogen bonding between bases remains intact), and that in the final form [double prime in Eq. (4)] the separation of the positive charge from the negative DNA phosphates is greater than in the intermediate form [single prime, Eq. (4)]. The rapidity and μ-dependence of the first stage indicates that the binding is external and the second stage was then reasonably interpreted as the transition of the externally bound proflavine to an internal site (i.e., intercalated; see Section 7), without any intermediary opening of the base-pairs. Therefore, in Eq. (4), (bound Pf)' is proflavine externally stacked and bound (bound Pf)'' is proflavine located in a more internal site, whose formation necessitates conformational changes in the DNA. At $\mu = 0.016$, the proportion of bound proflavine which was externally bound [single prime, Eq. (4)] was about 30% but decreased to 7% at $\mu = 0.2$, which accords with the conclusions drawn from the binding curves (see Section 2). It also explains why the increase in contour length deduced from hydrodynamic measurements (e.g., Fig. 2 and Ref. 89) is somewhat less than can be deduced for complete intercalation. Glucosylation of the DNA, as in that obtained from T2 bacteriophage, slowed down the penetration rate (k_{23}) and approximately doubled the proportion bound externally.[113] Variations in this mechanism seem to occur, since ethidium bromide is reported not to involve an intermediate external complex and its insertion, which is slower than that of proflavine (3,6-diaminoacridine), does require the opening of an adjacent base pair, whereas the 2,7-di-*tert*-butyl derivative of proflavine (16) binds only with a fast relaxation time, and so, only externally.[114]

6. Relation in the Complex of the Acridine Plane to the Base Plane and Helical Axis of the DNA

Studies on the polarized fluorescence of atebrin (4) bound to flowing DNA (at $r \simeq 0.015$) showed that the plane of the atebrin ring was nearer to that of the bases than that of the helical axis of DNA.[114] Since the orientation of the DNA molecules by flow could not be complete under the conditions employed, this evidence cannot go further than to affirm that, in the complex, the planes of the acridine rings are within $\pm 30°$ of those of the DNA bases.

Flow dichroism studies on the complexes of acridine derivatives with DNA show that the DNA itself is slightly more dichroic when the acridine is bound.[114, 115] Thus the DNA bases are not tilted from their usual relation to the helical axis and the dichroism of the acridine and of the DNA bases are similar, so that the planes of the bases appear to be approximately parallel to the planes of the bound acridine rings. But again, the actual experimental evidence and the incompleteness of orientation of the DNA by flow only allow the assertion that the plane of the acridine ring is within $\pm 30°$ of that of the DNA bases. Unfortunately, the values of r at which observations have been made are not always clear so that the relative extents of binding by I and II are uncertain. This is an important factor in the interpretation of such optical evidence; for example, the change of circular dichroism on flowing complexes of acridine orange and DNA indicated that the acridine orange cations were roughly halfway between being perpendicular and parallel to the DNA helix axis.[116] However, the value of r was 0.5, so that at least half the acridine orange was bound by process II and this evidence, therefore, could not give an unequivocal indication of the orientation of acridine orange with respect to DNA when bound by process I. Subsequent, more detailed studies of flow circular dichroism, nevertheless, still indicate that, when acridine orange is bound as a monomer (process I) to DNA, it is more perpendicular than parallel to the DNA axis but deviates sufficiently from the perpendicular to require some tilting of the bases to accommodate the acridine orange.[107]

That the transition from process I to process II binding of proflavine involves a progressive change from an orientation of proflavine approximately perpendicular to the DNA axis to a more disordered state is nicely demonstrated by the observation of Houssier and Fredericq that, when the DNA in nucleohistone is oriented in an electric field, there is a progressive loss in the negative dichroism of the bound proflavine as the value of r increases.[117]

7. The Structures of the Complexes

The binding curves clearly indicate the existence of a stronger (I) and weaker (II) mode of binding of the aminoacridines on nucleic acids, and the structure of each type of complex must be considered separately.

A. Binding by Process I

Process I has the following experimental characteristics:

1. It is strong, i.e., the binding energy is of the order of 6–10 kcal mole^{-1} of aminoacridine bound.

2. It is favoured by the possession by the acridines of three flat aromatic rings (ca. 38 $\mathring{A}^2$), which can interact with the nucleotide bases of the nucleic acid, although two such rings also appear to allow such binding.[59,83] These interactions can be broadly described as van der Waals forces and are dependent on the nature of the bases.

3. It is, at least in part, electrostatic, for the interaction increases with decreasing μ. Only the cations bind strongly.

4. The contour length of double-helical DNA is increased, and its mass per unit length is decreased on binding.

5. The contour length of denatured DNA is not increased on binding if μ is large enough to suppress polyelectrolyte effects.

6. The planes of the bound acridine cations are approximately (to within 30°), parallel to the nucleotide bases and perpendicular to the axis of double-helical DNA.

7. The reactivity of the amino groups of the aminoacridine is diminished on binding.

8. Binding by process I to denatured DNA occurs at 20–25°C, to the same extent and with approximately the same range of binding constants as on intact double-helical DNA.

9. Binding by process I occurs with approximately the same range of binding constants, although with fewer sites (lower n) on single-stranded RNA as on DNA.

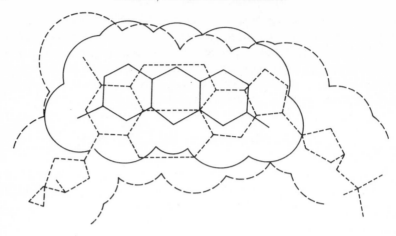

Fig. 4. A molecule of proflavine (solid-line hexagons) superimposed over a nucleotide base pair (lines in small dashes) with the deoxyribose phosphate chain in the extended configuration.[91]

10. Long side chains attached to the 9-amino position do not reduce binding in the 9-aminoacridine series.

11. There is an upper limit at $r \simeq 0.2$ to 0.25 to binding by process I.

Features (2) and (3) have been known since the first investigation of metachromasy of acridine dyes in cell staining reactions, and in 1947 Michaelis affirmed that the flat dye molecules might lie between different bases with the positive groups close to the negative phosphates of the DNA.[73] As additional evidence accrued, this view was reaffirmed by Oster[118] and, after the discovery of the double-helical structure of native DNA, it was suggested in 1955 and 1956 that the acridine rings could slip into a plane parallel to and between successive base pairs of this structure.[1, 52]

After the two types of binding had been distinguished and the strength of binding by process I clearly noted,[1] Lerman[2] made the initial observations on which characteristic (4) was based. This, and subsequent evidence for (6) and (7), led to his proposal of an intercalation model,[2, 91, 114] in which the amino-acridine cation is inserted between, and parallel to, successive base pairs of double-helical DNA, which has to untwist and extend in order to accommodate them (Fig. 3). In this model, the aminoacridine cation lies centrally over a base pair, so that the positive nitrogen atom is near to the central axis of the molecule and equidistant from the two polynucleotide chains (Fig. 4). Since there must be a limit to which the DNA can unwind in this way to accommodate acridine cations, the existence of a limit to binding by process I would be expected.

This model accounts adequately for the observations summarized in (1), (2), (3), (4), (6), and (11), but appeared to need modifying to account also for observations (5), (8), (9), and (10), which have emerged subsequently.

One modification (Figure 5) that has been suggested is that, when acridine cations are bound by process I, the acridine lies between successive nucleotide bases on the same polynucleotide chain, in a plane approximately parallel (6)

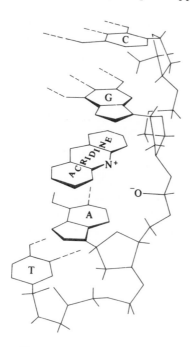

Fig. 5.   Modified Lerman model.

to the base planes, but at an angle (looking down the polynucleotide chain) such that the positive ring nitrogen is close to the polynucleotide phosphate group.[119] This condition is met equally well by native double-helical DNA, denatured disordered DNA, and single-stranded RNA [(8) and (9)]. Construction of this model shows that the 9-amino position is so placed that long chains attached at this point do not interfere with the structure (10), which is not true for the previous intercalation model[91] in which the acridines (and so the 9-position) are inside the centre of the double helix (Fig. 4). This modified intercalation model does not make the presence of intact double helices the condition of strong binding by process I, for such strong binding occurs when the nucleic acid structure is more mixed [(8) and (9)]; and it accords better with those structural requirements for strong binding of aminoacridines by polynucleotide structures that appear to be emerging from the various evidence (Section 3). It also explains why 9-amino-1,2,3,4-tetrahydroacridine (THA, **8**) which has one ring buckled by hydrogenation, can bind at all, although to an extent less than 9-aminoacridine. If intercalation between, and exactly over, the base pairs were necessary for strong binding, it is difficult to see how the THA structure (**8**) and chloroquine (**10**) could be inserted at all.[83]

In order to bind acridines through this modified intercalation, double helical DNA would have to untwist and extend, as observed (4), but denatured DNA and single-stranded RNA would already have open extended chains, like that in Figure 5, into which the acridine cation could be inserted without any further modification of the contour length of the polynucleotide. This accounts satisfactorily for observations (4) and (5), therefore. If acridines bind strongly *only* by intercalating into double helices, it is difficult to see why the viscosity of denatured DNA does not increase when they bind. Moreover the "hairpin" double helices of RNA would have to untwist, which would be sterically difficult.

This modified intercalation model has the added advantage of making it clearer how proflavine could maintain the spurious extension of the polynucleotide chains of DNA during transcription to RNA, when the double helix must open out, so that misreadings of base sequences can occur with their consequent genetic effects.[20]

This modified model (Figure 5) cannot represent the situation in all cases, however, for it does not explain why 2,7-di-*tert*-butyl proflavine (**16**) does not intercalate at all but only binds externally.[87b, 113] The modified model of Scheme 1 should allow it to intercalate readily, unless there were formidable energetic barriers influencing the rate, but not the overall free energy, of binding, whereas the original model (Figs. 3 and 4) should not allow it to do so. The work of Löber[84] has also revealed other structural relationships, and it now seems wisest to infer that the models of Figure 3 and Figure 5 represent the two extremes of a gamut of positions, for there seems to be a whole range of possible positions of bound acridine molecules all intercalated but varying in their positions relative to the helical axis and bases when viewed along the axis direction. (The position depicted in Fig. 4 is one limiting position; that in Figure 5 viewed along the chain direction, the other.)

Other models, in which the proflavine, for example, resides in the small groove of the DNA, and in which there are saltlike bonds between the proflavine amino groups and the DNA phosphates, have been canvassed but have not commanded general support and the broad features of the intercalation model for process I binding are now generally accepted.[120, 121] However, the position of any given acridine relative to the helical axis and the bases awaits determination in each particular case. Thus, in the case of the 3-aminoacridine derivatives including proflavine, a proposal has been made about their location, when intercalated, on the basis of the similarities to ethidium bromide in the variation of their induced circular dichroism with r (see Section 4). [75, 86]

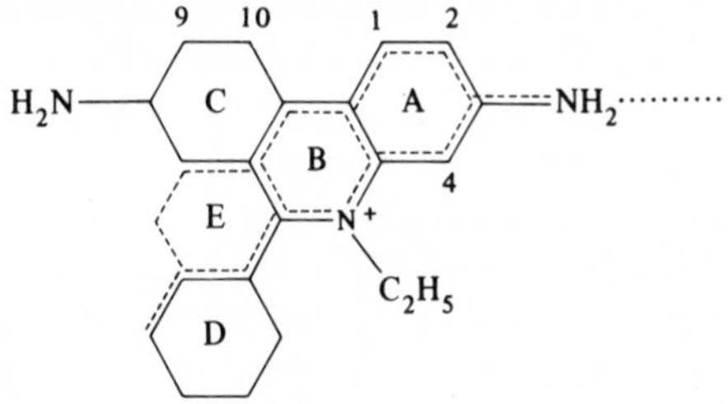

Fig. 6. A, B, C, and D—solid lines: ethidium bromide (numbers outside of rings). A, B, and E—broken lines: 3-aminoacridine and proflavine superimposed on the ethidium structure so that the 7-aminoquinoline rings (A, B) overlap two of those of the ethidium. The DNA base-pairs are regarded as lying under and over the rings A, B, and C with hydrogen bonds which join them parallel to the direction C-A, i.e., parallel to the C_9-C_{10} and C_1-C_2 bonds of the ethidium bromide (from ref. 86).

The basis of this proposal (see Figure 6) is that 3-aminoacridine derivatives and ethidium bromide possess in common the quinoline ring system with an amino group at the 7-position (numbered 3 in ethidium bromide, **9**). It is suggested that 3-aminoacridine derivatives and ethidium bromide intercalate, so that their 7-aminoquinoline portion (rings A and B, Figure 6) always bear the same relation to the DNA base pairs[75, 86]; thus A and B in Scheme 2 would be exactly over the H-bonded base pairs of the DNA with B central and the C_1—C_2 bond of the acridines, or the C_1—C_2 bond of ethidium, parallel to the base-pair hydrogen bonds (Fig. 4). If an acridine is intercalated in the modified way shown in Figure 5, part of the bound acridine is still accessible from the outside. Armstrong, Kurucsev, and Strauss have suggested that one of the two modes of strong binding, whose existence they have deduced (see Section 2), is the binding of another acridine molecule to one already intercalated in the modified manner of Figure 5, the other mode being the intercalation itself.[64]

Finally, and negatively, it is at least clear that "strong binding" must not be too readily identified with and taken as semantically equivalent to "intercalation" with all the features originally proposed.[2]

B. Binding by Process II

Binding by process II has the following experimental characteristics.

1. It is weak, of the order of, at most, a few kilocalories per mole of bound aminoacridine.

2. It involves interaction between bound aminoacridine molecules.

3. It is electrostatic and enhanced relatively more by decreases in ionic strength than is binding by process I.

4. Since it follows binding by process I, which is more internal and covers the available base rings, it is probably an external binding process.

5. The planes of the acridine rings are more disordered with respect to the planes of the base rings than in strong binding process I, although this does not exclude some degree of ordered stacking.

6. Denaturation of DNA enhances the tendency to bind by process II.

7. Binding by process II can proceed up to the limit prescribed by the condition for electroneutrality, i.e., to $r = 1.0$.

The general picture of process II binding which emerges is that of acridine cations attached approximately edgewise and externally to the double helix of DNA, with their positive ring nitrogen atoms close to the phosphate groups. When r is large enough, the acridine rings can stack upon each other in a direction parallel to the helix axis. However, when r is as large as this, the helix will be much disordered by strong binding process I with intercalation, so that this direction of the helical axis may be very ill-defined and no effect on the viscosity of DNA would be anticipated. Since the more open structures of denatured DNA would be expected to be more accommodating to this stacking and self-aggregation of the acridine cations attached to the phosphate groups, observation (6) is not surprising.

The mutual interactions between bound acridines can be sufficiently strong and directed in some cases, e.g., with acridine orange, that this type of binding may be described as the binding of dimers or aggregates, which are similar to those which exist in free concentrated solutions and give rise to similar spectral shifts.[102]

This stacking model is very feasible for acridine orange bound to DNA, since the forces between the ligand molecules are then relatively strong and are comparable with the electrostatic forces between the ligand and the DNA. However, the forces between bound proflavine molecules are much weaker (cf. its lower tendency to aggregate in solution) than for acridine orange. Therefore, although the picture of proflavine being bound externally by predominantly electrostatic forces in process II is probably substantially correct, this ought not always to be described as stacking, for the tendency of proflavine to interact with proflavine in a direction parallel to the DNA axis will be markedly less than with acridine orange. That this tendency is not absent is shown by the evidence that proflavine on nucleic acids can indeed exhibit an aggregate-type spectrum, with the accompanying loss of a good isosbestic point. But this tendency is markedly less than with acridine orange. For these reasons, it is unwise to assume that all binding by process II, which is weak in the energetic sense, can be described as a result of stacking, in the sense described above, although it is usually true that all stacked acridines are weakly bound, energetically speaking. It is confusing to describe either weak binding by process II or stacking simply as "electrostatic binding." Although electrostatic forces play a greater (probably dominant) role in process II compared with process I, and in stacking, yet attractive forces between bound aminoacridine molecules are also clearly involved—just as in binding by process I, attractive forces between aminoacridine and nucleotide rings must be involved, in addition to electrostatic forces. So binding by process I, by process II, and by stacking cannot be accurately described simply as "electrostatic binding," as if such forces were alone operative.

8. The Structure of the Complexes and the Biological Activity of the Acridines

In spite of the intricate and wide-ranging character of the experimental evidence, a reliable picture is emerging of the nature of the complex between nucleic acids and the aminoacridines, an important group of bacteriostatic and mutagenic agents. It would be tempting to attribute the entire biological effects of these aminoacridines to this interaction. Thus the ability of the aminoacridines to extend and stretch the polynucleotide chains of DNA is closely related to their ability to cause additions or deletions in the genetic code, and at least two plausible mechanisms whereby this could happen during the replication process have been proposed.[19, 114, 122] According to one of these,[19, 122] the acridine causes a miscopying of a single strand of the DNA by its intercalation into either the old or the new chain, causing, repectively, insertion or deletion of a base pair; the modified intercalation model (Figure 5) provides a molecular basis for such a mutagenic mechanism. Alternatively, an acridine molecule could cause insertion or deletion of a base pair by being intercalated in the initial DNA double helix, while it was undergoing recombination at another locus with a separate juxtaposed DNA double helix.[114] There is evidence that recombination is occurring during acridine mutagenesis, but this may be fortuitous.[123] Meanwhile genetic evidence[124] has shown that the + and − classes of mutants in the most intensively studied group of acridine-induced mutations (those in the rII region of T4 phage) do, in fact, correspond, respectively, to insertion and deletion of a base pair. The proof of the triplet nature of the genetic code (see Section 1) depended on the idea that, by causing the insertion or deletion of a base pair, acridines caused a misreading of the message decoded by a triplet reading "frame" from the base sequences of DNA and then translated into amino-acid sequences in a protein.[20] A third view attributes the mutagenic action of acridines to their ability to stabilize DNA against separation of its two chains at any point where they are intercalated.[125]

However, a word of caution is now necessary in case we too readily assume that all the biological effects of aminoacridines can be explained by their interaction with nucleic acids. A study of the mechanism of the inhibition of *E. coli* RNA-polymerase (a protein) by proflavine has shown that this inhibition occurs not only because of the interaction of the proflavine with the DNA primer, which is necessary for the reaction, but also by direct interaction between the proflavine and those binding sites on the RNA polymerase which interact with the nucleotide bases of the primer DNA and with the substrate nucleotide-tri-phosphates.[126]

Attention must be drawn to those interesting papers in which the ability of the flat ring compounds, ethidium bromide (9) and proflavine (3), to extend the contour length of DNA and to cause it to untwist has been utilized to determine the degree of supercoiling in circular mitochondrial and viral DNA.[127] Finally, it must be noted that most of the studies reported above have been made on systems in equilibrium. Under conditions *in vivo*, it may well be that it is the rate of formation of complexes which is of major significance, e.g., the residence time of an acridine cation on a nucleic acid chain, along which an enzyme is advancing and has been delayed.

References

1. A. R. Peacocke and N. J. H. Sherrett, *Trans. Faraday Soc.*, **52**, 261 (1956).
2. L. S. Lerman, *J. Mol. Biol.*, **3**, 18 (1961).
3. P. Ellinger and A. Z. Hirt, *Anat. Entwicklungsgeschichte*, **90**, 791 (1929).
4. F. Bukatsch and M. Haitinger, *Protoplasma*, **34**, 515 (1940).
5. (a) S. Strugger, *Jenaische Z. Naturwissenschaften*, **73**, 97 (1940); (b) *Can. J. Res.*, 26E, **229**, (1948).
6. P. P. H. de Bruyn, R. C. Robertson, and R. S. Farr, *J. Biol. Chem.*, **108**, 279 (1951).
7. A. Krieg, *Experientia*, **10**, 172 (1954).
8. J. A. Armstrong, *Exp. Cell. Res.*, **11**, 640 (1956).
9. J. Smiles and E. A. R. Taylor, *Nature*, **179**, 306 (1957).
10. D. Roth and M. L. Manjon, *Biopolymers*, **7**, 695 (1969).
11. (a) P. L. Pearson, M. Bobrow, and C. G. Vosa, *Nature*, **226**, 78 (1970); (b) P. Barlow and C. G. Vosa, *Nature*, **226**, 961 (1970).
12. C. G. Vosa, *Chromosoma* (Berlin), **30**, 366 (1970); **31**, 466 (1970).
13. R. I. De Mars, *Nature* **172**, 964 (1953).
14. R. Dulbecco and M. Vogt, *Virology*, **5**, 236 (1958).
15. S. Brenner, S. Benzer, and L. Barnett, *Nature*, **182**, 983 (1958).
16. E. Freese, *Proc. Nat. Acad. Sci.* U.S., **45**, 622 (1959).
17. A. Orgel and S. Brenner, *J. Mol. Biol.*, **3**, 762 (1961).
18. L. Astrachan and E. Volkin, *J. Amer. Chem. Soc.*, **79**, 130 (1957).
19. S. Brenner, L. Barnett, F. H. C. Crick, and A. Orgel, *J. Mol. Biol.*, **3**, 121 (1961).

20. E. Terzaghi, Y. Okada, G. Streisinger, J. Emrich, M. Inouye and A. Tsugita, *Proc. Nat. Acad. Sci. U.S.*, **56**, 500 (1966).

21. F. H. C. Crick, L. Barnett, S. Brenner, and R. J. Watts-Tobin, *Nature*, **192**, 1227 (1961).

22. R. A. C. Foster, *J. Bacteriol.*, **56**, 795 (1948).

23. R. I. de Mars, S. E. Lauria, H. Fisher, and C. Levinthal. *Ann. Inst. Pasteur*, **84**, 113 (1953).

24. M. Susman, M. M. Piechowski, and D. A. Ritchie, *Virology*, **26**, 163 (1965).

25. G. Dénes and L. Polgar, *Nature*, **185**, 386 (1960).

26. J. E. Hotchin, *J. Gen. Microbiol.*, **5**, 609 (1951).

27. D. Kay, *Biochem. J.*, **73**, 149 (1959).

28. J. D. Spikes and C. A. Ghiron, in *Physical Processes in Radiation Biology*, L. Angenstine, R. Mason, and B. Rosenberg, eds., Ademic Press, New York, 1964, p. 309.

29. F. L. Shaffer, *Virology*, **18**, 412 (1962).

30. (a) S. D. Rubbo, A. Albert, and M. Maxwell, *Brit. J. Exp. Pathol.*, **23**, 69 (1942); (b) A. Albert, S. D. Rubbo, R. J. Goldacre, M. E. Davey, and J. D. Stone, *Brit. J. Exp. Pathol.*, **26**, 160 (1945); (c) A. Albert and R. J. Goldacre, *Nature*, **161**, 95 (1948).

31. A. Albert, *The Acridines*, Arnold, London, 1966.

32. A. R. Peacocke and C. N. Hinshelwood, *J. Chem. Soc.*, 1235, 2290 (1948).

33. A. R. Peacocke, *Exp. Cell. Res.*, **7**, 498 (1954).

34. H. McIlwain, *Biochem. J.*, **35**, 1311 (1941).

35. (a) C. N. Hinshelwood, *The Kinetics of Bacterial Change*, Clarendon Press, Oxford, England, 1946; (b) A. C. R. Dean and C. N. Hinshelwood, *Growth Function and Regulation in Bacterial Cells*, Clarendon Press, Oxford, England, 1966.

36. E. M. Witkin, *Cold Spring Harbor Symp. Quant. Biol.*, **12**, 256 (1947).

37. L. Massart, G. Peeters, and A. Vanhoucke, *Arch. Int. Pharmacodyn, Ther.*, **75**, 210 (1947).

38. J. Deley, G. Peeters, and L. Massart, *Biochim. Biophys. Acta*, **1**, 393 (1947).

39. B. Ephrussi, "Nucleo-cytoplasmic relations" in *Microorganisms*, Clarendon Press, Oxford, England, 1953.

40. M. Nagao and T. Sugimura, *Biochim. Biophys. Acta*, **103**, 353 (1965).

41. (a) Y. Hirota and T. Iijima, *Nature*, **180**, 655 (1957); (b) Y. Hirota, *Proc. Nat. Acad. Sci. U.S.*, **46**, 57 (1960).

42. T. Watanabe, *Bacteriol. Rev.*, **27**, 87 (1936).

43. M. Steinert and S. van Assel, *J. Cell Biol.*, **34**, 489 (1967).

44. L. Simpson, *J. Cell Biol.*, **37**, 660 (1968).

45. F. D'Amato, *Caryologia*, **2**, 229; 3, 211 (1950).

46. I. Lasnitzki and J. H. Wilkinson, *Brit. J. Cancer*, **2**, 369 (1948).

47. P. Dustin, *Nature*, **159**, 794 (1947).

48. (a) G. Scatchard, *Ann. N.Y. Acad. Sci.*, **51**, 660 (1949); (b) I. M. Klotz, in *The Proteins*, H. Neurath and K. Bailey, eds.), Academic Press, New York, 1953; (c) F. Karush, *J. Amer. Chem. Soc.*, **72**, 2705 (1950).

49. A. Blake and A. R. Peacocke, *Biopolymers*, **6**, 1225 (1968).

50. (a) G. Schwarz, *Eur. J. Biochem.*, **12**, 442 (1970); (b) G. Schwarz, S. Klose, and W. Balthasar, *Eur. J. Biochem.*, **12**, 445 (1970); (c) G. Schwarz and W. Balthasar, *Eur. J. Biochem.*, **12**, 461 (1970).

51. P. H. Lloyd, R. N. Prutton, and A. R. Peacocke, *Biochem. J.*, **107**, 353 (1968).

52. H. G. Heilweil and Q. van Winkle, *J. Phys. Chem.*, **59**, 939 (1955).

53. G. Oster, *Trans. Faraday Soc.*, **47**, 660 (1951).

54. R. K. Tubbs, W. E. Ditmars, and Q. van Winkle, *J. Mol. Biol.*, **9**, 545 (1964).

55. J. C. Thomes, G. Weill, and M. Daune, *Biopolymers*, **8**, 1647 (1969).
56. N. F. Ellerton and I. Isenberg, *Biopolymers*, **8**, 767 (1969).
57. L. M. Chan and Q. van Winkle, *J. Mol. Biol.*, **40**, 491 (1969).
58. J. Chambron, M. Daune, and C. Sadron, *Biochim. Biophys. Acta*, **123**, 306, 319 (1966).
59. D. S. Drummond, V. F. W. Simpson-Gildemeister, and A. R. Peacocke, *Biopolymers*, **3**, 135 (1965).
60. S. Ichimura, M. Zama, H. Fujita, and T. Ito, *Biochim. Biophys. Acta*, **190**, 116 (1969).
61. D. Rogers and A. R. Peacocke, unpublished observations (1968); D. Rogers, Chemistry, Part II, Thesis, University of Oxford, England, 1968.
62. M. Liersch and G. Hartmann, *Biochem. Z.*, **343**, 16 (1961).
63. A. Blake and A. R. Peacocke, *Biopolymers*, **5**, 383 (1967).
64. R. W. Armstrong, T. Kurucsev, and U. P. Strauss, *J. Amer. Chem. Soc.*, **92**, 3174 (1970).
65. M. Daune, private communication, 1970.
66. M. Gilbert and P. Claverie, *J. Theor. Biol.*, **18**, 330 (1968); in *Molecular Associations in Biology*, B. Pullman, ed., Academic Press, New York and London, 1968, p 245.
67. N. F. Gersch and D. O. Jordan, *J. Mol. Biol.*, **13**, 138 (1965).
68. I. O. Walker, *Biochim. Biophys. Acta*, **109**, 585 (1956).
69. N. B. Kurnick, *J. Gen. Physiol.*, **33**, 243 (1950).
70. N. B. Kurnick and A. E. Mirsky, *J. Gen. Physiol.*, **33**, 265 (1950).
71. D. G. Dalgleish, H. Fujita, and A. R. Peacocke, *Biopolymers*, **8**, 633 (1969).
72. A. Blake, H. Fujita, and A. R. Peacocke, University of Oxford, England, unpublished observations, 1967.
73. L. Michaelis, *Cold Spring Harbor Symp. Quant. Biol.*, **12**, 131 (1947).
74. D. F. Bradley and M. K. Wolf, *Proc. Nat. Acad. Sci. U.S.*, **45**, 944 (1959).
75. A. R. Peacocke, *Studia Biophysica*, **24/25**, 213 (1971).
76. M. C. Feil and A. R. Peacocke, unpublished observations, (1971).
77. (a) M. J. Waring, *Biochim. Biophys. Acta*, **114**, 234 (1966); (b) *J. Mol. Biol.*, **13**, 269 (1965).
78. R. F. Beers and G. Armilei, *Nature*, **208**, 466 (1965).
79. V. Kleinwächter and J. Koudelka, *Biochim. Biophys. Acta*, **91**, 539 (1964).
80. G. Weill and M. Calvin, *Biopolymers*, **1**, 401 (1963).
81. G. Weill, *Biopolymers*, **3**, 567 (1965).
82. F. W. Northland, P. P. H. de Bruyn, and N. H. Smith, *Exp. Cell. Res.*, **7**, 201 (1954).
83. R. L. O'Brien, J. G. Olenick, and F. E. Hahn, *Proc. Nat. Acad. Sci. U.S.*, **55**, 1511 (1966).
84. (a) G. Löber, *Photochem. Photobiol.*, **4**, 607 (1965); (b) G. Löber, *Habilitationsschrift*, Jena, Germany, 1969; (c) G. Löber, *Studia Biophysica*, **3**, 113 (1967); **4**, 299 (1967); **8**, 99 (1968); (d) G. Löber, *Z. Chem.*, **7**, 252 (1969); (e) G. Löber and G. Achtert, *Biopolymers*, **8**, 595 (1969).
85. (a) R. M. Acheson, D. G. Dalgleish, C. Harvey, and A. R. Peacocke, unpublished observations 1970; (b) C. Harvey, D. Phil. Thesis, University of Oxford, England, 1970.
86. D. G. Dalgleish, A. R. Peacocke, G. Fey, and C. Harvey, *Biopolymers*, **10**, 1853 (1971).
87. (a) Proceedings of Vth Jena Molecular Biology Symposium on "Interactions in Biopolymers," August-September 1970, *Studia Biophysica*, **24/25** (1970). (b) G. Löber, p. 233; (c) W. Muller and D. M. Crothers, p. 279.
88. D. S. Drummond, N. S. Pritchard, V. F. W. Simpson-Gildemeister, and A. R. Peacocke, *Biopolymers*, **4**, 971 (1966).

89. G. Cohen and H. Eisenberg, *Biopolymers*, **8**, 45 (1969).
90. H. A. Scheraga and L. Mandelkern, *J. Amer. Chem. Soc.*, **75**, 179 (1953).
91. L. S. Lerman, *J. Cell. Comp. Physiol.*, *Suppl.* 1, **64**, 1 (1964).
92. (a) Proc. conference on "The Reaction of Antibiotics with Nucleic Acids and the Biological Consequences," *Hoppe-Seyler's Z. Physiol. Chemie*, **349**, 953 (1968); (b) M. Waring, *Nature*, **219**, 1320 (1968).
93. J. Cairns, *Cold Spring Harbor Symp. Quant. Biol.*, **27**, 311 (1962).
94. Y. Mauss, J. Chambron, M. Daune, and H. Benoit, *J. Mol. Biol.*, **27**, 579 (1967).
95. V. Luzzati, F. Masson, and L. S. Lerman, *J. Mol. Biol.*, **3**, 634 (1961).
96. G. Weill and M. Calvin, *Biopolymers*, **1**, 401 (1963).
97. W. Fuller and M. J. Waring, *Ber. Bunsenges. Phys. Chem.*, **68**, 805 (1964).
98. D. M. Neville and D. R. Davies, *J. Mol. Biol.*, **17**, 57 (1966).
99. D. Freifelder, P. F. Davison, and E. P. Geiduschek, *Biophys. J.*, **1**, 389 (1961).
100. L. S. Lerman, *J. Mol. Biol.*, **10**, 367 (1964).
101. J. Chambron, Ph.D. Thesis, University of Strasbourg, France, 1965.
102. (a) V. Zanker, *Z. Phys. Chem.*, **199**, 225 (1952); **200**, 250 (1952). (b) W. Appel and V. Zanker,*Z. Naturforsch.*, *B*, **13**, 126 (1958).
103. D. M. Neville and D. F. Bradley, *Biochim. Biophys. Acta*, **50**, 397 (1961).
104. (a) A. Blake and A. R. Peacocke, *Nature*, **206**, 1009 (1965); (b) *Biopolymers*, **4**, 1091 (1966); **5**, 383, 871 (1967).
105. (a) K. Yamaoka and R. A. Resnik, *J. Phys. Chem.*, **70**, 4051 (1966); (b) *Nature*, **213**, 1031 (1967).
106. V. I. Permogorov, Y. S. Lazurkin and S. Z. Shmurak, *Dokl. Biophys.*, **155**, 71 (1964).
107. B. J. Gardner and S. F. Mason, *Biopolymers*, **5**, 79 (1967).
108. H. J. Li and D. M. Crothers, *Biopolymers*, **8**, 217 (1969).
109. M. Zama and S. Ichimura, *Biopolymers*, **9**, 53 (1970).
110. W. Fuller and M. Waring, *Ber. Bunsenges. Phys. Chem.*, **68**, 805 (1964).
111. G. G. Hammes and C. D. Hubbard, *J. Phys. Chem.*, **70**, 2889 (1966).
112. H. J. Li and D. M. Crothers, *J. Mol. Biol.*, **39**, 461 (1969).
113. D. M. Crothers, *Studia Biophyscia*, **24/25**, 79 (1971).
114. L. S. Lerman, *Proc. Nat. Acad. Sci. U.S.*, **49**, 94 (1963).
115. C. Nagata, M. Kodama, Y. Tagashira, and A. Imamura, *Biopolymers*, **4**, 409 (1966).
116. S. F. Mason and A. J. McCaffrey, *Nature*, **204**, 468 (1964).
117. C. Houssier and E. Fredericq, *Biochim. Biophys. Acta*, **120**, 434 (1966).
118. G. Oster, *Trans. Faraday Soc.*, **47**, 660 (1951).
119. N. J. Pritchard, A. Blake, and A. R. Peacocke, *Nature*, **212**, 1360 (1966).
120. G. B. Gurski, *Biofizika*, **5**, 737 (1966); *Biophysics*, **11**, No. 5 [849], trans. (1966).
121. R. Rigler, *Ann. N.Y. Acad. Sci.*, **157**, 211 (1969).
122. L. E. Orgel, *Advan. Enzymol.*, **27**, 289 (1965).
123. J. W. Drake, *J. Cell. Comp. Physiol.*, *Suppl.* 1, **64**, 19 (1964).
124. F. H. C. Crick and S. Brenner, *J. Mol. Biol.*, **26**, 361 (1967).
125. G. Streisinger, et al., *Cold Spring Harbor Symp. Quant. Biol.*, **31**, 77 (1966).
126. B. H. Nicholson and A. R. Peacocke, *Biochem. J.*, **100**, 50 (1966).
127. (a) M. Gellert, *Proc. Nat. Acad. Sci. U.S.*, **57**, 148 (1967); (b) R. Radloff, W. Bauer, and J. Vinograd, *Proc. Nat. Acad. Sci. U.S.*, **57**, 1514 (1967); (c) L. V. Crawford and M. J. Waring, *J. Mol. Biol.*, **25**, 23 (1967); (d) *J. Gen. Virol.*, **1**, 387 (1967).

CHAPTER XV

Acridines and Enzymes

B. H. NICHOLSON

Department of Physiology and Biochemistry,
The University of Reading, England

Interest in the biological activity of the acridines has centered on their antibacterial, antimalarial and, more recently, mutagenic properties. Apart from a few isolated instances, little interest has been taken in their action on enzymes until the last 20 years or so. The stimulation of interest in this topic must stem largely from our increased understanding of cellular metabolism, which makes the significance of a particular interaction more readily appreciated; it now seems certain that some, if not all of the above properties are caused by the reaction of acridines with enzyme systems, which usually results in inhibition of the system. With this background it is not surprising that much

The literature review pertaining to this chapter was completed in November 1970. Abbreviations are as follows: FAD, flavin adenine dinucleotide; FMN, flavin mononucleotide; NAD, nicotinamide adenine dinucleotide; NADP, nicotinamide adenine dinucleotide phosphate; mRNA and tRNA, messenger and transfer ribonucleic acid; DNA, deoxyribonucleic acid; NMP, NDP, nucleoside mono and diphosphates; dAT, alternating copolymer of deoxyadenylate and deoxythymidylate residues. A, C, G, U, T and refer to the base adenine, cytosine, guanine, uracil, and thymine; the prefix d to the deoxyribotides of these bases.

of our information on the chemical and structural requirements for interaction has come from studies on the antibacterial and antimalarial properties.

The pioneering work of Albert[1] in the forties (see also Chapter XVI)
showed that antibacterial activity was dependent on the strength of the
acridine as a base, a high basicity being required to ensure that a considerable
proportion of the acridine is present as cation under normal physiological
conditions (pH 7.3), in order to react with the "receptors" of the cell. A
second requirement is that the area of the planar nucleus must be at least
38 Å². Thus 9-aminoacridine is antibacterial, 4-aminoquinoline is not. If
9-aminoacridine is hydrogenated in one ring, the antibacterial activity is lost,
since the ring is puckered and the flat area is only 28 Å².[1] The critical size of
the planar area must reflect the summation of a critical number of van der
Waals interactions, sufficient to become a strong binding force; without such
an interaction the electrostatic interaction between the acridine cation and
the receptor anion would be short-lived. Since the area required is planar, the
receptor must also be planar. A similar reaction between acridines and the
planar methylene blue forms the basis of a quantitative assay for acridines.[2]

These findings of Albert have been carried a stage further by Sharples and
Greenblatt,[3] who, since the pK_a of the acridines and the fraction in ionic
form are the outward manifestations of the electron distribution, consider
that the relative effectiveness of these acridines might show some correlation
with one or more of the values obtained by quantum-mechanical calculations.
They found a generally high correlation between the bacteriostatic index and
the energy of the lowest unoccupied molecular orbital (LUMO) for gram-
positive organisms.

Insufficient information is available to generalize on the effects of side
chains on interactions between acridines and enzymes, apart from those substituents, e.g., amino groups in the 3 or 9 position, necessary to raise the pK_a
of the acridine nucleus from 5.3 to 7.3 or higher. It may well be that the
modes of interaction are more numerous than those which dictate antibacterial or antimalarial properties. Consideration of these points, where
appropriate, is therefore left to the discussion of the particular enzyme or
group of enzymes. The division into groups is simply for convenience and does
not necessarily imply a common mechanism of interaction.

1. Enzymes with Nucleotide Prosthetic Groups

An interesting example of the effect of the side chain is shown with diamine
oxidase,[4] in which two homologous series of acridines, illustrated by structures
1 and **2**, analogous to atebrin (**3**), show increasing inhibitory power with the
increasing length of the side chain. The most efficient inhibitor, that is, the

NHCHMe(CH$_2$)$_x$NEt$_2$

NH(CH$_2$)$_x$NEt$_2$

OMe

OMe

Cl

Cl

N

N

1

2

minimum amount of acridine to give 50% inhibition for both series, occurs when four carbon atoms separate the two nitrogen atoms of the side chain.[5] The antimalarial activity also reaches a maximum with four separating carbons, but on increasing the number of carbon atoms in the side chains of both series, the level required for 50% inhibition remains constant, while the antimalarial activity of series 2 decreases.

Atebrin [mepacrine, quinacrine, 6-chloro-9-(4-diethylamino-1-methylbutyl-amino)-2-methoxyacridine], 3 has a marked affinity for diamine oxidase.

NHCHMe(CH$_2$)$_3$NEt$_2$

8
1
OMe

Cl

5
N
4

3

Using an acetone-dried preparation of porcine kidney,[6] it was shown that the uptake of oxygen during oxidation of cadaverine was depressed,[5] the inhibitions with 10^{-3}, 10^{-4}, and 10^{-5} M atebrin being, respectively, 92, 82, and 45%. Acridine antimalarials are efficient inhibitors of diamine oxidase because of the aliphatic diamine structure of the side chains. However, 9-amino-6-chloro-2-hydroxyacridine, i.e., atebrin demethylated and without the side chain, still inhibited the porcine diamine oxidase by 33% at $10^{-4}M$. Hellerman, Lindsay, and Bovarnick[6] have prepared D-amino acid oxidase (E.C. 1.4.3.3.) from lamb kidneys and removed the bound FAD cofactor (6). Atebrin was found to inhibit D-amino acid oxidase strongly when the concentration of FAD was low but had little effect in the presence of high concentrations. Furthermore, the inhibition could be reversed by increasing concentrations of FAD. At $10^{-3}M$ atebrin inhibition was 80% (FAD $= 1 \times 10^{-7}M$), and this dropped to 4% on adding $11.0 \times 10^{-7}M$ FAD. A similar observation had been made on tissue respiration by Wright and Sabine,[7] who found a lowering of the atebrin inhibition of tissue respiration and of D-amino oxidase activity by FAD. It was concluded that atebrin competed with FAD for the FAD binding site on the protein.[6] Recent studies of the FAD/protein binding have indicated that conformational changes of the protein are of importance in the formation of the catalytically active form of the enzyme.[8]

Webb[9] and Klein[10] have considered the structural requirements for effective inhibition of D-amino acid oxidase by inhibitor molecules. The advantages of having a positively charged nitrogen atom in the proper position, particularly with respect to the carboxyl group of pyridinecarboxylic acids, has been pointed out, as has the possibility of aromatic interactions with the enzyme-bound FAD.[11] A 50% inhibition of monoamine oxidase by 9-amino-1,2,3,4-tetrahydroacridine ($10^{-4}M$) has been reported, in which the main structural requirement for inhibition appeared to be the 4-aminoquinoline moiety.[12]

An FMN (5), containing cytochrome reductase, is 50% inhibited by $2 \times 10^{-4}M$ atebrin (80% at $10^{-3}M$). The inhibition is not an effect on cytochrome c, since this is unaffected by $10^{-3}M$ atebrin, but was attributed to a direct noncompetitive interaction with the FMN binding site on the protein. Greater inhibition was observed at 25° than at 0°, because of the temperature-induced dissociation of FMN from the enzyme. The interaction can be prevented by low concentrations of FMN (1 part per 500 of atebrin), but once it has occurred, the addition of excess FMN will not reverse it.[13]

Glucose-6-phosphate dehydrogenase was strongly inhibited by atebrin ($5 \times 10^{-4}M$). The addition of glucose-6-phosphate or the normal hydrogen acceptor NADP gave some protection against the inhibition, NADP being 1000 times more effective on a mole for mole basis, presumably because of the greater stability of the NADP enzyme complex ($K = 10^{-9}M$), compared to the substrate protein complex ($K = 10^{-3}M$).[13] Cytochrome oxidase was found to be unaffected. Acriflavine is a respiratory inhibitor in *Saccharomyces carlsberg* and *S. cerevisiae*[14]; Witt, Neufang, and Muller[15] mention that of all the glycolytic enzymes in yeast, only the NAD-dependent D-glyceraldehyde phosphate dehydrogenase (EC 1.2.1.12) is considerably inhibited by 0.4 or 0.8 mM proflavine. Inhibition was only observed when the enzyme was tested in the glycerol-1,3-diphosphate to glyceraldehyde-3-phosphate direction, proflavine having no effect in the reverse direction. The enzyme is inhibited faster when incubated, which is suggestive of the temperature-dependent dissociation mentioned earlier. Ten minutes after the addition of proflavine, the enzyme is 50% inhibited, but this can be completely reversed by adding cysteine or glutathione, suggesting an involvement of proflavine with the sulfhydryl groups of the enzyme. Atebrin has also been shown to inhibit hexokinase, and beef heart lactate dehydrogenase.[16, 17]

Most of the foregoing experiments were designed to find the site of inhibition responsible for the antimalarial action, rather than the particular features of acridine or enzyme structure responsible for the inhibition of a particular enzyme; thus we have little information on the effect of the side chains on a particular inhibition. In neutral solution the atebrin side chain carries a positively charged nitrogen distal to the ring; to form the dication the second proton is added to the ring nitrogen and the charge shared with the first nitrogen of the side chain (see also Chapter II). The greater the distance between the two side chain nitrogens, the less their mutual repulsion, and the less the depression of the basicity of the nucleus.

From the foregoing it will be seen that many, though not all, of the enzymes inhibited by atebrin contain either FMN or FAD prosthetic groups, both of which contain riboflavine (**4**).

4, Riboflavine, R = CH_2OH
5, FMN, R = $CH_2OPO(OH)_2$
6, FAD, R = $CH_2O-5'-AMP$

There is a similarity in the chemical formulas of riboflavine (**4**) with that of atebrin (**3**), and this has been the basis of suggestions that atebrin might be a structural analogue of riboflavine and compete for the same receptor, e.g., the prosthetic binding site of an enzyme. Albert[19] does not consider this idea to be well based, since (1) the pK_a values (7.9 and 10.4) of atebrin show it to be a doubly ionized cation in neutral solution, whereas riboflavine, with its pK values of -0.2 (cation) and 10.2 (anion) is completely uncharged under the same conditions; and (2) the three-ring nucleus of atebrin is completely conjugated and hence entirely flat, whereas in riboflavine the rings do not lie in the same plane because the central ring is not completely conjugated and he considers the dihedral angle between the left- and right-hand rings is nearly 90°. Albert considers it unlikely, therefore, that atebrin and riboflavine would be attracted by the same kind of receptor.

Hybrids of the two molecules were synthesized by Adams and found to have no antimalarial or antiriboflavine activity.[2-]. However, the pK_a values above refer only to the oxidized form of riboflavine, and for the free-radical or semiquinone intermediate these are 1.5 and 7.5, while the reduced flavin has a pK_a of 6.5. Although it is true that the flavin is most likely to bind in the oxidized state (indeed, kinetic analysis shows that the reaction is ordered and that the coenzyme is the leading substrate), since all these forms are bound by the enzyme, pK_a's may not be so relevant as the actual charge distribution. Also, resonance structures can be drawn for all three oxidation states in which the central ring of the isoalloxazine nucleus is conjugated, and hence planar, e.g., (**4**). It is possible that despite the specificity of the binding site, established by virtue of the restricted topography and charge distribution at the particular H-bonding and other binding positions, the two molecules, as seen by the binding sites, are sufficiently similar for binding to occur. For a particular molecule to be an inhibitor, it must have some points of difference with the true substrate or cofactor. It is clear that the differences in conformation of atebrin and flavin are not as large as has been suggested, and there must be some reason for the very large number of FMN or FAD enzymes inhibited by acridines. It may soon prove possible to answer this question for lactate dehydrogenase as studies on its tertiary structure are well under way.[21]

2. Enzymes Not Containing Nucleotide Prosthetic Groups

Quastel showed that acriflavin inhibited urease.[22] Purified preparations were inhibited more (up to 72% at 1/2000 concentration) than the crude soya bean extract (46% at 1/2000 concentration). Of all the dyestuffs examined, inhibitory activity was confined to basic dyes. It was postulated that as acidic dyes are inactive, there cannot be both acid and basic groups at the active site, even though the reaction is a hydrating one and symmetrical pH curves are obtained, although the latter might be explained by the effect of pH on the substrate.

So far, any deduction about the mechanism of inhibition of a given enzyme has had to rely almost exclusively on the kinetic data and/or the effect of structural modification to the original acridine inhibitor. Two enzymes can now be considered when we have detailed information on the nature of the active center: chymotrypsin, because its tertiary structure is now known,[23] and acetylcholinesterase, because of the large number of inhibitory studies made on the physiological importance of the enzyme. α-Chymotrypsin is known to be inhibited by many aromatic compounds. Wallace, Kurtz, and Niemann, using acetyl-L-valine methyl ester as a substrate, tested over 100 compounds for inhibition, in an attempt to determine the topography of the active center. The isomeric 1-, 2-, 3-, and 9-aminoacridines (pK_a 6.0, 5.9, 8.0, and 10.0, respectively) were all inhibitory, giving K_i (mM) of 0.34, 0.22, 0.23, and 00 at $0 - 5 \times 10^{-4}M$, that of 9-aminoacridine being so high as to be indeterminable, because of micelle formation. The high value of K_i obtained for atebrin (3) is provably a result of the bulky side chain, since 9-acetamido-acridine is also inhibitory, but the possiblility of an erroneous value due to micelle formation was not eliminated. The situation is complicated by the presence of both charged and uncharged forms because the pK_a of atebrin is close to that of the reaction mixture (pH 7.90). The stronger inhibition ($K_i = 0.08$ mM) by acriflavin (a mixture of proflavine and 3,6-diamino-10-methylacridinium salts) than by proflavine (3,6-diaminoacridine) (0.13 mM) would seem to indicate that the 3,6-diamino-10-methylacridinium cation is a more effective inhibitor than 1-acetyl-2-(L-tyrosyl)hydrazine ($K_i = 0.074$ mM) previously the most effective reversible inhibitor known.[25] Both proflavine (pK_a = 9.65) and 3,6-diamino-10-methylacridinium salts (pK_a > 12) are predominantly ionized at the pH of the assay.

Bernhard and Lee reported that proflavine underwent a spectral shift on binding chymotrypsin similar to that on binding to DNA, and that this binding could be prevented completely by the addition of the artificial substrate N-trans-cinnamoylimidazole.[26] Subsequently, it was reported that this shift could be used to study short-lived transients in the enzyme reaction,[27] since the displacement of dye by substrates could be followed spectrophotometrically by the disappearance of the 465 nm peak of the proflavine difference spectrum. This approach has been used to measure the dissociation constants of the proflavine complexes of both trypsin and chymotrypsin.[27, 28]

Further studies by Himoe, Brandt, DeSa, and Hess,[29-31] using the proflavine displacement technique, among others, have yielded detailed information on individual rate constants with a variety of substrates. Fluorescence depolarization and circular dichroism measurements indicate the formation of an asymmetric complex,[32] and temperature-jump methods show that the equilibrium of the binding reaction[33] is rapidly established (1–2 msec).

Bernhard, Lee, and Tashjian find that the inhibition constant of proflavine is $3.7 \times 10^{-5} M$, when determined against a specific substrate (acetyltyrosine ethyl ester).[34] This differs from the value reported by Wallace, Kurtz, and Niemann probably because the pseudosubstrate gives only a partially competitive inhibition.[24] They have also raised the possibility that proflavine is not a competitive inhibitor, since the acylation reaction can proceed with proflavine bound to the enzyme site. Once acylated, however, the enzyme does not bind dye, nor does the concentration of dye affect the rate of deacylation; they suggest that a change in conformation of the enzyme occurs, concomitant with release of the bound dye. However, Himoe et al. found the proflavine displacement method to agree well with data from two independent methods.[31]

As a result of studies mentioned earlier, Wallace, Kurtz, and Niemann were able to make certain postulates about the nature of the active center.[24] Some of these, which are relevant to acridine inhibitors, are summarized as follows:

1. The active center of α-chymotrypsin may contain both electron-rich and electron-deficient regions, and there is a negative charge in close proximity to at least one of the loci involved in interaction with aromatic molecules.

2. The binding of an aromatic molecule does not require a complementary coplanar surface, provided that the effect of steric obstructions on the complementary surface is less than the gain in binding energy on increasing the planar area of the inhibitor. Thus binding energy increased and hence K_i

decreased with derivatives of pyridine (17 Å²), quinoline or naphthalene (28 Å²), and acridines or azaphenanthrenes (38 Å²) in that order.

3. Orientation of the inhibitor at its binding site is governed in part by interaction of a polarized substituent at the site; where the presence of more than one polarized substituent gives rise to an asymmetric molecule, there will be a corresponding increase in modes of attachment.

4. The aromatic binding site, while predominantly flat, is considered to be curved along its length.

From the X-ray crystallographic studies of Birktoft, Matthews, and Blow[23] the structure of the active center is now known (see Fig. 1), and also the location of the pseudosubstrate formyl-L-tryptophan (for a detailed discussion of the structure and mechanism of action, see the review by Blow and Steitz[35]). This compound binds to the enzyme with the aromatic moiety in a hole, where it interacts with the peptide bonds of the protein that are parallel to the ring. The amide linkage is at the top of the hole and directed toward the active serine (serine 195). The carboxylate group of the adjacent aspartate 194 points away from the binding site and forms an ion pair with the alpha amino group of isoleucine 16. This free amino group is generated by the autolytic cleavage which occurs when α-chymotrypsin is formed from the inactive chymotrypsinogen precursor. It is essential for activity, since blocking by acetylation inactivates and reaction of irreversible inhibitors with the active site greatly increases its pK_a.[36, 37] It seems likely that the production of the free amino group causes the carboxyl group of aspartate 194 to move from an external position in contact with the solvent to its internal position, allowing the bond-breaking site to form.

The inhibition of trypsin with aminoacridines and the induction of difference spectra are of interest.[27, 28] The homology of trypsin with chymotrypsin is well established; the major point of difference in their active center appears to be the presence of an aspartate at the base of the hole, providing a negative charge that accounts for the specificity for arginine and lysine.[35] This residue may well be able to react with amino substituents of the acridine ring.

The similarity between the binding site determined by X-ray crystallography and the features postulated by Wallace, Kurtz, and Niemann[24] suggests that the 38 Å² planar surface of proflavine occupies the region of the hole in a manner similar to formyl-L-tryptophan (Fig. 2). In support of this, neither chymotrypsinogen nor the homologous enzyme elastase induce a difference spectrum in proflavine,[28] which they therefore presumably do not bind; nor does a hole form in elastase, since this region is filled by the replacement of two glycine residues by valine and threonine.[38] In chymotrypsinogen the ion pair no longer exists, and it seems likely[35] that the side chain of aspartate 194

Fig. 1. View of the active site of α-chymotrypsin, looking toward the interior of the molecule. On the far side (covered by ASP 194) is ILEU 16, whose free amino group is liberated
on activation of the zymogen to form an ion pair with ASP 194, at the same time swinging
SER 195 into its active site position. The outlined pseudosubstrate *N*-formyl-L-tryptophan
is held in the pocket by interaction with peptide bonds parallel to the plane of the indole
ring, and also some edge on van der Waals contacts. A charge-relay system allows electron
transfer from SER 195 to the ring δ nitrogen of HIS 57 and from the εN to the buried carboxyl group of ASP 102.

The peptide backbone is white and the colored side chains appear dark grey; the radius
of the atoms is one-third the van der Waals radius.

Fig. 2. The same view with 3,6-diaminoacridine in place of *N*-formyl-L-tryptophan. One end of the acridine lies in the pocket, while the other end covers the active site and places a positively charged nitrogen in the vicinity of the carboxyl group of ASP 102. The model was constructed from a kit supplied by Labquip, Reading, England.

768

is situated in a region near the substrate binding site, thereby interfering with substrate binding and also disrupting the bond-breaking site.[39] Thus, once an acridine is bound to chymotrypsin interactions of the positive charge dissipated over the three nitrogens with groups essential for activity, such as aspartate 102 and even aspartate 194 may then occur, thereby disrupting the hydrolytic site. The marked decrease in the affinity of proflavine for the enzyme below pH 4.0, with only a slight change in the environment as judged by the difference spectrum maximum, has been taken as evidence for a negative charge in close proximity to the proflavine binding site.[40]

Acetylcholinesterase is inhibited by a number of acridines: 1-, 2-, 3-, and 9-aminoacridine, and 9-amino-4-methylacridine, a 50% inhibition being reached at levels between 10^{-4} and $10^{-6} M$. It is interesting that the 4-aminoacridine and proflavine were inactive over this concentration range.[41] 9-Amino-1,2,3,4-tetrahydroacridine, in one of the few inhibitions reported with this compound,[12] gave 50% inhibition at $10^{-6} M$ and 9-butylamino-1,2,3,4-tetrahydroacridine was active at $10^{-5} M$. The lower activity of this latter compound was thought to be caused by steric hindrance from the N-butyl side chain. The entire structure of 9-amino-1,2,3,4-tetrahydroacridine appears to be essential for optimal inhibition, since 4-aminoquinoline and 4-aminopyridine were relatively poor inhibitors. The kinetics of this inhibition indicate that this may be the site of the pharmacological action of the drug, which is widely used as an antidepressant.[42] High doses of 9-amino-1,2,3,4-tetrahydroacridine inhibited cholinesterase of rat brain by 20–80%; and an even stronger inhibition of pseudocholinesterase (substrate butyrylcholine) indicated a degree of selectivity.[43] Unfortunately, the tertiary structure of acetylcholinesterase is not known, but because of its importance in nervous transmission, its inhibition by a variety of compounds has been widely studied. From the results of these studies, it is now known that there is an active serine in a sequence glutamate-serine-alanine at the active center,[44] which is acetylated during the course of the reaction (cf. serine 195 of chymotrypsin).

The function of an essential imidazole group is probably to assist in deacylation of the serine. Using inhibitors with a rigid structure, the presence of a binding site about 2.5 Å from the hydrolytic site was demonstrated.[45] This binding site is thought to be anionic in character but also partially hydrophobic, since 3,3-dimethylbutyl acetate is hydrolyzed just as readily as the positively charged acetylcholine, indicating the possibility of van der Waals interaction of the methyl groups surrounding the positively charged nitrogen with the binding site.[46] Butyryl cholinesterase has the same active site sequence.[47]

An unusual and interesting proflavine stimulation of the hydrolysis of benzoyl-L-arginine ethyl ester by the plant endopeptidase ficin (EC 3.4.4.12), has been studied in detail by Holloway.[48] The characteristic proflavine difference spectrum was observed on the interaction of ficin with proflavine and gave a dissociation constant of $3.5 \pm 1.8 \times 10^{-4}M$, compared to $1.55 \pm 0.61 \times 10^{-4}M$ found by equilibrium dialysis. The data were consistent with proflavine binding to a single site. Kinetic analysis gave a similar dissociation constant ($1.61 \times 10^{-4}M$) and showed that in the benzoyl-L-arginine ethyl ester/ficin reaction, the acylation step was rate-limiting, the only effect of proflavine binding being to accelerate the acylation of the enzyme by the substrate. With 4-nitrophenyl hippurate, a substrate in which the acylation is rapid owing to the good leaving properties of the 4-nitrophenoxy moiety, deacylation is the rate-limiting step; the addition of proflavine gave no activation. In support of this, an enhanced nucleophilicity of the essential thiol in the presence of proflavine was demonstrated by its increased reactivity toward N-methylmaleimide. Two explanations were advanced as to why the substrate is more rapidly hydrolyzed in the ternary dye-enzyme-substrate complex than in the binary enzyme-substrate complex. First, the dye may bind so as to become an integral part of the active site when the acceleration effect may then result from a group on the dye participating in the reaction. Second, proflavine may bind away from the active site and effect or facilitate a change in enzyme conformation in the acylation reaction.[48]

Not much is known about the active site of ficin. Two ionizing groups of pK_a[4] and 8.5[49] have been assigned to carboxyl and either the essential thiol, or by analogy to isoleucine 16 of chymotrypsin, an amino group that controls the conformation of the enzyme.[48] The similar plant endopeptidase papain, whose tertiary structure is known,[49, 50] has two similar pK_a's believed to correspond to aspartate 158 and cysteine 25. The carboxyl group is about 6–7 Å from the SH of cysteine 25 and the same distance from the essential histidine 159.[51] The enhanced reactivity observed for the linkage, two residues removed on the C-terminal side from an aromatic side chain, indicates the existence of an aromatic binding region.[52]

Crystallographic studies on the acridine-inhibited enzymes will be awaited with interest.

3. Acridines and Enzymes Associated with Nucleic Acids

A. Inhibition *in vivo* and in Cell Extracts

The inhibition of several cellular events has been ascribed to inhibition of enzymes acting on nucleic acids. Acriflavine at low concentrations inhibits the replication of kinetoplast DNA selectively by its localization in the kinetoplast of *Leichmania tarentolae*.[53] Proflavine has been widely used to inhibit messenger RNA synthesis, e.g., in experiments on β-galactosidase induction.[54] Soffer and Gros[55] have used proflavine to demonstrate the decay of pulse-labeled RNA by inhibiting RNA synthesis in *E. coli*.

The decrease in cytochrome content of mitochondria and the mitochondrial enzymes, NADH-succinic dehydrogenase, L-α-glycerophosphate dehydrogenase, and oxidases in *C. fasciculata*, grown in acriflavine for 3–4 days are also probably caused by the inhibition of mRNA synthesis.[56] Proflavine inhibition of mRNA synthesis has been used to demonstrate a light-induced mRNA for carotenoid synthesis.[57] The induction of catalase in *Rhodopseudomonas spheroides* by hydrogen peroxide was inhibited by $5 \times 10^{-4}M$ acridine, and by 5 μM acriflavine.[58] At $1 \times 10^{-4}M$ and 1.5 μM, respectively, they stimulated induction. The stimulatory effect was considered to be at transcriptional level (RNA synthesis), rather than translational (protein synthesis) level. The concentration of acridine that stimulated the induction of enzyme inhibited general protein synthesis in the cell under similar conditions. Also, acriflavine did not stimulate the induction of β-galactosidase in *E. coli* B; these authors suggest that acridine dyes may interact with the internal inducer-DNA complex, which leads to a higher rate of production of catalase mRNA.

Nontoxic doses of acriflavine ($3 \times 10^{-6}M$) were found to depress the unscheduled DNA synthesis of HeLa cells in the G1 growth period at low uv doses, but not at higher doses.[59] However, acriflavine at this concentration has little effect on survival at uv doses where the depression of unscheduled DNA synthesis is greatest.

The inhibition of RNA synthesis in rat liver mitochondria by low concentrations of acriflavine has been taken to indicate that mitochondrial DNA is involved in the process.[60] Atebrin is thought to inhibit cell division of *Tetrahymena* by interfering with nucleic acid synthesis,[61] and the inhibition of adenine incorporation into *Plasmodium berghei* parallels that of isolated RNA and DNA polymerases of *E. coli*.[62] 9-Aminoacridine and 9-(3-dimethylaminopropylamino)-1-nitroacridine (C 283) inhibits the synthesis of high-molecular-weight RNA more than that of low-molecular-weight RNA in regenerating rat liver and cell cultures.[63] Proflavine (167 μg ml^{-1}) has been shown to prevent the breakdown of pulse-labeled RNA in a thermophilic bacillus and to inhibit the function of the protein-synthesizing complex, believed to be due to complex formation with RNA that inhibits nuclease action.[64] At lower concentration (52 μg ml^{-1}), only RNA synthesis was inhibited, half the inhibition occurring in 75 sec.

Yeast cells exposed to acriflavine give rise to *petit* mutants. All such mutants lack cytochrome oxidase, a typical mitochondrial enzyme complex, and isolated mitochondria from acriflavine mutants are not capable of protein synthesis.[65] It has been demonstrated that the mutation is cytoplasmic in character, being inherited in a non-Mendelian fashion. Thus, when mutation is induced by acriflavine and the resultant *petit* mutants are crossed, all the progeny lack respiratory enzymes and are biochemically indistinguishable from the parent strain.[66] In other words, the mutational defects cannot be corrected either by complementation or crossing over. An observation that may be of some relevance here is that the rat liver mitochondrial DNA polymerase enzyme-template complex is more sensitive to inhibition by ethidium bromide and acriflavine than is the nuclear enzyme complex.[67] If such a sensitivity exists in yeast, it might be responsible for the remarkably specific effects of these drugs in producing *petit* mutations.

The phenomenon has been utilized to identify mitotic recombinants. Luzzati[68] produced *petit* mutants by culturing *S. cerevisiae* in acriflavine. Cells already present were *grande* and produced *grande* colonies after the removal of the acriflavine, so that recombinants occurring during growth in the presence of acriflavine could be distinguished from their off-spring. In this way, the clone stem cells of the mitotic recombinants could be identified.

A related phenomenon is the elimination of bacterial extrachromosomal DNA (episomes, plasmids), such as the F sex factor in *E. coli*, the R drug resistance factors, and Col, the colicine production factors, by treatment with acridines. The addition of 20–80 μg ml^{-1} of acridine orange (3,6-bis-dimethylaminoacridine) to cultures of *E. coli* results in the immediate inhibition of episomal replication.[69] Dyes have also been shown to produce cytoplasmic but not nuclear mutations in trypanosomes.[70]

A technique with interesting possibilities is the labeling of mitotic and meiotic chromosomes of various organisms with fluorescent acridine derivatives. Caspersson et al. have shown that fluorescent acridine mustards bind to well-defined regions in chromosomes of *Viva*.[71] They found that both" propyl-atebrin mustard" [6-chloro-9-(3-di-2'-chloroethylaminopropylamino)-2-methoxyacridine] and "atebrin mustard" [6-chloro-(4-di-2'-chloroethyl-amino-1-methylbutylamino)-2-methoxyacridine] gave the same result, whereas the noncovalently binding atebrin, proflavine, and acriflavine produced very much weaker fluorescence, although in approximately the same regions. Three "half-mustard" analogues of atebrin were also used. These monofunctional derivatives gave the same clear fluorescent regions, but the fluorescence intensities both locally and over the entire chromosomes were reduced. The reasons for the specific fluorescence are not clear: as alkylating agents, the acridine mustards probably attack the N-7 of guanine in the DNA, and hence may accumulate at G-C rich regions. Breaks in the DNA strands induced by a number of alkylating agents have been shown to occur preferentially at certain regions of the chromosomes.[72] By studying the DNA distribution pattern along several chromosomes of *Trillium*, Caspersson et al.

found a good correlation between regions reacting with several different fluorescent compounds and regions designated as hetero-chromatin by other criteria.[73]

The technique has been extended to human chromosomes at mitosis and meiosis, where a 5-min treatment of cell smears with 0.5% atebrin hydrochloride[74] or atebrin mustard[75] gives rise to a specific localized fluorescence in the Y chromosome. In contrast, spermatozoa stained with acridine orange or ethidium bromide, both of which are capable of forming fluorescent complexes with DNA, do not give localized fluorescent regions.[76] Since the presence or absence of the Y chromosome determines the sex of the individual, it may prove possible to identify the sex of the fetus from cell smears.[77] There are some difficulties, however. The normal Y chromosome has a long arm, the distal portion of which fluoresces clearly; but some otherwise normal males have short-arm chromosomes that lack this distal region and hence do not fluoresce on binding atebrin. This difference[78] or the possibility of damaged cells,[79] can therefore constitute a source of error in identifyng a male fetus.

In addition to inhibiting nucleic acid synthesis, acridines have been shown to inhibit protein synthesis. Proflavine ($6 \times 10^{-5}M$) inhibited endogenous leucine incorporation by 50% in a subcellular rat liver system,[80] and significant inhibition was also observed with pyrosin, methylene blue, azure B, and acridine orange, although actinomycin D was without effect. Quantitatively similar inhibitions were found when proflavine was added to intact reticulocytes, synthesizing haemoglobin. In the rat liver system, in the presence of polyuridylate which codes for phenylalanine, the inhibition was disproportionately greater for this amino acid than for leucine. The change in the leucine:phenylalanine incorporation ratio was thought to be a result of a change in the specificity of tRNA during translation, in addition to the overall inhibition of protein synthesis. This latter effect was the most marked and was probably a result of inhibition of the attachment of amino acids to tRNA. Proflavine was also found to inhibit the transfer of phenylalanine from tRNA, to which it binds strongly, to the nascent polypeptide chain.[81] Atebrin has been shown to inhibit protein synthesis, as well as DNA synthesis.[82] The synthesis of phenylalanyl-tRNA was reduced to 14% of control activity by proflavine ($1 \times 10^{-4}M$) but was completely protected from inhibition by a 10-fold increase in tRNA ($2 \times 10^{-3}M$, with respect to nucleotides).

Muscle preparations, treated with acridine orange, showed a decrease in adenosine triphosphatase activity.[83] Atebrin affects the photoswelling and photoshrinkage of isolated spinach chloroplasts[84] and has been shown to uncouple photophosphorylation associated with chloroplast movement.[85, 86] The effect could be reversed by ATP[85] and be prevented by bovine serum albumin or glutathione.[87]

B. Acridines and Purified Enzymes Acting on Nucleic Acids

In contrast to the proteolytic enzymes mentioned earlier, the situation with the enzymes that act on nucleic acids is complicated by the binding of acridines to the nucleic acids themselves. It is difficult, therefore, to distinguish whether an observed inhibition is caused by the acridine's interacting with the enzyme, or with the nucleic acid, or with both, either separately or as the enzyme nucleic acid complex.

In an early study, Beers, Hendley, and Steiner showed that acridine orange inhibits the rate of synthesis and phosphorolysis of polyadenylic acid by polynucleotide phosphorylase.[88] This enzyme catalyzes the reversible formation of polynucleotides from nucleoside diphosphates:

$$n\text{NDP} \rightleftharpoons (\text{NMP})_n$$

For the synthetic reaction, synthesis is initially slow unless a small oligonucleotide primer is present, since synthesis proceeds by terminal addition. The slope of the rate against the ADP concentration curve remained the same for all concentrations of acridine orange ($5.8 \times 10^{-5} - 3.3 \times 10^{-4}M$), however, and this was attributed to the removal of the oligo adenylate primer. The inhibition could be reversed by the addition of polyadenylate. In that sense, it is not a true enzyme inhibitor of the synthetic reaction. The phosphorolysis reaction, on the other hand, showed both relative and absolute decreases in inhibition. At high ionic strength, the rate of the synthetic reaction is increased, due to alteration in Km, but the percentage inhibition remains the same. At low concentrations of potassium chloride and acridine orange, the acridine could increase the polymerization rate by more than 100%, and this activation could be enhanced by the addition of magnesium. This activation was thought to be a result of the neutralization of internucleotide phosphates by the bound dye, analogous to the shielding effect of high concentrations of potassium chloride.

Several acridines have been reported to inhibit DNAase action. Atebrin is reported to inhibit DNAase I (E.C. 3.1.4.5 beef pancreas) by binding the DNA substrate.[89, 90] It was shown, using a plot of atebrin concentration vs. $i (1 - i)$, where i was the fractional inhibition, that inhibition was detectable from 3 μm ml^{-1} upward[90]; 50% inhibition occurring at about 12 μm ml^{-1}. One atebrin is bound per six nucleotides; this is believed to prevent the attachment of the enzyme to the DNA, thereby causing the inhibition.

Similar results have been reported with acridine orange and acriflavine.[91] The addition of either of these compounds, but not actinomycin D, markedly increased the viscosity of the DNA solution and also prolonged the time required for hydrolysis, as measured by the decrease in viscosity. The highest concentration of acridines used (0.3 mM) gave almost complete inhibition.

Analysis of the first-order kinetics by plotting inhibitor concentration against a function of the fractional inhibition indicated that the dye was competing with the enzyme for binding sites on the DNA. Extrapolation gave the average number of nucleotide pairs bound per dye molecule as 4.0 for acridine orange and 1.9 for acriflavine. Atebrin-bound 2.3[89] and 3.0[90] nucleotide pairs per dye molecule for inhibition. Since the saturation of DNA with intercalated dye occurs at a binding ratio of two base pairs per dye molecule, inhibition is maximal before saturation is reached. The variation in binding ratios for maximum inhibition has been interpreted as due to the positioning of the bulky dimethylamino group near the phosphate groups of the DNA, acridine orange having two, atebrin one, and acriflavine none.[91] A terminal amino group on the substituted 8-amino side-chain is required for strong binding of the antimalarial 8-aminoquinolines to DNA.[92]

C. Mutation, DNA and RNA Polymerases

One of the most important uses of acridines as biochemical tools in the investigation of living systems has been in the field of mutagenesis. The acridine mutants differ from most other mutants, in that they do not chemically modify a base in the DNA double helix but give rise to "addition" or "deletion" mutants in which a base pair, i.e., a base in one strand normally hydrogen-bonded to a complementary base in the other, is added or removed. Since the bases in one strand of DNA are translated sequentially three bases at a time (each group of three bases ultimately coding for one amino acid), starting from a defined point at the beginning of a cistron or genetic unit of DNA, the addition or deletion of a base pair results in an effective shift of the reading frame. This gives a grossly different translation of the message beyond the site of the mutation. If the deletion of a base is followed further along the strand by the insertion of a base, or vice versa, the reading frame, and the amino acid sequence of the polypeptide produced, might be altered only in the region between the mutations, and hence give rise to partially active protein and be pseudowild. Acridine-induced (usually proflavine, acridine orange, or acridine yellow) frameshift mutants actually gave the first proof of the triplet nature of the genetic code.

Although the existence of addition or deletion mutations is firmly established, therefore, it is not clear how the addition or deletion occurs in the first instance. Three theories have been proposed. The first was by Brenner, Barnett, Crick, and Orgel.[93] Following Lerman's report[94] that acridine binding to DNA can take place by an intercalation of the planar acridine ring between the planar base pairs (see Chapter XIV), they suggested that if this occasionally occurred between the bases on one strand of DNA but not on

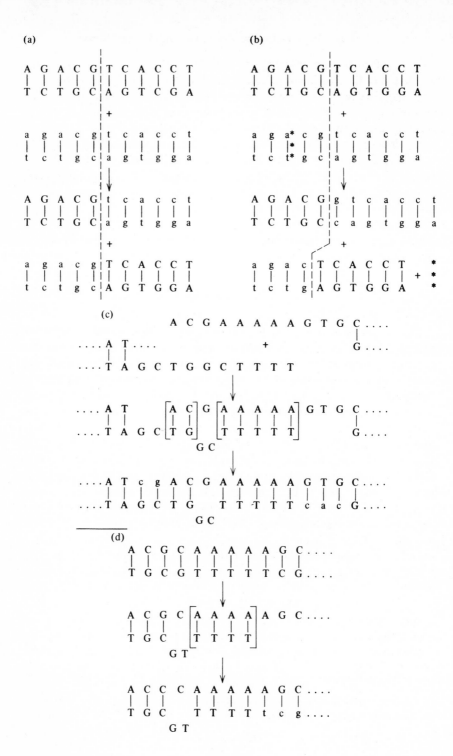

the other, then when the DNA polymerase moved along the DNA during replication, it might lead to the addition or subtraction of a base.

Subsequently, Lerman suggested that a mechanism could be devised on the simple intercalated model, where the base pairs are forced apart or "unscrewed" to 6.8 Å instead of the normal 3.4 Å, the extra gap being produced by the back rotation of the base pairs attached to the flexible sugar-phosphate backbone to a position above the preceding base pair.[95]

This second mechanism relies on recombination, which occurs frequently during phage multiplication. If two phage DNA duplexes or chromosomes are paired in perfect homology, the intercalation of an acridine in one DNA duplex will force it out of step with its neighbor. Enzyme action then breaks both chromosomes at the same position, side by side; in the subsequent repair process, the chromosomes may be reunited with an exchange of partners, resulting in the omission of one or more base pairs in one product and the corresponding duplication of one or more base pairs in the other product of the reunion (Fig. 3).

There are difficulties with both of these models, although neither can be rejected. Further elaboration of Lerman's model is required to account for the known role of heterozygotes in recombination, with the observation that acridine-induced mutations occur as heterozygotes,[96] while acridine binding to a single DNA chain, required by Brenner's model,[93] is now known to be very weak.[95] Therefore, Streisinger et al.[97] proposed that acridines are

---

Fig. 3. (a) A normal recombination between a pair of homologous double-stranded DNA molecules. The two molecules have identical base sequences and are distinguished by upper and lower case letters, so that the effect of recombination can be seen clearly. Cleavage of the phosphodiester backbone, exchange, and rejoining take place as indicated by the dashed line. The base sequence of the two molecules, and hence their information content, are exactly the same as before recombination. (b) Mechanism of acridine mutagenesis originally proposed by Lerman. The forces holding the molecules in register should be considered as acting on the left-hand ends. Temporarily and at the instant of exchange, an acridine ($\vdots$) intercalated in one of the DNA molecules moves the remaining right-hand end of the molecule one base pair to the right. Subsequent exchange and rejoining results in the insertion of an additional base pair in one molecule (addition mutant) and the removal of a base pair in the other (deletion mutant). (c) Frameshift mutation in a heterozygous region. Two parental DNA molecules with single-stranded ends of nearly identical base sequence join as shown. Complementary bases pair wherever possible, forcing a section of one strand out of the structure and thereby weakening it. Insertion of an acridine in a region indicated by a bracket would strengthen the structure and allow the missing bases at the ends of either strand to be added. The loop would remain, however, until replication of the DNA occurred, giving rise to a regular structure with an added or deleted base pair. (d) A mutation occurring at the end of a molecule, where limited endonuclease digestion leaves a single-stranded end of the DNA molecule. Partial rewinding of the double-helical region occurs with mispairing of bases. The structure is stabilized by an acridine intercalated in the region indicated by the bracket, while the digested strand is resynthesized.

mutagenic, not because they stretch the DNA molecule but rather because they stabilize it, as exemplified by the higher thermal denaturation temperature of DNA observed when acridines are intercalated. In this third mechanism, when a break occurs in one strand in a region of repeating bases or base pairs, the end of the broken strand may form a loop, the last two or so bases (the remainder of the strand) returning to pair with the wrong, but complementary, bases in the other strand. Because of the small number of H-bond contacts, such mispairing would be expected to be unstable and short-lived, but if acridines were intercalated between the base pairs they would stabilize the hydrogen bonding sufficiently to allow enzymatic synthesis to fill in the gap before the short region melts out, resulting in the addition or deletion of a base or bases.

In this theory the initial break can occur through the recombination mechanism, which proceeds via breaks in the two strands at different, not the same, places or at the end of the chains after exonuclease digestion of one of them.

Of the two alternatives, Streisinger et al. favor terminal mutations, partly on genetic grounds and partly because high concentrations of acridines greatly decrease the frequency of the recombinants.[97] However, this may have wider application in view of the recently discovered "discontinuous" replication of DNA in bacteria, in which short lengths of DNA are synthesized and subsequently joined.[98]

Neither Lerman's nor Streisinger's theories are entirely satisfactory. Proflavine and similar acridines are highly mutagenic in the T-even phages but have not been found to be mutagenic in several strains of bacteria. There are two pertinent differences between phage and bacteria: T-even phages contain hydroxymethylcytosine instead of cytosine in their DNA; more significantly, their replication is accompanied by extensive recombination. Acridines with half mustard or polyamine side chains are, however, highly mutagenic in bacteria and other organisms[99]; Streisinger et al. have suggested that their mechanism could apply at the site of a mutagen-induced break in the DNA.

A more serious objection is that substances that can intercalate in a similar manner to proflavine are not necessarily mutagenic. Lerman has shown that methylation of the ring nitrogen in acridines abolishes their mutagenic effect,[100] although they still stabilize the DNA, as indicated by increasing T_m, and can still bind, as shown by the hypochromicity at 260 nm and equilibrium dialysis.[101] The degree of binding and of stability decreased in the order 9-aminoacridine = 9-amino-10-methylacridinium chloride > 10-methylacridinium chloride > acridine orange. In view of the great difference in mutagenicity of 9-aminoacridine and 9-amino-10-methylacridinium chloride, it is believed that neither the intercalation (strong binding) nor the weak interaction are directly correlated with the mutagenic action of acridines.[100, 101]

It seems possible, therefore, that the specificity of acridine mutagenesis might result from the interaction of the acridine with the polynucleotide and the enzymes involved in polynucleotide synthesis. In this connection, McCarter, Kadohama, and Tsiapalis have reported some very interesting experiments,[102] in which they looked for mistakes in the incorporation of bases into polynucleotides synthesized by *E. coli* DNA polymerase (EC 2.7.7.7) in the presence of proflavine (0-1 $\times$ 10^{-4}M). The finding of Hurwitz et al.[103] that *E. coli* DNA polymerase was almost totally inhibited by proflavine (60 μM) was confirmed by these workers, who also showed that the incorporation of each base decreased in a parallel fashion with increasing proflavine concentration, when native calf thymus DNA was used as a primer. When an alternating copolymer of deoxyadenylate and deoxythymidylate (poly dAT) was used as a primer, and incubations were carried to minimum absorbency (corresponding to maximum polymer), the incorporation of ^{32}P-TTP increased to 176% of the control, up to 2 $\times$ 10^{-5}M proflavine, and then decreased to about 80% of the control at 5 $\times$ 10^{-5}M proflavine. The initial rise was attributed to the inhibition of exonuclease II, a degradative enzyme found in close association with DNA polymerase, and which when assayed directly is also inhibited by proflavine. In the poly (dG$_2$:dC) primed reaction, the incorporation of ^{32}P $-$ CTP was inhibited at all concentrations of proflavine. The amount of G or C incorporated in the dAT primed reaction was less than 0.2%, the limit of the assay; this was unaffected by proflavine, as would be expected since proflavine does not induce base substitutions of the sort G or C for A or T. However, when nearest neighbor analyses of the product of short incubations, corresponding to 10 replications of the former (where net synthesis was approximately 10 times the amount of primer present) were studied, changes were observed in the sequences formed in the dAT-primed reaction.

As the concentration of the proflavine increased, the proportion of ApA sequences increased such that the composition of the product varied from 1% ApA, 99% dTpA in the absence of proflavine, to 86% ApA, 14% dTpA at 6 $\times$ 10^{-5}M proflavine. Proflavine lengthened the lag period before synthesis of poly dA:dT in the unprimed (*de novo*) DNA polymerase reaction. Once started, however, synthesis was faster than in the primed reaction and proceeded at a concentration of proflavine, which is markedly inhibitory to poly dAT synthesis. Consequently, the results with the primed system could possibly be explained by a poly dA:dT trace impurity, which is replicated preferentially when poly dAT synthesis is blocked. Alternatively, repeating sequences of bases may arise through a reiterative replication almost certainly involving chain slippage. If the proflavine interferes with the copying mechanism of the template, one base may be copied more than once, thereby generating an insertion, or if repeatedly, a run of identical bases free to slip and be replicated. This possibility is of interest in view of the finding of Streisinger et al. that frame shift mutations were most often in, or adjacent to, runs of T's or A's alternating in DNA.[97] The recent finding of T. Okazaki and R. Okazaki that DNA replication proceeds discontinuously, short single strands being synthesized in the $5' \rightarrow 3'$ direction, which are then joined by a ligase, may well have some bearing on acridine mutagenesis.[104]

The related DNA-dependent polymerase of E. coli (EC 2.7.7.6), responsible for the production of messenger and other RNA species in the cell, has been shown to be partially inhibited by proflavine in several laboratories.[103, 105—107] Atebrin [6-chloro-9-(4-diethylamino-1-methylbutylamino)-2-methoxyacridine] (3) at the relatively high concentration of $3.3 \times 10^{-3}M$, inhibits the RNA polymerase of the crown gall tumor-inducing organism, Agrobacterium tumefaciens,[108] giving 99% inhibition, and at $1.7 \times 10^{-4}M$, 3.1% inhibition. At $8 \times 10^{-4}M$, it was reported to be strongly bactericidal for E. coli B, inhibiting DNA synthesis and RNA and protein synthesis strongly, but at $2 \times 10^{-4}M$ bacteria grew into long filaments, due to the inhibition of DNA synthesis.[109] At this bacteriostatic concentration, RNA synthesis was not affected. The inhibition of DNA replication at $8 \times 10^{-4}M$ paralleled the in vitro inhibition of the E. coli enzyme as the inhibition of RNA synthesis paralleled the in vitro RNA polymerase reaction.[110] A 50% inhibition of both DNA and RNA polymerase at $10^{-5}M$ and complete inhibition at $10^{-3}M$ was thought to be caused by the intercalated acridines' restraining the bases in their intrahelical positions. No specific binding groups are required, unlike chloroquine, which requires the 2-amino group of guanine. The inhibition of messenger RNA production probably accounted for the 40% reduction observed in the incorporation of ^{14}C-phenylalanine into protein.

Zillig's group reported a 93% inhibition of T2 DNA-primed E. coli RNA polymerase by $10^{-4}M$ proflavine.[110, 111] It is interesting that the single-stranded DNA of ϕX 174 gave similar control activity, but only 9% inhibition with the same concentration of proflavine; they consider that proflavine inhibits the elongation step (private communication). Richardson found that proflavine (5 μg and 10 μg ml^{-1}) retards initiation slightly[112]; once started, however, the rate of synthesis, although it is reduced, is constant for at least 10 min; he concludes that proflavine has a greater effect on initiation than on polymerization. By centrifuging a mixture of polymerase, DNA, and proflavine on a sucrose density gradient in 10 μg ml^{-1}, proflavine showed that only a small fraction of enzyme was bound to DNA, which under these conditions had a proflavine molecule per three base pairs. As Richardson pointed out, these experiments do not distinguish between an altered DNA unable to bind polymerase and a blocked enzyme unable to bind DNA.

Sentenac, Simon, and Fromageot, using a direct filtration technique to isolate DNA-bound *E. coli* RNA polymerase, showed that both synthesis and initiation are inhibited equally by up to $2.15 \times 10^{-4}M$ proflavine, when native thymus DNA is the template.[113] In the presence of denatured DNA, initiation is inhibited to the same extent, but RNA synthesis was less affected. The amount of initiation was found from the amount of γ-^{32}P nucleoside triphosphate (the first nucleotide of the synthesized RNA). Both γ-^{32}P-ATP and γ-^{32}P-GTP gave similar results. This is in contrast to a report by Maitra, Cohen, and Hurwitz, who found a greater inhibition by proflavine of initiation by ATP, using T4 DNA as the template.[114]

Scholtissek found that proflavine (50 μg ml^{-1}) strongly inhibited RNA synthesis in chick fibroblasts up to a maximum of 96% of the control activity.[115] The remaining RNA was isolated and shown to be labeled throughout its length with a nonrandom arrangement of bases, and to cover a range of molecular weights. The low-molecular-weight material (soluble RNA) contained up to 84% of pyrimidines in clusters; the high-molecular-weight material was rich in AMP and UMP, in contrast to the RNA of nontreated cultures. At high doses, all RNA fraction had a low GMP content; this correlated with a reduction in GC nearest neighbors, GA being relatively unaffected. He suggests a disturbance of template function.

Waring considers that the intercalation is likely to be responsible for the inhibition[116] and points out that proflavine and other inhibitors, e.g., ethidium bromide, inhibit both DNA and RNA synthesis about equally.[117] He also states that if, as has been suggested,[118] RNA polymerase "reads" its template from the narrow groove and DNA polymerase reads it from the wide groove of the DNA, the distortion of the DNA on acridine intercalation will appear much the same to both enzymes. However, actinomycin D, which probably also intercalates,[119] only inhibits DNA polymerase at a concentration sufficiently high to give a significant stabilization against thermal strand separation, whereas RNA polymerase is completely inhibited at 100-fold, lower concentration. The differential inhibition of the two enzymes can be explained by postulating an increased rate of dissociation of actinomycin DNA complex preceding the DNA polymerase[119]; in this connection it may be significant that the gene 32-protein of bacteriophage T4 may act to unwind and align the T4 template DNA ahead of the polymerase.[120]

Nicholson and Peacocke showed that 9-aminoacridine also inhibits the *E. coli* RNA polymerase to approximately the same extent as proflavine, both giving 60–80% inhibitions at high and low concentrations, respectively, of DNA with the same amount of enzyme.[121] 9-Amino-1,2,3,4-tetrahydroacridine, in which the hydrogenated ring is puckered, inhibited only slightly and bound weakly to DNA, so that qualitatively there appears a correlation between DNA binding and polymerase inhibition. However, a careful

quantitative analysis, in which the binding of proflavine and of 9-amino-
acridine to DNA was studied under the conditions of the enzyme assay,
showed that although the amount of the aminoacridine bound to DNA increases with inhibition, a stage is reached at which an increase in aminoacridine concentration still causes an increase in inhibition, with practically
no increase in the amount bound to DNA. Further, reciprocal plots of the
rate vs. DNA concentration or the concentration of all 4 nucleoside triphosphates were linear, and in each case had common intercepts on the
ordinate, suggesting competitive inhibition of the enzyme.

The dissociation constants were, for K_{NTP}, 2.2×10^{-4} mole l^{-1}; K_{DNA},
4.2×10^{-5} mole l^{-1}; K_i for proflavine, 5.4×10^{-9} moles l^{-1}; K_i for 9-amino-
acridine 2.4×10^{-9} mole l^{-1}. Hence DNA is more strongly bound than the
free nucleoside triphosphate. $(K_{\mathrm{NTP}}/K_{\mathrm{DNA}})$ is approximately 5, and the
dissociation constant of proflavine from the enzyme sites is twice that of
9-aminoacridine.

The recognition process required to produce RNA complementary to one
strand of the DNA implies that the DNA is opened out in some way. This is
confirmed by the facilitated exchange of tritium involved in hydrogen
bonding between the bases in the DNA duplex when it is transcribed by
RNA polymerase.[122] Once opened, its bases can form complexes with the
incoming nucleoside triphosphates. This presumably occurs near the active
center of the enzyme, which must contain a void sufficient to accommodate
the base, in a position to complex with the bases of the DNA, and, at the end
of the base distal to the DNA, the general incoming nucleoside recognition
sites (e.g., C-2'-hydroxyl) and the polymerization sites. More simply, the
DNA forms one side of the active center. It is conceivable that the similar-
sized flat ring of the acridine could compete for the site normally occupied by
the base of the nucleoside triphosphate and also interfere with the extension
of the DNA bases. This also makes it clear why the inhibitory activity of the
various acridines should parallel their binding to DNA (proflavine >
9-aminoacridine > 9-amino-1,2,3,4-tetrahydroacridine), since similar stereochemical factors would influence the ability to interact with the base rings of
DNA and with the enzyme sites capable of holding such rings.

The presence of an acridine ring in the vicinity of the residues involved in
the enzymatic action would be expected to protect these from inactivation by
chemical reagents. In experiments with diazonium-1H-tetrazole, which is
specific for histidine, tyrosine, and at high concentrations for amino groups,
the polymerase was completely inactivated when the first histidine and/or the
first tyrosine had been coupled.[123] The addition of 9-aminoacridine, sufficient
to give a 50% inhibition before the reagent, gave some slight protection
(approx. 17% of the initial activity remaining). In contrast, 9-amino-1,2,3,4-
tetrahydroacridine gave no protection. Albert has suggested that if the

antibacterial action of aminoacridines depends on blocking RNA polymerase, tyrosine (phenolic pK_a (10.0) may be the receptor.[124]

Whether the above mechanism can be applied to RNA polymerase, mutations, and to other enzymes remains to be seen. It is apparent that over the last decade, there has been a shift in viewpoint from a fairly straightforward, and to some extent mechanical, intercalation to the slightly more complex field of enzyme active centers.

D. Selective Photooxidation of Amino Acid Residues

Dye-sensitized photooxidation is being increasingly used as an approach to the elucidation of the relationship of structure to activity in proteins. 3,6-Diaminoacridines are often used as sensitizers, since they are easily photoreduced.[125] The procedure is to add the dye to the enzyme in the dark, and then irradiate with visible light from an incandescent bulb, under aerobic conditions. Only four amino acids appear to be affected. These are histidine, tyrosine, tryptophan, and methionine; other amino acid residues react very slowly or not at all.

The selectivity can be increased by irradiation in anhydrous formic acid, when methionine is oxidized to its sulfoxide, and tryptophan is converted to a variety of products but mainly to kynurenine. The following breakdown of tryptophan was proposed.[126] (Scheme 1.)

When this technique was applied to lysozyme, all six tryptophanyl residues were converted to kynurenine and both the methionines were converted to the sulfoxide.[127] Photooxidation caused an extensive inactivation of lysozyme. Enzyme activity dropped to 2% after 30 min of irradiation, when oxidation of tryptophanyl residues was not complete. This suggests that only some of these residues are at the active center, and this is supported by the tertiary structure determination, which shows only two tryptophans in this region.[128] The slow conversion of methionine to its sulfoxide was probably of little significance in this experiment.

Unfortunately, the use of formic acid must lead to denaturation in most and especially the larger proteins, the majority of which probably will not renature. Unless an alternative solvent can be found, the technique may prove to be limited, although it may provide a useful method of nonenzymatic cleavage of a polypeptide chain.[127]

References

1. A. Albert, *The Acridines*, 2nd ed., Arnold, London, 1966.
2. A. Bolliger, *Quart. J. Pharm.*, **13,** 1 (1940).
3. N. E. Sharples and C. L. Greenblatt, *Exp. Parasitol.*, **24,** 216 (1969).
4. S. F. Mason, *J. Chem. Soc.*, 351 (1950).
5. H. Blaschko and R. Duthie, *Nature*, **156,** 113 (1945).
6. L. Hellerman, A. Lindsay and M. R. Bovarnick, *J. Biol. Chem.*, **163,** 553 (1946).
7. C. I. Wright and J. C. Sabine, *J. Biol. Chem.*, **155,** 315 (1944).
8. V. Massey and B. Curti, *J. Biol. Chem.*, **241,** 3417 (1966).
9. J. L. Webb, *Enzymes and Metabolic Inhibitors*, Vol. 2, Academic Press, New York, 1966, p. 343.
10. R. J. Klein, *J. Biol. Chem.*, **205,** 725 (1953).
11. M. L. Fonda and B. M. Anderson, *J. Biol. Chem.*, **242,** 3957 (1967).
12. P. N. Kaul, *J. Pharm. Pharmacol.*, **14,** 237, 243 (1962).
13. E. Hass, *J. Biol. Chem.*, **155,** 321 (1944).
14. S. Toshiyuki, *Bull. Brew Sci.*, **12,** 35 (1966).
15. I. Witt, B. Neufang and H. Muller, *Biochem. Biophys. Acta*, **170,** 216 (1968).
16. D. M. Fraser and W. O. Kermack, *Brit. J. Pharmacol. Chemotherap.*, **12,** 16 (1957).
17. J. F. Speck and E. A. Evans, *J. Biol. Chem.*, **159,** 71, 83 (1945).
18. S. Mudd, *J. Bacteriol.*, **49,** 527 (1945).
19. A. Albert, *The Acridines*, 2nd ed., Arnold, London, 1966, p. 515.
20. R. R. Adams, C. A. Weisel, and H. S. Mosher, *J. Amer. Chem. Soc.*, **68,** 883 (1946).
21. M. G. Rossman, private communication.
22. J. H. Quastel, *Biochem. J.*, **26,** 1685 (1932).
23. J. J. Birktoft, B. W. Matthews, and D. M. Blow, *Biochem. Biophys. Res. Commun.*, **36,** 131 (1969).
24. R. A. Wallace, A. N. Kurtz, and C. Niemann, *Biochemistry*, **2,** 824 (1963).
25. A. N. Kurtz and C. Niemann, *Biochem. Biophys. Acta*, **53,** 324 (1961).
26. S. A. Bernhard and B. F. Lee, in Abstr. VI Int. Congress of Biochemistry, New York, 1964.
27. S. A. Bernhard and Gutfreund, *Proc. Nat. Acad. Sci. U.S.*, **53,** 1238 (1965).
28. A. N. Glazer, *Proc. Nat. Acad. Sci.*, U,S., **54,** 171 (1965).
29. A. Himoe, K. G. Brandt, and G. P. Hess, *J. Biol. Chem.*, **242,** 3963 (1967).
30. K. G. Brandt, A, Himoe, and G. P. Hess, *J. Biol. Chem.*, **242,** 3973 (1967).

31. A. Himoe, K. G. Brandt, R. J. DeSa, and G. P. Hess, *J. Biol. Chem.*, **244,** 3483 (1969).
32. R. J. Foster, *Federation Proc.*, **27,** 3200 (1968).
33. L. Faller, *Sci. American*, **220,** May 30, 1969.
34. S. A. Bernard, B. F. Lee, and Z. H. Tashjian, *J. Mol. Biol.*, **18,** 405 (1966).
35. D. Blow and T. A. Steitz, *Ann. Rev. Biochem.*, **39,** 84 (1970).
36. H. L. Oppenheimer, B. Labouesse, and G. P. Hess, *J. Biol. Chem.*, **241,** 2720 (1966).
37. C. Ghelis, J. Labouesse, and B. Labouesse, *Biochem. Biophys. Res. Commun.* **29,** 101 (1967).
38. D. M. Shotton and H. C. Watson, *Phil. Trans. Roy. Soc.*, *Ser. B*, **257,** 111 (1970).
39. P. B. Sigler, D. M. Blow B. W. Matthews and R. Henderson, J. Mol. Biol., **35,** 143 (1968).
40. G. Feinstein and R. E. Feeney, *Biochemistry*, **6,** 749 (1967).
41. F. H. Shaw and G. A. Bentley, *Aust. J. Exp. Biol. Med. Sci.*, **31,** 573 (1953).
42. I. S. de la Lande and R. B. Porter, *Aust. J. Exp. Biol.*, **41,** 149 (1963).
43. N. Rosic and M. P. Milosevic, *Jugoslav. Physiol. Pharmacol. Acta*, **3** (1), 43 (1967).
44. I. Wilson and E. Cabib, *J. Amer. Chem. Soc.*, **78,** 202 (1956).
45. S. Friess and H. Baldridge, *J. Amer. Chem. Soc.*, **78,** 2482 (1956).
46. V. Whittaker, *Physiol. Rev.*, **31,** 312 (1951).
47. R. A. Osterbahn and J. A. Cohen, *Structure and Activity of Enzymes*, T. W. Goodwin, B. S. Hartley, and J. I. Harris, eds., Academic Press, New York, 1964.
48. M. R. Holloway, *Eur. J. Biochem.*, **5,** 366 (1968).
49. B. R. Hammond and H. Gutfreund, *Biochem. J.*, **72,** 349 (1959).
50. J. Drenth, J. N. Jansonius, R. Koekoek, H. M. Swen, and B. G. Wolthers, *Nature*, **218,** 929 (1968).
51. J. Drenth, J. N. Jansonius, R. Koekoek, L. A. Sluyterman, and B. G. Wolthers, *Phil. Trans. Roy. Soc.*, *Ser. B*, **257,** 231 (1970).
52. I. Schechter and A. Berger, *Boochem, Biophys. Res. Commun.*, **27,** 157 (1967).
53. L. Simpson, *J. Cell. Biol.*, **37,** 660 (1968).
54. R. L. Perlman and I. Pastan, *J. Biol. Chem.*, **243,** 5420 (1968).
55. R. L. Soffer and F. Gros, *Biochem. Biophys. Acta*, **87,** 423 (1964).
56. G. C. Hill and W. A. Anderson, *J. Cell Biol.*, **41,** 547 (1969).
57. P. Batra and L. Storms, *Biochem. Biophys. Res. Commun.*, **33,** (5), 820 (1968).
58. K. T. Shanmugan and L. R. Berger, *Arch. Microbiol.*, **69,** 206 (1969).
59. R. G. Evans and B. Djordjevic, *Int. J. Radiat. Biol.*, **15,** 239 (1969).
60. C. Saccone, M. N. Gadaleta, and R. Gallerani, *Eur. J. Biochem.*, **10,** 61 (1969).
61. S. C. Chou, S. Ramanathan, and W. C. Cutting, *Pharmacology*, **1,** 60 (1968).
62. K. Van Dyke, C. Szustkiewicz, C. H. Lantz, and L. H. Saxe, *Biochem. Pharmacol.*, **18,** 1417 (1969).
63. J. Mendecki, Z. Wieckowska, and M. Chorazy, *Acta Biochem. Polon.*, **16,** 253 (1969).
64. J. Grinsted, *Biochem. Biophys. Acta*, **179,** 268 (1969).
65. S. Kuzela and E. Grecna, *Experentia*, **25,** 776 (1969).
66. H. Jakob, *Genetics*, **52,** 75 (1965).
67. R. N. Meyer and M. V. Simpson, *Biochem. Biophys. Res. Commun.*, **34,** 238 (1969).
68. M. Luzzati, Ph.D. Thesis, University of Paris, France, 1965.
69. S. Sesnowitz-Horn and E. A. Adelberg, *J. Mol. Biol.*, **46,** 17 (1969).
70. *M. Nageo* and T. Sugimara, *Biochem. Biphys. Acta*, **103,** 353 (1965).
71. T. Caspersson, L. Zech, E. J. Modest, G. E. Foley, U. Wagh, and E. Simonson, *Exp. Cell.*, **58,** 128 (1969).
72. B. A. Kihlman, *in Actions of Chemicals on Dividing Cells*, Prentice-Hall, Englewood Cliffs, N.J., 1966.

73. T. Caspersson, L. Zech, F. J. Modest, G. E. Foley, U. Wagh, and E. Simonson, *Exp. Cell. Res.*, **58**, 141 (1969).
74. P. L. Pearson and M. Bobrow, *Nature*, **226**, 959 (1970).
75. L. Zech, *Exp. Cell Res.*, **58**, 463 (1969).
76. P. Barlow and C. G. Vosa, *Nature*, **226**, 961 (1970).
77. L. B. Shettles, *Nature*, **230**, 52 (1970).
78. S. Borgaonkar and D. H. Hollander, *Nature*, **230**, 52 (1971).
79. A. Rook, L. Y. Hsu, M. Gertner, and K. Hirschhorn, *Nature*, **230**, 53 (1971).
80. I. B. Weinstein and I. H. Finkelstein, *J. Biol. Chem.*, **242**, 3757 (1967).
81. I. B. Weinstein and I. H. Finkelstein, *J. Biol. Chem.*, **242**, 3763 (1967).
82. A. K. Krey and F. E. Hahn, *Fed. Proc.*, **28**, 361 (1969).
83. R. G. Lyudkovskaya and L. P. Pevzner, *Biofizika*, **13**, 894 (1968).
84. K. Nishida, Comparative Biochemistry and Biophysics of Photosynthesis, University of Tokyo Press, Japan, 1967.
85. E. Schonbohm, *Z. Pflanzen. Physiol.*, **60**, 255 (1969).
86. J. R. Vose and M. Spencer, *Can. J. Biochem.*, **46**, (12), 1475 (1968).
87. C. D. Howes and A. I. Stern, *Plant Physiol.*, **44**, 1515 (1969).
88. R. F. Beers, D. D. Hendley, and R. F. Steiner, *Nature*, **182**, 242 (1958).
89. N. B. Kurnick and I. E. Radcliffe, *J. Lab. Clin. Med.*, **60**, 669 (1962).
90. J-B. Le Pecq, J-Y. Talaer, B. Festy, and R. Truhaut, *Acad. Sci., Paris*, **254**, 3918 (1962).
91. J. D. Leith, *Biochem. Biophys. Acta*, **72**, 643 (1963).
92. C. R. Morris, L. V. Andrew, L. P. Whichard, and D. J. Holbrook, *Mol. Pharmacol.*, **6**, 548 (1970).
93. S. Brenner, L. Barnett, F. H. C. Crick, and A. Orgel, *J. Mol. Biol.*, **3**, 121 (1961).
94. L. S. Lerman, *J. Mol. Biol.*, **3**, 13 (1961).
95. L. S. Lerman, *Proc. Nat. Acad. Sci.* (*U.S.*), **49**, 94 (1963).
96. J. W. Drake, *J. Cell. Comp. Physiol.*, *Suppl. 1*, **64**, 19 (1964).
97. G. Streisinger, et al., *Cold Spring Harbor Symp. Quant. Biol.*, **31**, 77 (1966).
98. R. Okazaki, et al., *Cold Spring Harbor Symp. Quant. Biol.*, **33**, 129 (1968).
99. B. N. Ames and H. J. Whitfield, *Cold Spring Harbor Symp. Quant. Biol.*, **31**, 221 (1966).
100. L. S. Lerman *J. Cell. Comp. Physiol. Suppl. 1*, **64**, 1 (1964).
101. S. C. Riva, *Biochem. Biophys. Res. Commun.* **23**, 606 (1966).
102. J. A. McCarter, N. Kadohama, and C. Tsiapalis, *Can. J. Biochem.*, **47**, 391 (1969).
103. J. Hurwitz, J. J. Furth, M. Malmy, and M. Alexander, *Proc. Nat. Acad. Sci. U.S.*, **48**, 1222 (1962).
104. T. Okazaki and R. Okazaki, *Proc. Nat. Acad. Sci., U.S.*, **64**, 1242 (1969).
105. I. H. Goldberg and M. Rabinowitz, *Science*, **136**, 315 (1962).
106. I. H. Goldberg, M. Rabinowitz, and E. Reich, *Proc. Nat. Acad. Sci., U.S.*, **48**, 2094 (1962).
107. M. J. Waring, *J. Mol. Biol.*, **13**, 269 (1965).
108. R. M. Hochster and V. M. Chang. *Can. J. Biochem. Physiol.*, **41**, 1503 (1963).
109. J. Ciak and F. E. Hahn, *Science*, **156**, 655 (1967).
110. R. L. O'Brien, J. G. Olenick, and F. E. Hahn, *Proc. Nat. Acad. Sci. U.S.*, **55**, 1511 (1966).
111. E. Fuchs, W. Zillig, P. H. Hofschneider, and A. Preuss, *J. Mol. Biol.*, **10**, 546 (1964).
112. J. J. Richardson, *J. Mol. Biol.*, **21**, 83 (1966).
113. A. Sentenac, E. J. Simon, and P. Fromageot, *Biochem. Biophys. Acta.*, **161**, (2), 299 (1968).

114. V. Maitra, S. N. Cohen, and J. Hurwitz, *Cold Harbor Symp. Quant. Biol.*, **31,** 113 (1966).

115. C. Scholtissek, *Biochem. Biophys. Acta.*, **103,** 146 (1965).

116. M. J. Waring, *Nature*, **219,** 1320 (1968).

117. M. J. Waring, *Symp. Soc. Gen. Microbiol.*, **16,** 235 (1966).

118. E. Reich, *Symp. Soc. Gen. Microbiol.*, **16,** 266 (1966).

119. W. Muller and D. M. Crothers, *J. Mol. Biol.*, **35,** 251 (1968).

120. B. M. Alberts and L. Frey, *Nature*, **227,** 1313 (1970).

121. B. H. Nicholson and A. R. Peacocke, *Biochem. J.*, **100,** 50 (1966).

122. R. Hewitt and B. H. Nicholson, *Biochem. J.*, **113,** 16P (1969).

123. F. Zaheer and B. H. Nicholson, *Biochem. Biophys. Acta*, **251,** 38 (1971).

124. A. Albert, *Selective Toxicity*, 3rd ed., John Wiley and Sons, New York, 1965.

125. F. Millich and G. Oster, *J. Amer. Chem. Soc.*, **81,** 1357 (1969).

126. C. A. Benassi, E. Scaffone, G. Galiazzo, and G. Jori, *Photochem. and Photobiol.*, **6,** 857 (1967).

127. G. Galiazzo, G. Jori, and E. Scaffone, *Biochem. Biophys. Res. Commun.*, **31,** 158 (1968).

128. C. C. F. Blake, G. A. Mair, D. C. Phillips, and B. R. Sarma, *Proc. Roy. Soc. Ser. B*, **167,** 365 (1967).

CHAPTER XVI

The Antibacterial Action of Acridines

A. C. R. DEAN

Physical Chemistry Laboratory,
University of Oxford, England

1. Introduction

Historically, interest in acridines as antibacterial agents stems from the pioneer work of Ehrlich and Benda, who observed that 3,6-diamino-2,7,10-trimethylacridinium chloride and 3,6-diamino-10-methylacridinium chloride possessed trypanocidal activity, particularly the latter, which was also called trypaflavine on account of its activity toward *Trypanosomona brucei*.[1] Almost simultaneously, Shiga[2] reported that trypaflavine was active in cholera infections, while Browning and Gilmour[3] showed that acridine yellow (3,6-diamino-2,7-dimethylacridine) and 3,6-diaminoacridine inhibited the growth of *Staphylococcus aureus*, *E. coli*, *Bacillus anthracis* and *Bacterium typhosum*. Browning and Gilmour's work stimulated much interest in the potentialities of acridines as wound disinfectants, since the antibacterial activity, in contrast to that of many other substances, was shown to be retained in the presence of body fluids and pus. Indeed, both 3,6-diaminoacridine (pro-flavine) bisulfate and 3,6-diamino-10-methylacridinium chloride (con-taminated with the unmethylated compound and known as acriflavine or trypaflavine) became widely used for this purpose.

Browning and his co-workers next investigated the action of a number of substituted 3,6-diaminoacridines, but none of the compounds showed any improvement on the original ones.[4] It became apparent, however, that changes in the pH of the media in which the acridines were examined caused marked variations in their antibacterial potency. The next important landmarks (cf. Ref. 5) were the synthesis and testing[6] of rivanol (3,9-diamino-7-ethoxy-acridine lactate) in 1921, of 7-alkoxy-9-amino-3-nitroacridines[7] in 1929, of flavazole (the salt of 3,6-diaminoacridine and sulphathiazole)[8] in 1946, and the realization, somewhat belatedly, during World War II, that 9-amino-acridine itself possessed antibacterial activity. The latter came directly from the extensive studies of Albert, Rubbo, and their associates, which, as will be seen later, contributed significantly to the elucidation of the mechanisms underlying the antibacterial actions of the acridines (Section 3).

Reviews of the toxicity and of the medical actions of the acridines are available.[5, 9] Today antibacterial acridines are used mainly in the treatment of burns and wounds, and in this area there has been a tendency to replace them by sulfonamides and antibiotics. This is, perhaps, unfortunate, since they are active against a wide range of Gram-positive and Gram-negative bacteria and the indiscriminate use of antibiotics has led to the widespread occurrence of antibiotic-resistant organisms. The troublesome pathogen *Pseudomonas aeruginosa*, however, is relatively resistant to acridines.[5] The more important acridine antibacterials will now be dealt with individually.

2. Acridine Antibacterials

A. 9-Aminoacridine

9-Aminoacridine (**1**) is a strong base and crystallizes from acetone as yellow prisms, mp 234°.

1

It is very sparingly soluble in water. Its preparation is described in Chapter II (p. 116). The monohydrochloride monohydrate is a potent antiseptic, known as aminacrine hydrochloride, monacrin, acramine, demacrine, and acramidine. At 20°, one part is soluble in 300 parts of water, 150 parts of alcohol, and 200 parts of normal saline. It is soluble in glycerol but almost insoluble in chloroform and ether. Aqueous solutions have a strong green-blue-fluorescence, changing to blue at great dilution. The pH of a 0.2% solution in water[10] is 5–6.5. 9-Aminoacridine can be estimated by precipitation as the dichromate or the ferricyanide, followed by volumetric titration of the excess precipitant[11a, b]; the latter is the more accurate. The fluorescence spectrum of 9-aminoacridine has also been used for its estimation.[12]

9-Aminoacridine is a more potent antibacterial than either proflavine or acriflavine. This will become apparent later (Sections 3 and 4). It is a constituent of many preparations for topical use and is the preferred acridine, since it is nonstaining. As with all aminoacridines, sodium alginate, soaps, and other anionic substances of high molecular weight, which form poorly diffusing ion pairs, counteract its antibacterial action.[13] In solution it is well-tolerated by tissues, although continual application tends to delay healing.[14, 15] 9-Aminoacridine can also be used in the solid state, but repeated applications may cause tissue damage.[16, 17]

B. Proflavine

3,6-Diaminoacridine (2) is almost universally known as proflavine, a term often used to describe the base, sulfate or acid sulfate. Its preparation is described in Chapter I (p. 31). Proflavine is now supplied as the neutral sulfate, called proflavine hemisulfate (3), although earlier the acid sulfate was the salt generally available.

At 20°, proflavine hemisulfate is soluble in 300 parts of water or 35 parts of glycerol; it is very slightly soluble in alcohol and insoluble in ether. A saturated solution in water is deep orange in color and gives a green fluorescence when freely diluted. Its pH is 6.0–8.0.[14] Proflavine may be estimated by the methods already described for 9-aminoacridine or colorimetrically after diazotization and coupling with N-(1-naphthyl)ethylenediamine.[18] Unlike

9-aminoacridine, solutions of proflavine stain the skin bright yellow and are decomposed by sunlight.

Both proflavine and acriflavine were used as wound disinfectants during both world wars, although later in World War II 9-aminoacridine was preferred.[9] The application of not more than 0.5 g of powdered proflavine acid sulfate at a time was recommended,[19] a treatment that gave good results in a number of intractable mixed infections. Other workers found that necrosis occurred and healing was delayed.[16, 17, 20] However, the type of wounds treated varied considerably. In general, it appears that the solid was advisable for old heavily infected wounds and did not then cause necrosis, but a dilute solution was best for lightly infected fresh wounds, where the solid had undesirable effects. Proflavine is a highly toxic substance in tissue culture[21]; a destructive action on leucocytes has also been reported,[22] but this has been contested.[5, 9] In contrast to acriflavine, proflavine caused very little damage to rabbit brain[23] at antibacterial concentrations and it also interfered less with the respiration of brain tissues.[24] Its toxicity to human tissues has been critically reviewed by Browning[5] and Albert.[9]

Treating tissues with proflavine acid sulfate subjects them to highly acid conditions, since aqueous solutions are strongly acid. The suggestions that it should be replaced by neutral salts of proflavine, which give aqueous solutions of pH about 6, were of much value[25, 26] and led to the introduction of neutral proflavine sulfate (3). Proflavine base, giving a strongly alkaline reaction, is as undesirable as the acid sulfate. The neutral sulfate (proflavine hemisulfate) is a good wound antiseptic, although, as pointed out earlier, 9-aminoacridine is even better on account of its higher antibacterial activity and its nonstaining properties. Proflavine and other acridines, in general, are strongly adsorbed by surgical dressings. However, Albert and Gledhill, in tests involving proflavine, 9-aminoacridine, and acriflavine, clearly demonstrated that this adsorption was readily reversed by water and particularly by serum, so that dressings can act as depots for the antibacterials.[27]

C. Acriflavine

As commercially supplied acriflavine is a mixture of 3,6-diamino-10-methylacridinium chloride (4) and proflavine (3,6-diaminoacridine) dihydrochloride, the latter being present to the extent of about one-third.

4

Similarly, neutral acriflavine is the corresponding mixture of 3,6-diamino-10-methylacridinium chloride and proflavine monohydrochloride. The preparation of acriflavine is described in Chapter VIII (p. 582). Pure 3,6-diamino-10-methylacridinium chloride[28] was obtained from the commercial material by treating a dilute aqueous solution with excess freshly precipitated silver oxide and allowing to stand overnight at 0°. The solution was then filtered, the filtrate adjusted to pH 7 with hydrochloric acid, and evaporated to dryness. The residual solid after several recrystallizations from methanol,[29] followed by water, gave pure 3,6-diamino-10-methylacridinium chloride as yellow prismatic needles, which became anhydrous at 120°. The hemihydrate is stable in air at room temperature, and its solubility in water at 20° is 0.4%. Commercial samples of acriflavine are much more soluble because of the presence of proflavine dihydrochloride and other impurities. Acriflavine may be estimated volumetrically using potassium ferricyanide[30] or dichromate,[31] or gravimetrically as the picrate.[32a, b] Aqueous solutions are rapidly decomposed in the presence of alkali[29] and by sunlight when photooxidation to the acridanone probably takes place.[33]

Following the discovery of its outstanding antibacterial properties in 1913, acriflavine[3] became widely used as a wound antiseptic, in spite of the many reports,[24, 25, 34] one of the earliest being in 1918,[35] that it is considerably more toxic but not more antiseptic than proflavine. The toxicity varies with the degree of purity of the sample[5]; chemically pure 3,6-diamino-10-methylacridinium chloride being 10 times as toxic as proflavine and 6 times as toxic as 9-aminoacridine.[9] It also causes sloughing of the tissues.[36] There appears to have been no attempt by manufacturers to produce a pure product for use as an antiseptic. Consequently, most of the biological work has been carried out with impure material of variable composition and is, therefore, difficult to assess. However, the availability of other more clinically acceptable acridines make this of little importance.

D. Other Acridines

A number of other acridines have had limited use as antibacterials. Rivanol (ethacridine), initially the hydrochloride and later the lactate of 6,9-diamino-2-ethoxyacridine (5), was first prepared in 1921, and has been used as a surgical antiseptic.[6]

5

It is of interest as being the first 9-aminoacridine used in wound therapy, but comparative studies of rivanol and acriflavine failed to show the superiority of either.[34, 37–39] 9-Amino-4-methylacridine (6), known as neomonacrin or salacrin, was used clinically as the hydrochloride, following a favorable *in vitro* examination.[40] It appears to be very similar in activity to 9-amino-acridine but is slightly less toxic and more soluble. 2,6-Diaminoacridine hydrochloride (7), also called diflavine or acramine red, was prepared by standard methods and shown to be a powerful antibacterial.[41] Clinically, it is not superior to 9-aminoacridine and causes a bright red stain.[16, 42, 43]

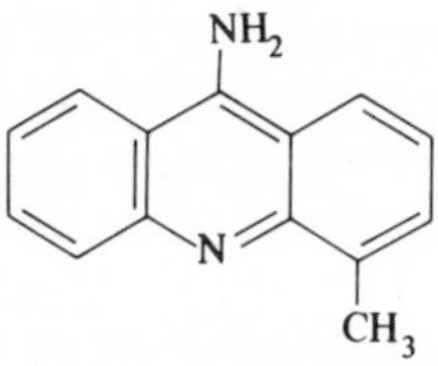

6

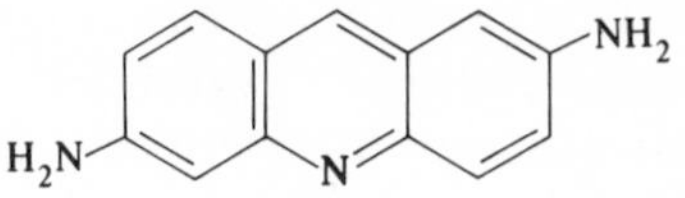

7

Two other acridines, sinflavine (3,6-dimethoxy-10-methylacridinium chloride)[44] and flavicid (3-amino-6-dimethylamino-2,7,10-trimethyl-acridinium chloride)[45, 46] were also discarded as wound disinfectants at a much earlier date. Entozon or nitroacridin 3582 (8) has been prepared[47a, b] in a manner similar to atebrin (see Chapter II, p. 125) and appears to be active against *streptococci* and *rickettsiae*.

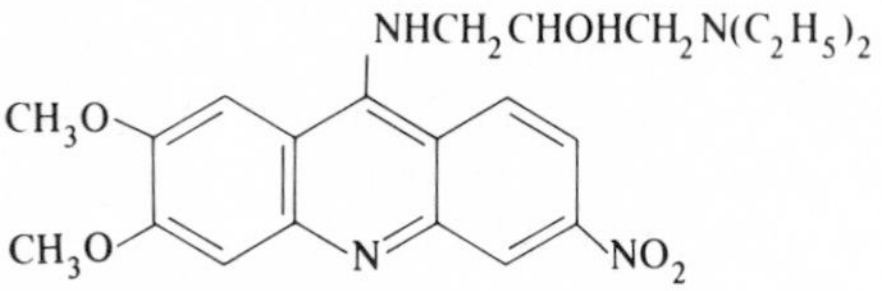

8

Earlier, Schnitzer and Silberstein[7] and Schnitzer[48] had pointed out the systemic antistreptococcal activity of substituted 9-amino-3-nitroacridines.

Salts of acridines and sulfonamides have also been made from acridine hydrochloride and sodium sulfonamide. These salts are not antagonized by *p*-aminobenzoic acid.[5] 3,6-Diaminoacridine sulfathiazole (flavazole),[49] when diluted to 2% with sulfathiazole powder, proved very useful in the treatment of wounds containing mixed infections; a chemotherapeutic action has been demonstrated in mice experimentally infected with *Clostridium septicum* and *Clostridium welchii*.[8] The corresponding sulfathiazole 9-aminoacridine salt is also highly antibacterial but was more irritant to tissues than the proflavine derivatives.[50] In contrast, acridines containing sulfonamido groups in the same molecule had no important antibacterial properties.[40, 51a, b] Salts of 9-aminoacridine with benzylpenicillin[52a-c] and with 4-hexylresorcinol have also been reported. The latter (akrinol, 9-aminoacridinium 4-hexylresorcinate) is fungicidal as well as antibacterial,[53] the fungicidal activity coming from the hexylresorcinol moiety. In general, acridines are not fungicidal.

3. Structure-Activity Relationships

Browning's work, about 50 years ago, demonstrated that the antibacterial properties of 3,6-diaminoacridine (proflavine) were dependent on both the presence of the amino groups and on complete acridine ring system. Acridine itself, 3,6-bisacetamidoacridine (9), 2-aminopyridine (10), quinoline and its 2-, 3-, and 4-amino derivatives (e.g., 11), and tetrahydroquinoline had very little activity.

Similar results were obtained when the heterocyclic nitrogen atom was alkylated. Of this class of compound, only 3,6-diamino-10-methylacridinium chloride (4) was strongly bacteriostatic. Subsequently, Albert, Francis, Garrod, and Linnell[54] showed that one amino group could suffice if present in the correct position (as in 9-aminoacridine) in the acridine nucleus and again the intact ring system was necessary.

Next a correlation between basic strength and antibacterial acivity was proposed by Albert, Rubbo, and their associates,[55, 56] and later confirmed by them in what must be one of the most extensive studies of this kind ever carried out.[40, 57-59] More than a hundred acridines were examined under comparable conditions against five different species of bacteria. The following general conclusions were drawn.[60] First, efficient bacteriostasis only occurs with compounds that are ionized to the extent of at least 50% as cations, those that ionize as anions or zwitterions being ineffective. Second, the chemical nature of the substituents in the acridine nucleus is seldom important, except insofar as it influences ionization. Third, the mode of action is most easily explained on the basis of a competition between acridine cations and hydrogen ions for a vital anionic site in the bacterial cell; this competition involves about 500 acridine cations for each hydrogen ion. Fourth, ionic linkages together with multiple van der Waals bonds fix the acridine cation to the vital receptor. This is made possible by the large flat area of the acridine molecule; the bacterial receptor is also assumed to be flat.

The results of Albert et al. are summarized in Table I, in which the compounds are arranged in alphabetical order. Minimum bacteriostatic concentrations were determined in a nutrient broth of pH 7.2–7.4. For this, the acridines were dissolved in broth at a concentration of 200 mg l^{-1}, this solution then being diluted with drug-free broth in a stepwise manner, so that the concentration in each dilution was one-half that in the dilution immediately preceding it (i.e., 200, 100, 50 mg l^{-1}). The tubes were innoculated with about 10^6 organisms, and the results read after 48-hr incubation at 37°. An arbitrary series of numbers termed "inhibitory indices" was used to signify the lowest concentration of drug inhibiting growth, an index of 1 corresponding to 200 mg l^{-1}, an index of 2 to 100 mg l^{-1}, an index of 3 to 50 mg l^{-1} and so on. Growth at 200 mg l^{-1}, the highest concentration used, was denoted by an index of 0.* The sum of the inhibitory indices for the five organisms tested with each acridine is stated to give a good indication of its antibacterial properties, a sum index of 15 or higher signifying a strongly antibacterial compound. The pK_a values in Table I are a measure of the basicities of the acridines. At 37°, a pK_a of 7.3 corresponds to the critical level of 50% cationic ionization. Nevertheless, the antibacterial activity was not proportional to the pK_a, which shows that once the critical level of cationic ionization has been reached, subsequent increases in pK_a do not affect the activity. The effect of substituents on the basicities of the acridines is that expected from the electronic character and position of the substituents (see Chapter II, p. 110). Examples of acridines that ionize almost completely as anions are acridine-2-sulfonic acid and acridine-2-carboxylic acid. 9-Aminoacridine-2- and -4-carboxylic acids and 9-amino-1-hydroxyacridine exist as zwitterions in water at pH 7.3 and 37°, while acridine itself and methyl acridine-9-carboxylate are un-ionized. The low antibacterial activity of these substances is apparent from Table I. Conversion of 9-aminoacridine-2-carboxylic acid to its methyl ester, which prevented the formation of the zwitterion and allowed only the formation of a cation, largely restored the antibacterial properties to the level expected from the parent 9-aminoacridine molecule. In general, antibacterial activity and systematic toxicity for mammals are not closely correlated.[4, 5] It is also of considerable interest that the order of sensitivity of the organisms given in Table I is nearly always the

---

*Albert et al. actually express their results in terms of dilutions, i.e., 1 in 5000, 1 in 10,000, but for conformity with what is reported later, these have been transformed into concentrations.

TABLE I. Antibacterial Properties of a Number of Acridine Derivatives

Compound	Cl. welchii	Strep. pyo- genes	Staph. aureus	E. coli	Pro- teus	Sum of in- dexes	Basic pK_a in water at 37°	% Ionized as cation at pH 7.3 and 37°
1-Acetamidoacridine	0	0	0	0	0	0	4.2*	0.05
2-Acetamidoacridine	4	2	1	1	1	9	4.5*	0.2
3-Acetamidoacridine	2	2	0	0	0	4	5.3*	1.0
4-Acetamidoacridine	2	1	0	0	0	3	2.8*	<0.01
9-Acetamidoacridine	0	0	0	0	0	0	4.2*	0.05
9-Acetamidobenz[a]acridine	0	0	0	0	0	0	5.0*	<1
12-Acetamidobenz[b]acridine	6	0	0	0	0	6	5.1*	<1
9-Acridanone	0	0	0	0	0	0	−0.1	<0.01
Acridine	3	1	1	0	1	6	5.3	1.0
Acridine-2-carboxylic acid	0	0	0	0	0	0	5.0, ca. 2[a]	99.3[b],0.7[d]
Acridine-4-carboxylic acid	0	0	0	0	0	0	7,4 ca. 2[a]	44,[b] 56[d]
Acridine-9-carboxylic acid	0	0	0	0	0	0	4.8, ca. 2[a]	99.5[b],**0.5**[d]
Acridine-9-carboxylic acid, methyl ester	0	0	0	0	0	0	3.2	<0.01, 100[c]
Acridine-2-sulfonic acid	0	0	0	0	0	0	4.5,ca. 1.5[a] 1.5[a]	**99.7**[b],0.2[d]
N,N'-Bis-(9-acridyl)ethylene diamine	5	3	1	0	0	9	8.7*	96
2-Aminoacridan	2	2	1	0	0	5	4.9*	0.6
1-Aminoacridine	5	2	1	1	0	9	5.7	2.6
2-Aminoacridine	5	2	1	0	0	8	5.6	2.2

(Table Continued)

797

Table I. (Continued)

Compound	Cl. welchii	Strep. pyogenes	Staph. aureus	E. coli	Proteus	Sum of indexes	Basic pK_a in water at 37°	% Ionized as cation at pH 7.3 and 37°
3-Aminoacridine	6	5	3	4	3	21	7.7	72
4-Aminoacridine	2	1	1	0	0	4	4.2	0.1
9-Aminoacridine	7	6	4	4	4	25	9.6	99
6-Aminoacridine-2-carboxylic acid	0	0	0	0	0	0	7.7, ca. 2[a]	28[b],72[d]
6-Aminoacridine-2-carboxylic acid, methyl ester	1	0	0	0	0	1	6.8	24,76[c]
9-Aminoacridine-2-carboxylic acid	0	0	0	0	0	0	9.6*, c.3.0[a]	1.0[b],99[d]
9-Aminoacridine-2-carboxylic acid, methyl ester	7	6	5	2	0	20	8.2*	89,11[c]
9-Aminoacridine-2-carboxamide	3	5	2	3	2	15	8.3*	91
9-Aminoacridine-4-carboxylic acid	0	0	0	0	0	0	13*, 2.7[a]	100[d]
3-Aminoacridine-7-sulfonic acid	1	0	0	0	0	1	7.3, ca 2[a]	50[b], 50[d]
3-Aminoacridine-7-sulfonamide	4	2	0	2	0	8	6.9*, ca. 11[a]	29,0.05[b] 0.02[d] 70.9[c]
9-Aminobenz[a]acridine	7	5	5	0	0	17	7.1	39
10-Aminobenz[a]acridine	0	0	0	0	0	0	5.4*	1
12-Aminobenz[a]acridine	7	6	5	3	2	23	8.6*	95
2-Aminobenz[b]acridine	2	0	0	0	0	2	6.1*,[e]	6[e]
12-Aminobenz[b]acridine	7	7	5	4	1	24	9.9*	100
7-Aminobenz[c]acridine	8	6	6	3	0	23	8.3*	91

Table I. (Continued)

Compound	Cl. welchii	Strep. pyogenes	Staph. aureus	E. coli	Proteus	Sum of indexes	Basic pK_a in water at 37°	% Ionized as cation at pH 7.3 and 37°
9-Aminobenz[c]acridine	0	0	0	0	0	0	4.5*	<1
10-Aminobenz[c]acridine	4	4	3	0	0	11	6.4	11
2-Amino-6-chloroacridine	0	0	0	0	0	0	4.3*	0.06
3-Amino-6-chloroacridine	4	4	3	2	0	13	6.8	24.0
3-Amino-9-chloroacridine	1	0	0	0	0	1	6.4	11.2
6-Amino-2-chloroacridine	5	4	3	1	0	13	6.7	20.0
9-Amino-1-chloroacridine	6	6	4	4	2	22	8.1*	86
9-Amino-2-chloroacridine	7	6	5	4	4	26	8.5*	94
9-Amino-3-chloroacridine	7	6	5	5	3	26	8.9	98
9-Amino-4-chloroacridine	6	5	4	3	2	20	8.0*	83
9-Amino-6-chloro-2-methoxy-acridine	8	6	5	2	0	21	8.5	94
9-Amino-2-cyanoacridine	7	5	4	4	3	23	7.7*	72
9-Amino-2,4-dimethylacridine	8	7	5	3	1	24	10.1*	100
9-Amino-4,5-dimethylacridine	8	7	5	4	3	27	9.0*	98
2-Amino-9,10-dimethylacridinium bromide	5	5	3	2	0	15	9.4	99
2-Amino-6-dimethylaminoacridine	6	5	3	1	0	15	8.2*	89
9-Amino-2-ethoxy-6-nitroacridine	7	8	4	0	0	19	7.3	50
9-(2-Aminoethyl)acridine	6	6	2	3	2	19	8.8*	97
9-Amino-4-ethylacridine	7	6	4	3	2	22	9.8*	99
9-Amino-1-hydroxyacridine	3	3	1	0	0	7	12.4, 5.3[a]	1.0, 99[d]
9-Amino-2-hydroxyacridine	4	6	2	0	0	12	>12, 7.4[d]	56, 44[d]

(Table Continued)

Table I. (Continued)

Compound	Cl. welchii	Strep. pyogenes	Staph. aureus	E. coli	Proteus	Sum of indexes	Basic pK_a in water at 37°	% Ionized as cation at pH 7.3 and 37°
9-Amino-3-hydroxyacridine	2	4	0	0	0	6	> 12, 6.3[a]	9, 91[d]
9-Amino-4-hydroxyacridine	5	5	3	3	3	19	> 12, 6.8[a]	24, 76[d]
9-Amino-1-methoxyacridine	7	6	4	3	2	22	9.8*	99
9-Amino-2-methoxyacridine	7	7	5	4	2	25	9.2	98
9-Amino-3-methoxyacridine	7	6	5	4	3	25	9.7*	99
9-Amino-4-methoxyacridine	7	6	4	3	2	22	9.5*	99
2-Amino-9-methylacridine	5	3	2	0	0	10	5.8*	3
9-Amino-1-methylacridine	6	6	4	4	4	24	9.7*	99
9-Amino-2-methylacridine	7	6	4	4	2	23	9.6*	99
9-Amino-3-methylacridine	6	6	4	3	2	21	9.8*	99
9-Amino-4-methylacridine	7	7	5	4	4	27	9.8*	99
1-Amino-10-methylacridinium bromide	3	3	2	0	0	8	Unstable	Very little
2-Amino-10-methylacridinium bromide	1	2	0	0	0	3	Unstable	Very little
3-Amino-10-methylacridinium bromide	6	7	1	5	0	19	11.8	100
9-Amino-10-methylacridinium chloride	6	6	4	4	3	23	10.7	100
9-Amino-1-nitroacridine	7	8	6	4	2	27	7.3*	50
9-Amino-2-nitroacridine	7	6	5	4	2	24	7.5*	61

800

Table I. (Continued)

Compound	Cl. welchii	Strep. pyogenes	Staph. aureus	E. coli	Proteus	Sum of indexes	Basic pK_a in water at 37°	% Ionized as cation at pH 7.3 and 37°
9-Amino-3-nitroacridine	6	10	6	5	3	30	7.5*	61
9-Amino-4-nitroacridine	6	7	4	4	3	24	7.5*	61
7-Amino-4-nitrobenz[c]acridine	6	1	0	0	0	7	7.1*e	39e
9-Amino-2-phenylacridine	8	7	6	3	0	24	9.4*	99
9-Amino-4-phenylacridine	5	5	3	2	0	15	8.9*	98
9-Amino-1,2,3,4-tetrahydroacridine	1	1	0	0	0	2	10.0	100
Benz[a]acridine	0	0	0	0	0	0	4.5*	<1
Benz[b]acridine	1	2	0	0	0	3	5.0*	<1
Benz[c]acridine	0	0	0	0	0	0	3.9*	<1
9-Butylaminoacridine	6	5	2	0	0	13	9.5*	99
6-Chloro-9-(4-diethylamino-1-methylbutyl)amino-2-methoxyacridine (atebrin, mepacrine)	5	4	1	0	0	10	10.0	100
9-Cyclohexylaminoacridine	6	5	3	0	0	14	9.0*	99
1,6-Diaminoacridine	7	7	4	4	4	26	8.7*	96
1,9-Diaminoacridine	5	5	3	3	2	18	10.6*	100
2,7-Diaminoacridine	5	3	1	0	0	9	5.9	4
2,9-Diaminoacridine	7	6	2	2	1	18	10.0*	100
3,5-Diaminoacridine	4	3	0	2	0	9	6.9*	28.5
3,6-Diaminoacridine (proflavine)	7	6	4	3	2	22	9.3	99
3,7-Diaminoacridine	7	6	4	5	4	26	7.8	76

(Table Continued)

801

Table I. (Continued)

Compound	Cl. welchii	Strep. pyogenes	Staph. aureus	E. coli	Proteus	Sum of indexes	Basic pK_a in water at 37°	% Ionized as cation at pH 7.3 and 37°
3,9-Diaminoacridine	7	6	2	2	0	17	11.1	100
4,5-Diaminoacridine	0	0	0	0	0	0	3.9	0.04
4,9-Diaminoacridine	5	5	4	4	4	22	9.0*	98
3,6-Diamino-9-acridyl 3,6-diamino-9-acridanyl ether	1	3	2	0	0	6	10*	100
4,7-Diaminobenz[c]acridine	7	7	6	3	2	25	8.6	95
3,6-Diamino-2,7-dichloroacridine	8	6	4	5	0	23	7.9*	88
3,6-Diamino-1,8-dimethylacridine	8	8	2	4	0	22	10.2*	100
3,6-Diamino-2,7-dimethylacridine	8	6	4	3	0	21	9.8*	99
3,6-Diamino-4,5-dimethylacridine	9	8	6	4	3	30	8.7*	96
3,9-Diamino-7-ethoxyacridine (rivanol)	6	7	4	3	0	20	11.1*	100
3,6-Diamino-10-methylacridinium chloride	7	8	3	3	1	22	>12	100
3,6-Dimethoxyacridine	5	4	0	0	0	9	5.9*	4.0
3-Dimethylaminoacridine	5	5	3	2	0	15	8.0*	83
3,6-bisDimethylaminoacridine	5	6	4	2	0	17	10.1*	100
9-Dimethylaminobenz[a]acridine	4	5	2	0	0	11	7.0	33
9-Dodecylaminoacridine	2	5	4	0	0	11	9.3*	99
9-Heptylaminoacridine	4	5	4	1	0	14	9.5*	99
9-Hexadecylaminoacridine	3	3	2	0	0	8	8.9*	98

Table I. (Continued)

Compound	Cl. welchii	Strep. pyogenes	Staph. aureus	E. coli	Pro-teus	Sum of in-dexes	Basic pK_a in water at 37°	% Ionized as cation at pH 7.3 and 37°
1,2,4,5,7,8-Hexamethylacridine	0	0	0	0	0	0	4.3*	0.06
1-Hydroxyacridine	0	0	0	0	0	0	5.5	1.4
2-Hydroxyacridine	0	0	0	0	0	0	5.4	1.0
3-Hydroxyacridine	3	2	1	0	0	6	5.0	0.5
4-Hydroxyacridine	5	4	3	0	0	12	5.2	0.8
9-Hydroxyacridine	See 9-Acridanone							
9-(2-Hydroxyethyl)aminoacridine	5	5	2	2	1	15	8.9*	98
2-Methoxyacridine	3	2	1	0	0	6	5.2	1.0
4-Methoxyacridine	3	0	0	0	0	3	5.0	0.6
2-Methylacridine	0	0	0	0	0	0	5.5	1.6
4-Methylacridine	2	0	0	0	0	2	5.4	1.2
9-Methylacridine	3	0	0	0	0	3	5.6	2.0
10-Methylacridinium bromide	0	0	0	0	0	0	Unstable	Very little
9-Methylaminoacridine	7	6	4	2	3	22	9.9*	99

Source: Refs. 40, 50-59

*Indicates that the pK_a was determined in 50% alcohol and corrected (approximately) to that which would be expected in water by adding a correcting figure of 0.5 pK_a unit in most cases.

a = the pK_a value of the acid ionization.

b = % ionized at pH 7.3 and 37° as anions.

c = % of compound present as the neutral molecule at pH 7.3 and 37°

d = % ionized at pH 7.3 and 37° as zwitterions.

e = at 20° instead of 37°.

803

same, and this does not appear to be related to their Gram-staining properties.[6]

Beside the compounds listed in Table I, a large number of acridines have been prepared and tested as antibacterials. The more important papers dealing with these are given in Ref. 62a–l; on the whole, the results are in agreement with the conclusions already given. Exceptions do occur, however, even among the compounds listed in Table I; for example, 4-hydroxyacridine and its 9-amino derivative, although not ionized to the extent of 50%, are antibacterial. This is thought to be a result of chelation with trace metals essential to the bacteria.[63] The four 9-amino-X-nitroacridines and 9-amino-2-ethoxy-6-nitroacridine are all highly antibacterial, despite their low pK_a values, which may reflect a specific effect of the nitro group.[40] On the other hand, 9-amino-1,2,3,4-tetrahydroacridine fulfills the requirement for cationic ionization and yet is a poor antibacterial. This is explained in terms of "dimensional factors," which can lead to an interference of the combination between the drug and receptor.[64] As already pointed out, the bacterial receptor is assumed to be flat and if the flat area of the drug molecule is reduced, it is conceivable that the interacting forces may be too weak to hold it on the adsorbing site. Converting 9-aminoacridine into 9-amino-1,2,3,4-tetrahydroacridine (12) reduces the flat area of the molecule, since the reduced ring is not planar. Support for this dimensional explanation comes from the finding that 12-amino-1,2,3,4-tetrahydrobenz[b]acridine (13), in which the area of flatness has increased, has substantial antibacterial activity.[64]

12

13

In other examples, N,N'-bis(9-acridinyl)ethylenediamine and 3,6-diamino-9-acridanyl ether, insufficient adsorption because of the very large bulk of the molecules was given as a possible explanation of low activity.[64]

4. Development of Resistance to Acridines

Although generally used when many compounds have to be screened for antibacterial properties, the minimum inhibitory concentration technique (Section 3) is open to criticism on strictly quantitative grounds. The cultures are scored as + or − after an arbitrarily chosen interval of time has elapsed, which can impose qualitative differences, when in fact what is present may be no more than quantitative changes. For example, at a given drug concentration, visible growth may be abundantly evident in a culture at, say, 48 hr, although complete inhibition may have persisted for 36 hr and at a lower drug concentration for 24 hr, and so on. Measuring the actual lags before growth provides a more strictly quantitative approach. This forms the basis of the determination of lag concentration relationships, which have been widely used by Hinshelwood and his associates in studying many drugs, including acridines.[65]

The lag concentration relationships obtained with *Klebsiella aerogenes* (NCIB 418, previously called *Aerobacter aerogenes* or *Bacterium lactis aerogenes*) and 3,6-diaminoacridine (proflavine) and 9-aminoacridine are shown in Figures 1 and 2. The experiments were carried out at 40° and pH 7.1, a chemically defined medium, aerated by a stream of sterile air and containing inorganic salts and glucose being used. The ΔL is the difference

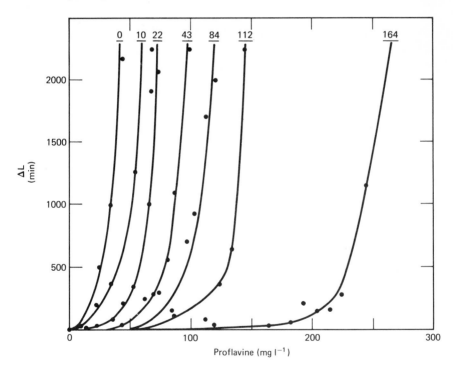

Fig. 1. Lag concentration relationships for *Klebsiella aerogenes* and proflavine. The numbers on the curves denote the concentration of drug to which the strains were trained. [ord]; ΔL (min); [abs]; proflavine (mgl^{-1}). From data given in Ref. 65.

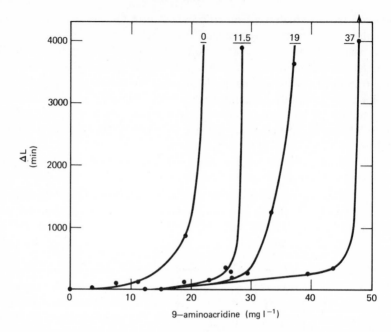

Fig. 2. Lag concentration relationships for *Klebsiella aerogenes* and 9-aminoacridine. The numbers on the curves denote the concentration of drug to which the strains were trained. [ord]; ΔL (min); [abs]; 9-aminoacridine (mgl⁻¹). From data given in Ref. 66.

between the lag observed in drug medium and that in drug-free medium. The latter was small and often negligible, since the inocula (ca. 10^6 cells) were taken from the logarithmic phase of the growth cycle of cultures fully adapted to drug-free medium. The curve at the extreme left of each figure shows the behavior of cells that have never been exposed to the drug, while the others refer to strains "trained" by serial subculture at the given concentration (preceded when necessary by growth at lower concentrations), until no further improvement occurred. When 3,7-diaminoacridine was used, the lag concentration relationships bore a close quantitative resemblance to the 9-aminoacridine set.[66]

Figures 1 and 2 show the superior bacteriostatic activity of 9-aminoacridine compared to proflavine. For example, using "untrained" cells, a value of ΔL of 1000 min occurred at a concentration of 19 mg l⁻¹ of the former, compared to 27 mg l⁻¹ of proflavine. For comparison, 2 mg l⁻¹ oxytetracycline, 2.5 mg l⁻¹ crystal violet, 28 mg l⁻¹ chloramphemicol, or 30 mg l⁻¹ methylene blue produced the same lag in analogous experiments. Figures 1 and 2 also clearly show that resistance to the two drugs develops and that the resistance of the trained strains is continuously graded to the concentration at which training has been carried out. The horizontal spacing of the curves is given by relations of the type $m_s = \overline{m} + $ constant, where $\overline{m}$ is the training concentration and m_s that required to produce large values of ΔL (i.e., 1000 min or longer).

Resistance to 9-aminoacridine at concentrations higher than 45–50 mg l⁻¹ could not, however, be achieved.[66] In contrast, strains of *K. aerogenes* fully resistant to 3000 mg l⁻¹ of proflavine have been obtained,[67] once more emphasizing the superiority of 9-aminoacridine. This difference is borne out in some experiments reported by Albert,[68] which show that although the maximum concentrations of proflavine or of 3,7- and 3,9-diaminoacridines tolerated by *S. aureus* increased four- to eightfold (depending on the drug) during the course of six subcultures in drug medium that of 9-aminoacridine or its 4-methyl, 2-phenyl or 3-chloro derivatives did not (cf. Ref. 69a,b). However, Hinge has obtained strains of *E. coli K12*, like *Klebsiella aerogenes*, resistant to 9-aminoacridine. Reports of proflavine resistance in *Micrococcus pyogenes*,[69a, b] of resistance to acriflavine,[70] and of resistance to an amino-nitroacridine[48] in *streptococci* have also appeared. The training of *Staphylococcus albus* to acriflavine[71] was lost on subculture in drug-free medium; in contrast, *Trypanosomona rhodiense*, which rapidly became resistant to this drug, retained its resistance after passage through 700 mice in a period of 7½ years[72a, b] Hinshelwood and his associates have studied the stability of the resistance of *K. aerogenes* to acridine antibacterials, particularly proflavine, intensively.[65] In general, they found that the longer a training had been impressed on the cells, the greater its stability. For example, the resistance of a strain, which had been trained by exposure to gradually increasing concentrations of drug until 195 mg l⁻¹ was reached, and then given 105 subcultures at this concentration, remained unchanged throughout 63 subsequent subcultures in a drug-free medium. More prolonged growth in the latter led to a loss of resistance, and this took place in a gradual manner. On the other hand, when cells that had undergone only a few generations of growth in drug were transferred to a drug-free medium, the resistance was rapidly lost.

Training *K. aerogenes* to proflavine led to a loss of sulfonamide adaptation but only if the cells were not first exposed to proflavine. If well-trained to proflavine first, both resistances were retained[73]; in an extension of this work,[74] a strain resistant to streptomycin, sulfanilamide, chloramphenicol, ampicillin, and 9-aminoacridine, both singly and in admixture, has been obtained. This multiple resistance is not episomic in character. Cross-resistance, indicating an action at the same stage of a reaction sequence in the cell, has also been observed with acridines. Thus training *K. aerogenes* to proflavine conferred resistance to 3,7-diaminoacridine and 9-aminoacridine. Other acridines were not tested, but in the three that were, the relations were reciprocal and almost quantitative.[65] The strains were also cross-resistant to methylene blue and propamidine (4,4′-diamidinodiphenoxypropane). Cross-resistance between proflavine and propamidine has also been observed in *staphylococci*.[69a, b]

Families of lag concentration curves, such as those obtained with proflavine and 9-aminoacridine (Figs. 1 and 2) and with other drugs, have been interpreted in terms of an adaptive theory of drug resistance. The detailed mathematical analysis, together with the reasons for advocating this mechanism rather than the selection of mutant types present in the initial population in very low frequency, is given by Dean and Hinshelwood.[65] (See also Refs. 75, 76.) This does not necessarily imply that some selection does not occur when untrained cells are first exposed to moderately high levels of drug, since, beside being bacteriostatic, acridines are also bactericidal. For example, exposing cells of *K. aerogenes* to 97 mg l^{-1} of proflavine in phosphate buffer (pH 7.1) at 40° led to a drop in the viable count from 6.3 × 10^8 ml^{-1} to 1.4 × 10^6 ml^{-1} in 1 hr.[77] At lower concentrations, the drop in viability was correspondingly reduced. It follows, then, that in media permitting growth (as used in lag concentration determinations), there could be a competition between the development of an adaptive process and the death of the cells. The physiological age of the cells innoculated into drug medium ranges from those that have just divided to those just about to divide.

In essence, the adaptive interpretation[65] envisages that resistance can come about by a change in the enzyme balance of the cells, and there is experimental evidence that the balance does change in a manner suggesting a coordinated pattern.[65] It is assumed that the drug interferes with the working but not with the reproduction of a particular enzyme, so that the supply of intermediate to the next enzyme in the sequence is reduced. Enzymes occurring later in the sequence are accordingly held up and cellular multiplication lags, while earlier enzymes continue to function, with the result that the inhibited enzyme increases in amount, until the drug is overcome. Equations of the type:

$$\Delta L = L - L_0 = \frac{A}{k} \left\{ \frac{1}{C_1 + \phi(\bar{m}) - \phi(m)} - \frac{1}{C_1 + \phi(\bar{m})} \right\} \qquad (1)$$

account for the families of curves obtained with the various drugs. The symbols L and L_0 are the lags in drug and in drug-free media, respectively; ΔL is set equal to A/R, where A is a constant (for a given type of organism and a given medium) and R the rate of operation of an enzyme process. The concentration, C, of the intermediate, which is the substrate for the enzyme, is related to R in a manner expressible approximately by a Langmuir isotherm, i.e., $R = kc/1 + bc$, where k and b are constants. In drug-free medium, C has the value C_1 in unadapted cells. The addition of the drug reduces it to $C_1 - \phi m$, where the function ϕm of the drug concentration (m) characterizes the drug. When adaptation is complete, C will have returned to C_1, so that if the cells are now transferred to drug-free medium, there will be an immediate increase to a value equal to $C_1 + \phi m$. Therefore, if cells adapted to a concentration $\bar{m}$ are transferred to a concentration m, the value will be given by $C = C_1 + \phi(\bar{m}) - \phi(m)$. When ΔL is fairly large, Eq. (1) reduces to

$$\Delta L = \frac{\text{constant}}{C_1 + \phi\bar{m} - \phi m} \tag{2}$$

Plots of the function ϕm against m are also informative.[65] A linear relation (as observed with proflavine) corresponds to a linear reduction in the rate of production of intermediate as m increases. In other examples, ϕm increased either more, or less rapidly than m, or even changed abruptly from a less than linear to a more than linar increase. A more than linear increase indicates cooperative effects between the drug molecules on the receptor, while a less than linear proportionality suggests that once a certain amount of drug has been taken up, the adsorption of more drug is impeded to a greater extent than that of a competing substrate. The latter could occur if the drug molecules were large and the substrate molecules small and is of particular interest, since the low antibacterial activity of some acridines has been attributed to their large size (Section 3). The condition for an abrupt change in the ϕm, m curves would be that the two effects just referred to are in competition with one another.

Equation (2) assumes a constant pH. Various workers have pointed out that the antibacterial activity of highly ionized acridines is reduced as the pH is lowered, and vice versa.[4, 5, 40, 59] If, then, the cells are trained at one pH value and tested at another, the families of lag concentration curves will be displaced horizontally either to the left or the right, depending on whether the test pH is higher or lower than the training pH. Equation (2) can be modified as follows to take account of this:

$$\Delta L = \frac{\text{constant}}{C_1 + \left[\dfrac{\phi\bar{m}}{[\bar{H}^+]^n}\right] - \left[\dfrac{\phi m}{[H^+]^n}\right]} \tag{3}$$

where $[H^+]$ is the test concentration and $[\bar{H}^+]$ the training concentration of hydrogen ion, n is the slope of the line relating the hydrogen ion concentration and the concentration of drug required to produce a standard value of ΔL. A simple kinetic treatment of the competition between hydrogen ions and acridine cations for sites on a cellular receptor, as proposed earlier (Section 3), predicts a linear relationship and a slope of unity. Slopes of 0.9 and 0.7 have been observed with *K. aerogenes* and 9-aminoacridine and proflavine, respectively,[65, 78] in tests carried out at pH's from 6.1 to 7.7. Over this pH range in which the bacteriostatic activity changed considerably, the degree of ionization of the drugs varied little. It should be pointed out, however, that reducing the pH can have quite a different effect on a feebly ionized acridine. If it is a poor antibacterial at neutral pH because of poor ionization, then reducing the pH should lead to an increase in activity if the degree of ionization (as cations) is thereby increased. This has been demonstrated, but obviously, for practical purposes such pH changes must fall within physiologically tolerable limits.[79]

Other mechanisms that could, in principle, affect the development of resistance must now be considered. Penetration of the drug does not appear to be a rate-limiting step in proflavine bacteriostasis.[80] No correlation was observed between oil-water solubility, surface activity or reduction protentials and antibacterial activity in a series of acridines.[81] Resistant cells of K. aerogenes did not bind less drug than sensitive cells when suspended in phosphate buffer containing proflavine,[82] implying that a loss of receptors does not accompany the development of resistance. However, the amount adsorbed did not vary much over a pH range in which the lag in growth media varies considerably, suggesting that the inhibitory action is exerted at specialized sites and not at every site where drug can be taken up. Similarly, resistant cells of E. coli B did not adsorb less proflavine from buffer than sensitive cells, and no differences could be detected between the amount taken up by the DNA or the cell envelopes of sensitive and resistant cells or in the binding sites. However, in these studies with E. coli B, it was maintained that a drug bound by resting cells in phosphate buffer could be released again by an active process involving metabolism, and that resistant and sensitive cells differ in the ease with which this takes place.[83] Thus, in a medium supporting growth, sensitive cells were found to contain 4 to 6 times as much proflavine as resistant cells. Nevertheless, the amount remaining bound to resistant cells, even during active growth, would be sufficient to inhibit sensitive cells. Adding metabolites or metabolizable sugars to suspensions of resistant cells in buffer also released drug from them. The presence of metabolic inhibitors prevented the release but not the passive uptake. Acid production is not thought to be involved, but the exact mechanism is not known. Observations of this sort are quite compatible with theories ascribing the development of resistance to changes in enzyme balance, and as already pointed out, such changes can indeed be demonstrated.

Acridines also inhibit division relative to the growth of the cells on the first exposure to concentrations just below the bacteriostatic level. With proflavine and K. aerogenes, filamentous cells several times longer than normal cells appeared.[84] This was ascribed to an inhibition of DNA synthesis, since if the filaments were regarded as the equivalent number of cells of standard size, their DNA content was low; but if counted as single cells it was normal. Proflavine has also been reported to inhibit DNA-dependent RNA synthesis[85] and some acridines free bacteria of infectious particles (episomes).[86] These findings might be taken to indicate that DNA is the "vital" receptor in the cell. However, a large proportion of the drug taken up is bound by other cellular structures, such as ribosomes (ribonucleoprotein particles),[87, 88] cell walls,[83] and enzymes[89], and until it is known definitely how much of the binding occurs at "sites of loss," the true identity of the vital receptor(s) must remain in doubt (cf. Ref. 90). The intercalation of acridines with nucleic acids is discussed in Chapter XIV (p. 727), but there is no reason whatsoever to correlate any mutations arising as a consequence of such an interaction with the development of resistance to acridines.[91, 92]

References

1. P. Ehrlich and L. Benda, *Chem. Ber.*, **46,** 1931 (1913).
2. K. Shiga, *Z. Immun. Forsch. Exp. Ther.*, **18,** 65 (1913).
3. C. G. Browning and W. Gilmour, *J. Pathol. Bacteriol.*, **18,** 144 (1913).
4. C. H. Browning, J. B. Cohen, R Gaunt, and R. Gulbransen, *J. Pathol. Bacteriol.*, **24,** 127 (1921).
5. C. H. Browning, in *Experimental Chemotherapy*, Vol. 2, R. J. Schnitzer and F. Hawking, ed., Academic Press, New York, 1964, p. 1.
6. J. Morgenroth, R. Schnitzer, and E. Rosenberg, *Deut. Med. Wschr.*, **47,** 1317 (1921).
7. R. Schnitzer and W. Silbertstein, *Z. Hyg. Infect Krankh.*, **109,** 519 (1929).
8. J. McIntosh and F. R. Selbie, *Brit. J. Exp. Pathol.*, **27,** 46 (1946).
9. A. Albert, *The Acridines*, 2nd ed., Arnold, London, 1966, pp. 403, 505 ff.
10. *British Pharmacopoeia*, Pharmaceutical Press, London, 1968, p. 33.
11. (a) E. Pedley, *Pharm. J.*, **155,** 148 (1945); (b) G. Amor and G. E. Foster, *Analyst*, **70,** 174 (1945).
12. J. G. Devi, M. L. Khorana, and M. R. Padhye, *Indian J. Pharm.*, **15,** 3 (1953).
13. G. Richardson and R. Woodford, *Pharm. J.*, **192,** 527 (1964).
14. *British Pharmaceutical Codex*, Pharmaceutical Press, London, 1968, p. 30.
15. *The Extra Pharmacopoeia* (*Martindale*), 28th ed., Pharmaceutical Press, London, 1968, p. 1176.
16. D. S. Russell and M. A. Falconer, *Lancet*, **1,** 580 (1943).
17. F. Hawking, *Lancet*, **1,** 710 (1943).
18. W. H. C. Shaw and G. Wilkinson, *Analyst*, **77,** 127 (1952).
19. G. A. C. Mitchell and G. A. H. Buttle, *Lancet*, **2,** 749 (1943).
20. J. Ungar and F. A. Robinson, *J. Pharmacol. Exp. Ther.*, **80,** 217 (1944).
21. F. Jacoby, P. G. Medawar, and E. N. Willmer, *Brit. Med. J.*, **2,** 149 (1941).
22. E. P. Abraham, et al., *Lancet*, **2,** 182 (1941).
23. D. S. Russell and M. A. Falconer, *Proc. Roy. Soc. Med.*, **33,** 494 (1940).
24. M. C. Manifold, *Proc. Roy. Soc. Med.*, **33,** 498 (1940); *Brit. Med. J.*, **1,** 631 (1940).
25. H. Berry, *Pharm. J.*, **146,** 181 (1941); *Quart. J. Pharm. Pharmacol.*, 14, **149** (1941).
26. A. Albert and W. S. Gledhill, *Pharm. J.*, **151,** 87 (1943).
27. A. Albert and W. S. Gledhill, *Lancet*, **1,** 759 (1944).
28. P. Gailliot, *Quart. J. Pharm. Pharmacol.*, **7,** 63 (1934); *Bull. Soc. Chim. Fr.*, **1,** 796 (1934).

29. J. Marshall, *Quart. J. Pharm. Pharmacol.*, **7**, 514 (1934).

30. G. F. Hall and A. D. Powell, *Quart. J. Pharm. Pharmacol.*, **10**, 486 (1937).

31. E. Pedley, *Pharm. J.*, **155**, 148 (1945).

32. (a) B. A. Ellis, *Analyst*, **67**, 226 (1942); (b) W. P. Chambers and R. M. Savage, *Quart. J. Pharm. Pharmacol.*, **18**, 227 (1945).

33. K. N. Mathur and S. S. Bhatnagar, *Indian J. Phys.*, **3**, 37 (1928).

34. D. S. Russell and M. A. Falconer, *Brit. J. Surg.*, **28**, 472 (1941).

35. C. H. Browning and R. Gulbransen, *Proc. Roy. Soc.*, *Ser. B*, **90**, 136 (1918).

36. H. Berry *Pharm. J.* **154**, 94 (1945).

37. M. Levrat and F. Morelon *C. R. Soc. Biol.*, **114**, 643 (1933).

38. M. Levrat and F. Morelon, *Bull. Sci. Pharmacol.*, **40**, 582 (1933).

38. G. R. Goetchius and C. A. Lawrence, *J. Lab. Clin. Med.*, **29**, 134 (1944).

40. A. Albert, S. D. Rubbo, R. J. Goldacre, M. E. Davey, and J. D. Stone, *Brit. J. Exp. Path.*, **26**, 160 (1945).

41. A. Albert and W. S. Linnell, *J. Chem. Soc.*, 1614 (1936).

42. G. A. C. Mitchell and G. A. H. Buttle, *Lancet*, **2**, 287 (1943).

43. N. H. Turnbull, *Aust., New Zealand J. Surg.*, **14**, 3 (1944).

44. R. Goldschmidt, *Z. Immunforsch. Exp. Ther.*, **54**, 442 (1927).

45. M. Levrat and F. Morelon, *Bull. Sci. Pharmacol.*, **40**, 582 (1933).

46. H. Langer, *Deut. Med. Wschr.*, **46**, 1015, 1143 (1920).

47. (a) C. S. Miller and C. A. Wagner, *J. Org. Chem.*, **13**, 891 (1948); B.I.O.S. Final Rep. No. 766; (b) C. A. Laurence, *Proc. Soc. Exp. Biol. Med.*, **52**, 90 (1943).

48. R. Schnitzer, *Medizin Chem.*, **3**, 34 (1936).

49. C. A. Hill and H. A. Stevenson, British Patent 573, 578; *Chem. Abstr.*, **43**, 2738 (1949).

50. J. McIntosh, et al., *Lancet*, **2**, 97 (1945).

51. (a) S. Rajagopalan and K. Ganapathi, *Proc. Indian Acad. Sci.*, **15A**, 432 (1942); (b) S. Singh and J. R. Chaudhri, *Indian J. Med. Res.*, **35**, 177 (1947).

52. (a) D. C. Brodie and E. Lowenhaupt, *J. Amer. Pharm. Assn.*, **38**, 498 (1949); (b) P. A. Kelly, *Aust. J. Pharm.*, **32**, 1142 (1951); (c) H. W. Rhodehamel, U.S. Patent 2,567,679; *Chem. Abstr.*, **46**, 221 (1952).

53. H. Seneca, *Antibiotics Chemother.*, **11**, 587 (1961).

54. A. Albert, A. E. Francis, L. P. Garrad, and W. S. Linnell, *British. J. Exp. Pathol.*, **19**, 41 (1938).

55. A. Albert, S. D. Rubbo, and R. J. Goldacre, *Nature*, **147**, 332 (1941).

56. S. D. Rubbo, A. Albert, and M. Maxwell, *Brit. J. Exp. Pathol.*, **23**, 69 (1942).

57. A. Albert and R. J. Goldacre, *J. Chem. Soc.*, 706 (1946).

58. A. Albert, S. D. Rubbo, and M. I. Burvill, *Brit. J. Exp. Pathol.*, **30**, 159 (1949).

59. A. Albert, *The Acridines*, 2nd ed., Arnold, London, 1966, p. 434.

60. A. Albert, *The Acridines*, 2nd ed., Arnold, London, 1966, p. 459.

61. A. Albert, *The Acridines*, 2nd ed., Arnold, London, 1966, p. 455.

62. (a) S. Singh, J. R. Chaudhri, and G. Singh, *J. Med. Res.*, **36**, 387 (1948); (b) J. H. Wilkinson and I. L. Finar, *J. Chem. Soc.*, 32 (1948); (c) K. Weise, *Z. Hyg. Infekt-Krankh.*, **97**, 56 (1922); (d) A. A. Goldberg and W. Kelly, *J. Chem. Soc.*, 637 (1947); (e) F. R. Bradbury and W. H. Linnell, *Quart. J. Pharm. Pharmacol.*, **11**, 240 (1938); (f) C. A. Lawrence, *Proc. Soc. Exp. Biol. Med.*, **52**, 90 (1943); (g) G. R. Goetchius and C. A. Lawrence, *J. Lab. Clin. Med.*, **30**, 145 (1945); (h) H. Berry, *Quart. J. Pharm. Pharmacol.*, **14**, 149 (1941); (i) W. H. Linnell and M. J. H. Smith, *Quart. J. Pharm. Pharmacol.*, **1**, 28 (1949); (j) R. C. Avery and C. B. Ward, *J. Pharmacol. Exp. Ther.*, **85**, 258 (1945); (k) W. Sharp, M. M. J. Sutherland and F. J. Wilson, *J. Chem. Soc.*,

344 (1943); (l) W. L. Glen, M. M. J. Sutherland, and F. J. Wilson, *J. Chem. Soc.*, 1482 (1936).

63. A. Albert, S. D. Rubbo, R. J. Goldacre, and B. G. Balfour, *Brit. J. Exp. Pathol.*, **28**, 69 1947).

64. A. Albert, *The Acridines*, 2nd ed., Arnold, London, 1966, p. 450.

65. A. C. R. Dean and C. Hinshelwood, *Growth, Function and Regulation in Bacterial Cells*, Clarendon Press, Oxford, England, 1966, pp. 136 ff.

66. J. M. Pryce and C. N. Hinshelwood, *Trans. Faraday Soc.*, **43**, 1 (1947).

67. D. S. Davies, C. N. Hinshelwood, and J. M. Pryce, *Trans. Faraday Soc.*, **41**, 778 (1945).

68. A. Albert, *The Acridines*, 2nd ed., Arnold, London, 1966, p. 466.

69. (a) J. R. McIntosh and F. R. Selbie, *Brit. J. Exp. Path.*, **24**, 246 (1943); (b) *Lancet*, **1**, 793 (1943); **2**, 224 (1943).

70. J. W. Howie, *J. Path. Bacteriol.*, **46**, 367 (1938).

71. V. Burke, C. Ulrich, and D. Hendrie, *J. Infect. Dis.*, **43**, 126 (1928).

72. (a) F. Murgatroyd and W. Yorke, *Amer. Trop. Med. Parasit.*, **31**, 165 (1937); (b) P. W. Schuler, *J. Infect. Dis.*, **81**, 139 (1947).

73. D. S. Davies, C. N. Hinshelwood, and J. M. Pryce, *Trans. Faraday Soc.*, **40**, 397 (1944).

74. P. G. Bolton, A. C. R. Dean, and P. J. Rodgers, *Antonie van Leeuwenhoek*, **33**, 274 (1967).

75. (a) A. C. R. Dean, in *Origins of Resistance to Toxic Agents*, ed. M. G. Sevag, R. D. Reid, and O. E. Reynolds, eds., Academic Press, New York, 1955, p. 42; (b) A. C. R. Dean and C. N. Hinshelwood, in Ciba Foundation Symposium on *Drug Resistance in Micro-organisms*, G. E. W. Wolstenholme and C. M. O'Connor, eds., Churchill, London, 1957, p. 4.

76. A. C. R. Dean, F. W. O'Grady, and R. F. Williams, in *Mechanism of Action and Development of Resistance to Antibiotics and Chemotherapeutic Agents*, *Proc. Roy. Soc. Med.*, **64**, 529, 534, 540 (1971).

77. A. A. Eddy, *Proc. Roy. Soc.*, *Ser. B*, **141**, 126, 137 (1953).

78. A. R. Peacocke and C. Hinshelwood, *J. Chem. Soc.*, 1235 (1948).

79. A. Albert, *The Acridines*, 2nd ed., Arnold, London, 1966, p. 445.

80. S. Jackson and C. N. Hinshelwood, *Trans. Faraday Soc.*, **44**, 527 (1948).

81. A. Albert, *The Acridines*, 2nd ed., Arnold, London, 1966, p. 462.

82. A. R. Peacocke and C. Hinshelwood, *J. Chem. Soc.*, 2290 (1948).

83. D. J. Kushner and S. R. Khan, *J. Bacteriol.*, **96**, 1103 (1968).

84. P. C. Caldwell and C. Hinshelwood, *J. Chem. Soc.*, 1415 (1950).

85. J. Hurwitz, J. J. Furth, M. Malamy, and M. Alexander, *Proc. Nat. Acad. Sci. U.S.*, **48**, 1222 (1962).

86. T. Watanabe, *Bateriol. Rev.*, **27**, 87 (1963).

87. M. J. Waring, *Symp. Soc. Gen. Microbiol.*, **16**, 235 (1966).

88. S. H. Miall and I. O. Walker, *Biochim. Biophys. Acta*, **145**, 81 (1967).

89. W. H. Vogel, R. Snyder, and M. P. Schulman, *J. Pharmacol. Exp. Ther.*, **146**, 66 (1964).

90. S. D. Silver, *Exp. Chemother.*, **4**, 505 (1966).

91. A. C. R. Dean and C. Hinshelwood, *Growth, Function and Regulation in Bacterial Cells*, Clarendon Press, Oxford, 1966, p. 151.

92. A. Albert, *The Acridines*, 2nd ed., Arnold, London, 1966, p. 501.

CHAPTER XVII

Carcinogenic and Anticarcinogenic Properties of Acridines

DAVID B. CLAYSON

Department of Experimental Pathology and Cancer Research,
School of Medicine, Leeds, England

1. Carcinogenesis

A. Introduction

Attempts to correlate chemical structure and carcinogenic activity have generally been unsuccessful. Substances capable of inducing cancer in man or laboratory animals are drawn from many chemical classes, including the polycyclic aromatic hydrocarbons and their heterocyclic analogues, aromatic amines and aminoazo compounds, nitrosamines, nitrosamides, hydrazines, triazines, biological alkylating agents (such as strained ring lactones, sulfur or nitrogen mustards, aziridines, epoxides), radiochemicals, metals or their derivatives and asbestos. The ability to induce cancer is also found in cetain miscellaneous compounds such as urethane, thioacetamide, or ethionine.[1]

Recent work suggests that many of these agents have a common mechanism of action. They are thought to be converted in the body to very reactive, positively charged ions (e.g., carbonium ions), which react with the major cellular macromolecules, DNA, RNA, and protein, with the result that cells are transformed to a cancerous state.[2] Because an effective carcinogen has to be absorbed into the body, transported to its site of activation, and metabolized, and the active metabolite must have an appropriate half-life to reach its site of action, the difficulties in elucidating meaningful structure-activity relationships are understandable. These practical difficulties are increased by the fact that the potential carcinogens may interact with systems other than those in which the tumor will ultimately appear. The immune system, for example, is believed to exercise a controlling function on the ability of some cancer cells to develop into clinically apparent tumors; the ability of a carcinogen to impair immunological reactivity, as well as to transform cells, may influence its apparent carcinogenic activity.

Relatively little work has been carried out on the carcinogenicity of simple acridine compounds. Lacassagne and his colleagues at the Institut de Radium in Paris[3] have made a major contribution to the activity of benz- and dibenz-acridines, but these chemicals have been but little used in the mainstream of biological and biochemical work on mechanisms of carcinogenesis. They probably resemble closely the polycyclic aromatic hydrocarbons in their mode of action. There are two major concepts about the mode of carcinogenic action of the latter compounds: (1) they require metabolic activation to a highly reactive, positively charged ionic form, which combines with nucleophilic groups in DNA, RNA, and protein,[4] or (2) they interact physically with DNA.[5] This controversy remains to be resolved although the weight of present evidence favors the former alternative.

B. Acridine Derivatives

Comparatively few simple acridine derivatives have been shown to induce cancer. Boyland reported that feeding acridine orange (3,6-bis-dimethyl-aminoacridine hydrochloride) in the diet led to malignant hepatomas and cholangiomas in all of 14 rats surviving more than 402 days.[6] He suggested that this might be due to the intercalation of the dye with DNA,[7] but it should also be noticed that this substance is an aromatic amine derivative that may undergo metabolic activation in the liver.[2] However, Van Duuren and his colleagues tested acridine orange for carcinogenic activity in mouse skin and by injection, after solution in tricaprylin, into the subcutaneous tissues of rats and mice.[8] On painting, the chemical had no effect on mouse skin but yielded 3 liver tumors in 20 mice. On injection, a single local sarcoma was obtained in both rats and mice. The reason for the comparative failure of the latter experiments to induce a significant yield of tumors may be in the smaller amounts of acridine orange presented to the animals by painting or local injection than by administration in the diet. Proflavine (3,6-diaminoacridine) hemisulfate on intradermal injection in ulcerative doses promoted carcinogenesis in mouse skin, which had previously been treated with subcarcinogenic doses of 7,12-dimethylbenz[a]anthracene.[9] On the other hand, Trainin, Kay, and Berenblum failed to demonstrate the initiation of skin tumors by intraperitoneal injection of acridine yellow (3,6-diamino-2,7-dimethylacridine hydrochloride) or acridine orange at levels approaching the median lethal dose.[10] Croton oil was used as the promoting agent. Repeated gastric instillations of 3,6-diamino-2,7-dimethylacridine induced mammary hyperplasia but not tumors in young female Sprague Dawley rats.[12] 9-Nitroacridine 10-oxide on subcutaneous injection in mice failed to induce local sarcomas, although its analogue 4-nitroquinoline 1-oxide was active under these conditions.[11]

C. Benzacridines and Dibenzacridines

Following the discovery[13] of the carcinogenic properties of dibenz[ah]-acridine and dibenz[aj]acridine (see p. 548 for nomenclature) in 1935, the majority of the systematic work on these substances has been carried out by Lacassagne and his colleagues.[3] Their review forms the foundation of this section because they reported testing 78 derivatives by skin painting and/or subcutaneous injection. For economic reasons, screening such large numbers of chemicals means that relatively few mice can be used for each compound. Therefore, comparison between the carcinognic potency of different chemicals is unreliable. Nevertheless, the results of the biological experiments may be roughly evaluated in terms of Iball's carcinogenic index:[14]

$$\text{carcinogenic index} = \frac{x}{NL} \times 100$$

where x is the number of animals bearing tumors, N the number of animals alive at the time of first tumor, and L the mean latent period in days. This gives an indication of relative carcinogenicity. Lacassagne found that repeated experiments generally gave similar values for this index.[3]

No benz[b]acridines have been reported to be carcinogenic; in general, benz[c]acridines are more potent than benz[a]acridines.

The apparent carcinogenic activity of certain compounds varied according to the test used. For example, 8,12-dimethyl-9-chlorobenz[a]acridine, 10-fluoro-9,12-dimethylbenz[a]acridine, 5,7,11-trimethylbenz[c]acridine, 7,8,11-trimethylbenz[c]acridine, 7,9,10-trimethylbenz[c]acridine, 7-methyl-9-ethylbenz[c]acridine and dibenz[ch]acridine were active against mouse skin on painting but failed to induce sarcomas on subcutaneous injection. Conversely, 7,11-dimethyl- and 7,10-dimethyl-benz[c]acridine induced a higher sarcoma than carcinoma index. The differences in the proportion of animals with tumors and the different induction times are, however, not impressive. These discrepancies are unlikely to be a result only of the small numbers of animals employed, as similar results have been reported following more extensive investigations in the polycyclic aromatic hydrocarbon series.[15] They illustrate the difficulties faced by those who wish to correlate fine differences in physical or chemical properties of carcinogens with their biological activity.

The most important point to emerge is that substitution on the *meso*-carbon atom (12 in benz[a]acridine, 7 in benz[c]acridine) appears to be almost essential for carcinogenic activity in the benzacridines but not in the dibenzacridines. In the benz[c]acridine series, every tested substance that was methylated at this position, with the exception of 5,7-dimethylbenz[c]acridine, was found to be carcinogenic.[3] The abnormal behavior of the latter was suggested to be caused by substitution on the phenanthrenoid bond (K region) and consequent steric hindrance militating against activity.

In general, substitution by large groups gave less carcinogenic compounds than substitution by small groups. Thus 7,11-dimethylbenz[c]acridine is more carcinogenic than 7-ethyl-11-methylbenz[c]acridine, and 10-fluoro-12-methyl-benz[a]acridine than 10-fluoro-12-ethylbenz[a]acridine. The only apparent exception, which Lacassagne et al. did not explain, was that 7-ethyl-9-methylbenz[c]acridine was carcinogenic, whereas 7-ethyl-9-fluorobenz[c]-acridine was not.[3]

It was suggested that substitution on the 10-carbon atom in benz[a]-acridine or the 9 position in benz[c]acridine (i.e., the position *para* to the nitrogen atom in the ring not carrying the angular benzene ring) was not of itself enough to make the molecule carcinogenic, but methyl- or fluoro-substitution at this position tended to augment the activity of derivatives that were already active. For example, 7,9-dimethylbenz[c]acridine was slightly more active than 7-methylbenz[c]acridine. If the number of animals used in the comparisons of carcinogenicity is considered, this correlation cannot be accepted as convincingly proved. Substitution at the ortho position to the nitrogen on the same ring (position 8 in benz[a]acridine and position 11 in benz[c]acridine) was suggested to diminish carcinogenic activity.

More recent work by Lacassagne and his collaborators showed that whereas hydrogenation of certain double bonds in derivatives of the benz[a]-pyrene series had little effect on the induction of sarcomas following subcutaneous injection, 1,2,3,4-tetrahydro-7-methyldibenz[*ah*]acridine and 1,2,3,4-tetrahydro-7-methyldibenz[*cj*]acridine were much less active in inducing sarcomas than the fully unsaturated 7-methyldibenz[*ah*]acridine and 7-methyldibenz[*cj*]acridine.[16] The same group remarked on the greater sensitivity of male mice to sarcoma induction by benzo[*a*]pyrene and dibenzo[*def,mno*]chrysene, but found that 7-methylbenz[*c*]acridine was able to induce sarcomas to a similar extent in both sexes.[17] These experiments were carried out at only one dose level in each case. It is, therefore, possible that the use of a smaller dose of the acridine would have permitted the influence of the sex of the animal to have been manifested. In a further communication, a series of benz[*a*]- and benz[*c*]-acridines were tested for their ability to induce sarcomas.[18] 7-Formyl-9-methylbenz[*c*]acridine and 7-formyl-11-methylbenz-[*c*]acridine were active and 10-aminobenz[*a*]acridine and 7-cyanobenz[*c*]-acridine possibly so. A further number of chemicals did not induce sarcomas.

Mashbits tested a series of benzacridine derivatives by injection into white mice.[19] He induced injection site tumors with 7-methylbenz[*c*]acridine and 7-methyldibenz[*ch*]acridine. Lung adenomas followed the injection of many other substances, but without a detailed account of their occurrence in mice that had been kept without treatment, their significance is extremely uncertain. Bergmann, Blum, and Haddow reported that tumors were induced by 5-fluoro-7,9-dimethylbenz[*c*]acridine and 5-fluoro-7,10-dimethylbenz[*c*]-acridine but did not report details of the test.[20]

The environmental importance of the carcinogenic dibenzacridines has been discussed.[21,22] Dibenz[*a,j*]- and dibenz[*a,h*]-acridine have been reported to occur in cigarette smoke in concentrations of 0.27 and 0.01 μg per 100 cigarettes smoked. These dibenzacridines can be formed by the pyrolysis of nicotine or pyridine at 750°C. Traces of dibenzacridines have also been reported in the urban atmosphere.

The most important and chemically unusual exposure to benzacridine carcinogens, suggested by Hakim,[23] concerns the use of argemone oil to adulterate vegetable cooking oils in India. This oil is derived from the seeds of the poppy *Argemone Mexicana Linn*, which was introduced into India about 250 years ago. It contains the alkaloid sanguinarine (**1**), which is reported to be metabolized by a series of, as yet, unproved stages to benz[*c*]-acridine (**2**).[24]

1   2

The metabolite was identified by the comparison of its uv and fluorescent spectra and its chromatographic and electrophoretic properties with genuine benz[c]acridine. This is a subject that requires further investigation. Hakim,[23] using much more satisfactory numbers of mice than previously employed,[3] showed that benz[c]acridine was weakly carcinogenic to mouse skin with an Iball's carcinogenicity index of 6. He presented evidence in rats to suggest that it was a directly acting bladder carcinogen by use of the bladder implantation test,[25] although this test has been considered unsatisfactory in rats.[26]

The substitution of further heteroatoms in the benzacridines and dibenzacridines may or may not lead to carcinogenic compounds. The most interesting of these from the viewpoint of carcinogenesis are the tricycloquinazoline derivatives, whose carcinogenic activity has been extensively examined by Baldwin and his colleagues.[27] As these substances are not strictly acridines, they will not be considered further.

D. Mechanism of Carcinogenic Action of Benzacridines

Robinson was the first to draw attention to the importance of the phenathrenoid bond, later called the K or Krebs region in the carcinogenic polycyclic aromatic hydrocarbons.[28] This type of bond occurs in both benz[a]- and benz[c]-acridines; following work initiated by Schmidt,[29] and Pullman,[30] attempts have been made to calculate the electron density of the bond and to correlate this parameter with the carcinogenicity of the molecule. Lacassagne et al., for example, indicated that the lesser carcinogenic activity of the benz[a]acridines (3), compared to the benz[c]acridines (4), might be caused by the nearness of the electron-attracting aza-atom in the [a] series, compared to the [c] series.[3]

3   4

Attempts to calculate the electron density at the K region at first suggested a correlation between the calculated value and the carcinogenic activity, but subsequent refinements in the method of calculation caused the difference in bond energy between inactive and active compounds to diminish from more than 0.01 electron to less than 0.001 electron.

The hypothesis of Dipple, Lawley and Brookes that hydrocarbon activation occurs at either a methyl substituent on the mesocarbon atom (5) or at the phenathrenoid bond (6) suggests that a study of the ease of conversion of the methyl substituent to a carbonium ion might be profitable.[31]

One of the consequences of such an activation, the combination of the activated molecule with cellular proteins, has been much less comprehensively investigated with benzacridines than with polycyclic aromatic hydrocarbons. Nevertheless, the potently carcinogenic acridines bind to mouse skin protein much more strongly than those that are weakly carcinogenic or apparently noncarcinogenic.[32] Similarly, the more potent a carcinogen, the more readily it competes with a different radio-labeled carcinogen for binding sites in mouse skin protein.[33]

Recently, effort has been devoted to showing that carcinogens are both electron acceptors and electron donors.[34-36] An extensive study in which the ability of a considerable number of carcinogens and noncarcinogens to form complexes with iodine, chloranil, trinitrobenzene, and acridine, and their photodynamic activity composed, failed to develop any significant correlations.[37] This approach to the prediction of carcinogenic activity is, therefore, not valid. This is unfortuaate because adequate carcinogenicity tests in animals may cost several thousands of dollars for each compound, and a chemical method of distinguishing carcinogens from noncarcinogens would be valuable.

There have been several other attempts to link the carcinogenic activity of benzacridines to their physical properties. The actual or calculated partition of any aromatic hydrocarbon or benzacridine between a nonpolar solvent and water is loosely correlated to its biological activity.[38] Carcinogens have greater partition coefficients in aqueous solutions of DNA than noncarcinogens.[39] The number of molecules of very potent benzacridine carcinogens in the asymmetric unit crystal tended to be greater than for weak carcinogens in the six examples quoted.[40] Molecular optical anisotropy, measured by Rayleigh-depolarized diffusion, appeared to be abnormal with carcinogenic hydrocarbons and benzacridines.[41] Similarly, attention has been drawn to the fact that there is a relation between carcinogenicity and the ability of benzacridines to absorb light[42] at a wavelength greater than 3545 Å. In a consideration of the electronic properties of a large series of methylated benzacridines, Chalvet and Sung suggested there is (1) a rough parallelism between the ionization potential at the phenanthrenoid bond and carcinogenicity, and (2) a better agreement between carcinogenic activity and the calculated amount of the carcinogen that is fixed to protein.[43] The significance of all these findings will be easier to appreciate when the carcinogenic reactions of benzacridines are understood at the molecular level.

2. Antineoplastic Agents

A. Introduction

The aim of cancer chemotherapy is the selective poisoning of tumor cells, while leaving the host relatively unharmed. Antibacterial chemotherapy has been successful because there are considerable differences in essential intermediary metabolism between bacterial and mammalian cells. Tumor cells, however, derive from the normal cells of the host, and metabolic differences are usually more subtle. The problem is made even more difficult because the properties and behavior of individual tumors may differ from each other. In consequence, the problems in finding successful antineoplastic agents are great.

There is no certain way in which to test antineoplastic agents. Three-stage testing schedules are often employed. Substances are generally examined first *in vitro* by determining whether they are toxic to cell cultures; then those that pass successfully through this screen are tried against transplanted tumors in animals. Only occasionally, because of the expense involved in maintaining large numbers of animals to the tumor-bearing age, is it possible to use tumors that have grown spontaneously in the primary hosts, despite the fact that this model is much closer to the situation in man.

After an extensive pharmacological investigation to ensure that the side effects of these drugs are not likely to be worse than the disease they are supposed to cure, the successful substances are subjected to clinical trial, and ultimately a very few may pass into more general clinical use. It is debatable whether the money invested in developing antineoplastic agents over the past 30 years has justified itself, except in the limited areas of the treatment of lymphoma and leukemia. Reference to a current prescribing handbook shows that no acridine derivative is recommended in the United Kingdom as a clinically acceptable antineoplastic agent.[44] This section is, therefore, concerned with past failures and present attempts to establish a clinically effective acridine antineoplastic agent.

There are several different reasons why acridine derivatives have been examined for antineoplastic activity. It was realized at quite an early stage that the carcinogenic polycyclic aromatic hydrocarbons were able to inhibit, as well as to evoke, tumors.[45] This was applied to carcinogenic and non-carcinogenic derivatives of acridines.[46, 47] Because of the general belief that cancer tissue proliferates more rapidly than normal tissue, a belief which is only a partial truth, efforts to attack cancer have centered around methods of killing dividing cells through the use of biological alkylating agents[48] or other substances which interfere with DNA metabolism.[49] A few substances with miscellaneous actions have also to be mentioned. There do not appear to be any acridine derivatives that act as metabolic antagonists against tumor cells by mimicking the structure of a naturally occurring intermediate in the essential metabolism of the cell nor do there appear to be any acridine-containing synthetic steroids with the ability to control tumor growth.

B. Carcinogenic and Other Acridine Derivatives

It was shown that a number of carcinogenic methylated benz[a]acridines inhibited the growth of the Walker 256 carcinoma.[50] Using either transplanted mouse sarcomas or mammary tumors, Lewis and Goland investigated 331 acridine derivatives, of which 26 did not possess an amino group and 35 were diamino compounds.[47] The compounds without an amino group did not stain the tumor or retard its growth. Sixteen of the compounds with one or more amino groups, including atebrin (see Chapter II, p. 125), inhibited tumor growth to 2.5–5.0% of the normal level.

The most extensive work on the antineoplastic properties of the acridines is due to Ledochowski and his collaborators. Their compounds were substituted in the *meso* position with dialkylaminoalkylamino groups, and in other positions with methyl, methoxy, and halogen.[51–53] The inhibition of growth of sarcoma 180 was used as the primary screen and inhibition of sarcoma 37, Ehrlich ascites carcinoma, ascitic lymphoma, and Walker carcinoma 256 as secondary screens. In the course of a large synthetic program, they showed that 9 of 151 compounds were active against the primary screen. These were characterized by a methoxy, but not ethoxy, group in position 7 in **7**, usually dimethylaminobutylamino but sometimes dimethyl-amino-propyl or -ethylamino in position 9, and bromine but usually not chlorine in position 1 or 3.

7

The extension of this work led to investigations of the tumor-inhibiting properties of nitroacridines against *in vitro* systems.[54] 1-Nitro- and 3-nitro-acridines were most effective. The most promising of these substances (7, 9-dimethylaminopropylamino-1-nitro and 9-dimethylaminobutylamino-1-nitro), were investigated more fully, and the former was proposed for clinical trial. Both substances were active against sarcoma 180, Ehrlich sarcoma, exudative leukaemia; the former was active by vascular perfusion against Walker 256 carcinoma.[55]

It is suggested that 7 (9-dimethylaminopropylamino-1-nitro) acts in a similar manner to proflavine by forming complexes with DNA and thus inhibiting RNA synthesis.[56] It does not resemble the antineoplastic, biological alkylating agents that inhibit the growth of the sensitive form of the Yoshida sarcoma to a much greater extent than the resistant form. This acridine derivative is significantly active against the alkylating agent-resistant but not the sensitive solid form of this tumor. On the other hand, when the ascites form of the tumor was used, the alkylating agent-sensitive form was slightly inhibited but not the resistant form.[57]

Cain, Atwell, and Seelye suggested that the group of coplanar fully ionized cationic agents, which demonstrate high activity against the L1210 leukemia, may lodge temporarily in the minor groove of the polynucleotide helix.[49] A subsequent intercalated form of binding leads to cell death. From a consideration of these sequential binding sites, a number of nonquaternary bis-bases active against the L1210 have been prepared (8–12), in contrast to the inactive derivatives based on 13.

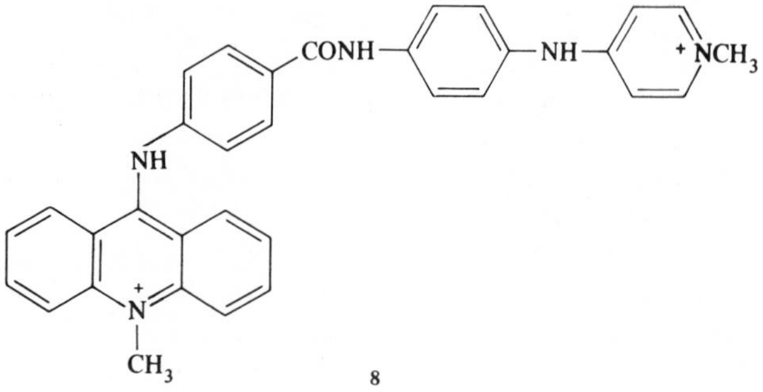

8

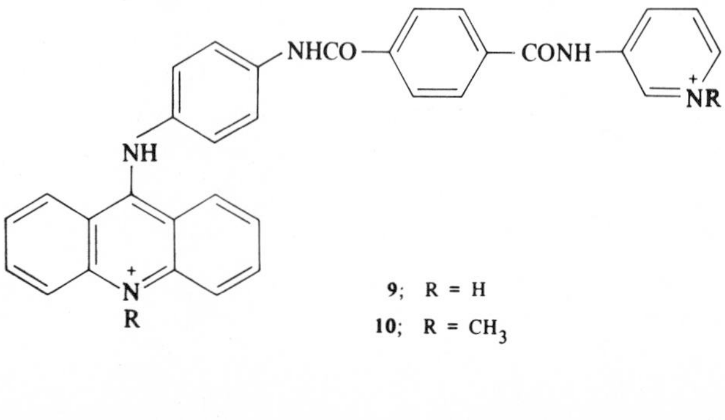

9;   R = H
10;   R = CH₃

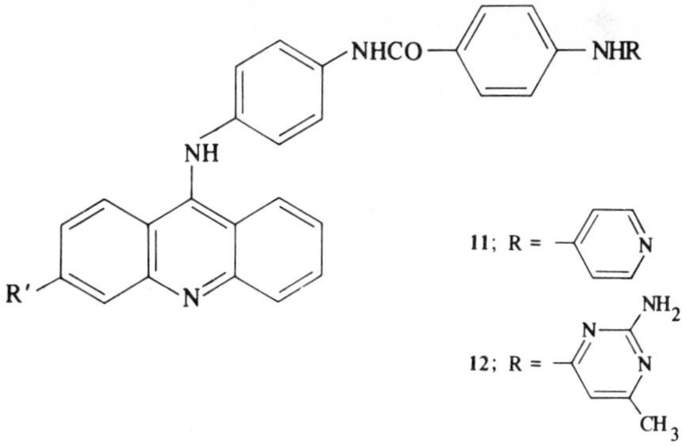

11; R = (structure)   R' = NO$_2$

12; R = (structure)   R' = H

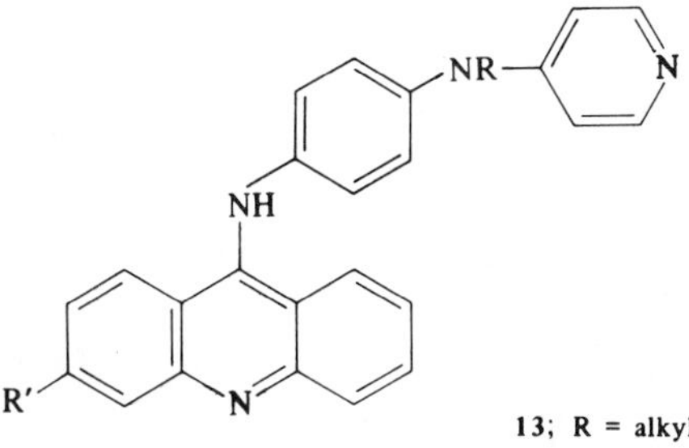

13; R = alkyl   R' = H

It was deduced that to permit it to enter the cell, the molecule must be capable of existing in a nonionic form at pH 7.3, and for antileukemic activity there should be a separation between the positive charges of 20 Å to permit it to form ionic bonds with phosphate groups in DNA.

Moszew and Adamczyk showed that analogues of the aza-substituted acridine **14** were active against Ehrlich ascites tumor *in vivo*.[58]

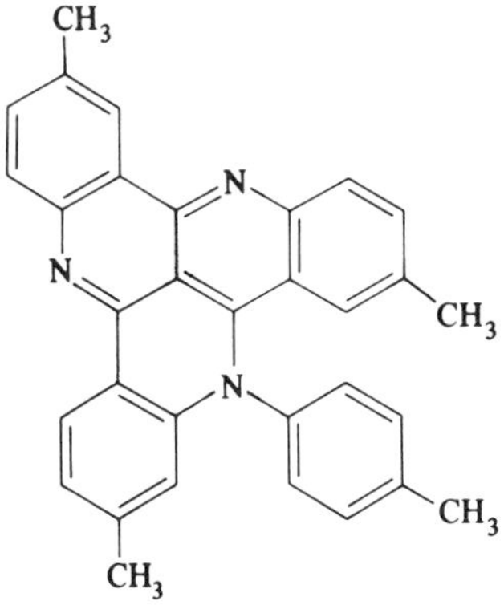

14

Shortridge, Turner, and Green found 7-aminobenz[c]acridine to be active against the Furth rat leukemia, but 21 analogues were inactive.[59] They suggested that the correct degree of ionization, intramolecular charge distribution, molecular shape, and surface area were necessary for activity.

C. Biological Alkylating Agents

Peck, O'Connell, and Creech examined the effect on the Ehrlich ascites tumor in mice of a variety of heterocyclic nuclei linked to monofunctional sulfur mustard or nitrogen mustard through an aminoalkyl side chain.[48] The sulfur mustard derivatives were generally more effective. Compounds **15, 16,** and **17** were among those found to be active. The biological side of this work was particularly well done, between 100 and 200 mice being used for each chemical.

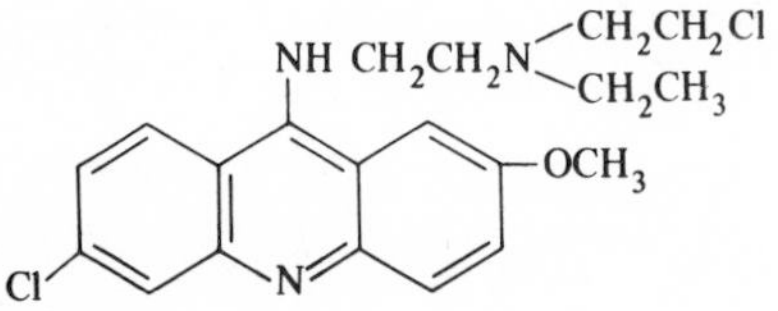

15

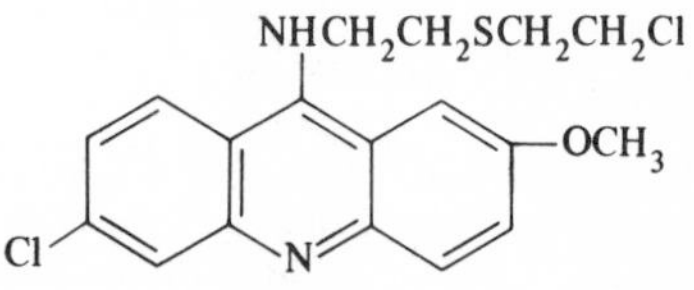

16

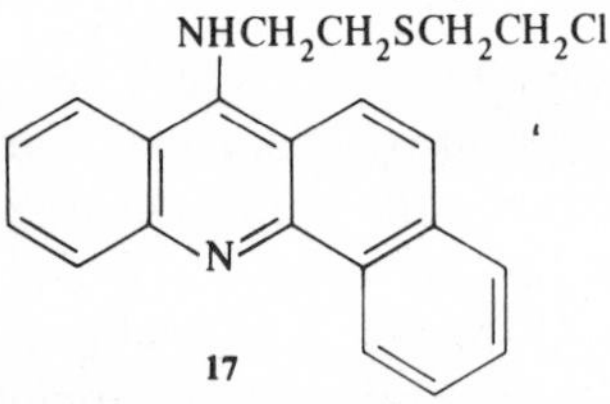

17

D. Miscellaneous Compounds

Davis and Soloway[60] described the preparation of a series of boron-containing acridines, which were tested for their ability to act as neutron-capture agents and to localize in tumors, for use in the neutron radiotherapy of cancer. Limited success was obtained with compounds of the nature of **18.**

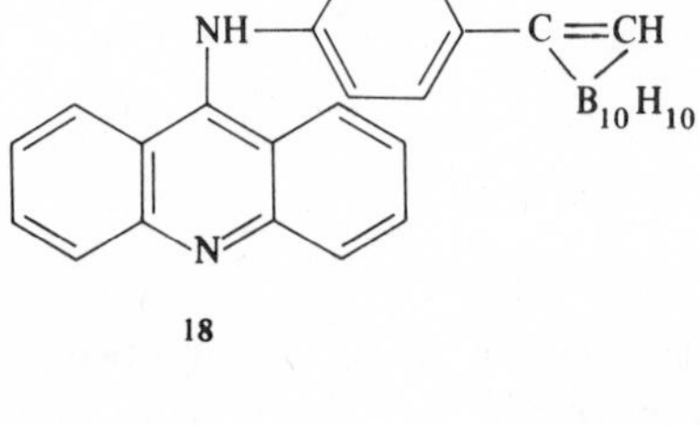

18

References

1. D. B. Clayson, *Chemical Carcinogenesis*, Churchill, London, 1962.
2. J. A. Miller, *Cancer Res.*, **30**, 559 (1970).
3. A. Lacassagne, N. P. Buü-Hoï, R. Daudel, and F. Zajdela, *Advan. Cancer Res.*, **4**, 315 (1956).
4. P. Brookes, *Biochem. Pharmacol.*, **20**, 999 (1971).
5. H. Marquandt, A. Bendich, F. S. Philips, and D. Hoffmann, *Chem.-Biol. Interact.*, **3**, 1 (1971).
6. E. Boyland, *Proc. Roy. Soc. Med.*, **60**, 93 (1967).
7. S. F. Mason, and A. J. McCafferty, *Nature*, **204**, 465 (1964).
8. B. L. van Duuren, A. Sivak, C. Katz, and S. Melchionne, *Brit. J. Cancer*, **23**, 587 (1969).
9. M. H. Salaman, and O. M. Glendenning, *Brit. J. Cancer*, **11**, 434 (1957).
10. N. Trainin, A. M. Kay, and I. Berenblum, *Biochem. Pharmacol.*, **13**, 263 (1964).
11. W. Nakahara, Arzneim. Forsch. **14**, 482 (1964).
12. D. P. Griswold, A. E. Casey, E. K. Weisburger, and J. H. Weisburger, *Cancer Res.*, **28**, 924 (1968).
13. G. Barry et al., *Proc. Roy. Soc., Ser. B*, **117**, 318 (1935).
14. J. Iball, *Amer. J. Cancer*, **35**, 188 (1939).
15. M. J. Shear, *Amer. J. Cancer*, **33**, 499 (1938).
16. A. Lacassagne, N. P. Buü-Hoï, F. Zajdela, and P. Jacquignon, *C. R., Acad. Sci., Paris*, **251**, 1322 (1960).
17. A. Lacassagne, F. Zajdela, and N. P. Buü-Hoï, *C. R. Soc. Biol.*, **152**, 1312 (1958).
18. A. Lacassagne et al., *C. R. Acad. Sci., Paris, Ser. D*, **267**, 981 (1968).
19. F. D. Mashbits, *Vop. Onkol.*, **1**, 52 (1955); *Chem. Abstr.*, **51**, 2999 (1957).
20. E. D. Bergmann, J. Blum, and A. Haddow, *Nature*, **200**, 480 (1963).
21. B. L. van Duuren, J. A. Bilbao, and C. A. Joseph, *J. Nat. Cancer Inst.*, **25**, 53 (1960).
22. E. Sawicki, G. P. McPherson, T. W. Stanley, J. Melker, and W. C. Elbert *Air Water Pollution* **9**, 515 (1965).
23. S. A. E. Hakim, *Indian J. Cancer*, **5**, 183 (1968).
24. S. A. E. Hakim, V. Mijovic, and J. Walker, *Nature*, **189**, 201 (1961).
25. J. W. Jull, *Brit. J. Cancer*, **5**, 328 (1951).
26. G. M. Bonser, D. B. Clayson, J. W. Jull, and L. N. Pyrah, *Brit. J. Cancer*, **7**, 456 (1953).
27. R. W. Baldwin, G. J. Cummingham, A. T. Davey, M. W. Partridge, and H. J. Vipond, *Brit. J. Cancer*, **17**, 226 (1963).
28. R. Robinson, *Brit. Med. J.*, **(i)**, 943 (1946).
29. O. Schmidt, *Naturwissenschaften*, **29**, 146 (1941).
30. A. Pullman and B. Pullman, *Advanc. Cancer Res.*, **3**, 117 (1955).
31. A. Dipple, P. D. Lawley, and P. Brookes, *Eur. J. Cancer*, **4**, 493 (1968).
32. P. Daudel et al., *Bull. Soc. Chem. Biol.*, **42**, 135 (1960).
33. P. Daudel, G. Prodi, A. Fabel, and M. C. Morniche, *C. R. Acad. Sci., Paris*, **253**, 593 (1961).
34. A. C. Allison, M. E. Peover, and T. A. Gough, *Nature* **197**, 764 (1963).
35. E. Szent-Gyorgui, I. Isenberg, and S. L. Baird, *Proc. Nat. Acad. Sci. U.S.*, **46**, 1444 (1960).
36. A. Szent-Gyorgyi and J. McLaughlin, *Proc. Nat. Acad. Sci. U.S.*, **47**, 1397 (1961).
37. S. S. Epstein, I. Bulon, J. Koplan, M. Small, and L. Mantel, *Nature*, **204**, 750 (1964).
38. C. Hansch and T. Fujita, *J. Amer. Chem. Soc.*, **86**, 1616 (1964).

39. J. Brigando, *Bull. Soc. Chim. Fr.*, 1797 (1956).
40. R. Mason, *Naturwissenschaften*, **43**, 252 (1956).
41. P. Bothorel et al., *Bull. Soc. Chim. Fr.*, 2920 (1966).
42. S-S. Sung, *C. R. Acad. Sci.*, *Paris*, **257**, 1425 (1963).
43. O. Chalvet and S-S. Sung, *C. R. Acad. Sci.. Paris* **251**, 2092 (1960).
44. F. J. Wilson, Ed., *MIMS Annual Compendium* 1971 Haymarket Publishing Co., London, 1971, p. 92.
45. A. Haddow, and A. M. Robinson, *Proc. Roy. Soc.*, *Ser. B*, **122**, 442 (1937).
46. N. P. Buü-Hoï, A. Lacassagne, J. Lecocq, and G. Fudali, *C. R. Soc. Biol.*, **139**, 485 (1945).
47. M. R. Lewis and P. P. Goland, *Amer. J. Med. Sci.*, **215**, 282 (1948).
48. R. M. Peck, A. P. O'Connell, and H. J. Creech, *J. Med. Chem.*, **9**, 217 (1966).
49. B. F. Cain, G. J. Atwell, and R. N. Seelye, *J. Med. Chem.*, **14**, 311 (1971).
50. G. M. Badger, L. A. Elson, A. Haddow, C. L. Hewett, and A. M. Robinson, *Proc. Roy. Soc. Ser.*, *B*, **130**, 255 (1942).
51. C. Radzikowski, Z. Ledochowski, and A. Ledochowski, *Acta Unio Int. Cancer*, **187**, 222 (1962).
52. Z. Ledochowski, A. Ledochowski, and C. Radzikowski, *Acta Unio Int. Cancer*, **20**, 122 (1964).
53. C. Radzikowski et al., *Arch. Immunol. Ther. Exp.*, **15**, 241 (1967).
54. J. Konopa, A. Ledochowski, A. Matuszkiewicz, and E. Jereczek-Morawska, *Neoplasma*, **16**, 171 (1969).
55. C. Radzikowski et al., *Arch. Immunol. Ther. Exp.*, **15**, 126 (1967).
56. L. S. Lerman, *J. Cell Comp. Physiol.*, **64**, 1 (1964).
57. C. Kwasniewska-Rokicinska and A. Winkler, *Arch. Immunol. Ther. Exp.*, **17**, 376, (1969).
58. J. Moszew and B. Adamczyk, *Zesw. Nauk. Uniw. Jagiellon Pr. Chem.* **11**, 27 (1966); *Chem. Abstr.*, **68**, 105048 (1968).
59. D. Shortridge, R. Turner and H. N. Green, *Brit. J. Cancer*, **23**, 825 (1969).
60. M. A. Davis and A. H. Soloway, *J. Med. Chem.*, **10**, 730 (1967).

CHAPTER XVIII

Acridine Antimalarials

DAVID W. HENRY

*Department of Bio-Organic Chemistry, Stanford Research Institute,
Menlo Park, California*

1. Introduction

Research on acridine antimalarials has now spanned over 40 years, rendering it one of the more venerable topics intertwining chemistry with medicine. As a result, the subject has been reviewed a number of times through the years.[1-6] To avoid repetition, this chapter discusses in detail only the more recent developments in the field. For a broader discussion and more complete referencing to the earlier literature on this subject, the reader is referred especially to the excellent review by Albert.[4]

In addition to acridines associated with malaria, this chapter includes acridines with inhibitory properties against some other diseases that are not noted elsewhere. This broadening of the subject is easily justified, as the major acridine drug, quinacrine [4, originally called atebrin and systematically named 6-chloro-9-(4-diethylamino-1-methylbutylamino)-2-methoxyacridine], possesses a variety of useful biological attributes in addition to its prominent antimalarial properties. For example, a class of potent antiamebic drugs are closely related to quinacrine, and this drug has also served as the starting point for an extensive synthetic program directed at antitumor drugs.

829

The latter topic has been considered in Chapter XVII. Where pertinent, benzologs, isosteres, or other compounds closely related to acridines are also included in the discussion.

The initial surge in research on acridines as antimalarial agents occurred in the early 1930s. This work was initiated in the laboratories of I. G. Farbenindustrie where the search for a synthetic substitute for quinine (1) had been proceeding for some years. The starting point for these investigations was the early observation by Ehrlich that methylene blue (2), a phenothiazine dye, exhibited some antiplasmodial activity.[7] These early efforts led initially to pamaquine (3), which was introduced in 1926 and was the first representative of the still important, causally prophylactic, 8-aminoquinoline group of antimalarial drugs.[8] Because of toxicity and the relatively low degree of effectiveness of pamaquine against the acute stage of malaria infections, research continued[9, 10] and led to the preparation of the acridine derivative, quinacrine (4), in 1930. A key feature of this research was the discovery of the importance of a basic side chain on the aromatic nucleus. This structural moiety, no doubt initially investigated because of its presence in quinine, has provided a dominant theme in antimalarial drug research since that time.[5, 9]

Although quinacrine had been available for some years, its status as a clinically useful drug was uncertain in the Allied countries at the beginning of World War II. During this conflict, because it involved many malarious areas of the world, large amounts of antimalarial drugs were in demand at a time when quinine supplies were denied to the Allied countries by the hostilities. After initial uncertainties about the toxicity of quinacrine were dispelled, it became the mainstay of malaria prophylaxis and treatment throughout the war.[11, 12] Although quinacrine was satisfactory, it was far from ideal, and a massive search for an improved agent began in both the United States and England during and immediately after World War II.[12-14] Among the several drugs that evolved from this effort, chloroquine (5), whose close relationship to quinacrine is apparent, had the greatest impact.

1

2

$$CH_3$$
$$HNCH(CH_2)_3NEt_2$$

MeO

N

NHCH(CH$_2$)$_3$NEt$_2$

3 CH$_3$

MeO

N

Cl

4

$$CH_3$$
$$HNCH(CH_2)_3NEt_2$$

Cl

N

5

Chloroquine proved to be more potent and less expensive and to exact fewer side effects than quinacrine. As a result, it largely replaced quinacrine as the drug of choice for treatment and prophylaxis during the 1950s.

At this point, many believed that a combination of appropriate public health measures with the effective drugs and residual insecticides that had become available during and immediately following the war years could lead to total eradication of malaria. Indeed, a World Health Organization program directed toward that end has made great strides.[15-17] However, a side effect of this attitude was a simultaneous, almost complete collapse of all phases of malaria research.[18] Thus the 1961 discovery of a strain of *Plasmodium falciparum* in Colombia that resisted nontoxic doses of chloroquine was an unpleasant surprise. (See Ref. 19 and references therein.) Subsequent experience, especially in Southeast Asia, revealed strains of falciparum malaria resistant to all available drugs.[19] Thus the same problem that shattered the complacency surrounding the availability of penicillin for bacterial infections arose with antimalarial therapy as well. As a result of this recurring threat, malaria research experienced a renaissance during the mid-1960s, with special emphasis on drugs to combat chloroquine-resistant *P. falciparum*. This effort is continuing and, among many other results, has reinstated synthetic research devoted to acridine-derived antimalarials.

2. Biological Aspects*

Malaria is caused by infection with a protozoan parasite from the genus *Plasmodium.* Four species—*P. falciparum, P. vivax, P. malariae,* and *P. ovale*—

*For a more detailed discussion of the medical and biological aspects of malaria, see Ref. 20.

are disease agents in man, but many other species infect a variety of other vertebrate hosts. The plasmodia are characterized by a complex life cycle that requires alternate passage through a vertebrate species and an intermediate mosquito host of the genus *Anopheles*. An abbreviated version of the cycle is presented in Figure 1. This cycle is of considerable importance in considering malaria chemotherapy because individual drugs typically affect the various stages differentially. The mosquito ingests male and female forms of the parasite, the gametocytes, while taking a blood meal from an infected vertebrate host. Fertilization takes place in the mosquito gut, and the result of this union eventually matures to an oocyst. The oocyst, upon rupturing, releases large number of sporozoites; these migrate to the salivary glands of the mosquito and are injected into the vertebrate host during feeding. The sporozoites quickly infect liver cells and undergo preerythrocytic development.

The end result of the development stage is the release of greatly increased numbers of plasmodia, now in the form of merozoites. These enter the bloodstream, where they individually penetrate red blood cells to form trophozoites. The trophozoites are asexual multiplicative forms that mature into structures (schizonts) containing greatly increased numbers of merozoites. At maturity, the infected red cell ruptures and the newly released merozoites invade additional red blood cells, and the cycle is repeated. The coordinated release of merozoites into the blood at characteristic intervals is responsible for the onset of the periodic fever and chills for which malaria is so well known. During the blood phase of the disease, a small fraction of the merozoites are diverted from asexual multiplication and develop into male and female gametocytes, thus completing the parasite life cycle.

Some species of plasmodia, notably *P. vivax* and *P. malariae* among the

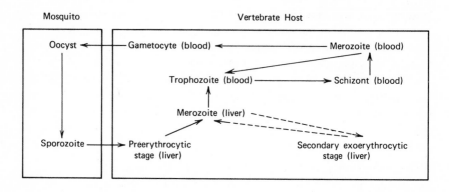

Fig. 1. Simplified plasmodial life cycle.

human parasites, persist in a dormant form after the acute phase of the illness has passed. This is due to reinvasion of liver cells by merozoites released during the preerythrocytic stage. The resulting secondary exoerythrocytic forms can remain inactive until conditions are suitable for reoccurrence of the disease, typically when natural immunity has subsided or suppressive drug treatment is stopped.[21]

The diseases caused by *P. vivax* and *P. falciparum* are the most common throughout the world. Except for *P. malariae*, for which the incubation period is about 1 month, the onset of symptoms occurs about 2 weeks after infection. Initial manifestations may be headache, lassitude, or nausea. This may be followed by the classical paroxysm, progressing through chills and shivering, pallor, gradual increase of body temperature to 104–106°F (40–41°C), profuse sweating, and finally abrupt return of the temperature to normal or subnormal. Without treatment, this is repeated more or less regularly, but with lessening severity, for several weeks or longer. The intervals between attacks vary from 1 to 3 days, depending on the infecting species. This pattern, although common, is not universal, and individual cases, especially with *P. falciparum*, may differ markedly. Of the four species of malaria parasitic in man, only *P. falciparum* is normally considered to be life-threatening. In infections with this organism, up to 10% of the red blood cells may be parasitized in severe cases; a 2% level of parasitemia is unusual for the other species.[20]

The asexual blood forms of the parasite are the most easily influenced by drug treatment. All types of drugs in clinical use inhibit growth at this point of the life cycle, although they vary widely in the rate at which the effect is demonstrable. Quinine, quinacrine, and especially chloroquine act rapidly and are therefore valuable in treating acute attacks; taken prophylactically, they also are (or have been) widely used to suppress the disease in endemic areas of the world. However, as they have no effect on the preerythrocytic or secondary exoerythrocytic phases, the cessation of treatment may result in the reoccurrence of overt disease. The 8-aminoquinolines, i.e., pamaquine and its descendents, are the only class of drug capable of destroying these tissue forms of the plasmodia.[19, 20]

A great many methods for evaluating potential antimalarial drugs have been developed over the years. Until about 1950, almost all primary screening studies employed avian species of plasmodia in various bird hosts. Pamaquine and quinacrine were developed on the basis of the Roehl assay, the first test system that provided comparative information for a series of drug candidates.[22] This assay employed *P. relictum* or *P. cathemerium* in canaries; subsequent systems employed *P. gallinaceum* in chicks and *P. lophurae* in ducks.[23] Wiselogle lists 38 suppressive tests employing these host-parasite combinations that were used in the U.S. World War II Malaria Program.[12] In most of these tests, a depressed level of blood parasitemia, compared with untreated controls, indicates antimalarial properties for a candidate drug. Not surprisingly, assays employing parasites and hosts only distantly related to the human disease were less than perfect as predictors of drug activity in man. The discovery in 1948 of *P. berghei*,[24] a parasite of small rodents from Central Africa, provided a better model.[25] This organism, which adapted readily to common laboratory rodents, has largely displaced the avian species of plasmodia for many types of experimental studies. In addition to being a better biological model, *P. berghei* infections in the mouse do not require time-consuming blood examinations to ascertain drug activity. For example, an increase in survival time of treated animals over that of untreated controls provides an accurate assessment in the assay used by Rane in the current U.S. Army Research Program on Malaria.[26] Another feature of *P. berghei* that has proved very valuable is its susceptibility to the induction of drug resistance. As reviewed by Schmidt,[19] this organism relatively easily acquires resistance to chloroquine, with simultaneous cross resistance to quinacrine and, to a lesser extent, to quinine. Many earlier attempts to develop chloroquine-resistant strains of avian or simian plasmodia were essentially unsuccessful. The cross-resistance phenomenon displayed by chloroquine-resistant *P. berghei* is also characteristic of drug-resistant *P. falciparum* infections in man. The use of *P. berghei* in chemotherapeutic studies has been recently summarized.[27]

Secondary examination of promising drugs from primary screening systems has often employed the simian species, *P. cynomolgi* and *P. knowlesi*, in conjunction with appropriate monkey hosts.[23, 28] Both therapeutic and prophylactic studies are possible in these systems. *P. knowlesi*, in addition to its major role as an *in vivo* model of human malaria, has also been used in an automated *in vitro* test system that has promise as a mass screening technique.[29] Another area of important recent developments is the successful adaptation of human malaria species to simian hosts. For example, *P. falciparum* and *P. malariae* infections have been established in splenectomized chimpanzees, and all human species except *P. ovale* have been found adaptable to the owl monkey.[30] *P. falciparum* appears to be especially versatile; it has recently also been successfully introduced into Cebus monkeys[31] and gibbons.[32] The use of these new hosts for human plasmodia can be expected to substantially advance research on new drug development and other phases of malaria control.

3. Quinacrine, Structure-Activity Relationships

Up to and during World War II, major structure-activity studies on acridine antimalarials were undertaken in Germany, Russia, England, and

the United States. Although occasional compounds from each of these pro-
grams were as effective as quinacrine in experimental test systems or clinical
trials, none succeeded in displacing it. In general, all these efforts reached
similar conclusions regarding the parameters determining maximum potency. However, it is very difficult to make direct comparisons because the
groups commonly used different systems for assessing their drugs. As these
studies have recently been reviewed in depth,[4] only selected relationships will
be discussed here; for complete details, the reader is referred to the original
literature.[1-6]

In the most simple terms, acridine is converted to an antimalarial drug of
the quinacrine class by the insertion of a diamine side chain into the 9-position
to give structures encompassed by the general formula **6**.

6

When the 9-amino group bears an additional substituent, the compounds are
inactive. This has been ascribed to their chemical instability, as 9-dialkyl-
aminoacridines readily undergo hydrolysis to 9-acridanones.[3, 4, 6, 33] (See
also Chapter II, p. 124.) The substitution of oxygen or sulfur for nitrogen in
the 9-position results in derivatives completely devoid of activity.[34] Similarly,
deletion of the terminal amine[12] or replacement by another polar group
(e.g., hydroxyl) results in inactive compounds.[35]

The nature of Y, the linkage between the two amino functions in the side
chain, is probably the least demanding element of structure-activity relationships in this class of compounds. A chain length of two to nine carbon atoms
is consistent with significant activity, as illustrated by compounds **4**, **7**, and
8, shown in Table I. The chain may be entirely aliphatic or largely aromatic
(e.g., **9**). The replacement of carbon atoms in the chain by heteroatoms is
entirely consistent with activity, as in **10** and **11**. Cyclic structures (e.g., **12** and
13) also provide highly active compounds. Branches of various types are also
well tolerated in the Y segment of the side chain. The α-methyl branch
characteristic of the quinacrine side chain may be successfully omitted or
replaced with a variety of other groups (e.g., **14–20**) of varying steric and
polar characteristics. Moving the α-phenyl substituent of **18** to the β-position,
as in **21**, had no effect on activity. More polar branches are also acceptable;
hydroxy derivative **22** and triamine **23** possess significant antimalarial properties. Triamine **24**, however, was inactive. Hydroxy compound **22**, known as

TABLE I. Structure-Activity Relationships of Quinacrine Analogues: Side Chain Variants[a]

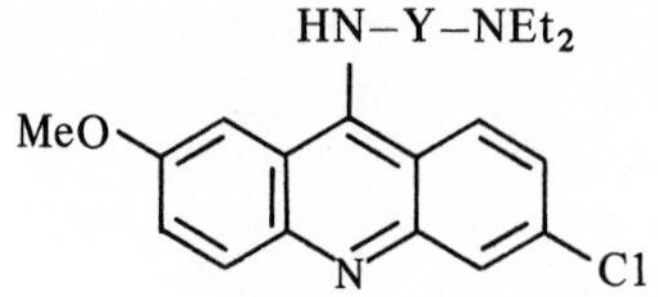

No.	Y	Anti-malarial activity[b]
4	CH_3 $-CHCH_2CH_2CH_2-$ (quinacrine)	3
7	CH_3 $-CHCH_2-$	1.5
8	CH_3 $-CH(CH_2)_7-$	2
9	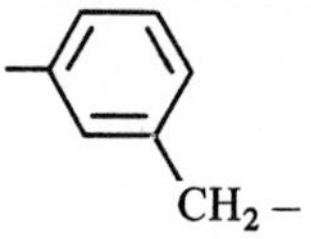	2
10	$-CH_2)_3-O-(CH_2)_2-$	0.6[c]
11	$-(CH_2)_2-S-(CH_2)_2-$	2
12		3
13	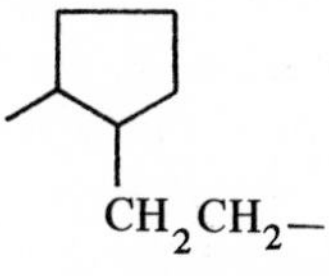	3
14	$-CH_2CH_2CH_2CH_2-$	2
15	CH_2CH_3 $-CHCH_2CH_2CH_2-$	3

(Table Continued)

Table I. (Continued)

No.	Y	Anti-malarial activity[b]
16	CH_3CHCH_3 $-\overset{|}{C}HCH_2CH_2CH_2-$	3
17	$(CH_2)_5CH_3$ $-\overset{|}{C}HCH_2CH_2CH_2-$	3
18	⬡ $-\overset{|}{C}HCH_2CH_2CH_2-$	3
19	$COOEt$ $-\overset{|}{C}HCH_2CH_2CH_2-$	1.5
20	⬡N $-\overset{|}{C}HCH_2CH_2CH_2-$	1.5
21	⬡ $-CH_2\overset{|}{C}HCH_2CH_2-$	3
22	OH $-CH_2\overset{|}{C}HCH_2-$ (acranil)	2
23	CH_2NEt_2 ⬡ CH_2-	2^c
24	CH_2NMe_2 $-\overset{|}{C}HCH_2NMe_2$	$<0.4^d$

[a]All examples taken from Ref. 12. [b]Expressed as quinine equivalents, the ratio of the minimum effective therapeutic does of quinine divided by that of the test drug. Unless otherwise noted, the test system is *P. lophurae* in the duck. [c]*P. gallinaceum* in the chick. [d]The "less than" symbol means that the drug was inactive up to the toxic dose.

acranil, has been used clinically to treat giardiasis (lambliasis), an intestinal disease caused by the protozoan parasite, *Giardia lamblia*.[36]

More recent work on the side chain by Elslager et al.[37, 38] has established that incorporation of the hydrazino or hydroxylamino moiety into the Y segment of the side chain, either at the proximal or distal end, is generally deleterious to antimalarial activity. However, some of the compounds prepared in these studies possessed significant antiamebic and anthelmintic properties.[37]

As in the Y segment of the side chain, a wide variety of structural features in the terminal *N*-substituents are compatible with substantial activity. Selected examples are given in Table II. The terminal amine may be primary (**25**), secondary (**26** snd **27**), or tertiary (**28** and **29**), although the majority of structure-activity work on this portion of the side chain has dealt with tertiary amines. In general, with aliphatic hydrocarbon substituents, a total of up to seven or eight carbon atoms on the terminal amine is compatible with significant activity. The substituents may be linear, branched, or cyclic, as illustrated by compounds **27–35**. Diethylamino appears optimum and was used most frequently by the majority of investigators, but the advantage is not always marked. Occasionally, small structural differences in the alkyl groups cause substantial effects on activity (e.g., compare **34** with **36** and **30** with **37**). As the total number of carbon atoms in the terminal aliphatic substituents reaches or exceeds eight, toxicity begins to appear (e.g., **38** and **39**). The acceptability of unsaturated hydrocarbon substituents (i.e., diallylamino) was established very early.[39, 40] Although relatively little investigation has been done on them, heteroatoms in the terminal groups are allowable under some circumstances but do not enhance activity (e.g., **40–44**). The attachment of almost any type of aromatic function to the terminal nitrogen is highly detrimental to activity, as illustrated by **45** and **46**. The only exception to this generalization is compound **47** in which the aromatic substituent is 6-methoxy-8-quinolyl; this aromatic moiety is the same as that from the pamaquine series of 8-aminoquinoline antimalarials and renders interpretation of the result difficult.

Approximately 50 different substituted acridine nuclei were investigated during the structure-activity studies associated with quinacrine-related antimalarials, and the majority provided active compounds in at least one assay system. Table III lists some examples of interest from early studies by the German group.[39] The completely unsubstituted nucleus (**49**) provided only marginal activity in these studies, using *P. relictum* and *P. cathemerium* in canaries. However, when evaluated against *P. lophurae* in the duck, compound **49** was approximately half as active as quinacrine.[12] Independent incorporation of the methoxy and chloro substituents of quinacrine (**50** and **51**) clearly indicates that the chlorine is providing most of the activity enhancement.

TABLE II. Structure-Activity Relationships of Quinacrine Analogues: Terminal Amine Variants[a]

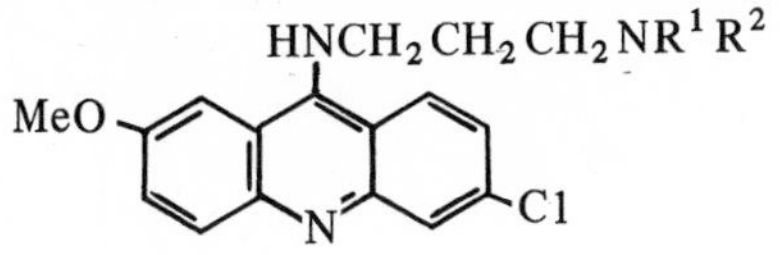

$$HNCH_2CH_2CH_2NR^1R^2$$

No.	$-NR^1R^2$	Antimalarial activity[b]
25	$-NH_2$	1.5
26	$-NHCH_3$	2
27	$-NHCH_2CH(CH_3)_2$	2
28	$-N\big(CH_2CH(CH_3)_2\big)_2$	3
29	$-N(CH_2CH_3)_2$	$3,4,1.0^d$
30	$-N(CH_2CH_2CH_3)_2$	2
31	$-N\big(CH_3\big)\big(CH(CH_2CH_2CH_3)_2\big)$	1
32	$-N\big(CH_3\big)\big(CHCH(CH_3)_2\ |\ CH_3\big)$	1
33	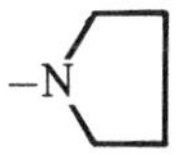 $-N$ (pyrrolidine)	1
34	$-N$ (2-methylpiperidine, CH_3)	2
35	$-N$ (cyclohexyl, CH_3)	2
36	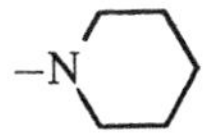 $-N$ (piperidine)	0.4

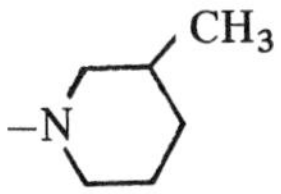

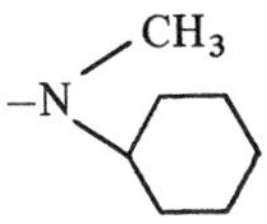

(Table Continued)

839

Table II. (Continued)

No.	$-NR^1R^2$	Anti-malarial activity[b]
37	$-N\begin{smallmatrix} CH_2CH_2CH_3 \\ CH(CH_3)_2 \end{smallmatrix}$	0.4
38	$-N(CH_2CH_2CH_2CH_3)_2$	$< 2^c$
39	$-N\begin{smallmatrix} CH_2CH_2CH(CH_3)_2 \\ CH_2CH_2CH(CH_3)_2 \end{smallmatrix}$	$<2^c$
40	$-N\begin{smallmatrix} CH_2CH_3 \\ CH_2CH_2OCH_3 \end{smallmatrix}$	1.0
41	$-N\begin{smallmatrix} CH_2CH_3 \\ CH_2CH_2OH \end{smallmatrix}$	$\overline{1.0}$
42	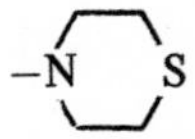 $-N\underset{\smile}{\frown}S$	0.2^e
43	$-NH(CH_2)_3N(CH_2CH_3)_2$	0.2^e
44	$-N\begin{smallmatrix} (CH_2)_3N(CH_2CH_3)_2 \\ (CH_2)_3N(CH_2CH_3)_2 \end{smallmatrix}$	0.6^e
45	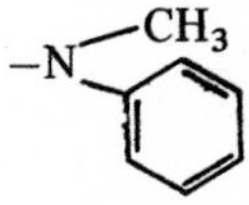 $-N\begin{smallmatrix} CH_3 \\ C_6H_5 \end{smallmatrix}$	0.02^f
46	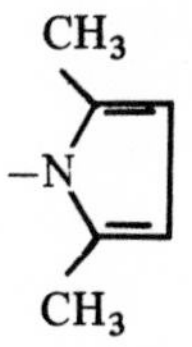 $-N$ (2,5-dimethylpyrrole) $CH_3 \ldots CH_3$	$<0.04^{c,f}$

(Table Continued)

Table II. (Continued)

No.	$-NR^1R^2$	Anti-malarial activity[b]
47	−NH− (quinoline with OMe substituent)	0.6, 0.8, 1.5[d]

[a] All examples taken from Ref. 12. [b] Expressed as quinine equivalents, the ratio of the minimum effective therapeutic dose of quinine divided by that of the test drug. Unless otherwise noted, the test system is *P. lophurae* in the duck. [c] The "less than" symbol means that the drug was inactive up to the toxic dose. [d] Results from three different drug regimens. [e] *P. cathemerium* in the canary. [f] *P-gallinaceum* in the chick.

Relatively few other monosubstituted nuclei have been reported, and only one provided more favorable activity than that of **51**; Topchiev and Bekhli[41] reported that the 1-chloro acridine nucleus gave a corresponding compound substantially more active than quinacrine.

Disubstituted nuclei were the most commonly prepared, and series **52–59** illustrates the effect of replacing the methoxy of quinacrine with various other substituents. The sequence of increasingly larger alkoxy groups (e.g., **55**, *n*-hexyloxy) provides compounds of the same potency as quinacrine, despite the pronounced steric and hydrophobic effects that would be expected to accompany their incorporation. The primary member of this series, the quinacrine derivative with methoxy cleaved to phenolic hydroxy, has been reported by Russian investigators to retain the activity of quinacrine but to have increased toxicity.[42] The deletion of the alkoxy and replacement by methylthio, methyl, or chloro (**57–59**) affected activity only slightly in this group.

A similar series of quinacrine analogues, in which the methoxy is retained and the 6-position substituent varied, is provided by **60–65**, as shown in Table III. It is apparent that the 6-chloro substituent has only a slight advantage over 6-fluoro, 6-bromo, and 6-methyl, and essentially has no advantage over 6-cyano. The corresponding iodo- and nitro-bearing drugs were clearly deactivated. The rationalization of these results in terms of the chemical or

TABLE III. Structure-Activity Relationships of Quinacrine Analogues Aromatic Substituent Variations[a]

$$CH_3$$
$$HNCHCH_2CH_2CH_2NEt_2$$

No.	Aromatic substituents	METD	TI
48	6-Cl-2-MeO (quinacrine)	0.33	30
49	None	5.0	1
50	2-MeO	2.5	2
51	3-Cl	0.33	15
52	6-Cl-2-EtO	0.67	8
53	6-Cl-2-iPrO	0.67	8
54	2-BuO-6-Cl	0.33	15
55	6-Cl-2-nC$_6$H$_{13}$O	0.33	15
56	6-Cl-2-nC$_{12}$H$_{25}$O	1.25	8
57	6-Cl-2-MeS	0.33	15
58	6-Cl-2-Me	0.33	15
59	2,6-Cl$_2$	0.67	15
60	6-F-2-MeO	0.67	15
61	6-Br-2-MeO	0.67	15
62	6-I-2-MeO	1.25	4
63	2-MeO-6-Me	0.67	15
64	2-MeO-6-NO$_2$	10.0	1
65	6-CN-2-MeO	0.3	23[b]
66	6-NH$_2$-2-MeO	–	–
67	7-NH$_2$-6-Cl-2-MeO (aminoacrichine)	–	–

[a] Data from Ref. 39, unless otherwise noted. Assay systems employed *P. relictum* or *P. cathemerium* in canaries. METD = minimum effective therapeutic dose in mg. TI = therapeutic index (maximum tolerated dose divided by METD).

[b] Data from Ref. 43.

electronic properties of the substituents has proved frustrating, as noted by Albert.[4]

The 6-cyano analogue (65) was sufficiently promising to merit clinical trials,[43] but the skin discoloration associated with its use proved unacceptable.[3] An unresolved conflict in this series of 2,6-disubstituted acridine nuclei is represented by 6-amino derivative 66; this drug was independently found inactive by one group[39] but fully active by another.[3] Numerous substitution patterns other than 2,6- were examined during studies on disubstituted, quinacrine-related drugs of this type, but none provided compounds so consistently active.

Comparatively few trisubstituted quinacrine congeners have been prepared, and only one provided interesting activity. This compound, aminoacrichine (67, Table III), proved as active as quinacrine and was significantly less toxic.[44, 45] Compound 67 has found clinical application in the U.S.S.R.[45]

The effect of partially reducing the acridine nucleus in quinacrine-type antimalarials was investigated by two groups[46—48]; both found that 1,2,3,4-tetrahydroacridines of general structure 68 were virtually without activity.

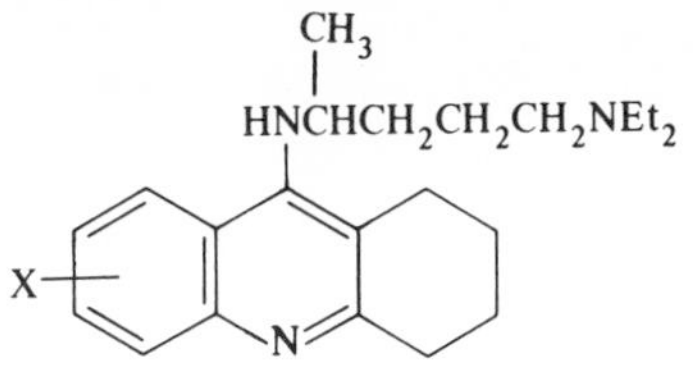

$$CH_3$$
$$HNCHCH_2CH_2CH_2NEt_2$$

68   X = Cl, NO_2, OMe

The benzologs are another type of nuclear variation that has received considerable attention. All three of the possible benzacridines[49—53] and four pyridoacridines[52, 54, 55] have been fitted with appropriate diamine side chains for examination as antimalarials. Benz[a]acridine[12, 49] and pyrido[2,3-a]-acridine[55] provided compounds (e.g., 69 and 70) with activity approaching that of quinacrine in avian test systems; the remaining heterocycles gave compounds with little or no antiplasmodial properties.

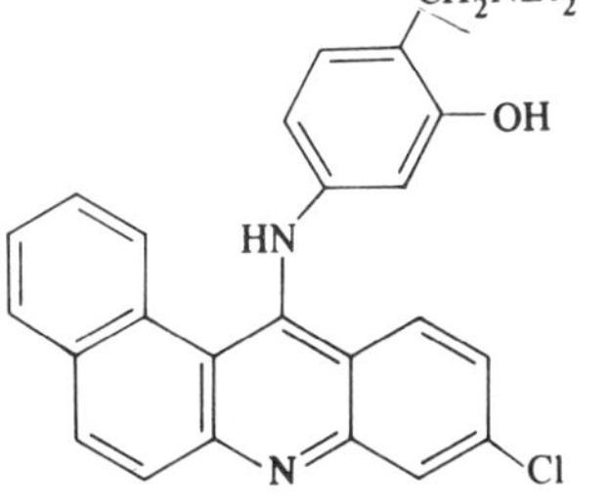

69

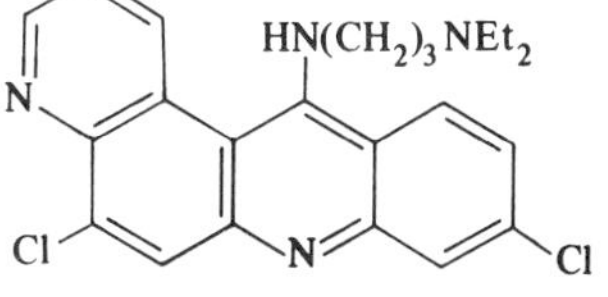

70

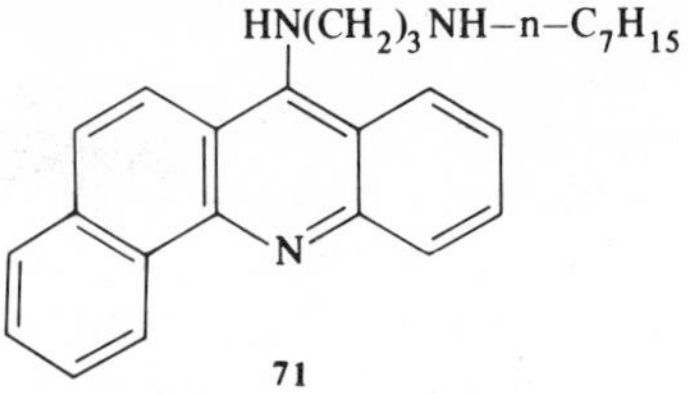

$$HN(CH_2)_3NH-n-C_7H_{15}$$

71

Although lacking antimalarial activity, the benz[c]acridines proved quite interesting as antiamebic agents. Elslager and collaborators have extensively investigated this area[56]; one derivative from this work, **71**, was found to be quite effective against intestinal and hepatic amebiasis in a clinical trial.[57]

Another type of nuclear substituent that proved advantageous with the 9-aminoacridines is the 10-oxide. Elslager et al. prepared an extended series of quinacrine analogues with this functional group and found many that were more active than their desoxy parents.[58] The most promising compound, the desmethoxy derivative **72**, was four times more potent than quinacrine in chicks having *P. lophurae* infections and twice as active in rhesus monkeys infected with *P. cynomolgi*.[59]

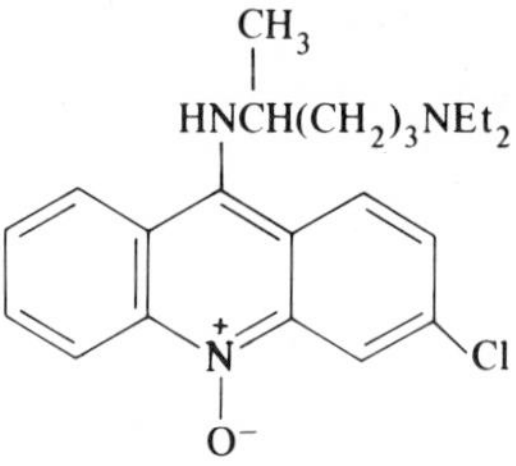

$$CH_3$$
$$HNCH(CH_2)_3NEt_2$$

72

Compound **72** also lacked the tissue-staining properties of quinacrine, a highly desirable improvement. In clinical trials against the human malarias, *P. falciparum* amd *P. malariae*, the activity of **72** was about equivalent to that of the widely used 4-aminoquinoline antimalarial, amodiaquine, and provided no apparent advantage.[60, 61] Oxidation of the 1-nitrogen atom of 4-aminoquinoline antimalarials is also reported to provide enhanced activity in experimental assay systems.[62]

The effect of substituting nitrogen for one of the ring carbons of the acridine moiety has been considered by several investigators. The initial effort in this area yielded azacrin (**73**), the 1-aza analogue of quinacrine. This drug, prepared independently in two laboratories, combines structural features of the causally prophylactic 8-aminoquinolines with those of the 9-aminoacridines.[63, 64] It proved to be approximately twice as active as quinacrine in the chick and mouse[64] and also displayed somewhat improved properties in the clinic, compared with quinacrine. However, it possessed no advantage over the 4-aminoquinolines.[65-67] A more recent study has revealed that oxidation of azacrin to the 5-oxide increases potency nearly fourfold against *P. berghei* in the mouse.[68] This same 5-oxide is also effective for treating experimentally induced filariasis in the gerbil.[68] A closely related isomeric aza-acridine derivative, **74**, possessed only marginal activity in the mouse-*P. berghei* system.[69] Another aza-acridine variant with the side chain in the 4-position (**75**) was completely inactive.[70]

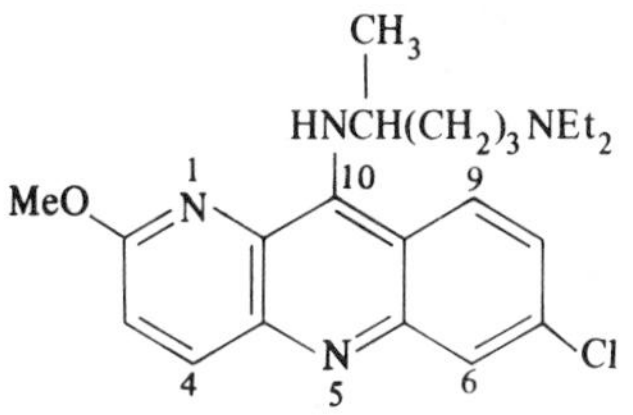

73

Azacrin

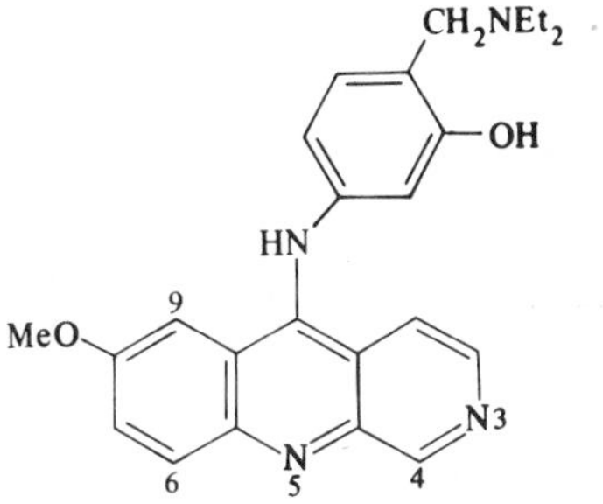

74

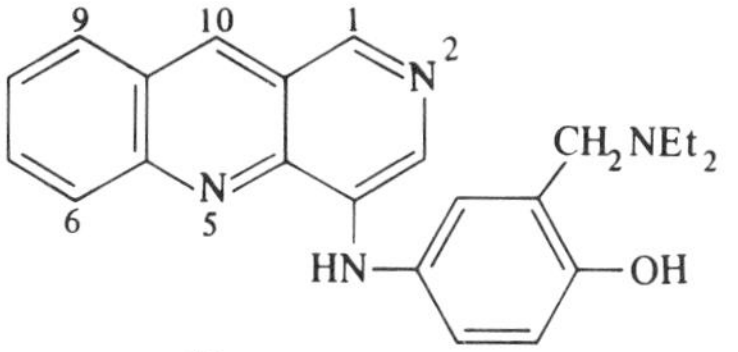

75

Elslager, Tendick, and Werbel have prepared a series of derivatives of quinacrine (e.g., **76**), chloroquine, and other antimalarial drugs designed to introduce repository antimalarial action to their respective classes.[71]

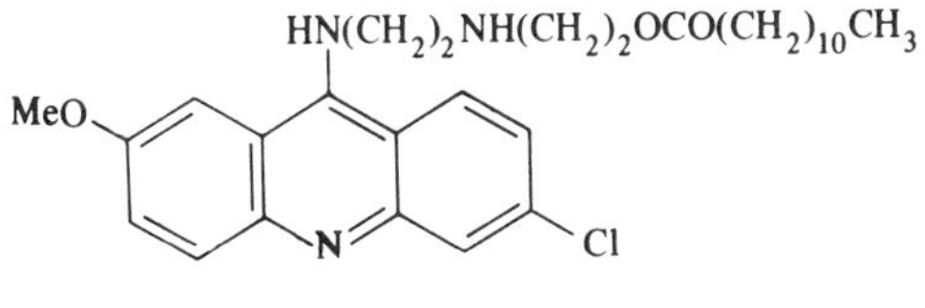

76

The stimulus for this work was the success of the repository drug, cycloguanil pamoate [4,6-diamino-1-(4-chlorophenyl)-1,2-dihydro-2,2-dimethyl-1,3,5-triazine 2:1 molar complex with 4,4'-methylenebis-(3-hydroxy-2-napthoic) acid], in protecting animals and man from plasmodial infection for periods of several months after a single injection.[5, 21] Unfortunately, the effort was unsuccessful in the acridine group.

4. Quinacrine, Pharmacological and Biochemical Aspects

For the treatment of acute clinical attacks of malaria, the usual dose of quinacrine is 200 mg every 4 to 6 hr for the first day, followed by 100 mg three times daily for 6 days.[72] For suppressive use in areas where malaria is endemic, 100 mg per day is the recommended dose, treatment being started 2 weeks in advance of exposure. Although a relatively benign drug, extensive use over many years has shown that quinacrine may provoke a variety of side effects. Generally, these involve the skin (yellow discoloration, pruritis), the gastrointestinal tract (nausea, vomiting, diarrhea, abdominal cramps), and the central nervous system (toxic psychosis).[1, 72] These effects, however, are transient, or reversible on cessation of treatment. Because of these difficulties, quinacrine has largely been superseded by other drugs for malaria treatment.[72] It is, however, still used as a remedy for tapeworm infestation.[73]

On ingestion, quinacrine is widely dispersed among the body tissues, especially on chronic administration when it progressively accumulates at sites such as the liver, spleen, lungs, and adrenal glands. The fingernails and hair also retain the drug; in the former, it may be detected by its brilliant, greenish yellow fluorescence under uv illumination.[72] The drug has a high affinity for body tissues and is rapidly adsorbed; after intravenous injection, for example, 95% of the drug leaves the circulatory system within 3 min.[1] Tissue concentrations may be thousands of times higher than that in the blood plasma. It is rapidly absorbed from the gut when taken orally but is excreted only slowly and is accompanied by a variety of metabolites.[1, 2, 4, 6]

Numerous speculations and hypotheses concerning the mode of action of quinacrine on plasmodia have been advanced during past years; these are discussed in some detail by Albert.[4] Most recently, attention has focused on the action of the drug on nucleic acid metabolism, and it appears that inhibition of DNA and/or RNA synthesis may be a primary effect. Kurnick and Radcliffe noted that quinacrine forms a molecular complex with DNA *in vitro*,[74] and subsequent work by Lerman[75] and others[76] has defined this complex as occurring through the very interesting intercalative binding mechanism (see Chapter XIV). *In vitro* enzyme studies have shown that quinacrine inhibits the DNA and RNA polymerases from *E. coli*[76, 77] and, in a parallel manner, suppresses DNA and RNA synthesis in the intact microorganism.[78] In work with *P. berghei*, Van Dyke and collaborators[79–81] have shown that the drug inhibits incorporation of radioactively labeled adenosine triphosphate into plasmodial RNA and, to a lesser extent, DNA. These investigators have also observed that intact parasite nuclei display the characteristic fluorescence associated with intercalative binding after parasitized erythrocytes are treated with quinacrine.[81] Quinacrine may also act by interfering with the absorption of nutrients essential to *P. berghei* metabolism, as it inhibits uptake of adenosine by infected erythrocytes.[81] Similarly, the inhibition of amino acid uptake by the drug has been demonstrated in another protozoan, *Tetrahymena pyriformis*.[82] In much of the cited work dealing with the mode of action of quinacrine, parallel experiments were performed with chloroquine and quinine. Although differences exist, especially with quinine, all these drugs seem to act by similar mechanisms.[76]

5. Quinine-related Acridine Antimalarials

Although antimalarial acridines are usually associated solely with 9-amino derivatives related to quinacrine, a significant effort has also been applied to acridine methanols formally derived from quinine (**1**). Perrine has commented on unsuccessful efforts in the 1930s to obtain such 9-(α-hydroxy-β-aminoethyl)-acridines.[83] Acridans (9,10-dihydroacridines) with this type of side chain are completely inactive,[13, 84] as are 1,2,3,4-tetrahydroacridines with similar 7-position side chains.[13, 85] However, a 6-position side chain on the tetrahydro nucleus (e.g., **77**) provided significant antimalarial activity (about one-third that of quinine) against *P. gallinaceum* in the chick.[13, 85] A four-carbon α-hydroxy side chain in the 9-position of a fully aromatic acridine (**78**) produced a compound equal to quinine in potency and about two-thirds as active as quinacrine in the *P. gallinaceum*-chick test.[14, 86] Very recently, Rosowsky et al.[87] have prepared related benz[c]acridines (e.g., **79**) that are fully curative in *P. berghei*-infected mice at 40 mg kg^{-1}. Quinacrine is not curative in this test in doses up to 320 mg kg^{-1}, the maximum nontoxic dosage.

References

1. G. M. Findlay, *Recent Advances in Chemotherapy*, Vol. 2, Churchill, London, 1951.
2. J. Hill, in *Experimental Chemotherapy*, Vol. 1, R. J. Schnitzer and Frank Hawking, eds., Academic Press, New York, 1963, pp. 542 ff.
3. O. Y. Magidson, "The Synthesis of Antimalarial Drugs," World Health Organization Working Document No. 54, 1960.
4. A. Albert, *The Acridines*, 2nd ed., Arnold, London, 1966, pp. 469-492.
5. R. M. Pinder, in *Medicinal Chemistry*, 3rd ed., Alfred Burger, ed., Wiley-Interscience, New York, 1970, p. 508.
6. R. M. Acheson, *Acridines*, Interscience, New York, 1956, pp. 362-382.
7. P. Ehrlich and P. Guttman, *Klin. Wochenschr.*, **28**, 953 (1891).
8. P. Mühlens, *Naturwissenschaften*, **14**, 1162 (1926).
9. W. Schulemann, *Proc. Roy. Soc. Med.*, **25**, 897 (1931).
10. H. Mauss and F. Mietzsch, *Klin. Wochenschr.*, **12**, 1276 (1933).
11. A. Albert, *The Acridines*, 2nd ed., Arnold, London, 1966, pp. 424-425.
12. F. Y. Wiselogle, *A Survey of Antimalarial Drugs*, 1941-1945, J. W. Edwards, Ann Arbor, Mich., 1946.
13. G. R. Coatney, W. C. Cooper, N. B. Eddy, and J. Greenberg, "Survey of Antimalarial Agents," Public Health Monograph No. 9, Public Health Service Publication No. 193, U.S. Govt. Printing Office, 1953.
14. F. L. Rose, *J. Chem. Soc.*, 2770 (1951).
15. *Chron. World Health Org.*, **13**, 341 (1959).
16. *Chron. World Health Org.*, **20**, 286 (1966).
17. L. J. Bruce-Chwatt, *Trans. Roy. Soc. Trop. Med. Hyg.*, **59**, 105 (1965).
18. C. G. Huff, *Amer. J. Trop. Med. Hyg.*, **14**, 339 (1965).
19. L. H. Schmidt, *Ann. Rev. Microbiol.*, **23**, 427 (1969).
20. L. J. Bruce-Chwatt, in *Textbook of Medicine*, 12th ed., P. B. Beeson and W. McDermott, eds., W. B. Saunders and Co., London, 1967, pp. 350-362.
21. F. A. Neva, *New England Med. J.*, **277**, 1241 (1967).
22. W. Roehl, *Naturwissenschaften*, **14**, 1156 (1926).
23. D. G. Davey, in *Experimental Chemotherapy*, Vol. 1, F. Hawking and R. J. Schnitzer, eds., Academic Press, New York, 1963, pp. 487-511.
24. I. H. Vincke and M. Lipps, *Ann. Soc. Belg. Med. Trop.*, **28**, 97 (1948).
25. J. P. Thurston, *Brit. J. Pharmacol.*, **5**, 409 (1950).
26. T. S. Osdene, P. B. Russell, and L. Rane, *J. Med. Chem.*, **10**, 431 (1967).
27. D. M. Aviado, *Exp. Parasitol.*, **25**, 399 (1969).
28. F. C. Goble, in *Annual Reports in Medicinal Chemistry, 1969*, C. K. Cain, ed., Academic Press, New York, 1970. pp. 116 ff.

29. C. J. Canfield, L. B. Altstatt, and V. B. Elliot. *Am. J. Trop. Med. Hyg.*, **19,** 905 (1970).
30. R. L. Hickman, *Mil. Med.*, **134,** 741 (1969).
31. M. D. Young and D. C. Craig, *Mil. Med.*, **134,** 767 (1969).
32. R. A. Ward, J. H. Morris, D. J. Gould, A. F. C. Bourke, and F. C. Cadigan, *Science*, **150,** 1604 (1965).
33. A. F. Bekhli, *J. Gen. Chem.* (*USSR*), **28,** 1943 (1958).
34. R. O. Clinton and C. M. Suter, *J. Amer. Chem. Soc.*, **70,** 491 (1948).
35. F. E. King, R. J. S. Beer, and S. G. Waley, *J. Chem. Soc.*, 92 (1946).
36. M. A. R. Ansari, *Pakistan J. Health*, **4,** 175 (1954).
37. E. F. Elslager and D. F. Worth, *J. Med. Chem.*, **12,** 955 (1969).
38. E. F. Elslager, F. H. Tendick, L. M. Werbel, and D. F. Worth, *J. Med. Chem.*, **12,** 970 (1969).
39. C.I.O.S. Report, "Pharmaceuticals at the I. G. Farbenindustrie Plant, Elberfield, Germany," No. XXV-54, Item 24, 1946.
40. H. Mauss and F. Mietzsch (I. G. Farbenindustrie), German Patent 553072 (1930); *Chem Abstr.*, **26,** 4683 (1932).
41. K. S. Topchiev and A. F. Bekhli, *C. R. Acad. Sci. Russia*, **55,** 629 (1947); *Chem. Abstr.*, **41,** 6986 (1947).
42. O. Yu. Magidson, A. M. Grigorovskii, and E. P. Galperin., *J. Gen. Chem.* (*USSR*), **8,** 56 (1938); *Chem. Abstr.*, **32,** 5405 (1938).
43. O. Yu. Magidson and A. I. Travin, *Chem. Ber.*, **69B,** 537 (1936).
44. A. M. Grigorovskii and Y. M. Terentyeva, *J. Gen. Chem.* (*U.S.S.R*), **17,** 517 (1947); *Chem. Abstr.*, **42,** 910 (1948).
45. A. M. Grigorovskii and F. A. Veselitskaya, *J. Gen. Chem.* (*U.S.S.R.*), **26,** 491 (1956); *Chem. Abstr.*, **51,** 2778 (1957).
46. O. Yu. Magidson and A. I. Travin, *J. Gen. Chem.*, **8,** 842 (1937); *Chem. Abstr.*, **31,** 5800 (1937).
47. L. J. Sargent and L. Small, *J. Org. Chem.* **11,** 359 (1946).
48. L. J. Sargent and L. Small, *J. Org. Chem.*, **12,** 567 (1947.
49. D. P. Spalding, E. C. Chapin, and H. S. Mosher, *J. Org. Chem.*, **19,** 357 (1954).
50. G. B. Bachman and G. M. Picha, *J. Amer. Chem. Soc.*, **68,** 1599 (1946).
51. G. B. Bachman and J. W. Wetzel, *J. Org. Chem.*, **11,** 454 (1946).
52. J. Dobson, W. C. Hutchison, and W. O. Kermac, *J. Chem. Soc.*, 123 (1948).
53. G. B. Bachman and F. M. Cowen, *J. Org. Chem.*, **13,** 89 (1948).
54. J. Dobson and W. O. Kermac, *J. Chem. Soc.*, 150 (1946).
55. W. C. Hutchison and W. O. Kermac, *J. Chem. Soc.*, 678 (1947).
56. E. F. Elslager, in *Medicinal Chemistry*, 3rd ed., Alfred Burger, ed., Wiley-Interscience, New York, 1970, pp. 543-544.
57. R. A. Radke, *Gastroenterology*, **36,** 590 (1959).
58. E. F. Elslager, R. E. Bowman, F. H. Tendick, D. J. Tivey, and D. F. Worth, *J. Med. Pharm. Chem.*, **5,** 1159 (1962).
59. P. E. Thompson, J. E. Meisenhelder, H. H. Najarian, and A. Bayles, *Amer. J. Trop. Med. Hyg.*, **10,** 335 (1961).
60. World Health Org. Tech. Rep. Ser., No. 324, 1966.
61. E. T. Reid, R. J. Fraser, and F. B. Wilford, *Central African J. Med.*, **9,** 478 (1963).
62. E. F. Elslager, E. H. Gold, F. H. Tendick, L. M. Werbel, and D. F. Worth, *J. Heterocycl. Chem.*, **1,** 6 (1964).
63. F. Takahashi, S. Shimada, and K. Hayase, *J. Pharm. Soc. Jap.*, **64,** 2A, 5 (1944); *Chem. Abstr.*, **45,** 8531 (1951).
64. D. M. Besley and A. A. Goldberg, *J. Chem. Soc.*, 2448 (1954).

65. L. J. Bruce-Chwatt and H. M. Archibald, *Brit. Med. J.*, **1,** 539 (1953).

66. J. O. W. Ang'Awa and N. R. E. Fendall, *J. Trop. Med. Hyg.*, **57,** 59 (1954).

67. J. F. B. Edeson, *Ann. Trop. Med. Parasitol.*, **48,** 160 (1954).

68. E. F. Elslager, S. C. Perricone, and D. F. Worth, *J. Heterocycl. Chem.*, **7,** 543 (1970).

69. P.-L. Chien and C. C. Cheng, *J. Med. Chem.*, **11,** 164 (1968).

70. N. D. Heindel and S. A. Fine, *J. Med. Chem.*, **13,** 760 (1970).

71. E. F. Elslager, F. H. Tendick, and L. M. Werbel, *J. Med. Chem.*, **12,** 600 (1969).

72. I. M. Rollo, in *The Pharmacological Basis of Therapeutics* 3rd ed. L. S. Goodman and A. Gilman, eds., Macmillan, New York, 1965, pp. 1090-1091.

73. I. M. Rollo, in *The Pharmacological Basis of Therapeutics*, 3rd. ed., L. S. Goodman and A. Gilman, eds., Macmillan, New York, 1965, pp. 1072-1074.

74. N. B. Kurnick and I. E. Radcliffe, *J. Lab. Clin. Med.*, **60,** 669 (1962).

75. L. S. Lerman, *Proc. Nat. Acad. Sci. U.S.*, **49,** 94 (1963).

76. F. E. Hahn, R. L. O'Brien, J. Ciak, J. L. Allison, and J. G. Olenick, *Mil. Med.*, (*Suppl.*), **131,** 1071 (1966).

77. R. L. O'Brien, J. G. Olenick, and F. E. Hahn., *Proc. Nat. Acad. Sci. U.S.*, **55,** 1511 (1966).

78. J. Ciak and F. E. Hahn, *Science*, **156,** 655 (1967).

79. K. Van Dyke and C. Saustkiewicz, *Mil. Med.*, **134,** 1000 (1969).

80. K. Van Dyke, C. Szustkiewicz, C. H. Lantz, and L. H. Saxe, *Biochem. Pharmacol.*, **18,** 1417 (1969).

81. K. Van Dyke, C. Lantz, and C. Szustkiewicz, *Science*, **169,** 492 (1970).

82. K. A. Conklin and S. C. Chou, *Science*, **170,** 1213 (1970).

83. T. D. Perrine, *J. Org. Chem.*, **25,** 1516 (1960).

84. L. J. Sargent and L. F. Small, *J. Org. Chem.*, **13,** 447 (1948).

85. L. J. Sargent and L. F. Small, *J. Org. Chem.*, **19,** 1400 (1954).

86. T. D. Perrine and L. F. Small, *J. Org. Chem.*, **14,** 583 (1949).

87. A. Rosowsky, M. Chaykovsky, S. A. Yeager, R. A. St. Amand, M. Lin, and E. J. Modest, *J. Med. Chem.*, **8,** 809 (1971).

Index

This is a combined subject and compound index. Compo[...] are not detailed in the index, but may be found by referring [...] of compound. Tables are indicated by the letter "T" before [...] in heavy type refer to major discussions; in italics to main [...]